HANDBOOK OF MAMMALIAN VOCALIZATION

AN INTEGRATIVE NEUROSCIENCE APPROACH

HANDBOOK OF MAMMALIAN VOCALIZATION

AN INTEGRATIVE NEUROSCIENCE APPROACH

Edited by

STEFAN M. BRUDZYNSKI

ELSEVIER

AMSTERDAM • BOSTON • HEIDELBERG • LONDON
NEW YORK • OXFORD • PARIS • SAN DIEGO
SAN FRANCISCO • SINGAPORE • SYDNEY • TOKYO
Academic Press is an imprint of Elsevier

Academic Press is an imprint of Elsevier

32 Jamestown Road, London NW1 7BY, UK
30 Corporate Drive, Suite 400, Burlington, MA 01803, USA
525 B Street, Suite 1900, San Diego, CA 92101-4495, USA

First edition 2010

British Library Cataloguing-in-Publication Data
A catalogue record for this book is available from the British Library

Library of Congress Cataloging-in-Publication Data
A catalog record for this book is available from the Library of Congress

ISBN: 978-0-12-374593-4

For information on all Academic Press publications
visit our website at elsevierdirect.com

Typeset by Macmillan Publishing Solutions
www.macmillansolutions.com

Printed and bound in United States of America

10 11 12 13 14 15 10 9 8 7 6 5 4 3 2 1

Contents

Section 11. Vocal Communication Systems in Mammals: Semiotic Codes in Vocalization

About the Authors

Steven M. Barlow, Ph.D., speech physiologist, neuroscientist and Professor of Speech-Language-Hearing, Neuroscience, Human Biology, and Bioengineering at the University of Kansas. He works on the central patterning of ororhythmic and respiratory movements in preterm and term infants at risk for neurodevelopmental outcome, and has engineered a new orocutaneous entrainment system undergoing randomized clinical trials for the prohabilitation of suck and feeding disorders in premature infants.

Roozbeh Behroozmand received his BSc degree from the Department of Electrical and Computer Engineering at the University of Tehran. He received the MSc degree in Biomedical Engineering at Amirkabir University of Technology. He is currently a Ph.D. student in the Department of Communication Sciences and Disorders at Northwestern University, Evanston, Illinois, USA. His primary research interest is focused on neural processing of speech and hearing, audio–vocal integration, speech motor control and biomedical signal processing.

Gerald Berke, MD, laryngeal physiologist, otolaryngologist, head and neck surgeon. Presently, Professor of Surgery and Chief of the Division of Head & Neck Surgery at the University of California Los Angeles and Director of the UCLA Voice Center for Medicine and the Arts. He is a member of the Board of Directors of the American Board of Otolaryngology and a past President of the American Laryngology Association. He is clinically involved with patients with voice disorders, nasal and sinus diseases and esophagal disorders and is currently working on requirements for organ and nerve preservation for laryngeal transplantation in humans and measures of laryngeal function.

Rekha Bhatt, Ph.D., Research Associate in the Department of Neurology and Neuroscience, UMD-New Jersey, New Jersey Medical School, Newark New Jersey, USA. Research interest: neurobiology of aggressive behavior; neuroimmunology.

Suresh Bhatt, DVM, Research Associate in the Department of Neurology & Neuroscience, UMD-New Jersey, New Jersey Medical School, Newark New Jersey, USA. Research interest: neurobiology of aggressive behavior.

D. Caroline Blanchard, Ph.D., biopsychologist and ethologist, Research Professor, Pacific Biosciences Research Center, and Professor, Department of Genetics and Molecular Biology, John A. Burns School of Medicine, University of Hawaii at Manoa, Honolulu, Hawaii, USA. Her main research interest is in aggression, defense and social behaviors in mammals, and the brain systems that support these patterns.

Robert J. Blanchard, Ph.D., biopsychologist and ethologist, Professor of Psychology and Neuroscience at the University of Hawaii at Manoa, Honolulu, Hawaii, USA. His main research interest is in aggression, defense, and social behaviors in mammals, and the brain systems that support these patterns.

George S. Borszcz, Ph.D., behavioral neuroscientist, Associate Professor of Behavioral and Cognitive Neuroscience in the Department of Psychology at Wayne State University, Detroit, Michigan, USA. His main research interest is in the neurobiology of pain processing and analgesia, and particularly the brain circuitry responsible for the affective response to pain.

Igor Branchi, Ph.D., neuroscientist and psychobiologist, Researcher at the Department of Cell Biology and Neuroscience, Istituto Superiore di Sanità, Rome, Italy. His major field of interest is the role of genetic and epigenetic factors in shaping neurobehavioral development and adult vulnerability to mental dysfunction.

Stefan M. Brudzynski, Ph.D., D.Sc., neurophysiologist, neuroscientist and biopsychologist, presently Professor of Psychology and Neuroscience at Brock University, St. Catharines, Ontario, Canada. His main research interest is in neural substrate of animal behavior, neuropsychopharmacology, control of locomotor behavior, vocal communication in animals, and particularly central control of ultrasonic vocalization and communication in rodents.

Susan A. Brunelli, Ph.D., researcher, Developmental Neuroscience, Department of Psychiatry, Columbia College of Physicians and Surgeons, New York, NY. Her research interests lie broadly in the social, behavioral and affective development of young organisms. Research involves the selective breeding of two lines of rats that differ in the infant expression of anxiety and serve as a powerful model system to experimentally analyze genetic predisposition and effects of early experience to create vulnerability to affective disorders over the life span.

Jeffrey Burgdorf, Ph.D., neuroscientist, biopsychologist and molecular biologist. He is an Assistant Professor in the Department of Biomedical Engineering at the Falk Center for Molecular Therapeutics at Northwestern University, Evanston, Illinois, USA. His main research interest is in the molecular underpinnings of emotional states, using ultrasonic vocalizations as a model system and translating these findings into novel therapeutics to treat psychiatric disorders.

Yale E. Cohen, Ph.D., cognitive neuroscientist, neurophysiologist, behavioral physiologist. Associate Professor in the Department of Psychological and Brain Sciences and the Center for Cognitive Neuroscience, Dartmouth College, Hanover, New Hampshire, USA. His research focuses on the neural and behavioral mechanisms underlying auditory cognition with an emphasis on ethological stimuli and social cognition.

Francesca R. D'Amato, D.Sc., ethologist and psychobiologist working in the Psychobiology and Psychopharmacology Department of the Institute of Neuroscience of the National Research Council in Italy. Her main research concerns the role of early environment on the development of social, emotional and cognitive capabilities in rodents. In particular, she has focused on the role of the mother and maternal behaviors as early programming factor.

Eva Maria Eberl, Ph.D., researcher and lecturer at Graz University, Austria, in neuroscience, acoustic phonetics and language acquisition. Her research concerns the evolutionary and ontogenetic development of speech, language and gestural communication, in particular the study of manual gestures as precursors to language. She is currently investigating the pooling of resources in gesture cognition as the foundation for the development of verbal language in the human infant.

Günter Ehret, Ph.D., neurobiologist, behavioral physiologist and ethologist, Professor of Neurobiology at the Institute of Neurobiology, University of Ulm, Germany. His research interests are the neural substrates of animal behavior, sound processing in the auditory system to perception and recognition, sound communication, hormonal regulation of parental behavior, psychoacoustics, evolutionary bases of speech perception, genetics of hearing and vocalizing in the mouse.

Meredith E. Estep, ABD, neuroscience, University of Kansas, Kansas, USA. Her main research interests are in the neurodevelopment and sensorimotor control of human oromotor behaviors including speech, suck and mastication.

Camille L. Finck, MD, Ph.D., otorhinolaryngologist, phonetician and microsurgeon; she is Chef de Clinique in the Department of Otorhinolaryngology of the Centre Hospitalier Universitaire of Liège, Belgium. Her main research interest is in microsurgical revalidation of biomechanically altered vocal folds, and especially in developing regenerative microsurgery. She also works on clinical evaluation of voice. She teaches the anatomo-physiology of phonation at the Faculté de Psychologie et des Sciences de l'Education of the University of Mons and at the Faculté de Psychologie of the University of Liège, Belgium.

Neville H. Fletcher, Ph.D., DSc, FAA, now retired after 20 years as Professor of Physics at the University of New England in Australia, and five years as Director of Physical Sciences for Australia's national research organization CSIRO, is a full-time Visiting Fellow at the Australian National University. His research interests include semiconductor materials and devices, cloud physics, ice physics, musical acoustics and biological acoustics.

Roland Frey, Ph.D., evolutionary morphologist, comparative anatomist, member of the Research Group "Reproduction Management" at the Institute for Zoo and Wildlife Research (IZW), Berlin, Germany. His main research topics are mammalian vocal tract adaptations and sexual selection, head anatomy of ruminants, nasal structure in particular, evolutionary history and macroscopic anatomy of the male genital organs and copulation modes in mammals.

Michael Garstang, Ph.D., is a meteorologist specializing in surface-to-atmosphere interactions with associated cross-discipline studies. He is presently Emeritus Professor and Distinguished Research Professor in the Department of Environmental Sciences at the University of Virginia, USA. He has contributed to ecosystem sustainability and to long-range, low-frequency elephant communication.

Alban Gebler, Dipl. Phys., Dipl. Biol., is a physicist and biologist at the Institute of Theoretical Biology, Humboldt University of Berlin, Germany, and coordinator for the airborne earth observation mission

GEOHALO at Earth's Magnetic Field Group, Helmholtz Centre Potsdam, GFZ German Research Centre for Geosciences. In addition, he works as a self-employed consultant for computer and laboratory technology. In the field of bioacoustics, he focuses mainly on sound analysis and nonlinear effects of animal vocalization.

Francisco Gonzalez-Lima, BSc (Biology), BA (Psychology), Ph.D. (Anatomy and Neurobiology, University of Puerto Rico School of Medicine), holds the first endowed chair named after a Hispanic professor in the USA, the George I. Sanchez Centennial Professorship at the University of Texas at Austin, Department of Neuroscience and Psychology. He has been a Humboldt fellow and visiting scientist in Germany, England and Canada. His research interests include neuroscience, neuroanatomy, neurobiology, physiological psychology, psychobiology, learning and memory, and neural mechanisms of behavior.

Eva Gruber-Dujardin, Dr of Veterinary Medicine, veterinarian, neuropathologist and neuroscientist, after Junior Lecturer position at the Veterinary Faculty of Utrecht University, is presently Veterinary Pathologist in the Department of Infectious Pathology in the German Primate Centre, Göttingen, Germany. Her main research interest is neuroanatomy and neuropathology with a special interest in neuroanatomy of primate vocalization system, as well as neural substrates of animal behavior, communication and vocalization.

Steffen R. Hage, MSc, Ph.D., neurophysiologist, neuroscientist, neuroethologist, presently Group Leader in the Department of Animal Physiology at the University of Tübingen, Germany. His main research interests are audio–vocal integration processes, vocal–motor control mechanisms and acoustic communication in mammals.

Myron A. Hofer, MD, psychiatrist and behavioral neuroscientist, is the Sackler Professor and Director of the Sackler Institute of Developmental Psychobiology, at Columbia University, New York, NY, USA. Through an experimental analysis of physiological and behavioral interactions that enmesh the infant rat and its mother, he and his colleagues have discovered novel regulatory processes that have become the basis for a new understanding of the early origins of attachment, the dynamics of separation responses and the shaping of later development in adulthood. He has had a long-standing interest in evolution, the nature of development and the relationship between the two.

Joseph F.Y. Hoh, BSc (Med), MB, BS, Ph.D., DSc, is a muscle physiologist, Honorary Associate Professor of Physiology, formerly Reader, Discipline of Physiology, University of Sydney, Sydney, Australia. His main research interest is in myosin isoforms in skeletal and cardiac muscles in relation to contractility and comparative physiology, neurogenic and myogenic regulation of muscle function, and thyroid and adrenergic regulation of cardiac muscle.

Gareth Jones, Ph.D., evolutionary biologist, bioacoustics scientist and bat researcher, Professor of Biological Sciences at the University of Bristol, UK. His research is on the echolocation, molecular ecology and evolution, and conservation biology of bats.

Kate E. Jones, Ph.D., biodiversity scientist, Senior Research Fellow at the Institute of Zoology, Zoological Society of London, London, UK. Her main research interest is in understanding how evolutionary processes produce past, present and future global biodiversity patterns. She works on a number of projects ranging from the evolution of echolocation in bats to disease macroecology in humans.

Christoph Kayser, Ph.D., is a neurophysiologist, neuroscientist and Max Planck Junior Group Leader at the Max Planck Institute for Biological Cybernetics, Tübingen, Germany. His main research interests are the neural basis underlying the integration of sensory information, the neural codes used for the encoding of sensory information and the adaptation of sensory systems to properties of their natural sensory inputs.

Arlette Kolta, Ph.D., neuroscientist and Professor at the Faculty of Dentistry at the Université de Montréal. She is also an accredited member of the Physiology Department of the same institution and Adjunct Professor of the Faculty of Dentistry, McGill University, Canada. She uses electrophysiological and neuroanatomical approaches to investigate the cellular circuits and mechanisms involved in generating the rhythmic motor pattern for mastication and the potential role of neuron–glia interactions in rhythmogenesis. She is also interested in the role of spindle primary afferents in the development of chronic orofacial muscle pain.

Koji Kuraoka, Ph.D., neurophysiologist and cognitive neuroscientist, Research Fellow of the Japan Society for the Promotion of Science, Japan. His main research interest is in neural substrate of emotion and social cognition, physiological reaction to emotional stimuli and particularly single neuron activity in relation to facial and vocal communication in primates.

Simone Kurt, Ph.D., neurophysiologist, electrophysiologist and neuroscientist, Research Associate at the Institute of Neurobiology, University of Ulm, Germany. Her main research interests are sound processing in the auditory system to perception and recognition, auditory learning and cortical processing, neuropharmacology, genetics of learning behavior and knowledge transfer in mice.

Jeffrey T. Laitman, Ph.D., anatomist and physical anthropologist, Distinguished Professor, Professor of Anatomy and Functional Morphology, and Professor of Otolaryngology, at Mount Sinai School of Medicine, New York, NY, USA. He is an international expert on the evolution and development of the upper respiratory tract, and has made strides in understanding clinicopathologies such as sudden infant death syndrome (SIDS), gastro-esophageal reflux disease and choking. His main research interests are in human upper respiratory tract and skull base anatomy, comparative vocal tract anatomy, evolution of human vocal tract, laryngeal development and anatomy.

Charles R. Larson, Ph.D., received a Ph.D. in Speech and Hearing Science from the University of Washington in 1975. He is currently a Professor and Chairman of the Department of Communication Sciences and Disorders at Northwestern University, Evanston, Illinois, USA. His primary research interest is focused on neural mechanisms relating auditory feedback to vocalization. He teaches courses in anatomy and neurobiology of speech and vocal motor control.

Jung Hoon Lee, Ph.D., Physicist, postdoctoral researcher, Psychological and Brain Sciences at Dartmouth College, Hanover, New Hampshire, USA. He is mainly interested in studying the mechanism of information processing in the brain and developing computational models to explain brain's higher functionalities, such as categorization and memory.

Lionel C. Lejeune, MD, is an intern in the Department of Otorhinolaryngology at the Centre Hospitalier Universitaire of Liège, Belgium. He is also researcher in the Laboratory of Connective Tissues Biology of the University of Liège. His research interest is in cellular and molecular biology of the extracellular matrix of the vocal fold's lamina propria.

Gregory F. Lewis, Ph.D. candidate in the Department of Bio-Engineering at the University of Illinois at Chicago, Illinois, USA. His research focuses on the adaptive features of the neural regulation of the middle ear muscles and how dysfunction of middle ear muscles may lead to clinical features of hyperacusis and language delay.

Yoav Litvin, Ph.D., biopsychologist and ethologist, Researcher, University of Hawaii at Manoa, Honolulu, Hawaii, USA. His interests are in defense, aggression, stress, arousal, learning and memory, and the brain systems that support the related behavioral patterns.

Hanjun Liu, Ph.D., worked as a postdoc in speech physiology laboratory at Northwestern University and is now a professor of rehabilitation medicine at Sun Yat-sen University, P.R. China. His main research interest is normal and pathological speech production and perception, speech motor control and neural processing of speech and hearing, and particularly neural mechanisms underlying the online control of speech and voice.

Nikos K. Logothetis, Ph.D., neurophysiologist, neuroscientist and Professor and Director at the Max Planck Institute for Biological Cybernetics, Tübingen, Germany. His main research interests are in the neural mechanisms of perception, object recognition, cortical plasticity and neuromodulation. Over the last decade developments and research in the Logothetis laboratory have contributed to our understanding of the neurophysiological mechanisms underlying the functional magnetic resonance imaging (fMRI) signal.

Jennifer L. Long, MD, Ph.D., is an Otolaryngology-Head and Neck Surgery resident at University of California-Los Angeles, CA, USA. Her special interests are in clinical laryngology and voice, vocal fold development and function, and tissue engineering approaches to treating laryngeal disorders.

James P. Lund, BDS, Ph.D., FCAHS, dentist and neuroscientist and Professor in the Faculty of Dentistry and Alan Edwards Centre for Research on Pain at McGill University, Montréal, Canada. He is also a member of the Groupe de Recherche sur le Système Nerveux Central of the Université de Montréal. He works on the central patterning of rhythmical jaw movements and their modification by sensory inputs, particularly pain. He is also part of a group that carries out clinical trials to improve function in people who have lost teeth.

Alanna Maltby, BSc, is a Ph.D. student at the Institute of Zoology, Zoological Society of London and University College London, UK, researching the evolution and diversity of echolocation calls in bats.

Marta B. Manser, Ph.D., MSc, zoologist, Professor of Animal Behaviour at Zürich University, Zürich, Switzerland. Her main research interest is in communication and cognition in social mammals in their natural habitat, in particular a comparative approach on vocal communication, group decision-making and spatial orientation in several mongoose species with different social structures, but living in similar ecological habitats.

Carolyn McGettigan, MA (Cantab), Ph.D. (London), neuroscientist and Research Associate at the Institute of Cognitive Neuroscience, University College London, London, UK. Her work combines behavioral and neuroimaging techniques to investigate speech input and output processes in the human brain. She is particularly interested in the role of rhythm in speech comprehension, and in the cognitive factors that underlie individual differences in speech processing.

Walter Metzner, Ph.D., neurobiologist, Professor at the University of California, Los Angeles, USA. His main research interest is in the neural control of animal behavior, particularly bioacoustics of echolocation and vocal communication in bats, neuronal processing in the vertebrate auditory system and in related senses, vocal motor control and auditory-motor feedback, as well as electrolocation and communication in electric fish.

Joseph R. Moskal, Ph.D., Professor in the Department of Biomedical Engineering and Director of the Falk Center for Molecular Therapeutics in the McCormick School of Engineering and Applied Sciences at Northwestern University, Evanston, IL, USA. The mission of the Falk Center is to use gene discovery technologies, such as focused microarray and quantitative polymerase chain reaction technologies, in concert with appropriate animal models to identify novel targets for drug discovery for central nervous system diseases, including learning and memory disorders and depression.

Jeff Muller, Ph.D., clinical psychologist and Research Scientist at the New York State Psychiatric Institute and Psychiatry Department of the College of Physicians and Surgeons of Columbia University, New York, NY. His general interest is in the neural basis of emotional phenomena in animal models of anxiety and depression, and he works currently on a mouse model of the neural basis of the vulnerability to depression in stressed individuals carrying a low-expressing allele of the serotonin transporter gene.

Katsuki Nakamura, Ph.D., cognitive neuroscientist, neurophysiologist and experimental psychologist, Professor of Primate Research Institute, Kyoto University, Japan. His main research interest is in neural substrate of social and communicative behavior, emotion, and visual recognition and memory in humans and non-human primates.

John D. Newman, Ph.D., neuroethologist, primatologist, Head, Unit on Developmental Neuroethology, Laboratory of Comparative Ethology, Eunice Kennedy Shriver National Institute of Child Health and Human Development, National Institutes of Health, Poolesville, Maryland, USA. His main research interests are primate bioacoustics, neuroethology of auditory communication and the evolution of the neural systems mediating auditory communication.

John Nyby, Ph.D., behavioral neuroscientist/behavioral neuroendocrinologist, Professor of Biology at Lehigh University, Bethlehem, Pennsylvania, USA. His interests are in the endocrine, pheromonal and neurological regulation of animal behavior and animal communication. Research over the years has focused on the pheromonal and endocrine regulation of reproductive behaviors in mice, with a special interest in mouse ultrasonic communication.

Diego Oddi, Ph.D. student working in the Psychobiology and Psychopharmacology Department of the Institute of Neuroscience of the National Research Council in Italy. His scientific activity is centered on the communicative role of ultrasonic vocalizations as a tool to investigate the role of environmental and genetic factors on mice emotionality and sociability.

Michael J. Owren, Ph.D., Associate Professor of Psychology at Georgia State University, Atlanta, GA, USA. His main research interest is in animal bioacoustics and in vocal communication in humans, including speech and nonverbal vocalizations, such as laughter.

Jaak Panksepp, Ph.D., behavioral and affective neuroscientists, is Baily Endowed Professor of Animal Well-Being Science at the College of Veterinary Medicine at Washington State University and Distinguished Research Professor of Psychobiology, emeritus, at Bowling Green State University, as well as Head of Affective Neuroscience Research at the Falk Center for Molecular Therapeutics of Northwestern University, Evanston, IL, USA. His main research interest has been the understanding the neurobiological basis of affective states through a focus on the evolutionary foundations of human emotional and motivational systems as studied in animals models. He is also interested in new insights into and treatments of psychiatric disorders, with a current emphasis on depression, autism, addictions and post traumatic stress disorder.

Christopher I. Petkov, Ph.D., comparative neuroscientist and Senior Lecturer (Assistant Professor) at the Institute of Neuroscience at Newcastle University in the United Kingdom. His main research interests are the comparative study of communication in humans and other primates using noninvasive neuroimaging, and the link between neuroimaging and the study of the mechanisms of cognition and perception at the neuronal level.

Stephen W. Porges, Ph.D., neuroscientist, Professor of Psychiatry and Bio-Engineering, and Director of Brain-Body Center at the University of Illinois at Chicago, IL, USA. His research interests are in the study of relations between the autonomic nervous system and adaptive behaviors, and how atypical neural regulation of physiological state promotes maladaptive behaviors in clinical populations.

Joy S. Reidenberg, Ph.D., comparative anatomist, Associate Professor of Anatomy and Functional Morphology at Mount Sinai School of Medicine, New York, NY, USA. She is an international expert on cetacean anatomy and development. Her main research interests are in aquatic animal anatomy, growth and development, particularly anatomical adaptations to environmental extremes, vocalization mechanisms, respiratory tract (including larynx) anatomy and development, and musculo-skeletal anatomy.

Drew Rendall, Ph.D., Associate Professor of Psychology at the University of Lethbridge, Lethbridge, Alberta, Canada. His main research interest is in the functions and mechanisms of vocal communication in non-human primates and their potential connections to vocal communication in humans, including language.

Brian E. Russ, Ph.D., cognitive neuroscientist, neurophysiologist and behavioral psychologist, Postdoctoral Fellow at Havard University, Cambridge, Massachusetts, USA. His research focuses on the neural and behavioral mechanisms involved in higher-order cognitive processes, with particular interest in the neural processing of ethological signals related to social cognition in primates.

Gillian D. Sales, Ph.D., AKC, Dip TCDHE, MIBiol, Cbiol., currently Curator of the Museum of Life Sciences, King's College London, Guy's Campus, London, UK, where she was previously 34 years as Lecturer and Senior Lecturer in Zoology. Research interests in animal behavior, communication, especially ultrasonic communication, physiology and ecology of small mammals, animal welfare, effect of sound in the animal environment and the effects of sound on captive animals.

Disa A Sauter, Ph.D., psychologist, post-doctoral researcher at the Max Planck Institute for Psycholinguistics in Nijmegen, The Netherlands. Her research investigates the vocal signals that humans use to communicate emotional states. She is particularly interested in nonverbal emotional vocalizations, and has investigated these signals using a range of methodologies including experimental psychology, fMRI, ERPs, acoustic analysis and cross-cultural comparisons.

Maria Luisa Scattoni, Ph.D., neuroscientist and developmental psychobiologist, Researcher at Istituto Superiore di Sanità, Rome, Italy. Her main research interest is translational neuroscience, with a focus on neurodevelopmental disorders, such as autism. She is studying ultrasonic vocalizations in altricial rodents to evaluate their social communication role and to develop ethological models for the study of normal and pathological neurodevelopment.

Gerd Schuller, Dipl. Phys., Dr. rer. nat. (Ph.D.), neuroscientist and neurophysiologist, Professor of Vertebrate Neuroscience at Ludwig-Maximilians-University, Munich, Germany. His main research interest is in neural mechanisms of animal behavior in echolocating bats, functional neuroanatomy of hearing and vocalization in bats and bioacoustics, and particularly the organization of the auditory feedback to the vocal control in echolocation calls and production of communication calls in bats.

Rainer K.W. Schwarting, Prof. Dr. rer. nat. (Ph.D.), psychologist, Professor in the Section for Experimental and Biological Psychology, at the Philipps-University of Marburg, Germany. His main research interests are emotion, motivation and procedural learning in rats, and their neural bases, including neuroanatomy (basal ganglia, limbic system), neuropharmacology (psychostimulants, cytokines) and neurochemistry (dopamine, serotonin, acetylcholine).

Christine Schwartz is a Ph.D. graduate student in the Department of Biology at Texas A&M University, College Station, Texas, USA. Her thesis research focuses on mapping forebrain vocal motor pathways in echolocating free-tailed bats.

Sophie K. Scott, Ph.D., Group Leader of the Speech Communication at the Institute of Cogntitve Neuroscience at University College London, London, UK. She spent several years as a post-doctoral researcher at the MRC CBU in Cambridge UK. Her research uses functional imaging to investigate the cortical basis of human vocal communication, including speech perception and production, and vocal emotional processing. She applies models from primate auditory processing to the neural basis of human perception.

Harry N. Shair, Ph.D., Research Scientist in the Division of Developmental Neuroscience at the New York State Psychiatric Institute in New York City, New York, as well as an Associate Professor of Clinical Psychology in the Psychiatry Department of Columbia University, College of Physicians and Surgeons, New York, NY. His long-term interest in the effects of environmental influence on the developmental trajectory of young organisms has led to a model of the formation of early life social bonds.

Ari Daniel Shapiro, Ph.D., biological oceanography, former Postdoctoral Investigator in the Department of Biology, Woods Hole Oceanographic Institution, Woods Hole, MA, USA. His research focused on marine mammal behavior, communication and acoustics. He now works as an independent science radio and multimedia producer (www.aridanielshapiro.com).

Keisuke Shiba, MD, Ph.D., is an otolaryngologist, head and neck surgeon, and neurophysiologist, presently Director of Hikifune ORL and Day Surgery Clinic, Tokyo, Japan. His main research interest is in brainstem mechanism of upper airway reflexes and vocalization.

Allan Siegel, Ph.D., Professor of Neurology & Neuroscience and Psychiatry, presently in the Department of Neurology and Neuroscience, UMD-New Jersey, New Jersey Medical School, Newark, NJ, USA. His research interest includes neurobiology of aggressive behavior, neuropharmacology, neuroanatomy and behavioral pharmacology.

Michael S. Smotherman, Ph.D., Assistant Professor of Biology and Neuroscience at Texas A&M University, College Station, Texas, USA. His research exploits echolocating bats as a model system for behavioral, pharmacological and electrophysiological studies of vocal control in mammals, which serves the larger goal of understanding the neurobiology and evolution of the acoustic communication in vertebrates.

Catherine A. Spuz, MA, behavioral neuroscientist, doctoral candidate of Behavioral and Cognitive Neuroscience in the Department of Psychology at Wayne State University, Detroit, Michigan, USA. Her main research interest is in the contribution of NMDA receptors in the central nucleus of the amygdala to the generation and suppression of affective responses to pain.

Josef Syka, Ph.D., D.Sc., neurophysiologist, neuroscientist, Professor of Physiology at Charles University in Prague, Head of the Department of Auditory Neuroscience at the Institute of Experimental Medicine, Academy of Sciences, Prague, Czech Republic. His main research interest is in investigation of the processing of acoustical signals including vocal communications in the auditory system of animals, function of descending pathways in the auditory system, changes in the inner ear and central auditory system elicited by noise exposure and during aging, as well as the development of the auditory system.

Joji Tsunada, DVM, Ph.D., neuroscientist, neurophysiologist and primatologist. Post-doctoral Fellow, Department of Psychological & Brain Sciences, Dartmouth College, Hanover, New Hampshire, USA. His research interest is understanding the neural mechanism of social cognition in primates.

Markus Wöhr, Dr. rer. nat. (Ph.D.), psychologist, is working in the Section for Experimental and Biological Psychology, at the Philipps-University of Marburg, Germany. His main research interest is in social behavior in rats and mice, particularly in their ultrasonic communication. Using rodent ultrasonic vocalizations, he is studying the neuroanatomical, neurochemical and genetic mechanisms underlying fear and anxiety, as well as autism.

Steven S. Zalcman, Ph.D., Associate Professor, Department of Psychiatry, UMD-New Jersey, New Jersey Medical School, Newark, NJ, USA. His research interest is in neuroimmunology and behavior, and the role of cytokines in normal and abnormal behavior.

Elke Zimmermann, Dipl.-Biol., Ph.D., zoologist, behavioral and evolutionary biologist, primatologist, Professor of Zoology at the University of Veterinary Medicine, the Leibniz University and the Center of Systems Neuroscience, Hannover, Germany. Her main research interest is in bioacoustics, speciation and adaptation of mammals (lemurs and tree shrews in particular), evolution of communication and cognition, affective and evolutionary neuroscience.

Klaus Zuberbühler, Ph.D., is a comparative psychologist and Professor of Psychology at the University of St Andrews in Scotland, UK. His main research interest is in human and non-human primate cognition and communication. Most of his research is with free-ranging primates in their natural habitats.

Preface

The *Handbook of Mammalian Vocalization* represents a collective effort of over 80 authors representing 13 countries from 5 continents in creating this extensive monograph. The central subject, the mammalian vocalization, has been approached from a multitude of angles originating both from essential as well as side disciplines. The main, leading concept of the *Handbook* comes from the perspective of integrative neuroscience, which cements together many views and approaches and offers the best synthetic grasp of the phenomenon. In order to achieve this goal, however, many chapters from disciplines other than neuroscience appeared to be necessary to broaden the general knowledge of all important aspects of vocal behavior from molecular, cellular, pharmacological, organ and systems approaches on one side, to zoological, developmental, genetic, evolutionary, ecological, ethological and communication issues on the other. At the same time, the *Handbook of Mammalian Vocalization* presents a rich, although not complete, array of neuroscience issues ranging from neurophysiology, neuroanatomy and neurobiology to neurobehavioral, neuroethological, biophysical, neurological, neurocognitive, biopsychological and neuropsychiatric aspects of animal and human vocalization. Modern development of biomedical science demands constant broadening of our horizons and integrating and inter-relating more and more scientific subdisciplines. The history of the recent half a century illustrates this process very well, when single disciplines began joining forces to form interdisciplinary studies, and then progressed to multidisciplinary approaches and finally reached an equivalent of global integration of all possible science approaches. This trend is important not from the point of view of dramatically increasing our extensive knowledge, but from the point of view of synthesizing and understanding the basic tenets of the subject. Researchers in each subfield, subdiscipline and even individual research groups have different strategies, different ideas, somewhat different understanding and a different order of importance of many of the science issues. Thus, the best global understanding depends on studying the multitude of views, results and approaches from the perspective of any available subject, and this *Handbook* was created with this notion in mind. Although a complete account of any topic might not be possible, any significant step forward in this direction would be constructive. The reader will find most aspects of mammalian vocalization mentioned in more than one chapter of the *Handbook*, but always from different perspectives, with different details and always with an exhaustive bibliography. This slightly overlapping composition of the *Handbook* was intentional. I would like to encourage any reader, from beginners to advanced academicians, not only to read selected chapters, but also to thoroughly and extensively study this compendium. The editorial goal of this enterprise will be achieved if the reader finds the information both enriching and providing incentive for new ideas in further research.

Stefan M. Brudzynski
(Editor)

Acknowledgments

The editor would like to express his sincere thanks to Joe Huston, a Series Editor for the Elsevier *Handbook of Behavioral Neuroscience*, who first suggested the idea of writing this *Handbook*, and provided the initial guidance and continuous support. The editor also wishes to thank all persons from Elsevier involved in the publication process, and particularly Johannes Menzel, Clare Caruana and Maureen Twaig. Finally, the editor would like to express his gratitude to numerous persons who helped as reviewers and consultants. There are too many of them to list all of the names. Among the coauthors of the *Handbook* who critically read the manuscripts of their colleagues, a particular acknowledgment should be directed to Neville Fletcher, Gillian Sales, Allan Siegel, Jeffrey Burgdorf and Mike Smotherman for their time and meticulous and helpful comments. Among external reviewers, many thanks are directed to Uwe Jürgens for his comments and critical reading of some of the chapter drafts. Finally, the editor would like to thank numerous colleagues and students from Brock University for their help in reading the manuscripts, particularly Sid Segalowitz, Cheryl McCormick, Joffre Mercier, John Mitterer, Kathy Mondloch, Stan Sadava, and Melanie Komadoski.

SECTION 1

Introduction

CHAPTER 1.1

Vocalization as an ethotransmitter: introduction to the *Handbook of Mammalian Vocalization*

Stefan M. Brudzynski

Department of Psychology, Brock University, St. Catharines, Ontario, Canada

I. Introduction

Animal vocal communication has received much attention from researchers for many decades. Over the last 70 years, attention has been focused on different aspects of the process of sending and receiving signals. Some approaches were centered on the mechanisms of emitting vocalizations, regardless of the recipient, and others represented a unidirectional process of producing and sending vocal signals by the signaler to the receiver. Adequate responses by the receiver were evidence of the communicatory process. Also, significant research attention has been focused on receivers with regard to the behavioral consequences of perceived species-specific vocal signals. Finally, vocal communication was also approached as an exchange of vocalizations between or among animals in a process of bi- or multidirectional communication, where the main biological function was centered not on sending or receiving specific information, but on testing the responsiveness of the receiver or influencing the receiver's behavior. In larger social groups, this type of communication could be seen as a social negotiation or social decision-making process.

The fundamental and common element of all of these different approaches is vocalization as a biological phenomenon, and this is the main subject of this handbook.

I.A. Mammalian vocalization as a particularly well-developed communication system

Despite numerous modes of mammalian communication, the production of vocalizations in mammals remains one of the most widespread behavioral and physiological processes, requiring highly complex coordination and integration of numerous subsystems, including the central nervous system, autonomic and endocrine systems, and peripheral organs. There are several biological reasons why vocalization is found ubiquitously among mammalian species regardless of their habitat (subterrean, terrestrial, aerial, or aquatic). Acoustic signals can propagate for considerable distances, and their reception is dependent neither on daylight, nor on good visibility. Vocalizations may be produced and terminated almost instantly (with negligible latency), and the discontinued signals do not leave any permanent trace. Finally, the combination of different acoustic parameters within calls provides an immensely rich combination of features for encoding signals. As a consequence, all mammalian vocalizations represent a theoretically unlimited repertoire of biological signals, which may differ from each other just as individual mammalian organisms differ within any species. Finally, vocal signals have also been used for rapid echolocation (as self-communication) in aerial and some aquatic mammals. There is no other communication method (visual, olfactory, thermal, or electromagnetic) which can combine all these features together and which has been subjected to such a phylogenetic variation. One of the goals of this handbook is to recognize this uniqueness of vocalizations as mammalian communication signals.

During intraspecific interactions, all mammals use combinations of communication modes (e.g., vocal, visual and olfactory, or visual, vocal and tactile), and all these signals have the capability of affecting the behavior of recipients. Communication based on direct contact, both mechanical and/or chemical, is

Stefan M. Brudzynski (Ed.)
Handbook of Mammalian Vocalization
ISBN 978-0-12-374593-4

DOI: 10.1016/B978-0-12-374593-4.00001-2

phylogenetically older, while communication based on telereception appeared later in evolution. Among telereception modes, electroreception and distant chemoreception (pheromones) seem to have appeared earlier in evolution than visual and vocal communication, which can convey reliable signals for the longest distances. Among these last two communication methods, vocalization has reached the highest sophistication in mammals as a group. Vocal signals have reached considerable levels of complexity, and can usually travel for longer distances than those determined by visual range. Thus, all mammalian communicatory signals may be classified according to their evolutionary history, with a gradually increasing range of operation and signal complexity. This observation leads to the conclusion that vocalization is the most recent and advanced mode of communication. Vocalization also has its own evolutionary history within the mammalian taxon, from simple vocal displays, through complex vocalization signaling to semiotically- and referentially-organized vocal communication. Examples of all these stages of mammalian vocal communication are represented in this handbook. The evolutionary development of vocal communication may be documented not only for mammals, but also for birds, which are known to produce sophisticated vocalizations.

I.B. The neuroscience approach in studies of mammalian vocalization

Direct studies of brain mechanisms involved in producing vocalizations, and of the central mechanisms of reception, recognition and analysis of species-specific calls, provide additional approaches to the topic and reveal many aspects of mammalian vocal communication. Since Charles Darwin published *The Expression of the Emotions in Man and Animals* in 1872 (Darwin, 1872/1889), behavioral sciences, including comparative psychology, have struggled for almost 140 years to explain clearly the biological nature and origin of vocal communication in mammals. One of the obstacles was, and still is, our inability to know the subjective feelings and intentions of animals when they emit vocalizations. One of the ways to gain some insight into this issue is to use a comparative and multidisciplinary approach, as is richly illustrated in this handbook. This includes using a comparative approach to identify functions of the brain structures and systems which are directly involved in the production of vocalizations. The high degree of homology in transmitters, brain structures and neural regulation underlying the production of vocalizations among different groups of animals, and among animals and humans, represents the most powerful evidence of a common origin of internal states, particularly affective states, and common biological reasons for communicating. The subtitle of this handbook, *An Integrative Neuroscience Approach*, illustrates well one of the main goals of the book: to gain a better understanding of why the vocal communication system evolved in mammals, how vocal production is controlled by the brain, what neural factors influence production and reception of vocalizations, and what physiological, behavioral, affective and cognitive processes are involved in this process.

II. The concept of ethotransmitter

All modes of communication may be regarded as behavioral initiators, modulators, or inhibitors, depending on the situation, state of the organism and structure of the signal. Although all communicating signals, regardless of their mode, may fulfill the criteria of behavioral transmitters, vocalization may serve as a particularly good example. A behavioral transmitter may be defined as a relatively weak physical or chemical stimulus originating from one organism, which is capable of inducing energetically costly responses in another organism through elevated activity of the central nervous system and activation of the somatic, autonomic and endocrine systems.

The function of behavioral transmitters is analogous to the action of neurotransmitters in the brain. While the latter operates among individual cells, the former operates among individual organisms. Binding of a neurotransmitter molecule to its receptor can activate the entire neuron, central nervous system, or the entire organism, where the energy of the resulting process (from utilization of ATP) initiates a response with much higher output energy than the initial energy of the stimulus molecule. Both situations, at the molecular and organism levels, have an initial biological stimulus (a signaling molecule or vocal signal), which can affect specialized receptors (pharmacological or physiological), involve action of the central nervous system, and cause an amplified response of neuronal systems or an entire organism, respectively. The action of these transmitters is based on the "lock-and-key" principle, where the transmitter is the key and the lock is the receptor capable of initiating an amplified response.

Using the analogy to neurotransmitters, behavioral transmitters could be termed ethotransmitters (from the Greek *ethos* meaning "character," "habit," "behavior" plus the Latin *transmissio*, meaning "sending across"). Which communicatory signals should be termed ethotransmitters, and at what evolutionary level they should act, are undoubtedly debatable points. Nevertheless, mammalian vocalization represents the best developed example of the postulated ethotransmitter, and it deserves special attention. This term not only signifies a characteristic biological phenomenon, but also provides a new framework for the understanding of the evolution of vocalization and vocal communication, and their biological roles in social animals. Moreover, a better understanding of mammalian vocalization as ethotransmitter may provide us with a basis for a full understanding of the origin of human speech. A provisional definition of ethotransmitter is suggested below.

The "lock-and-key" metaphor has been mentioned frequently for neurotransmitter action for more than a century, and was also noted a long time ago for behavioral transmission. This metaphor was used by Konrad Lorenz in his studies of bird behavior. The signaling stimuli were termed releasers, which were capable of releasing specific behavior in the receivers by acting on the central nervous system and activating the innate releasing mechanism (Lorenz, 1935). The concept of the "innate releasing mechanism" (Tinbergen, 1951) or "inborn releasing scheme" (Uexküll von, 1934) emphasized the specific relationship between the character of a particular stimulus and a particular stereotypic response released. The ethological studies which followed, however, were trapped within the narrow framework of these concepts by trying to generalize the behavior of all animals from invertebrates to vertebrates within this conceptual scheme, and by assuming that the released responses are instinctive, i.e., innate and largely mechanistic in nature. This introduction does not allow for a detailed description of the history of these concepts in the abundant ethological literature over the last 70 years. However, it might not be an essential topic for this handbook, because the best known classical ethological textbooks summarizing substantial ethological knowledge paid little attention to neural mechanisms of behavior, and marginal attention to mammalian vocalization (Hinde, 1966; Eibl-Eibesfeldt, 1970; Immelmann, 1980). As mentioned previously, studies of the role of the central nervous system in the process of communication are essential for understanding this phenomenon, and to shape our understanding of vocalization as an ethotransmitter.

An ethotransmitter produced during mammalian social behavior should fulfill the following criteria:

1. It should have evolved to play an adaptive role in regulating behavior, particularly in highly social animals;
2. It should be produced and released by specialized tissues and organs;
3. Production and emission of the transmitter should be controlled by the central nervous system;
4. It should be predominantly targeted at other organisms of the same species, and emission of the ethotransmitter should occur in a way that maximizes its reception and behavioral effects on the recipients;
5. It should have characteristic structure suitable for transmission from one organism to another;
6. It should be selectively recognized by receptive mechanisms of the recipient organisms;
7. It should have properties of activating the central nervous system of the recipient and amplifying its responses (both in the sense of behavioral excitation or inhibition);
8. It should have the capability of inducing differential responses in the recipients, depending on the transmitter structure and mode of release.

Mammalian vocalization fulfills all of these criteria. The *Handbook of Mammalian Vocalization* is organized in such a way that successive sections of the book discuss evidence for one or more of the points listed above. The main sections of the handbook deal with the evolution and variability of vocalizations, production and emission, transmission and reception of calls, as well as responses of the recipients, mechanisms of generation of amplified responses, generation of sound and semiotic content of emitted vocalizations.

III. Overview of the content of this handbook

The content of the handbook is organized in ten sections, with 4–6 individual chapters in each section. The order of sections roughly follows the criteria listed for an ethotransmitter, except for Section 1, which forms an introduction.

III.A. Section 2

Section 2 provides exemplary studies of evolution of the vocal system and vocalization. The larynx is the chief organ for the production of vocalizations in mammals, and the evolution of its specialized muscles is described by Joseph Hoh. Some other aspects of larynx evolution are also mentioned in Section 10, dealing with the mechanism of sound production. Parallel to the evolution of the components needed for the physical production of vocalizations, a complex neural control needed to evolve. An overview of the core neural system in the mammalian brain responsible for such control is outlined by John Newman. The evolution of particular mammalian groups may differ significantly, depending on the environment in which the organisms live, their biological adaptations and social organization. Rodents represent an interesting group which has developed a fast reproduction cycle with infants being born at a relatively undeveloped (altricial) stage. It is interesting that these altricial infants can communicate by ultrasonic vocalization with their mother. Myron Hofer discusses the evolution of infant rat separation calls and their adaptive role. Other details of infant rat vocalizations and their adaptations are provided in Section 6. As another example, Alanna Maltby and co-workers from Gareth Jones' laboratory explain the evolutionary origin of bat echolocation calls. Other aspects of evolution and specific functions of the larynx are provided in Section 10.

III.B. Section 3

Mammalian vocalizations have evolved in many directions. Examples of this diversity are provided in Section 3. One general rule is that the sound frequency range of basic vocalizations produced by vibration of the vocal folds remains in proportion to the body mass of the vocalizing organism. The larger the body mass, the lower the sound frequency. This relationship is described by Neville Fletcher. In general, mammalian vocalizations may range from infrasounds to ultrasounds (i.e., below and above human hearing range). While infrasounds evolved for long-distance communication, ultrasonic calls serve for short-range communication. This is discussed in chapters written by Michael Garstang about elephant vocalizations and by Neville Fletcher with my coauthorship about rat vocalizations. These chapters explain how vocalizations are structured and emitted in order to maximize their reception by other members of the species. The largest number of mammalian species is represented by rodents. Almost all of them have developed an effective system of ultrasonic communication with a considerable acoustic repertoire. Examples of a very high variability of these calls are described in two chapters, by Gillian Sales for adult wild and wild-type rodent species, and by Maria Luisa Scattoni and Igor Branchi for mouse infants of different laboratory strains.

III.C. Section 4

One of the criteria listed for an ethotransmitter is its selective recognition by recipients. Section 4 summarizes evidence that the mammalian brain has mechanisms responsible for the selective recognition of species-specific vocalizations. The selective perception and recognition of vocal signals are summarized by Günter Ehret and Simone Kurt. Selective responses to vocal signals may be found at different levels of the central nervous system, from subcortical regions to the neocortex. Markus Wöhr and Rainer Schwarting, working on rats, report that although the semiotic content of calls may require learning, there is an innate preparedness to do so. Josef Syka describes activation of other subcortical structures in guinea pigs, and concludes that the detection of species-specific vocalizations is not based on "call detectors" but on encoding the spectrotemporal acoustic patterns of vocalization by specialized circuits. Finally, Christopher Petkov and his colleagues, from the laboratory of Nikos Logothetis, describe a specific cortical region in the primate brain which is activated by species-specific vocalizations and is homologous to the human voice recognition region.

III.D. Section 5

Section 5 presents evidence that species-specific vocalizations activate the central nervous system and behavior. Yoav Litvin, from the laboratories of Caroline Blanchard and Robert Blanchard, describes the initiation of specific defensive responses of rats to alarm vocalizations. Markus Wöhr and colleagues, working in the laboratory of Francesca D'Amato, discuss the effects of altricial rodent pup vocalizations on the behavior of their mother. A subsequent chapter provides further evidence that species-specific vocalizations are effective triggers of affective behavior in recipients, as reported by Koji Kuraoka and Katsuki Nakamura in the monkey. Finally, vocalizations may be used by animals as tools influencing affect and behavior of others, as described by Drew Rendall and Michael Owren. The human brain is not different in this respect. Sophie Scott and her colleagues present evidence that human non-verbal vocalization can activate human brain areas specifically involved in the perception of certain emotions.

III.E. Section 6

The following four sections provide details of the brain structures and systems responsible for the initiation of vocalizations. Section 6 reviews experimental evidence suggesting that the vast majority, if not all, of mammalian vocalizations are generated from emotional causes by limbic structures of the brain. Jaak Panksepp argues that vocalizations represent basic emotional states and are the consequences of social-affective interactions. Vocalizations signaling affective states are reported in rats by Jeffrey Burgdorf and Joseph Moskal, who describe a specific subtype of 50 kHz calls as an index of positive affective state in this species. The expression of a defined affective state is also reported for vocalizations of prosimian mouse lemurs by Elke Zimmermann. There is also significant evidence that altricial rat infants may be able to express their infantile emotionality by emitting isolation calls, as reviewed by Jeff Muller from the laboratory of Susan Brunelli and Harry Shair.

III.F. Section 7

Chapters in Section 7 are concentrated on mechanisms within the brain structures associated with functions of the hypothalamus and related diencephalic and mesencephalic limbic regions in the initiation of species-specific vocalizations. The circuits responsible for the generation of vocalization characteristics of the so-called "rage response," with an integrated pattern of somatic, autonomic and endocrine components in cats, are summarized by Allan Siegel and his co-workers. The phylogenetic shift in the neural regulation of the autonomic component has consequences for the production of vocalizations. This issue is viewed from the perspective of the Polyvagal Theory by Steven Porges and Gregory Lewis. The hypothalamic control of pain vocalization and the affective dimension of pain are described by George Borszcz and Catherine Spuz, while structures of the brain activated during electrically-induced pain-like vocalizations from the brainstem are described by Francisco Gonzalez-Lima. The hypothalamus is also associated with functions of the rhinencephalon and endocrine system. John Nyby reviews results of studies of male mouse ultrasonic vocalization appearing as part of the multisensory communication in reproductive behavior, and provides evidence for hormonal and pheromonal regulation of mouse behavior and calling.

Hypothalamic and limbic structures of the brain are controlled by the ascending reticular activating system, which is capable of changing the state of the organism and initiating affective states with concomitant vocalization. These ascending systems may be viewed as the true initiators of affective states in mammals. One of the tegmental ascending projections, the ascending mesolimbic cholinergic system, and the induction of aversive vocalizations is reviewed and discussed on the basis of pharmacological studies in cats and rats by Stefan Brudzynski.

III.G. Sections 8 and 9

Sections 8 and 9 provide neuroanatomical and cellular evidence for the functions of the brain stem neurons responsible for activation of the central pattern generators for vocalization, orchestrating this activity with other relevant ongoing functions (e.g., respiration, swallowing), and activating the output to specific muscles in the larynx and other associated structures for the production of calls. Eva Gruber-Dujardin describes the role of the periaqueductal gray in the triggering and gating expression of vocalization, while Steffen Hage discusses localization of the central pattern generators for vocalization in the pontine reticular formation and the whole network of descending connections for vocalization to the nucleus ambiguus and retroambiguus as the motor nuclei for the laryngeal muscles. Steven Barlow and his colleagues review the central pattern generation of orofacial movements, closely associated with vocal functions.

Other chapters (in Section 9) provide descriptions of the integrated functions of the larynx in breathing, vocalization, and airway protective reflexes (e.g., coughing and sneezing) by Keisuke Shiba. The vocal–respiratory interactions in the parabrachial nucleus are further reported by Michael Smotherman and his colleagues. Motor output for vocalization is also modulated by audio–vocal feedback interactions, which are discussed by Hanjun Liu and colleagues from the laboratory of Charles Larson. Vocal control of bat echolocation signals developed some additional neural regulatory adaptations, which are reflected in the organization of the motor system for bat echolocation signals, as reported by Walter Metzner and Gerd Schuller.

III.H. Section 10

Specific neural output from the brainstem structures activates an intricate system of muscles within the larynx for the generation of sound. Section 10 is devoted to this topic. Gerald Berke and Jennifer Long describe functions of the mammalian larynx in the production of sound, while Camille Finck and Lionel Lejeune provide detailed evidence that vocal folds have a specific histological and macromolecular structure which supports their oscillatory function.

Modifications of the acoustic parameters of vocalizations produced by the larynx may occur by means other than direct regulation of the laryngeal musculature. Roland Frey and Alban Gebler describe a broad range of mechanisms responsible for the modification of the vocal outcome in roaring-like vocalizations in many mammalian species. Finally, Joy Reidenberg and Jeffery Laitman describe, in a unique chapter, a diverse array of laryngeal and extralaryngeal structures for production of vocalizations in marine mammals. Of particular interest are mechanisms which allow underwater vocalization without releasing air, by a mechanism involving internal air reservoirs which can capture airflow from the lungs during vocalization.

III.I. Section 11

The last section, Section 11, is devoted to development of semiotic codes and complex features conveying different categories of information (e.g., motivational, referential) in mammalian vocalizations, and the transition to the semantics of human language. Eva Eberl discusses a close relationship between gestural and vocal communication in monkeys. Marta Manser summarizes information about functionally referential and motivational vocal signals, while Yale Cohen and co-workers present a summary of studies of auditory categories in non-human primates. These studies provide evidence that differential responses of recipients depend on the structure, and hence semiotic content, of the vocalizations. The recognition of individuals within a social group based on their signature vocalizations is reviewed by Ari Shapiro. Finally, Klaus Zuberbühler reviews the evolutionary development of the semiotic content of vocalizations and the semantics of human language.

IV. Summary

All of the material accumulated in this handbook provides convincing evidence that mammalian vocal communication represents a particularly well-developed system, in which vocalization fulfills all the criteria of an ethotransmitter. The production of vocalization in mammals is not an accidental or trivial activity without significant consequences for receivers. The importance of the generation of vocalizations may be illustrated by the fact that the vocalizing animal needs to modify its respiration, and thus oxygen supply, and as a consequence modify its heart rate and autonomic functions, just to be able to emit a call. Vocalizations play an important role in mammalian social behavior which is, in turn, regulated to a large degree by vocal signals. The evolution of the larynx and its highly complex neural regulation, and the evolution of complex forms of vocalizations with the capacity to deeply affect and modify the functions of individual organisms and even groups of organisms, strongly suggests that vocalization as a signaling entity may be treated as ethotransmitter, i.e., a signal transmitted for a distance from organism to organism with biologically significant consequences observed at all levels, from the cellular level, through the organ level, to the organismal level. The important point is that an ethotransmitter is capable of causing neurophysiological and neurobehavioral responses in receivers. In other words, it is a biologically active and neuroactive stimulus which is transmitted by air or water between organisms. Although there are a large number of signaling agents with similar properties (e.g., pheromonal or scent molecules, light emission, electric discharges, changes of coloration, visual displays, etc.), none of them has reached such a level of sophistication, none of them was subjected to such a degree of evolutionary modifications, and none of them progressed in evolution to human language. Although the evidence suggests that mammalian vocalizations are true ethotransmitters, the question of what other signaling agents may be, and may not be, included into this category is a matter for further research, discussion and definition.

References

Darwin, C., 1872/1998. The Expression of the Emotions in Man and Animals, 3rd edn. Oxford University Press, New York, NY.

Eibl-Eibesfeldt, I., 1970. Ethology. The Biology of Behavior. Holt, Rinehart and Winston, New York, NY.

Hinde, R.A., 1966. Animal Behavior. A Synthesis of Ethology and Comparative Psychology. McGraw-Hill Book Company, New York, NY.

Immelmann, K., 1980. Introduction to Ethology. Plenum Press, New York, NY.

Lorenz, K.Z., 1935. Der Kumpan in der Umwelt des Vogels. J. Ornithol. 83, 137–214.

Tinbergen, N., 1951. The Study of Instinct. Clarendon Press, Oxford, UK.

Uexküll von, J., 1934. Streifzüge durch die Umwelten von Tieren und Menschen. Springer, Berlin, Germany.

SECTION 2

Evolution of Vocal System and Vocalization

The mammalian larynx evolved initially as a protective valve for the respiratory system and then as a chief organ with specialized musculature for production of vocalizations. Larynx, together with the brain core structures controlling its function, forms the mammalian communication system.

CHAPTER 2.1

Laryngeal muscles as highly specialized organs in airway protection, respiration and phonation

Joseph F.Y. Hoh*

Discipline of Physiology and the Bosch Institute, School of Medical Sciences, Sydney Medical School, The University of Sydney, NSW, Australia

Abstract: Laryngeal muscles are a distinct allotype, differing from other muscles with respect to embryological origin and initiators (upstream activators) of myogenesis. They are endowed with a capacity to express a wider range of myosin heavy chains (MyHCs) than limb muscles. This includes extraocular (EO or 2L), 2B, 2X, 2A and slow MyHCs, generating fibers with decreasing speeds of contraction. MyHC expression and fiber-type profiles in laryngeal muscles show within-species and between-species variations. Within species, the vocal fold adductor, thyroarytenoid muscle (Ta), including the vocalis division that modulates vocal fold function, has faster fibers than the posterior cricoarytenoid muscle (PCA), the abductor, which has faster fibers than the vocal fold tensor, cricothyroid muscle (CT). Between species, small animals have faster muscle fibers than larger animals: EO MyHC is expressed in the Ta of rat and rabbit, but not in cat, dog, baboon, human, horse or cattle, while 2B MyHC is expressed in laryngeal muscles of cat and dog, but not their limb muscles. In the baboon, human, horse and cattle, only fibers expressing 2X, 2A and slow MyHCs are found. The CT only has fiber types found in limb muscles of the same species. Scaling considerations suggest that between-species variations in laryngeal muscle fiber types are evolutionary adaptations in response to changes in body mass and respiratory frequency. Within-species variations in fiber types ensure that protective closure of the glottis is always faster than movements regulating airflow during respiration. In common with muscles of other allotypes, laryngeal muscles are subjected to neural and hormonal regulation.

Keywords: larynx; laryngeal muscle; myosin heavy chain; muscle fiber types; muscle allotype; contraction; vocalization; respiration; scaling; comparative physiology

Abbreviations: CT: cricothyroid muscle; EO: extraocular (myosin heavy chain); eo: extraocular (muscle fiber type); IA: interarytenoid muscle; LCA: lateral cricoarytenoid muscle; MyHC: myosin heavy chain; PCA: posterior cricoarytenoid muscle; SDS-PAGE: sodium dodecyl sulfate polyacrylamide gel electrophoresis; Ta: thyroarytenoid muscle; Ta-V: vocalis division of thyroarytenoid muscle; Ta-X: external division of thyroarytenoid muscle

I. The laryngeal muscles and their physiological functions

Laryngeal muscles and their functions have been reviewed recently (Hoh, 2005), and a brief overview is presented here. In mammals there are five intrinsic laryngeal muscles: (1) thyroarytenoid (Ta); (2) lateral cricoarytenoid (LCA); (3) interarytenoid (IA); (4) posterior cricoarytenoid (PCA); and (5) cricothyroid (CT). The CT is innervated by the superior laryngeal nerve, while all the others are innervated by the recurrent laryngeal nerve. These muscles move the vocal fold to mediate three types of functions: airway protection; respiratory control; and phonation. The laryngeal muscles are classically divided into three functional groups: adductors of the vocal fold, comprising Ta, LCA, and IA, which close the glottis; the abductor, comprising of only PCA, which opens the glottis; and the vocal fold tensor, the CT. The Ta is comprised of two divisions: the external division (Ta-X), which adducts the vocal fold; and the vocalis division (Ta-V), which modulates sound quality.

*Corresponding author. E-mail: joeh@physiol.usyd.edu.au

Stefan M. Brudzynski (Ed.)
Handbook of Mammalian Vocalization
ISBN 978-0-12-374593-4

DOI: 10.1016/B978-0-12-374593-4.00002-4

Airway protection is phylogenetically the oldest and the most important function of the larynx. This involves coughing, sneezing and closing the glottis during deglutition. High speed of contraction is probably of paramount importance in this function.

Laryngeal muscles control respiration by adjusting the resistance to airflow to match respiratory frequency. The PCA dilates the glottis during inspiration to reduce the load on respiratory muscles, while the adductors contract during expiration to regulate expiratory airflow. During deep breathing, the CT increases the anteroposterior diameter of the larynx, thus increasing the size of the glottal airway.

Vocalization involves complex deployment of the various laryngeal and respiratory muscles. Sustained adduction of the vocal folds is necessary to build up subglottic pressure for vocalization, while controlling the pitch and amplitude of the voice requires varying the vibrating mass and the stiffness of the vocal folds. The vertical compartment of the PCA helps to stabilize the position of the vocal process during phonation. The Ta-V is a specialized muscle for vocalization. It lies subjacent to the vocal ligament, and its contraction increases the tension and stiffness of the non-ligamentous portion of the vocal fold. Contraction of Ta-V fibers may reduce the mass of the vocal fold taking part in its vibration during vocalization, or modulate the viscoelastic properties of the vibrating vocal fold. In conjunction with the CT, which tightens the vocal fold, Ta-V is important in modulating sound frequency and sound amplitude during phonation.

II. Molecular physiology of myosins

The molecular basis of muscle contraction is the interaction of myosin and actin, whereby chemical energy stored in ATP is transduced into mechanical energy as ATP is hydrolyzed. Myosin plays a dominant role in controlling the kinetics of this process, and consequently the dynamic properties of muscle fibers. The maximal speed of shortening of a fiber is correlated with the ATPase activity of its myosin (Bárány, 1967). Structurally, myosin is a hexamer consisting of a pair of heavy chains (MyHCs) and two pairs of light chains. Ten major striated muscle MyHC genes exist in the mammalian genome (Schiaffino and Reggiani, 1996; Qin et al., 2002). These encode MyHC isoforms that encompass a wide range of kinetic characteristics. By transducing energy from ATP at different rates, these MyHCs control the speed and power of muscle contraction. The selective expression of MyHCs and other genes in different muscles enables organisms to meet the wide range of functional demands on muscles in different parts of the body. For example, in limb muscles, fibers expressing slow, 2A, 2X and 2B MyHCs have increasing speeds and power, in the order listed (Bottinelli et al., 1991), while expression of EO MyHC in the extraocular muscle makes it faster than any limb muscle (Asmussen et al., 1994). Thus, MyHC composition is advantageously used for the classification of muscle fibers, as the mechanical profile of a muscle is reflected in its fiber type composition. In this classification system, fibers expressing slow, 2A, 2X, 2B and EO MyHCs are respectively designated as slow, 2a, 2x, 2b and eo fibers. Within species, these fibers have increasing speed of contraction in the listed order.

The functional unit of a muscle is the motor unit, defined structurally as a motoneuron together with all the muscle fibers innervated by it. Motor units in laryngeal muscles are characteristically small, containing few fibers and producing small forces. Studies in rats (Hinrichsen and Ryan, 1982) and humans (Neto and Marques, 2008) showed that the CT motor units had an average of 8 and 20 fibers in these species, respectively, while those in other laryngeal muscles had 4 and 10 fibers in rats and humans, respectively. In contrast, slow and 2a motor units in the rat soleus muscle have, on average, 110 and 142 fibers, respectively (Chamberlain and Lewis, 1989). Laryngeal muscles are thus capable of extremely fine-tuned movements.

III. Laryngeal muscles as a unique allotype

While limb and trunk muscles are used mainly for locomotion and postural maintenance, head and neck muscles are used for visceral functions, such as airway protection, mastication and moving the eyes. Each muscle group has evolved characteristics appropriate to its specific function. Laryngeal, extraocular and masticatory muscles differ from limb and trunk muscles in their capacities to express specific isoforms of MyHC. Viewed across different species, laryngeal muscles do express MyHC isoforms found in limb muscles, but in some species may also express other isoforms, such as the EO isoform, which is otherwise specific to extraocular muscles (DelGaudio et al., 1995; Lucas et al., 1995; Briggs and Schachat, 2000). The neonatal MyHC, normally expressed only during myogenesis, has recently been reported in cattle laryngeal muscles (Toniolo et al., 2005). The pattern of MyHC expression and other distinctive qualitative differences between muscle groups characterize their muscle allotype (Hoh, 1991; Lucas et al., 1995; Fraterman et al., 2007). Table 1 lists the known range of MyHC isoforms potentially expressed in adult limb, laryngeal, masticatory and extraocular muscle allotypes across various species.

Muscles of different allotypes also differ in the pattern of expression of many other genes than those for MyHCs (Sciote et al., 2003; Fraterman et al., 2007), and are restricted in phenotypic options when subjected to neural control (Hoh, 1991). Extraocular and laryngeal muscle fibers are more richly endowed than limb muscles with satellite cells, which play prominent roles in muscle growth, hypertrophy and regeneration (McLoon et al., 2007). Further, the facts that laryngeal muscles in humans are less severely affected than limb muscles in the muscular dystrophies (Manolides and Baloyannis, 1981), and that laryngeal muscles except the CT are spared in the mouse model of Duchenne muscular dystrophy (mdx mouse, Marques et al., 2007), also reflect allotypic differences between laryngeal and limb muscles. The cellular basis for the different muscle allotypes appears to be of the diverse developmental origins of their myogenic precursor cells, and the signaling pathways they go through to become myoblasts (Hoh, 2005).

IV. Laryngeal muscle fiber types

Laryngeal muscle fibers were thought to be indistinguishable from limb muscle fibers until the Ta in rabbits (Lucas et al., 1995) and rats (DelGaudio et al., 1995) were found to express EO or 2L MyHC. The more recent finding in cattle Ta of neonatal MyHC (Toniolo et al., 2005), another isoform expressed in adult extraocular but not limb muscle fibers, serves to distinguish laryngeal muscles from limb muscles in this species. Immunohistochemical evidence for fibers expressing slow-tonic MyHC, also expressed in extraocular muscle, was reported to occur in the human vocalis division of Ta (Han et al., 1999), but this was not confirmed in a more rigorous study (Sokoloff et al., 2007).

The presence of EO MyHC in laryngeal muscle myosin is associated with a higher ATPase activity than limb fast muscle myosin (Syrovy and Gutmann, 1971). Laryngeal muscle expressing EO MyHC is expected to contract at a higher speed than limb muscle in the same species, according to the Bárány relationship, but no direct comparison has been made. Although mechanical analysis of single laryngeal muscle fibers containing EO MyHC has not been reported, such analysis on single extraocular muscle fibers of the rabbit showed higher rates of cross-bridge kinetics compared with rabbit limb fast fibers (Li et al., 2000). Analysis of human vocalis fibers has indicated that laryngeal fibers containing the same MyHC as limb fibers contract at the same speed (D'Antona et al., 2002).

Fiber types in laryngeal muscles have been studied by MyHC immunohistochemistry and single fiber MyHC SDS-PAGE. Results from a number of species revealed wide variations in fiber types between laryngeal muscles of different function within species, and between homologous muscles of different species (Wu et al., 2000a,b,c; Rhee et al., 2004; Rhee and Hoh 2008; Rhee et al., 2009). This is illustrated by the fiber type distributions of Ta-X and Ta-V

Table 1. Myosin heavy chain isoforms potentially expressed in adult muscles of different allotypes across various species

Myosin heavy chain isoform	Limb/trunk allotype	Laryngeal allotype	Masticatory allotype	Extraocular allotype
EO (2L)		X		X
2B	X	X	X	X
2X	X	X	X	X
2A	X	X	X	X
Masticatory			X	
Neonatal		X	X	X
Embryonic				X
α-cardiac			X	X
β-cardiac (slow)	X	X	X	X
Slow-tonic				X

An "X" indicates that the isoform in question is expressed within the allotype in some species. Within an allotype, the isoform(s) actually expressed can vary between muscles, and between homologous muscles in different species.

(Table 2) and of PCA and CT (Table 3) in rat, rabbit, cat and baboon. Within species, the Ta, an adductor, has faster fiber type profiles than the abductor, PCA, which has faster fibers than the vocal fold tensor, the CT. Between species, a given laryngeal muscle generally has a faster fiber type profile in smaller animals than in larger ones. Thus, variations between muscles within species appear to be related to their specific functional roles, while variations between species appear to be influenced by body mass. A considerable proportion of laryngeal fibers in some species expressed multiple MyHC isoforms in the same fiber (Wu et al., 1998, 2000c). This is not a universal feature of laryngeal muscles (Rhee and Hoh, 2008), but may be associated with the diverse functional roles each laryngeal muscle is called upon to play (Hoh, 2005).

IV.A. Fiber types in Ta and PCA muscles and their functional significance

For all species studied, Ta-X has the fastest fiber type or MyHC profile among laryngeal muscles, but what constitutes the fastest fiber type varies with the species. For rat (Wu et al., 2000a; Rhee et al., 2004) and

Table 2. Distribution of fiber types in Ta-X and Ta-V (vocalis) muscles of the rat, rabbit, cat, and baboon

Muscle fiber type	Rat		Rabbit		Cat		Baboon	
	Ta-X	Ta-V	Ta-X	Ta-V	Ta-X	Ta-V	Ta-X	Ta-V
Slow			0.6	7.4	4.8	19.0	18.9	58.0
Slow/2a						0.3	0.9	1.5
2a				1.6	2.3	27.0	45.9	20.5
2a/2x				0	0	1.5	4.3	4.1
2x		50.5		61.0	14.5	51.9	30.0	15.9
2x/2b		5.7		1.0	0.4	0.3		
2b		4.3		0.2	78			
2x/2b/eo		0.8	9.9	5.2				
2b/eo	100.0	37.0	89.5	3.3				
eo				19.0				

Data for the rat are from Rhee et al. (2004) and those for other animals are from Rhee and Hoh (2008). Fiber types are classified according to their MyHC composition, e.g., those expressing 2B/EO MyHCs are designated 2b/eo. They are listed in the order of increasing speed of contraction. Note that pure eo fibers are present only in rabbit Ta-V. Values are percentages of total fibers. For rat Ta-V, values are averages for rostral and caudal regions of the muscle.

Table 3. Distribution of fiber types in PCA and CT muscles of the rat, rabbit, cat, and baboon

Muscle fiber type	Rat		Rabbit		Cat		Baboon	
	PCA	CT	PCA	CT	PCA	CT	PCA	CT
Slow	7.4	19.0	36.0	23.9	29.8	38.7	42.5	43.0
Slow/2a			0.1	0.9	0.7	0.8	0.3	11.2
2a	16.6	12.5	21.6	34.2	44.7	60.5	31.6	45.8
2a/2x	0.6	0.3	0.7	0.5	20.7		3.3	
2x	29.0	61.2	23.1	40.5	3.7		22.3	
2x/2b		2.1	18.5		0.4			
2b	44.6	4.9						
2x/2b/eo								
2b/eo	1.9							

Data for the rat are from Rhee et al. (2004) and those for other animals are from Rhee and Hoh (2008). Fiber types are classified according to their MyHC composition, and are listed in the order of increasing speed. Values are percentages of total fibers.

rabbit (Rhee and Hoh, 2008), the fastest fibers express EO MyHC, whereas for cat (Rhee and Hoh, 2008) and dog (Wu et al., 2000c), the fastest fibers express 2B MyHC. In baboon (Rhee and Hoh, 2008), human (Wu et al., 2000b; Li et al., 2004) and horse (Rhee et al., 2009), however, the fastest fibers express 2X-MyHC. There is thus a shift towards a slower fiber type or MyHC profile with increasing body size, a feature also seen in the PCA and the CT.

Tables 2 and 3 show that, in the rat and rabbit, the vast majority of fibers in the Ta-X are 2b/eo, co-expressing EO and 2B MyHCs, while EO MyHC expression is virtually absent in the PCA. In the cat, 2B MyHC is strongly expressed in the Ta, but virtually absent in the PCA. Furthermore, in most species, the fiber type profile of the PCA is faster than that of the CT. These immunohistochemical results are confirmed by analysis of MyHC composition in the dog laryngeal muscles (Wu et al., 2000c). In the human, only adductors express 2X MyHC, the fastest isoform expressed in laryngeal muscles in this species (Li et al., 2004). Consistent with these observations, the contraction times of the isometric twitch for the PCA in several species are intermediate between those of the Ta and the CT (Hoh, 2005).

The high speed of contraction for the adductor Ta relative to the abductor PCA and the CT in the same species may have a survival value to the animal, by preventing foreign bodies from entering the lungs during inspiration. To be effective, the adductor needs to be faster relative to the abductor, which is active during inspiration (Green and Neil, 1955). The higher speed of the adductor may represent an evolutionary adaptation to ensure that protective closure of the glottis is always faster than movements regulating airflow during respiration.

The Ta-V in each species has a slower and wider fiber type profile than the Ta-X (Table 2). The Ta-V muscles in rats (Rhee et al., 2004) and rabbits (Rhee and Hoh, 2008) have substantial proportions of slower fibers (2x, 2a and slow) in contrast to the high content of 2b/eo fibers in the Ta-X. The broader range of fiber types in the Ta-V is also evident in the dog, in which this muscle is anatomically less well-defined. The region corresponding to the Ta-V is rich in 2x, 2a and slow fibers, whereas the rest of the muscle is rich in the faster 2x and 2b fibers (Bergrin et al., 2006). The slower and wider range of fiber types may enable the vocalis to modulate the mechanical properties of the vocal fold, and hence the sound quality, over a wide range of characteristics.

IV.B. Fiber types in the CT and their functional significance

Studies on the CT in the rat (Wu et al., 2000a; Rhee et al., 2004), human (Wu et al., 2000b), rabbit, cat and baboon (Rhee and Hoh, 2008) are in general agreement that this muscle expresses only isoforms of MyHC found in their limb muscles. EO MyHC, absent in rat and rabbit limb muscles, is also absent in rat and rabbit CT, and 2B MyHC, absent in cat limb muscle (Lucas et al., 2000), is also absent in cat CT. However, dog CT was reported to have 2% of fibers expressing 2B MyHC (Wu et al., 2000c), whereas dog limb muscles do not express this isoform (Latorre et al., 1993). This exceptional finding in so few fibers needs confirmation. Functionally consistent with the general similarity in fiber type profiles in CT and limb fast muscles, the isometric contraction time of the CT in several species is similar in value to that of fast limb muscle of the same species (Hoh, 2005).

The difference in the pattern of MyHC expression between the CT and other laryngeal muscles, together with their different developmental origins and innervation, suggests that they may have allotypic differences. The fact that the CT is not spared in the dystrophic mdx mouse (Marques et al., 2007) supports this notion. However, baboon (Rhee and Hoh, 2008) and human (Li et al., 2004) Ta and PCA muscles do not express EO or 2B MyHC, making them indistinguishable from limb muscle when MyHC expression is used as a criterion for allotypic differences. An alternative hypothesis for the pattern of MyHC expression in the CT is simply that an abundance of the slowest fibers is necessary in all species to support the functions of the CT, including the development of sustained tension in the vocal fold without fatigue during phonation.

Increasing body mass also shifts the CT towards slower fiber types. Type 2b fibers are present only in the rat CT, and 2x fibers, which are abundant in the rat and rabbit CT, are absent in cat and baboon CT. The proportion of slow fibers is smallest in the rat CT and highest in the baboon CT (Table 3).

V. Phylogenetic plasticity (adaptability) of laryngeal muscles and functional significance

Phylogenetic plasticity refers to adaptive changes in muscle properties in response to changes in functional requirements during phylogeny. Variations in fiber types between homologous laryngeal muscles in animals of different body size are of this kind. Interspecies variations in contractile properties of a muscle can be brought about through two distinct mechanisms: changing the ATPase activity of its MyHC isoforms by mutation; and changing its fiber type or MyHC profile. Both these mechanisms are used in limb (Pellegrino et al., 2003) and cardiac (Hoh et al., 2007) muscles during phylogeny in response to changes in body mass.

In laryngeal muscles, the two mechanisms for changing contractile characteristics with body size are similarly employed, as shown in Tables 2 and 3. Changing isoforms are exemplified by the expression in laryngeal muscles of EO MyHC, the fastest MyHC isoform, only in small animals (rat, rabbit), and of 2B MyHC, the fastest among limb MyHC isoforms, in cat and dog, but not in the larger baboon, horse (Rhee et al., 2009) and human (Li et al., 2004). Changing fiber type profiles in laryngeal muscles with body size is also seen in Tables 2 and 3. Slow and 2a fibers are more abundant in larger animals. Earlier literature also showed that the percentages of slow fibers in Ta, PCA and CT muscles in 11 species were low in small but high in large animals (Table 2, Hoh, 2005). The MyHC profiles of these muscles in rat, dog and human showed a shift towards faster isoforms in smaller animals (Table 3, Hoh, 2005).

It has been hypothesized that the variations in fiber types in laryngeal muscles between species represent an adaptation to optimize the control of airway resistance in relation to the respiratory frequency and metabolic rate of the animal (Rhee and Hoh, 2008). The basal metabolic rate of eutherian and marsupial mammals is strongly influenced by body mass, being scaled approximately to (body mass)$^{0.75}$ (Kleiber, 1961; Dawson and Hulbert, 1970; Schmidt-Nielson, 1984). It follows that specific metabolic rate, or metabolic rate per unit body mass, scales to (body mass)$^{-0.25}$. The supply of oxygen, and consequently the ventilation of the lungs, also needs to change proportionally. As lung volumes do not change as a fraction of body size (Tenney and Remmers, 1963), the respiratory system can cope with changes in specific metabolic rate if respiratory frequency is scaled to the same exponent. In close agreement, empirical data show that respiratory frequency scales to (body mass)$^{-0.26}$ (Schmidt-Nielson, 1984). This means that small animals should have higher respiration rates than large animals. Laryngeal muscles of small animals consequently have less time to regulate airflow during each breath, and should therefore be faster contracting. In conforming to this expectation, it has been shown (Cooper et al., 1994) that, for animals over a wide range of body mass values, the contraction time of the CT was a small, constant fraction of the duration of their respiratory cycle. As body mass decreases, laryngeal muscles must increase in speed by shifting their fiber type profile towards faster fibers, or expressing new, faster myosins, in conformity with the allometric equation relating respiratory frequency to body mass.

Since some MyHC isoforms expressed in laryngeal muscles are also expressed in cardiac (slow MyHC) and limb (slow, 2A, 2X, 2B MyHCs) muscles, the use of these massive muscles exerts a considerably greater impact on the energy economy of the animal. It is thus not surprising that the kinetic properties of each of these isoforms are greatly influenced by adaptations to the functional requirements of skeletal (Rome et al., 1990) and cardiac (Hoh et al., 2007) muscles in relation to body size. The exponent of the allometric equation relating speed of muscle fiber contraction to body mass depends on the fiber type or MyHC isoform, being -0.18 for slow muscle fibers (Rome et al., 1990), -0.175 for 2a fibers, -0.098 for 2x fibers and -0.048 for 2b fibers (Pellegrino et al., 2003). These values fall short of the -0.26 needed to meet requirements for respiratory control, a problem that is particularly acute for 2b fibers. The more moderate situation for slow fibers, with an exponent closer to -0.26, is likely due to adaptation of the slow MyHC to cardiac functional requirements (Hoh et al., 2007).

The above discussion implies that, as body mass decreases, the increase in speed of laryngeal fiber types achieved by changing myosin ATPase activity of each isoform alone in conformity with the allometric relation to body mass would not be fast enough for respiratory control. New and faster fiber types would thus be necessary. In animals the size of dogs and cats, 2x fibers, the fastest found in their limb muscles, are apparently no longer fast enough for respiratory control, thus requiring the expression of 2B MyHC in laryngeal muscles in these species. As body mass decreases further in rabbits and rats, even 2b fibers are apparently not fast enough, thus necessitating the coexpression of the fastest known MyHC isoform, EO MyHC, with 2B MyHC. With further decrease in body mass, slow and 2a fibers would be expected to disappear from laryngeal muscles, as in limb muscles of the shrew (Savolainen and Vornanen 1995), while pure eo fibers would predominate in all laryngeal muscles, with the possible exception of the CT, if it is different allotypically from other laryngeal muscles.

VI. Physiological plasticity of laryngeal muscles

Physiological plasticity refers to changes in muscle fiber type properties that occur during the lifetime of the animal. Extensive work on limb muscles has shown that muscle fibers show a considerable degree of plasticity, which enables their phenotypic properties to adapt to changes in functional load or hormonal state (Pette and Staron, 1997). Recent studies on laryngeal muscles reveal that they also show physiological plasticity, responding to neural and hormonal stimuli, just as limb muscles do (Hoh, 2005).

VI.A. Neural regulation

The neural hypothesis for regulating muscle fiber types, based primarily on studies done on limb muscle, states that different patterns of motor nerve activity promote selective changes in gene expression to establish the specialized characteristics of various fiber types. However, muscle fibers of a given allotype can only change within the phenotypic options available to the allotype (Hoh, 1991). Nerves supplying slow fibers carry sustained low frequency impulses, while those to fast fibers carry short bursts of high frequency. Stimulation of fast limb fibers with sustained low frequency impulses converts limb fast fibers into slow fibers via a calcineurin-dependent pathway (Chin et al., 1998; Tothova et al., 2006).

The earliest report suggesting that laryngeal muscles are subject to neural regulation came from the observation that the normal checkerboard pattern of fast and slow fibers in apparently normal horse laryngeal muscles was disrupted by the occurrence of patches of contiguous fibers of the same fiber type (Gunn, 1972). Such fiber type grouping was thought to arise from the denervation and reinnervation of muscle fibers due to recurrent laryngeal neuropathy, a common condition in horses (Rhee et al., 2009). Further support for neural regulation of laryngeal muscles was the observation that sustained low frequency stimulation of denervated sheep PCA muscle led to an increase in slow fibers in this muscle (Carraro et al., 1988). Work is needed to verify that this change also comes via a calcineurin-dependent pathway. Histochemical evidence showing that slow laryngeal fibers in the dog can undergo transformation into fast fibers was provided by cross-innervation of PCA and LCA muscles with the hypoglossal nerve (Paniello et al., 2001). The possibility that various subtypes of fast laryngeal muscle fibers are also under neural regulation was demonstrated by immunohistochemical analysis of the rat Ta-X, which is normally composed of purely 2b/eo fibers. Following recurrent laryngeal nerve section and reinnervation of laryngeal muscles, there was a progressive transformation of 20% of these 2b/eo fibers into pure 2x fibers, suggesting that nerve fibers normally innervating 2x muscle fibers cross-innervated 2b/eo fibers and transformed them into 2x fibers (Rhee et al., 2004). Although pure eo fibers are virtually absent in rat Ta, rabbit Ta-V contained 19% of these fibers (Table 2). The existence of such a high proportion of pure eo fibers shows that EO MyHC is not always coregulated with 2B MyHC in 2b/eo fibers, but that a distinct signal, presumably a unique impulse pattern, exists to cater for the exclusive expression of the EO MyHC.

The occurrence of neural regulation in laryngeal muscles implies that the variations in fiber type profiles between muscles within species is at least partly due to the impulses these muscles receive from their motor nuclei. It is believed that laryngeal fiber types would change in response to altered patterns of use, such as in voice training.

VI.B. Hormonal regulation

In limb muscles, hyperthyroidism shifts fiber types in the direction of fast fibers, while hypothyroidism has the reverse effect (Caiozzo and Haddad, 1996). Analysis of MyHC isoforms in rat laryngeal muscle fibers showed that modulation of the thyroid state also shifts isoform distribution in a similar way, the effects are seen primarily to involve changing the relative proportions of 2X and 2B isoforms (Wu et al., 2000a). There appears to be a surprising lack of literature on the effects of sex hormones on mammalian laryngeal muscles.

VII. Conclusions

In conclusion, laryngeal muscles are highly specialized organs for airway protection, respiratory control and vocalization. They represent a distinct muscle allotype with specialized fiber type characteristics.

Airway protection and respiratory control played dominant roles in molding their fiber types during evolution. Mammals have made use of these structures for vocalization and have further evolved a specialized compartment of the Ta muscle, the vocalis division (Ta-V), to enhance this function.

References

Asmussen, G., Beckers-Bleukx, G., Marechal, G., 1994. The force-velocity relation of the rabbit inferior oblique muscle: influence of temperature. Pflugers Arch. 426, 542–547.

Bárány, M., 1967. ATPase activity of myosin correlated with speed of muscle shortening. J. Gen. Physiol. 50 (Suppl), 197–218.

Bergrin, M., Bicer, S., Lucas, C.A., Reiser, P.J., 2006. Three-dimensional compartmentalization of myosin heavy chain and myosin light chain isoforms in dog thyroarytenoid muscle. Am. J. Physiol. Cell Physiol. 290, C1446–C1458.

Bottinelli, R., Schiaffino, S., Reggiani, C., 1991. Force-velocity relations and myosin heavy chain isoform compositions of skinned fibers from rat skeletal muscle. J. Physiol. 437, 655–672.

Briggs, M.M., Schachat, F., 2000. Early specialization of the superfast myosin in extraocular and laryngeal muscles. J. Exp. Biol. 203, 2485–2494.

Caiozzo, V.J., Haddad, F., 1996. Thyroid hormone: modulation of muscle structure, function, and adaptive responses to mechanical loading. Exer. Sport Sci. Rev. 24, 321–361.

Carraro, U., Catani, C., Saggin, L., Zrunek, M., Szabolcs, M., Gruber, H., Streinzer, W., Mayr, W., Thoma, H., 1988. Isomyosin changes after functional electrostimulation of denervated sheep muscle. Muscle Nerve 11, 1016–1028.

Chamberlain, S., Lewis, D.M., 1989. Contractile characteristics and innervation ratio of rat soleus motor units. J. Physiol. 412, 1–21.

Chin, E.R., Olson, E.N., Richardson, J.A., Yang, Q., Humphries, C., Shelton, J.M., Wu, H., Zhu, W., Bassel-Duby, R., Williams, R.S., 1998. A calcineurin-dependent transcriptional pathway controls skeletal muscle fiber type. Genes. Dev. 12, 2499–2509.

Cooper, D.S., Shindo, M., Sinha, U., Hast, M.H., Rice, D.H., 1994. Dynamic properties of the posterior cricoarytenoid muscle. Ann. Otol. Rhinol. Laryngol. 103, 937–944.

D'Antona, G., Megighian, A., Bortolotto, S., Pellegrino, M.A., Marchese-Ragona, R., Staffieri, A., Bottinelli, R., Reggiani, C., 2002. Contractile properties and myosin heavy chain isoform composition in single fiber of human laryngeal muscles. J. Muscle Res. Cell Motil. 23, 187–195.

Dawson, T.J., Hulbert, A.J., 1970. Standard metabolism, body temperature, and surface areas of Australian marsupials. Am. J. Physiol. 218, 1233–1238.

DelGaudio, J.M., Sciote, J.J., Carroll, W.R., Escalmado, R.M., 1995. Atypical myosin heavy chain in rat laryngeal muscle. Ann. Otol. Rhinol. Laryngol. 104, 237–245.

Fraterman, S., Zeiger, U., Khurana, T.S., Wilm, M., Rubinstein, N.A., 2007. Quantitative proteomics profiling of sarcomere associated proteins in limb and extraocular muscle allotypes. Mol. Cell Proteomics 6, 728–737.

Green, J.H., Neil, E., 1955. The respiratory function of the laryngeal muscles. J. Physiol. 129, 134–141.

Gunn, H.M., 1972. Histochemical observations on laryngeal skeletal muscle fibers in "normal" horses. Equine Vet. J. 4, 144–148.

Han, Y., Wang, J., Fischman, D.A., Biller, H.F., Sanders, I., 1999. Slow tonic muscle fibers in the thyroarytenoid muscles of human vocal folds; a possible specialization for speech. Anat. Rec. 256, 146–157.

Hinrichsen, C.F., Ryan, A., 1982. The size of motor units in laryngeal muscles of the rat. Experientia 38, 360–361.

Hoh, J.F.Y., 1991. Myogenic regulation of mammalian skeletal muscle fibers. News Physiol. Sci. 6, 1–6.

Hoh, J.F., 2005. Laryngeal muscle fiber types. Acta Physiol. Scand. 183, 133–149.

Hoh, J.F., Kim, Y., Lim, J.H., Sieber, L.G., Lucas, C.A., Zhong, W.W., 2007. Marsupial cardiac myosins are similar to those of eutherians in subunit composition and in the correlation of their expression with body size. J. Comp. Physiol. (B) 177, 153–163.

Kleiber, M., 1961. The Fire of Life. An Introduction to Animal Energetics. Wiley, New York, NY.

Latorre, R., Gil, F., Vazquez, J.M., Moreno, F., Mascarello, F., Ramirez, G., 1993. Skeletal muscle fiber types in the dog. J. Anat. 182, 329–337.

Li, Z.B., Lehar, M., Nakagawa, H., Hoh, J.F., Flint, P.W., 2004. Differential expression of myosin heavy chain isoforms between abductor and adductor muscles in the human larynx. Otolaryngol Head Neck Surg. 130, 217–222.

Li, Z.B., Rossmanith, G.H., Hoh, J.F., 2000. Cross-bridge kinetics of rabbit single extraocular and limb muscle fibers. Invest. Ophthalmol. Vis. Sci. 41, 3770–3774.

Lucas, C.A., Rughani, A., Hoh, J.F.Y., 1995. Expression of extraocular myosin heavy chain in rabbit laryngeal muscle. J. Muscle Res. Cell Motil. 16, 368–378.

Lucas, C.A., Kang, L.H., Hoh, J.F., 2000. Monospecific antibodies against the three mammalian fast limb myosin heavy chains. Biochem. Biophys. Res. Commun. 272, 303–308.

Manolides, L., Baloyannis, S., 1981. Histological and ultrastructural findings in the vocal muscles of patients suffering from muscular dystrophies. Arch. Otorhinolaryngol. 230, 181–188.

Marques, M.J., Ferretti, R., Vomero, V.U., Minatel, E., Neto, H.S., 2007. Intrinsic laryngeal muscles are spared from myonecrosis in the mdx mouse model of Duchenne muscular dystrophy. Muscle Nerve 35, 349–353.

McLoon, L.K., Thorstenson, K.M., Solomon, A., Lewis, M.P., 2007. Myogenic precursor cells in craniofacial muscles. Oral Dis. 13, 134–140.

Neto, H.S., Marques, M.J., 2008. Estimation of the number and size of motor units in intrinsic laryngeal muscles using morphometric methods. Clin. Anat. 21, 301–306.

Paniello, R.C., West, S.E., Lee, P., 2001. Laryngeal reinnervation with the hypoglossal nerve. I. Physiology, histochemistry, electromyography, and retrograde labeling in a canine model. Ann. Otol. Rhinol. Laryngol. 110, 532–542.

Pellegrino, M.A., Canepari, M., Rossi, R., D'Antona, G., Reggiani, C., Bottinelli, R., 2003. Orthologous myosin isoforms and scaling of shortening velocity with body size in mouse, rat, rabbit and human muscles. J. Physiol. 546, 677–689.

Pette, D., Staron, R.S., 1997. Mammalian skeletal muscle fiber type transitions. Int. Rev. Cytol. 170, 143–223.

Qin, H., Hsu, M.K., Morris, B.J., Hoh, J.F., 2002. A distinct subclass of mammalian striated myosins: structure and molecular evolution of "superfast" or masticatory myosin heavy chain. J. Mol. Evol. 55, 544–552.

Rhee, H.S., Hoh, J.F., 2008. Immunohistochemical analysis of myosin heavy chain expression in laryngeal muscles of the rabbit, cat, and baboon. J. Histochem. Cytochem. 56, 929–950.

Rhee, H.S., Lucas, C.A., Hoh, J.F., 2004. Fiber types in rat laryngeal muscles and their transformations after denervation and reinnervation. J. Histochem. Cytochem. 52, 581–590.

Rhee, H.S., Steel, C.M., Derksen, F.J., Robinson, N.E., Hoh, J.F., 2009. Immunohistochemical analysis of laryngeal muscles in normal horses and horses with subclinical recurrent laryngeal neuropathy. J. Histochem. Cytochem. 57, 787–800.

Rome, L.C., Sosnicki, A.A., Goble, D.O., 1990. Maximum velocity of shortening of 3 fiber types from horse soleus muscle – implications for scaling with body size. J. Physiol. 431, 173–185.

Savolainen, J., Vornanen, M., 1995. Fiber types and myosin heavy chain composition in muscles of common shrew (*Sorex araneus*). J. Exp. Zool. 271, 27–35.

Schiaffino, S., Reggiani, C., 1996. Molecular diversity of myofibrillar proteins: gene regulation and functional significance. Physiol. Rev. 76, 371–423.

Schmidt-Nielson, K., 1984. Scaling: Why Animal Size is So Important. Cambridge University Press, Cambridge, UK.

Sciote, J.J., Horton, M.J., Rowlerson, A.M., Link, J., 2003. Specialized cranial muscles: how different are they from limb and abdominal muscles? Cells Tissues Organs 174, 73–86.

Sokoloff, A.J., Li, H., Burkholder, T.J., 2007. Limited expression of slow tonic myosin heavy chain in human cranial muscles. Muscle Nerve 36, 183–189.

Syrovy, I., Gutmann, E., 1971. ATPase activity of two rabbit laryngeal muscles. Experientia 27, 248.

Tenney, S.M., Remmers, J.E., 1963. Comparative quantitative morphology of the mammalian lung: diffusing area. Nature 197, 54–56.

Toniolo, L., Maccatrozzo, L., Patruno, M., Caliaro, F., Mascarello, F., Reggiani, C., 2005. Expression of eight distinct MHC isoforms in bovine striated muscles: evidence for MHC-2B presence only in extraocular muscles. J. Exp. Biol. 208, 4243–4253.

Tothova, J., Blaauw, B., Pallafacchina, G., Rudolf, R., Argentini, C., Reggiani, C., Schiaffino, S., 2006. NFATc1 nucleocytoplasmic shuttling is controlled by nerve activity in skeletal muscle. J. Cell Sci. 119, 1604–1611.

Wu, Y.Z., Baker, M.J., Crumley, R.L., Blanks, R.H., Caiozzo, V.J., 1998. A new concept in laryngeal muscle: multiple myosin isoform types in single muscle fibers of the lateral cricoarytenoid. Otolaryngol. Head Neck Surg. 118, 86–94.

Wu, Y.Z., Baker, M.J., Crumley, R.L., Caiozzo, V.J., 2000a. Single-fiber myosin heavy-chain isoform composition of rodent laryngeal muscle: modulation by thyroid hormone. Arch. Otolaryngol. Head Neck Surg. 126, 874–880.

Wu, Y.Z., Crumley, R.L., Armstrong, W.B., Caiozzo, V.J., 2000b. New perspectives about human laryngeal muscle: single-fiber analyses and interspecies comparisons. Arch. Otolaryngol. Head Neck Surg. 126, 857–864.

Wu, Y.Z., Crumley, R.L., Caiozzo, V.J., 2000c. Are hybrid fibers a common motif of canine laryngeal muscles? Single-fiber analyses of myosin heavy-chain isoform composition. Arch. Otolaryngol. Head Neck Surg. 126, 865–873.

CHAPTER 2.2

Evolution of the communication brain in control of mammalian vocalization

John D. Newman*

Laboratory of Comparative Ethology, Eunice Kennedy Shriver National Institute of Child Health and Human Development, National Institutes of Health, Poolesville, MD, USA

Abstract: The evolution of the brain structures mediating vocal production in mammals is outlined. The structures together form a system. Central to this system is an ancient core that is found in all vertebrates. Mammals have added components in the cerebral cortex, consisting of a limbic midline cortex, as well as neocortical areas that influence laryngeal movement. Together, the elements of this system regulate the broad functional categories of vocalizations, along with the subtle nuances that reflect variation in motivational state, social context, and developmental status of the vocalizer. Despite the great variety of acoustic qualities and species differences found in mammalian vocalizations, the brain pathways and circuits regulating their expression appear to have changed very little over the course of evolution.

Keywords: evolution of vocal system; limbic system; cingulate gyrus; brainstem; periaqueductal gray

I. Introduction

Communication by sound is a widespread behavior found in all vertebrate classes. The neural pathways mediating auditory communication have been worked out to some extent in fish (Goodson and Bass, 2002; Bass and McKibben, 2003), anuran amphibians (Emerson and Boyd, 1999), reptiles (Kennedy, 1975, 1981), birds (Brown, 1971; Jarvis, 2007), and mammals (references in this chapter). In non-mammalian vertebrates, auditory communication serves, for the most part, in a limited set of contexts, largely centered round reproduction. It is true that many bird species, having established more complex social groupings, use sound communication in other contexts. However, with the exception of the pathways related to singing (which have been extensively studied), the exact neural pathways mediating vocalization in these other contexts are poorly known. In mammals, communicating by sound is associated with many social contexts and, while establishing the neural pathways mediating vocal communication is correspondingly more of a challenge, considerable progress has been made. The goal of this chapter is to provide a summary of the relevant brain structures from a comparative and evolutionary perspective. The evidence suggests that, despite the great variety of vocalizations across mammalian groups, there is a limited set of neural structures that appear to underlie their production.

One issue that separates mammalian vocalization from auditory communication in non-mammalian vertebrates is the role of emotions in activating and modulating mammalian vocal behavior. In fact, going back to Darwin, it has been assumed that emotions govern vocal expression in mammals. More recently, growing evidence suggests that cognition also plays a role in the vocal behavior of some mammals. Referential signals in non-human primates are one example (see Manser, Chapter 11.2 in this volume).

A number of conceptual models for the evolution of mammalian vocalization have been created. Among these is the "motivation–structural rules" model of Morton (1977). While based on considerable data, these conceptual frameworks generally leave out the underlying neural pathways responsible. It seems likely that broad schema of vocal behavior, particularly those based on shared acoustic structure, have common pathways in terms of vocal production.

*Corresponding author. E-mail: jdnewman@helix.nih.gov

ISBN 978-0-12-374593-4
DOI: 10.1016/B978-0-12-374593-4.00003-6

It may be, following Jürgens (2002), that a limited number of brainstem structures are responsible for all vocalizations, and that what differentiates the various classes of calls are differing circuits higher in the brain that are the basis of the differing motivation underlying a specific vocalization. There must still be differing subcircuits at some level of the brainstem circuitry that control production of specific calls, given that differing sequences of laryngeal actions and respiratory actions are required. Looking at the broad overview of vertebrate communication by sound, one can find a thread suggesting shared pathways across groups. Perhaps the most widely shared structures for sound production lie in the caudal brainstem (Bass et al., 2008). In the midbrain the same region, called the periaqueductal gray matter (PAG) in mammals, is important for sound production across a wide range of vertebrate classes. Of particular interest is the fact that this region is involved in sound communication regardless of the peripheral end organ responsible for making the sound. The piscine equivalent of the PAG is even found to be involved in communicating by an entirely different medium, namely weakly electrical signals, in certain fish (Heiligenberg, 1988).

II. Approaches to studying the neural basis of communication in mammals

One approach is to identify a widely shared vocal subtype, and examine the evidence for underlying neural substrates. Newman (2007) took this approach with respect to the infant isolation call ("infant cry") in tracing the evidence for a "mammalian cry circuit." A similar approach has been taken for certain rodent vocalizations, in particular the 50kHz vocalizations associated with juvenile play in rats (Burgdorf et al., 2007). Microchiropteran bats have come under study by neuroscientists, due to their well-developed systems of echolocation. While not, strictly speaking, communication, the vocalizations that develop into echolocation sounds have been shown to develop out of infant isolation calls. Also, the neural substrates mediating the production of certain echolocation sounds are intimately related to neural systems used to elaborate non-echolocation communication sounds (cf. Gooler and O'Neill, 1988; Schuller and Radtke-Schuller, 1988).

Sound-producing (non-seismic) motor behavior and brain mechanisms have evolved at least five different times over the course of vertebrate evolution. First, there is a central core pathway running from the preoptic area of the forebrain back through the PAG of the midbrain (torus semicircularis in non-mammalian vertebrates) to the motor nuclei of the caudal brainstem innervating the muscles responsible for sound production. Second, there are the specialized parts of the striatum in the forebrain that are involved in the production of song in songbirds. Third, there are the specialized regions of the brain, mainly in the cerebral cortex, underlying speech production. Fourth, there are the specialized areas responsible for echolocation in microchiropteran bats. And fifth, there is sound production through "phonic lips" by odontocete cetaceans. As this is a chapter in a volume mostly devoted to laryngeal communication in mammals (for underwater sound communication, see Reidenberg and Laitman, Chapter 10.4 in this volume), I will focus mainly on the brain mechanisms mediating vocal communication in terrestrial mammals. An important corollary of the first system, found in many vertebrates, is the evolution of new, forebrain structures centered round the anterior parts of the limbic system. In particular, the so-called limbic cortex, the cerebral part of the "thalamo–cingulate" subdivision of MacLean's "limbic system" (MacLean, 1990), needs to be considered in some detail.

III. Vocal repertoires

Although some general patterns emerge, mammalian vocalizations are characterized by diversity with respect to the acoustic patterns employed. This diversity is no doubt fostered by the variety of ecological adaptations, degree of sociability and tremendous range of body size represented among living mammals. There is also strong evidence for species-specificity in some classes of vocalization, particularly those related to finding and distinguishing potential mates in an acoustically rich environment. One might reasonably expect a corresponding diversity in underlying mechanisms. Insufficient data prevent generalizations, but the reoccurrence of brain substrates mediating vocalization across a number of mammalian orders suggests that a common plan prevails.

IV. Brain systems associated with vocalization: an overview

IV.A. Limbic system

The limbic system of MacLean (1952) has frequently been identified as an important part of the underlying neural basis for mammalian vocalizations. Evidence for this varies according to the specific structure in question. In contrast to the widely held view of the importance of limbic structures in mediating vocal production, the literature has sparse reference to such an association with respect to humans (e.g., MacLean, 1990).

However, Cummings et al. (1983), in a single case study, concluded that loss of both emotional and volitional phonation in a human was a consequence of damage to both limbic and neocortical structures. Some evidence also suggests that infarcts in and around the anterior cingulate gyrus result in mutism or loss of prosodic components of speaking (Brown, 1988). Most of the evidence implicating limbic structures in vocal production has come from experimental studies using animals (see below and Siegel et al., Chapter 7.1 in this volume).

IV.B. Cerebral cortex

It is generally accepted that the neocortex plays an important role in human speech, but is not an essential part of non-human vocal communication (e.g., Lieberman, 2006). Nevertheless, electrical stimulation of a portion of the ventral motor cortex causes movement of the vocal folds in macaques (Hast et al., 1974) and squirrel monkeys (Jürgens, 1974). Hast et al. (1974) stimulated the surface of the inferolateral motor cortex in macaques. They were able to differentially activate the thyroarytenoid, cricothyroid and extrinsic laryngeal muscles. Heart rate changes occurred only from the area principally devoted to responses of the cricothyroid muscle. Jürgens and Zwirner (2000) found evidence for asymmetrical representation of the larynx on one or the other cerebral hemisphere. Jürgens (1974) stimulated the region of motor cortex from which vocal fold movements could be produced. No vocalization or other sound was produced. The author concluded that this cortical region functioned to modulate sound production, but was not the structure where coordination of the vocal act arose. Jürgens et al. (1982) performed deep-reaching lesions in the cortical face area of the squirrel monkey (matching an infarct in a human patient rendered aphonic) and found that there was no change in vocalization.

One part of the cerebral cortex has consistently been shown to be important in vocal production. The cingulate gyrus of the cerebral cortex has proven to be an important part of the circuitry regulating vocalization. This was first discovered by W.K. Smith (1941), who stimulated the brain of macaques under light ether anesthesia. He found that electrical stimulation of the rostral part of Brodmann's area 24 produced vocal responses: "identical to those which the animal makes under ordinary conditions" (p. 451). He also found that bilateral extirpation of the excitable area did not result in loss of the ability to vocalize. MacLean (1988) argued that part of the limbic system, the so-called "thalamo–cingulate" division, was unique to mammals and, as a consequence, likely played a role in regulating behavior that was found in mammals, but not well-represented in other vertebrates. One of these behaviors was mother–infant communication, in particular the infant cry. In a lesion study using adult squirrel monkeys, MacLean and Newman (1988) found that bilateral destruction of the anterior cingulate gyrus (including the portion below the genu of the corpus callosum) resulted in long-lasting elimination of the isolation call (the species-specific cry in infancy), but not other vocalizations. A previous study (Sutton et al., 1974) likewise found that destruction of this region in macaques eliminated the adult version of the "coo" vocalization, although the authors concluded that the deficit was restricted to conditioned vocalizations. Although an earlier electrical brain stimulation study had found that vocalizations could be evoked in subgenual cingulate regions (Jürgens and Ploog, 1970), a more dorsal area became known as the "cingular vocalization area" (Müller-Preuss and Jürgens, 1976). This region lies around the cingulate sulcus at the level of the genu of the corpus callosum. These same authors found that this region projected into many areas from which electrical stimulation evoked vocalization, including the periaqueductal gray matter of the midbrain. Interestingly, a projection to auditory association cortex in the temporal lobe also arises from this area (Müller-Preuss et al., 1980). Activation of the cingular vocalization area suppressed spontaneous activity of neurons in the auditory association cortex, as well as inhibiting auditory responses.

IV.C. Brainstem

Vocal pathways have been extensively studied in the macaque monkey and squirrel monkey. In both, virtually the entire motor pathway from the caudal brainstem areas, from which respiration and laryngeal muscle activation are directly controlled, through the midbrain, hypothalamus, preoptic area, limbic structures up to the cerebral cortex have been explored. As with other studies, the midbrain PAG has been implicated in phonation in macaques. Larson (1985) recorded from single neurons within the PAG and found an increase in firing rate prior to onset of laryngeal EMG activity or vocalization. Microstimulation in the same region elicited phonation similar to natural calls.

Dressnandt and Jürgens (1992) determined which brain structures are involved in motor coordination of vocalization in squirrel monkeys. Brainstem sites in which electrical stimulation caused a deterioration of ongoing vocalization were found. In the rostral brainstem, there were two sites: one was in the pyramidal tract and the other in the caudoventral PAG. In the caudal brainstem, sites in the nucleus ambiguus (the motor nucleus of the larynx) produced stimulation-bound changes in the frequency range of the call. In the area around the nucleus of the solitary tract, the sensory relay from the larynx, stimulation caused changes in vocalization indicating that somato-motor feedback is essential for vocal motor control. A third group of sites lie scattered in the reticular formation, where motor neurons connected to several important motor structures (larynx, respiratory muscles, cranial nerves, etc.), indicating that this region is one of the motor coordination regions for vocalization, i.e., an area in which the activities of the different muscles involved in phonation are integrated into a specific vocal pattern (see Hage, Chapter 8.2 in this volume).

A classic study by Magoun et al. (1937) using macaques, normal cats, and decerebrated cats (just behind the diencephalon) under a light dose of Nembutal (pentobarbital), found that responses to electrical stimulation in the monkey consisted of vocalization occurring rhythmically with each expiratory phase; most commonly cries of a wide variety of intensity and pitch; also barking or chattering; occasionally, cooing or whistling sounds. As a whole, the calls resembled those which are expressions of unpleasant affective states. Facial muscle activity also resembled that exhibited by normal aversive emotional expression. Vocal responses were obtained by electrical stimulation from a large part of the rostrocaudal extent of the central gray matter of the aqueduct (PAG) from the dorsal part of the tegmentum of the midbrain, together with parts of the midbrain and pons, and from the reticular formation of the medulla. In cats, cries and spitting were produced; the responsive area extending from the rostral part of the central gray surrounding the aqueduct (PAG) through the dorsal part of the midbrain tegmentum to its lateral part. A reactive region was also found in the anterior hypothalamus (not tested in monkeys). The same responses as found in the intact cats were found in decerebrated cats. A subsequent study by Kelly et al. (1946) confirmed the importance of the midbrain for evoked vocalization in cats.

In a landmark study, Jürgens and Ploog (1970) electrically stimulated sites throughout the brain of awake squirrel monkeys and linked each stimulation site to the specific vocalizations evoked. By doing this, they were able to construct the course of general pathways through the brain mediating specific call types. It should be emphasized, however, that the electrical stimulation method did not allow for the obtaining of precise neuroanatomical results or the direction of neuronal projection of studied pathways.

The pathway for cackling calls (also known as "chucks" and "keckers") runs from the caudal end of the PAG and adjacent parabrachial nuclei throughout the midbrain and diencephalon. At the level of the inferior thalamic peduncle, the system branches into three components: the first follows the inferior thalamic peduncle dorsally towards the anteriomedial thalamic nucleus; the second follows the inferior thalamic peduncle ventrolaterally into the central and basal nuclei of the amygdala and further through the external capsule and uncinate fasciculus to the rostroventral temporal cortex; the third component follows the anterior thalamic radiation along the ventromedial border of the internal capsule into the ventromedial orbital cortex and precallosal cingulate gyrus. Except for electrode positions in the precallosal cingulate gyrus, structures evoking growling calls (also known as "purr," "err" and "churr") form a continuous system. In the midbrain, the responsive areas follow the PAG, but only a small part of it. At the level of the dorsal tegmental nucleus of Gudden, one branch leaves the periaqueductal component and runs in a ventrolateral direction through the central tegmentum and ventral tegmental area (Tsai) into the lateral hypothalamus. In the posterior hypothalamus this tegmental component joins the periaqueductal component, and both continue up the anterior border of the diencephalon; one branch follows the same path up the inferior thalamic peduncle as the cackling sites; the third component takes its course through the preoptic area, bends around the anterior commissure, and joins the stria terminalis, reaching the amygdala. The positions yielding chirping and trilling (also known as "twitters" and "peeps") do not represent a continuous system. Trilling occurs along the precommissural fornix along a small section of the spinothalamic tract, while chirping can be elicited in five separate areas, namely, the gyrus subcallosus-medioventral caudatum region; the midline thalamus; rostral hippocampus; caudal PAG; and along the caudal spinothalamic tract.

In a study using single unit recording in the medulla oblongata, Lüthe et al. (2000) determined that vocalization-related cells exist in great numbers in the nucleus ambiguus, nucleus retroambiguus, in the reticular formation, solitary tract nucleus and spinal trigeminal nucleus. The authors proposed that initiation of vocalization takes place via the parvocellular reticular formation; vocal pattern control is mainly brought about by the parvocellular reticular formation, nucleus ambiguus, solitary tract nucleus and spinal trigeminal nucleus, expiratory control and respiratory–laryngeal coordination is carried out by the retroambiguus nucleus and nucleus ambiguus, whereas vocalization-specific postural adjustments are carried out via the vestibular and cuneate nuclei.

Larson (1991) examined the relationship of PAG neurons to laryngeal and respiratory muscles during vocalization in macaque monkeys. Most neurons were related to single muscles. However, a significant number were related to functionally-related groups of two or more muscles. The author concluded that the PAG determines qualitative aspects of vocalization by multisynaptic action.

Zealear and Larson (1988) used microelectrodes to investigate laryngeal motor neurons in the nucleus ambiguus of awake vocalizing macaque monkeys. Monkeys were trained to produce coos and barks to obtain a food reward. Neurons innervating the thyroarytenoid muscles responded to laryngeal phasic activity such as swallowing and vocalizing. Neurons firing during inspiration were not active during swallowing or vocalizing.

Larson and Kistler (1984) recorded neuronal activity in the PAG that was associated with laryngeal muscular activity and vocalization in awake macaques. The neuronal activity began prior to the onset of vocalization, and most times prior to EMG activity in the laryngeal muscles. The vocalizations were barks and shrieks. It appears that the role of PAG may be related to initiation of vocalizations (see Gruber-Dujardin, Chapter 8.1 in this volume).

It is important to add that mapping of this limited set of subcortical structures associated with the induction of vocalization by electrostimulation can be largely replicated by chemical stimulation of the same brainstem structures and pathways, which elicit comparable vocalization in non-primate species (see Brudzynski, Chapter 7.3 in this volume).

V. Conclusions

Evidence suggests that a limited number of brain structures are involved in vocal production in mammals. With the possible exception of some cortical areas, these structures are in the subcortical forebrain, particularly limbic structures and the hypothalamus, and midbrain, particularly the PAG, leading to the motor neurons that innervate the larynx – the peripheral organ of vocal production. Coordination of laryngeal activity and respiratory activity likewise resides in the caudal brainstem. A major unanswered question is how the detailed circuitry in this limited set of neural structures accounts for the great diversity of vocal subtypes found in mammalian communication.

Neuroscience has found it useful to organize the mammalian brain around "systems" or "circuits" that conceptualize organizational schema according to certain unifying themes. There is the "reward system," the "limbic system," the "visual system," etc. Communication has lacked a unifying schema, although several overviews have been presented for single species. I have proposed the term "communicating brain" (Newman, 2003) to refer to those brain structures commonly linked to vocal production in a number of mammalian species. As vocal communication requires both a sender and receiver to be evolutionarily adaptive, we must also include structures generally known to be part of the auditory pathways as part of a "communicating brain" concept.

Acknowledgments

This chapter was written with support from the intramural research program of the Eunice Kennedy Shriver National Institute of Child Health and Human Development, NIH.

References

Bass, A.H., McKibben, J.R., 2003. Neural mechanisms and behaviors for acoustic communication in teleost fish. Prog. Neurobiol. 69, 1–26.

Bass, A.H., Gilland, E.H., Baker, R., 2008. Evolutionary origins for social vocalization in a vertebrate hindbrain-spinal compartment. Science 321, 417–421.

Brown, J.L., 1971. An exploratory study of vocalization areas in the brain of the Red-winged blackbird (*Agelaius phoeniceus*). Behaviour 39, 91–127.
Brown, J.W., 1988. Cingulate gyrus and supplementary motor correlates of vocalization in man. In: Newman, J.D. (Ed.), The Physiological Control of Mammalian Vocalization. Plenum Press, New York NY, pp. 227–243.
Burgdorf, J., Wood, P.L., Kroes, R.A., Moskal, J.R., Panksepp, J., 2007. Neurobiology of 50 kHz ultrasonic vocalizations in rats: electrode mapping, lesion, and pharmacological studies. Behav. Brain Res. 182, 274–283.
Cummings, J.L., Benson, D.F., Houlihan, J.P., Gosenfeld, L.F., 1983. Mutism: loss of neocortical and limbic vocalization. J. Nerv. Ment. Dis. 171, 255–259.
Dressnandt, J., Jürgens, U., 1992. Brain stimulation-induced changes of phonation in the squirrel monkey. Exp. Brain Res. 89, 549–559.
Emerson, S.R., Boyd, S.K., 1999. Mating vocalizations of female frogs: control and evolutionary mechanisms. Brain, Behav. Evol. 53, 187–197.
Goodson, J.L., Bass, A.H., 2002. Vocal-acoustic circuitry and descending vocal pathways in teleost fish: convergence with terrestrial vertebrates reveals conserved traits. J. Comp. Neurol. 448, 298–322.
Gooler, D.M., O'Neill, W.E., 1988. The control of biosonar signal production in bats demonstrated by microstimulation of anterior cingulate cortex in the echolocating bat, *Pternonotus parnelli parnelli*. In: Newman, J.D. (Ed.), The Physiological Control of Mammalian Vocalization. Plenum Press, New York, NY, pp. 153–183.
Hast, M.H., Fischer, J.M., Wetzel, A.B., Thompson, V.E., 1974. Cortical motor representation of the laryngeal muscles in *Macaca mulatta*. Brain Res. 73, 229–240.
Heiligenberg, W., 1988. Neural mechanisms of perception and motor control of a weakly electric fish. Adv. Study Behav. 18, 73–98.
Jarvis, E.D., 2007. Neural systems for vocal learning in birds and humans: a synopsis. J. Ornithol. 148 (Suppl. 1), S35–S44.
Jürgens, U., 1974. On the elicitability of vocalization from the cortical larynx area. Brain Res. 81, 564–566.
Jürgens, U., 2002. Neural pathways underlying vocal control. Neurosci. Biobeh. Rev. 26, 235–258.
Jürgens, U., Ploog, D., 1970. Cerebral representation of vocalization in the squirrel monkey. Exp. Brain Res. 10, 532–554.
Jürgens, U., Zwirner, P., 2000. Individual hemispheric asymmetry in vocal fold control of the squirrel monkey. Behav. Brain Res. 109, 213–217.
Jürgens, U., Kirzinger, A., von Cramon, D., 1982. The effects of deep-reaching lesions in the cortical face area on phonation. A combined case report and experimental monkey study. Cortex 18, 125–139.
Kelly, A.H., Beaton, L.E., Magoun, H.W., 1946. A midbrain mechanism for facio-vocal activity. J. Neurophysiol. 9, 181–189.
Kennedy, M.C., 1975. Vocalization elicited in a lizard by electrical stimulation of the midbrain. Brain Res. 91, 321–325.
Kennedy, M.C., 1981. Motoneurons that control vocalization in a reptile: an HRP histochemical study. Brain Res. 218, 337–341.
Larson, C.R., 1985. The midbrain periaqueductal gray: a brainstem structure involved in vocalization. J. Speech Hear. Res. 28, 241–249.
Larson, C.R., 1991. On the relation of PAG neurons to laryngeal and respiratory muscles during vocalization in the monkey. Brain Res. 552, 77–86.
Larson, C.R., Kistler, M.K., 1984. Periaqueductal gray neuronal activity associated with laryngeal EMG and vocalization in the awake monkey. Neurosci. Lett. 46, 261–266.
Lieberman, P., 2006. Toward an Evolutionary Biology of Language. Belknapp Press, Harvard University, Cambridge, MA.
Lüthe, L., Häusler, U., Jürgens, U., 2000. Neuronal activity in the medulla oblongata during vocalization. A single-unit recording study in the squirrel monkey. Behav. Brain Res. 116, 197–210.
MacLean, P.D., 1952. Some psychiatric implications of physiological studies on frontotemporal portion of limbic system (visceral brain). Electroencephalogr. Clin. Neurophysiol. 4, 407–418.
MacLean, P.D., 1988. Evolution of audio-vocal communication as reflected by the therapsid-mammalian transition and the limbic thalamocingulate division. In: Newman, J.D. (Ed.), The Physiological Control of Mammalian Vocalization. Plenum Press, New York, NY, pp. 185–201.
MacLean, P.D., 1990. The Triune Brain in Evolution. Role in Paleocerebral Functions. Plenum Press, New York, NY.
MacLean, P.D., Newman, J.D., 1988. Role of midline frontolimbic cortex in production of the isolation call of squirrel monkeys. Brain Res. 450, 111–123.
Magoun, H.W., Atlas, D., Ingersoll, E.H., Ranson, S.W., 1937. Associated facial, vocal and respiratory components of emotional expression: an experimental study. J. Neurol. Psychopathol. S1–17, 241–255.
Morton, E.S., 1977. On the occurrence and significance of motivation-structural rules in some bird and mammal sounds. Am. Nat. 111, 855–869.
Müller-Preuss, P., Jürgens, U., 1976. Projections from the "cingular" vocalization area in the squirrel monkey. Brain Res. 103, 29–43.
Müller-Preuss, P., Newman, J.D., Jürgens, U., 1980. Anatomical and physiological evidence for a relationship between the "cingular" vocalization area and the auditory cortex in the squirrel monkey. Brain Res. 202, 307–315.
Newman, J.D., 2003. Vocal communication and the triune brain. Physiol. Behav. 79, 495–502.
Newman, J.D., 2007. Neural circuits underlying crying and cry responding in mammals. Behav. Brain Res. 182, 155–165.
Smith, W.K., 1941. Vocalization and other responses elicited by excitation of the Regio cingularis in the monkey. Am. J. Physiol. 133, P451–P452.
Schuller, G., Radtke-Schuller, S., 1988. Neural control of vocalization in bats at peripheral to midbrain levels. In: Newman, J.D. (Ed.), The Physiological Control of Mammalian Vocalization. Plenum Press, New York, NY, pp. 67–85.
Sutton, D., Larson, C., Lindeman, R.C., 1974. Neocortical and limbic lesion effects on primate phonation. Brain Res. 71, 61–75.
Zealear, D.L., Larson, C.R., 1988. A microelectrode study of laryngeal motoneurons in the nucleus ambiguus of the awake vocalizing monkey. In: Fujimora, O. (Ed.), Vocal Physiology: Voice Production. Mechanisms and Functions. Raven Press, New York, NY, pp. 29–37.

CHAPTER 2.3

Evolution of the infant separation call: rodent ultrasonic vocalization

Myron A. Hofer*

The Sackler Institute for Developmental Psychobiology at Columbia University, College of Physicians and Surgeons, New York, NY, USA

Abstract: This chapter describes research that shows how mammalian infant vocalization is likely to have evolved as a component of the neuro-behavioral response of infants to separation from mother, littermates and home cage nest. The following questions are asked: how might evolution of the larynx as a protective valve during food and water ingestion and thermoregulatory physiology that was central to evolution of mammalian brain function have played a role in the evolution of infant vocal responses? And how might other selection pressures, such as the risk of detection by predators and variation in the likelihood of maternal responses to infant calls, have further shaped neural regulation of the infants' vocal response within the cognitive/emotional state elicited by separation? Finally, the results of experimental selection over many generations for high and low levels of infant vocalization provide laboratory tests of some of the hypotheses generated by this research.

Keywords: ultrasonic vocalization; USV; rats; infants; social isolation; development; selection; larynx; ontogenetic adaptation

I. Introduction

Infants of virtually every species of mammal, as well as birds, vocalize when separated from their home nest, conspecifics and most clearly their mothers. The survival value of this form of sound production and its communicative "meaning" are generally thought to reflect the dependency of the young on a particular set of environmental conditions, and of parental care and protection. For altricial species, such as the Norway rat, this also includes maintenance of the body temperature, nutrition and protection from climate extremes, as well as predators. The sound functions as a signal that the infant's survival is at risk, and strongly activates searching and retrieval behavior from the mother and sometimes in closely-related conspecifics. The name "separation call" reflects the general sense of these contingencies, but does not imply that the call is emitted by the infant for the *purpose* of summoning its mother. Like many other vocalizations, it is a component of the animal's state, and becomes shaped during evolution by selection for its signal value in promoting survival of the infant, while the effectiveness of the mother's sound perception, her searching and her retrieval behavior in response to that signal are shaped by the increased survival of the more responsive mother's genes (carried by her offspring) in the next generation.

One of the most interesting questions about these infant vocalizations is how they became included as a component of the physiological/behavioral state induced by the condition of social separation. We know that evolution can produce novel traits, by using some that evolved in one set of conditions as the basis for creating a new structure or function that is adaptive in another set of conditions. Can we trace back this process to find out how vocalization may have become a part of the infants' response to separation? And what further adaptations may have taken

*Corresponding author.
Phone: (212) 543-5284; Fax: (212)-543-6922

ISBN 978-0-12-374593-4

DOI: 10.1016/B978-0-12-374593-4.00004-8
Revised from Acta Paediatrica 2002, Vol.91 No.5,

place as a result of other selection pressures, such as the risk that calling may attract potential predators instead of closely-related conspecifics? Finally, can we test some of the hypotheses we construct from evolutionary principles in a laboratory experiment using repeated selective breeding over generations?

II. Evolution of the larynx

Before I consider the evolution of infant isolation calling, I'd like to step back and consider the evolution of the organ for vocalization, the larynx. Negus (1931), a distinguished British anatomist and physiologist, published a monumental work in which he compared the larynx in over 1,500 different species, ranging from primitive vertebrates to primates. He concluded that the larynx evolved initially as a valve to protect the delicate membranes of early air-breathing fish, to ensure that food on its way to the esophagus, or other particulate matter entering the mouth, did not damage the epithelium for gas exchange. As evolution progressed, air breathing became a more prominent and important physiologic function, and muscular development of the larynx began. Some muscles developed to keep the larynx open and promote airflow, and others worked to close it for protection. Later, cartilage developed to make the structure more rigid, to prevent airway collapse and further improve airflow (see Berke and Long, Chapter 10.1 in this volume). Thus, although we recognize the importance of the larynx in vocalization, it is important to keep in mind that it evolved initially as a valve to optimize airflow.

Furthermore, the anatomic structure of the larynx varies widely among modern mammals. Negus observed tremendous structural variation, ranging from a simple tube in the ox to an elaborate structure with delicate valve leaflets, actually vocal cords, in a primitive primate with a limited vocal repertoire, the lemur. Negus came to the interesting conclusion that across species the degree of relative complexity of the larynx relates *not* to vocal capacity, but rather to the capacity for powerful and precise upper arm movements. One functional example of this can be noted in humans. After each serve in tennis there is often an audible grunt, or expulsion of air. During the serve, very precise upper arm movements are required, so the laryngeal valves close in order to stabilize the thorax as a platform for the arm. When the serve is completed, the valves open and air is expelled.

Thus, while the larynx originally evolved to protect gas exchange structures and later to decrease airway resistance, the vocal cords evolved as an adaptation within the larynx that improved upper limb function. There are likely to be other derivative functions of different aspects of laryngeal anatomy. For example, vocalization involves closing the larynx and using thoracic and abdominal muscles to force air through the vocal cords. This alters intrathoracic pressure, modulating central blood circulation and pulmonary gas exchange that may play a role in the regulation of body temperature. As requirements for communication and the exchange of information became critical to more complex social species, functional adaptations associated with vocalization may have played a role in further shaping the evolution of the larynx. For example, a mother's need to detect accurately the state or condition of her infant at a distance may have facilitated the evolution of an association between crying and emotional states that fulfill the "signal" function of crying.

III. The connection between crying and cold

Homeothermy, or the maintenance of stable body temperature, was a critical capacity in the evolution of early mammals. Young offspring needed to stay in close proximity to the nest and to the mother to help maintain their body temperature, and needed a means to signal their mothers to retrieve them from the cold, if displaced from the nest. Infant crying appears to have evolved to play that role. Evolutionary adaptation of certain reptilian anatomic features in mammals resulted in the development of inner ear bones and an extra set of sensory hair cells, allowing the early mammals to hear much higher sounds than reptiles. This allowed mammals to communicate with ultrasound at frequencies that their reptile predators were unable to hear. This trait has been preserved in the ultrasonic frequency of many rodent vocalizations, which are in the range of 20 to 100 kHz.

The evolution of infant crying may have had an earlier connection with the evolution of mechanisms for endogenous heat production in early mammals. In recent years, physiologists have established that hypothermia in small infant mammals results in cessation of heartbeat and respiration, leaving animals essentially in a state of "suspended animation" from which they can recover readily, due to their capacity to generate heat from specialized fat called brown adipose tissue. The high surface area/body mass ratio of infant rats allows rapid cooling in ice water and is used as a highly effective anesthetic in the laboratory. Harry Shair and I found to our surprise that hypothermic infant rats vocalize in the ultrasonic range while still comatose (Hofer and Shair, 1992). In our laboratory, we measure ultrasonic infant rat vocalizations (USV) using portable bat echolocation detectors that transduce ultrasounds into the audible range. The mean USV rate of unresponsive hypothermic rat pups, at about 20°C body temperature, approximates the USV rate for normal rat pups that have just been separated from their mother. There is initially no change in the USV rate of hypothermic rat pups placed in contact with their mother. As the pups warm beyond 25°C, however, they become responsive to sensory cues from her. This results in a decline and cessation of pup USV that is not seen when the gradually warming pups are alone. This contact quieting was likely to be a function of the increasing level of consciousness that we documented in a study of the recovery of reflex motor acts and instinctive responses after induced hypothermia.

A few years earlier, Blumberg and Alberts (1990) had hypothesized that the USV of normal infant rats in response to isolation (in relatively cool ambient temperatures) had evolved because it facilitated pulmonary oxygen transport to brown adipose tissue (BAT) and restored core temperature, rather than being related to social isolation *per se*. BAT acts as an internal heater, it envelops all of the major blood vessels and, when activated, has a very high oxygen requirement. Blumberg and Alberts noted that all vocalization results in an increase in intrathoracic pressure as the larynx closed against the escaping breath. This phenomenon is known as "laryngeal braking," and under some conditions enhances pulmonary oxygenation of the blood. For example, laryngeal braking, in the form of grunting respiration, is known to play a role in ameliorating respiratory distress syndrome in premature human infants.

To determine whether the USV production we had discovered in hypothermic rat pups was related functionally to oxygenation of BAT during recovery from hypothermia, as predicted by the laryngeal braking hypothesis, Harry Shair and I measured the temperature in intrascapular brown fat (BAT) and at a control site, in intact rats pups and in those who were unable to use laryngeal braking (e.g., denervated larynx or with tracheostomy) and could not emit USV (Hofer and Shair, 1993). We found no differences in the temperature of BAT during recovery in the two groups of pups. However, we did observe that the core body temperature of intact animals rose faster than in devocalized animals. Approximately 20% of devocalized animals ultimately did not recover from hypothermia, and at autopsy we found pulmonary edema. What could account for these findings? A human analogy may be found in mountaineers and skiers who fall into crevasses and are subsequently rescued by helicopter. These hypothermic individuals often develop acute pulmonary edema during warming, evidently an effect of extreme cold on pulmonary vascular permeability. Clinically, this is treated with positive end-expiratory pressure ventilation (PEEP) which increases the alveolar pressure and reduces the development of lung edema. We observed that, with each ultrasonic vocalization in the hypothermic rat pups, intrathoracic pressure increased significantly, producing a self-administered form of PEEP that would similarly minimize the development of pulmonary edema. Our rat pups that could not vocalize were unable to use this mechanism, and this may well have accounted for the slower thermal recovery and high incidence of pulmonary edema and death in this group.

Thus, hypothermic vocalization may have evolved to serve dual purposes. In addition to modulating tracheal pressure to prevent pulmonary edema after severe cold exposure, hypothermic USV could also serve a communicative function, stimulating the mother to leave the nest to look for the comatose but vocalizing pup, and retrieve it to the nest. Indeed, Susan Brunelli, Harry Shair and I found that rat mothers hearing these vocalizations leave their nests and direct their searching behavior accurately toward the pup, even though hypothermic USVs are lower-pitched, longer and have greater intervals between calls than typical isolation USVs (Brunelli et al., 1994). We can infer that rat pup ultrasonic calls may have evolved first as the product of a physiological adaptation facilitating recovery from hypothermia, and their role in communication was likely to have evolved later, under repeated selection for this derivative function.

IV. From communication to emotion

In exploring the early communicative function of infant calls we found that infant rats provide a good animal model to study early attachment and also the separation response. Very young infant rats learn to identify their own mother by smell. They readily vocalize in response to separation from the mother, a response that had been attributed simply to handling and cold stress. But our research led us to another process: the separation-induced withdrawal of specific sensorimotor regulators of USV that are intrinsic to the mother–pup interaction. First, we observed that moving a rat pup from the nest to a test chamber where it is alone results in a vigorous outburst of ultrasonic calling, but if an anesthetized mother or littermates were in the test chamber, the pup immediately reduced and then stopped the calling (Hofer and Shair, 1978). This quieting is due simply to contact with the dam or littermates, and is effective despite the novel environment. We then deconstructed the pup's contact experience into many individual stimuli, to attempt to identify the particular factors responsible for its down-regulating pup vocalizations so effectively. When we studied pup responses to various part surrogates for the mother, such as an anesthetized littermate, a flat piece of synthetic fur (with and without nest odor), or a clay model the shape of a littermate (either warmed or unwarmed), we learned that there are multiple sensory functions of the pup that are involved in mediating maternal regulation of pup vocalization. This regulation is a central component of a complex homeostatic system, including both mother and pup. In it there are at least three maternal regulators, including thermal, tactile and olfactory stimuli. When these regulators are withdrawn, the pup vocalizes, eliciting retrieval by the mother, and completing the circle (Hofer, 1996).

The central brain pathways for pup ultrasound can similarly be analyzed into their component neurotransmitter mechanisms. Interestingly, the profile of neurotransmitters involved in separation vocalization in infant rats is very similar to the neurotransmitter pattern that has been shown to mediate human anxiety and presumably separation anxiety in children (for a review see Brunelli and Hofer, 2001). Apparently, there is evolutionary conservation of the neural mechanisms by which the mammalian infant separation cry communicates information about the affective state of the infant. Benzodiazepines are anxiolytics in humans, and they powerfully and selectively inhibit rat pup vocalizations. In contrast, substances such as pentylenetetrazol or beta carbolines that produce disintegrative anxiety in humans by acting on the GABA receptor complex drastically increase vocalizations in infant rats. Oxytocin, a hormone with several roles in attachment, reduces pup calling rate. Opioid receptor agonists, serotonin, adrenergic agonists and peptides such as CRF and substance P are also involved in the regulation of both human affective states and rodent ultrasonic calls.

V. Evolutionary "trade-offs" may have shaped other crying responses

Subsequently, we found that if an isolated pup is allowed to interact briefly with a lactating female, its vocalization rate in isolation after removal of the dam increases far beyond that observed when the pup was first isolated (Hofer et al., 1996). This "maternal potentiation" effect is specific to interaction with a lactating female and is more intense after contact with an active, rather than an anesthetized dam. Repeating the same experiment with littermates or a virgin female instead of the dam results in the same degree of quieting during contact, but USV rates only return to the pups' initial isolation level after their removal. Maternal potentiation shows that the pup is sensitive to a specific context, its dam, immediately *prior* to separation, and demonstrates that features of that context can regulate USV rates even after the context is no longer present. This response begins to resemble the complexity of human infant crying responses, such as the sudden outcry typical of toddlers if their own mother returns to the day-care center unexpectedly (after having left her child earlier that morning), and then departs again after a brief interaction.

A very different form of vocal response involves the suppression of isolation calling (Takahashi, 1992). Unfamiliar male rats are predators for infant rats. Males from outside the colony can maximize their "evolutionary fitness" (their genetic transmission to future generations) by killing unfamiliar young rat pups and mating with the mother. The odor of an unfamiliar adult male silences isolated infant rats, and the inhibition of USV continues for many minutes after pups are removed to another chamber far from the male, the opposite of maternal potentiation (Shair et al., 1999). In contrast, prepubescent unfamiliar males or familiar adult males do not elicit any inhibition of vocalization. This vocal response to a highly specific threat stimulus may represent an early developing state of fear, as contrasted to the anxiety-like state elicited by the far less well-defined dangers inherent in the isolation situation. If this is so, the opposite direction of this effect on crying (isolation calls) may be one of the clearest experimental distinctions yet observed between the closely-related emotional states of fear and anxiety.

Both the inhibition of vocalization after the male is removed and the potentiation of isolation calling when the dam is removed are examples of vocalization patterns in response to prior contexts. In other studies (Hofer et al., 1998) we have investigated the time course of development of these context-sensitive effects. In the first two days after birth, the isolation calling and the contact quieting responses first develop. In contrast, the maternal potentiation and the male suppression effects develop later, after the first postnatal week, possibly related to maturation of the pups' ability to regulate the intensity of their responses according to the specific environmental contexts prior to the events.

Can we better understand the nature and origins of these complex vocal responses by thinking about them from an evolutionary perspective? If a pup has detected the odor of an unfamiliar male, it would be evolutionarily advantageous to inhibit all vocalizations to avoid detection by such a predator. The strength of the inhibition response would ideally be regulated by the proximity of the potential predator, and thus selection pressure may have favored evolution of the olfactory signal that we found to elicit the response. On the other hand, if a pup is with an active dam and is then suddenly parted from her, it is highly likely that the mother will not have gone far and pup vocalizations will result in immediate retrieval. In this case, the risk of attracting a predator is much less than the likelihood of maternal response. Therefore, maximal calling rates are relatively safe. Further, we have found that vocalization rates following separation from an active mother were higher than if the dam had been anesthetized (Hofer et al., 1996). One could hypothesize that an unresponsive mother is equivalent to a sleeping one, and this context signals a lower likelihood of rapid maternal response relative to predator risk. A less than maximal calling rate would be most adaptive in this situation, and this is precisely what occurs. If pups have been with littermates and the mother absent for more than five minutes, pup calling rates are even lower, consistent with a further reduced likelihood of maternal response relative to predator risk, but not yet the adverse risk–benefit ratio signaled by the odor of the predator itself, which causes an absolute suppression of calling.

VI. A genetic approach: "evolution" in the laboratory

Infant crying responses thus appear to be adaptively calibrated to the relative risk–benefit ratio, signaled by the prior contexts in which they take place. We next wondered whether we could test these evolutionary ideas by asking whether differential selection repeated over generations could indeed influence the characteristic USV rates of the infant isolation vocalization response in a population. To explore this, Susan Brunelli and I began a program of selective breeding of adult rats based on their level of USV response to isolation as 10-day-old infants (reviewed in Brunelli and Hofer, 2001; Muller et al., Chapter 6.4 in this volume). We used a genetically-heterogeneous strain from NIH and formed two lines, in addition to randomly-bred controls. The line selected for the highest USV responses represented selection in an environment with low risk of predators and high likelihood of maternal response. The line we selected for the lowest USV response to isolation as infants would represent the effects of rearing in an environment with high predator risk and low likelihood of their mother or other responsive adults being nearby.

What we found was that infants' calling during isolation was strikingly elevated in high line pups, starting as early as the response develops at three days of age, a week before the age at which selection had been carried out. Low line pups, in contrast, were moderately decreased at this early age. The greatest difference between high and low lines was at the age of repeated selection, 10 days postnatal. Response differences were much less evident at 14 days, and the lines converged as the response ceased to occur at 18–20 days postnatal. Thus selection had resulted in high line pups showing retention (and exaggeration) of the high response rates of newborn pups throughout their development, right up to and including the age of selection (10 days) and in the low line pups, a more rapid decline than normal occurred from 3 to 10 days of age. Thus, selection at 10 days of age appeared to be acting on the whole early developmental trajectory of the vocal response to separation, shifting it in time: either delaying or hastening the normal developmental decline in response intensity with age.

In addition, at 18 days of age, high line animals showed increased defecation/urination and a sympathetically mediated tachycardia during isolation testing, whereas lows showed low levels of defecation/urination and profound vagal withdrawal in the same test situation (Brunelli and Hofer, 2007). This adds further evidence to the conclusion that isolation calling is an expression of an underlying anxiety-like state, and one shaped by selection for the consequences of calling on pup survival. Following the two lines into adulthood, Susan Brunelli has recently found that selection during infancy has resulted in adult phenotypes that resemble different temperaments (Brunelli and Hofer, 2007). The high line adults were slower to emerge into an open test box and showed a much more quiet, passive response in the Porsolt swim test, whereas low line adults showed much diminished adolescent play behaviors, were active in open field and Porsolt swim test, and 70% of adult males engaged in aggressive behavior versus 30% of controls. These results constitute a particularly clear example of the organization of development into complex "paths" with different timing patterns (heterochrony).

VII. Summary and perspective

I have approached infant crying as an evolving trait. From classic studies of the larynx, the organ of crying, we find that while the larynx developed initially to regulate airflow, its structure became more complex as a mechanism in mammals to stabilize the thorax for upper arm movement, rather than to facilitate vocalization. In our experiments, we have found an unexpected adaptation still present today, hypothermic vocalization, in which the laryngeal–thoracic act of calling itself functions to maintain intra-thoracic pressure and avoid pulmonary edema as rat pups warm after severe hypothermia. Crying in a communicative role appears to have been the most recent evolutionary step. Yet, even in a relatively simple mammal such as the laboratory rat, it is a highly-regulated response that is controlled by many different environmental and social signals, cues that connect crying with the risks and benefits of this behavior in the environments within which mammals evolved. Recent research is beginning to elucidate the neurochemical pathways by which ultrasonic infant calling responses are transduced in the brain of the infant, and to shed new light on the nature of the emotional state underlying the infant's vocal response to isolation, one that appears to be highly similar to human anxiety.

Two other more complex forms of infant crying (maternal potentiation and inhibition by unfamiliar male scent) are described as adaptations within a set of opposing evolutionary selection pressures. The intensity of infant crying may thus be regulated by contextual cues representing the relative risk or benefit of the behavior present at that time. A laboratory selective breeding study has provided us with a model of how natural selection may have shaped the evolution of this most interesting early behavior, and linked it with other traits into a complex pattern of emotional behavior or temperament developing over time.

Acknowledgments

The work described in this chapter was supported by project grants and a Research Scientist Award from NIMH, and by the Sackler Institute for Developmental Psychobiology at Columbia University.

My colleagues Harry Shair and Susan Brunelli played crucial roles in carrying out and interpreting the results of the research.

Permission was granted by Blackwell Publishing to reproduce portions of the paper, "Unexplained infant crying: an evolutionary perspective" by Myron A. Hofer (published in Acta Pediatrica, 91: 491–496 (2002)).

References

Blumberg, M.S., Alberts, J.R., 1990. Ultrasonic vocalizations by rat pups in the cold: an acoustic by-product of laryngeal braking? Behav. Neurosci. 104, 808–817.

Brunelli, S.A., Hofer, M.A., 2001. Selective breeding for an infantile phenotype (isolation calling): a window on developmental processes. In: Blass, E. (Ed.), Handbook of Behavioral Neurobiology. Plenum Press, New York, NY, pp. 433–482.

Brunelli, S.A., Hofer, M.A., 2007. Selective breeding for infant rat separation-induced ultrasonic vocalizations: developmental precursors of passive and active coping styles. Behav. Brain Res. 182, 193–207.

Brunelli, S.A., Shair, H.N., Hofer, M.A., 1994. Hypothermic vocalizations of rat pups (*Rattus norvegicus*) elicit and direct maternal search behavior. J. Comp. Psychol. 108, 298–303.

Hofer, M.A., 1996. Multiple regulators of ultrasonic vocalization in the infant rat. Psychoneuroendocrinol. 21, 203–217.

Hofer, M.A., Shair, H., 1978. Ultrasonic vocalization during social interaction and isolation in 2-week-old rats. Dev. Psychobiol. 11, 495–504.

Hofer, M.A., Shair, H.N., 1992. Ultrasonic vocalization by rat pups during recovery from deep hypothermia. Dev. Psychobiol. 25, 511–528.
Hofer, M.A., Shair, H.N., 1993. Ultrasonic vocalization, laryngeal braking, and thermogenesis in rat pups: a reappraisal. Behav. Neurosci. 107, 354–362.
Hofer, M.A., Brunelli, S.A., Masmela, J., Shair, H.N., 1996. Maternal interactions prior to separation potentiate isolation-induced calling in rat pups. Behav. Neurosci. 110, 1158–1167.
Hofer, M.A., Masmela, J.R., Brunelli, S.A., Shair, H.N., 1998. The ontogeny of maternal potentiation of the infant rats' isolation call. Dev. Psychobiol. 33, 189–201.
Negus, V.E., 1931. The Mechanism of the Larynx. The C.V. Mosby Company, St. Louis, MO.
Shair, H.N., Masmela, J.R., Hofer, M.A., 1999. The influence of olfaction on potentiation and inhibition of ultrasonic vocalization of rat pups. Physiol. Behav. 65, 769–772.
Takahashi, L.K., 1992. Ontogeny of behavioral inhibition induced by unfamiliar adult male conspecifics in preweanling rats. Physiol. Behav. 52, 493–498.

CHAPTER 2.4

Understanding the evolutionary origin and diversification of bat echolocation calls

Alanna Maltby[1], Kate E. Jones[2] and Gareth Jones[3*]

[1]*Department of Genetics, Evolution and Environment, University College London, London, UK*
[2]*Zoological Society of London, Regent's Park, London, UK*
[3]*School of Biological Sciences, University of Bristol, Bristol, UK*

Abstract: The production of sound and the interpretation of the resulting echoes (echolocation) is used by some birds, dolphins and many bats. However, the complexity, sophistication and diversity of the design of echolocation calls are unparalleled within bats. Most bats echolocate by emitting frequencies between 20 and 60 kHz (ranging from 9 to 212 kHz) and the specific frequencies they use are a compromise between the costs associated with the likelihood of target detection and call attenuation. Bats have evolved a variety of designs, from simple broadband clicks to constant frequency calls. Interpretation of how different call designs have evolved has been hampered by the lack of a robust phylogenetic framework. However, recent improvements in our understanding of bat evolutionary relationships suggest that laryngeal echolocation either evolved once and was subsequently lost in non-echolocating bats, or evolved independently twice in different bat lineages. Reconstructing ancestral call designs is challenging, but it is suggested that early echolocating bats used calls that were tonal, short, broadband and multi-harmonic. There is some suggestion that flight evolved before echolocation, and that coevolution with insects and plants has shaped the evolution of call designs in bats. The key to understanding call evolution may be in understanding the genetic basis of echolocation, and this seems an active and important area for future research.

Keywords: ancestral calls; bats; call design; coevolution; echolocation; evolution; genes

I. Introduction

Echolocation, or biosonar, is the active use of calls and interpretation of their echoes to detect, localize, and classify objects (Griffin, 1944; Jones, 2005). An animal produces a sound and then listens to and analyzes echoes that reflect from the surroundings. Auditory cues relating to the direction, timing and spectral composition of the objects allow the animal to then perceive, pinpoint and recognize potential prey and obstacles (Pollak and Casseday, 1989; Thomas et al., 2004).

Echolocation has evolved in environments where vision is of limited use; the nocturnal niche of bats reduces the efficacy of vision. In dolphins, the turbidity of the water and the tendency to dive to depths where very little light permeates have contributed to the evolution of echolocation, and bird species which echolocate do so only to orientate in the dark caves in which they roost. The echolocation signals of non-bat echolocators are clicks, unlike the usually longer duration and more complex calls often produced in the larynx by bats (Thomas and Jalili, 2004). Although echolocation confers bats with remarkable abilities to perceive objects in the dark, there are times when it is ineffective. For example, bats such as the Greater Mouse-eared Bat (*Myotis myotis*) almost cease echolocating when searching for prey hidden under complex surfaces such as leaf litter and rely on listening for prey-generated sounds instead (Arlettaz et al., 2001).

The evolution of sophisticated echolocation probably came at a cost. The small body size of bats constrains the evolution of large brains, and given that the brain may consume up to 20% of circulating oxygen

*Corresponding author. E-mail: Gareth.Jones@bristol.ac.uk

Stefan M. Brudzynski (Ed.)
Handbook of Mammalian Vocalization
ISBN 978-0-12-374593-4

DOI: 10.1016/B978-0-12-374593-4.0005-X

and glucose (Martin, 1981; Armstrong, 1983), specialized enlargement of brain regions associated with auditory processing may only be possible if reduction in the size of areas associated with visual processing occurs (Harvey and Krebs, 1990). Hence, the evolution of sophisticated visual capabilities in bats may be limited because of trade-offs associated with auditory specialization.

II. Echolocation call frequencies

Bat echolocation calls tend to be in the ultrasonic frequency range (sound above the upper limit of human hearing, standardized at 20 kHz), although some echolocation calls are audible to humans. The lowest frequencies used in echolocation are emitted by the spotted bat (*Euderma maculatum*, Vespertilionidae), which uses frequencies containing most energy around 9 kHz (Fullard and Dawson, 1997). In contrast, the highest frequencies reach about 212 kHz in Percival's short-eared trident bat (*Cloeotis percivali*, Hipposideridae) (Fenton and Bell, 1981). Ultrasonic frequencies are not a requirement for echolocation, although there are several advantages to using them. Sounds reflect most clearly from objects larger than the wavelength of the sound, and because high-frequency sounds have short wavelengths, they allow strong echoes to be generated from small objects such as flying insects (Houston et al., 2004). High frequencies are also directional (e.g., Surlykke et al., 2008; see also Brudzynski and Fletcher, Chapter 3.3 in this volume) and can limit the spread of echolocation calls so that objects other than the target of interest are not detected. While bats may avoid using low frequencies in echolocation because of the need to detect small targets, extremely high frequencies are also avoided because excess atmospheric attenuation limits the range over which echolocation is effective (Lawrence and Simmons, 1982; Brudzynski and Fletcher, Chapter 3.3 in this volume). Most bats utilize "compromise" frequencies to avoid the costs associated with very high and very low frequencies, and call between 20 and 60 kHz (Fenton et al., 1998).

III. Diversity in call design

Bats show considerable diversity in call design (Fig. 1; Table 1). Some bats do not echolocate at all, i.e., most of the bats within the family Pteropodidae. Although species in a given family often tend to have similar call designs to one another, selective pressures imposed by the environment have overridden phylogenetic constraints in many cases. For example, the common moustached bat *Pteronotus parnellii* (Mormoopidae) has independently evolved a very similar call design to bats in the Rhinolophidae family, and even compensates for Doppler shifts caused by varying flight speeds (Schnitzler, 1972). In addition, there is much

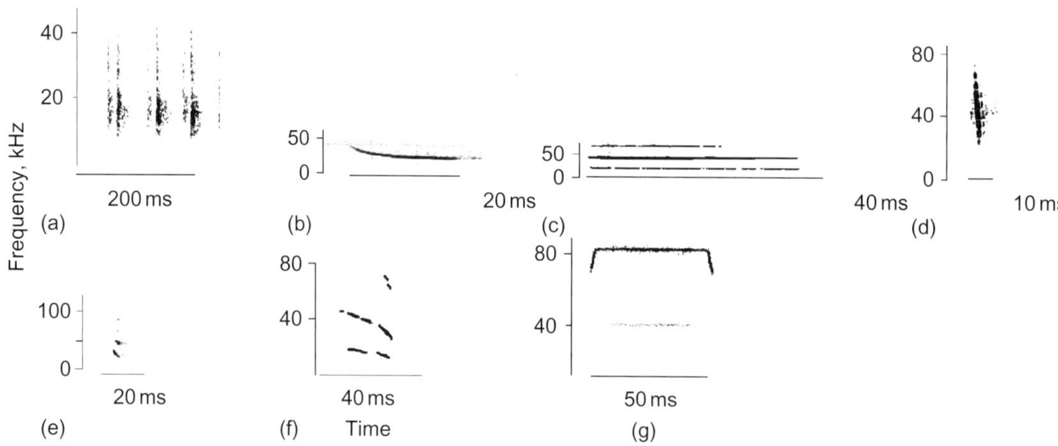

Fig. 1 Variation in signal design among different species of echolocating bats. (a) is a brief, broadband tongue click from *Rousettus aegyptiacus*; (b) is a narrowband call dominated by the fundamental harmonic, from *Eptesicus fuscus*; (c) is a narrowband multiharmonic signal from *Rhinopoma hardwickii*; (d) is a short broadband signal dominated by the fundamental harmonic from *Myotis daubentonii*; (e) is a short broadband multiharmonic call from *Mystacina tuberculata*; (f) is a long broadband multiharmonic signal from *Myzopoda aurita*; (g) is a constant frequency signal from *Rhinolophus ferrumequinum*. Based on an illustration in Jones and Teeling (2006).

intraspecific variation in call design, as different designs are suited to different perceptual challenges (e.g., Kalko and Schnitzler, 1993). This plasticity often makes it more difficult to classify taxa by the echolocation call type. However, with some generalization, it is possible to categorize the call types of bats. For example, there are several aspects of call design that define a bat's call. The first is the duration of the call – calls may be extremely brief (<1 ms) as in the broadband tongue clicks of the only echolocating genus in the Pteropodidae (*Rousettus*) relatively short (<10 ms) as in the calls of some bats in the family Vespertilionidae (e.g., many species in the genus *Myotis*); or long (>10 ms and up to 60 ms) such as calls emitted by many bats in the families Myzopodidae, Rhinolophidae and Hipposideridae (Fig. 1; Table 1).

A further defining aspect is the bandwidth of the call. Calls may be very broadband, exceeding 120 kHz in some *Kerivoula* species (Vespertilionidae) (Kingston et al., 1999), or they may be very narrowband, as in some species in the Rhinopomatidae and Vespertilionidae (e.g., Habersetzer, 1985) (Fig. 1; Table 1). Frequencies change across the time domain, as in the calls of the Rhinolophidae, which start with a brief upward broadband sweep, followed by a long constant frequency (CF) component, before ending with a downward broadband sweep (e.g., Jones and Rayner, 1989).

Many bats also use several harmonics in their calls, although usually with most of the call's energy concentrated into either the first (the fundamental) or the second harmonic (Jones and Teeling, 2006). Each subsequent harmonic is a multiple of the fundamental frequency.

IV. Phylogenetic context

In order to understand how different call types have evolved, for example, whether similar call designs have evolved convergently in phylogenetically distant taxa, it is necessary to place these calls into a context of the evolutionary history of bats. However, any

Table 1. Diversity of bat echolocation signals across families

Family	Call type	Clade size
Craseonycteridae (Hog-nosed bat)	4	1
Emballonuridae (Sheath-tailed bats)	4	51
Furipteridae (Smoky bats)	no data	2
Hipposideridae (Old World leaf-nosed bats)	8	81
Megadermatidae (False vampire bats)	6	5
Miniopteridae (Long-fingered bats)	3	19
Molossidae (Free-tailed bats)	3	100
Mormoopidae (Naked-backed bats)	4,8	10
Mystacinidae (Short-tailed bats)	6	2
Myzopodidae (Sucker-footed bats)	7	1
Natalidae (Funnel-eared bats)	6	8
Noctilionidae (Bulldog bats)	5,8	2
Nycteridae (Slit-faced bats)	6	16
Phyllostomidae (New World leaf-nosed bats)	6	160
Pteropodidae (Old World fruit bats)	1, 2	186
Rhinolophidae (Horseshoe bats)	8	77
Rhinopomatidae (Mouse-tailed bats)	4	4
Thyropteridae (Disk-winged bats)	4	3
Vespertilionidae (Vesper bats)	3, 5, 6	388

Call types are categorized as follows: 1, no echolocation; 2, brief broadband tongue clicks; 3, narrowband dominated by fundamental harmonic; 4, narrowband multiharmonic; 5, short broadband dominated by fundamental harmonic; 6, short broadband multiharmonic; 7, long broadband multiharmonic; 8, constant frequency. Taxonomy following Simmons (2005) and Miller-Butterworth et al. (2007) and echolocation call data from Jones and Teeling (2006).

such inferences are dependent on the accuracy of the reconstruction of bat evolutionary relationships. Bat evolutionary history has been, and remains, a hotly debated topic, with disagreements at all levels of the phylogenetic tree, both within the order and regarding bats' relationships to other mammals, making inferences about the evolution of call designs challenging. To place this into context we briefly review the most important issues to shape our understanding of the evolutionary relationships of bats.

Bats (Order Chiroptera) are the second most speciose mammalian order after the Rodentia, comprising around 1,116 species in 202 genera and 18 families (Simmons, 2005), although there may be as many as 19 families according to the latest taxonomic revisions (Gunnell and Simmons, 2005; Miller-Butterworth et al., 2007; Simmons et al., 2008). Following morphological studies, bats were previously placed within the group Archonta, which considered the flying lemurs, primates and tree shrews as possible sister taxa (Gregory, 1910). More recent molecular studies place Chiroptera as a basal or near-basal member of the Laurasiatheria, which contains the pangolins, carnivores, cetaceans, artiodactyls, perissodactyls and eulipotyphlans (such as shrews and moles) (Murphy et al., 2001). Understanding the possible sister taxa to bats has important implications to our understanding of the evolution of echolocation in bats. For example, as other taxa in Laurasiatheria are known to echolocate (e.g., cetaceans and perhaps shrews), it is possible to infer how primitive or derived this characteristic is in bats.

V. The chiropteran monophyly debate

In the 1980s and 1990s, despite previously widespread acceptance that all bats shared a common ancestor, a debate began around the suggestion that bats were diphyletic rather than monophyletic. Pettigrew (1986) discovered that the system of neural connections between the midbrain and the retina of pteropodids (Pteropodidae, formerly classified in the suborder Megachiroptera, see below) was similar to that of primates, whereas other bats (the former suborder Microchiroptera) showed the putatively "primitive" system of connections, in common with all other mammals. This evidence, in association with other aspects of morphology (Smith, 1977; Smith and Madkour, 1980; Pettigrew, 1986; Kennedy et al., 1987; Pettigrew et al., 1989; Buhl and Dann, 1991) and some early molecular evidence (Kleinschmidt et al., 1988) suggested that the megachiropterans evolved from a shared ancestor with primates, in a separate lineage from that which led to the evolution of microchiropterans. The bat diphyly hypothesis proposed that flight evolved on two independent occasions in bats. However, as molecular techniques developed and became more widely used, support for monophyly has proven overwhelming (e.g., Teeling et al., 2000; Murphy et al., 2001; Teeling et al., 2005; although see Pettigrew, 2008). If we accept that bats are monophyletic, then echolocation calls can be understood to have evolved from a common bat ancestor and diversifying into different designs across the clade.

VI. The microchiropteran monophyly debate

Within the 19 currently recognized extant and seven extinct bat families, familial interrelationships have also been debated. Bats had traditionally been placed in two monophyletic suborders, the Microchiroptera and the Megachiroptera (Dobson, 1875) (Fig. 2). Megachiroptera comprise one family, the Old World fruit bats (Pteropodidae) the majority of which do not use echolocation (although some use broadband clicks), whereas Microchiroptera includes all the other families (which all use some form of echolocation) (see Table 1). However, the monophyly of the Microchiroptera is now disputed, as molecular evidence suggests that several of the microchiropteran families are more closely related to Pteropodidae than to the remaining families (Fig. 2). The most widely used division of the order places the Pteropodidae, Rhinolophidae, Hipposideridae, Megadermatidae, Craseonycteridae, and the Rhinopomatidae in a new suborder, initially termed the Yinpterochiroptera. This was formed by the concatenation of "Yinochiroptera" introduced by Koopman (1984) and "ptero" by Springer et al. (2001), and leaves the remaining families in the suborder Yangochiroptera (Fig. 2). Yangochiroptera was originally named by Koopman (1984), although at that time it excluded Nycteridae and Emballonuridae (Springer et al., 2001; Gunnell and Simmons, 2005). These clades are now commonly referred to as Pteropodiformes and Vespertilioniformes for Yinpterochiroptera and Yangochiroptera respectively (following Hutcheon and Kirsch, 2006), and we follow this convention throughout the rest of this chapter.

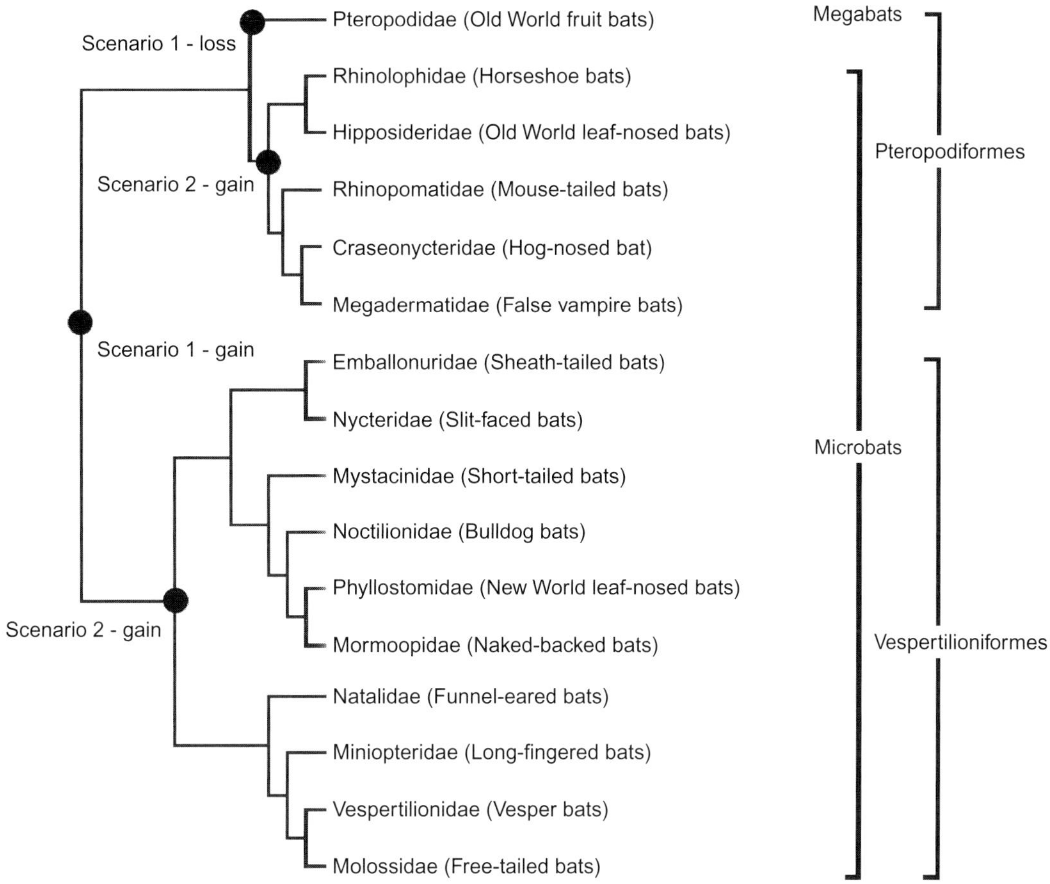

Fig. 2 Phylogenetic relationships among 19 bat families following Teeling et al. (2005). Circles indicate where echolocation may have evolved, either once (Scenario 1) or twice (Scenario 2).

The new understanding of the phylogenetic relationships between families suggests that the most parsimonious explanation under the current phylogeny is that laryngeal echolocation evolved once, at the root of the bat lineage, and was later lost in the Pteropodidae. This would suggest that the pteropodid genus *Rousettus* then independently evolved their broadband clicks. Alternatively, echolocation evolved at least twice: once in the ancestors of echolocating Pteropodiformes and independently in Vespertilioniformes (Fig. 2). However, whether laryngeal echolocation evolved once or more than once in bats remains unresolved: molecular scaffolds that incorporate fossil taxa support a single evolutionary event, while analyses of genes associated with hearing (Li et al., 2008) suggest that separate evolutionary events may have occurred.

Reconstructions of the ancestral states of echolocation calls are problematic, given the difficulties imposed by convergence in signal design across phylogenetically distant taxa. Mapping of call characteristics onto phylogenetic trees suggests that the use of calls dominated by the fundamental harmonic (as in many members of the Vespertilionidae and Molossidae) is a derived state, and that multiharmonic signals are ancestral (see Fig. 3) (Jones and Teeling, 2006). Schnitzler et al. (2004) suggested that early echolocating bats used calls that were tonal, low intensity, short, broadband and multiharmonic. Conversely, Eick et al. (2005) hypothesized that the ancestral protobat emitted high intensity calls.

Some features of bat echolocation have evolved independently on several occasions. We have already described the independent evolution of constant frequency calls and Doppler shift compensation in rhinolophids (Pteropodiformes) and in *Pteronotus parnellii* (Vespertilioniformes). Furthermore, nasal emission of calls has evolved in some vespertilionids

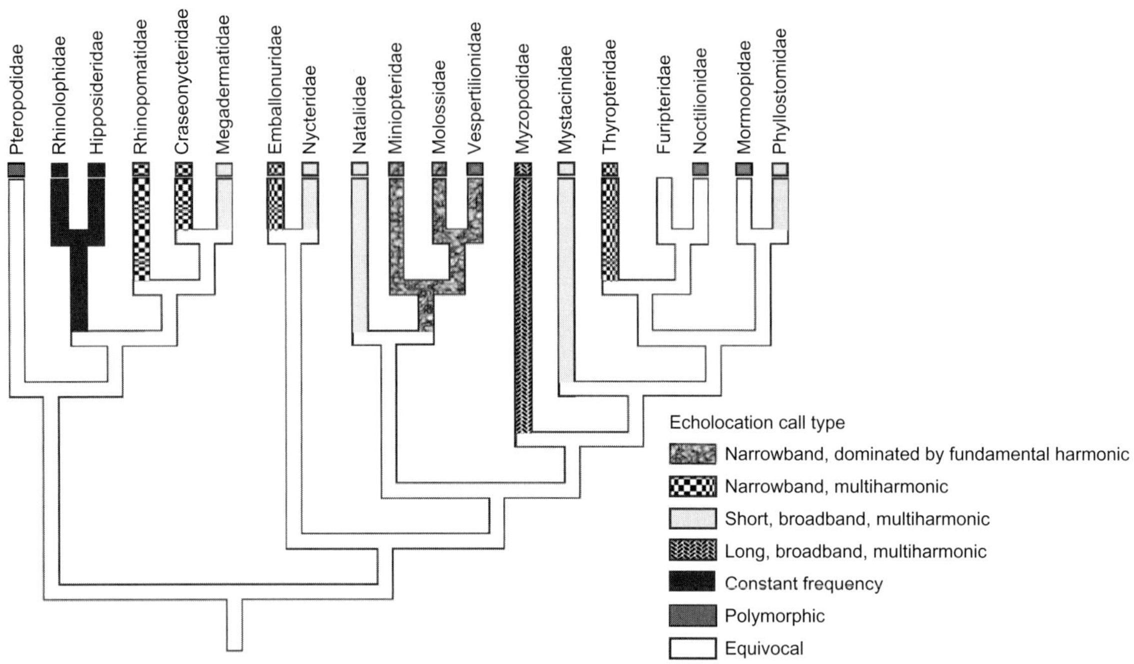

Fig. 3 Linear parsimony ancestral reconstructions of eight echolocation call types using the family-level molecular phylogeny of Teeling et al. (2005) using MacClade v.3.8 (Maddison and Maddison, 1992). Adapted from Jones and Teeling (2006). Call types were categorized as in Table 1. Different patterns represent either the call type of the family (patterns in the terminal squares at the end of the branches) or ancestral reconstructions of the call types (branch patterns). Some families have more than one call type and this is denoted as polymorphic (i.e., Pteropodidae has call types 1 and 2; Noctilionidae 5 and 8; Mormoopidae 4 and 8; and Vespertilionidae 3, 5 and 6). Branches are denoted as equivocal where it is not possible to estimate the call type ancestral condition. Adapted from Jones and Teeling (2006).

(e.g., Rafinesque's big-eared bat *Corynorhinus rafinesqui*) (Griffin, 1958), nycterids and phyllostomids within the Vespertilioniformes, and in megadermatids in the Pteropodiformes (Eick et al., 2005).

VII. Which came first: flight or echolocation?

Bats are thought to have evolved around 55 to 65 million years ago (Teeling et al., 2005), with the oldest known fossils dated to the Eocene about 53 million years ago (Simmons and Geisler, 1998). Whether bats evolved flight before echolocation or vice versa has been hotly debated (summarized in Speakman, 2001). Briefly, there are three hypotheses that have been posited: echolocation first; flight first; and tandem (where the ancestral bat uses echolocation only for communication and this develops with the ability to fly). The strongest evidence to date to test these hypotheses is from a recent fossil *Onychonycteris finneyi*, dated at around 52.5 million years ago (Simmons et al., 2008). Cranial features of this fossil, together with its relatively small cochlea, suggest that bats may have evolved powered flight prior to evolving the ability to echolocate. However, whether this bat lacked echolocation entirely, or whether it was capable of basic echolocation that did not involve morphological specialization, is open to debate. Prior to the discovery of *Onychonycteris*, there was no clear evidence in support of any of the three hypotheses for the evolution of echolocation and flight (Speakman, 2001).

VIII. Coevolution with insects

There is some evidence that bat call designs may also have been influenced by interactions with other taxa, for example, the arms race between some bat species and their tympanate moth prey (which can hear ultrasound) (Fullard, 1998). Some bats have evolved the use of low echolocation frequencies that are below the range of moth hearing, and this enables them to prey on moths. For example, the Mexican free-tailed bat (*Tadarida teniotis*) calls as low as 11 kHz and the spotted bat (*Euderma maculatum*) calls at around 9 kHz. Both feed mainly on tympanate insects (Rydell and Arlettaz,

1994; Fullard and Dawson, 1997). Hipposiderid and rhinolophid bats that emit high frequencies tend to eat more moths than low-frequency congeners (Jones, 1993; Bogdanowicz et al., 1999). Hence, bats may be able to catch tympanate prey by using echolocation calls at allotonic frequencies (i.e., by calling at frequencies outside the range of moth hearing) (Fenton and Fullard, 1979; Schoeman and Jacobs, 2003). However, whether bats evolved allotonic frequencies primarily to exploit tympanate prey, or initially for other reasons associated with improving echolocation performance (e.g., for detecting smaller targets by using higher frequencies) is open to question. In response to the echolocation calls of moth-eating bats, some insects in turn have evolved tympanal organs (ears) which can detect frequencies with highest sensitivity roughly between 20 and 50 kHz, and react to these sounds by changing direction or folding their wings and dropping out of the flight path of the bat (Jones and Rydell, 2003). It is clear that hunting by echolocating bats is an important selection pressure shaping insect hearing, because sensitivity to ultrasound has evolved in at least six insect orders, including several times independently in moths (Hoy, 1998).

IX. Coevolution with plants

Similarly, we may be able to gain insight into the influence that flowering, fruiting and nectar-producing plants have had on the bat echolocation call design and vice versa. For example, around 1,000 plant species in the Neotropics are bat-pollinated by bats in the family Phyllostomidae. Adaptations of the plants to the bats include outward-facing flowers; large, sturdy petals with exposed stamens; inconspicuously colored, strong-smelling, night-opening flowers; high protein content pollen; and large volumes of nectar, released in small doses (Dobat and Peikert-Holle, 1985). There is some evidence that bats have tailored their echolocation calls to the recognition of the plants they feed on, and it is possible that plants have evolved structures, shapes and textures that help bats to locate them and to feed (Von Helversen et al., 2003).

X. Genes and echolocation

With complete genome sequences soon available from two bat species, the little brown bat (*Myotis lucifugus*) and the greater horseshoe bat (*Rhinolophus ferrumequinum*), several implications for better understanding the evolution of echolocation exist. Comparisons with the genomes of other mammals will elucidate some of the genetic changes responsible for the evolution of echolocation. Sequencing studies of candidate genes selected because of a likely influence on echolocation have already been illuminating. For example, FOXP2 is a transcription factor involved in the neural control of orofacial coordination. Mutations in the *FOXP2* gene are responsible for impaired coordination in human speech, and in the comprehension of language (Fisher and Marcus, 2006). Mutations in *FoxP2* may have been important in the evolution of language, with positive selection resulting in two amino acid changes in *FoxP2* since humans shared a common ancestor with chimpanzees (Enard et al., 2002). Infant *FoxP2* knockout mice fail to produce ultrasonic isolation calls (Shu et al., 2005).

FoxP2 has traditionally been viewed as being highly conserved in mammals, but recent studies on *FoxP2* evolution in bats have changed this view, and have shown accelerated evolution of the *FoxP2* gene in bats compared with other vertebrates (Li et al., 2007). *FoxP2* is extremely diverse in echolocating bats, especially at exons 7 and 17. Exon 7 is of interest because its mutations have been linked to language impairment in humans (Fisher and Marcus, 2006). Exon 17 was invariant across all other eutherian mammals surveyed by Li et al. (2007), with the exception of one non-synonymous mutation in the pig. Variation in bats was substantial, with up to 8 non-synonymous mutations recorded.

FoxP2 is implied to be involved in vocal learning, as well as in orofacial coordination, but specific mutations linked to vocal learning abilities have not been pinpointed (Webb and Zhang, 2005; Li et al., 2007). Li et al. (2007) argued that the high variability in *FoxP2* in bats was more probably associated with sensorimotor demand than with vocal learning, despite bats showing evidence for vocal learning of both social (Boughman, 1998) and echolocation (Jones, 1993; Hiryu et al., 2006) calls. The high levels of *FoxP2* sequence diversity in bats may be involved in the mobilization of downstream genes or genetic cascades associated with complex sensorimotor coordination necessary for echolocation. Bats emit calls at rates of up to 200 pulses per second when attacking insects (Jones and Rydell, 2003), and rapid feedback from individual calls must represent a demanding challenge (Moss and Sinha, 2003). The diversity of echolocation signals used by echolocating bats is high, and perhaps diversity in *FoxP2* shapes some of this extensive variability and influences neural circuits that underpin echolocation. Information on the expression patterns of *FoxP2* in bat brains will be valuable for comparisons with data already available for other mammals. *FoxP2* is strongly expressed in the inferior colliculus of mice (Ferland et al., 2003), a part of the brain that shows enlargement and morphological specialization in bats (Glezer et al., 2004), and which is fundamental in bat echolocation (Pollak and Casseday, 1989). A deeper understanding of the ways in which *FoxP2* interacts with downstream targets will also be valuable.

Another gene of interest to understanding the evolution of echolocation is that coding for the motor protein Prestin, which drives amplification in the outer hair cells (OHCs) of the mammalian cochlea. Its expression pattern correlates with the appearance of electromotility in thc OHCs (Zheng et al., 2000). Prestin differs from classic motors that require ATP hydrolysis, and instead appears to work by direct conversion of voltage into force. Prestin can therefore work several orders of magnitude faster than cellular motor proteins, and targeted gene disruptions revealed a >100-fold (40 dB) loss of auditory sensitivity in homozygote *Prestin*-knockout mice (Liberman et al., 2002). Prestin appears crucial for auditory sensitivity and selectivity, and perhaps also for high-frequency hearing in mammals (Liberman et al., 2002), and so the evolution of the *Prestin* gene in mammals was identified as being of interest (Li et al., 2008).

Prestin belongs to a superfamily of anion transporter genes collectively known as *SLC26* genes. A phylogeny of *SLC26* genes showed positive selection during the evolution of *Prestin* on the lineage leading to mammals, but gene conservation suggested strong purifying selection after mammals evolved (although only four species were studied) (Franchini and Elgoyhen, 2006). Evolutionary analyses of *Prestin* sequences from a wide range of bat species and from additional mammal taxa revealed several interesting points (Li et al., 2008). First, positive selection was identified in the lineage leading to the evolution of rhinolophoid bats that use long constant frequency components in their echolocation calls, and which use full or partial Doppler shift compensation in echolocation. These bats possess an "auditory fovea" (Schuller and Pollak, 1979) and show extremely sharp auditory tuning, with Q10 dB values 10–30 times higher than those observed in other mammals (Kössl and Vater, 1995). It seems likely that further modifications in *Prestin* may be associated with exquisite frequency selectivity in rhinolophid bats (Li et al., 2008).

A phylogeny of bats based on *Prestin* sequences (Fig. 4) supports monophyly in taxa that use laryngeal echolocation, i.e., it recovers the traditional placement of Megachiroptera and Microchiroptera as monophyletic clades. However, other extensive sequencing studies, often using genes that may be neutral in function, support paraphyly in laryngeal echolocators, i.e., the Pteropodiformes and Vespertilioniformes clades

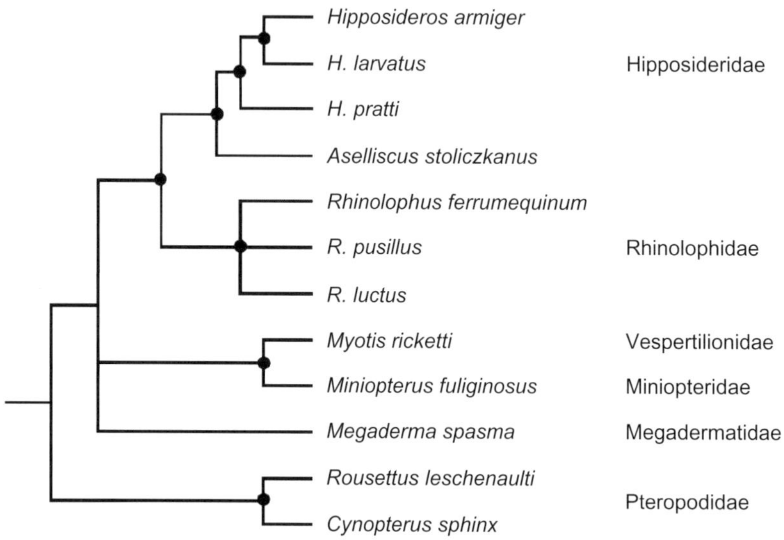

Fig. 4 Gene tree based on the complete *Prestin* coding sequence in bats. Adapted from Li et al. (2008).

(e.g., Eick et al., 2005; Teeling et al., 2000, 2005; Van Den Bussche and Hoofer, 2004). It is therefore likely that the phylogenetic history of *Prestin* arises from convergence at the genetic level, rather than reflecting the true evolutionary history of bat species. That sequences from genes such as *Prestin* reunite echolocating bats cautions against the use of putatively functional genes in molecular phylogenetic analyses. Such genes, like many morphological traits, may be subject to convergence and may misrepresent evolutionary histories (Li et al., 2008). The *Prestin* phylogeny also suggests that there has been independent evolution of echolocation across bats.

Genome comparisons have already illuminated some of the key areas of genetic divergence between chimpanzees and humans (Clark et al., 2003). The capacity for genomic comparisons to allow a better understanding of genes that might have undergone positive selection in the evolution of echolocation is vast. We anticipate this as an area of major research interest in the coming years. Using genetic data to better resolve phylogenies is also crucial. More in-depth phylogenies will allow a better understanding of constraints on signal evolution. The combination of phylogenetics and better categorization and documentation of signal diversity in bats bodes for exciting developments in research on the evolution of bat echolocation.

Acknowledgments

This work was financially supported through a NERC studentship to AM, the Zoological Society of London (AM & KEJ) and the Darwin Initiative and BBSRC (GJ).

References

Arlettaz, R., Jones, G., Racey, P.A., 2001. Effect of acoustic clutter on prey detection by bats. Nature 414, 742–745.

Armstrong, E., 1983. Relative brain size and metabolism in mammals. Science 220, 1302–1304.

Bogdanowicz, W., Fenton, M.B., Daleszczyk, K., 1999. The relationships between echolocation calls, morphology and diet in insectivorous bats. J. Zool. (London) 247, 381–393.

Boughman, J., 1998. Vocal learning by greater spear-nosed bats. Proc. Royal Soc., London, B 265, 227–233.

Buhl, E.H., Dann, J.F., 1991. Cytoarchitecture, neuronal composition and entorhinal afferents of the flying fox hippocampus. Hippocampus 1, 32–53.

Clark, A.G., Glanowski, S., Nielsen, R., Thomas, P.D., Kejariwal, A., Todd, M.A., Tanenbaum, D.M., Civello, D., Lu, F., Murphy, B., Ferriera, S., Wang, G., Zheng, X., White, T.J., Sninsky, J.J., Adams, M.D., Cargill, M., 2003. Inferring nonneutral evolution from human–chimpanzee–mouse orthologous gene trios. Science 302, 1960–1963.

Dobat, K., Peikert-Holle, T., 1985. Blüten und Fledermäuse. Verlag Waldemar Kramer, Frankfurt am Main, Germany.

Dobson, G.E., 1875. Conspectus of the suborders, families and genera of Chiroptera arranged according to their natural affinities. Annals of the Magazine of Natural History Series 4, 345–357.

Eick, G.N., Jacobs, D.S., Matthee, C.A., 2005. A nuclear DNA phylogenetic perspective on the evolution of echolocation and historical biogeography of extant bats (Chiroptera). Mol. Biol. Evol. 22, 1869–1886.

Enard, W., Przeworski, M., Fisher, S.E., Lai, C.S.L., Wiebe, V., Kitano, T., Monaco, A.P., Pääbo, S., 2002. Molecular evolution of *FOXP2*, a gene involved in speech and language. Nature 418, 869–872.

Fenton, M.B., Fullard, J.H., 1979. The influence of moth hearing on bat echolocation strategies. J. Comp. Physiol. 132, 77–86.

Fenton, M.B., Bell, G.P., 1981. Recognition of species of insectivorous bats by their echolocation calls. J. Mammal. 62, 233–243.

Fenton, M.B., Portfors, C.V., Rautenbach, I.L., Waterman, J.M., 1998. Compromises: sound frequencies used in echolocation by aerial-feeding bats. Can. J. Zool. 76, 1174–1182.

Ferland, R.J., Cherry, T.J., Preware, P.O., Morrisey, E.E., Walsh, C.A., 2003. Characterization of Foxp2 and Foxp1 mRNA and protein in the developing and mature brain. J. Comp. Neurol. 460, 266–279.

Fisher, S.E., Marcus, G.F., 2006. The eloquent ape: genes, brains and the evolution of language. Nature Rev. Genetics 7, 9–20.

Franchini, L.F., Elgoyhen, A.B., 2006. Adaptive evolution in mammalian proteins involved in cochlear outer hair cell electromotility. Mol. Phylogenet. Evol. 41, 622–635.

Fullard, J.H., 1998. The sensory coevolution of moths and bats. In: Hoy, R.R., Popper, A.N., Fay, R.R. (Eds.), Comparative Hearing: Insects. Springer, Berlin, Germany, pp. 279–326.

Fullard, J.H., Dawson, J.W., 1997. The echolocation calls of the spotted bat *Euderma maculatum* are relatively inaudible to moths. J. Exp. Biol. 200, 129–137.

Glezer, I.L., Hof, P., Morgane, P.J., Fridman, A., Isakova, T., Joseph, D., Nair, A., Parhar, P., Thengampallil, A., Thomas, S., Venugopal, R., Jung, G.H., 2004. Chemical neuroanatomy of the inferior colliculus in brains of echolocating and nonecholocating mammals: immunocytochemical study. In: Thomas, J.A., Moss, C.F., Vater, M. (Eds.), Echolocation in Bats and Dolphins. University of Chicago Press, Chicago, Il, pp. 161–172.

Gregory, W.K., 1910. The orders of mammals. Bull. Am. Museum Nat. Hist. 27, 1–524.

Griffin, D.R., 1944. Echolocation by blind men, bats and radar. Science 100, 589–590.

Griffin, D.R., 1958. Listening in the Dark. Yale University Press, New Haven, CT.

Gunnell, G., Simmons, N.B., 2005. Fossil evidence and the origin of bats. J. Mammal. Evol. 12, 209–246.

Habersetzer, J., 1985. Vergleichende flügelmorphologische Untersuchungen an einer Fledermausgesellschaft in Madurai. In: Nachtigall, W. (Ed.), In Biona Report 5. Bat Flight – Fledermausflug. Gustav Fischer, Stuttgart, Germany, pp. 75–106.

Harvey, P.H., Krebs, J.R., 1990. Comparing brains. Science 294, 140–146.

Hiryu, S., Katsura, K., Nagato, T., Yamazaki, H., Lin, L.-K., Watanabe, Y., Riquimaroux, H., 2006. Intra-individual variation in the vocalized frequency of the Taiwanese leaf-nosed bat, *Hipposideros terasensis*, influenced by conspecific colony members. J. Comp. Physiol., A 192, 807–815.

Houston, R.D., Boonman, A.M., Jones, G., 2004. Do echolocation signal parameters restrict bats' choice of prey? In: Thomas, J.A., Moss, C.F., Vater, M. (Eds.), Echolocation in Bats and Dolphins. University of Chicago Press, Chicago and London, pp. 339–349.

Hoy, R.R., 1998. Acute as a bug's ear: an informal discussion of hearing in insects. In: Hoy, R.R., Popper, A.N., Fay, R.R. (Eds.), Comparative Hearing: Insects. Springer-Verlag, New York, NY.

Hutcheon, J.M., Kirsch, J.A.W., 2006. A moveable face: deconstructing the Microchiroptera and a new classification of extant bats. Acta Chiropterol. 8, 1–10.

Jones, G., 1993. Bats vs. moths: studies on the diets of rhinolophid and hipposiderid bats support the allotonic frequency hypothesis. In: Horáček, I.,Vohralík, V. (Eds.), Prague Studies in Mammalogy. Charles University Press, Prague, Czech Republic, pp. 87–92.

Jones, G., 2005. Echolocation. Current Biol. 15, R484–R488.

Jones, G., Rayner, J.M.V., 1989. Foraging behavior and echolocation of wild horseshoe bats *Rhinolophus ferrumequinum* and *R. hipposideros* (Chiroptera, Rhinolophidae). Behav. Ecol. Sociobiol. 25, 183–191.

Jones, G., Rydell, J., 2003. Attack and defense: interactions between echolocating bats and their insect prey. In: Kunz, T.H., Fenton, M.B. (Eds.), Bat Ecology. Chicago University Press, Chicago, IL, pp. 301–345.

Jones, G., Teeling, E.C., 2006. The evolution of echolocation in bats. Trends Ecol. Evol. 21, 149–156.

Kalko, E.K.V., Schnitzler, H.-U., 1993. Plasticity in echolocation signals of European pipistrelle bats in search flight – implications for habitat use and prey detection. Behav. Ecol. Sociobiol. 33, 415–428.

Kennedy, W.D., Pettigrew, J.D., Calford, M.B., 1987. Cells of origin of the corticospinal tract in the little red flying fox. Proc. Australian Physiol. Pharmacol. 18, 102.

Kingston, T., Jones, G., Zubaid, A., Kunz, T.H., 1999. Echolocation signal design in Kerivoulinae and Murininae (Chiroptera: Vespertilionidae) from Malaysia. J. Zool., London 249, 359–374.

Kleinschmidt, T., Sgouros, J.G., Pettigrew, J.D., Braunitzer, G., 1988. The primary structure of the hemoglobin from the gray-headed flying fox (*Pteropus poliocephalus*) and the black flying fox (*P. alecto*, Megachiroptera). Biol. Chem. Hoppe-Seyler 369, 975–984.

Koopman, K.F., 1984. Bats. In: Anderson, S., Jones, J.K. (Eds.), Orders and Families of Recent Mammals of the World. John Wiley & Sons, New York, NY, pp. 145–186.

Kössl, M., Vater, M., 1995. Cochlea. In: Popper, A.N., Fay, R.R. (Eds.), Hearing by Bats. Springer-Verlag, New York, NY, pp. 191–234.

Lawrence, B.D., Simmons, J.A., 1982. Measurements of atmospheric attenuation at ultrasonic frequencies and the significance for echolocation by bats. J. Acoust. Soc. Am. 71, 585–590.

Li, G., Wang, J., Rossiter, S.J., Jones, G., Zhang, S., 2007. Accelerated *FoxP2* evolution in echolocating bats. PLoS ONE 2, e900.

Li, G., Wang, J., Rossiter, S.J., Jones, G., Cotton, J.A., Zhang, S., 2008. The hearing gene *Prestin* reunites echolocating bats. Proc. Nat. Acad. Sci. USA 105, 13959–13964.

Liberman, M.C., Gao, J., He, D.Z.Z., Wu, X., Jia, S., Zuo, J., 2002. Prestin is required for electromotility of the outer hair cell and the cochlear amplifier. Nature 419, 300–304.

Maddison, W.P., Maddison, D.R., 1992. MacClade: Analysis of Phylogeny and Character Evolution. Version 3.08. Sinauer Associates, Sunderland, MA.

Martin, R.D., 1981. Relative brain size and basal metabolic rate in terrestrial vertebrates. Nature 293, 57–60.

Miller-Butterworth, C.M., Murphy, W.J., O'Brien, S.J., Jacobs, D.S., Springer, M.S., Teeling, E.C., 2007. A family matter: conclusive resolution of the taxonomic position of the long-fingered bats, *Miniopterus*. Mol. Biol. Evol. 24, 1553–1561.

Moss, C.F., Sinha, S.R., 2003. Neurobiology of echolocation in bats. Curr. Opin. Neurobiol. 13, 751–758.

Murphy, W.J., Eizirik, E., O'Brien, S.J., Madsen, O., Scally, M., Douady, C.J., Teeling, E., Ryder, O.A., Stanhope, M.J., de Jong, W.W., Springer, M.S., 2001. Resolution of the early placental mammal radiation using Bayesian phylogenetics. Science 294, 2348–2351.

Pettigrew, J.D., 1986. Flying primates – megabats have the advanced pathway from eye to midbrain. Science 231, 1304–1306.

Pettigrew, J.D., 2008. Primate-like retinotectal decussation in an echolocating megabat, *Rousettus aegyptiacus*. Neurosci. 153, 226–231.

Pettigrew, J.D., Jamieson, B.G.M., Robson, S.K., Hall, L.S., McAnally, K.I., Cooper, H.M., 1989. Phylogenetic relations between microbats, megabats and primates (Mammalia, Chiroptera and Primates). Philos. Trans. Royal Soc. London, B. Sci. 325, 489–559.

Pollak, G.D., Casseday, J.H., 1989. The Neural Basis of Echolocation in Bats. Springer-Verlag, Berlin, Germany.

Rydell, J., Arlettaz, R., 1994. Low-frequency echolocation enables the bat *Tadarida teniotis* to feed on tympanate insects. Proc. Royal Soc. London, B. Sci. 257, 175–178.

Schnitzler, H.-U., 1972. Control of Doppler shift compensation in the greater horseshoe bat, *Rhinolophus ferrumequinum*. J. Comp. Physiol. 82, 79–82.

Schnitzler, H.-U., Kalko, E.K.V., Denzinger, A., 2004. Evolution of echolocation and foraging behavior in bats. In: Thomas, J.A., Moss, C.F., Vater, M. (Eds.), Echolocation in Bats and Dolphins. University of Chicago Press, Chicago, IL, pp. 331–339.

Schoeman, M.C., Jacobs, D.S., 2003. Support for the allotonic frequency hypothesis in an insectivorous bat community. Oecologia 134, 154–162.

Schuller, G., Pollak, G., 1979. Disproportionate frequency representation in the inferior colliculus of Doppler-compensating greater horseshoe bats: evidence for an acoustic fovea. J. Comp. Physiol. A 132, 47–54.

Shu, W.G., Cho, J.Y., Jiang, Y.H., Zhang, M.H., Weisz, D., Elder, G.A., Schmeidler, J., De Gasperi, R., Gama Sosa, M.A., Rabidou, D., Santucci, A.C., Perl, D., Morrisey, E., Buxbaum, J.D., 2005. Altered ultrasonic vocalization in mice with a disruption in the *FoxP2* gene. Proc. Nat. Acad. Sci. USA 102, 9643–9648.

Simmons, N.B., 2005. Order Chiroptera. In: Wilson, D.E., Reeder, D.M. (Eds.), Mammal Species of the World: A Taxonomic and Geographic Reference. Smithsonian Institution Press, Washington, DC, pp. 312–529.

Simmons, N.B., Geisler, J.H., 1998. Phylogenetic relationships of *Icaronycteris, Archaeonycteris, Hassianycteris* and *Palaeochiropteryx* to extant bat lineages, with comments on the evolution of echolocation and foraging strategies in Microchiroptera. Bull. Am. Museum Nat. Hist. 235, 1–182.

Simmons, N.B., Seymour, K.L., Habersetzer, J., Gunnell, G.F., 2008. Primitive Early Eocene bat from Wyoming and the evolution of flight and echolocation. Nature 451, 818–821.

Smith, J.D., 1977. Chiropteran evolution. In: Baker, R.J., Jones, J.K., Carter, D.S., Robert, J. (Eds.), Biology of Bats of the New World Family, Phyllostomidae I Special Museum Publications. Texas Tech University, Lubbock, TX, pp. 49–69.

Smith, J.D., Madkour, G., 1980. Penial morphology and the question of chiropteran phylogeny. In: Wilson, D.E., Gardner, A.L. (Eds.), Proceedings Fifth International Bat Research Conference. Texas Tech Press, Lubbock, TX, pp. 347–365.

Speakman, J.R., 2001. The evolution of flight and echolocation in bats: another leap in the dark. Mammal Rev. 31, 111–130.

Springer, M.S., Teeling, E.C., Madsen, O., Stanhope, M.J., de Jong, W.W., 2001. Integrated fossil and molecular data reconstruct bat echolocation. Proc. Nat. Acad. Sci. USA 98, 6241–6246.

Surlykke, A., Pedersen, S.B., Jakobsen, L., 2008. Echolocating bats emit a highly directional sonar sound beam in the field. Proc. Royal Soc., B. 276, 853–860.

Teeling, E.C., Scally, M., Kao, D.J., Romagnoli, M.L., Springer, M.S., Stanhope, M.J., 2000. Molecular evidence regarding the origin of echolocation and flight in bats. Nature 403, 188–192.

Teeling, E.C., Springer, M.S., Madsen, O., Bates, P., O'Brien, S.J., Murphy, W.J., 2005. A molecular phylogeny for bats illuminates biogeography and the fossil record. Science 307, 580–584.

Thomas, J.A., Jalili, M.S., 2004. Echolocation in insectivores and rodents. In: Thomas, J.A., Moss, C.F., Vater, M. (Eds.), Echolocation in Bats and Dolphins. The University of Chicago Press, Chicago, IL, pp. 547–564.

Thomas, J.A., Moss, C.F., Vater, M., 2004. Echolocation in Bats and Dolphins. University of Chicago Press, Chicago, IL.

Van Den Bussche, R.A., Hoofer, S.R., 2004. Phylogenetic relationships among recent chiropteran families and the importance of choosing appropriate out-group taxa. J. Mammal. 85, 321–330.

Von Helversen, D., Holderied, M.W., Von Helversen, O., 2003. Echoes of bat-pollinated bell-shaped flowers: conspicuous for nectar-feeding bats? J. Exp. Biol. 206, 1025–1034.

Webb, D.M., Zhang, J., 2005. *FoxP2* in song-learning birds and vocal-learning mammals. J. Hered. 96, 212–216.

Zheng, J., Shen, W., He, D.Z.Z., Long, K.B., Madison, L.D., Dallos, P., 2000. Prestin is the motor protein of cochlear outer hair cells. Nature 405, 149–155.

SECTION 3

Diversity of Vocalizations

Mammalian vocalizations evolved in many directions with sound frequencies ranging from infrasounds to ultrasounds. The vocal system allows for a high variability of produced sounds and suitable adaptations, for example, for long-range (infrasound) or short-range (ultrasound) communication.

CHAPTER 3.1

A frequency scaling rule in mammalian vocalization

Neville H. Fletcher[*]

Research School of Physics and Engineering, Australian National University, Canberra, ACT 0200, Australia

Abstract: The main biological purpose of mammalian vocalization is communication with other members of the same species, and it is generally beneficial to maximize the distance over which this can be done, though in some instances short-range confidential communication is desired. Considering the anatomical and acoustical parameters involved, a scaling rule for maximized communication distance is predicted, with frequency proportional to body-mass to a power of about −0.4, and this agrees well with observation over a very large size range. The communication distance varies about as body mass to the power 0.6.

Keywords: frequency of vocalization; frequency scaling; animal size; communication distance; vocalization and body mass; hearing; underwater communication

I. Introduction

Animals produce sounds mainly for communication with other members of the same species for a variety of purposes, such as to define territory, attract a mate, warn of predators, or signal the existence of a food source. For most of these purposes it is an advantage to be heard over as long a distance as possible. In just a few situations, however, it is desirable to have communications that cannot be detected by predators. These two styles of communication are analogous to normal speech or loud shouting in human communication on the one hand, and to whispering on the other. In this section we shall be concerned mostly with the former type of vocalization. Mention should also be made of sounds produced by whistling, as in the traditional whistled languages of some African tribes and in the ultrasonic calls of rats, the frequencies in each case being around five times that of the usual vocalization frequency. These will not be included in the discussion of the present section, but are treated in the following chapters. Land-based mammals will be the main concern, with a small section towards the end devoted to the very different rules for those that live in the sea.

A detailed treatment of acoustic communication in a wide variety of animals has been given by Stebbins (1983) and by Bradbury and Vehrenkamp (1998), and also in two collections of papers edited by Busnel (1963) and by Lewis (1983). The acoustics underlying the subject has been described in detail by Fletcher and Thwaites (1979) and by Fletcher (1992), while there are many papers on the acoustic behavior of individual animal species. The matter of frequency scaling has been discussed by Fletcher (2004), and it is on this paper that the present section is largely based.

The initial discussion in what follows will be based on the simple assumption of sound propagation in the open air, which gives a surprisingly good prediction of the variation of vocalization frequency with animal mass. To be more realistic, however, it must be recognized that animals live in environments that contain vegetation of many kinds, including grassy surfaces or dense woodlands. The latter part of the section therefore examines the effects these have on the conclusions reached earlier.

[*]Corresponding author. E-mail: neville.fletcher@anu.edu.au

Stefan M. Brudzynski (Ed.)
Handbook of Mammalian Vocalization
ISBN 978-0-12-374593-4

DOI: 10.1016/B978-0-12-374593-4.00006-1

II. Sound production and transmission

While communication through vibrations produced by stamping feet or similar impacts is used by some animals such as elephants, the major communication medium is by vocal sounds produced by pressurized air released through a vibrating valve – the vocal folds – in the larynx. In any vibrating system, the frequency scales inversely with the linear size, provided all dimensions are scaled similarly and density and elastic moduli remain unchanged. If we take L to be a measure of the length or linear size of the animal and assume all parts of its anatomy to scale similarly, then the obvious scaling rule is that the dominant vocalization frequency f is proportional to $1/L$ or equivalently to $M^{-1/3}$ where M is the mass of the animal. This rule was proposed by Bradbury and Vehrenkamp (1998) and, as will be discussed below, requires only minor modification to take proper account of other acoustic matters.

The acoustic power of the radiated sound depends on the lung pressure, the oscillating area of the vocal fold aperture, and the frequency of the vocal fold vibrations. If A_V is the amplitude of oscillation of the vocal fold area and p is the lung pressure, then the oscillating airflow amplitude U is given by:

$$U = \left(\frac{2p}{\rho}\right)^{1/2} A_V, \tag{1}$$

where ρ is the density of air. Since both the mouth opening and the head of the animal are small compared with the dominant wavelength of the animal's call, the radiated power P at the call frequency f can easily be shown to be (Morse, 1984; Fletcher, 1992):

$$P = \frac{\pi \rho f^2 U^2}{2c} \tag{2}$$

where c is the speed of sound in air.

As the sound propagates away from the animal, its intensity decreases because it is spread over a larger area, giving a decrease as r^{-2} where r is the distance from the source. In addition, there is a decrease in intensity because of energy absorption by the air, caused by its viscosity and thermal conductivity. This absorption depends in a complex way upon temperature, pressure, humidity and frequency, but overall it increases with frequency about as αf^n with n close to 1.5 and α about $3.6 \times 10^{-8}\,\mathrm{m^{-1}Hz^{-1.5}}$ under typical atmospheric conditions (Bass et al., 1995; Sutherland and Daigle, 1997). This gives a sound attenuation due to atmospheric absorption of about 0.5 dB per 100 m at 1 kHz, with smaller absorption at lower frequencies.

Combining the spreading effect with the absorption, the sound intensity $I(r)$ at a distance r from a source of frequency f and acoustic power P is therefore:

$$I(r) \approx \frac{P}{4\pi r^2} \exp(-\alpha f^n r). \tag{3}$$

This relationship for the assumed values of α and n is valid only under normal conditions in an open environment. Even in such an open environment, however, propagation can be changed by inversion layers of low temperature air near the ground, which tend to capture the sound waves and confine them to the thickness of the layer, thus increasing propagation distance. Wind will generally have little effect, except for the noise it produces, since wind speed is small compared with the speed of sound in air. In an environment such as a forest, of course, the value of the attenuation coefficient α may be increased significantly, as is discussed later.

III. Sound reception

The acoustic signal provided to the cochlea of a listening animal of the same species at a distance r is:

$$S(r) = I(r) A_E \tag{4}$$

where A_E is the cross-sectional area of the outer ear or pinna. The pinna amplifies the pressure signal (Fletcher and Thwaites, 1988) and feeds it through the tympanic membrane and a bony link to the cochlea, where hair cells ultimately translate it into action potentials. The acoustics and physiology of this transduction process are both complex to model, but experimental results (Fay, 1988, 1997) show that most animals have about the same auditory threshold S^* within about ± 10 dB, although the frequency of this optimal response and the upper and lower frequency limits of hearing may differ greatly between species, becoming higher for smaller animals. This accords with a scaling model in which the neural output is proportional to the angular deflection of the auditory hair-cells.

The requirement for audibility of a call from another animal at distance r can be deduced from Equations 1–4 to be that:

$$S(r) \equiv \frac{pf^2 A_V^2 A_E}{4cr^2} \exp(-\alpha f^n r) \geq SZ^* . \quad (5)$$

Fig. 1 shows how this relation operates at various frequencies (f). Remembering that the maximum lung pressure p is fixed and that A_V and A_E are determined by the overall size of the animal, the concern is with the way in which the acoustic stimulus intensity S varies with distance as a function of frequency. At small distances where atmospheric attenuation is not important, the animal can produce a louder signal if the frequency is raised, as detailed in Equation 2. A high frequency signal, however, is more steeply attenuated with propagation distance than a low frequency signal, as detailed in Equation 3. There is therefore a particular frequency f^* at which the signal strength exceeds the threshold value S^* for a maximum distance, and it is to be presumed that this is the call frequency that will have evolved for the animal species concerned.

To determine the optimal frequency for maximum communication distance we simply require that $dr/df = 0$ in Equation 5, which gives:

$$\alpha f^n r = \frac{2}{n}, \quad (6)$$

and when this is substituted back into Equation 5 this gives the optimal frequency as:

$$f^* = \left(\frac{16cT}{\alpha^2 n^2 A_V^2 A_E p} \right)^{1/(2n+2)} . \quad (7)$$

III.A. Relation to animal size

It is now simple to apply the resulting Equation 7 to deduce how the optimal vocalization frequency should vary with animal size. If L is a measure of the linear size of the animal and all anatomical parts scale similarly, then both the vocal valve area A_V and the external ear area A_E will vary as L^2, so that Equation 7 predicts that:

$$f^* \propto L^{-3/(n+1)} \propto M^{-1/(n+1)}, \quad (8)$$

where M is the mass of the animal. Inserting the value $n = 1.5$ as a good approximation to the frequency dependence of sound attenuation in the atmosphere gives the result that the optimal call frequency f^* should be proportional to the animal length L to the power -1.2 or mass M to the power -0.4.

It is interesting to see how well this scaling rule agrees with observations of the calls of mammals of different sizes. Fig. 2 assembles data collected from a wide

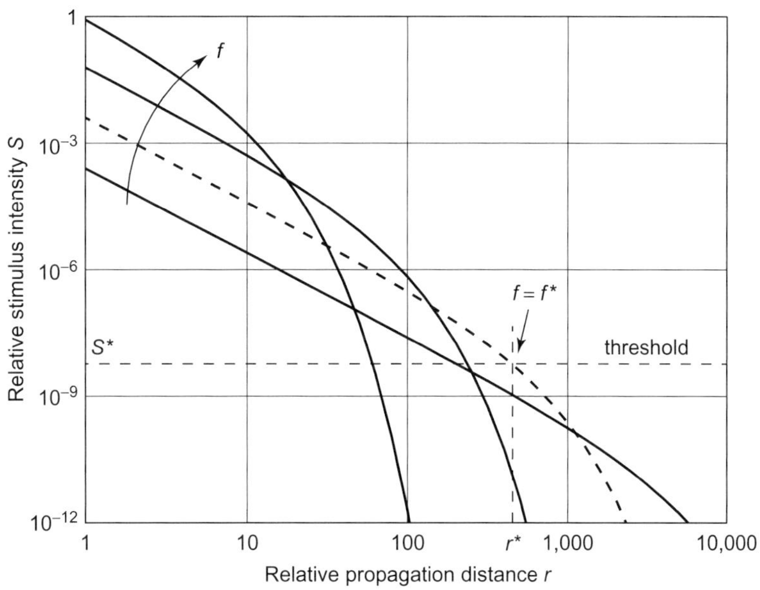

Fig. 1. Radiated sound intensity as a function of distance for a range of sound frequencies f, assuming other parameters to be constant. As shown by the broken curve, reception distance r for a threshold sensitivity S^* is maximized to the value r^* at frequency f^* (Fletcher, 2004).

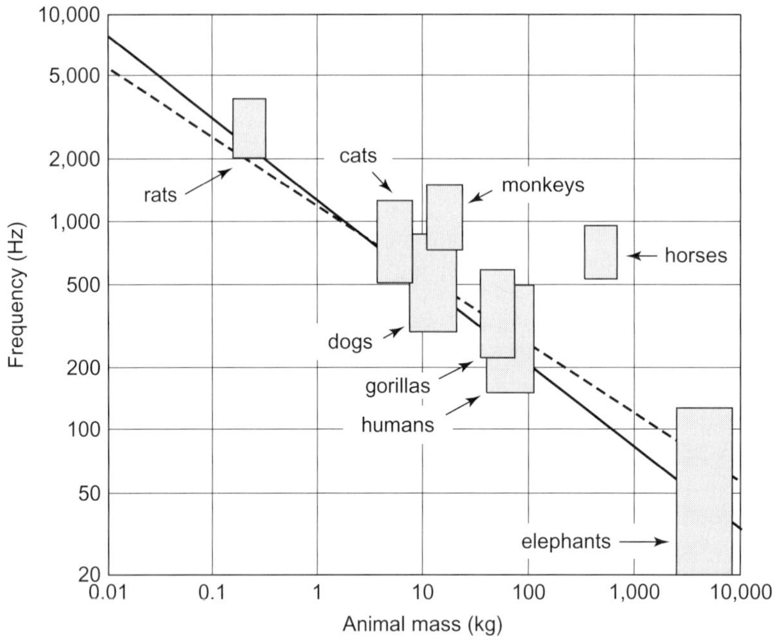

Fig. 2. Typical fundamental vocalization frequencies for a large range of mammals. The ultrasonic calls of rats and bats and the trumpetings of elephants are not included. The full regression line is for the relation $f \propto M^{-0.4}$ and the broken line is for $M^{-0.33}$ (modified from Fletcher, 2004).

range of sources and covers a mass range of six orders of magnitude, or a factor 10^6. The overall agreement is very good, although there are significant deviations in the case of animals such as horses and monkeys. Rats and mice, and of course bats, also vocalize in the ultrasonic range above 20kHz (see Chapter 3.3), and elephants have "trumpetings" at higher frequencies than shown, but it is likely that these all involve vocal mechanisms different from those of ordinary calls.

It is now also interesting to see how the maximum communication distance r^* varies with animal size. This follows from Equations 6 and 8 which together give the relation:

$$r^* \propto M^{n/(n+1)} \propto M^{0.6}, \tag{9}$$

where the final result comes from inserting the value $n = 1.5$ for the measured behavior of the atmospheric attenuation, as discussed before.

III.B. The effects of habitat

As mentioned in the introduction, the discussion above really only applies to sound transmission in the open air, such as might be possible from a cliff top or high rock. In reality, however, most communication takes place nearer to the ground and often through an environment such as a forest. Surely this will have a considerable impact on vocal behavior.

First consider the case of an open landscape with a nearly flat floor of sand or grass. As well as the sound wave propagating directly through the air, there will be another one that has been reflected from the ground and a third that is essentially "captured" by the ground impedance. The reflected wave will be weaker than the direct wave because of absorption during the reflection, and it will also be delayed a little in time because its path is longer. At the listening point these two waves may thus either reinforce or partially cancel each other, depending on phase difference and thus on the height of singer and listener above the ground. The contribution of the "ground wave" must also be added in, making a quite complex calculation even for an ideally flat surface (Embleton et al., 1976; Rasmussen, 1981; Embleton, 1996). For flat ground and a typical animal height of about 0.5m there is a reinforcement of up to 6dB at frequencies below about 200Hz, then attenuation increasing to a local maximum of about 15dB at 1000Hz at a distance of about 20m. At higher frequencies or larger distances the attenuation increases and the maxima are less pronounced.

If the ground surface is not ideally flat, as will generally be the case in nature, then things become even more complicated and it is impossible to give any reliable figures for the attenuation to be expected. Broadly viewed, however, the attenuation is greater at frequencies above about 500 Hz than at lower frequencies, and is greater though less oscillatory for a source close to the ground – a small animal – than for an elevated source or large animal. For a typical grassland surface, the attenuation is much greater than that for atmospheric absorption.

The other environment of interest is woodland, which may be either evergreen or deciduous. Here, in addition to absorption, there are scattering phenomena to be considered and perhaps even resonances with leaves or branches. Measurements (Embleton, 1963) for propagation distances over about 50 m show a nearly constant absorption for the forests studied of about 0.2 dB m^{-1} between about 200 Hz and 2 kHz and a steady increase for higher frequencies. This is again much greater than the free atmospheric absorption in this frequency range, but shows a similar general increase with increasing frequency.

An interesting example of the effect of habitat on vocalization occurs in the case of wild cats of the genus *Felis*. A study of several species of this genus by Peters et al. (2009) shows a surprising direct rather than inverse correlation between body mass and dominant frequency in the range 1.3–10 kg and 700–1000 Hz. While these ranges are both small, the explanation appears to lie in the fact that the smaller cats inhabit an open environment where absorption near 1 kHz is high, while the larger cats live in woodland and can benefit by using as high a formant frequency as possible, given their vocal anatomy.

Since the animals which we are interested in here are land-dwelling mammals, which live either in forests or else relatively close to the ground in grasslands, the high attenuation in both these environments calls into question some of the analysis above relating to propagation attenuation. Fortunately, this does not have any great influence on the final conclusions if we simply assume that, over a large range of animal sizes, anatomical scaling results in a vocalization frequency that is proportional to animal length or to its mass to the power $-1/3$ as suggested by Bradbury and Vehrenkamp (1998). The broken line in Fig. 2 shows the expected correlation on this assumption.

IV. Noise, bandwidth and close communication

Vocal communication between animals does not, of course, take place in an ideally quiet environment, instead there is competing noise produced by wind and by other animals. As a general rule, such noise tends towards a $1/f$ frequency distribution, which is equivalent to saying that it has the same average intensity per octave across the spectrum, so that the sound energy per hertz is much greater at low frequencies than at high. Animal vocal calls, however, and the related hearing sensitivity of animals, generally have a bandwidth that is about proportional to the central frequency involved, which means that the noise intensity obscuring the signal in the perception band is about constant, independent of the call frequency. What this implies, therefore, is that the analysis given above is still valid, except that the threshold sensitivity S^* should be interpreted to mean the detectable signal, which is again approximately constant because the detection bandwidth shrinks as the frequency is lowered and the noise level per hertz increases.

There is another interesting feature of the frequency-scaling rule for both call frequency and bandwidth, and this is that the rate of information transfer by the vocal signal decreases about as $M^{-0.4}$ in the same way as does the call frequency. While this might appear to be a disadvantage for large animals, it accords with the fact that the rate of relative motion of animals is about inversely proportional to their linear size – a mouse, for example, can jump around and run ten times its body length much more rapidly than an elephant can.

The analysis above is based on the assumed use of vocalization for conspecific communication over the largest possible distance, the signals being territorial, warning, or collaborative. In some cases, however, vocalization serves a different purpose and the aim is to communicate over relatively short distances and to avoid attracting predators. An example is the cry of baby animals to their mothers. These short-distance vocalizations use frequencies much higher than those for adult long-range communication, and the advantage of this is clear from Fig. 1. High-frequency signals are produced more efficiently, but are much more rapidly attenuated at large distances, achieving just the desired result.

V. Underwater communication

Some mammalian species such as whales and dolphins live under water but still use sound as a means of communication, so it is interesting to see what modifications are necessary to the discussion above. The two main differences are that the surrounding medium is water, with a density 1,000 times that of air and an acoustic absorption coefficent α that is 100 times smaller than in air. Also, while the water medium is three-dimensional at close range, it becomes effectively two-dimensional once the range considered is large compared with the water depth.

Aquatic mammals also produce sound by passing air through a vibrating valve, as in the larynx, and this air may either be vented through the nose or else stored temporarily in an air sac. Sound radiation mostly takes place through transmission of acoustic vibrations through the body tissues into the surrounding water, since these tissues have nearly the same acoustic properties as water. Analysis of the acoustic behavior follows much the same path as for animals in the air, except that r^{-2} must be replaced by r^{-1}. The result is a figure that is qualitatively like Fig. 1 except that the general slope of the curves is reduced by a factor of 2, and the values of r are increased by about a factor of 100. The conclusion is that aquatic animals should use much higher frequencies for communication than those living in air, and this is borne out by observation, quite large whales having communication frequencies in the 1–2 kHz range. The communication range of such large aquatic mammals is extremely long because of the low attenuation and $1/r$ spreading of the signal.

VI. Conclusion

This section has provided a wide-ranging theoretical background against which the acoustic communication behavior of mammals can be judged. Given this background, interest centers on deviations from the predicted norm for particular species, the reason for these deviations, and the effects that this has on their individual and social behavior.

References

Bass, H.E., Sutherland, L.C., Zuckerwar, A.J., Blackstock, D.T., Hester, D.M., 1995. Atmospheric absorption of sound: recent developments. J. Acoust. Soc. Am. 97, 680–683.

Bradbury, J.W., Vehrenkamp, S.L., 1998. Principles of Animal Communication. Sinauer, Sunderland, MA.

Busnel, R.G. (Ed.), 1963. Acoustic Behavior of Animals. Elsevier, New York.

Embleton, T.F.W., 1963. Sound propagation in homogeneous deciduous and evergreen woods. J. Acoust. Soc. Am. 35, 1119–1125.

Embleton, T.F.W., 1996. Tutorial on sound propagation outdoors. J. Acoust. Soc. Am. 100, 31–48.

Embleton, T.F.W., Piercy, J.E., Olson, N., 1976. Outdoor sound propagation over ground of finite impedance. J. Acoust. Soc. Am. 59, 267–277.

Fay, R.R., 1988. Hearing in Vertebrates: A Psychophysics Data-book. Hill-Fay Associates, Winnetka, IL.

Fay, R.R., 1997. Hearing and sound perception among vertebrate animals. In: Crocker, M.J. (Ed.), Encyclopedia of Acoustics. John Wiley and Sons, New York, pp. 1789–1797.

Fletcher, N.H., 1992. Acoustic Systems in Biology. Oxford University Press, New York.

Fletcher, N.H., 2004. A simple frequency-scaling rule for animal communication. J. Acoust. Soc. Am. 115, 2334–2338.

Fletcher, N.H., Thwaites, S., 1979. Physical models for the analysis of acoustic systems in biology. Quart. Rev. Biophys. 12, 26–66.

Fletcher, N.H., Thwaites, S., 1988. Obliquely truncated simple horns: idealized models for vertebrate pinnae. Acustica 65, 194–204.

Lewis, B. (Ed.), 1983. Bioacoustics: A Comparative Approach. Academic Press, London, UK.

Morse, P.M., 1984. Vibration and Sound. Acoustical Society of America, New York, NY, chap. 7.

Peters, G., Baum, L., Peters, M.K., Tonkin-Leyhausen, B., 2009. Spectral characteristics of intense mew calls in cat species of the genus *Felis* (Mammalia: Carnivora: Felidae). J. Ethol. (in press).

Rasmussen, K.B., 1981. Sound propagation over grass covered ground. J. Sound Vibr. 78, 247–255.

Stebbins, W.C., 1983. The Acoustic Sense of Animals. Harvard University Press, Cambridge, MA.

Sutherland, L.C., Daigle, G.A., 1997. Atmospheric sound propagation. In: Crocker, M.J. (Ed.), Encyclopedia of Acoustics. Wiley, New York, pp. 341–365.

CHAPTER 3.2

Elephant infrasounds: long-range communication

Michael Garstang*

University of Virginia, Department of Environmental Sciences, Charlottesville, VA, USA

Abstract: Infrasound in the range of 1 to 20 Hz may be generated and/or detected by elephants over distances in excess of 10 km. Complex sounds generated by conspecifics can be interpreted over ranges of kilometers to aid in social cohesion, definition of territory, reproduction, resource utilization and avoidance of predators. Abiotic sounds may be detected over distances of a few hundred kilometers, further aiding in the animals survival. Physical conditions at the surface and in the atmosphere can inhibit or enhance these processes, and evidence is presented to suggest that elephants respond, if not directly, then indirectly to these physical constraints.

Keywords: elephant calling; infrasound; acoustic signaling; atmospheric conditions and vocalization; long-range vocalization; abiotic sounds

I. Introduction

Near infrasound lies in frequencies below human hearing between 1 and 20 Hz, with wavelengths between 350 and 17 m. A number of terrestrial animals generate, detect and respond to infrasonic signals. The African elephant, *Loxodonta africana*, stands out among those as a mammal known to use loud, low-frequency, long-range communication as an integral part of its behavior. Elephants can generate sounds with frequencies below 10 Hz and can detect sounds as low as 1 Hz. The focus of this chapter will be on the African elephant and its use of long-range infrasonic communication.

The concept of acoustic range, when applied to an animal, is a behaviorally-dependent variable. Range is functionally dependent on physical conditions, such as habitat and atmospheric state. Similarly, characterizing range as "long" is not subject to simple definition. For the purposes of elephant infrasonic communication, long range may be considered as beyond visual detection and out of high-frequency acoustic range. The maximum range over which the loudest low-frequency call of an animal can be transmitted and detected by a conspecific is a desirable number since it has implications for territory, reproduction, resources utilization and predation.

Communication can be defined as an association between a sender and a receiver's behavior as a consequence of a signal (Wiley and Richards, 1978). The information contained in a signal at its source, referred to as the broadcast information, always equals or exceeds the received information. Harmonics of infrasonic fundamental frequencies extend into the audible range, and may contain important information (McComb et al., 2003). These higher frequencies will be the first to be attenuated over distance, thus complicating the concept of range. Nevertheless, atmospheric conditions most conducive to the transmission of the fundamental infrasonic signal will mean that the associated higher frequency harmonics will also travel the greatest distance. Long-range, low-frequency communication must therefore be considered both from a theoretical point of view, as well as a practical acoustic and behavioral response.

Sound pressure level (SPL) is measured with respect to a specified level of 20 micropascals and is stated in decibels (dB) relative to that level. For sound sources it is often specified at a fixed distance of 1 m (Pierce, 1981). The distance over which a given animal call can be detected depends on the intensity or loudness of the call and the threshold of hearing of the animal receiving the call. Both numbers are poorly known for animals other than humans. The intensity of a call by an adult elephant is thought to

*Corresponding author. E-mail: mxg@swa.com

Stefan M. Brudzynski (Ed.)
Handbook of Mammalian Vocalization
ISBN 978-0-12-374593-4

DOI: 10.1016/B978-0-12-374593-4.00007-3

approach 120 dB between 14 and 35 Hz (Poole et al., 1988). Heffner and Heffner (1980, 1982) measured the threshold of hearing of a 7-year-old Asian elephant (*Elephus maximus*) of 60 dB SPL calling at 17 Hz and 65 dB SPL at 16 Hz. The uncertainty in numbers describing the intensity of a call and the threshold of hearing for elephants probably results in an uncertainty in the determination of the range over which a given loud, low-frequency call can be heard is at least ±10%. This uncertainty is amplified when translated into area.

Elephants can certainly hear and respond to abiotic infrasonic sounds. Ocean waves breaking on shore lines generate sounds around 1 Hz which can travel thousands of kilometers in the atmosphere (Bedard and Georges, 2000; Boermer, 2006; Bass et al., 2007). Large thunderstorm clouds contain convective eddies which generate infrasound that travels hundreds of kilometers (Garstang, 2004). Helicopter blades similarly produce infrasound that can be heard by animals at ranges of 100 km (Payne, personal communication). Low-frequency elephant vocalizations may be transmitted as seismic waves in the surface substrata (Günther et al., 2004). Equally, elephants may respond to low-frequency sounds produced by other mammals, such as lion (*Panthera leo*) and hippopotamus (*Hippopotamus amphibius*) (Barklow, 2004).

In the sections below we will explore some of the specific adaptations which allow elephants to communicate over long distances at low frequencies, the biological benefits of such long-range communication and the physical conditions of elephant habitat which promote or inhibit communication.

II. Specific adaptations

Although the vocal and auditory characteristics of an elephant are by no means unique in the mammalian world, they both exhibit features which support the ability to generate and detect infrasonic sounds. Probicidians evolved within forests over a period of many millions of years. By the time of the beginning of the Miocene (24 million years ago), forests receded and savannas appeared and most of the morphological features of present-day elephants were well-established (Shoshani, 1998). Large lungs and vocal folds with their own mass and elasticity act as self-oscillating mechanical vibrators generating acoustic energy and fundamental frequencies which are at least as low as 15 Hz.

The supralaryngeal vocal tract of an adult African elephant, extending from the larynx to the tip of the trunk, can measure close to 5 m in length. Elephants have only 5 bones in the hyoid opporatus, as opposed to 10 in most mammals. The more flexible space filled by muscles, tendons, and ligaments favors the production of low-frequency sound. Elephants are capable of extending the larynx (laryngeal descent) as well as extending the trunk (Fitch and Reby, 2001; Reby and McComb, 2003). The pharangeal pouch (just behind the tongue), used to store water, as well as the nasal cavity in the forehead, may further change the acoustic characteristics of the vocal tract. The length of this vocal tract may be equated to a column of air in a tube (such as an organ pipe) which is closed at one end and is equal in length to one-quarter the wavelength measured in meters. For a sound with a frequency of 15 Hz and in air with a temperature of 5°C, the wavelength is approximately 20 m and the length of the tube is 5 m.

The air column within the vocal tract has elasticity and mass which will vibrate preferentially at certain frequencies, termed normal modes or resonances. The vocal tract will shape the final form of oscillations originating in the larynx. The form of the vocal signal will contain the fundamental frequency (FØ), harmonics of this frequency and selectively amplified parts of the signal referred to as formants can be clearly seen in the sonogram shown in Fig. 1.

The vocal tract length governs formant spacing, which is a better predictor of body size than the size of vocal folds or larynx (Fitch, 2000; Fitch and Hauser, 2002; Reby and McComb, 2003). Lower temperatures, because they slow down the speed of sound, favor the production of lower frequencies for the same vocal tract length. On the open savannas, the temperature of inhaled air by an elephant can exceed 50°C during the day and drop below 5°C at night. There is, thus, a bias towards the generation of lower frequencies under cold night-time conditions (Garstang, 2004).

Hearing is interpreted in terms of a behavioral response to sounds which may be used to obtain and interpret information about an animal's environment, including the ability to identify and localize a sound source. In general, the largest terrestrial animals are most sensitive to sound frequencies below 10,000 Hz. The African elephant has large, mobile pinnae and a large skull. The external ears can be raised and extended outwards, behavior noted when elephants are listening (Moss, 1988). The elephant has a large tympanic membrane (ear drum), and the size of the

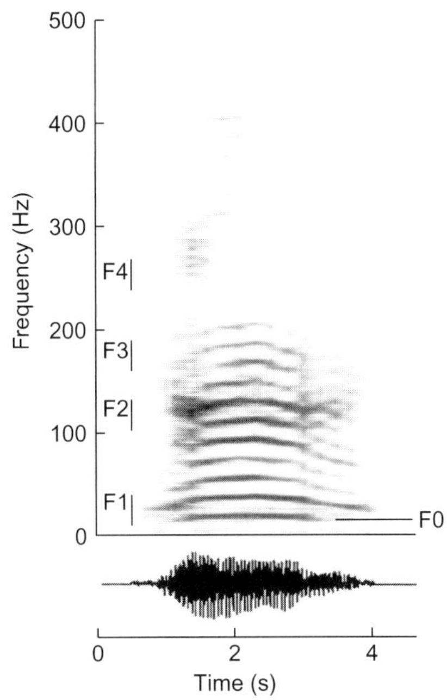

Fig. 1. Waveform of a female contact call showing the fundamental frequency (FØ) and harmonics, and the position of the first four formants (F1–F4). Frequency bandwidth: 8.74 Hz (lower trace); FFT size: 1,024 points; overlap: 50%. After Garstang (2004); from McComb et al. (2003).

ossicles and the shape and size of cochlea, which in *Loxodonta* has two spiral turns, are all conducive to the detection and transmission to the brain of low-frequency sounds (von Békésy, 1960, pp. 506–509).

Localization of a sound source depends on the difference in wave form phases creating a time delay between the two ears (McAlpine et al., 2001). A phase difference, $\Delta\phi$ for a given frequency, f, corresponds to an interaural time difference of Δt where $\Delta t = \Delta\phi/2\ \pi f$ (Hartman, 1999). A 90 degree or pi/2 radius phase shift at a frequency (f) of 15 Hz yields a time delay of 0.017 s. Lower frequencies favor localization and elephants have been shown to be unable to locate sounds with frequencies $\geq$4,000 Hz (Heffner et al., 1982).

The path length of an acoustic wave traveling around the head and ears of an adult African elephant from one ear to the other probably exceeds 150% of the actual perimeter (Kuhn, 1977, 1987; Brown, 1994, pp. 64–69). Heffner and Heffner (1982, 1984) showed that elephants can localize low-frequency sounds within an azimuth angle of one degree. Langbauer et al. (1991), using low-frequency playback calls of an estrous African elephant, demonstrated that males located the sound source over a distance of 2 km.

III. Biological benefits of long-range, low-frequency communication

In African elephants the closest social relationships are between members of the family unit and between different family units called bond groups (Moss and Poole, 1983). Family units are composed of adult females that are matrilineal relatives and their immature offspring. Bond groups are groups of family units that have frequent contact and exhibit mutual recognition. Individual members of the family unit, over a given 24 hour period, may range over a number of kilometers, especially in arid habitats at the end of the dry season. The family unit will, however, assemble, usually near sunset before proceeding to water. Family units move over even larger areas, and depending on resources, tend not to coalesce (Moss and Poole, 1983). McComb et al. (2000) estimate that calls can be recognized by as many as 14 different families, and that individuals within these families can identify 100 other adult females. They further found that the adult females can discriminate between the calls of less frequent associates and identify strangers whom they might regard as representing a threat to their unit. Age is a crucial factor in the ability of the elephant to retain and store information about those whom they encounter, and can affect the social knowledge of the unit as a whole.

These abilities to communicate over long distances have far reaching consequences to the health and survival of elephant populations. McComb et al. (2001) have shown that the ability of the matriarch to retain information that allows her a wide range of recognition materially improves the reproductive success of the group. Social cohesion of the group and avoidance of unnecessary stress brought on by false alarms or failure to recognize the source of the sounds all affect the fitness of the herd.

Elephants use their long-distance calls to maintain separation and so optimize use of scarce resources. Langbauer (2000) has shown that elephant herds maintain a separation for days if not weeks at a time, one herd never crossing the path of another. Garstang et al. (1997) have suggested the concept of a temporal range which is governed by the distance (area) over which an elephant's loud, low-frequency call can be heard by another elephant. Payne (1998, p. 224), delineates areas in which elephants in the Sengwa Reserve in Zimbabwe spend at least 50% of their time. These areas typically have an equivalent radius of 3–4 km, and are within the potential daytime range of a loud, low-frequency call.

Females in estrous produce frequent loud, low-frequency calls. Forrest and Raspet (1994) and Leong et al. (2003) found no distinct acoustic characteristics in the ovulatory follicular phase of the estrous cycle. The rate of low-frequency calls, however, was highest in the first period of follicular growth potentially attracting males prior to ovulation and resulting in both male–male competition and female choice. Payne et al. (2003) noted a similar result in the wild, showing that females in estrous make long sequences of powerful low-frequency calls. They also noted that there is a relationship between elephant numbers, social complexity, vocalization and calling rates.

Cold temperatures typical of early evening, night and early morning conditions over the dry savannas of Africa and Asia produce pervasive calm conditions with a strongly inverted temperature lapse rate. A cold layer of air in contact with the surface forms a sound channel or duct. Spherical spreading is replaced by the ducting of the low-frequency signal, which now travels over distances potentially greater than 10 km (Garstang et al., 1995; Larom et al., 1997a,b). A loud infrasonic call which during the middle of the day might travel no more than 1 km and be heard over an area of 3 km^2 can now be heard over an area of 300 km^2. No males may be present in the first area, while a significant number will be present in the second, ensuring competition and selection.

Adult elephants drink as much as 200 liters of water a day. As water supplies diminish in the dry season lions and spotted hyenas (*Crocuta crocuta*) will frequent water holes, therefore posing a threat to young elephants (Joubert, 2006; Loveridge et al., 2006). Family units of elephants, as well as bond groups, will assemble together before going to a water hole thus protecting their young by weight of numbers. Social cohesion, definition of territory, reproduction, resource utilization and predation, are all a function of low-frequency communication taking place between elephants over distances equal to or greater than 1 km.

Elephants react to a wide range of low-frequency abiotic sounds. These include infrasound produced by natural phenomena such as thunderstorms and those produced by human activity such as helicopters (Garstang, 2004). It is possible that elephants can detect and interpret these sounds over distances of up to at least a few hundred kilometers.

IV. Physical conditions

The nature of the terrain, vegetation, and the state of the atmosphere in and over habitat occupied by elephants can have a marked effect on both the theoretical distance a given infrasonic call can be transmitted and detected and on a wide range of acoustic characteristics that are attached to the low-frequency signal and that extend into higher frequencies.

Conversely, neglect of fundamental physical factors which influence both the transmission and detection of near infrasonic frequencies can lead to errors of up to an order of magnitude in range, and up to two orders of magnitude in the area ensonified. In the complex field of animal communication where the detection and interpretation of the signal must be measured in terms of the receiver's response, definitive conclusions must be treated with caution.

Sound propagating away from a point source into an unlimited environment (three-dimensional propagation) is subject to an attentuation of 6 dB for every doubling of the distance from the source. Infrasound propagates outwards in a spherical wave in all directions from a source. As the distance or the radius from the source doubles, the surface area of the spherical wave is increased four-fold and the sound pressure level reduced by 10 log 4 or 6 dB. The level of the low-frequency signal emitted must be high enough for the received signal to exceed this attenuation and be above the receiver's threshold of hearing if the signal is to be detected.

In practice, the vertical gradients of temperature and wind above the ground control the propagation of low-frequency sounds which can now be described in two-dimensional cylindrical coordinates with a numerical solution of the Helmholtz form of the acoustic wave equation (Raspet et al., 1985; Franke and Swenson, 1989; Garstang et al., 1995; Larom, 1996).

Temperatures and winds above the earth's surface change with height under virtually all conditions. The sign of the change or lapse rate in both temperature and wind in the atmospheric boundary layer is governed primarily by the presence or absence of heating of the surface by the sun (Garstang and Fitzjarrald, 1999). The dry, cloud-free atmosphere typical of the subtropical savannas making up much of the elephants' habitat leads to rapid daytime heating and high surface temperatures (45–50°C), and rapid night-time cooling and low surface temperatures (10–20°C).

A direct consequence of this diurnal surface heating and cooling is a diurnal cycle in wind speed

(Greco et al., 1992). Daytime heating produces turbulence, mixing higher winds aloft to the surface. Night-time cooling eliminates mixing, producing calm or low-wind speeds at the surface.

In the atmosphere near the surface described above, sound is refracted and sound waves are "bent" upwards, away from the surface during the day and downwards towards the earth's surface at night. Garstang et al. (1995) and Larom (1996) using model calculations based on a simplified subtropical atmosphere (Fig. 2), representing typical daytime, night-time and transitional conditions, show that for a 117 dB signal at 15 Hz and an assumed threshold of hearing of 50 dB, the range changes from near 2 km in the day to near 10 km at night.

When examining the rapid cooling which takes place in the 2 hours before sunset, Garstang et al. (1995) found that the range of the signal expands three-fold, and the area over which the call can be heard changes from about 15 km^2 in the middle of the day to over 300 km^2 soon after sunset. An empirical method which can be applied in the field to determine the presence and depth of a nocturnal inversion and its consequent effect on the range of low-frequency signal is presented in Appendix A.

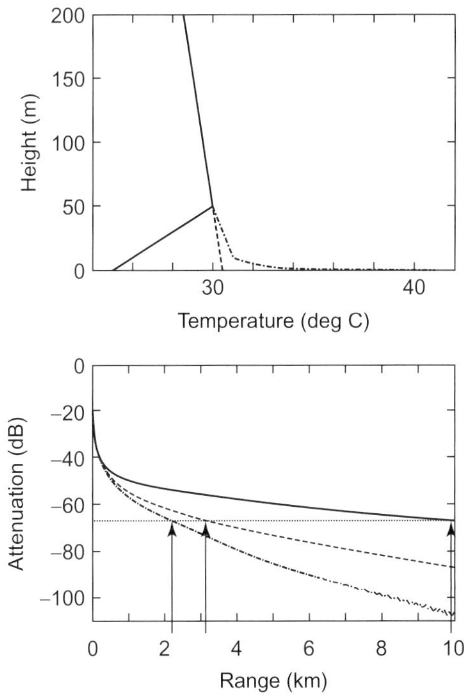

Fig. 2. Idealized temperature profiles (upper graph) and the 15 Hz attenuation profiles predicted from them (lower graph). The temperature profiles show: midday super-adiabatic lapse (dash-dot line); late afternoon, adiabatic lapse (dashed line) and an evening inversion (solid line). The elephant low-frequency calling range based on a −67 dB difference between call strength and hearing threshold is shown by a horizontal line in the lower figure. After Garstang (2004); from Larom et al. (1997).

Atmospheric conditions do not remain static overnight. The stronger winds overlying the cold surface air accelerate over sloping terrain, generating a nocturnal jet and causing episodic incursions of higher winds that penetrate the inversion layer and reach the surface. Acoustic conditions improve in the early morning as the nocturnal jet decays and surface winds once again approach calm conditions (Greco et al., 1992; Garstang et al., 2005).

Turbulence, with eddy sizes ranging from meters to a few hundred meters in diameter, is one of the few processes in the atmosphere that seriously attenuate infra- and near-infrasonic signals. Molecular absorption of infrasound in ambient conditions is essentially negligible. All but the softest surfaces, such as thick forest humus, are excellent acoustic reflectors of low-frequency sounds.

Reverberation and scattering can attenuate infrasound in the presence of obstacles such as trunks, limbs and leaves in a forest. In closed canopy forests or at night over open savannas stratification of the surface air eliminates most, if not all, of the turbulent fluctuations. Wiley and Richards (1978, p. 69), suggest that frequency-dependent attenuation does not differ among major classes of terrestrial habitats. The premise is that non-stationary heterogeneous turbulence increases as vegetation decreases, producing scattering in open habitats equal to that in forests. This is not true in the stably stratified surface atmosphere of the open savannas at night.

Wind is directly related to turbulence and will attenuate a signal along its path, as well as creating flow noise at the elephant's ear, effectively elevating the threshold of hearing and reducing the ability of the animal to detect or interpret the signal.

Conditions within a closed canopy forest are significantly different from those over the open savannas. Scattering by leaves, limbs and trunks affects frequencies mostly above 3,000 Hz. Reverberations in a forest are found to be least between 2,000–8,000 Hz, but may also significantly decrease below 200 Hz (Richards and Wiley, 1978). Tree trunks cause multiple reflections and scattering, and can reduce a 100 Hz signal by 5–6 dB for each doubling of range (Heimann, 2003). For frequencies below 100 Hz, attenuation becomes progressively negligible.

Studies now indicate that African elephants comprise two species: the savanna (*Loxodonta africana*) and the forest (*Loxodonta cyclotis*) elephant (Roca and O'Brien, 2005; Roca et al., 2007). Thompson et al. (2008) have found low-frequency calls of forest elephants between 7 and 8 Hz with a mean frequency at 19 Hz; of 423 forest elephant calls recorded, 95% were below 32 Hz.

The state of the atmosphere in a closed canopy equatorial rain forest reflects the steady-state conditions of an equatorial climate. Day-to-night and season-to-season changes in temperature are extremely small (less than a few °C), with change coming mainly in the form of wet and dry periods or seasons. Similarly, the wind speeds are very low and there is little change in wind regimes. Fig. 3 shows the vertical structure and diurnal changes in temperature, specific humidity and wind speed in a closed canopy forest. Because of high moisture contents in the rain forest, virtual temperature is used instead of temperature. Virtual temperature is the temperature that a parcel of air would have if, at the same pressure and density, all the latent heat contained in the water vapor were converted to sensible heat (AMS, 2000, p. 820).

Both day and night show inversion conditions within the canopy, with the floor of the forest always being colder than the top of the canopy. Wind speeds are low, essentially calm at the forest floor, and only 2 m/s at the top of the canopy. Conditions which favor ducting persist throughout the 24 hours, and may be slightly better during the day (stronger inversion of temperature) than at night. The marked changes in temperature and wind which occur on a regular diurnal basis over the savannas and which dramatically change transmission and reception of sound, are absent in the forest. Thompson (personal communication) found in the forests of the Central African Republic (CAR) that calling rates remained linear with increasing numbers of elephants, regardless of day or night.

Thompson et al. (2008) have measured the detection of calls on four recorders at increasing distances from calling elephants in the forest around the Dzanga bai in the CAR, finding an average range of just over 800 m. More preliminary results from their work suggest that powerful forest elephant calls might travel comparable distances through the forest as those of the savanna elephants do over the open plains under optimal acoustic conditions of ducting and no wind.

Acoustic conditions for the transmission of infrasound in intermediate habitat between closed canopy forests and open savannas are undocumented, but should not differ substantially from those pertaining to the open savannas. Acoustic conditions in broken terrain are complex, due to differential enhancement and attenuation (Piercy et al., 1977; Canard-Caranna et al., 1990; Heimann and Gross, 1999).

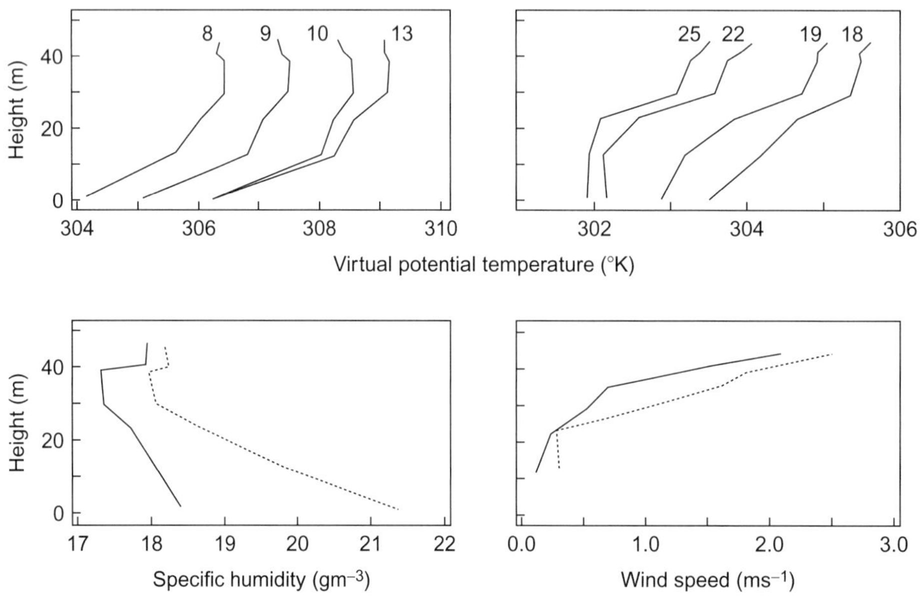

Fig. 3. Mean profiles of virtual potential temperature, (°K), specific humidity q (g/m) (night is solid, day is dotted), and wind speed U (m/s) (night is solid, day is dotted) within a 45 m rainforest. The virtual potential temperature profiles are identified by the hour of day over which they were averaged (12 = 1200; 25 = 0100, local time). After Garstang and Fitzjarrald (1999).

V. Infrasound and elephant behavior

Central to the discussion of infrasound and elephant behavior is the question of the distance or range over which an infrasonic call can be transmitted, detected and interpreted by a conspecific. McComb et al. (2003) have emphasized that a wide range of acoustic characteristics associated with infrasonic calls, but at higher frequencies, may carry important social information. This information, because it is at higher frequencies, is lost over distances shorter than that traveled by the fundamental infrasonic signal.

The frequencies attached to the fundamental infrasonic call are mostly below 300 Hz with wavelengths in excess of 1 m. Such signals under favorable atmospheric acoustic conditions (ducting and no wind) suffer little attenuation with distance. Conversely, these same signals are seriously attenuated in the presence of refraction and scattering (by turbulence), conditions characteristic of the open tropical savannas during the day. It would, therefore, appear appropriate to determine whether elephants show a behavioral response to these pronounced diurnal changes in atmospheric acoustic conditions.

Garstang et al. (2005) recorded loud, low-frequency calls from eight fixed microphones in an array around an isolated water hole in eastern Etosha National Park. The hourly distribution of 1,650 low-frequency calls recorded over seven days is shown in Fig. 4. The calls recorded cluster in the hours following sunset and sunrise (1800 and 0600). The highest number of calls occurs in the two hours after sunrise (184 and 196, or 14% and 15% respectively). The largest number of detected calls in contiguous hours occurs in the three hours following sunset (558 or 42%). Thus, just over 70% of all calls recorded fall in the five hours following sunset and sunrise. Of the remaining 30% of the detected calls, 24% were recorded at night and only 6% recorded during the day.

Model results based on atmospheric measurements made in Etosha National Park (Larom et al., 1997b) are also shown in Fig. 4. The −67 dB attenuation contours of calling range were calculated using the Fast Field Program (FFP) for a 15 Hz call at 117 dB and a threshold of hearing of 50 dB. The model results follow the clear diurnal cycle in calls recorded, and reflect the early evening maximum. Model results

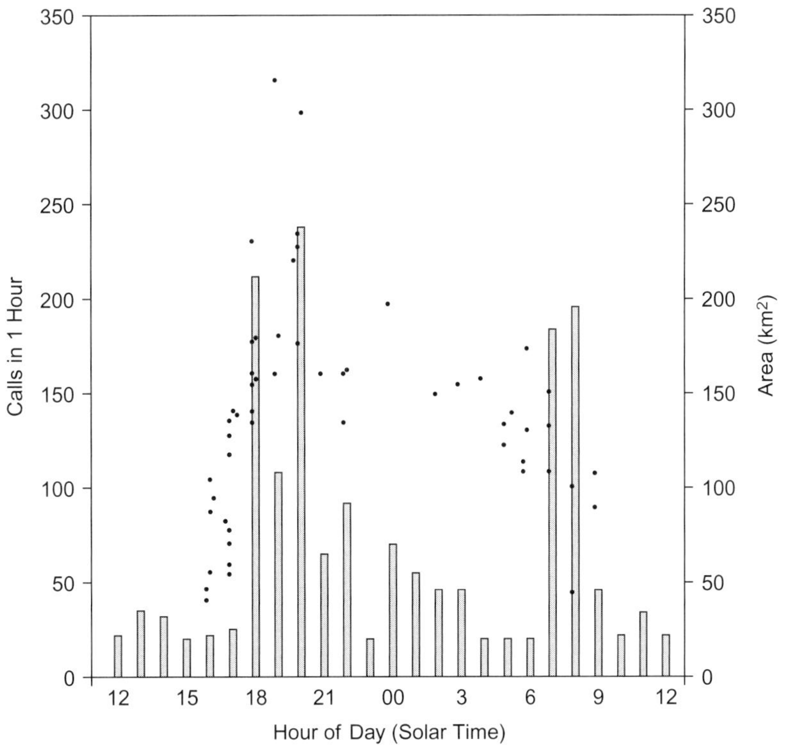

Fig. 4. Total number in each hour of calls recorded (bars: left ordinate) from 8 fixed microphones and the model calculated area (km^2) covered by the −67 dB attenuation contour as a function of time of day (dots: right ordinate).

simulate the increase in calls recorded near sunrise, followed by a marked decline over daytime hours.

Fig. 5 shows that heat is being lost by the atmosphere to the surface nearly one hour before sunset (1800). This negative flux persists for more than one hour after sunrise (0600). Both factors prolong the cold surface night-time conditions, and are reflected in the time of formation and desolation of the nocturnal inversion (Fig. 5, heavy line).

The number of calls detected at the Mushara water hole reflects three dynamic factors: proximity; range of detection; and rate of calling. Elephants were most often observed at Mushara near and following sunset. No breeding herds were seen at the water hole during the day. Proximity clearly contributes to the large number of calls recorded in the early evening, and possibly to the low number of calls recorded during the day.

The pronounced increase in the number of recorded calls following sunrise is not explained by proximity. Tracking by aircraft found no elephants within 3 km of the water hole in the 2 hour period after sunrise. The early morning maximum in calls recorded must depend on changes in detection range or changes in calling rates, or both.

Improvement in acoustic range would result in an increase in the number of recorded calls. Detection of more calls by elephants may trigger more calling, thus a feedback effect results from the improved detection conditions leading to a significant increase in the number of calls recorded. Soltis et al. (2005) found that in captive elephants, females did not produce rumbles at random, but were nearly twice as likely to produce rumbles if rumbles were heard from other group members.

VI. Conclusions

Knowledge of the distance over which an animal's call is likely to travel is essential in any study of animal communication. The basic state of the atmosphere must be known before conclusions can be drawn about range or interpretation of signals. Atmospheric conditions change the area over which a call can be detected, influence the content of the call, and can change the pattern and rate of calling.

Because acoustic conditions over most of the elephant's habitat are poor during the day, field work faces a practical problem of being required to pay particular attention to atmospheric state and to replicate many experiments near and after sunset and before sunrise. At the very least, basic atmospheric measurements must be taken to characterize the acoustic state of the fluid in which the signal is being transmitted in both captive and free-ranging situations.

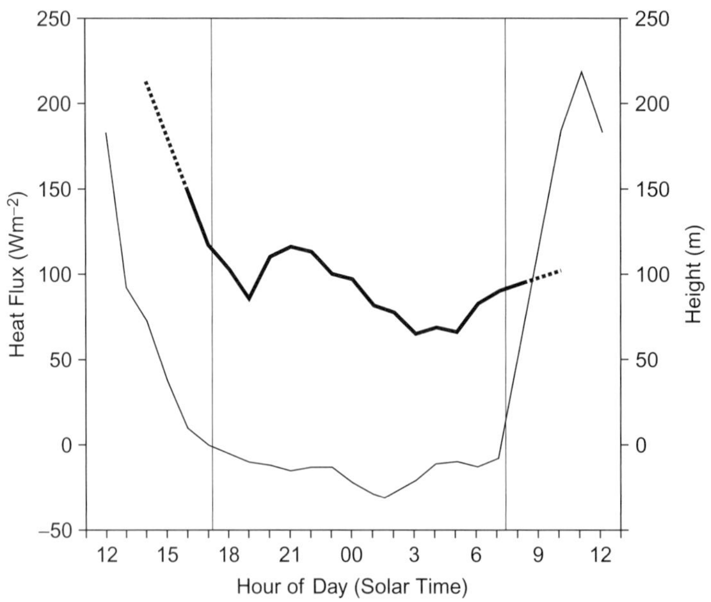

Fig. 5. Sensible heat flux (light black line and lefthand ordinate) at the surface based on eddy flux median values for a three-week period in Etosha National Park, Namibia, and the height of the base of the nocturnal inversion as measured by a vertical pointing sodar and by tethered balloon soundings (heavy black line and righthand ordinate). The vertical lines delineate the times of neutral stability partitioning the 24 hours into stable night-time and unstable daytime conditions. After Garstang et al. (2005).

Transmission and reception of infra- and near-infrasound in a closed canopy forest is in need of particular attention. While near steady-state conditions of temperature, wind and the gradients of these variables suggest that the large changes in the distance these sounds can be transmitted and detected experienced over a diurnal cycle will be absent in a forest, the actual distances traveled are poorly known. Over both the savannas and in the forest, but particularly in the forest, rain or storm conditions may totally disrupt communication.

Acknowledgment

I would like to thank Mary Morris who assisted in the preparation of the manuscript, including the compilation of the bibliography. This, and the support of the University of Virginia in producing the figures, is gratefully acknowledged.

Appendix A: Graphical determination of the presence and depth of a nocturnal inversion

Two sets of lines are shown in Fig. 6: (1) the dry adiabatic lapse rate, DALR, of temperature (heavy black line), cooling at $-1°C/100\,m$; and (2) a characteristic positive lapse rate of $+4.2°C/100\,m$ occurring over the tropical and subtropical savannas on clear nights (light black line).

Three measurements are required: (1) two air temperatures, T_1 and T_6, at 1 m and 6 m above the ground; and (2) wind speed at 1 or 6 m. The onset (or decay) of the nocturnal inversion is signaled when $T_1 = T_6$ designated T_0. T_0 determines the reference DALR. The example shown is where $T_0 = 30°C$ and the reference DALR line is shown as a heavy black line.

The growth of nocturnal inversion will occur as the 1 m temperature, T_1, continues to drop below T_o. The height (depth) of the inversion is determined by the intersection of the observed T_1 temperature following the positive lapse rate line (light or dashed black line) upwards to intersect the DALR, as shown by the example where $T_1 = 23°C$ and the inversion height $h_1 = 135\,m$.

Optimum atmospheric acoustic conditions for the transmission of low-frequency sounds exist when the height of the inversion lies between 50 and 200 m and surface winds are less than 2 m/s. Model calculations show that under these conditions a loud, low-frequency elephant call can be detected by another elephant at a range of approximately 10 km.

(Working nomograms to determine inversion presence and height are available at http://www.swa.com/members/publications/HandbookMammalianVocalizationChap3.2nomogram.pdf)

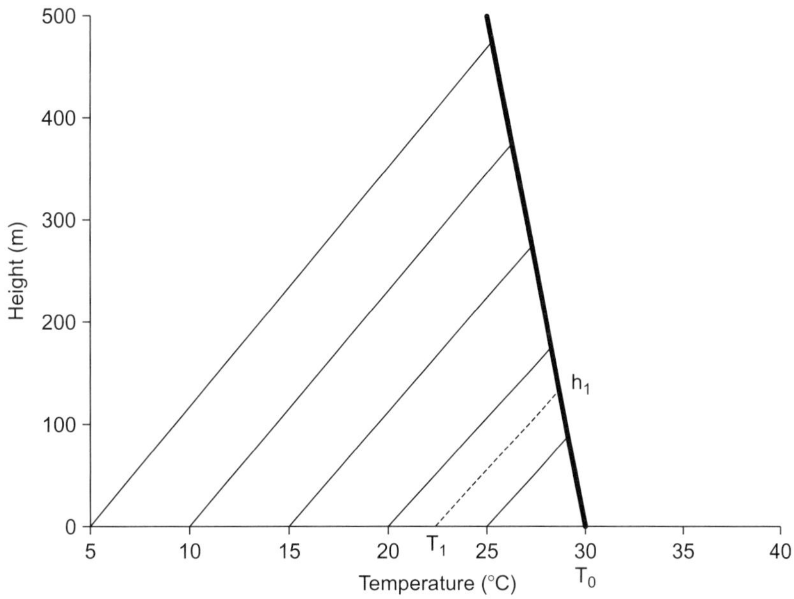

Fig. 6. Height (h_1) of nocturnal inversion, where T_0 = equilibrium temperature following the dry adiabatic line (heavy black line) to intersect with the current temperature at 1 m, T_1, following the characteristic lapse rate (light and dashed black line).

References

AMS (American Meteorological Society), 2000. Glossary of Meteorology. American Meteorological Society, Boston, MA, p. 820.

Barklow, W.E., 2004. Low-frequency sounds and amphibious communication in Hippopotamus amphibious (A). J. Acoust. Soc. Am. 115, 2555.

Bass, H.E., Hetzer, C.H., Raspet, R., 2007. On the speed of sound in the atmosphere as a function of altitude and frequency. J. Geophy. Res. 112, D15110.

Bedard Jr., A.J., Georges, T.M., 2000. Atmospheric infrasound. Phys. Today 53, 32–37.

Boermer, W.M., 2006. How infrasonic imaging, HF-surface radar and HF-OTHR and GPS technology can favorably be implemented for detecting the onset of tsunamis and the real-time imaging of its spreading. Proceedings MIKON 2006 Internat. Conf. Microwaves, Radar and Wireless Communications, May, pp. 47–52.

Brown, C.H., 1994. Sound localization. In: Fay, R.R., Popper, A.N. (Eds.), Comparative Hearing: Mammals. Springer, Berlin, Heidelberg, New York, pp. 57–96.

Canard-Carauna, S., Lewy, S., Vermorel, J., Parmentier, G., 1990. Long-range sound propagation near the ground. Noise Control Eng. 34, 111–119.

Fitch, W.T., 2000. Skull dimensions in relation to body size in non-human mammals: the causal bases for acoustic allometry. Zool. 103, 40–58.

Fitch, W.T., Reby, D., 2001. The descended larynx is not uniquely human. Proceedings R. Soc. London, B Biol. Sci. 268, 1669–1675.

Fitch, W.T., Hauser, M.D., 2002. Unpacking "honesty:" vertebrate vocal production and the evolution of acoustic signals. In: Simmons, A., Fay, R.R., Popper, A.N. (Eds.), Acoustic Communication, Vol. 16, Springer Handbook of Auditory Research. Springer-Verlag, Berlin, Heidelberg, New York.

Forrest, T.G., Raspet, R., 1994. Models of female choice in acoustic communication. Behav. Ecol. 5, 293.

Franke Jr., S.J., Swenson, G.W., 1989. A brief tutorial on the fast field program (FFP) as applied to sound propagation in the air. Appl. Acoust. 27, 203–215.

Garstang, M., 2004. Long-distance, low-frequency elephant communication. J. Comp. Physiol. A 190, 791–805.

Garstang, M., Fitzjarrald, D.R., 1999. Observations of Interactions Between the Tropical Surface and Atmosphere. Oxford University Press, New York, NY.

Garstang, M., Larom, D., Raspet, R., Lindeque, M., 1995. Atmospheric controls on elephant communication. J. Exp. Biol. 198, 939–951.

Garstang, M., Larom, D., Payne, K., Lindeque, M., 1997. Defining territory with sound. Paper presented at the AMS Internat. Conf. So. Hemisphere Meteorol. and Oceanogr., Pretoria, South Africa, April.

Garstang, M., Fitzjarrald, D.R., Fristrup, K., Brain, C., 2005. The daily cycle of low-frequency elephant calls and near-surface atmospheric conditions. Earth Interac. 9 (14).

Greco, S., Ulanski, S., Garstang, M., Houston, S., 1992. Low-level nocturnal wind accelerations over the Central Amazon Basin. Bound. Layer Meteor. 58, 91–115.

Günther, R.H., O'Connell-Rodwell, C.E., Klemperer, S.L., 2004. Seismic waves from elephant vocalizations: a possible communication mode? Geophys. Res. Lett. 31, L11602.

Hartman, W.M., 1999. How we localize sound. Phys. Today 52, 24.

Heffner, R., Heffner, H., 1980. Hearing in the elephant (*Elephas maximus*). Science 208, 518–520.

Heffner, R., Heffner, H., 1982. Hearing in the elephant (*Elephas maximus*): absolute sensitivity, frequency discrimination, and sound localization. J. Comp. Physiol. 96, 926–944.

Heffner, R., Heffner, H., 1984. Sound localization in large mammals: localization of complex sounds by horses. Behav. Neurosci. 98, 541–555.

Heffner, R., Heffner, H., Stichman, N., 1982. Role of elephant pinna in sound localization. Anim. Behav. 30, 628–629.

Heimann, D., 2003. Numerical simulations of wind and sound propagation through an idealised stand of trees. Acta Acustica 89, 779–788.

Heimann, D., Gross, G., 1999. Coupled simulation of meteorological parameters and sound level in a narrow valley. Appl. Acoust. 56, 73–100.

Joubert, D., 2006. Hunting behaviour of lions (*Panthera leo*) on elephants (*Loxodonta africana*) in the Chobe National Park, Botswana. Am. J. Ecol. 44, 279–281.

Kuhn, G.F., 1977. Model for the interaural time differences in the azimuthal plane. J. Acoust. Soc. Am. 62, 157–167.

Kuhn, G.F., 1987. Physical acoustic and measurements pertaining to directional hearing. In: Yost, W.A., Gourevitch, G. (Eds.), Directional Hearing. Academic Press, New York, NY.

Langbauer Jr., W.R., 2000. Elephant communication. Zool. Biol. 19, 425–445.

Langbauer Jr., W.R., Payne, K., Charif, R., Rapport, L., Osborne, F., 1991. African elephants respond to distant playback of low-frequency conspecific calls. J. Exp. Biol. 157, 35–46.

Larom, D.L. 1996. Meteorological controls on the infrasonic communication of the African elephant. Ph.D. dissertation, University of Virginia, VA.

Larom, D.L., Garstang, M., Payne, K., Raspet, R., Lindeque, M., 1997a. The influence of surface atmospheric conditions on the range and area reached by animal vocalizations. J. Exp. Biol. 200, 421–431.

Larom, D.L., Garstang, M., Lindeque, M., Raspet, R., Zunckel, M., Hong, Y., Brassel, K., O'Beirne, S., Sokolic, F., 1997b. Meteorology and elephant infrasound at Etosha National Park, Namibia. J. Acoust. Soc. Am. 101, 1710–1717.

Leong, K.M., Ortolani, A., Burks, K.D., Mellen, J.D., Savage, A., 2003. Quantifying acoustic and temporal characteristics of vocalizations for a group of captive African elephants, *Loxodonta africana*. Bioacoustics 13, 213–231.

Loveridge, A.J., Hunt, J.E., Murindagomo, F., Macdonald, D., 2006. Influence of drought on predation of elephant (*Loxodonta africana*) calves by lions (*Panthera leo*) in an African wooded savanna. J. Zool. 270, 523–530.

McAlpine, D., Jiang, D., Palmer, A.R., 2001. A neural code for low-frequency sound localization in mammals. Nature Neurosci. 4, 396–401.

McComb, K., Moss, C., Sayialel, S., Baker, L., 2000. Unusually extensive networks of vocal recognition in African elephants. Anim. Behav. 59, 1103–1109.
McComb, K., Moss, C., Durant, S.M., Baker, L., Sayialel, S., 2001. Matriarchs as repositories of social knowledge in African elephants. Science 292, 491–494.
McComb, K., Reby, D., Baker, L., Moss, C., Sayialel, S., 2003. Long-distance communication of acoustic cues to social identity in African elephants. Anim. Behav. 65, 317–329.
Moss, C., 1988. Elephant Memories: Thirteen Years in the Life of an Elephant Family. Ballantine Books, New York, NY.
Moss, C.J., Poole, J.H., 1983. Relationships and social structure in African elephants. In: Hinde, R.A. (Ed.), Primate Social Relationships: An Integrated Approach. Blackwell Scientific Publications, Oxford, UK, pp. 315–325.
Payne, K., 1998. Silent Thunder: In the Presence of Elephants. Simon and Schuster, New York, NY.
Payne, K., Thompson, M., Kramer, L., 2003. Elephant calling patterns as indicators of group size and composition: the basis for an acoustic monitoring system. Afr. J. Ecol. 41, 99–107.
Pierce, A.D., 1981. Acoustics: An Introduction to Its Physical Principles and Applications. McGraw-Hill, New York, NY, p. 61.
Piercy, J., Embleton, J., Sutherland, L., 1977. Review of noise propagation in the atmosphere. J. Acoust. 61, 1403–1418.
Poole Jr., J.H., Payne, K., Langbauer, W.R., Moss, C., 1988. The social contexts of some very low-frequency calls of African elephants. Behav. Ecol. Sociobiol. 22, 385–392.
Roca, A.L., O'Brien, S.J., 2005. Genomic inferences from Afrotheria and the evolution of elephants. Curr. Opin. Genet. Dev. 15, 652–659.
Roca, A.L., Georgiadis, N., O'Brien, S.J., 2007. Cytonuclear genomic dissociation and the African elephant species question. S. J. Quaternary Internat. 169, 4–16.
Raspet, R., Lee, S.W., Kuester, E., Chang, D.C., Richards, W.F., Gilbert, R., Bong, N., 1985. Fast-field program for a layered medium bounded by complex impedance surfaces. J. Acoust. Soc. Am. 77, 345–352.
Reby, D., McComb, K., 2003. Anatomical constraints generate honesty: acoustic cues to age and weight in roars of red deer stags. Anim. Behav. 65, 519–530.
Shoshani, J., 1998. Understanding proboscidean evolution: a formidable task. Trends Ecol. Evol. 13, 480–487.
Soltis, J., Leong, K., Savage, A., 2005. African elephant vocal communication, I: Antiphonal calling behaviour among affiliated females. Anim. Behav. 70, 579–587.
Thompson, M.E., Schwager, S.J., Payne, K.P., Turkalo, A.K. 2008. Acoustic estimation of wildlife abundance: methodology for vocal mammals in forested habitats. Afr. J. Ecol. accepted.
von Békésy, G. 1960. Experiments in Hearing. (Translated and edited by Werer, E.G.) McGraw-Hill, New York, NY, pp. 506–509.
Wiley, R.H., Richards, D.B., 1978. Physical constraints on acoustic communication in the atmosphere: implications for the evolution of animal vocalizations. Behav. Ecol. Sociobiol. 3, 69–94.

CHAPTER 3.3

Rat ultrasonic vocalization: short-range communication

Stefan M. Brudzynski[1*] and Neville H. Fletcher[2]

[1]*Department of Psychology, The Centre of Neuroscience, Brock University, St. Catharines, Ontario, Canada*
[2]*Research School of Physics and Engineering, Australian National University, Canberra, Australia*

Abstract: Rodents are subjected to a significant environmental pressure as prey for a large number of carnivore predators. Ultrasonic vocalization is one of the defensive adaptations which minimize the chances of being detected by a predator. Two mechanisms of ultrasound production in the larynx are discussed, with a whistle mechanism being the most probable one. Physical features of ultrasounds, such as greater directionality, greater attenuation, greater scattering, decreased localizability than vocalizations audible to humans and suitability for communication in underground burrows, make ultrasound a superior alternative to sonic vocalization for short-range communication, particularly in emergencies.

Keywords: defensive adaptation; ultrasonic vocalization; evolution of ultrasound calls; conspecific communication; alarm calls; whistle mechanism; rodents; rat

I. Introduction

The vast majority of rodents are prey for numerous carnivores belonging to different groups of vertebrates (Lack, 1946; Ryszkowski et al., 1973; Andersson and Erlinge, 1977). Many species of predators have specialized in hunting and killing rodents, and rely on them as their main source of food (e.g., owls, buzzards, weasels, wildcats and some species of snakes). They are referred to as specialists. Probably the most effective predators within this group are birds of prey which, after depleting the local population of rodents, will undertake migrations in search of their prey in other areas. Weasels represent one of the most specialized hunter species and their slender and elongated body allows them to capture prey in burrows and hidden nesting chambers. Thus, rodents are not only endangered in open spaces, but the safety of their nest areas may also be ineffective against this predator and as a result adult rodents of both sexes, their young and infants are all subject to predation.

The other category of rodent predators, termed generalists, may consume a large range of different foods, but they will feed on small rodents when these are available (Andersson and Erlinge, 1977). Foxes, martens, polecats, domestic cats, badgers and hawks are examples of these generalists. All the predators together have a powerful impact on populations of small rodents. Some recent studies reported predators having up to 95% predation impact on rodent population as studied within a three year period in the wild (Jędrzejewski and Jędrzejewska, 2007). It is not surprising that predators are believed to be one of the main factors influencing not only population dynamics, but also behavior and evolution of rodent species (Sundell, 2006).

As in most other rodent species, the position of rats in the food chain and the constant environmental pressure shaped rat evolution for millions of years, and resulted in various antipredator adaptations. This chapter is focused on one of the most complex adaptations, ultrasonic communication among rats, and particularly on ultrasonic alarm calls, which cannot be heard by many (reptiles and birds) but not all predators, and provide warning to the entire social group.

[*]Corresponding author. E-mail: sbrudzyn@brocku.ca

Stefan M. Brudzynski (Ed.)
Handbook of Mammalian Vocalization
ISBN 978-0-12-374593-4

DOI: 10.1016/B978-0-12-374593-4.00008-5

Alarm calls probably significantly contribute to survival of the rat colony.

II. Development of ultrasonic vocalization in rats

For the purpose of this chapter, "ultrasonic" will be taken to mean frequencies above 20 kHz, and calls of lower frequency will be called "sonic." While these words derive from human hearing, they may also apply to some extent to the hearing of rodent predators. Although the evolution of vocalization and auditory communication in vertebrates has a long phylogenetic history stemming from fish (Bass et al., 2008; Margoliash and Hale, 2008), ultrasonic calling in rats, as a defensive adaptation, seems to be a relatively recent development in evolution. Mammalian radiation began about 65 million years ago (Easteal, 1999). The myomorph rodents (suborder Myomorpha, which includes Muroidea, to which the rat belongs) are a group of mammals particularly rich in species, with about 40 million years of phylogenetic history (Catzeflis et al., 1992). Since the genera of the mouse (*Mus*) and rat (*Rattus*) have emerged as separate groups, probably about 16–23 million years ago (Catzeflis et al., 1992; Springer et al., 2003), and both these groups use ultrasound for communication, the mechanisms for ultrasonic calling would appear to have arisen between 20 and 40 million years ago. One of the prerequisites for ultrasonic communication was development of auditory sensitivity reaching far into the ultrasonic frequency spectrum. This sensitivity would be stimulated by a nocturnal life style, when the visual system is unhelpful in avoiding night predators (e.g., owls).

As suggested by Newman and by Hofer (see Newman, Chapter 2.2 and Hofer, Chapter 2.3 in this volume), ultrasonic vocalization probably appeared first in mother–infant interactions. Altricial rat infants rely entirely on maternal help for survival, and calling mother (e.g., after falling out of the nest, which, based on laboratory observations, may happen relatively frequently and is a life-threatening event) was probably a necessary development. The ultrasonic nature of infant vocalizations may arise from increased air pressure in the thorax, together with constricted vocal folds, both of which evolved in the respiratory system in response to cold (outside of the nest, see Hofer, Chapter 2.3 in this volume). Air escaping through a very small orifice of initially closed vocal folds (laryngeal breaking) could create high-pitched ultrasonic sound (see Section III). Rat pups maintain this way of sound production for about the first 20 days of their life, up to weaning. During that time, young rats demonstrate an extremely rich repertoire of different calls, with sound components reaching from audible to humans to high ultrasonic frequencies. In a sonographic study, 10–17 day old rat pups emitted almost all their calls as frequency modulated vocalizations with minimum frequency as low as 1.9 kHz and maximum frequency as high as 125 kHz (Brudzynski et al., 1999). This variety of vocalizations indicates that pups use different forms of vocalizations, and can readily make the functional transition from sonic to ultrasonic calls, which happens sometimes even within a single-frequency sweep beginning as an audible sound and ending in a high ultrasonic range of frequencies. Since rat pups are born at a very immature state, the early developmental stages right after their birth may represent a brief recapitulation of the evolutionary history and developmental pattern of the mechanisms needed for communication with ultrasonic vocalization, in an analogous way to the biogenetic law of Haeckel, which applies to embryogenesis. Adaptive advantages of vocalization in the ultrasonic range will be discussed later in this chapter.

The infantile pattern of vocalization in the rat (isolation calls) changes abruptly at weaning into the adult forms of vocalization. Adult rats have two forms of vocalization: an audible (to humans) or sonic form of calls with fundamental frequency between 2–4 kHz and rich harmonic components (Nitschke, 1982); and an ultrasonic form with much higher frequency, 20–70 kHz, and possibly as high as 100 kHz. Recent behavioral studies have shown that these sonic calls (squeals) may be used by rats in direct confrontation with a predator as a short-distance warning signal directed to the predator (Litvin et al., 2007), but would also be detectable over a larger area by conspecifics. These defensive threat vocalizations (heterospecific communication) increase with predator proximity, and they are thought to inform the predator that the targeted rat is ready for a defensive attack (Litvin et al., 2007). On the other hand, the ultrasonic calls are well-suited for communication at short distances and exclusively for communication with conspecifics within the social group.

There are then two major questions to be answered: what is the mechanism used for producing these ultrasonic calls, and why are they favored for short-range conspecific communication? Here we mostly confine our attention to laboratory Norway rats, since these have been most often studied, and attempt to arrive at conclusions that apply more generally to rodent species.

III. Ultrasound production

III.A. The whistle mechanism

The vocal anatomy of rodents, as shown in Fig. 1, is similar to that of other mammals and consists of a pair of vocal folds at the top of the trachea. During the production of most audible sounds the folds vibrate together under the influence of raised lung pressure, and thereby create a pulsating airflow into the upper vocal tract, the frequency of vibration being determined by the mass and tension of the vocal folds (see Berke and Long, Chapter 10.1 in this volume). Because the vocal folds close once in each vibration, their valve behavior is nonlinear and the airflow consists of a fundamental frequency together with a relatively strong admixture of many harmonic components. The acoustic influence of this flow is modified by resonances of the upper vocal tract to produce emphasized frequency bands, or vocal formants, like those distinguishing different vowel sounds in human speech. The object of this usual vocalization appears to be conspecific communication at large distances, and the fundamental frequency is related to animal size as discussed in Chapter 3.1 and is typically in the range 2–4 kHz for rats with body mass around 250 g.

There is, however, much uncertainty about the way in which ultrasonic calls are produced. For rats of this size the frequency range for conspecific alarm calls is typically 20 to 30 kHz (see Fig. 2), and the sound consists of a nearly pure tone with very little acoustic power in higher harmonics (Roberts, 1975a; Brudzynski and Holland, 2005). A functioning larynx is essential for this to occur, as indicated by laryngeal denervation studies (Roberts, 1975b; Nunez et al., 1985). These and other observations have led to the speculation that the tones are actually produced by some sort of whistle mechanism in the vocal tract, rather than by vibration of the vocal folds.

Roberts (1975a) has provided support for this hypothesis by simulating the rat calls using a "bird whistle" device in which air is blown through two aligned circular holes in two parallel plates separated

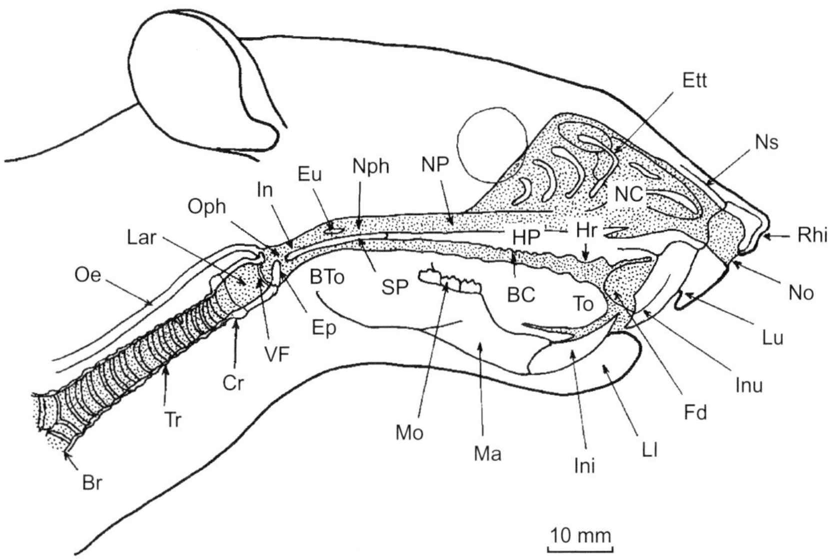

Fig. 1. The cross-section of the upper respiratory tract (stippled) with outlines of selected head and neck structures of an adult male Wistar rat, 300 g body weight. Abbreviations: Bc: buccal (oral) cavity; Br: bronchus; BTo: base of the tongue; Cr: cricoid cartilage of the larynx; Ep: epiglottis; Ett: ethmoid bones and nasal turbinates; Eu: opening into eustachian tube; Fd: fold of skin covering diastema; HP: hard palate; Hr: hard palate rugae; In: opening of internal nostrils (choana); Ini: lower jaw incisor; Inu: upper jaw incisor; Lar: larynx; Ll: lower lip; Lu: upper lip; Ma: lower fragment of the mandible; Mo: three molar teeth in the lower jaw (the upper ones are omitted); NC: nasal cavity; No: external nostrils; NP: nasal passage; Nph: nasopharynx (above soft plate); Ns: nasal bones of the skull (fragment); Oe: esophagus; Oph: oropharynx (behind the soft palate); Rhi: rhinarium; SP: soft palate; To: rostral part of the tongue; Tr: trachea; VF: vocal folds. For other anatomical details omitted in this diagram, see Rowett (1960) and Wells (1964).

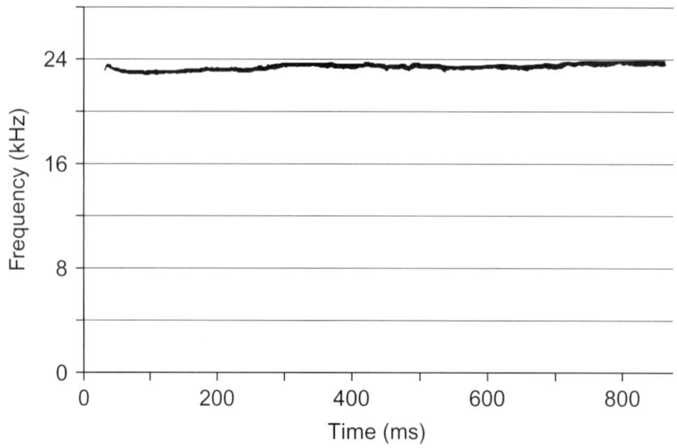

Fig. 2. A sonogram of a single typical alarm call emitted by an adult Wistar rat. Peak frequency of this call is 22.7 kHz.

by about 1.5 mm and sealed with a Perspex ring. The blowing pressures used were said to be "within the physiological range," although no details were given. This whistle could be made to produce sounds in the frequency range 20 to 80 kHz that closely resembled the vocalizations of rodents in both frequency and temporal pattern.

If this hypothesis is correct, then the one thing that remains to be identified is the anatomical structure responsible for the sound production mechanism. Studies on rats in which either the mouth or the nose were blocked during vocalization (Roberts, 1975b) established that the source of the sound was below the junction between the nasal cavities and the buccal cavity, and the opening of larynx (i.e., in the oropharynx; see Fig. 1). The vocal folds provide one of the two required apertures, and the evidence for that was provided by a direct observation of vocal folds during production of ultrasounds (Sanders et al., 2001). Sanders and colleagues reported that vocal folds were tightly opposed during emission of ultrasounds in a rat and an opening of 1–2 mm in diameter was observed in the back of the folds (dorsal direction). It was also noticed that the vocal cords did not vibrate during production of ultrasounds. Thus, the question remains: where is the other opening of the "whistle" located?

It is conceivable that the other opening of the whistle is created in the oropharynx, which is roughly 3–4 mm in diameter (these dimensions may change depending on the position of the soft structures around them). The space of the oropharynx is surrounded by movable elements (see Fig. 1), such as the epiglottis, the end of the soft palate around the internal nostrils, and the muscular base of the tongue. It is possible that the base of the tongue may push up against the tip of the soft palate while shaped in a form of a trough. This would create a second small opening between the tongue and soft palate. The second opening would be leading to the mouth cavity (or predominantly to the mouth), which could explain Roberts' observations that rat ultrasounds are emitted mainly through the mouth (Roberts, 1975b).

The other possibility is that the epiglottis could be stabilized in a semi-closed position, in this way creating the second opening above the vocal folds at the epiglottal level of the larynx. The rodent epiglottis is reinforced by an epiglotic cartilage joined with the thyroid cartilage and probably could be rigidly stabilized (Roberts, 1975c). The distance between these two openings would then be about 1–1.5 mm. This arrangement would explain why Sanders et al. (2001) have observed the opening in the vocal folds toward the dorsal side of folds. In this way, the two openings (in vocal folds and between the epiglottis and the top of the larynx) would be aligned. These mechanisms would be almost identical as to the location, size of the openings and the distance between them, to that found in Roberts' (1975a) simulation experiment. The option that two openings could be created by vocal folds (inferior folds), and superior folds (false folds) is unlikely, because false folds are not well-described in the rat and their distance from inferior folds would be too small.

It is helpful to have a general understanding of the operation of such a device (Chanaud, 1970; Wilson et al., 1971; Fletcher, 1992; Fletcher and Rossing, 1998). Air from the lungs forms a jet that emerges from the first aperture and is aligned so that it passes out through the second aperture. Such jets are, however, subject to instabilities and the one of interest here is often referred to as a "varicose," with the diameter of the jet varying. If the jet is too wide to pass through the exit aperture, this causes an increase in the pressure in the enclosed region between the apertures which, in turn, influences the diameter of the jet emerging from the entry aperture. If the propagation delay of the wave along the jet between the apertures is such that there is about one wavelength along the jet between them, this causes a reinforcement of the wave and the oscillating flow from the exit aperture generates a sound wave at that frequency.

If the lung pressure is p, then the speed of the jet is $v = (2p/\rho)^{1/2}$, where ρ is the density of air, and the speed of the varicose wave varies from v if the wavelength is long compared with the jet diameter, down to $v/2$ if the wavelength is less than the jet diameter (Fletcher and Rossing, 1998). Inserting some figures, we might take the lung pressure p to be about 1 kPa (10 cm water pressure) and, since $\rho \approx 1.2\,\text{kg/m}^3$ this gives a jet speed of about 40 m/s, and so a wave speed of 20–40 m/s. If the distance between the apertures is about 1 mm, then the frequency of the oscillating airflow emerging from the exit will be 20–40 kHz, which accords well with what is observed and with the mechanism proposed above.

There are two ways in which the frequency of the sound could be controlled: either by changing the lung pressure or by changing the distance between the two apertures, both of which seem possible. Further support for this whistle mechanism comes from studies by Roberts (1975a), who recorded the effect on the frequency of the ultrasonic calls of young rats when air was replaced by a heliox ($He–O_2$ mixture). The frequency of the calls was found to increase in proportion to the increase in sound speed, and thus $1/\rho^{1/2}$ as expected for a whistle-generated sound, since there must be about one wavelength of the disturbance along the jet.

III.B. Frequency-shift mechanism

While this whistle mechanism is plausible, no direct evidence of a suitable opening structure above the larynx has been observed, so alternatives must be considered. The obvious one is something analogous to the human singing of counter-tenors, who have two clearly defined vocal ranges, a low-frequency one about 100–300 Hz for normal speech, and a high-frequency range about 400–1,000 Hz for singing – a frequency difference of about a factor of three between the two registers, which is similar to that in some rats. Something similar is heard in the voices of male children reaching puberty, where there can be uncontrolled jumps between the two registers. Men and women in some cultures can also produce "yodels" in which there are sudden jumps between two registers quite widely separated in frequency. The physiological difference relates to the way in which the vocal folds vibrate, the low registers involving vibration of the whole fold while the high register operates with vibration confined to a soft surface layer (see Finck and Lejeune, Chapter 10.2 in this volume). In some versions of the high-register technique the vocal folds do not touch each other during their vibration, with the result that the airflow is much more nearly sinusoidal and the resulting sound has only very weak upper harmonics. It has not yet been established whether something similar might occur in rodents. The frequency-shift evidence of Roberts (1975a), however, would support either the simple jet mechanism or a mechanism involving excitation of a cavity resonance by a jet, as in some musical wind instruments, but not a mechanism relying solely on vocal fold vibrations, the frequency of which is determined by the mass and elasticity of the vibrating structures and is affected very little by the properties of the surrounding atmosphere.

III.C. Acoustic power of ultrasonic calls

There is very little information in the literature about the acoustic power typically produced by rodents during their ultrasonic calls. Roberts (1975a), however, gives a value for sound pressure level of up to 100 dB re 20 μPa at a distance of 10 cm for the loudest calls of the rodents he studied. This figure is equivalent to 80 dB at 1 m, which would be about 1 mW of sound power if the signal were to spread uniformly over a sphere. As will be discussed later, however, the ultrasonic signal is quite directional once the sound wavelength becomes comparable with the mouth diameter, so that the actual radiated sound power is probably closer to 0.1 mW. For a typical young adult rat the body mass is about 200 g, so that this corresponds to about 0.5 mW/kg. This sound production level per unit of body mass is comparable with that of birds, but much greater than that of larger mammals.

A study of auditory sensitivity in rodents (Brown, 1973) has shown that the *Mus musculus* species shows a main peak in cochlear microphonic output at about 15 kHz and a subsidiary peak at about 50 kHz. This corresponds roughly to hearing sensitivity as shown by behavioral studies, and indicates that these rodents and presumably most other murid rodents have evolved hearing abilities tuned to both their sonic vocalizations, and also to their ultrasonic calls. A study of audibility in rats using operant conditioning provided similar results of audibility within the frequency range from 10 to 50 kHz with the band of greatest auditory sensitivity being approximately one octave in width and located around 40 kHz (Gourevitch and Hack, 1966). Bats, unsurprisingly, show similarly tuned hearing abilities.

IV. Ultrasonic vocalization as adaptation for a short-range communication

The cost of a complex laryngeal control during emission of ultrasounds, the need for significantly prolonged exhalations and increased lung pressure during production of alarm calls, as well as the use of ultrasonic vocalizations for rat communication in both appetitive and aversive behavioral situations (Brudzynski, 2007) indicate a significant adaptive value of high sound frequency in minimizing chances of being detected by potential predators. As compared to sonic vocalization, these features could be summarized as directionality, increased attenuation, deflection and scattering, and decreased localizability of the source. Most of these features are important for the role of ultrasonic vocalization as a short-range communication system.

IV.A. Directionality

One advantage of using ultrasounds for communication is their directionality. Thus, rats may have, to some extent, control over emitting their vocalizations in a desired general direction of conspecific recipients and away from the suspected or real location of a predator.

Biological ultrasound behaves rather differently in the environment from the sounds audible to humans, because the wavelengths involved (typically in the range 5–15 mm) are small compared with physical features of the immediate surroundings. This is apparent in the directionality of the sound emitted from the rodent mouth. Even if for simplicity we take the animal mouth to be a circular aperture of radius a, which is set in a sphere representing the animal's head, then the analysis is still very complicated (Morse, 1981). A more drastic approximation, in which the mouth is set in a rigid plane as though the animal were against a tree or a bank, is more readily soluble and leads to the conclusion that the radiated sound is concentrated in a beam with an angular half-width equal to about 20 λ/a degrees, where λ is the sound wavelength and the beam width is taken to be that beyond which its intensity has declined by more than about 10 dB. For a rodent with external mouth diameter of 10 mm, this gives a half-width of about 50° at 20 kHz and 20° at 70 kHz.

IV.B. Attenuation and scattering

As discussed in Chapter 3.1 in this volume, the attenuation of sound due to atmospheric absorption is greater at high frequencies than at low frequencies, the attenuation coefficient increasing about as frequency to the power 1.5 in an open environment. This could be an incentive to use ultrasound for conspecific short-range communication, as atmospheric attenuation has little effect at close range. The directionality of ultrasound is also an advantage, since the rat's head can be turned away from the predator and towards other members of its community.

More important than attenuation is probably the effect of scattering of the sound from surrounding obstacles in the environment, such as rocks and large, leafy plants, which obscure information about the location of its source. This scattering becomes extreme when the scattering objects are larger in size than the sound wavelength, which in this case is typically about 1 cm, and of course varies with the number of scattering objects involved.

The use of a simple ultrasonic cry with fixed frequency and duration certainly limits the amount of information it can convey, although sometimes the duration and frequency are varied. This limitation is probably not of great importance, however, since only a small amount of information needs to be encoded in an alarm call.

IV.C. Localizability of the vocalizing rat

The ultrasonic alarm vocalizations produced by rats are generally of a very simple form, as shown in Fig. 2, and consist of a repetitive series of approximately fixed-frequency pulses each lasting for about 1 second, and typically arranged in groups of 3 to 5 calls (Brudzynski and Holland, 2005), although shorter pulses of 30–60 ms duration can be generated by rats and are observed in some species (Sewell, 1967). As noted before, these tones have very little harmonic development. The call frequency is very nearly constant, except for an initial adjustment in which the frequency of the first call in the sequence may begin a little higher and then fall to the dominant frequency within the series.

Since the biological role of these vocalizations is largely that of alarm calls, the objective would be to distribute the signal strongly over a local area without giving too many clues as to its place of origin. Multiple local reflections would help to achieve this aim as though the source is in the sonic analog of a hall of mirrors. It is also observed that most of the calls begin and end in a gradual way, and the rat maintains an almost constant frequency during the entire call. These acoustic features make localization of sound source even more difficult.

IV.D. Underground communication

It is well-justified from an acoustical point of view that vocalizations in the ultrasonic range of frequencies represent an effective adaptation in rats, minimizing the danger of being detected and attacked by predators. It seems, however, that rats safely hidden in underground burrows would not need to communicate in the ultrasonic range and could return to the regular mammalian sonic range. This could even extend the range of communication within the tunnels. Rats, however, appear not to do that and keep the ultrasonic form of communicating underground.

Rodents spend much of their time in underground burrows, and it is relevant to examine the transmission of sound and ultrasound through such environments. Although there is the normal attenuation due to molecular losses in the air, the attenuation underground is primarily due to losses to the burrow walls (Fletcher, 1992). This causes the sound intensity to decrease with distance x along the burrow as $\exp(-2\alpha x)$, and the attenuation coefficient α at frequency f in a smooth-walled pipe of radius r is approximately $3 \times 10^{-5} f^{1/2}/r$. For a rough-walled burrow, the attenuation will be greater by a large factor. The rate of increase of attenuation with frequency in a burrow is, thus, much less than for sound in the open air where attenuation varies about as $f^{3/2}$, and of course there is no extra effect of spherical spreading of the sound. It is therefore advantageous for rodents to use high frequencies for communication within a burrow, since they can produce these with greater source power than low frequencies (see Fletcher, Chapter 3.1 in this volume). Because of their higher information content, however, rapidly varying sonic vocalizations may be preferred for other communication.

V. Conclusions

The position of rodents in the food chain created a substantial environmental pressure on them. Rodents developed vocal communication in the ultrasonic range of frequencies as part of their antipredator adaptations. Particularly, ultrasonic alarm calls appeared to be an effective social defensive behavior. Physical features of ultrasounds and acoustic features of rat vocalizations provide congruent evidence that the ultrasonic form of vocalization serves for short-range communication. Ultrasonic calls have probably evolved to facilitate conspecific communication on the one hand, and to minimize chances of being detected by predators, on the other. Ultrasonic vocalizations, which may perhaps be produced with greater acoustic power than lower-frequency sonic vocalizations, are suitable for communication both inside and outside the burrows when only simple information such as an alarm call needs to be conveyed.

References

Andersson, M., Erlinge, S., 1977. Influence of predation on rodent populations. Oikos 29, 591–597.

Bass, A.H., Gilland, E.H., Baker, R., 2008. Evolutionary origins for social vocalization in a vertebrate hindbrain – spinal compartment. Science 321, 417–421.

Brown, A.M., 1973. High frequency peaks in the cochlear microphonic response of rodents. J. Comp. Physiol. 83, 377–392.

Brudzynski, S.M., 2007. Ultrasonic calls of rats as indicator variables of negative or positive states. Acetylcholine-dopamine interaction and acoustic coding. Behav. Brain Res. 182, 261–273.

Brudzynski, S.M., Holland, G., 2005. Acoustic characteristics of airpuff-induced 22-kHz alarm calls in direct recordings. Neurosci. Biobehav. Res. 29, 1169–1180.

Brudzynski, S.M., Kehoe, P., Callahan, M., 1999. Sonographic structure of isolation-induced ultrasonic calls of rat pups. Dev. Psychobiol. 34, 195–204.

Catzeflis, F.M., Aguilar, J.-P., Jaeger, J.-J., 1992. Muroid rodents: phylogeny and evolution. TREE 7, 122–126.
Chanaud, R.C., 1970. Aerodynamic whistles. Sci. Am. 222, 40–46.
Easteal, S., 1999. Molecular evidence for the early divergence of placental mammals. BioEssays 21, 1052–1058.
Fletcher, N.H., 1992. Acoustic Systems in Biology. Oxford University Press, New York, NY, Chapters 6, 7, 10, 14.
Fletcher, N.H., Rossing, T.D., 1998. The Physics of Musical Instruments, second ed. Springer, New York, NY, Chapter 16.
Gourevitch, G., Hack, M.H., 1966. Audibility in the rat. J. Comp. Physiol. Psychol. 62, 289–291.
Jędrzejewski, W., Jędrzejewska, B., 2007. Predation on rodents in Białowieża primeval forest, Poland. Ecography 16, 47–64.
Lack, D., 1946. Competition for food by birds of prey. J. Anim. Ecol. 15, 123–129.
Litvin, Y., Blanchard, D.C., Blanchard, R.J., 2007. Rat 22 kHz ultrasonic vocalizations as alarm cries. Behav. Brain Res. 182, 166–172.
Margoliash, D., Hale, M.E., 2008. Vertebrate vocalizations. Science 321, 347–348.
Morse, P.H., 1981. Vibration and Sound. second ed., reprinted 1976. New York, NY: Acoustical Society of America, pp. 323–332.
Nitschke, W., 1982. Acoustic Behavior in the Rat. Research, Theory, and Applications. Praeger Publishers, New York, NY.
Nunez, A.A., Pomerantz, S.M., Bean, N.J., Youngstrom, T.G., 1985. Effects of laryngeal denervation on ultrasound production and male sexual behavior in rodents. Physiol. Behav. 34, 901–905.
Roberts, L.H., 1975a. The rodent ultrasound production mechanism. Ultrasonics 13, 83–88.
Roberts, L.H., 1975b. Evidence for the laryngeal source of ultrasonic and audible cries of rodents. J. Zool. Lond. 175, 243–257.
Roberts, L.H., 1975c. The functional anatomy of the rodent larynx in relation to audible and ultrasonic cry production. Zool. J. Linn. Soc. 56, 255–264.
Rowett, H.G.Q., 1960. The Rat as a Small Mammal, second ed., John Murray, London, UK.
Ryszkowski, L., Groszczyński, J., Truszkowski, J., 1973. Trophic relationships of the common vole in cultivated fields. Acta Theriol. 18, 125–165.
Sanders, I., Weisz, D.J., Yang, B.Y., Fung, K., Amirali, A., 2001. The mechanism of ultrasonic vocalization in the rat. Soc. Neurosci. Abstr. 27, 88.19.
Sewell, G.D., 1967. Ultrasound in adult rodents. Nature 215, 512.
Springer, M.S., Murphy, W.J., Eizirik, E., O'Brien, S.J., 2003. Placental mammal diversification and the Cretaceous-Tertiary boundary. PNAS 100, 1056–1061.
Sundell, J., 2006. Experimental tests of the role of predation in the population dynamics of voles and lemmings. Mammal. Rev. 36, 107–141.
Wells, T.A.G., 1964. The Rat. A Practical Guide. Heinemann Educational Books Ltd, London, UK.
Wilson, T.A., Beavers, G.S., DeCoster, M.A., Holger, D. K., Regenfuss, M.D., 1971. Experiments on the fluid mechanics of whistling. J. Acoust. Soc. Am. 50, 366–372.

CHAPTER 3.4

Ultrasonic calls of wild and wild-type rodents

Gillian D. Sales*

Department of Anatomy and Human Sciences, King's College London, Guy's Campus, London, England

Abstract: Acoustic design of rodent ultrasonic signals and evolutionary selection pressures involved in this process are considered through a comparative approach to ultrasonic signal structure. Frequency characteristics of calls of wild and wild-type rodents are reviewed, together with factors that may affect call design. These include the mechanism of ultrasound production, phylogeny, ecology and habits of the signaler and recipient, the function of calls and the sensory abilities of recipients. There are no species-specific frequency bands or patterns, but there are some phylogenetic differences. The previously proposed whistle mechanism of ultrasound production can account for some differences in calling in different situations. There is some evidence for matching of call structure to habitat, function and receiver sensory abilities. Areas that deserve further study are highlighted.

Keywords: wild rodents; ultrasonic calls; acoustic communication; signal structure; phylogenetic factors; ecological factors; rodent hearing ability

I. Introduction

Communication by rodents using ultrasonic vocalization (USV) that is above the nominal upper limit of human hearing (taken to be 20 kHz) has received increasing attention over the last 30–40 years. Calls have been detected in a range of different situations including isolation of infants from the nest, aggressive and sexual interactions, encounters between females, from mothers deprived of their litters, lone animals separated from the colony, animals when alarmed and during exploratory behaviour (Sales and Pye, 1974; Francis, 1977; Hoffmeyer and Sales, 1977; Smith and Sales, 1980; Wilson and Hare, 2004; Costantini and D'Amato, 2006; Portfors, 2007). We now know much of the ontogeny of calling and its relationship to the development of homeothermy (e.g., Okon, 1972; Blake, 1992) and the different situations and states associated with ultrasound emission (e.g., Sales and Pye, 1974; Gyger et al., 1986; Brudzynski, 2005; Costantini and D'Amato, 2006; Portfors, 2007). The brain centers, pathways and systems involved (e.g., Floody and O'Donohue, 1980; Brudzynski, 2007) and the vocal mechanism and genes implicated in calling (Roberts, 1975; Shu et al., 2005) have also been studied, as have the auditory sensitivity of rodents to ultrasound (e.g., Brown and Pye, 1975; Ehret and Haack, 1981; Fay, 1988) and the effect of various signals on recipient behavior (e.g., Sewell 1970a; Barfield and Thomas, 1986; Cherry, 1989; Sales, 1991; Ehret, 2005).

Less attention has been focused on how factors, such as the vocal mechanism, signal function, habits and habitat of signaler and recipient, as well as sensory abilities of intended recipients might affect call design. This aspect of ultrasonic communication requires a comparative consideration of USVs of wild and wild-type rodents, the calls of which will not be affected by selective breeding as those of laboratory strains such as rats and mice may be (Price, 1987). Most studies on rodents have been laboratory-based and we know little of where calls are emitted in the wild. However, innovative application of new technology has recently allowed USVs to be recorded in the field (Wilson and Hare, 2004; Kalcounis-Rueppell et al., 2006, 2008) which should provide valuable information. This chapter reviews information on the physical characteristics of USVs of wild and wild-type rodents, somewhat arbitrarily defined here as rodents

*Corresponding author. E-mail: gillian.sales@kcl.ac.uk

Stefan M. Brudzynski (Ed.)
Handbook of Mammalian Vocalization
ISBN 978-0-12-374593-4

DOI: 10.1016/B978-0-12-374593-4.00009-7

which have not been selectively bred sufficiently to produce distinctive strains and factors that do and may affect their call design, with reference to laboratory rats and mice where relevant. Coverage is not exhaustive; a range of examples is used for illustration.

II. Diversity of ultrasonic calls of rodents

The Order Rodentia is the largest taxon of mammals, and is comprised of a diverse group of animals ranging from about 7 g to 80 g in body weight. Rodents occur in every continent except Antarctica, and live in habitats as diverse as Arctic tundra, forests and hot, dry deserts. They are mainly terrestrial, often burrowing, forms but they include arboreal and semi-aquatic animals. There are approximately 2,280 species of rodents within five suborders, 33 families and 481 genera (Wilson and Reeder, 2005). This group therefore provides an excellent opportunity for making multiple comparisons among taxa of varying relationship. USVs have been detected from at least 50 species in 30 genera, all from two suborders, the Sciuromorpha (two families) and Myomorpha (eight families). Apparently no USVs have been detected from suborders Anomaluromorpha, Castorimorpha or Hystricomorpha, although it is not clear whether this is due to their genuine absence, a lack of studies or reporting, or difficulties in finding such reports. Most of the USVs reported are 25–200 ms in duration, with a total range of ca. 2 ms to more than 3,000 ms; they are generally within the frequency range 20–80 kHz, although frequencies can reach 140 kHz (Sales and Pye, 1974). Many USVs consist of a single frequency component (Fig. 1a,b,c,f), but calls with more harmonics or harmonic series are also produced (Fig. 1e,g) (Sewell, 1970b; Smith and Sales, 1980). Some calls recorded from golden hamsters, *Mesocricetus auratus*, and gerbils, *Gerbillus*, have harmonically unrelated components within the ultrasonic range with either a narrow (5–15 kHz; Fig. 1d) or broad (up to 50 kHz or more) frequency range. Within each call the frequency may change little (slow frequency change; Fig. 1a,d) or show one or more changes which can occur as rapid changes (sometimes, but not always, defined as >1 kHz/ms; Fig. 1b), as instantaneous steps (Fig. 1c) or as sweeps in frequency upwards and/or downwards (Fig. 1e). Continuous, repeated sweeps give rise to "U," reversed "U" ("n"), or "W" shaped changes or frequency modulations (Fig. 1e,f,h). Seven frequency patterns have been identified in calls of infant field voles, *Microtus agrestis* (Mandelli and Sales, 2004) that are similar to those described for infant laboratory rats (Brudzynski et al., 1999) and adult mice (Holy and Guo, 2005; Portfors, 2007).

There are no species-specific frequency bands or patterns, but some species are more similar in their calls than others (see below). In many cases calls of infants of the subfamily Arvicolinae, family Cricetidae (genera *Myodes*, *Microtus*, *Dicrostonyx* and *Lagurus* in Fig. 1), appear more restricted in frequency than those of other members of this family, and within Family Muridae, calls of infants in the genera *Acomys* through to *Mus* tend to have higher frequency calls than those of the genera *Thamnomys* through to *Rattus* in Fig. 1. In adults, frequency modulations appear more common in murid than in cricetid rodents. However, these differences may, to a certain extent, reflect the methods of recording and reporting of the calls. Harmonic components appear more common among Cricetidae than Muridae, although not exclusively so (Sewell, 1970b; Smith and Sales, 1980), and also occur in calls of juvenile flying squirrels, *Glaucomys volans*, which have a fundamental frequency of 22–29 kHz (L.M. Gilley, personal communication). Some calls of species such as *Peromyscus maniculatus*, the deer mouse, and *Meriones unguiculatus*, the Mongolian gerbil, have one or more harmonic series, the fundamental of which may drift from above to below the nominal lower limit of ultrasound (Smith and Sales, 1980). This structure has rarely been reported from the Family Muridae, although it has been found in juvenile *Meriones* (Sales and Pye, 1974) and seven harmonics occur in USVs of murid genera *Notomys* and *Pseudomys* (Watts, 1975, 1976).

Particular call types may be emitted only in specific situations, or in several different ones. In laboratory rats short "50 kHz" USVs with varying degrees of frequency change are associated with sexual and aggressive interactions, while the longer "22 kHz" calls have been detected in a variety of situations including submission, post-copulatory periods and alarm (see Costantini and D'Amato, 2006 and Portfors, 2007 for reviews). Frequency modulated calls, often described as chirps, chirrups or twitters, have been described for a number of other species, generally in heterosexual encounters (Table 1), while less marked frequency changes are associated with aggressive interactions and have been reported for fewer species (Table 2). Lower frequency calls similar to the 22 kHz calls of laboratory rats have been reported

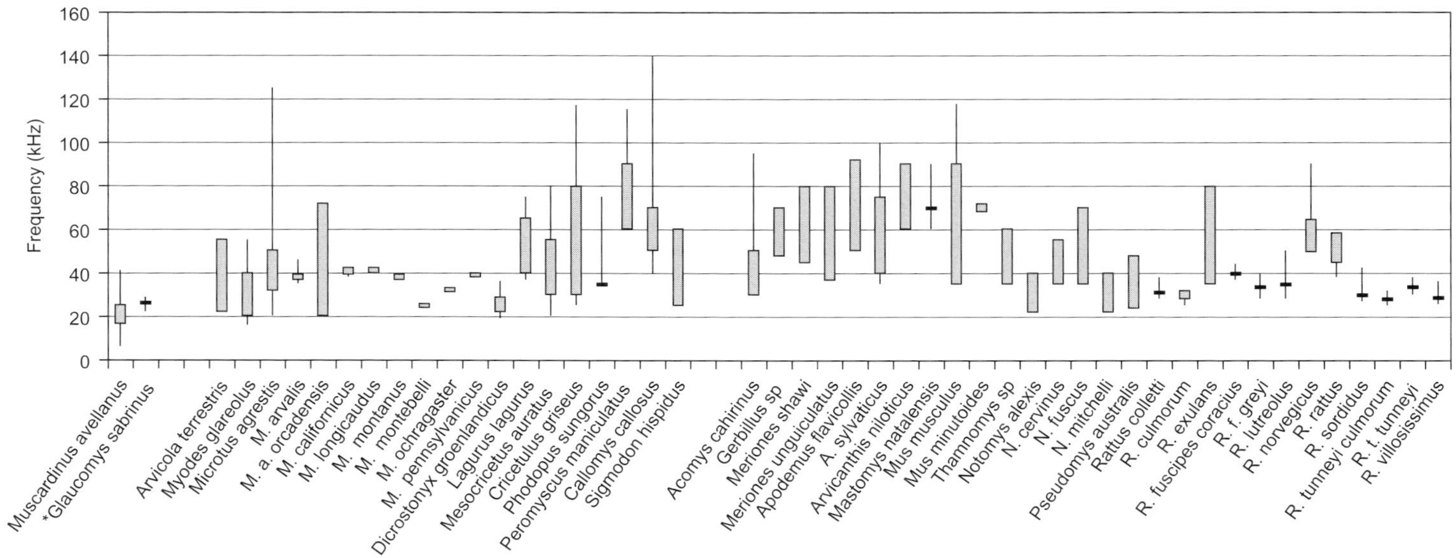

Fig. 1. Frequency ranges of ultrasonic calls emitted by infant rodents of 49 species. Values are given for general ranges or for means or medians (±2 kHz) together with extreme ranges as available. Three rodent families represented are separated by gaps in the horizontal axis. These are: Family Gliridae (genus *Muscardinus*). *Data are for juvenile rather than infant *Glaucomys sabrinus* but are included for comparison), Cricetidae (*Myodes* to *Peromyscus*) and Muridae (*Acomys* to *Rattus*). Data are taken from Brooks and Banks, 1973; Colvin, 1973; Schüpbach, 1974; Zippelius, 1974; Watts, 1976; Sales and Smith, 1978; Smith and Sales, 1980; Price, 1987; Yoshinaga et al., 1997; Motomura et al., 2002; Mandelli and Sales, 2004; Blake personal communication; M.L. Gilley personal communication.

Table 1. Characteristics of ultrasonic calls emitted in heterosexual encounters

	Duration	Frequency (within 10 kHz bands)*									Bandwidth							Source of data
	ms	<20	20	30	40	50	60	70	80	90+	kHz	fc	sd	sw	st	rc	fm	
Family Gliridae																		
Glaucomys sabrina	130	•	•								7.82	h					•	Gilley (pers. communication)
G. volans	137	•	•								8.87	h					•	Gilley (pers. communication)
Family Cricetidae																		
Microtus agrestis	15–68	•	•	•	•	•	•				8–36	f		•	•			Mandelli, 1998
M. arvalis	67		•										•					Kapusta et al., 2007
M. montanus				•														Pierce et al., 1989
M. ochrogaster				•														Pierce et al., 1989
Myodes glareolus	2–25		•	•	•	•					1–5	f, nh	•					Sales, 1972b
Dicrostonyx groenlandicus	c70	•	•	•								f		•			•	Brooks and Banks, 1973
Lagurus lagurus	5–60						•	•	•	•	10–30	f			•	•	•	Sales, 1972b
Allocricetulus eversmanni	112		•	•	•						16.8	h	•					Kapusta et al., 2006
Cricetulus migratoria	134		•	•	•	•					28	h	•					Kapusta et al., 2006
Mesocricetus auratus (female present)	99		•	•	•	•					1–5	f, nh	•		•			Sales, 1972b; Floody and Pfaff, 1977a
M. auratus (female removed)	77		•	•	•	•						f, nh	•		•			Floody and Pfaff, 1977a
Phodopus sungorus	36				•	•					4–10	f	•			•		Sales and Lof (unpublished)
P. cambellii	17–92	•	•	•	•	•	•	•			3–40	f	•		•	•		Sales and Rhodes (unpublished)
Peromyscus maniculatus	2–50			•	•	•	•				6–20	f	•			•		Sales, 1972b

Continued

Table 1. (Continued)

	Duration	Frequency (within 10 kHz bands)*									Bandwidth							Source of data
	ms	<20	20	30	40	50	60	70	80	90+	kHz	fc	sd	sw	st	rc	fm	
Family Muridae																		
Acomys caharinus	100–300		•	•	•	•	•	•	•		1–50	h			•		•	Sales, 1972b
Gerbillurus. paeba. exilis	31–175				•	•	•					f		•				Dempster and Perrin, 1991
G. p. paeba	31–150					•	•	•				f		•				Dempster and Perrin, 1991
G. setzeri	50–262	•	•	•	•	•	•					f		•				Dempster and Perrin, 1991
G. tytonis sweep call	37–118				•	•	•					f		•				Dempster and Perrin, 1991
G. vallinus sweep call	25–75		•	•	•							f		•				Dempster and Perrin, 1991
G. vallinus whistle call	131–950	•	•	•								f	•					Dempster and Perrin, 1991
Tatera brandtsii sweep call	43–243	•	•									f		•				Dempster and Perrin, 1991
Tatera brandtsii whistle call	268–806	•	•	•	•							f	•					Dempster and Perrin, 1991
Meriones unguiculatus pre-copulation	3–200		•	•								f		•			•	Holman, 1980
M. unguiculatus post-copulation	40–314		•									f	•					Holman, 1980
Apodemus sylvaticus	8–150					•	•	•			2–65	f	•	•	•		•	
Mus minutoides	2–30								•	•	1–10	f				•		Sales, 1972b
Rattus norvegicus pre-copulation	5–60					•	•	•	•		10–20	f				•	•	Kaltwasser, 1985
R. norvegicus post-copulation	600–1500		•								−5	f	•					Kaltwasser, 1985
R. rattus pre-copulation	3–180				•	•	•	•	•		10–20	f				•	•	Kaltwasser, 1985, 1990
R. rattus post-copulation	200–1700		•									f	•					Kaltwasser, 1985, 1990
R. villosisimus				•	•				•			f						Begg, 1976
Mastomys natalensis	5–30							•	•		2–5	f						Sales, 1972b

*lower limit of each band given except for <20 when upper limit given. See Table 2 for abbreviations.

Table 2. Characteristics of ultrasonic calls emitted in aggressive encounters between male rodents

	Duration	Frequency (within 10 kHz bands)*									b-w						Source of data
	ms	<20	20	30	40	50	60	70	80	90+	kHz	fc	sd	sw	st	rc	
Family Cricetidae																	
Microtus agrestis	67				•								•				Kapusta et al., 2007
M. arvalis	68			•									•				Kapusta et al., 2007
Myodes glareolus	10–64		•	•	•	•					2–5		•		•		Sales and Pye, 1974/ pers obs
Allocricetulus eversmanni	136		•	•	•	•					22		•				Kapusta et al., 2006
Cricetulus migratoria	211	•	•	•	•						31		•	•			Kapusta et al., 2006
Phodopus sungorus	17–106		•	•							4–10	f	•		•		Sales and Lof (unpub. observation)
Family Muridae																	
Gerbillus sp.						•						f					Sewell, 1969
Meriones shawii	10–40			•	•	•	•				10–40	f	•		•		Sales and Pye, 1974
Apodemus flavicollis	4–45			•	•	•	•				1–12	f					Hoffmeyer and Sales, 1977
A. sylvaticus	20–79					•	•	•			2–20	f	•		•	•	Sales and Pye, 1974
Thamnomys sp.	20–60			•	•	•					5–15	f			•	•	Sales and Pye, 1974
Rattus norvegicus	5–60					•	•	•	•		−10	f				•	Kaltwasser, 1985
R. rattus	30–70			•	•	•	•				−20+	f		•		•	Kaltwasser, 1985
R. villosisimus	5–7			•	•							f	•				Begg, 1976

*lower limit of each band given except for <20 when upper limit given; b-w: bandwidth; fc: frequency composition; f: fundamental; h: one or more harmonics present; nh: non-harmonic components present; sd: slow drifts; sw: one or two sweeps in frequency upwards and/or downwards; st: instantaneous "steps" in frequency; rc: rapid frequency changes; fm: repeated frequency modulation upwards and downwards.

Values for duration and bandwidth have been rounded to the nearest whole number and where necessary have been extrapolated from published figures.

after aggressive encounters for at least 11 species of *Rattus* (Watts, 1980; Kaltwasser, 1985, 1990) and after mating for wild Norway rats, *Rattus norvegicus*, and black rats, *R. rattus* (Kaltwasser, 1985), golden hamsters (Cherry, 1989) and Mongolian gerbils (Holman, 1980). In field voles, *Microtus agrestis*, shorter, higher frequency calls have been detected during attack (Table 2), while a few longer, 100–125 ms, lower frequency, 15–18 kHz, calls were detected during approach (Gilligan and Sales, unpublished observations).

III. Factors affecting the characteristics of ultrasonic calls

III.A. The nature of the sound production mechanism

The use of ultrasound by rodents is, in part, a function of their size. Low frequencies require large body structures; for small animals, therefore, ultrasound is easier to produce, transmit and receive than lower sounds (Pye, 1979). However, in rodents, the frequency of ultrasonic calls does not necessarily appear to be related to body size, as it is in some other animals including birds and frogs (Bradbury and Vehrencamp, 1998) and, in general, in other mammals (see Fletcher, Chapter 3.1 in this volume). Exploratory calls of both male and female wood mice, *Apodemus sylvaticus*, are higher in frequency than those of the larger yellow-necked mouse, *A. flavicollis* (Hoffmeyer and Sales, 1977; Sales and Green, unpublished observations); however, USVs emitted during male–male encounters in field voles have higher fundamental frequencies than those of smaller common voles, *Microtus arvalis* and bank voles (Kapusta et al., 2007). Rodents also emit lower frequency, broadband "audible" calls (audible to humans, i.e., below 20 kHz) that may extend into the ultrasonic range (Begg, 1976; Watts, 1980). Both of these types of call are produced by the larynx, but in different ways (Roberts, 1975; Nunez et al., 1985). For audible calls the vocal folds vibrate as in the human voice, but the USVs of rodents (but not of bats) appear to be produced by a mechanism similar to a Rayleigh whistle in which sound is produced by air passing through apertures in two parallel plates (Roberts, 1975). Roberts found that in a whistle-like model of the rodent larynx, changing the air pressure by even small amounts could lead to marked changes in frequency and amplitude and that, while mostly a pure tone was produced, certain pressures resulted in complex waveforms. Roberts' suggestion that rodents could emit USVs by the vocal folds being held rigid with a small aperture rather than vibrating has been confirmed by direct observation (Sanders et al., 2001; for further discussion see Brudzynski and Fletcher, Chapter 3.3 in this volume). The structure of rodent USVs can be related to use of the larynx in this way. The various frequency changes described above could reflect small changes in pressure during sound emission, and the increasing complexity of call structure with age in infants could result from an increased ability to produce higher pressures. It is perhaps not surprising that frequency changes and modulated calls are associated with active states, such as aggression and mating, when thoracic and abdominal pressures could be high, while the lower frequency calls with no or minor frequency changes detected after such behavior could reflect a reduction in activity and lower pressures.

III.B. Phylogeny

As sound is produced by anatomical structures, it might be expected that rodent ultrasonic calls would vary somewhat with phylogeny (Fig. 1). Recordings in the wild show that the USVs of California mice, *Peromyscus californicus*, appear to differ in frequency from those of sympatric but not closely-related brush mice, *P. boylii* (Kalcounis-Rueppell et al., 2008). Calls of infant *Microtus pennsylvanicus*, *M. montanus* and *M. californicus* are of shorter duration, higher frequency and are more structurally complex than those of infant *M. ochragaster* (Colvin, 1973). These differences reflect a lack of close relationship of *M. ochrogaster* to the other species (Colvin, 1973; Conroy and Cook, 2000). Similarly, in nine species of Australian rats, *Rattus* (Watts, 1980), and six taxa of South African rodents from the genera *Gerbillurus* and *Tatera* (Dempster and Perrin, 1991), calls of more closely-related species are more similar than those of less close relatives. Phlogeny, therefore, does appear to play some role in the physical nature of the calls, although differences may reduce with further analyses of larger numbers of species. Comparative studies of vocal tracts would be interesting here. Not all interspecific differences are phylogenetic: the calls of the closely-related field and common voles differ more from each other than either do to the less closely-related bank voles (Kapusta et al., 2007).

III.C. Function of calls and ecology of callers and recipients

It is not known to what extent calls are emitted in burrows or what effect a burrow would have on sound transmission. The diameters of burrows of small rodents, such as wood mice, are several times the wavelengths of ultrasonic calls which in air are approximately 7 mm at 50 kHz. In the open, ultrasounds are attenuated more rapidly than lower frequencies, so where they are used for communication over some distance they should be at the lower end of the ultrasonic range. The relatively constant and high-frequency (approximately 48 kHz) alarm calls of *Spermophilus richardsonii*, Richardson's ground squirrel, are therefore somewhat surprising (Wilson and Hare, 2004). As such calls would not travel far or be easy to locate, the authors suggest they are aimed at nearby philopatric kin, while remaining undetected by more distant predators.

Ultrasounds are reflected and scattered from small objects. In woods with sparse undergrowth there is little scattering, but in grass or wheat ultrasounds are rapidly attenuated and scattered (Smith, 1979), so although within the hearing range of several predators (Fay, 1988), they would be difficult to detect and to localize from a distance. In addition, complex signals are more degraded than simpler calls by repeated reflection (reverberation) from objects such as vegetation (Bradbury and Vehrencamp, 1998). In vegetation, lower frequencies and simpler calls would therefore be more effective for communication over distances than higher frequencies and complex calls. This may be why many calls of grassland rodents such as voles show relatively little frequency modulation and are at the lower end of the ultrasonic scale (Kapusta et al., 2006), and why calls emitted during male–female encounters of *Tatera brantsii*, a southern African grassland rodent, tend to have lower frequencies than those of related species living in more open country (Dempster and Perrin, 1991). USVs of deer mice and brush mice recorded in the wild are also in the lower ultrasonic range (Kalcounis-Rueppell et al., 2006).

Calls which must be located by conspecifics should be relatively low in frequency, stereotyped, repetitive and cover a broad bandwidth or have rapid changes in frequency (Bradbury and Vehrencamp, 1998). This is so for the ultrasonic songs of pygmy mice, *Baiomys musculus* and *B. taylori*, and singing mice, *Scotinomys teguina* and *S. xerampelinus*, which consist of stereotypic repeated, rapid downward frequency sweeps and may be used to maintain contact with conspecifics and facilitate location of the caller (Miller and Engstrom, 2007). Locatability may also account for differences in the mating calls of golden hamsters. During oestrus females, which are aggressive outside this period, have precopulatory calls that are more complex than those of males, and are more likely to include rapid changes in amplitude and frequency. They may therefore operate over greater distances to signal female receptivity (Floody and Pfaff, 1977a,b). Male calls emitted after separation from an oestrus female have wider bandwidths and more rapid frequency changes than those emitted in the presence of the female, and so could facilitate resumption of contact between separated animals.

One function of communication is to transmit certain information about the caller to the recipient(s) (Bradbury and Vehrencamp, 1998). In rats and mice there is good evidence that motivational state can be coded in the calls (Brudzynski 2005; Costantini and D'Amato, 2006) and this is probably so for other rodents. Information about the size of the caller could be obtained from the inverse relationship between size and call frequency that is seen in a number of species, including many primates (Hauser, 1993; Bradbury and Vehrencamp, 1998). However, as indicated above, this is not necessarily the case in rodents. Other features of the vocal mechanism must therefore determine the characteristics of the calls. Lower frequency, harsher signals have been linked to larger more aggressive animals, while higher-pitched more tonal signals have been associated with more amicable species and especially with friendly, escape or appeasing behavior: the motivation–structural rule of Morton (1977). This rule appears to apply to a range of mammalian taxa including marsupials, rodents, carnivores and ungulates, but not to all primates (Morton, 1977; Hauser, 1993). Rodent USVs, however, do not easily fit this scheme; the calls emitted during aggression are often at similar frequencies to those emitted during mating, and those emitted by submissive animals are of lower frequency than those emitted during fighting. Audible calls are emitted during aggressive encounters, but often by the attacked animal.

There is evidence that USVs emitted during mating are attractive to the opposite sex and promote reproductive behavior in recipients (e.g., Barfield and Thomas, 1986; Floody and Bauer, 1987). It might be expected that where related species are sympatric and have a chance of meeting, their mating calls would differ more than those of closely-related species that are unlikely to meet or those of less closely-related species: a phenomenon known as character displacement (Brown and Wilson, 1956). This occurs in a range of species (Bradbury and Vehrencamp, 1998), but there are apparently no examples from mammals. Calls of male California and brush mice recorded in the wild in an area of sympatry appear to emit calls at different frequencies (Kalcounis-Rueppell et al., 2008) and calls emitted during male–female encounters of field voles are higher in frequency than those of either common voles or bank voles. These vole species are sympatric, but their microhabitats differ: field voles prefer to live at the forest edge so could meet with both bank voles, which tend to live in forests, and with common voles which live in grassy areas away from the forest (Kapusta and Sales, 2009). This does not seem to be a case of character displacement *per se* and the ultrasonic calls of these and other species in areas of both sympatry and allopatry deserve further study.

III.D. Sensory ability of recipients

Communication signals must be detectable by recipients and discriminable from other signals. There is good evidence that rodents can hear the ultrasonic signals of their own species and many have a peak of auditory sensitivity in the ultrasonic range (Brown and Pye, 1975). Less is known about the ability of their auditory systems to discriminate different aspects of ultrasonic signals and the subtle call differences between species, or to pick out relevant features such as frequency modulation, although temporal responses of single brain neurons to 22 kHz calls have been recorded in rats (Allen et al., 2007). Lactating female mice can discriminate narrow-band sounds of frequencies of 37.5 kHz or higher from tones of 20 kHz and preferentially respond to the higher and to narrower-band sounds. They prefer tones of 50 kHz and of 30 ms or longer to shorter tones of similar frequency, but only if the two tones differ by at least 20–25 ms in length (Ehret and Haack, 1981; Ehret, 1992). This auditory selectivity matches the characteristics of infant calls, but the extent to which one has driven the other is unclear. More studies are needed on neural and behavioral responses of rodents and their ability to discriminate different types of signal, as well as subtle differences between signals, for example of related species (Fig. 2).

IV. Concluding comments

The ecology and evolution of USVs and their design has been under-represented in studies of ultrasonic calling so far and offers many potentially rich areas of study as indicated above. A problem with comparing studies of rodent calls is lack of agreement in terms used for describing calls and, sometimes, of information given on the exact equipment and conditions used. Different methods of displaying the calls visually can lead to differences in interpretation: what appears to be a slow change in frequency on an expanded time base can appear a rapid change on a shorter time base, and may lead to terminological differences. In addition, ultrasonic calls are often considered in isolation from the rest of the acoustic repertoire of rodents. Rodent calls within our hearing range (i.e., calls <20 kHz) are often excluded from such studies, and yet some are similar in acoustic structure to ultrasonic calls but are of lower frequency. For example, Begg (1976) described two whistle-type calls with a single frequency component at about 11–14 kHz in *Rattus villosissimus* that were otherwise similar in structure to USVs of other *Rattus* species, a demonstration of the artificial nature of the human-determined boundary of ultrasonic calls. Consideration of ultrasonic calls within the wider acoustic repertoire, and indeed within the whole range of communication signals of a species, should be valuable.

Finally, it is important to note that ultrasonic communication signals are also found in mammalian taxa other than rodents. For example, purely ultrasonic calls have been detected from the bush baby, *Galago senegalensis*, the slow loris, *Nycticebus coucang*, the mouse lemur, *Microcebus murinus* and the fat-tailed dwarf lemur, *Cheirogaleus medius* (Zimmerman, 1981; Cherry et al., 1987; see also Zimmermann, Chapter 6.3 in this volume), and some calls in the audible range emitted by small primates such as pygmy marmosets, *Cebuella pygmaea* (Pola and Snowdon, 1975), show structural similarities to some rodent USVs. Knowledge of how such signals are produced would be of interest. The two different mechanisms of rodent, and possibly primate, ultrasound production can perhaps be likened to the vocal registers of humans; it is tempting to compare the ultrasonic range to the human whistle register that is used to produce extremely high pitches. In that register, vocal cords are fully stretched but do vibrate along their entire length, although in one plane rather than the usual two (Keilmann and Michek, 1993).

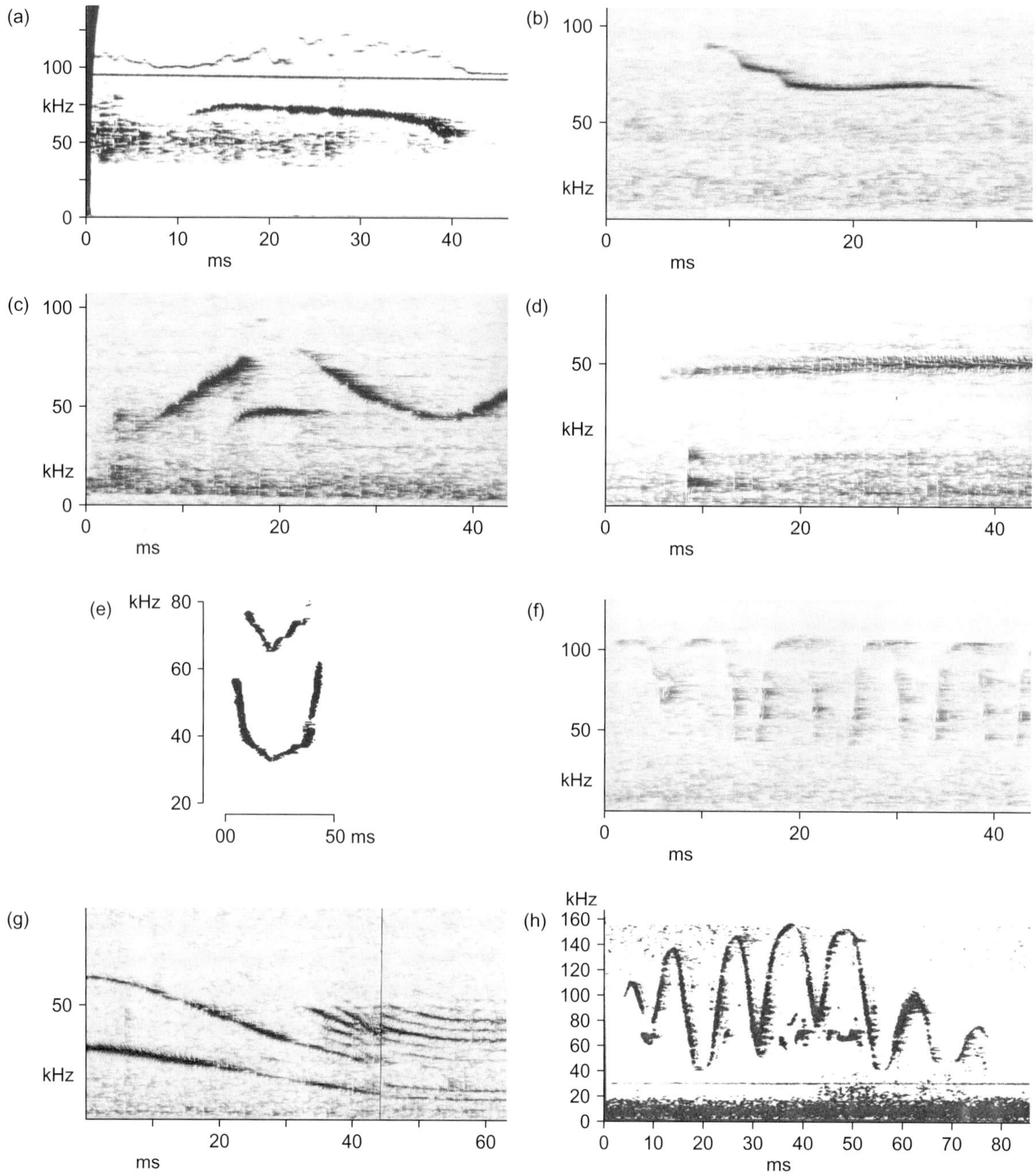

Fig. 2. Frequency patterns in the ultrasonic calls of rodents. (a) Call with slow frequency drift and amplitude modulation emitted by *Apodemus sylvaticus* during aggression; (b) rapid frequency changes emitted by *Calomys callosus* during sexual behavior; (c) frequency sweeps and steps emitted by *Lagurus lagurus* during sexual behavior; (d) narrow-band, non-harmonic components produced by *Gerbillus* sp. during aggression; (e) "U" shaped call with harmonic component emitted during sexual behavior in *Microtus agrestis*; (f) train of frequency modulations and "n" shaped changes emitted during sexual behavior by *Apodemus sylvaticus*; (g) call that begins in the ultrasonic range and decreases to the audible range with additional harmonic components, emitted by juvenile *Meriones shawii* – the vertical line at about 44 ms is where two sonagrams are abutted to show the complete call; (h) extensive frequency modulation produced by infant wild *Rattus norvegicus*. In all graphs time is represented in ms on the horizontal axis (every 10 ms) and frequency in kHz on the vertical axis.

The mammalian vocal mechanism is clearly flexible in its modes of operation, and rodents at least exploit a mode that most other mammals apparently do not. The evolution of this mode of operation of the larynx and of the various sounds it produces certainly deserves further study.

Acknowledgments

I am deeply grateful to Dr Barbara Blake for much help in providing information, contacts, advice and support; to her and Ms L. Michelle Gilley for permission to use unpublished data and to Dr Matina Kalcounis-Roupell for comments on her work. I am indebted to Professor David Pye for improving my understanding of ultrasound production and I thank Professor Stefan M. Brudzynski for advice and support during the writing process.

References

Allen, A.A., Furtrak, S.C., Brown, T.H., 2007. Single-unit responses to 22 kHz ultrasonic vocalizations in rat perirhinal cortex. Behav. Brain. Res. 182, 327–336.

Barfield, R.J., Thomas, D.A., 1986. The role of ultrasonic vocalizations in the regulation of reproduction in rats. Ann. N.Y. Acad. Sci. 474, 33–43.

Begg, R.J., 1976. Ultrasound production by the long haired rat of Australia (*Rattus villosissimus*). Ultrasonics 14, 177–182.

Blake, B.H., 1992. Ultrasonic vocalization and body temperature maintenance in infant voles of three species (Rodentia: Arvicolidae). Dev. Psychobiol. 25, 581–596.

Bradbury, J.W., Vehrencamp, S.L., 1998. Principles of Animal Communication. Sinauer Associates Inc., Sunderland, MA.

Brooks, R.J., Banks, E.M., 1973. Behavioural biology of the collared lemming (*Dicrostonyx groenlandicus* (Traill)): an analysis of acoustic communication. Anim. Behav. Mongr. 6, 1–83.

Brown, A.M., Pye, J.D., 1975. Auditory sensitivity at high frequencies in mammals. Adv. Comp. Physiol. Biochem. 6, 1–73.

Brown Jr., W.L., Wilson, O.E., 1956. Character displacement. Syst. Zool. 5, 49–64.

Brudzynski, S.M., 2005. Principles of rat communication: quantitative parameters of ultrasonic calls in rats. Behav. Genet. 35, 85–92.

Brudzynski, S.M., 2007. Ultrasonic calls of rats as indicator variables of negative or positive states: acetylcholine-dopamine interactions and acoustic coding. Behav. Brain Res. 182, 261–273.

Brudzynski, S.M., Kehoe, P., Callahan, M., 1999. Sonographic structure of isolation-induced ultrasonic calls of rat pups. Dev. Psychobiol. 34, 195–204.

Cherry, J.A., 1989. Ultrasonic vocalization by male hamsters: parameters of calling and effects of playback on female behaviour. Anim. Behav. 38, 138–153.

Cherry, J.A., Izard, M.K., Simons, E.L., 1987. Description of ultrasonic vocalizations of the mouse lemur (*Microcebus murinus*) and the fat-tailed dwarf lemur (*Cheirogaleus medius*). Am. J. Primatol. 13, 181–185.

Colvin, M.A., 1973. Analysis of the acoustic structure and function in ultrasounds of neonatal *Microtus*. Behaviour 44, 234–263.

Conroy, C.J., Cook, J.A., 2000. Molecular systematics of a holarctic rodent (*Microtus*: Muridae). J. Mammal. 81, 344–359.

Costantini, F., D'Amato, F.R., 2006. Ultrasonic vocalizations in mice and rats: social contexts and functions. Acta Zool. Sinn. 52, 619–633.

Dempster, E.R., Perrin, M.R., 1991. Ultrasonic vocalizations of six taxa of southern African gerbils (Rodentia: Gerbillinae). Ethology 88, 1–10.

Ehret, G., 1992. Categorical perception of mouse-pup ultrasounds in the temporal domain. Anim. Behav. 43, 409–416.

Ehret, G., 2005. Infant rodent ultrasounds – a gate to the understanding of sound communication. Behav. Genet. 35, 19–29.

Ehret, G., Haack, B., 1981. Categorical perception of mouse pup ultrasound by lactating females. Naturwissenschaften 68, 208.

Fay, R.R., 1988. Hearing in Vertebrates: A Psychophysics Data Book. Hill-Fay Associates, Winnetka, IL.

Floody, O.R., Bauer, G.B., 1987. Selectivity in the responses of hamsters to conspecific vocalizations. Horm. Behav. 21, 522–527.

Floody, O.R., Pfaff, D.W., 1977a. Communication among hamsters by high-frequency acoustic signals: I. Physical characteristics of hamster calls. J. Comp. Physiol. Psychol. 91, 794–806.

Floody, O.R., Pfaff, D.W., 1977b. Communication among hamsters by high-frequency acoustic signals: III. Responses evoked by natural and synthetic sounds. J. Comp. Physiol. Psychol. 91, 820–829.

Floody, O.R., O'Donohue, T.L., 1980. Lesions of the mesencepahalic central gray depress ultrasound production and lordosis by female hamsters. Physiol. Behav. 24, 79–85.

Francis, R.L., 1977. 22-kHz calls by isolated rats. Nature 265, 236–238.

Gyger, M., Pontet, A., Schenk, F., 1986. Le canal acoustique chez les Rongeurs: les émissions à haute fréquence. Rev. Suisse Zool. 93, 623–640.

Hauser, M.D., 1993. The evolution of nonhuman primate vocalizations: effects of phylogeny, body weight and social context. Am. Nat. 142, 528–542.

Hoffmeyer, I., Sales, G.D., 1977. Ultrasonic behaviour of *Apodemus sylvaticus* and *A. flavicollis*. Oikos 29, 67–77.

Holman, S.D., 1980. Sexually dimorphic ultrasonic vocalizations of Mongolian gerbils. Behav. Neural Biol. 28, 183–192.

Holy, T.E., Guo, Z., 2005. Ultrasonic songs of male mice. PloS Biol. 3, e286.

Kalcounis-Rueppell, M.C., Metheny, J.D., Vonhof, M.J., 2006. Production of ultrasonic vocalizations by *Peromyscus* mice in the wild. Front. Zool. 3, 3.

Kalcounis-Rueppell, M.C., Carney, C., Briggs, J.R., Petric, R., 2008. Individual context of ultrasonic vocalizations produced by wild deer mice (*Peromyscus*). Book of

Extended Abstracts, Second International Conference on Acoustic Communication by Animals, pp. 114–115.

Kaltwasser, M.-Th., 1985. Vergleichende Untersuchungen zum akustischen Kommunikationsverhalten und zur Orientierung bei *Rattus rattus* und *Rattus norvegicus*. Dissertation zur Erlangung des Grades eines Doktors der Naturwissenschaften der Fakultät für Biologie der Eberhard-Karls-Universität, Tübingen, Germany.

Kaltwasser, M.-Th., 1990. Acoustic signalling in the Black rat (*Rattus rattus*). J. Comp. Psychol. 104, 227–232.

Kapusta, J., Sales, G., 2009. Male–female interactions and ultrasonic vocalization in three sympatric species of voles during conspecific and heterospecific encounters. Behaviour 146, 939–962.

Kapusta, J., Szentgyörgyi, H., Surov, A., Ryurikov, G., 2006. Vocalization of two palaearctic species of hamster: Eversmann hamster *Allocricetulus eversmanni* and grey hamster *Cricetulus migratorius*. Bioacoustics 15, 315–330.

Kapusta, J., Sales, G.D., Czuchnowski, R., 2007. Aggression and vocalization of three sympatric vole species during conspecific and heterospecific same-sex encounters. Behaviour 144, 283–305.

Keilmann, A., Michek, F., 1993. Physiology and acoustic analysis of whistle voice of the woman. Folia Phoniatr. 45, 247–255.

Mandelli, M.-J., 1998. Aspects of social behaviour, sexual dimorphism and ultrasonic vocalizations in the field vole, Microtus agrestis. Unpublished Ph.D. Thesis, University of London, UK.

Mandelli, M.-J., Sales, G., 2004. Ultrasonic vocalizations of infant short-tailed field voles, *Microtus agrestis*. J. Mammal. 85, 282–289.

Miller, J.R., Engstrom, M.D., 2007. Vocal stereotypy and singing behaviour in baoimyine mice. J. Mammal. 88, 1447–1465.

Morton, E.S., 1977. On the occurrence and significance of motivational–structural rules in some bird and mammal sounds. Am. Nat. 111, 855–869.

Motomura, N., Shimizu, K., Shimizu, M., Aoki-Komori, S., Taniguchi, K., Serizawa, I., Saito, T.R., 2002. A comparative study of isolation-induced ultrasonic vocalization in rodent pups. Exp. Anim. 51, 187–190.

Okon, E.E., 1972. Factors affecting ultrasound production in infant rodents. J. Zool. 168, 139–148.

Nunez, A.A., Pomerantz, S.M., Bean, N.J., Youngstrom, T.G., 1985. Effects of laryngeal denervation on ultrasound production and male sexual behavior in rodents. Physiol. Behav. 34, 901–905.

Pierce, J.D., Sawrey, D.K., Dewsbury, D.A., 1989. A comparative study of rodent ultrasonic vocalizations during copulation. Behav. Neural Biol. 51, 211–221.

Pola, Y.V., Snowdon, C.T., 1975. The vocalizations of pygymy marmosets (*Cebuella pygmaea*). Anim. Behav. 23, 826–842.

Portfors, C.V., 2007. Types and functions of ultrasonic vocalizations in laboratory rats and mice. J. Lab. Anim. Sci. 46, 28–34.

Price, F.E.F., 1987. Aspects of the ontogeny and function of ultrasonic calling by the Norway rat, *Rattus norvegicus* Berkenhout. Ph.D. Thesis, University of London, UK.

Pye, J.D., 1979. Why ultrasound? Endeavour 3, 57–62.

Roberts, L.H., 1975. The rodent ultrasound production mechanism. Ultrasonics 13, 83–88.

Sales, G.D., 1972. Ultrasound and mating behaviour in rodents with some observations on other behavioural situations. J. Zool. 168, 149–164.

Sales, G.D., 1991. The effect of 22 kHz calls and artificial 38 kHz signals on activity in rats. Behav. Processes 24, 83–93.

Sales, G.D., Pye, J.D., 1974. Ultrasonic Communication by Animals. Chapman and Hall, London, UK.

Sales, G.D., Smith, J.C., 1978. Comparative studies of the ultrasonic calls of infant murid rodents. Dev. Psychobiol. 11, 595–619.

Sanders, I., Weisz, D.J., Yang, B.Y., Fung, K., Amirali, A., 2001. The mechanism of ultrasonic vocalization in the rat. Soc. Neurosci. Abs. 27, 241.

Schüpbach, U., 1974. [Ethological investigation with regard to acoustic communication between dormice (*Muscardinus avellarius*. L)]. Licentiate paper, Berne University, Switzerland (in German).

Sewell, G.D., 1970a. Ultrasonic communication in rodents. Nature 227, 410.

Sewell, G.D., 1970b. Ultrasonic signals in rodents. Ultrasonics 8, 26–30.

Shu, W., Cho, J.Y., Jiang, Y., Zhang, M., Weisz, D., Elder, G.A., Schmeidler, J., De Gasperi, R., Gma Sosa, M.A., Rabidou, D., Santucci, A.C., Perl, D., Morrisey, E., Buxhaum, J.D., 2005. Altered ultrasonic vocalization in mice with a disruption of the *Foxp2* gene. Proc. Natl. Acad. Sci. USA 102, 9643–9648.

Smith, J.C., 1979. Factors affecting the transmission of rodent ultrasounds in natural environments. Am. Zool. 19, 432–442.

Smith, J.C., Sales, G.D., 1980. Ultrasonic behavior and mother-infant interactions in rodents. In: Smotherman, R., Bell, R. (Eds.), Maternal Influences and Early Behavior. Spectrum Publications Inc., New York, NY, pp. 103–133.

Watts, C.H.S., 1975. Vocalizations of Australian hopping mice (Rodentia: *Notomys*). J. Zool. 117, 247–263.

Watts, C.H.S., 1976. Vocalizations of the plains rat *Pseudomys australis* Gray (Rodentia: Muridae). Aust. J. Zool. 24, 95–103.

Watts, C.H.S., 1980. Vocalization of nine species of rat (*Rattus*; Muridae). J. Zool. Lond. 191, 531–555.

Wilson, R.W., Hare, J.F., 2004. Ground squirrel uses ultrasonic alarms. Nature, Lond. 430, 523.

Wilson, D.W., Reeder, D. (Eds.), 2005. Mammal Species of the World: A Taxonomic and Geographic Reference. Johns Hopkins University Press, Baltimore, MD.

Yoshinaga, Y., Ohnol, W., Shiraishi, S., 1997. Postnatal growth, development and ultrasonic vocalization, of young Japanese voles, *Microtus montebelli*. Mamm. Study 22, 53–70.

Zimmerman, E., 1981. First record of ultrasound in two prosimian species. Naturwissenschaften 68, 531.

Zippelius, H.-M., 1974. Ultraschall-Laute Nestjunge Mäuse. Behaviour 49, 197–204.

CHAPTER 3.5

Vocal repertoire in mouse pups: strain differences

Maria Luisa Scattoni[1*] and Igor Branchi[2]

[1]*Section of Neurotoxicology and Neuroendocrinology*
[2]*Section of Behavioural Neurosciences, Department of Cell Biology and Neurosciences, Istituto Superiore di Sanità, Rome, Italy*

Abstract: Ultrasonic calls emitted by mouse pups are whistle-like sounds with frequencies between 30 and 90 kHz. These signals play an important communicative role in mother–offspring interactions, since they elicit in the dam a prompt caregiving response. Both physical and social factors modulate ultrasound emission. Recently, an approach based on sonographic analysis has allowed a deeper investigation of this behavior, providing a detailed characterization of the features of ultrasonic call structure. Overall, studies on mouse pup ultrasonic calls have shown that the rate of emission of these signals follows an ontogenetic profile, the different sonographic types are emitted according to the environmental context, and different mouse strains emit specific repertoires of sonographic types.

Keywords: ultrasound; ultrasonic vocalization; ultrasonic call; sonographic analysis; spectrographic analysis; sonographic type; development; ontogeny; mouse strains

I. Introduction

In recent genetic, physiological and pharmacological studies, researchers have developed particularly large numbers of laboratory mice strains. These strains were obtained by mating brothers and sisters generation after generation (inbreeding), creating a strain that is genetically unique, or mating between strains (outbreeding) to increase the genetic variation of the resulting strain. They both represent a good model for studying genetic differences and the effect of particular genes on physiological functions and behavior. There have been a particularly large number of behavioral studies performed on these strains. Among many behavioral parameters studied, mouse ultrasonic vocalization has recently focused the attention of researchers.

Ultrasonic vocalizations (USVs) emitted by mouse pups are whistle-like sounds (see Brudzynski and Fletcher, Chapter 3.3 in this volume) with frequencies ranging from 30 to 90 kHz, thus above the upper limit of human hearing range, which is approximately 20 kHz. In this species, as well as in other altricial rodent species, ultrasonic signals play an important communicative role in the mother–offspring interaction, since they elicit a prompt care-giving response in the dam. Both physical and social parameters modulate the USV emission in the infant rodent (Hofer, 1996; Branchi et al., 2001; Hahn and Schanz, 2002; Ehret, 2005; Hahn and Lavooy, 2005). The study of this ethologically and ecologically relevant behavior represents a reliable and effective tool for assessing the neurobehavioral development of the mouse.

The USV parameter most frequently used for measuring vocal behavior is vocalization rate, although duration and amplitude of vocalizations are also considered important. In addition, other acoustic parameters, obtained through a spectrographic (sonographic) analysis (i.e., displays of sound frequency on the y-axis, over time on the x-axis and with the amplitude on the color-coded z-axis), have been exploited to study vocal behavior in mouse pups. These parameters, taken from different sonographic displays, include call duration, sound frequency and the overall sonographic structure.

[*]Corresponding author. E-mail: marialuisa.scattoni@iss.it

Stefan M. Brudzynski (Ed.)
Handbook of Mammalian Vocalization
ISBN 978-0-12-374593-4

DOI: 10.1016/B978-0-12-374593-4.00010-3

II. General strain differences

The USV production differs among mouse strains. First of all, these differences concern the amount of USVs emitted. Some mouse strains spontaneously emit a high number of USVs, while others emit a very low number. For instance, NZB and C57BL/6J strains emit a lower number of vocalizations than other inbred mice, such as BALB/c, DBA, A/J, BTBR and FVB/NJ (Bell et al., 1972; Robinson and D'Udine, 1982; Cohen-Salmon et al., 1985; Roubertoux et al., 1996; Scattoni et al., 2008a). Also, differences in other call characteristics such as call duration, sound frequency and bandwidth are found to be dependent on the genetic background (Hahn et al., 1987; Roubertoux et al., 1996; Hahn et al., 1997; Hahn et al., 1998; Hahn and Schanz, 2002; Thornton et al., 2005). The importance of the role played by genes in affecting USV emission is corroborated by studies carried out on knockout mice, showing that deletion or insertion of selected genes markedly modifies the vocal behavior (Bolivar and Brown, 1994; Brunner et al., 1999; Winslow et al., 2000; Weller et al., 2003; El-Khodor et al., 2004; Moles et al., 2004; Fride et al., 2005; Shu et al., 2005; Picker et al., 2006; Scattoni et al., 2008b).

A recent study analyzed the contributions of genetic background, gender and early environmental factors in detail on vocalizing behavior in C57BL/6JOlaHsd and C57BL/6NCrl mice by using an embryo-transfer procedure (Wöhr et al., 2008). The results showed that, while such features as call frequency and amplitude are primarily dependent on the genotype, the differences concerning the amount of emitted ultrasonic calls were dependent mainly on the nest environment, particularly on the dyadic interaction between mother and the pup.

USV production can differ not only in the rate, duration, and amplitude of calls, but also in the way these parameters change day after day during the first postnatal weeks. Indeed, the rate of USVs emitted by mouse pups follows the ontogenetic profile, increasing during the first days of life, reaching a peak during the first postnatal week, then starting to decrease and completely disappearing when the pups open their eyes (Elwood and Keeling, 1982). However, such ontogenetic profiling shows important strain-dependent differences (Branchi et al., 2006; Scattoni et al., 2008a). For example, C57BL/6 and BALB/c strains show peak USV rates very early, at around postnatal day 3 (PND 3) (Bell et al., 1972; Robinson and D'Udine, 1982) as compared either to other inbred strains, such as BTBR, FVB/NJ, 129X1, DBA, A/J, SEC, C3H, or to outbred strains, such as CD-1, which show the peak around PND 7 (Hennessy et al., 1980; Elwood and Keeling, 1982; Branchi et al., 1998; Scattoni et al., 2008a).

Finally, strain differences in vocalization rate have also been found in the maternal potentiation response. This phenomenon occurs when, after the first isolation period and a subsequent brief contact with the dam, the isolated infant mouse produces USVs at a rate significantly higher than during the first isolation period (Shair et al., 2005). In the CD-1 Swiss-derived albino outbred strain, no maternal potentiation has been reported (Branchi et al., 2004). By contrast, inbred strains, particularly C57BL/6J mice, show maternal potentiation, significantly increasing their USV production during the second isolation period, when tested on PND 8 and PND 12 (Scattoni et al., 2009). Recent data indicate that the induction of maternal potentiation in mice depends on the experimental protocol used. Differing from the paradigm originally developed for rat species, the paradigm most effective in mouse species requires the reunion in the home cage with both mother and littermates, not only with the mother itself, in a novel and unfamiliar cage. In the home cage, time spent by the dam in contact with the experimental pup is significantly increased, enhancing the contrast experienced by the pups between the maternal reunion and the second isolation period (Scattoni et al., 2009).

III. Strain differences in vocal repertoire: new insights from the spectrographic analysis

III.A. Infant mice

The development of digital sound spectrographic analysis, which occurred in the last 5–10 years due to progress in computer technology, allows further insight into both environmental and genetic factors shaping the USV responses, and provides an additional and more detailed source of information about USV structure compared to analysis based only on vocalization rate, which has so far been the most utilized method (Branchi et al., 2001; Scattoni et al., 2008a, 2009).

A decade ago, Branchi and his colleagues showed for the first time that USVs with different sonographic structures were emitted by neonatal laboratory mice (see Fig. 1). These differently-shaped

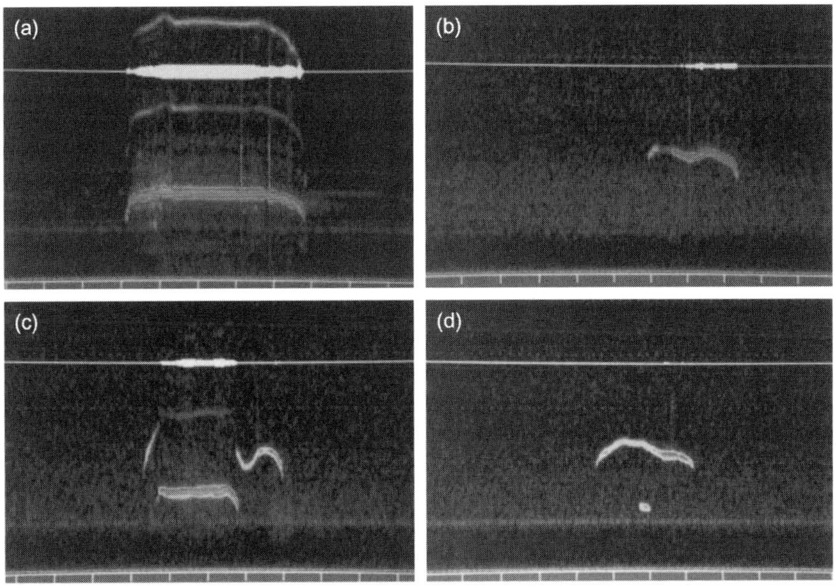

Fig. 1. Sonograms (frequency in kHz; time in ms; amplitude in dB – seen as a gray scale) of ultrasonic calls of infant CD-1 mice illustrating four call categories. (a) Ultrasonic vocalization of the constant frequency; (b) modulated-frequency; (c) frequency steps; and (d) composite call category. A waveform trace describing amplitude is superimposed on the spectrogram (white line). From Branchi et al. (1998) Dev. Psychobiol., 33: 249–256.

USVs were emitted in different numbers according to the environmental (social and physical) conditions to which the pup was exposed, which were odor from the nest, social isolation, low isolation temperature, tactile stimulation and odor of an unfamiliar conspecific adult male. The greatest changes in parameters of acoustic signals involved mainly a gradual frequency modulation, and particularly the occurrence of frequency steps. For instance, isolated mouse pups produced more frequency steps but an equal number of gradually modulated frequency calls as compared to mouse pups exposed to the odor of a conspecific unfamiliar adult male (Branchi et al., 1998).

More recently, spectrographic analysis has been used to assess the noxious effects of the nest environment. For instance, analysis of the spectrographic call patterns revealed significant effects of prenatal malnutrition on USV response that were not evident in the number of emitted calls (Tonkiss et al., 2003). Detailed spectrographic analysis showed that the reduction in the USV number following neonatal ethanol exposure was not a general reduction of all call types, but rather a selective reduction in certain acoustic forms (rising frequency, U-shaped and three sweep calls; Barron and Gilbertson, 2005).

Further analyses using playback are needed to clarify whether these selective changes have a specific impact on dam response and, thus, consequences for effective pup care and survival. As a first step, a recent investigation has already determined that calls with different sonographic types activated the auditory cortical response of the dam differently (Liu et al., 2006). In particular, cortical responses associated with USV detection and discrimination were elicited by the sonographic types characteristic for the natural pup vocalizations, but not by those of non-communicative artificial sound ensembles (Liu and Schreiner, 2007).

Recently, in search of strain differences, Scattoni and colleagues (Scattoni et al., 2008a) analyzed and classified every call emitted by four strains of mouse pups (BTBR, C57BL/6J, 129X1 and FVB/NJ) into ten categories. All patterns of the separation calls collected from every studied strain at PND 8 were subjected to an extensive spectrographic analysis. Each call was identified as one of ten distinct categories (see Fig. 2) (Scattoni et al., 2008a), defined according to the internal frequency changes, duration and sonographic shape, using previously published call categorizations (Sales and Smith, 1978; Branchi et al., 1998; Brudzynski et al., 1999; Panksepp et al., 2007). USVs were classified as *Complex calls* if one sound component contained two or more changes in frequency with varying direction and rate of changes, each $\geq$6.25 kHz. A second USV category, *Composite calls*,

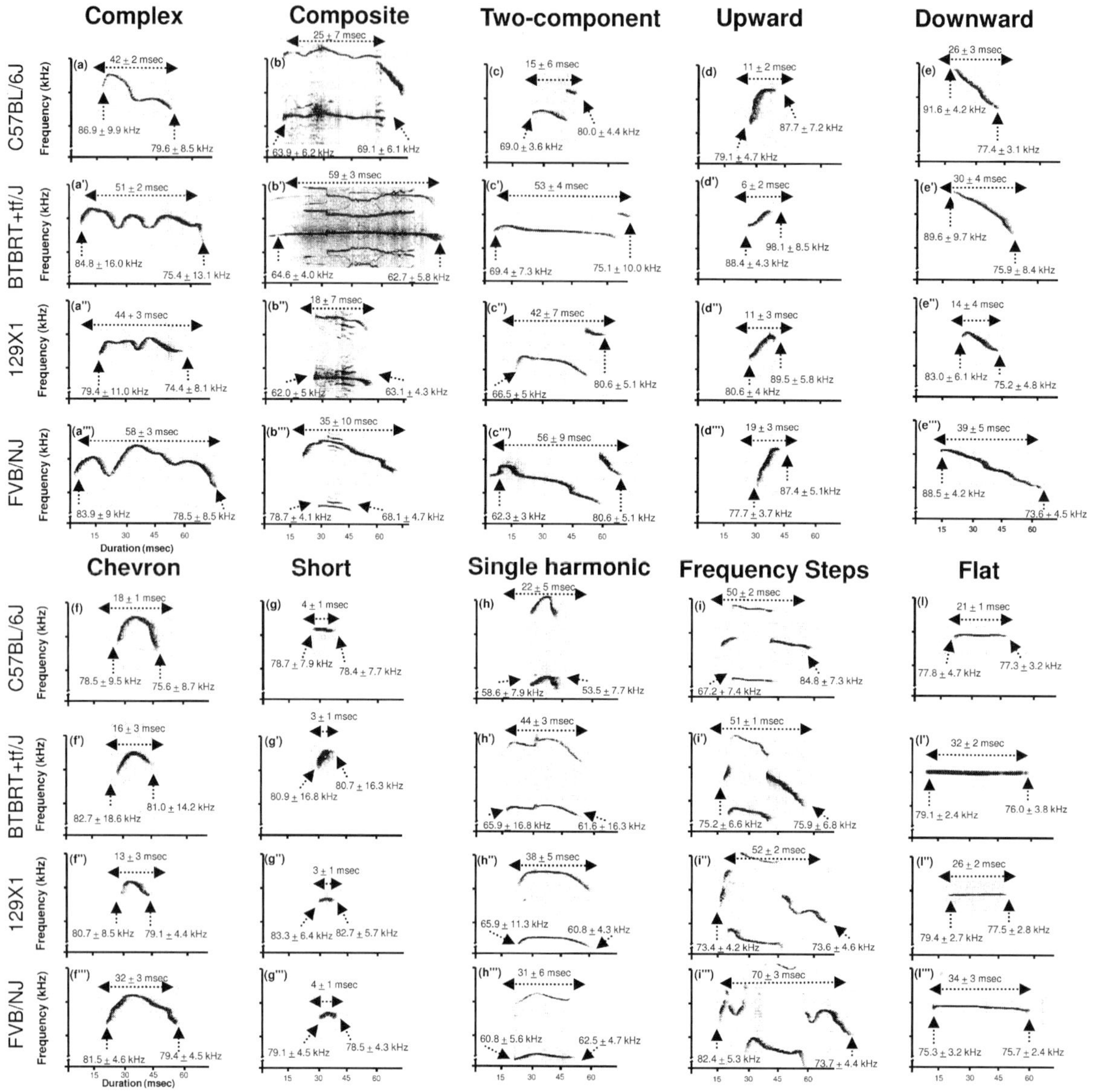

Fig. 2. Typical sonograms of ultrasonic vocalizations, classified into ten distinct categories of calls emitted by (a–l) B6, (a′–l′) BTBR, (a″–l″) 129X1, and (a‴–l‴) FVB/NJ mouse strains. The strains are labeled on the left-hand side; the first five categories are in the top panel and the remaining five categories in the lower panel. Sound frequency (y-axis) is in kHz, and time (x-axis) in ms. Duration of each call, as well as the initial and terminal frequencies are indicated on each graph by arrows. From Scattoni et al. (2008) PloS ONE, 3: e3067.

was characterized by one main call, resembling the complex call, and additional harmonic components of varying intensity and duration. *Two-component* calls consisted of two components: a main call (with a flat or downward frequency change) with an additional short component of higher frequency towards the end of the call. *Upward-modulated calls* exhibited a continuous increase in frequency that was ≥12.5 kHz, with a terminal dominant frequency at least 6.25 kHz more than the frequency at the beginning of the vocalization. *Downward-modulated calls* exhibited a continuous decrease in frequency that was ≥12.5 kHz, with a terminal dominant frequency at least 6.25 kHz less than the initial frequency of the vocalization. The general sonographic profile of *Chevron-like calls* resembled an "inverted-U shape," which was identified by a continuous increase in frequency ≥12.5 kHz followed by a decrease that was ≥6.25 kHz. Some calls had equal arms and others were asymmetric, with a shorter falling arm. *Short calls* were shorter than 5 ms. *Single harmonic calls* were formed by two parallel lines consisting of a continuous sound with a minor frequency modulation and a single higher amplified harmonic component. *Frequency steps* were instantaneous frequency changes appearing as a vertically discontinuous "step" on the sonographic display, but with no gaps on the time scale. *Flat calls* presented a constant frequency, including the initial and terminal frequency with changes ≤ 3 kHz.

The categories described above were studied as to the probability of their occurrence in vocalizations of different mouse strains. Significant differences among strains were found in the distribution of the call types. Particularly, 8-day-old C57BL/6J, FVB/NJ and 129X1 pups emitted a wide repertoire of calls, including high numbers of frequency steps and complex calls. By contrast, BTBR pups had a narrower repertoire of calls, which included high levels of harmonics and composite calls, but minimal numbers of chevron-shaped, upward, downward and short calls (see Fig. 2). When analyzing the data for the probability of vocalization type occurrence, a strain-dependent effect was found in the probability of producing specific types of calls for nine of the ten call categories (expressed as the number of calls in each category for each subject/total number of calls analyzed in each subject). The probability of emitting flat calls did not differ across strains. BTBR pups emitted more harmonics than C57BL/6J, 129X1 and FVB/NJ pups, and more composite and two-component calls than C57BL/6J pups. BTBR pups emitted less complex and frequency step calls than 129X1 and FVB/NJ pups, and less upward calls than C57BL/6J and FVB/NJ pups. C57BL/6J pups emitted more downward, chevron-like, and short calls than the other strains. Moreover, C57BL/6J pups emitted less two-component calls than the other strains.

III.B. Juvenile and adult mice

USVs are emitted by mice not only during the infancy, but throughout their lifespan (Nyby, 2001; Scattoni et al., 2009). The USV emission of adult mice has been primarily reported in reproductive contexts (Nyby, 2001), with males being responsible for most of the calls (Whitney and Nyby, 1979; Maggio et al., 1983). At variance with rats, adult mouse USVs were not detected during antagonistic encounters in laboratory settings (Nyby, 2001). Only recently, however, mouse USVs have been recorded in adolescent C57BL/6J and Balb/cJ mice of both sexes during social interactions, after weaning and in a repeated schedule of social housing (four days) followed by one-day social isolation (Panksepp et al., 2007). These vocalizations were remarkably complex, with a significant effect of genotype (strain) on each distinct USV category that was classified. Detailed qualitative analysis revealed significant strain differences in the distribution of call types, with the prevalence of downward and complex sonogram types in C57BL/6J mice, and the prevalence of upward and inverted U-shaped sonograms in BALB/cJ mice (Panksepp et al., 2007).

Holy and Guo first reported that vocalizations of adult male mice (C57BL/6 × DBA/2J F1 cross) were more complex than previously known, and shared some characteristics with birdsong. These vocalizations consisted of different components (syllable types), whose temporal sequencing included repeated phrases (Holy and Guo, 2005).

Comparing the results from the previously-described studies, it is possible to draw some further conclusions as to how USV features change according to strain and age in mice. For instance, it is possible to notice that 8-day-old C57BL/6J mice showed similar percentages of upward- and downward-modulated calls as compared to 30-day-old C57BL/6J mice (Panksepp et al., 2007), but had a higher percentage of chevron-like and complex calls. Differences in age, social setting, analysis parameters and the use of five versus ten distinct call categories may explain the discrepancies in the percentages reported by Panksepp and co-workers (Panksepp et al., 2007) and our recent infant data (Scattoni et al., 2008a). Moreover, the mean call duration and peak frequencies of these types of calls are not identical in the 8- and 30-day-old C57BL/6J mice, most likely due to the different ages of the subjects and differences in the test conditions. Branchi and co-workers (Branchi et al., 1998) previously reported a spectrographic characterization of 8-day-old outbred CD-1 mice isolated from their mothers and littermates. CD-1 pups emitted a wide spectrum of USVs classified into five categories (flat, complex, frequency steps, short and composite). As compared to the results obtained with C57BL/6J, 129X1 and FVB/NJ, the 8-day-old CD-1 emitted a higher percentage of frequency steps and complex calls, but low numbers of flat, short and complex calls. Thus, vocalization categories and their components in C57BL/6J, F1 BXD, BALB/cJ and

CD-1 mice seem to be generally similar in each strain, to such an extent that comparisons of results can be made across different laboratories.

IV. Conclusions

The growing literature on USVs in mice suggests that the analysis of this behavioral response is likely to be useful to characterize communication between the pup and the mother, and between adults of the same or opposite sex in some inbred strains and some lines of mutant mice. It appears that a quantitative, as well as a qualitative (spectrographic/sonographic analysis), approach may be useful in understanding the communicative value of calls emitted by mice, by comparing call categories and their components from different strains or genotypes. It is interesting to speculate that variability in the types of categories emitted by each strain may represent different "dialects" specific to each inbred strain. Further experiments using a wider number of inbred mouse strains, at different ages and in different environmental situations, is warranted to make statements about differences in strain genotype variations and complex vocal repertoires.

References

Barron, S., Gilbertson, R., 2005. Neonatal ethanol exposure but not neonatal cocaine selectively reduces specific isolation-induced vocalization waveforms in rats. Behav. Genet. 35, 93–102.

Bell, R.W., Nitschke, W., Zachman, T.A., 1972. Ultrasounds in three inbred strains of young mice. Behav. Biol. 7, 805–814.

Bolivar, V.J., Brown, R.E., 1994. The ontogeny of ultrasonic vocalizations and other behaviors in male jimpy (jp/Y) mice and their normal male littermates. Dev. Psychobiol. 27, 101–110.

Branchi, I., Santucci, D., Vitale, A., Alleva, E., 1998. Ultrasonic vocalizations by infant laboratory mice: a preliminary spectrographic characterization under different conditions. Dev. Psychobiol. 33, 249–256.

Branchi, I., Santucci, D., Alleva, E., 2001. Ultrasonic vocalisation emitted by infant rodents: a tool for assessment of neurobehavioural development. Behav. Brain Res. 125, 49–56.

Branchi, I., Campolongo, P., Alleva, E., 2004. Scopolamine effects on ultrasonic vocalization emission and behavior in the neonatal mouse. Behav. Brain Res. 151, 9–16.

Branchi, I., Santucci, D., Alleva, E., 2006. Analysis of ultrasonic vocalizations emitted by infant rodents. In: Costa, L.G., Hodgson, E., Lawrence, D.A., Reed, D.J. (Eds.), Current Protocols in Toxicology. John Wiley and Sons, Inc., Hoboken, NJ.

Brudzynski, S.M., Kehoe, P., Callahan, M., 1999. Sonographic structure of isolation-induced ultrasonic calls of rat pups. Dev. Psychobiol. 34, 195–204.

Brunner, D., Buhot, M.C., Hen, R., Hofer, M., 1999. Anxiety, motor activation, and maternal-infant interactions in 5HT1B knockout mice. Behav. Neurosci. 113, 587–601.

Cohen-Salmon, C., Carlier, M., Roubertoux, P., Jouhaneau, J., Semal, C., Paillette, M., 1985. Differences in patterns of pup care in mice. V. Pup ultrasonic emissions and pup care behavior. Physiol. Behav. 35, 167–174.

Ehret, G., 2005. Infant rodent ultrasounds – a gate to the understanding of sound communication. Behav. Genet. 35, 19–29.

El-Khodor, B.F., Dimmler, M.H., Amara, D.A., Hofer, M., Hen, R., Brunner, D., 2004. Juvenile 5HT(1B) receptor knockout mice exhibit reduced pharmacological sensitivity to 5HT(1A) receptor activation. Int. J. Dev. Neurosci. 22, 405–413.

Elwood, R.W., Keeling, F., 1982. Temporal organization of ultrasonic vocalizations in infant mice. Dev. Psychobiol. 15, 221–227.

Fride, E., Suris, R., Weidenfeld, J., Mechoulam, R., 2005. Differential response to acute and repeated stress in cannabinoid CB1 receptor knockout newborn and adult mice. Behav. Pharmacol. 16, 431–440.

Hahn, M.E., Lavooy, M.J., 2005. A review of the methods of studies on infant ultrasound production and maternal retrieval in small rodents. Behav. Genet. 35, 31–52.

Hahn, M.E., Schanz, N., 2002. The effects of cold, rotation, and genotype on the production of ultrasonic calls in infant mice. Behav. Genet. 32, 267–273.

Hahn, M.E., Hewitt, J.K., Adams, M., Tully, T., 1987. Genetic influences on ultrasonic vocalizations in young mice. Behav. Genet. 17, 155–166.

Hahn, M.E., Hewitt, J.K., Schanz, N., Weinreb, L., Henry, A., 1997. Genetic and developmental influences on infant mouse ultrasonic calling. I. A diallel analysis of the calls of 3-day olds. Behav. Genet. 27, 133–143.

Hahn, M.E., Karkowski, L., Weinreb, L., Henry, A., Schanz, N., Hahn, E.M., 1998. Genetic and developmental influences on infant mouse ultrasonic calling. II. Developmental patterns in the calls of mice 2–12 days of age. Behav. Genet. 28, 315–325.

Hennessy, M.B., Li, J., Lowe, E.L., Levine, S., 1980. Maternal behavior, pup vocalizations, and pup temperature changes following handling in mice of 2 inbred strains. Dev. Psychobiol. 13, 573–584.

Hofer, M.A., 1996. Multiple regulators of ultrasonic vocalization in the infant rat. Psychoneuroendocrinology 21, 203–217.

Holy, T.E., Guo, Z., 2005. Ultrasonic songs of male mice. PLoS Biol. 3, e386.

Liu, R.C., Schreiner, C.E., 2007. Auditory cortical detection and discrimination correlates with communicative significance. PLoS Biol. 5, e173.

Liu, R.C., Linden, J.F., Schreiner, C.E., 2006. Improved cortical entrainment to infant communication calls in mothers compared with virgin mice. Eur. J. Neurosci. 23, 3087–3097.

Maggio, J.C., Maggio, J.H., Whitney, G., 1983. Experience-based vocalization of male mice to female chemosignals. Physiol. Behav. 31, 269–272.

Moles, A., Kieffer, B.L., D'Amato, F.R., 2004. Deficit in attachment behavior in mice lacking the mu-opioid receptor gene. Science 304, 1983–1986.

Nyby, J.G., 2001. Auditory communication among adults. In: Willott, J.F. (Ed.), Handbook of Mouse Auditory Research: From Behavior to Molecular Biology. CRC Press, New York, NY, pp. 3–18.

Panksepp, J.B., Jochman, K.A., Kim, J.U., Koy, J.J., Wilson, E.D., Chen, Q., Wilson, C.R., Lahvis, G.P., 2007. Affiliative behavior, ultrasonic communication and social reward are influenced by genetic variation in adolescent mice. PLoS ONE 2, e351.

Picker, J.D., Yang, R., Ricceri, L., Berger-Sweeney, J., 2006. An altered neonatal behavioral phenotype in Mecp2 mutant mice. Neuroreport 17, 541–544.

Robinson, D.J., D'Udine, B., 1982. Ultrasonic calls produced by three laboratory strains of *Mus musculus*. J. Zool. Lond. 197, 383–389.

Roubertoux, P.L., Martin, B., Le Roy, I., Beau, J., Marchaland, C., Perez-Diaz, F., Cohen-Salmon, C., Carlier, M., 1996. Vocalizations in newborn mice: genetic analysis. Behav. Genet. 26, 427–437.

Sales, G.D., Smith, J.C., 1978. Comparative studies of the ultrasonic calls of infant murid rodents. Dev. Psychobiol. 11, 595–619.

Scattoni, M.L., Gandhy, S.U., Ricceri, L., Crawley, J.N., 2008a. Unusual repertoire of vocalizations in the BTBR T + tf/J mouse model of autism. PLoS ONE 3, e3067.

Scattoni, M.L., McFarlane, H.G., Zhodzishsky, V., Caldwell, H.K., Young, W.S., Ricceri, L., Crawley, J.N., 2008b. Reduced ultrasonic vocalizations in vasopressin 1b knockout mice. Behav. Brain Res. 187, 371–378.

Scattoni, M.L., Crawley, J., Ricceri, L., 2009. Ultrasonic vocalizations: a tool for behavioural phenotyping of mouse models of neurodevelopmental disorders. Neurosci. Biobehav. Rev. 33 (4), 508–515.

Shair, H.N., Brunelli, S.A., Hofer, M.A., 2005. Lack of evidence for mu-opioid regulation of a socially mediated separation response. Physiol. Behav. 83, 767–777.

Shu, W., Cho, J.Y., Jiang, Y., Zhang, M., Weisz, D., Elder, G.A., Schmeidler, J., De Gasperi, R., Sosa, M.A., Rabidou, D., Santucci, A.C., Perl, D., Morrisey, E., Buxbaum, J.D., 2005. Altered ultrasonic vocalization in mice with a disruption in the Foxp2 gene. Proc. Natl. Acad. Sci. USA 102, 9643–9648.

Thornton, L.M., Hahn, M.E., Schanz, N., 2005. Genetic and developmental influences on infant mouse ultrasonic calling. III. Patterns of inheritance in the calls of mice 3–9 days of age. Behav. Genet. 35, 73–83.

Tonkiss, J., Bonnie, K.E., Hudson, J.L., Shultz, P.L., Duran, P., Galler, J.R., 2003. Ultrasonic call characteristics of rat pups are altered following prenatal malnutrition. Dev. Psychobiol. 43, 90–101.

Weller, A., Leguisamo, A.C., Towns, L., Ramboz, S., Bagiella, E., Hofer, M., Hen, R., Brunner, D., 2003. Maternal effects in infant and adult phenotypes of 5HT1A and 5HT1B receptor knockout mice. Dev. Psychobiol. 42, 194–205.

Whitney, G., Nyby, J., 1979. Cues that elicit ultrasounds from adult male mice. Amer. Zool. 19, 457–463.

Winslow, J.T., Hearn, E.F., Ferguson, J., Young, L.J., Matzuk, M.M., Insel, T.R., 2000. Infant vocalization, adult aggression, and fear behavior of an oxytocin null mutant mouse. Horm. Behav. 37, 145–155.

Wöhr, M., Dahlhoff, M., Wolf, E., Holsboer, F., Schwarting, R.K., Wotjak, C.T., 2008. Effects of genetic background, gender, and early environmental factors on isolation-induced ultrasonic calling in mouse pups: an embryo-transfer study. Behav. Genet. 38, 579–595.

SECTION 4

Vocalizations as Specific Stimuli: Selective Perception of Vocalization

Mammalian vocalizations can be selectively recognized by members of the same species. There are subcortical detection mechanisms based on spectrotemporal acoustic patterns of calls, and the neocortex is also involved. Monkey neocortex showed a region specifically activated by vocalizations and homologous to a human cortical voice recognition region. Although the semiotic content of calls may require a learning process, there is an innate preparedness to perceive them.

CHAPTER 4.1

Subcortical responses to species-specific vocalizations

Josef Syka*

Institute of Experimental Medicine, Academy of Sciences of the Czech Republic, Prague, Czech Republic

Abstract: Acoustical signals are processed successively in several relay nuclei from the inner ear up to the auditory cortex. It is generally believed that the auditory system is optimized for processing biologically significant signals that occur in the natural environment, such as species-specific vocalizations. Most information about the processing of animal vocalizations, and about the relevant responses of neurons in the auditory cortex, has only been accumulated recently. Before reaching the auditory cortex, information about the acoustical signals is first coded in several subcortical auditory nuclei. This chapter summarizes recent information about the processing of vocalizations in two major subcortical auditory nuclei: the inferior colliculus (IC); and the medial geniculate body (MGB). Most experimental data are from the guinea pig, a species with a significant repertoire of vocalization signals. The results of these studies provided evidence for the encoding of the spectrotemporal acoustic patterns of vocalizations by ensembles of IC neurons. MGB neurons, in contrast to IC units, showed less precise representation of the temporal envelope in the case of long calls (such as whistle), but not in the case of short calls or calls consisting of many short components. Also, the MGB neurons, as compared to the IC neurons, showed a less precise rate representation of the spectral envelope for the low-frequency calls, but not for the broadband calls.

Keywords: species-specific vocalizations; inferior colliculus; medial geniculate body; call detectors; neuronal population; spectrotemporal responses; auditory nuclei

Abbreviations: A1: primary auditory cortex; AM: amplitude modulation; CF: characteristic frequency; CNIC: central nucleus of the inferior colliculus; DCIC: dorsal cortex of the inferior colliculus; ECIC: external cortex of the inferior colliculus; FM: frequency modulation; IC: inferior colliculus; MGB: medial geniculate body; PSTH: peristimulus time histogram

I. Introduction

Understanding how the auditory system of the brain processes vocalization is one of the most challenging problems of contemporary neuroscience. Perception of species-specific communication sounds is an important aspect of the auditory behavior of many animal species, and is crucial for mammalian social interactions, reproductive success and survival (Wang, 2000). Lewicki (2002) suggested that the auditory system is adapted for the efficient representation of a broad range of sounds in the natural environment, including animal vocalizations. According to this author, the code that is used in the auditory system depends on the class of sound detected. This coding may resemble Fourier transformation when optimized for animal vocalizations, or a wavelet transformation when optimized for non-biological environmental sounds. The initial stage of auditory processing could have differentially evolved to fit the different statistical profiles optimally of these two important groups of natural sounds. A unique specialization of the auditory system in detecting vocalizations

*Corresponding author. E-mail: syka@biomed.cas.cz

Stefan M. Brudzynski (Ed.)
Handbook of Mammalian Vocalization
ISBN 978-0-12-374593-4

DOI: 10.1016/B978-0-12-374593-4.00011-5

has particularly developed in bats (Suga, 1965). It is, however, beyond the scope of this chapter to cover the extensive information available about the processing of echolocating signals in the central auditory systems of bats (for a recent review see, for example, Covey, 2005; see also Maltby et al., Chapter 2.4 and Metzner and Schuller, Chapter 9.4 in this volume).

The first attempts to clarify the issue of vocalization coding in the auditory system of mammals, excluding bats, date to around 1970 when the first series of papers appeared, mostly based on studies of single unit responses in the auditory cortex of monkeys to species-specific vocalizations (Wollberg and Newman, 1972; Newman and Wollberg, 1973; Winter and Funkenstein, 1973; Newman and Symmes, 1974; Newman, 1978). The prevailing hypothesis at that time was that species-specific vocalizations were represented by the activity of specific neurons: the call detectors. Call detectors were considered to be specialized neurons that extract specific acoustic features, similar to visual cortex neurons performing feature extraction. Almost all of these studies were focused on the auditory cortex, with a few exceptions concerning the subcortical auditory nuclei. The first such study was performed by Watanabe and Katsuki (1974), who were interested in the contribution of neuronal inhibition in the responses of individual auditory neurons of the cat to species-specific vocalizations. The call was presented unaltered or after filtering through bandpass and band-rejection filters. According to their data, a mutual inhibitory interaction, which was exerted by different frequency components of the call, was clearly found in the cochlear nucleus and became more evident at higher auditory centers, reaching up to the medial geniculate body (MGB). In another study, Symmes et al. (1980) investigated the responses of the MGB neurons in unanesthetized squirrel monkeys to species-specific vocalizations and to artificial acoustic stimuli (tone and noise bursts and clicks). Parameters of the responses, such as absolute thresholds, the response vigor, rate-level functions, binaural interactions and the spectral composition of the responses, were compared with both types of stimuli, leading to the conclusion that MGB neurons process vocalizations and other acoustic stimuli in a similar way. Since the authors had previously collected representative data from the auditory cortex in the same species, they suggested that the processing of communication calls in the MGB is relatively simple and represents a preliminary step, in comparison with the processing in the auditory cortex. In contrast to the results of studies performed in anesthetized animals, MGB neurons in the study by Symmes et al. (1980) did not respond to tone, noise bursts and to vocalizations predominantly with an onset response, but rather discharged spikes throughout all the stimulus duration. The selectivity for individual types of calls was low; most neurons responded to all types of vocalizations. There were no significant differences found in responsiveness to vocalizations among neurons of the ventral, anterodorsal and magnocellular parts of the MGB. Similarly results obtained by Smolders et al. (1979) in the cat cochlear nucleus led only to a general conclusion that, at the cochlear nuclear level, the response to natural stimuli could be understood in terms of the neuronal responses to tones and the acoustic spectrum of the call. Such an interpretation, however, was inadequate to explain the MGB responses to different sounds fully.

The existence of specific cells in the auditory system that function as call detectors was not proven in all of these studies, and therefore the idea was ultimately accepted that the pattern discrimination of a complex sound can only be accomplished by a functional ensemble (circuit) of many neurons (Pelleg-Toiba and Wollberg, 1991). The call-detector neuron hypothesis was replaced in the 1990s by the idea of encoding complex sounds by the discharge patterns of neuronal populations, already proposed in 1980 (Creutzfeld et al., 1980). These authors observed that, in the MGB of unanesthetized guinea pigs, all MGB and cortical neurons responded to a variety of natural calls from the same or other species. The MGB neurons responded to more components of a call than cortical cells. Fast frequency modulation within a call, such as those of the fundamental frequency, could still be identified in the response of some MGB neurons, but not in those of cortical neurons. Although the onset type of responses in both structures to artificial acoustical stimuli and vocalizations prevailed, cortical responses were found to be more phasic than those of the MGB neurons.

The idea that animal calls are preferentially processed by populations of auditory neurons led to a new approach in the studies of auditory processing first introduced by Wang et al. (1995). According to these authors, and based on the results of studies of cortical plasticity, cortical responses to sounds that do not have behavioral relevance to the animal were coded differently from the responses produced by biologically and behaviorally relevant sounds. The features of complex sounds which arrive from the periphery to the auditory system are converted in the primary auditory cortical field A1 into a synchronization-based population code. Auditory nerve fibers accurately encode the fine temporal details of complex acoustical stimuli with responses phase-locked to the fundamental frequency and envelope of a complex stimulus. MGB neurons follow the amplitude modulated (AM) and frequency modulated (FM) stimuli at much faster rates than do cortical neurons, therefore the essential change of the neuronal processing occurs in the A1. This implies that the spectrotemporal acoustic pattern of the vocalization is coded by the spectrotemporal discharge pattern of A1 neurons with a reduced temporal complexity. Thus, the detailed representations of features of complex sounds from the periphery appear to be transformed into a more abstract, synchronization-based population code in the cortical field A1.

Wang et al. (1995) proposed that A1 representation of different stimuli does not have to be composed of isolated cortical zones, each corresponding to an individual stimulus, but rather that it consists of a collection of cortical patches that may partially overlap each other. This may represent a common strategy of the cortex to convert temporal information into spatial information as a requirement for processing complex sensory input patterns. Such theoretical conclusions were supported by the results of experiments in the study by Wang et al. (1995), in which representation of complex species-specific calls by populations of neurons in the primary auditory cortex (A1) was analyzed in a small New World monkey, the common marmoset. The spectrotemporal discharge pattern of spatially distributed neuronal populations in A1 was found to be correlated with the spectrotemporal acoustic pattern of a vocalization. This was demonstrated by recordings of population responses of A1 neurons organized on the basis of the occurrence of discharges of individual units over time, which were aligned according to their characteristic frequency. The group of A1 neurons responded selectively to natural vocalizations, in comparison to a poor responsiveness to synthetic sounds that had the same spectral, but different temporal, characteristics (they were time-compressed, time-expanded or time-reversed). These groups of neurons were distributed widely across the A1 cortex, with only a partial overlap. Discharges of the A1 neuronal populations were found to be phase-locked to specific stimulus events, but not to their spectral contents. As a consequence, a reduction in temporal complexity and an increase in the cross-population response synchronization occurred. In the spectral profile of the A1 responses, individual spectral peaks and troughs were clearly represented in the rate-characteristic frequency (CF) profiles, but the details of the spectrum were not represented.

The important role of the cross-population response synchronization in the coding of species-specific vocalizations in the auditory cortex was later confirmed by Gehr et al. (2000), who found significantly enhanced correlations among firing neurons at different recording sites of the A1 in the cat during presentation of a species-specific vocalization – the cat's meow. The enhanced correlation was, however, present only in peak-tracking units, the firing of which occurred throughout the whole period of vocalization. The approach of Wang et al. (1995), which was developed with the aim of analyzing the responses of cortical neuronal ensembles to vocalizations, initiated several studies on the subcortical processing of vocalizations. The results of these studies, focused on the IC and the medial geniculate body (MGB), will be discussed below. It must also be emphasized that substantial work has been done to understand the responses of neurons at the lower levels of the auditory pathway (such as the auditory nerve and cochlear nucleus) to complex acoustical stimuli, but with a few exceptions (e.g., Smolders et al., 1979); speech or speech-like stimuli were used in these experiments and not species-specific vocalizations. The results of experiments with speech stimuli were reviewed by Sachs (1984) and recently by Young (2008).

II. Inferior colliculus

II.A. Responsiveness of individual inferior colliculus neurons to vocalizations

The inferior colliculus (IC) is a midbrain auditory structure that integrates information from several ascending auditory pathways, descending corticotectal projections, intercollicular pathways and extra-auditory sources, particularly from the somatosensory system. Three subdivisions may be distinguished in the IC, both morphologically and on the basis of different inputs and outputs. The processing of information is different in each of the three main IC subdivisions: the central nucleus (CNIC); the dorsal cortex (DCIC); and the external cortex (ECIC). Response properties to tonal and noise stimuli distinguish neurons of the CNIC from those in the DCIC and ECIC. For example, in the guinea pig (Syka et al., 2000), CNIC neurons are characterized by a sharper frequency tuning, a lower average threshold, a shorter average first-spike latency of response to tones at the characteristic frequency, a higher occurrence of non-monotonic rate/level functions and a higher rate of spontaneous activity than neurons in the two other subdivisions. The CNIC represents the main part of the ascending auditory system, and plays a role similar to the ventral nucleus in the medial geniculate body. The roles of the DCIC and ECIC lie, in particular, in the integration of influences which descend to the IC from the cortex via corticotectal pathways and which also arrive from other sensory systems.

Only a few studies have been designed to investigate the responses of neurons in the IC to vocal stimuli (Aitkin et al., 1994; Syka et al., 1997; Šuta et al., 2003). The aim of the study by Aitkin et al. (1994), performed in anesthetized cats, was to gain information about the differential coding properties of neurons in three subdivisions of the IC: the central nucleus, and the external and dorsal cortices. Four types of feline vocalizations (kitten calls, screech, adult meow and low-meow) were used in this study, in addition to tone bursts and noise bursts. In general, the average thresholds to CF stimuli were lower in the IC than those to noise and vocal stimuli. As to the selectivity of responses, there appeared to be a continuum of response preferences, ranging from a complete lack of response to any of the complex stimuli used, to apparently unselective responding to all presented stimuli. Feline vocal stimuli were found to be more effective in inducing higher firing rates than white noise or pure tone stimuli at the CF in the ECIC and DCIC in comparison with the CNIC. There were no units that responded exclusively to one vocal stimulus, but a high proportion of units in the ECIC responded strongly to broadband stimuli, and some of these showed a clear preference for one vocal stimulus over another. Interestingly, the reactivity of neurons to pure tones and to noise bursts was different from that observed in response to animal calls. In the CNIC, neurons responded to tone and noise bursts more frequently with a sustained firing pattern than those in the DCIC and ECIC and non-monotonic rate-level functions were more common in the CNIC than in the DCIC and ECIC as well. What functional meaning may the fact that vocal stimuli were more effective than pure tone stimuli in the ECIC have? It is possible that the efferent projections of the call-preferring neurons in the IC make synaptic contacts at the premotor areas necessary for the initiation of motor responses to certain vocalizations. Such projections are more closely associated with the ECIC than with the CNIC, and include connections to the superior colliculus (Syka and Straschill, 1970; Hashikawa and Kawamura, 1983) and dorsolateral pontine nuclei (Aitkin and Boyd, 1978).

In the study by Aitkin et al. (1994), the authors paid great attention to the responses of IC neurons to suprathreshold stimuli. A dominant observation arising from this study was that, throughout the IC, particularly in the external nucleus, neurons which were clearly tuned to the tone-burst stimuli responded vigorously to the bandpass stimuli (noise and vocalizations). Neither the rate-level functions nor the peristimulus histograms suggested that strong inhibitory interactions existed between CF and off-CF frequencies for the majority of the units. For some units with non-monotonic rate-level functions to CF stimuli, monotonic functions were demonstrated using noise and vocal stimuli (Fig. 1). This observation means that lateral inhibition is not evoked in the IC units by off-CF components. The widespread effectiveness of vocal stimuli on cellular responses in the IC may also relate to the amplitude modulation (AM) and frequency modulation (FM) of components within each call. The majority of neurons recorded in the IC by Rees and Moller (1987) responded effectively to modulation rates up to 200 Hz. In cats, Langner and Schreiner (1988) reported even higher rates; over half of their sample responded to AM stimuli at frequencies up to 100 Hz, with some units still sensitive at 1,000 Hz. For further review of the subcortical neural coding mechanisms for auditory temporal processing, see, for example, Frisina (2001).

The most extensive study to date of the responses of the IC to vocalizations was performed in guinea pigs (Syka et al., 1997; Šuta et al., 2003). Guinea pigs represent a species where voluminous knowledge about the structure and function of the inner ear has been accumulated over the twentieth century, with less information being available about the structure and function of the central part of the auditory pathway. Guinea pigs produce expressed vocalization signals in different behavioral situations. Their vocalization repertoire contains at least 11 distinct calls (Berryman, 1976), fundamentally different in their spectrotemporal features (Syka et al., 1997). The four most recognized guinea pig calls are purr, whistle, chutter and chirp (Fig. 2). Purr consists of a series of regular low-frequency acoustical pulses with a fundamental frequency around 300 Hz; whistle is a long-lasting frequency and amplitude modulated sound consisting of many harmonics over a wide frequency range; chutter is a sequence of irregular noise bursts;

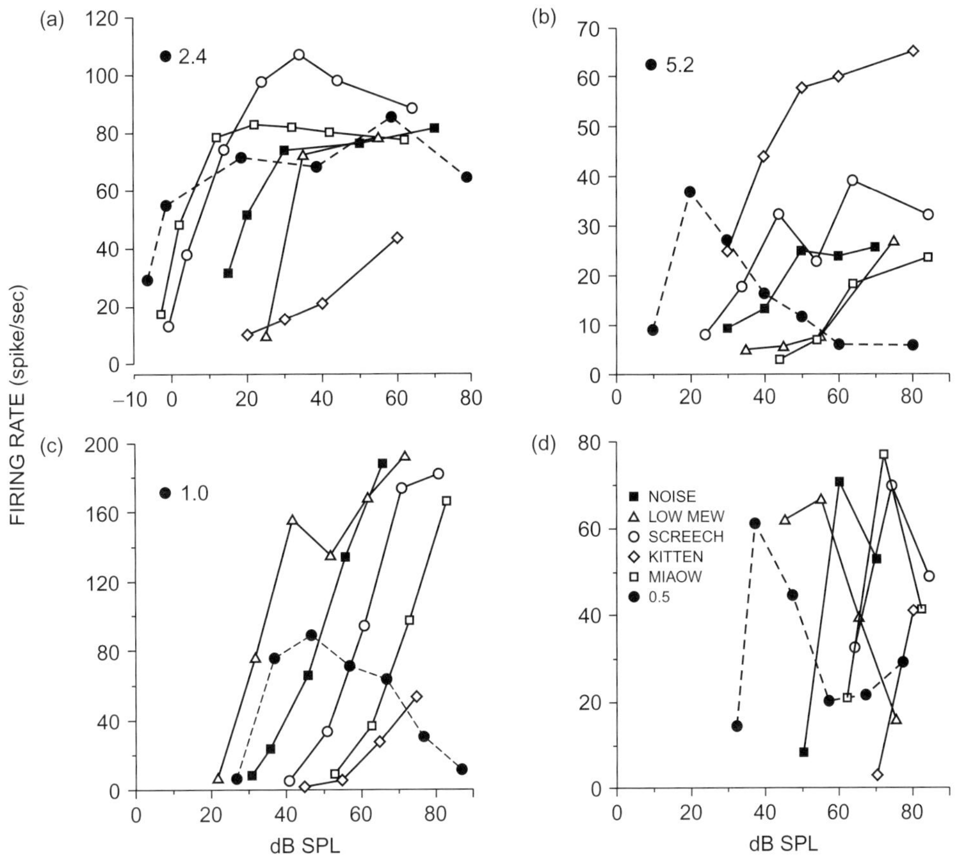

Fig. 1. Rate/level (R/L) functions for four units from the central nucleus of the inferior colliculus in the cat. Each R/L function plots the firing rate (spikes/s) during the stimulus, averaged for 20 repetitions of the stimulus, against the sound level in dB SPL. The R/L function to the characteristic frequency (CF) stimulus is shown as a dashed line joining filled circles. Other symbols: filled squares: noise; diamond: kitten calls; open circle: screech; triangle: low-meow; open square: meow. (a) Unit 91–2–4, CF 2.4 kHz; (b) unit 91–1–9, CF 5.2 kHz; (c) unit 91–1–4, CF 1 kHz; (d) unit 90–18–9, CF 0.5 kHz. From Aitkin et al. (1994), Fig. 6, p. 58.

and chirp is an isolated brief acoustic impulse with a harmonic structure.

The selectivity of neurons in the IC of the guinea pig to individual vocalizations was found to be relatively low. In general, 55% of neurons responded to all four types of vocalizations, 23% responded to three vocalization types, 16% to two and only a small portion of neurons responded to only one call type (3%) or did not respond to any call (3%). There were small differences among the responses of neurons to individual type of call – each call evoked a response in approximately 80% of recorded neurons. The strength of the response, evaluated as the firing rate over the duration of the stimulus, was greater to vocalizations than to tones or broadband noise in 37% of neurons. The responsiveness to individual types of vocalizations also corresponded with the characteristic frequency of the recorded neurons, i.e., their tuning. The low-frequency call, the purr, produced excitation in only a small number of high-frequency neurons; similarly, whistle, with its strong representation of high sound frequencies, elicited a response in a few low-frequency neurons. In approximately 90% of units, vocalizations evoked excitatory responses; inhibition was present throughout stimulation with vocalization only in 10% of neurons, most frequently in the response of neurons with a low CF to presentation of whistle.

Since the 1970s, one of the frequently used criteria for recognizing a selective response of a neuron to an animal call (call detector) was a difference between the response to a temporally reversed call and the response to a proper call. Such a procedure changes the temporal features of a sound, but preserves its spectral characteristics. In 45% of units evaluated in the IC of the

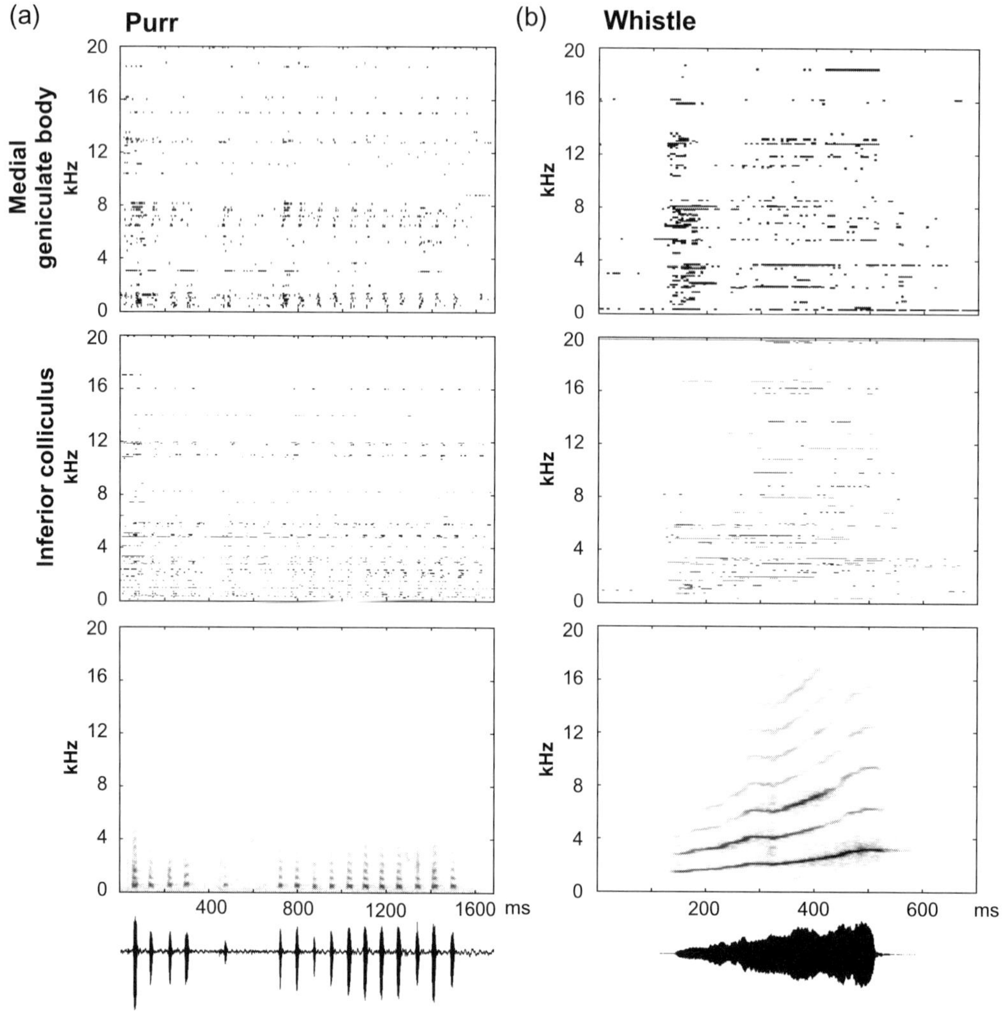

Fig. 2. Population response maps for purr (a); whistle (b); chutter (c); and chirp (d) in the medial geniculate body (top) and inferior colliculus (middle) accompanied by a call spectrogram with waveform (bottom). Each dot represents a discharge of 80 spikes/s at the indicated time (abscissa) from a neuron with a CF as shown on the ordinate. No averaging or interpolation was used. Bin width: 5 ms (abscissa) and 0.05 kHz (ordinate). From Šuta et al. (2003), Fig. 5, p. 3799 and Šuta et al. (2007), Fig. 1, p. 379.

guinea pig, the responses to reversed whistle were significantly different from those to the normal call; in general, the responses to whistle were stronger than to reversed whistle (Šuta et al., 2003). The possibility that the weaker response to temporally manipulated sounds is attributable to their lack of behavioral relevancy is supported by the study of Wang and Kadia (2001). This study showed that cortical neurons in the cat responded similarly to marmoset natural and time-reversed communication sounds, but that marmoset cortical units preferred the naturally patterned call over the time-reversed call. Further details were studied by Gehr et al. (2000), who found that in feline cortical neurons the onset part of the response, as well as the whole response to forward meows (naturally patterned), was significantly larger than for time-reversed meows. Interestingly, the sustained part of the response to the time-reversed meows was slightly larger than that for the forward meows. There is a possibility that the preference for natural whistle could be based on the preference for rising amplitude versus falling amplitude, or for rising frequency over falling frequency of the whistle versus temporally reversed whistle. This idea is supported by studies of neuronal responses to amplitude or frequency modulated sounds. For example, Chiu and Poon (2000) observed that more than half of the cells in the IC of rats could be considered as "AM sensitive," with a preference for the rising phase of the AM sound.

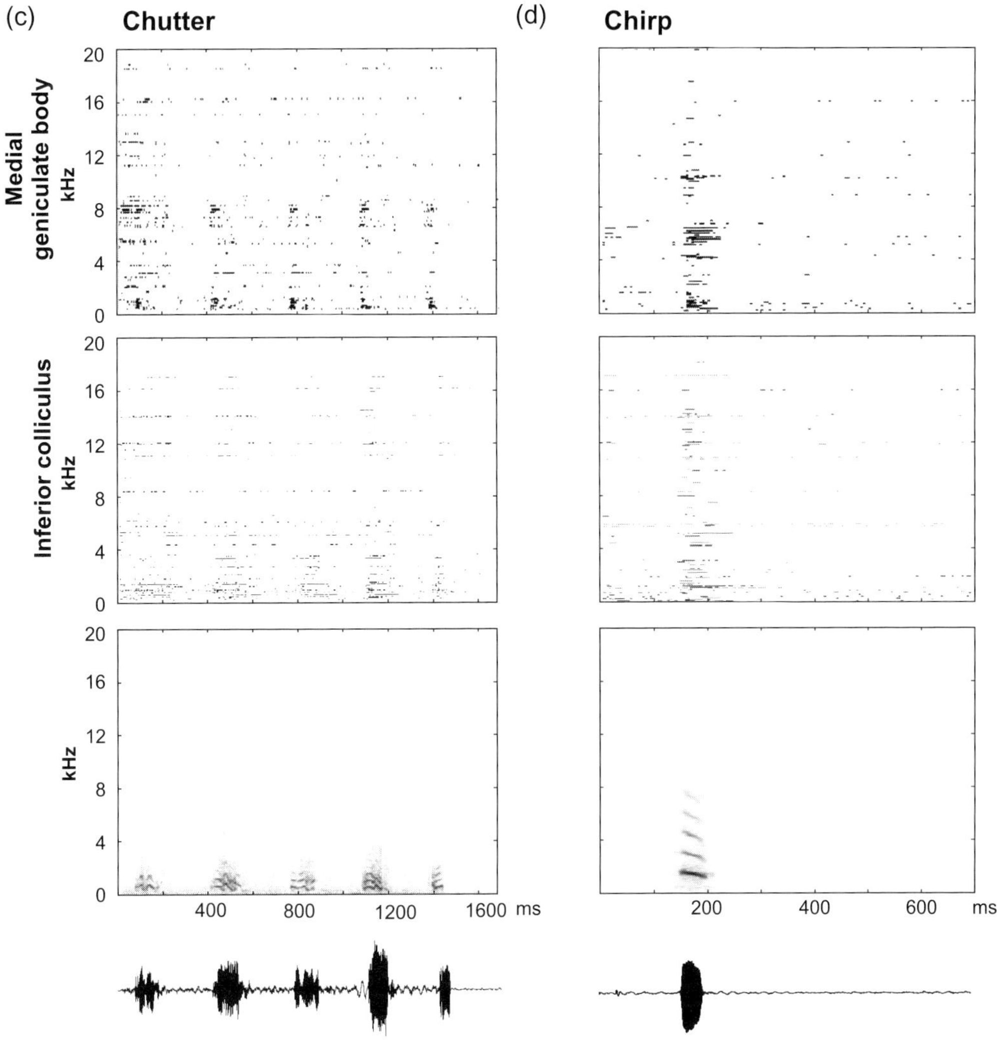

Fig. 2. Cont'd

Similarly, Neuert et al. (2001) reported an asymmetry in the responses to damped and ramped sinusoids in the IC of anesthetized guinea pigs, where all units displayed significant asymmetry in the discharge rate for at least one time constant of the AM.

II.B. Responsiveness of populations of inferior colliculus neurons to vocalization

The spectrotemporal features of vocalizations are processed in populations of auditory neurons; therefore the evaluation of population response maps, which are constructed from peristimulus time histograms (PSTHs) of individual neurons, may contribute significantly to our understanding of this processing. Population response maps of IC neurons in the guinea pig to all four types of vocalizations are displayed in Fig. 2 and compared with the spectrograms of vocalization sounds. A black dot in the population response map represents an elevated discharge of a unit with a given CF (ordinate) occurring during or after vocalization presentation (abscissa). It is evident from Fig. 2 that the areas of elevated activity of IC neurons match the areas of energy in the spectrograms. The match is, however, not complete since the CF range of the discharging units is wider than the corresponding frequency range within the stimulus. Since the vocalization stimuli were presented at 75 dB SPL, the response occurred in more units than only those with a corresponding CF according to the shape of the tuning curves. The bias towards discharging to a wider frequency range was more prominent in the responses of high-CF neurons to low-frequency sounds, such as purr and chutter, than in the responses of low-CF neurons to sounds of higher frequencies, such as whistle. This pattern is evidently associated with the shape of the tuning curves at higher intensities, where the low-frequency tail enables responses of high-CF neurons to low-frequency sound content, whereas the sharp slope of the high-frequency border of the tuning curve does not allow the response of low-CF units to sounds of higher frequencies.

The representation of the temporal features of vocalization sounds was evident when the population PSTHs (i.e., average PSTHs) was calculated. A comparison of population PSTHs and sound envelopes (Fig. 3) showed a good agreement in their shapes for short calls and calls composed of many short components, i.e., purr, chutter and chirp. Neuronal responses to these calls reflected the energy of the sound, and the magnitude of the peaks in the population PSTHs was approximately proportional to the peak magnitude in the sound envelope. The similarity between the population PSTH and the sound envelope for the three identified calls was very high – with a correlation coefficient ranging from 0.71 to 0.93. In many individual neurons, the correlation coefficients between their PSTH and sound envelope for chirp and chutter showed even higher values. These findings suggested that information about the peaks in the envelope of the calls could be available at the level of single units or small neuronal pools. Fig. 3 shows that a different situation was observed in the case of the whistle. The shape of the population PSTH was markedly different from the monotonically increasing envelope of the call, as the population PSTH was multimodal with two main peaks, one at the beginning of the sound and the second at approximately 320 ms. There were different sources of the peaks, as was shown by comparing the subpopulation PSTH divided according to the CF. The early peak represents the onset activity of units with a CF lower than 7 kHz, while the second peak reflects the onset of activity of units with CFs higher than 8 kHz. These high-CF units responded weakly or remained silent until about 320 ms after the stimulus onset, because there was almost no energy in the appropriate frequency component in whistle during this interval. The neurons began to respond at about 320 ms, when the frequency spectrum of the stimulus expanded above 8 kHz (Fig. 2).

To understand what information about the spectral features is coded by the firing of IC neurons in the guinea pig, analysis of the neuronal representation of the spectral characteristics using a rate versus CF profile was performed and compared with the short-term sound spectrum. Fig. 4 shows rate versus CF profiles for the whistle at three different times and for chirp, purr and chutter. Fig. 4a shows the situation at the beginning of the whistle (150–260 ms from the onset of the call). The peaks in the rate versus CF profile indicate the positions of the two main spectral peaks. The magnitudes of the rate peaks do not exactly fit the spectrum of the sound because the rates of the units with a CF corresponding to the first harmonic (fundamental) frequency are slightly lower than that of the second harmonic. This is probably caused by inhibition occurring in a part of the low-CF units. The range of the third and fourth harmonics is represented by one peak, and there is a near-zero level of driven activity in the high CF region. This fact corresponds with the absence of a high-frequency component in the sound at this time. A different picture is present in the middle part of whistle (260–370 ms, Fig. 4b). The magnitudes of the peaks do not correspond with the magnitudes of the spectral peaks. The strongest response is at the frequency of the fourth harmonic (about 8 kHz), and the high-CF area is also elevated more than would be expected from the short-term spectrum of the sound. The discrepancy could be explained by the onset response of high-CF neurons, which exceeds the sustained activity of lower-CF neurons. In the latest phase of the whistle (430–540 ms, Fig. 4c), the maximal peak in the rate–CF profile indicates the dominant spectral peak of the fundamental frequency in the short-term sound spectrum. The rate–CF profile for chirp (Fig. 4e) indicates the positions of the main spectral peaks, with isolated peaks of the fundamental and first harmonic frequencies and one peak for the third and fourth harmonics. A somewhat different situation is present for the other two sounds (Fig. 4d,f) where some local spectral peaks are intensified in the rate–CF profile and create dominant peaks. The correlation in these two cases between the sound spectrum and the rate–CF profile is the highest among all the mentioned vocalization sounds.

The response of IC neurons to vocalization sounds seems to omit the slow modulation of the sound envelope. This phenomenon is mainly present in the response to whistle, in which the sustained character of the response indicates just the presence of energy, but the slow changes in the sound envelope are not reflected in the modulation of the firing rate. This inability to follow slow fluctuations in the envelope corresponds well with the weak synchronization between neuronal discharge in the IC and sound envelope as described, for example, by Rees and Moller (1983) in the rat, or later by Rees

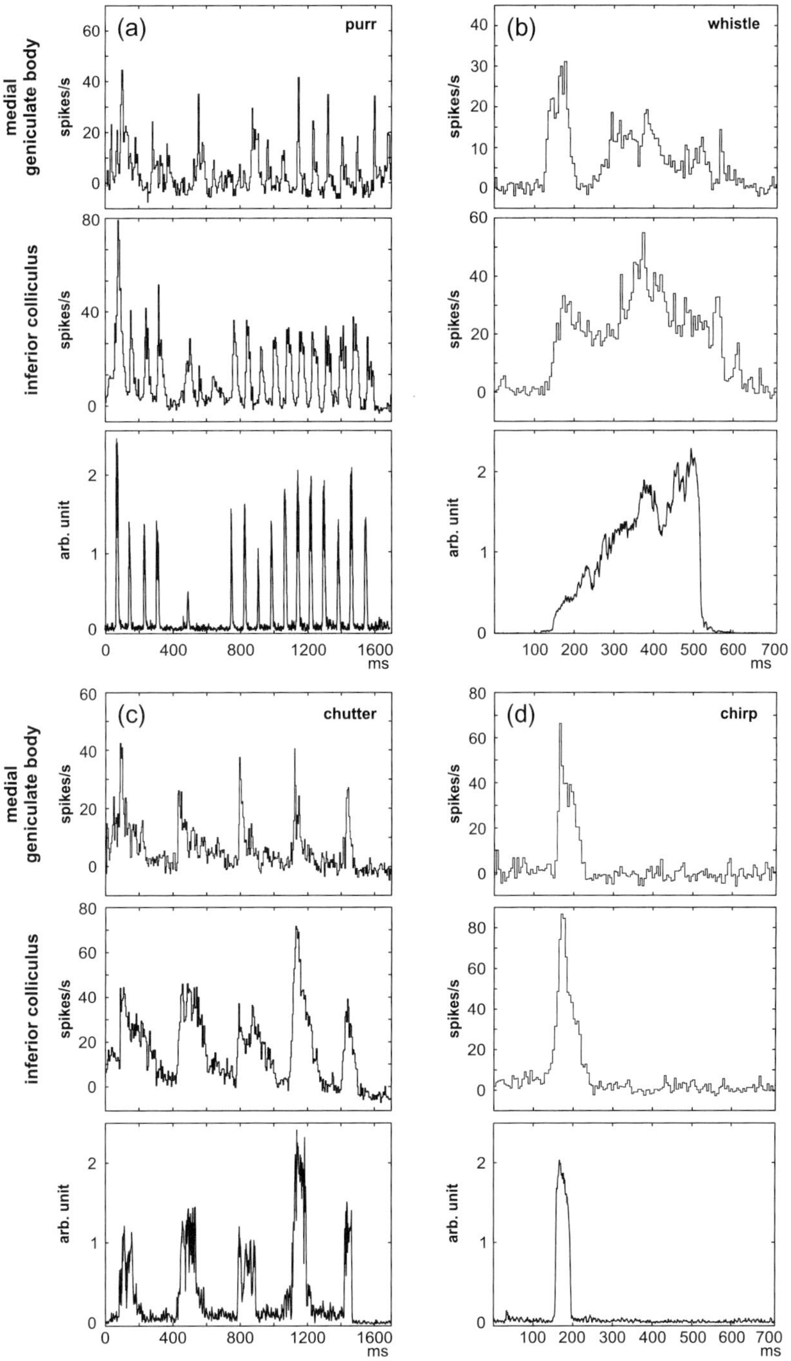

Fig. 3. Relationship between the temporal pattern of calls and the response of inferior colliculus and medial geniculate body neurons. A comparison of the sound envelopes (bottom) and the population peristimulus time histogram (PSTHs) in the inferior colliculus and in the medial geniculate body is shown for all four calls: purr (a); whistle (b); chutter (c); and chirp (d). Each population PSTH is calculated as the average PSTH of all recorded units. Bin width: 5 ms. From Šuta et al. (2003), Fig. 6, p. 3800 and Šuta et al. (2007), Fig. 4, p. 383.

and Palmer (1989) in the guinea pig. According to their findings, the modulation transfer functions of IC units typically have a bandpass character similar to units in the auditory cortex of the marmoset (Nagarajan et al., 2002), where degradation in the temporal envelope, performed by lowpass filtering of the temporal envelope at 4 and 10 Hz, dramatically diminished the synchronized response.

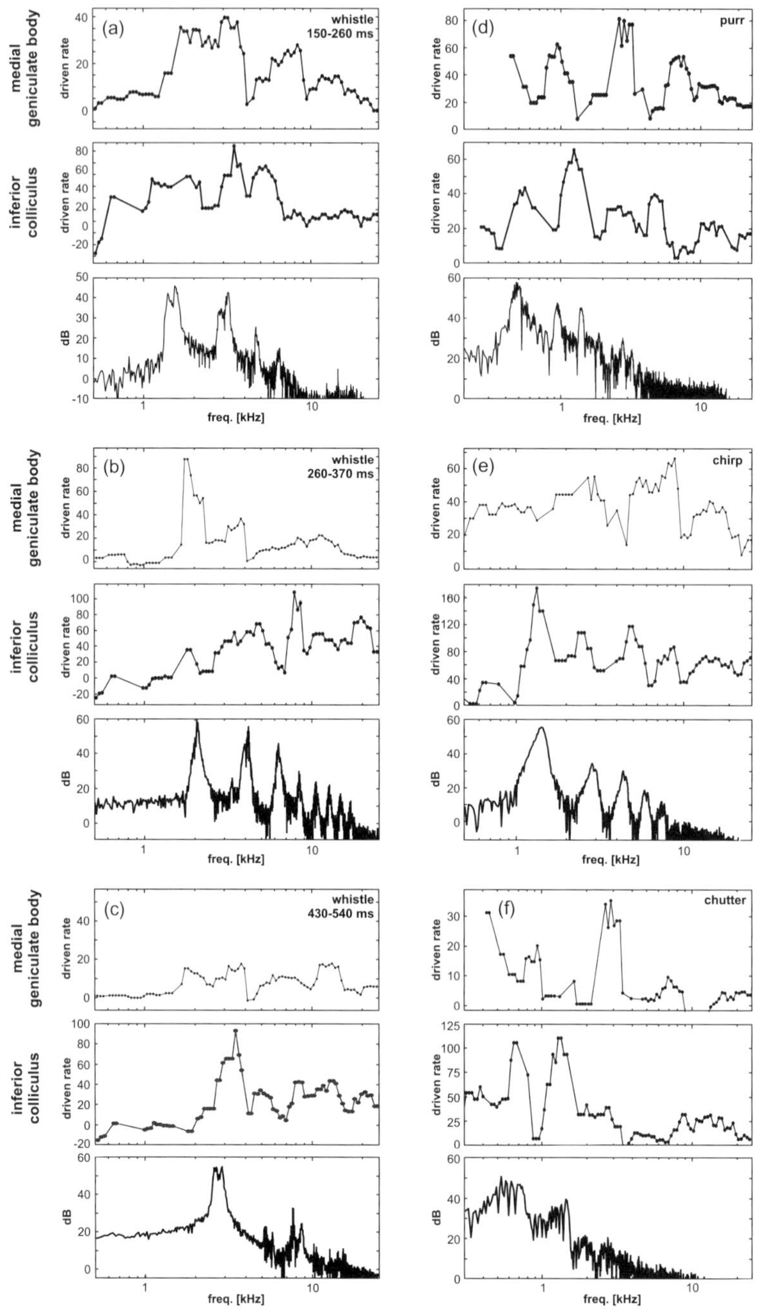

Fig. 4. Comparison of rate–CF profiles of inferior colliculus neurons, medial geniculate body neurons and the short-term spectra of the call (bottom) for three consecutive parts of whistle (a, b, c): the appropriate interval is indicated above every rate–CF profile; for purr (d): data calculated over the first phase containing four elementary phrases; for chirp (e); and for chutter (f): calculated over the first phrase of the sound. From Šuta et al. (2003), Fig. 13, p. 3804 and Šuta et al. (2007), Fig. 7, p. 385.

Another aspect of the processing of vocalization signals in the IC was described recently by Lesica and Grothe (2008). They demonstrated in the gerbil that the addition of ambient noise to vocalization stimuli evoked dramatic changes in the temporal processing strategy of IC neurons, which served to increase the information content conveyed by the neural response. They found that the onset of the ambient noise evoked a change in the receptive field dynamics of IC neurons that corresponded to a change from bandpass to lowpass temporal filtering. These effects are, to some extent, similar to those described previously by Rees and Moller (1987) and Rees and Palmer (1989), who observed that the addition of a broadband noise masker to an AM pure tone stimulus increased the temporal response of neurons in the IC of the rat or guinea pig to low modulation frequencies. Recently, Lesica and Grothe (2008) rightly point out that their results suggest that, because of differences in the modulation spectra of the "signal" and "noise" in their stimuli (vocalizations and ambient environmental noises), the changes in the modulation tuning they observed increased the information content in the neural response. This increase in information was not evident in the results of Rees and colleagues.

III. Medial geniculate body

III.A. Responsiveness of individual medial geniculate body neurons to vocalizations

The major target of projections from the IC is the medial geniculate body (MGB), which, similarly to the IC, consists of several subdivisions: the ventral; dorsal; and medial parts (Morest, 1964; Winer, 1985; Anderson et al., 2007). The ventral division of the MGB has the same role as the central nucleus of the IC, and represents the major part of the primary ascending auditory pathway. Similarly to the IC, a large percentage of MGB neurons in the guinea pig respond to individual types of vocalizations (Syka et al., 1997). Forty-one percent of MGB neurons in the guinea pig were able to follow all four types of calls, 20% three call types, 25% two call types, 11% one call type and only 5% of MGB neurons did not respond to any of the vocalization sounds. The selectivity for calls was even lower than in the IC, particularly for purr (only 50% of neurons responded to stimulation by purr) and for chutter (57% of neurons produced a response). Very rarely was a neuron found in the MGB that was inhibited by vocalization (only in 2% of neurons), whereas inhibition was more common in the IC (9% of neurons were inhibited by a call), particularly among low-CF neurons in response to whistle. Similarly to the IC, there was a greater responsiveness of MGB neurons to presented unchanged whistle than to time-reversed whistle (by 13% on average, Šuta et al., 2007). However, in the IC, the difference in responsiveness to the two acoustical stimuli was more pronounced (24%).

III.B. Responsiveness of populations of medial geniculate body neurons to vocalizations

The neuronal responses in the MGB of the guinea pig are, in many aspects, similar to those in the IC; however, there are also some striking differences present (Šuta et al., 2007). A comparison of the spectrograms of the calls with the spectrotemporal response maps (only neurons in the ventral part of the MGB were analyzed) revealed, similarly to the IC, that the patterns present in the population response maps corresponded with many features of a call's acoustical pattern (Fig. 2). However, some striking dissimilarities were also present. Fig. 2b, for example, shows that high-CF MGB neurons (CF $\geq$ 8 kHz) display enhanced activity at the beginning of the whistle (140–250 ms after the onset of the stimulus) even though there is no energy present in this frequency range. Another difference is evident from a comparison of the spectrotemporal map of chutter (Fig. 2c) and the spectrogram of this call, where the duration of the response to the individual bouts of the call is shorter than the duration of the bouts themselves.

Population PSTHs of MGB neurons reflect the sound temporal envelopes, as shown in Fig. 3. When the sound envelope and the population PSTH are correlated, the highest correlation is present for chirp ($r = 0.94$), less for purr ($r = 0.65$) and for chutter ($r = 0.57$). A specific relationship is evident when the sound envelope of a whistle is compared with its population PSTH. The dominant element in the response is the onset response, which contrasts with the weak energy at the beginning of the whistle. The low correlation between the call envelope and the population PSTH is, in this structure, mainly due to the phasic character of the neuronal response of MGB neurons, which is present even with simple acoustical stimuli such as tone or noise bursts (Kvašňák et al., 2000). The early onset response in the population PSTH of whistle is produced by the response of high-frequency neurons to the low-frequency part of the whistle spectrum (Šuta et al., 2007). When high-CF MGB neurons were stimulated by lowpass filtered whistle, they responded vigorously in comparison with a weak response to highpass filtered whistle. Therefore, the early onset response is really a response to the low-frequency component of the whistle, even though this frequency band is well below the neuronal CF.

The relationship between the spectral features of calls and the firing of MGB neurons is evident from the comparison of the rate–CF profiles with the short-term sound spectrum. Similarly to the case of IC neurons, rate–CF profiles are first shown in Fig. 4 for the three consecutive parts of whistle (a–c), for purr (d), chirp (e) and chutter (f). In the first 110 ms of whistle, the correlation coefficient between the short-term spectrum and the relevant rate–CF profile is 0.48. The rate–CF profile reflects the dominant spectral component of the first harmonic (fundamental) frequency and the second harmonic (1–4 kHz), but these two frequencies are not separated. A second peak appears in the range of 5–9 kHz, even though there is almost no energy in the sound above 7 kHz at this time. In the middle part of whistle the correlation coefficient is 0.56. The dominant first harmonic (fundamental) and the second harmonic frequencies produce the strongest firing of neurons with corresponding CFs. Higher harmonic frequencies are reflected in a flat and weak rate profile. The late part of whistle shows the weakest correlation coefficient ($r = 0.28$); the rate–CF profile is flat because of weak neuronal activity at the end of whistle. For purr, chirp and chutter (Fig. 4d–f), some local spectral peaks are intensified in the rate–CF profile and create dominant elements, whereas other frequencies are not reflected at all. In general, the values of the correlation coefficients between the sound spectra and rate–CF profiles are low (purr: $r = 0.10$; chirp: $r = 0.30$; chutter: $r = 0.11$).

A direct comparison of the IC and MGB data suggests that the representation of the spectral features found at the level of the IC is preserved in the MGB for wideband calls (whistle, chirp); however, this representation is less precise for the low-frequency calls (chutter, purr). Several studies have suggested that the coding of a vocalization spectrum cannot be described by a simple linear model. For example, Yeshurun et al. (1989) concluded from their study in awake squirrel monkeys that some of the responses of MGB cells to natural vocalizations could be predicted by assuming a linear transformation function, whereas other responses could be predicted by nonlinear (second order) kernels. Tanaka and Taniguchi (1991) similarly concluded that a majority of neurons in the MGB of the guinea pig showed discharge patterns not predictable from the spectral energy of the call near the neuronal CF. An interesting contribution to the problem of vocalization processing in the auditory system of mammals was offered by Philibert et al. (2005), who compared the responses of MGB neurons in the guinea pig and rat. They found that neurons in the guinea pig and rat MGB displayed similar response strengths to guinea pig vocalizations, and did not exhibit a preference for the natural pattern over the time-reversed versions of the calls in either species. This finding led them to the conclusion that, in mammals, selectivity for the natural version of species-specific vocalizations is prominent only at the cortical level.

As suggested by Creutzfeld et al. (1980), significant differences in the coding of vocalizations exist between the MGB and the auditory cortex. Thalamo–cortical transformation in the temporal and spectral domains was studied by Miller et al. (2002). They simultaneously recorded the responses of neurons in the ventral division of the medial geniculate body and in the primary auditory cortex in the cat. The spectro-temporal receptive fields of single units were derived in their approach by dynamic ripples. They found a similar spectral integration, as measured by excitatory bandwidth and spectral modulation preference, while temporal modulation rates were lower in the cortex by a factor of two (with the upper cut-off of temporal modulation function in the thalamus being 62.9 Hz and in the cortex 37.4 Hz). There was no evident correlation between the spectral and temporal integration properties, suggesting that the excitatory–inhibitory interactions are largely independent.

IV. Conclusions

Mammalian brain possesses mechanisms for processing behaviorally significant acoustic signals, such as species-specific vocalizations. Results of the contemporary research in auditory neuroscience indicate that the mechanisms of vocalization detection are not based on specific "call detectors," but rather on encoding of the spectrotemporal acoustic patterns of vocalizations by ensembles (circuits) of neurons. Before being ultimately recognized in the auditory cortex, the vocalization signals undergo significant processing in the subcortical auditory nuclei such as the inferior colliculus and medial geniculate body.

Acknowledgments

I am thankful to D. Šuta for his comments on the manuscript. The study was supported by the following grants: Grant Agency of the Czech Republic 309/07/1336, Academy of Sciences of the Czech Republic AVOZ50390512 and the Ministry of Education of the Czech Republic LC554.

References

Aitkin, L., Boyd, J., 1978. Acoustic input to the lateral pontine nuclei. Hear. Res. 1, 67–77.

Aitkin, L., Tran, L., Syka, J., 1994. The response of neurons in subdivisions of the inferior colliculus of cats to tonal, noise an vocal stimuli. Exp. Brain Res. 98, 53–64.

Anderson, L.A., Wallace, M.N., Palmer, A.R., 2007. Identification of subdivisions in the medial geniculate body of the guinea pig. Hear. Res. 228, 156–167.

Berryman, J.C., 1976. Guinea-pig vocalizations: their structure, causation and function. Z. Tierpsychol. 41, 80–106.

Chiu, T.W., Poon, P.W.F., 2000. Basic response determinants of single neurons to amplitude modulation in the auditory midbrain. Exp. Brain Res. 39, 87–104.

Covey, E., 2005. Neurobiological specializations in echolocating bats. Anat. Rec. A. Discov. Mol. Cell. Evol. Biol. 287, 1103–1116.

Creutzfeld, O., Hellweg, F.-C., Schreiner, Chr., 1980. Thalamocortical transformation of responses to complex auditory stimuli. Exp. Brain Res. 39, 87–104.

Frisina, R.D., 2001. Subcortical neural coding mechanisms for auditory temporal processing. Hear. Res. 158, 1–27.

Gehr, D.D., Komiya, H., Eggermont, J.J., 2000. Neuronal responses in cat primary auditory cortex to natural and altered species-specific calls. Hear. Res. 150, 27–42.

Hashikawa, T., Kawamura, K., 1983. Retrograde labeling of ascending and descending neurons in the inferior colliculus. A fluorescent double-labeling study in the cat. Exp. Brain Res. 49, 457–461.

Kvašňák, E., Popelář, J., Syka, J., 2000. Discharge properties of neurons in subdivisions of the medial geniculate body of the guinea pig. Physiol. Res. 49, 369–378.

Langner, G., Schreiner, C.E., 1988. Periodicity coding in the inferior colliculus of the cat. I. Neuronal mechanisms. J. Neurophysiol. 60, 1799–1822.

Lesica, N.A., Grothe, B., 2008. Efficient temporal processing of naturalistic sounds. PLoS ONE 3, e1655.

Lewicki, M.S., 2002. Efficient coding of natural stimuli. Nat. Neurosci. 5, 356–363.

Miller, L.M., Escabí, M.A., Read, H.L., Schreiner, C.E., 2002. Spectrotemporal receptive fields in the lemniscal auditory thalamus and cortex. J. Neurophysiol. 87, 516–527.

Morest, D.K., 1964. The neuronal architecture of the medial geniculate body of the cat. J. Anat. 98, 611–630.

Nagarajan, S.S., Cheung, S.W., Bedenbaugh, P., Beitel, R.E., Schreiner, C.E., Merzenich, M.M., 2002. Representation of spectral and temporal envelope of twitter vocalizations in common marmoset primary auditory cortex. J. Neurophysiol. 87, 1723–1737.

Neuert, V., Pressnitzer, D., Patterson, R.D., Winter, I.M., 2001. The responses of single units in the inferior colliculus of the guinea pig to damped and ramped sinusoids. Hear. Res. 159, 36–52.

Newman, J.D., 1978. Perception of sounds used in species-specific communications: the auditory cortex and beyond. J. Med. Primatol. 7, 95–105.

Newman, J.D., Symmes, D., 1974. Arousal effects of unit responsiveness to vocalizations in squirrel monkey auditory cortex. Brain Res. 78, 125–138.

Newman, J.D., Wollberg, Z., 1973. Multiple coding of species-specific vocalizations in the auditory cortex of squirrel monkeys. Brain Res. 54, 287–304.

Pelleg-Toiba, R., Wollberg, Z., 1991. Discrimination of communication calls in the squirrel monkey: "call detectors" or "cell ensembles"? J. Basic Clin. Physiol. Pharmacol. 2, 257–272.

Philibert, B., Laudanski, J., Edeline, J.-M., 2005. Auditory thalamus response to guinea-pig vocalizations: a comparison between rat and guinea-pig. Hear. Res. 209, 97–103.

Rees, A., Moller, A.R., 1987. Responses of neurons in the inferior colliculus of the rat to AM and FM tones. Hear. Res. 10, 301–330.

Rees, A., Palmer, A.R., 1989. Neuronal responses to amplitude-modulated and pure-tone stimuli in the guinea pig inferior colliculus, and their modification by broadband noise. J. Acoust. Soc. Am. 85, 1978–1994.

Sachs, M.B., 1984. Neural coding of complex sounds: speech. Ann. Rev. Physiol. 46, 261–273.

Smolders, J.W.T., Aertsen, A.M.H., Johannesma, P.I.M., 1979. Neural representation of the acoustic biotope. Biol. Cybern. 35, 11–20.

Suga, N., 1965. Responses of cortical auditory neurones to frequency modulated sounds in echo-locating bats. Nature 206, 890–891.

Šuta, D., Kvašňák, E., Popelář, J., Syka, J., 2003. Representation of species-specific vocalizations in the inferior colliculus of the guinea pig. J. Neurophysiol. 90, 3794–3808.

Šuta, D., Popelář, J., Kvašňák, E., Syka, J., 2007. Representation of species-specific vocalizations in the medial geniculate body of the guinea pig. Exp. Brain Res. 183, 377–388.

Syka, J., Straschill, M., 1970. Activation of superior colliculus neurons and motor responses after electrical stimulation of the inferior colliculus. Exp. Neurol. 28, 384–392.

Syka, J., Popelář, J., Kvašňák, E., Šuta, D., Jilek, M., 1997. Processing of species-specific vocalizations in the inferior colliculus and medial geniculate body of the guinea pig. In: Syka, J. (Ed.), Acoustical Signal Processing in the Central Auditory System. Plenum Press, New York, NY, pp. 431–441.

Syka, J., Popelář, J., Kvašňák, E., Astl, J., 2000. Response properties of neurons in the central nucleus and external and dorsal cortices of the inferior colliculus in guinea pig. Exp. Brain Res. 133, 254–266.

Symmes, D., Alexander, G.E., Newman, J.D., 1980. Neural processing of vocalizations and artificial stimuli in the medial geniculate body of squirrel monkey. Hear. Res. 3, 133–146.

Tanaka, H., Taniguchi, I., 1991. Responses of medial geniculate neurons to species-specific vocalized sounds in the guinea pig. Jap. J. Physiol. 41, 817–829.

Wang, X., 2000. On cortical coding of vocal communication sounds in primates. Proc. Natl. Acad. Sci. USA, 97, 11843–11849.

Wang, X., Kadia, S.C., 2001. Differential representation of species-specific primate vocalizations in the auditory cortices of marmoset and cat. J. Neurophysiol. 86, 2616–2620.

Wang, X., Merzenich, M.M., Beitel, R., Schreiner, C. E., 1995. Representation of a species-specific vocalization in the primary auditory cortex of the common

marmoset: temporal and spectral characteristics. J. Neurophysiol. 74, 2685–2706.
Watanabe, T., Katsuki, Y., 1974. Response patterns of single auditory neurons of the cat to species-specific vocalizations. Jap. J. Physiol. 24, 135–155.
Winer, J.A., 1985. The medial geniculate body of the cat. Adv. Anat. Embryol. Cell Biol. 86, 1–98.
Winter, P., Funkenstein, H.H., 1973. The effect of species-specific vocalization on the discharge of auditory cortical cells in the awake squirrel monkey (*Saimiri sciureus*). Exp. Brain Res. 18, 489–504.
Wollberg, Z., Newman, J.D., 1972. Auditory cortex of squirrel monkey: response patterns of single cells to species-specific vocalizations. Science 175, 212–214.
Yeshurun, Y., Wollberg, Z., Dyn, N., 1989. Prediction of linear and nonlinear responses of MGB neurons by system identification methods. Bull. Math. Biol. 51, 337–346.
Young, E.D., 2008. Neural representation of spectral and temporal information in speech. Philos. Trans. R. Soc. Lond. B. Biol. Sci. 363, 923–945.

CHAPTER 4.2

Activation of limbic system structures by replay of ultrasonic vocalization in rats

Markus Wöhr* and Rainer K.W. Schwarting

Experimental and Physiological Psychology, Philipps-University, Marburg, Germany

Abstract: Since the first observation of ultrasonic vocalizations in rats, several hypotheses have been proposed regarding a possible function of such calls as echolocation, temperature regulation and others. Today, however, it is well-established that ultrasonic vocalizations in rats serve as important social signals. Rat behavioral and neuronal responses to 22 kHz calls indicate that this call type serves as an alarm call. Although 22 kHz calls are not innately recognized as alarm calls, they can reach alarm signal value as a consequence of associative learning, which is facilitated by a biological preparedness to associate 22 kHz calls with aversive events. The perirhinal cortex might be at least part of the "neural template" responsible for such a biological preparedness. On the other hand, behavioral and neuronal responses to 50 kHz calls indicate that this call type serves as a contact call. Social approach displayed in response to 50 kHz calls is paralleled by an activation of frontal cortex and the nucleus accumbens, which might be related to the appetitive value of 50 kHz calls.

Keywords: ultrasonic vocalization; social communication; alarm call; contact call; periaqueductal gray; amygdala; nucleus accumbens; rat; c-Fos

I. Introduction

Since the first observation of ultrasonic vocalizations in rats (Anderson, 1954), several hypotheses have been proposed regarding a possible function of such calls as echolocation (Rosenzweig et al., 1955), temperature regulation (Blumberg and Sokoloff, 2001), and other purposes (for some early hypotheses see Nyby and Whitney, 1978; Smith, 1979). Today, however, it is well-established that ultrasonic vocalizations in rats serve as important social signals. Thus, by performing playback experiments, several studies have demonstrated that pup ultrasonic vocalizations can induce maternal search and retrieval behavior (Allin and Banks, 1972; Brunelli et al., 1994; Wöhr and Schwarting, 2008a). In juvenile and adult rats a wealth of evidence has been gathered indicating that 22 kHz calls which are emitted in aversive situations, such as fear conditioning (Jelen et al., 2003; Wöhr et al., 2005; Borta et al., 2006; Wöhr and Schwarting, 2008a,b) and startle (Kaltwasser, 1990, 1991), seem to signal alarm (Blanchard et al., 1991), while 50 kHz calls, which are emitted in appetitive situations such as play (Knutson et al., 1998), mating (Sales, 1972; Thomas and Barfield, 1985), and "tickling" (heterospecific play; Panksepp and Burgdorf, 1999, 2000, 2003; Schwarting et al., 2007), serve as social contact calls in sexual (Geyer et al., 1978; McIntosh et al., 1978) and non-sexual (Wöhr and Schwarting, 2007) situations.

II. 22 kHz vocalization: behavioral responses

The hypothesis that 22 kHz calls serve as alarm calls to warn conspecifics about external danger was borne out in studies by Blanchard and colleagues (Blanchard et al., 1991), who showed that the production of 22 kHz calls in response to a predator was dependent on the presence of conspecifics, i.e., on the presence of an audience. This indicates that the production of

*Corresponding author. E-mail: markus.woehr@staff.uni-marburg.de

Stefan M. Brudzynski (Ed.)
Handbook of Mammalian Vocalization
ISBN 978-0-12-374593-4

DOI: 10.1016/B978-0-12-374593-4.00012-7

vocal signals is not only sensitive to specific eliciting stimuli, but also to the caller's social context, i.e., the presence of an identified group of listeners. In the study by Blanchard and colleagues, the group of listeners was a naturally-established colony in a visible burrow system (Blanchard et al., 1991).

In line with a putative alarm function, it was also shown that the emission of 22 kHz calls in response to a cat could induce a profound and long-lasting set of defensive behaviors in conspecifics that had not seen the cat themselves (Blanchard et al., 1990). The experimental evidence that actually the production of 22 kHz calls, and not an unknown causal third factor, was crucial for the induction of these behavioral changes in the recipients was obtained from playback studies. It was demonstrated that the presentation of natural 22 kHz calls or 20 kHz sine wave tones to naïve rats could activate the fight/flight/freeze system. The studies carried out so far indicated that the behavioral response could be strain dependent. Thus, in response to the replay, Wistar rats (Sales, 1991; Brudzynski and Chiu, 1995; Commissaris et al., 2000; Neophytou et al., 2000; Burman et al., 2007; Wöhr and Schwarting, 2007) and Sprague–Dawley rats (Endres et al., 2007) showed reduction in locomotor activity and limited freezing response (behavioral inhibition), whereas Lister hooded rats showed bursts of running and jumping (behavioral excitation), which is a characteristic feature of an active defensive behavior (Beckett et al., 1996, 1997; Commissaris et al., 1998, 2000; Voits et al., 1999; Neophytou et al., 2000; Finn et al., 2004; Nicolas et al., 2007).

However, when critically reviewing the literature, the evidence in favor of obvious behavioral responses to 22 kHz calls is weak. It has to be noted that the effects were only clearly evident when rather loud and artificial continuous sine wave tones were used (Beckett et al., 1996, 1997; Commissaris et al., 1998, 2000; Voits et al., 1999; Neophytou et al., 2000; Finn et al., 2004; Nicolas, et al., 2007), whereas they were rather weak in the case of natural stimuli (Sales, 1991; Brudzynski and Chiu, 1995; Burman et al., 2007; Endres et al., 2007; Wöhr and Schwarting, 2007). In the first study using natural 22 kHz calls as playback stimuli, Sales (1991) was able to show that they could induce a modest locomotor inhibition. In particular, rats entered floor squares about 70 times when noise was presented, but only about 55 times during playback of 22 kHz calls. Shortly after this study, Brudzynski and Chiu (1995) performed a similar experiment, but did not observe acute effects during playback. Rats reached about 500 activity counts per observation time, both before and during playback. Only a slight, but significant, decrease in locomotor activity to about 350 activity counts was observed after cessation of the playback. Furthermore, three recent studies also observed only weak effects of 22 kHz calls. In a study performed by Burman and colleagues (Burman et al., 2007) only one out two natural 22 kHz stimuli was effective in increasing the latency to emerge from the test box – the other was without any effect. Also, Endres and colleagues (Endres et al., 2007) observed only a non-significant and modest increase in the time rats spent freezing when exposed to 22 kHz calls (about 25% versus about 8% during silence). Similarly, Wöhr and Schwarting (2007) observed only a non-significant and modest decrease in the distance traveled during playback of 22 kHz calls (about 10 cm/min versus about 40 cm/min during silence). Finally, some studies did not detect any behavioral response to playback of natural 22 kHz calls (Lindquist et al., 2004; Bang et al., 2008; Sadananda et al., 2008).

Moreover, in all the studies, when the responses to natural 22 kHz calls were compared with the responses to other ultrasonic stimuli, a clear differential response to 22 kHz calls as an unconditioned stimulus was not observed (Sales, 1991; Endres et al., 2007; Bang et al., 2008). Thus, Sales (1991) detected an inhibition of locomotor activity during playback of 22 kHz calls similar to that in response to an artificial 38 kHz stimulus. In the study by Endres and colleagues, the response to playback of 22 kHz calls was compared to various other stimuli, such as 50 kHz calls, 22 kHz sine wave tones, 22 kHz calls shifted to about 45 kHz and white noise in the range from 17 to 27 kHz (Endres et al., 2007). Comparisons between responses to 22 kHz calls and all the other stimuli demonstrated that there was no specific response of naïve rats to the playbacks of 22 kHz calls. However, when the different stimuli with acoustic characteristics close to 22 kHz were pooled together, a moderate increase in freezing was observed during and after the presentation of the stimulus. Such an induction of freezing was not seen in the other pooled groups. Bang and colleagues used a similar set of stimuli, including 22 kHz calls, 50 kHz calls, continuous tones, or temporally matched discontinuous tones, and obtained no differences in freezing responses between 22 kHz calls and any of the other stimuli (Bang et al., 2008).

Furthermore, in playback studies with artificial stimuli it was found that animals showed a stronger behavioral response to 7 kHz or 12 kHz sine wave tones than to 20 kHz sine wave tones (Commissaris et al., 2000). This observation indicates that the behavioral effects in these studies were clearly not related to the communicative value of the 22 kHz calls. More likely, these effects were caused by rather high sound pressure levels used in the experiments. In some studies, artificial stimuli were presented with more than 100 dB SPL (Voits et al., 1999; Commissaris et al., 2000), which is much higher than the usual sound pressure level of natural 22 kHz calls (approximately 60–80 dB SPL from 20–30 cm distance) (Wöhr et al., 2005; Wöhr and Schwarting, 2008a,b). A critical role of sound pressure in inducing behavioral changes was shown by Commissaris and colleagues, who found that the maximum velocity of the stimulus-induced locomotor activity was much higher when stimuli were presented at higher intensities (Commissaris et al., 2000).

Due to the fact that weak behavioral changes (Sales, 1991; Brudzynski and Chiu, 1995; Burman et al., 2007; Endres et al., 2007; Wöhr and Schwarting, 2007) or no changes (Lindquist et al., 2004; Bang et al., 2008; Sadananda et al., 2008) were obtained in the studies where natural 22 kHz calls were presented as the unconditioned stimulus, fear conditioning studies were performed in which 22 kHz calls served as the conditioned stimulus (CS) (Endres et al., 2007; Bang et al., 2008). During fear conditioning, a formerly neutral stimulus gains the efficacy to elicit fear-related conditional responses (CRs), such as freezing, after being paired with an aversive unconditioned stimulus (US).

Endres and colleagues (Endres et al., 2007) addressed the question of whether recognition of 22 kHz calls as alarm calls can be learned, and whether this learning is facilitated by a preparedness to acquire defensive behavioral patterns in response to such stimuli. They showed that rats quickly learned to associate an aversive event with 22 kHz calls, retained this information longer in memory, and were more reluctant to extinguish this memory than in the case of association of aversive events with other types of ultrasonic stimuli as artificial 22 kHz sine wave tones. The authors therefore concluded that "rats are predisposed to acquire adaptive defensive behavior in response to alarm calls," and that "better encoding of such learning in rats leads to a stable memory which better resists extinction" (Endres et al., 2007, p. 69).

Evidence in support of a predisposition to associate 22 kHz calls with aversive events was also obtained by Bang and colleagues (Bang et al., 2008) using a differential fear conditioning paradigm. Here, 22 kHz calls, 50 kHz calls and various artificial stimuli that were created to deconstruct the 22 kHz calls into simpler acoustic features, as frequency changes, frequency and amplitude modulation, and temporal patterning, were used. During differential fear conditioning, one of these stimuli (CS+) always co-terminated with a foot-shock, which served as a US, while another stimulus (CS−) was explicitly unpaired with the US. As in the study by Endres and colleagues, the 22 kHz calls did not differ from the other ultrasonic stimuli in terms of the unconditional elicitation of freezing behavior, but after pairing 22 kHz calls and foot-shocks, 22 kHz calls induced freezing behavior, suggesting "that freezing to 22 kHz calls is not innate, but instead, emerges as a consequence of associative learning" (Bang et al., p. 213). In contrast to the study by Endres and colleagues, however, Bang and colleagues found that 22 kHz calls serving as the CS+ were not more effective than 50 kHz calls or the artificial stimuli in supporting conditional freezing. However, in favor of a biological preparedness to associate 22 kHz calls with aversive events, an asymmetrical stimulus generalization was discovered. In particular, when 22 kHz calls served as the CS+, there was less generalization of fear to the CS− than when 22 kHz calls served as the CS−. Under the latter circumstances, rats failed to discriminate between the CS+ and CS−. By contrast, they clearly did discriminate when 22 kHz calls served as the CS+. Thus, the amount of stimulus generalization (from CS+ to CS−) depended on which stimuli served as the CS+ or the CS−. Despite the differences between the findings reported by Endres et al. (2007) and Bang et al. (2008), both studies indicated that the behavioral response to 22 kHz calls is not innate, but rather emerges through associative learning which, in turn, is facilitated by a predisposition. The existence of biological preparedness to associate certain stimuli over others was demonstrated before, for instance for taste and illness (Garcia and Koelling, 1966; for review see Seligman, 1970).

Apart from the fact that results of recent studies put in question the innate power of 22 kHz calls in consistently inducing defensive behavior (Sales, 1991; Brudzynski and Chiu, 1995; Lindquist et al., 2004; Burman et al., 2007; Endres et al., 2007; Wöhr and Schwarting, 2007; Bang et al., 2008; Sadananda et al., 2008), the observation that the presence of conspecifics can potentiate 22 kHz calling (Blanchard et al., 1991) was also not found consistently (Wöhr and

Schwarting, 2008b). In this last study, a conventional fear conditioning paradigm was used, which was shown to induce 22 kHz calling in previous experiments (Borta et al., 2006; Wöhr et al., 2005; Wöhr and Schwarting, 2008a). Experimental rats were tested in one of three experimental conditions, i.e., either alone, with an anesthetized conspecific, or with an active conspecific in an adjacent chamber. In line with the hypothesized alarm function of 22 kHz calls, one would expect that the presence of a conspecific would potentiate the production of 22 kHz calls. In contrast, however, it was found that the emission of 22 kHz calls during fear conditioning was not potentiated by the close presence of another conspecific. The presence of another rat might even have a mild attenuating effect, if any, on the call rate. It has to be emphasized that this does not rule out the possibility of an audience effect in other conditions. It appears to be possible that the audience effect in the study by Blanchard et al. (1991) was based on a naturally-established rat colony with a dominant alpha male. This may critically decide about the appearance of the audience effect. Interestingly, however, although no evidence for an audience effect on ultrasonic calling was provided in the study by Wöhr and Schwarting (2008b), some evidence in favor of the alarming function of 22 kHz calls was still demonstrated. Thus, a high positive correlation between 22 kHz calls emitted by the experimental rat which underwent fear conditioning and immobility of the recipient was observed. This result supports the notion that 22 kHz calls can act as alarm signals, even if no audience effect is evident. This result may indicate that such an alarm function does not require that the sender has full cognitive control over call production, i.e., the sender may not willingly produce such calls to warn conspecifics about danger. It is in line with this assumption that rats can also emit 22 kHz calls in a number of non-social aversive situations. Thus, individually tested rats can emit 22 kHz calls during fear conditioning (Jelen et al., 2003; Wöhr et al., 2005; Borta et al., 2006; Wöhr and Schwarting, 2008a,b), when startled in a cage (Kaltwasser, 1990, 1991), or when handled for the first time by an experimenter (Brudzynski and Ociepa, 1992). Most importantly, Francis (1977) has also shown that social isolation can act as an acute inducer of 22 kHz calling.

In summary, the studies on behavioral responses to 22 kHz calls accumulated so far indicate that: (1) the likelihood of emitting 22 kHz calls is largely independent of the social context, but in some conditions of established colonies may be potentiated by the presence of companions; (2) 22 kHz calls are not innately recognized as alarm calls; but (3) they can obtain alarm signal value as a consequence of associative learning; (4) which is facilitated by a biological preparedness to associate 22 kHz calls with aversive events.

III. 22 kHz vocalization: neuronal responses

The behavioral changes induced by playback of 22 kHz calls or artificial 20 kHz sine wave tones are accompanied by brain activity implicated in the regulation of anxiety and fear. By means of immunohistochemical assessment of the immediate early gene c-Fos, which is a marker for neuronal activity, Beckett and colleagues (Beckett et al., 1997) have demonstrated that the locomotor hyperactivity in Lister hooded rats in response to playback of artificial 20 kHz sine wave tones was paralleled by increased neuronal activity in the periaqueductal gray (PAG), amygdala, hypothalamus and thalamus. In more detail, the dorsal, but not ventral, part of the rostrocaudal PAG was activated. Within the amygdala, the medial, basolateral, central and lateral nuclei showed an increase in neuronal activity. Within the hypothalamus, activity was observed in the dorsomedial nucleus, but not the anterior and ventromedial nuclei. Finally, the stria terminalis and the paraventricular nucleus of the thalamus were also activated. No activation was observed in the entorhinal cortex.

In a subsequent study, which addressed the question of whether strain differences in behavioral responses to artificial 20 kHz sine wave tones are paralleled by differences in brain activity, the general pattern of neuronal activation was confirmed for Lister hooded rats (Neophytou et al., 2000). As compared to Lister hooded rats, however, Wistar rats showed much less pronounced neuronal activation to playback of 20 kHz sine wave tones. Increased activity was observed in the basolateral amygdala, stria terminalis and entorhinal cortex. Most importantly, the pattern of activation in the PAG differed from that of Lister hooded rats, and paralleled substantial behavioral differences. In Lister hooded rats, playback of vocalization induced locomotor hyperactivity, which was associated with neuronal activity preferentially in the dorsal region of the rostral and caudal PAG, whereas in Wistar rats the same playback induced locomotor hypoactivity and freezing responses, which were paralleled by activation in the ventral region of the caudal PAG. The PAG, which presumably represents the final common pathway in the behavioral expression of aversive states (Vianna and Brandão, 2003), is organized in functional circuits and it was shown that electrical or chemical stimulation of the dorsal part of the PAG elicited fleeing, while stimulation of the ventral parts of the PAG produced freezing (Depaulis et al., 1992; Morgan et al., 1998). Furthermore, inactivation of dorsal parts of the PAG increased fear-induced freezing, while inactivation of ventral parts disrupted this behavior (De Oca et al., 1998). In addition to the up-regulation of activity in the brain areas implicated in the regulation of fear and anxiety, a reduction in activity was observed in the paraventricular nucleus of the thalamus and the raphé nuclei in Wistar rats.

Overall, these studies have shown that 22 kHz call-like stimuli activated brain regions implicated in the regulation of fear and anxiety. In support of this conclusion, it was also shown that the behavioral response to 22 kHz sine wave tones can be attenuated efficiently by anxiolytics (Beckett et al., 1996; Nicolas et al., 2007).

Apart from these studies using artificial stimuli, the expression of c-Fos induced by natural 22 kHz calls was examined, only in one study so far (Sadananda et al., 2008) (see Fig. 1). In this study, the overall immunohistochemical staining pattern was similar to that obtained in studies using artificial signals (Beckett et al., 1997; Neophytou et al., 2000). Specifically, an increase in neuronal activity induced by 22 kHz calls was observed in the PAG, particularly in its rostral part. Also, in the amygdala activation was observed. The increase of activity observed there adds the 22 kHz calls to the group of other motivationally-relevant and negatively-valenced stimuli capable of increasing amygdalar activity, such as foot-shock, restraint, and others (Duncan et al., 1996; Kovács, 1998). Based on these and other findings, it was postulated that the amygdala is a key structure in emotional information processing, and fear has been the function most closely associated with it (Fendt and Fanselow, 1999; LeDoux, 2000; Maren and Quirk, 2004). The fact that 22 kHz calls yielded an increase in the basolateral, but not in the central, amygdala, points to the functional importance of intra-amygdaloid circuits (Pitkänen et al., 1997). The basolateral part is generally considered as the sensory gateway into the amygdala. It is the major site receiving inputs from all sensory systems, including the auditory one. The central amygdala, however, is viewed as an important output region. A number of relay stations are situated between these main input and output regions, where information coming from long-term storage systems may be integrated into the stimulus representation (Pitkänen et al., 1997). Considerable evidence has been accumulated in support of the hypothesis that the basolateral amygdala is critically involved in fear conditioning, i.e., in the acquisition of fear-related CRs (Fendt and Fanselow, 1999; LeDoux, 2000; Maren and Quirk, 2004). In most situations, incoming novel stimuli elicit responses which rapidly habituate, as long as they are not presented together with biologically-significant stimuli. However, pairing an initially neutral stimulus (later becoming CS) with a biologically-significant stimulus (US) leads to changes in synaptic plasticity in the basolateral amygdala in a way that the CS alone is able to flow through the intra-amygdaloid circuits and to activate the central amygdala. The central amygdala then orchestrates responses appropriate to cope with the detected biologically-significant event. In the case of threat, for instance, the output connections of the central amygdala to the brainstem, particularly PAG, induce freezing. The lack of increase in neuronal activity in the central amygdala in response to 22 kHz calls (Sadananda et al., 2008) is therefore in accordance with the observation that 22 kHz calls induce only a subtle amount of freezing (Sales, 1991; Brudzynski and Chiu, 1995; Burman et al., 2007; Endres et al., 2007; Wöhr and Schwarting, 2007), or no freezing at all (Lindquist et al., 2004; Bang et al., 2008; Sadananda et al., 2008). However, the clear increase in c-Fos activity in the basolateral part of the amygdala may indicate initiation of synaptic changes in this area, which might reflect a learning process or provide prerequisites to it (Sadananda et al., 2008). This assumption is also supported by results of a recent study, showing that infusion of the GABA-agonist muscimol into the basolateral amygdala prior to fear conditioning impaired the acquisition of fear to 22 kHz calls (Allen et al., 2008).

Furthermore, neuronal activation was also evident in the perirhinal cortex (Sadananda et al., 2008), which is laterally adjacent to and reciprocally connected with the basolateral amygdala (Pitkänen et al., 1997). First evidence that the perirhinal cortex is implicated in the processing of 22 kHz calls was obtained in a lesion study. Results of this study have shown that perirhinal lesions, performed prior to training, severely impaired delay fear conditioning to a 22 kHz call or artificial 22 kHz call-like stimuli as CS,

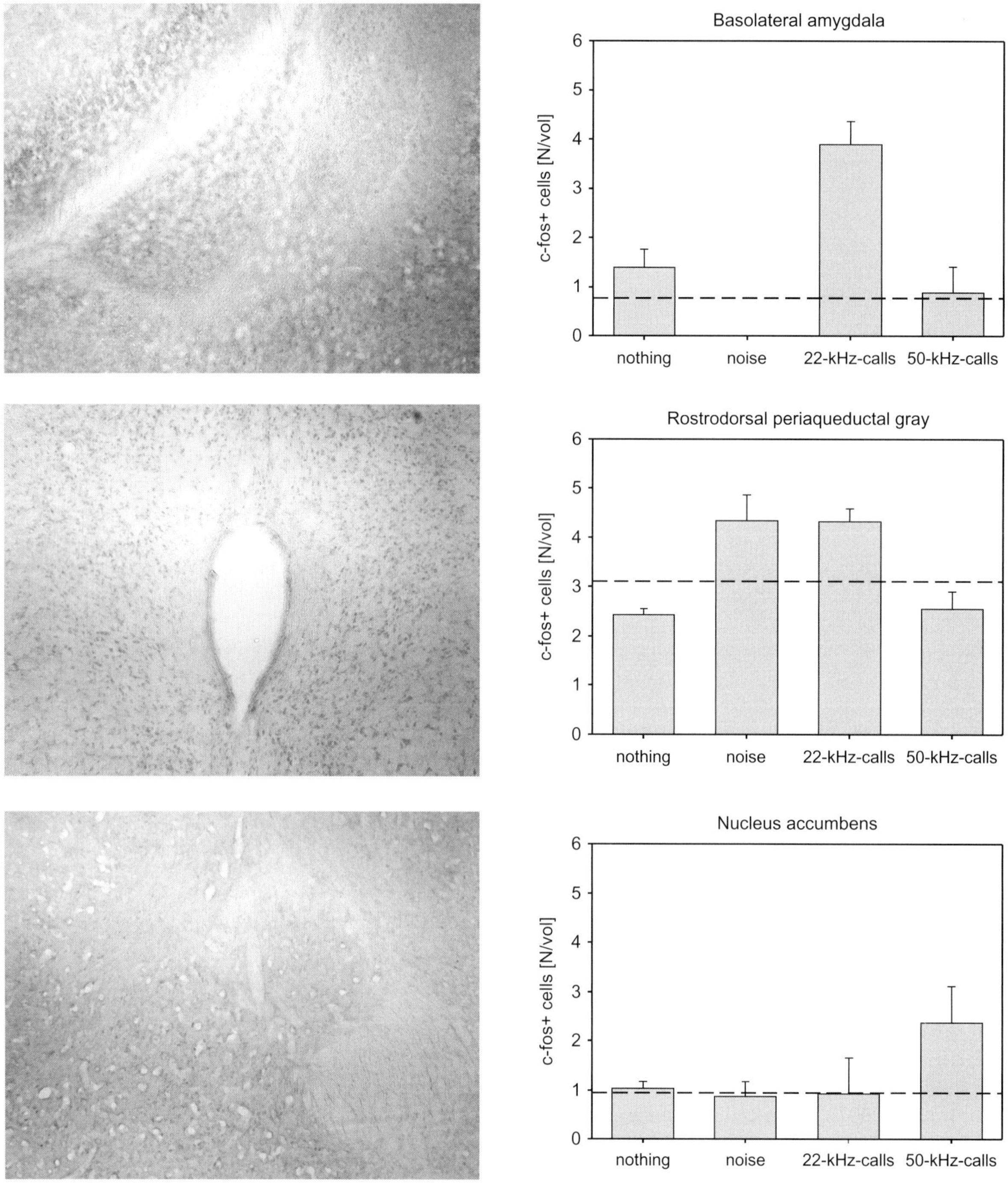

Fig. 1. On the left-hand side, photomicrographs of Fos-like immunoreactivity in basolateral amygdala (upper panel), periaqueductal gray (middle panel) and nucleus accumbens (lower panel). On the right-hand side, bar graphs depicting the number of c-Fos-positive cells (mean ± S.E.M.) in response to no acoustic stimulation ("nothing"), background noise ("noise"), 22 kHz ultrasonic vocalization ("22 kHz calls") and 50 kHz ultrasonic vocalization ("50 kHz calls") observed in basolateral amygdala (upper panel), periaqueductal gray (middle panel) and nucleus accumbens (lower panel). Cells were counted within 0.25 mm × 0.25 mm squares. Data from Sadananda et al. (2008).

while such lesions were ineffective when the CS was a continuous tone of the same or a lower frequency (Lindquist et al., 2004). This observation was replicated in a recent study, which has additionally shown that lesion of the perirhinal cortex also impaired the delay fear conditioning to artificial 22 kHz calls without frequency modulation, when the frequency and temporal pattern of this stimulus matched natural 22 kHz calls (Kholodar-Smith et al., 2008a). This is an important finding, since lesions of the perirhinal cortex did not affect fear conditioning to a continuous 22 kHz sine wave tone (Lindquist et al., 2004).

Based on these findings, it was suggested that the discontinuous nature of 22 kHz calls, i.e., its bout structure, is at least part of the reason why normal fear conditioning to 22 kHz calls requires cortical processing, whereas cortical processing is not necessary for conditioning to continuous tones: "Cortical processing may be required to integrate these discontinuous auditory stimuli across time, in order for normal fear conditioning to occur" (Allen et al., 2007, p. 335). In contrast to the delay fear conditioning, the perirhinal cortex is required in trace fear conditioning to discontinuous and continuous tones (Kholodar-Smith et al., 2008b). The difference between both paradigms is that in delay fear conditioning, the US is presented at the end of the CS, whereas in trace fear conditioning the CS is followed by a trace interval, which is terminated by the US. This result indicates that the role of the perirhinal cortex in trace fear conditioning is distinct from its more perceptual functions in delay fear conditioning.

To address the question of whether neurons in the perirhinal cortex are tuned to respond to 22 kHz calls as an ethologically-important natural stimulus, single-unit responses in the perirhinal cortex to natural 22 kHz calls and various control stimuli, namely frequency and temporally matched discontinuous tones, and continuous tones with the same or lower frequencies, were recorded in freely behaving rats (Allen et al., 2007). When comparing the number of units responding to the auditory stimuli, no difference between 22 kHz calls and control stimuli was observed. Overall, about 40% of the units responded to one or more of the auditory stimulus types used, out of which 69% responded to 22 kHz calls. Unlike the continuous tones, however, the 22 kHz calls sometimes elicited call-related firing patterns, which consisted of a transient increase in the firing frequency that was triggered by the onset, or less often the offset, of each of the successive calls in a bout of calls. Remarkably, similar firing patterns were elicited by the discontinuous tones. Therefore, Allen and colleagues (Allen et al., 2007) concluded that the naturally-occurring frequency modulation associated with individual calls does not affect firing pattern or the overall level of responsiveness.

In a subsequent study, Furtak and colleagues (Furtak et al., 2007) used a classical fear conditioning paradigm, in which 22 kHz calls or a continuous 22 kHz sine wave tone served as CS, and examined fear conditioning-induced changes in single-unit firing elicited in the perirhinal cortex. Firing changes were observed in about 70% of the recorded units in response to 22 kHz calls or a continuous 22 kHz sine wave tone after pairing the stimuli with a footshock (US). Conditioning caused widespread changes in neuronal firing, regardless of whether 22 kHz calls or a 22 kHz sine wave tone served as a cue. Remarkably, about 30% of units that were initially CS-unresponsive became CS-responsive after conditioning. Despite these general changes, however, two differences between single-unit responses elicited by the 22 kHz calls and those elicited by the 22 kHz sine wave tone were evident. First, about 10% of the units recorded from the rat group, which was conditioned to 22 kHz calls, displayed a precisely timed increase in firing rate during the interval, in which the US occurred during conditioning. This response pattern was unique to this group, and was not seen in rats conditioned to a 22 kHz sine wave tone. Second, before conditioning the neurons started firing to both CS after about 55 ms; after conditioning, however, neurons started firing to the 22 kHz sine wave tone after about 25 ms, whereas conditioning to 22 kHz calls had no effect on the firing latency. Based on these findings, the authors suggested that firing to both CS was mediated by cortical rather than subcortical pathways to the perirhinal cortex before conditioning, but that subcortical pathways gained control of firing through conditioning to the 22 kHz sine wave tone, but not to the 22 kHz calls (Furtak et al., 2007).

These studies are in line with the assumption of Endres and colleagues (Endres et al., 2007) who have suggested a "neural template," which better encodes the calls of conspecifics than other auditory stimuli. This conclusion was based on their behavioral study, where rats quickly learned to associate an aversive event with 22 kHz calls, retained this information longer in memory, and were more reluctant to extinguish this memory than in the case of other types of ultrasonic stimuli. It seems likely that the perirhinal cortex is at least part of this "neural template," which is responsible for the predisposition observed when 22 kHz calls are used as CS.

In summary, the studies on neuronal responses to presentation of 22 kHz calls accumulated so far indicate that: (1) 22 kHz calls activate brain regions implicated in the regulation of fear and anxiety in the receivers; (2) at least part of these regions are required for fear conditioning to 22 kHz calls; and (3) that among the activated structures, the perirhinal cortex might be at least part of the "neural template" responsible for the biological preparedness to associate 22 kHz calls with aversive events.

IV. 50 kHz vocalization: behavioral responses

The hypothesis that 50 kHz calls serve as contact calls is mainly based on behavioral observations in the sexual context. During mating, 50 kHz calls are emitted by both males and females, primarily while approaching and investigating the partner (Sales, 1972; Thomas and Barfield, 1985).

A functional effect of male 50 kHz calls on female proceptivity was indicated by devocalization and playback studies. When adult males were devocalized by resection of the laryngeal nerves and mated with estrous females, a reduced rate of darting and ear-wiggling was observed in these females in comparison to other females mated with sham-operated controls (Thomas et al., 1981, 1982). Furthermore, although playback of male 50 kHz calls into a cage with solitary estrous females had no obvious effect on their behavior (Geyer et al., 1978), females showed an increased level of proceptive behavior if they were exposed to males immediately after playback (Geyer et al., 1978). Playback of male 50 kHz calls could even restore proceptive behavior in estrous females when mated with devocalized males (Geyer and Barfield, 1978; McIntosh et al., 1978; White and Barfield, 1990). Finally, and most intriguingly, when male 50 kHz calls were presented while the females were in contact with a castrated non-mating male, the females showed elevated solicitation behavior and even lordosis responses without being mounted (McIntosh et al., 1978).

Apart from these behaviors, playback of male 50 kHz calls also elicited ultrasonic calling from females (White et al., 1993). Females' 50 kHz calls themselves appear to be important for mating, since devocalized females received fewer intromissions, despite displaying enhanced darting and approaches towards the partner, while normal mating activity was restored when tape-recorded female ultrasonic calls were presented in such pairs (White and Barfield, 1987, 1989).

More recently, it has become apparent that 50 kHz calls also serve communicative purposes in non-sexual contexts, since Panksepp and colleagues (Panksepp et al., 2002) observed that rats spent more time with rats, which vocalize a lot, rather than with conspecifics, which displayed less calling behavior. This observation suggested that 50 kHz calls are used as contact calls to (re)establish or to maintain contact among conspecifics. This view is supported by different lines of research. First, Siviy and Panksepp (1987) showed that deafening or devocalizing rats can affect juvenile play, a situation where 50 kHz calls are typically emitted (Knutson et al., 1998). Second, Brudzynski and Pniak (2002) found that rats emitted 50 kHz calls when exposed to odor of other conspecifics, and the number of emitted calls was directly proportional to the number of rats leaving their odor. This result indicated that the production of 50 kHz calls is driven by a potential social contact. Third, 50 kHz calls were also detected after separation of cage mates from each other during short social isolation in the animal's own, or in an unfamiliar, soiled, or fresh housing cage (Schwarting et al., 2007; Wöhr et al., 2008). Remarkably, the propensity to call differed depending on the time-point of the last social contact, i.e., rats mostly emitted 50 kHz calls initially after separation from their cage mate (Wöhr et al., 2008). Finally, it was found that not only the animal which was isolated in a new housing cage emitted 50 kHz calls, but the calls were also emitted by the cage mate that remained alone in the home cage after the removal of the test rat (Wöhr et al., 2008).

In order to prove the communicative value of 50 kHz calls in non-sexual contexts, a playback study was conducted where overt (i.e. visible) and calling behavior of the receiver was measured (Wöhr and Schwarting, 2007). The results clearly demonstrated that 50 kHz calls can induce approach behavior and ultrasonic calling in non-sexual contexts, supporting the hypothesis that 50 kHz calls are used to (re)establish or to maintain contact among conspecifics (Siviy and Panksepp, 1974; Brudzynski and Pniak, 2002; Panksepp et al., 2003; Schwarting et al., 2007; Wöhr et al., 2008). Exposure of rats to 50 kHz calls induced a three-fold increase in locomotor activity in adult rats in comparison to that seen in test phases without acoustic playback or phases where background noise was presented. Furthermore, the induced locomotor activity was directed towards the stimulus source (i.e. the loudspeaker). During the 60 second period of playback of 50 kHz calls, adult rats spent on average less than 5 seconds away from the speaker, and more than 10 seconds near to it. Such a preference was not observed during the phases of the experiment without playback of calls, or with the playback of background noise. This preference was more clearly evident in juvenile rats. Similar to the adults, they displayed a three-fold increase in their locomotor activity, which was also directed to the loudspeaker. Compared to adults, the preference towards the speaker was much more pronounced, and juveniles spent more than 40 seconds of the 60 second playback period near to the speaker. Overall, the male approach response observed during playback of 50 kHz calls resembled that of a mother searching for her pups in response to isolation-induced pup calls (Allin and Banks, 1972; Brunelli et al., 1994; Wöhr and Schwarting, 2008a).

The finding that playback of 50 kHz calls can induce social approach in non-sexual situations indicates that 50 kHz calls are an important element in the evolution and maintenance of social relationships in rats. Similar conclusions can be drawn for mice, where ultrasonic vocalizations were reported during social exploration (Moles and D'Amato, 2000; Moles et al., 2007; Panksepp et al., 2007). Moreover, Panksepp and colleagues observed that high-frequency calling in mice is positively correlated with social investigation (Panksepp et al., 2007). Therefore, ultrasonic vocalization in mice appears to reflect the level of social motivation. Supporting this notion, Moles and D'Amato (2000) have shown that social investigation and the number of ultrasonic calls can be modulated by manipulating the attractiveness of the test partner. They have suggested, therefore, that ultrasonic calls facilitate proximity between animals, which helps to acquire relevant social information. In line with this suggestion is also the fact that a reduced level of gregariousness among older mice, indicated by a decrease in the time spent investigating the partner, was paralleled by a decrease in the emission of ultrasonic vocalizations (Moles et al., 2007). Remarkably, a decrease in social interest as a function of ageing was also seen in rats in response to playback of 50 kHz calls (Wöhr and Schwarting, 2007), which is in accordance with observations that 50 kHz calls are emitted in the highest numbers in juvenile rats (Panksepp and Burgdorf, 1999).

In summary, the studies on behavioral responses to 50 kHz calls accumulated so far indicate that: (1) 50 kHz calls of both males and females appear to facilitate mating; (2) but they also have communicative value in non-sexual contexts, where they appear to serve to (re)establish or to keep contact among conspecifics.

V. 50 kHz vocalization: neuronal responses

So far, very little is known about the neuronal responses to 50 kHz calls. Sadananda and colleagues have shown that, in comparison to 22 kHz calls, the presentation of 50 kHz calls led to only few changes in c-Fos labeling in the brain (Sadananda et al., 2008) (see Fig. 1). More interestingly, the majority of the changes indicated a decreased neuronal activation compared to the non-stimulated controls. The reason for the decrease in labeling in the 50 kHz group is unclear. Presumably, general experimental factors led to neural activation in the control group, such as handling by the experimenter and being exposed to the test apparatus. These activations were probably prevented by 50 kHz stimulation and some of its consequences. In detail, decreased numbers of labeled cells were observed in the central amygdala, the lateral habenula, and the dorsal raphe nuclei. Such decreases were also observed in the 22 kHz group (non-significantly in the lateral habenula); therefore, they were probably not specific to the stimulus valence, i.e., aversive versus appetitive. The fact that 50 kHz calls led to similar changes in the central amygdala as the 22 kHz calls might appear surprising in the light of a bulk of evidence relating this nucleus to the expression of learned fear (Fendt and Fanselow, 1999; LeDoux, 2000; Maren and Quirk, 2004). However, there is other evidence showing that the central amygdala is also involved in appetitive functions, at least in relation to conditioned behavior (Knapska et al., 2008).

The lateral habenula is considered part of the network, which provides a downstream cross-talk between limbic forebrain sites (such as septum, lateral preoptic area, amygdala, nucleus accumbens/ventral pallidum) and ascending modulatory pathways (i.e., dopaminergic, serotonergic, cholinergic pathways) (Geisler and Trimble, 2008). Through these connections, the lateral habenula seems to affect arousal, attention and emotional state setting (Geisler and Trimble, 2008). At least some of its targets are regulated in such a way that habenula activation inhibits them. Since ultrasonic stimulation led to a decrease of habenular activity, the inhibitory outcomes in its efferent targets were probably reduced. These target effects, in turn, might allow for the presentation of motivationally relevant ultrasonic stimuli to lead to arousal and appropriate actions. Again, these links seem not to be valence-selective, which may explain similar neuronal changes after presentation of both 50kHz and 22kHz calls. This conclusion may also hold true in relation to the dorsal raphe nuclei, where decreased c-Fos labeling was also observed with either stimulus type.

Compared to the decreases, an increased c-Fos labeling after 50kHz calls was observed only in the frontal cortex, and non-significantly in the nucleus accumbens (Sadananda et al., 2008). The cortical neuronal activation was measured in the secondary motor cortex and was specific to the 50kHz group. This activation was probably associated with the observation that only this type of ultrasonic stimulation was effective in inducing pronounced behavioral activation, which was mainly directed towards the source of the stimulation (Sadananda et al., 2008; Wöhr and Schwarting, 2007). Such activation and social approach might also explain why the nucleus accumbens showed some sign of activation (it did not reach the significance level, probably because of the small sample sizes used in this study). The nucleus accumbens is well-known for its critical role in motivated behavior, where it is thought to serve as an "interface between motivation and action" (Mogenson et al., 1980). The nucleus accumbens is important for locomotor activity and approach behavior, which is critically modulated by its dopaminergic input. Therefore, one can assume that dopaminergic activation in the nucleus accumbens was necessary for approach towards appetitive 50kHz calls. Besides, this brain area is also efficient in eliciting 50kHz calls, but not 22kHz calls, for example, by local administration of the catecholaminergic agonist, amphetamine (Burgdorf et al., 2001; Thompson et al., 2006). Thereby, the nucleus accumbens may serve to close the functional link between mechanisms of detection and production of 50kHz calls, which seems to be especially relevant in appetitive social and reciprocal communicatory signals. Such a link might also help to explain why juvenile play, which is accompanied and maintained by increased emission of 50kHz calls, also leads to enhanced c-Fos labeling in the nucleus accumbens (Gordon et al., 2002).

In summary, the study on neuronal responses to 50kHz calls indicates that: (1) 50kHz calls decrease activation in a large number of brain areas compared to non-stimulated controls; however, (2) there are two brain regions, frontal cortex, specifically secondary motor cortex, and to a lesser degree the nucleus accumbens, where neuronal activity was increased by presentation of 50kHz calls. Neuronal activation in the former region is probably due to the fact that 50kHz calls induced pronounced behavioral activation, while activation of the nucleus accumbens might be related to the appetitive value of 50kHz calls.

VI. Conclusion

Behavioral and neuronal responses to 22kHz calls indicate that this call type serves as an alarm call. However, 22kHz calls seem not to be innately recognized as alarm calls, but can also obtain alarm signal value as a consequence of associative learning. This learning is facilitated by a biological preparedness to associate 22kHz calls with aversive events. The perirhinal cortex might be at least part of the "neural template" responsible for such a biological preparedness. Behavioral and neuronal responses to 50kHz calls indicate that this call type serves as an appetitive contact call. Social approach, displayed in response to 50kHz calls, is paralleled by an activation of frontal cortex and the nucleus accumbens, which might be related to the appetitive value of 50kHz calls. Therefore, it can be concluded that 22kHz and 50kHz vocalizations represent two behaviorally opposite states associated with aversive/agonistic responses and affiliative/appetitive responses, respectively, which activate dissimilar brain regions.

References

Allen, T.A., Furtak, S.C., Brown, T.H., 2007. Single-unit responses to 22kHz ultrasonic vocalizations in rat perirhinal cortex. Behav. Brain Res. 182, 327–336.

Allen, T.A., Narayanan, N.S., Kholodar-Smith, D.B., Zhao, Y., Laubach, M., Brown, T.H., 2008. Imaging the spread of reversible brain inactivations using fluorescent muscimol. J. Neurosci. Methods 171, 30–38.

Allin, J.T., Banks, E.M., 1972. Functional aspects of ultrasound production by infant albino rats (*Rattus norvegicus*). Anim. Behav. 20, 175–185.

Anderson, J.W., 1954. The production of ultrasonic sounds by laboratory rats and other mammals. Science 119, 808–809.

Bang, S.J., Allen, T.A., Jones, L.K., Boguszewski, P., Brown, T.H., 2008. Asymmetrical stimulus generalization following differential fear conditioning. Neurobiol. Learn. Mem. 90, 200–216.

Beckett, S.R.G., Aspley, S., Graham, M., Marsden, C. A., 1996. Pharmacological manipulation of ultrasound induced defence behaviour in the rat. Psychopharmacology 127, 384–390.

Beckett, S.R.G., Duxon, M.S., Aspley, S., Marsden, C.A., 1997. Central c-fos expression following 20 kHz/ultrasound induced defence behaviour in the rat. Brain Res. Bull. 42, 421–426.

Blanchard, R.J., Blanchard, D.C., Rodgers, J., Weiss, S.M., 1990. The characterization and modelling of antipredator defensive behavior. Neurosci. Biobehav. Rev. 14, 463–472.

Blanchard, R.J., Blanchard, D.C., Agullana, R., Weiss, S.M., 1991. Twenty-two kHz alarm cries to presentation of a predator, by laboratory rats living in visible burrow systems. Physiol. Behav. 50, 967–972.

Blumberg, M.S., Sokoloff, G., 2001. Do infant rats cry? Psychol. Rev. 108, 83–95.

Borta, A., Wöhr, M., Schwarting, R.K.W., 2006. Rat ultrasonic vocalization in aversively motivated situations and the role of individual differences in anxiety-related behavior. Behav. Brain Res. 166, 271–280.

Brudzynski, S.M., Chiu, E.M.C., 1995. Behavioural responses of laboratory rats to playback of 22 kHz ultrasonic calls. Physiol. Behav. 57, 1039–1044.

Brudzynski, S.M., Ociepa, D., 1992. Ultrasonic vocalization of laboratory rats in response to handling and touch. Physiol. Behav. 52, 655–660.

Brudzynski, S.M., Pniak, A., 2002. Social contacts and production of 50-kHz short ultrasonic calls in adult rats. J. Comp. Psychol. 116, 73–82.

Brunelli, S.A., Shair, H.N., Hofer, M.A., 1994. Hypothermic vocalizations of rat pups (*Rattus norvegicus*) elicit and direct maternal search behavior. J. Comp. Psychol. 108, 298–303.

Burgdorf, J., Knutson, B., Panksepp, J., Ikemoto, S., 2001. Nucleus accumbens amphetamine microinjections unconditionally elicit 50-kHz ultrasonic vocalizations in rats. Behav. Neurosci. 115, 940–944.

Burman, O.H.P., Ilyat, A., Jones, G., Mendl, M., 2007. Ultrasonic vocalizations as indicators of welfare for laboratory rats (*Rattus norvegicus*). Appl. Anim. Behav. Sci. 104, 116–129.

Commissaris, R.L., Beckett, S.R.G., Marsden, C.A., 1998. Strychnine effects on ultrasound-elicited behaviours in Lister hooded rats. Psychopharmacology (Berl.) 136, 162–171.

Commissaris, R.L., Palmer, A., Neophytou, S.I., Graham, M., Beckett, S.R.G., Marsden, C.A., 2000. Acoustically elicited behaviours in Lister hooded and Wistar rats. Physiol. Behav. 68, 521–531.

De Oca, B.M., DeCola, J.P., Maren, S., Fanselow, M.J., 1998. Distinct regions of the periaqueductal grey are involved in the acquisition and expression of defensive responses. J. Neurosci. 18, 3426–3443.

Depaulis, A., Keay, K.A., Bandler, R., 1992. Longitudinal neuronal organization of defensive reactions in the midbrain periaqueductal gray region of the rat. Exp. Brain Res. 90, 307–318.

Duncan, G.E., Knapp, D.J., Breese, G.R., 1996. Neuroanatomical characterization of Fos induction in rat behavioral models of anxiety. Brain Res. 713, 79–91.

Endres, T., Widmann, K., Fendt, M., 2007. Are rats predisposed to learn 22 kHz calls as danger-predicting signals? Behav. Brain Res. 185, 69–75.

Fendt, M., Fanselow, M.S., 1999. The neuroanatomical and neurochemical basis of conditioned fear. Neurosci. Biobehav. Rev. 23, 743–760.

Finn, D.P., Jhaveri, M.D., Beckett, S.R.G., Kendall, D.A., Marsden, C.A., Chapman, V., 2004. Cannabinoids modulate ultrasound-induced aversive responses in rats. Psychopharmacology (Berl.) 172, 41–51.

Francis, R.L., 1977. 22-kHz calls by isolated rats. Nature 265, 236–238.

Furtak, S.C., Allen, T.A., Brown, T.H., 2007. Single-unit firing in rat perirhinal cortex caused by fear conditioning to arbitrary and ecological stimuli. J. Neurosci. 27, 12277–12291.

Garcia, J., Koelling, R., 1966. Relation of cue to consequence in avoidance learning. Psychon. Sci. 4, 123–124.

Geisler, S., Trimble, M., 2008. The lateral habenula: no longer neglected. CNS Spectr. 13, 484–489.

Geyer, L.A., Barfield, R.J., 1978. Influence of gonadal hormones and sexual behavior on ultrasonic vocalization in rats. I. Treatment of females. J. Comp. Physiol. Psychol. 92, 436–446.

Gordon, N.S., Kollack-Walker, S., Akil, H., Panksepp, J., 2002. Expression of c-*fos* gene activation during rough and tumble play in juvenile rats. Brain Res. Bull. 57, 651–659.

Jelen, P., Soltysik, S., Zagrodzka, J., 2003. 22-kHz ultrasonic vocalization in rats as an index of anxiety but not fear: behavioral and pharmacological modulation of affective state. Behav. Brain Res. 141, 63–72.

Kaltwasser, M.T., 1990. Startle-inducing acoustic stimuli evoke ultrasonic vocalization in the rat. Physiol. Behav. 48, 13–17.

Kaltwasser, M.T., 1991. Acoustic startle induced ultrasonic vocalization in the rat: a novel animal model of anxiety? Behav. Brain Res. 43, 133–137.

Kholodar-Smith, D.B., Allen, T.A., Brown, T.H., 2008a. Fear conditioning to discontinuous auditory cues requires perirhinal cortical function. Behav. Neurosci. 122, 1178–1185.

Kholodar-Smith, D.B., Boguszewski, P., Brown, T.H., 2008b. Auditory trace fear conditioning requires perirhinal cortex. Neurobiol. Learn. Mem. 90, 537–543.

Knapska, E., Walasek, G., Nikolaev, E., Neuhäusser-Wespy, F., Lipp, H.-P., Kaczmarek, L., Werka, T., 2008. Differential involvement of the central amygdala in appetitive versus aversive learning. Learn. and Mem. 13, 192–200.

Knutson, B., Burgdorf, J., Panksepp, J., 1998. Anticipation of play elicits high-frequency ultrasonic vocalizations in young rats. J. Comp. Psychol. 112, 65–73.

Kovács, K.J., 1998. c-fos as a transcription factor: a stressful (re)view from a functional map. Neurochem. Int. 33, 287–297.

LeDoux, J.E., 2000. Emotion circuits in the brain. Annu. Rev. Neurosci. 23, 155–184.

Lindquist, D.H., Jarrard, L.E., Brown, T.H., 2004. Perirhinal cortex supports delay fear conditioning to rat ultrasonic social signals. J. Neurosci. 24, 3610–3617.

Maren, S., Quirk, G.J., 2004. Neuronal signaling of fear memory. Nature Rev. Neurosci. 5, 844–852.

McIntosh, T.K., Barfield, R.J., Geyer, L.A., 1978. Ultrasonic vocalisations facilitate sexual behaviour of female rats. Nature 272, 163–164.

Mogenson, G.J., Jones, D.L., Yim, C.Y., 1980. From motivation to action: functional interface between the limbic system and the motor system. Prog. Neurobiol. 14, 69–97.

Moles, A., D'Amato, F.R., 2000. Ultrasonic vocalization by female mice in the presence of a conspecific carrying food cues. Anim. Behav. 60, 689–694.

Moles, A., Constantini, F., Garbugino, L., Zanettini, C., D'Amato, F.R., 2007. Ultrasonic vocalization emitted during dyadic interactions in female mice: a possible index of sociability? Behav. Brain Res. 182, 223–230.

Morgan, M.M., Whitney, P.K., Gold, M.S., 1998. Immobility and flight associated with antinociception produced by activation of the ventral and lateral/dorsal regions of the rat periaqueductal grey. Brain Res. 804, 159–166.

Neophytou, S.I., Graham, M., Williams, J., Aspley, S., Marsden, C.A., Beckett, S.R.G., 2000. Strain differences to the effects of aversive frequency ultrasound on behaviour and brain topography of c-fos expression in the rat. Brain Res. 854, 158–164.

Nicolas, L.B., Klein, S., Prinssen, E.P., 2007. Defensive-like behaviours induced by ultrasound: further pharmacological characterization in Lister-hooded rats. Psychopharmacology (Berl.) 194, 243–252.

Nyby, J., Whitney, G., 1978. Ultrasonic communication of adult myomorph rodents. Neurosci. Biobehav. Rev. 2, 1–14.

Panksepp, J., Burgdorf, J., 1999. Laughing rats? Playful tickling arouses high frequency ultrasonic chirping in young rodents. In: Hameroff, S., Chalmers, C., Kazniak, A. (Eds.), Toward a Science of Consciousness III. MIT Press, Cambridge, MA, pp. 231–244.

Panksepp, J., Burgdorf, J., 2000. 50-kHz chirping (laughter?) in response to conditioned and unconditioned tickle-induced reward in rats: effects of social housing and genetic variables. Behav. Brain Res. 115, 25–38.

Panksepp, J., Burgdorf, J., 2003. "Laughing" rats and the evolutionary antecedents of human joy? Physiol. Behav. 79, 533–547.

Panksepp, J., Gordon, N., Burgdorf, J., 2002. Empathy and the action-perception resonances of basic socio-emotional systems of the brain. Behav. Brain Sci. 25, 43.

Panksepp, J., Jochman, K.A., Kim, J.U., Koy, J.J., Wilson, E.D., Chen, Q., Wilson, C.R., Lahvis, G.P., 2007. Affiliative behavior, ultrasonic communication and social reward are influenced by genetic variation in adolescent mice. PLOS One 2, e351.

Pitkänen, A., Savander, V., LeDoux, J.E., 1997. Organization of intra-amygdaloid circuitries in the rat: an emerging framework for understanding functions of the amygdala. Trends Neurosci. 20, 517–523.

Rosenzweig, M.R., Riley, D.A., Krech, D., 1955. Evidence for echolocation in the rat. Science 121, 600.

Sadananda, M., Wöhr, M., Schwarting, R.K.W., 2008. Playback of 22-kHz and 50-kHz ultrasonic vocalizations induces differential c-fos expression in rat brain. Neurosci. Lett. 435, 17–23.

Sales, G.D., 1972. Ultrasound and mating behaviour in rodents with some observations on other behavioural situations. J. Zool. 168, 149–164.

Sales, G.D., 1991. The effect of 22kHz calls and artificial 38kHz signals on activity in rats. Behav. Processes 24, 83–93.

Schwarting, R.K.W., Jegan, N., Wöhr, M., 2007. Situational factors, conditions and individual variables which can determine ultrasonic vocalizations in male adult Wistar rats. Behav. Brain Res. 182, 208–222.

Seligman, M.E.P., 1970. On the generality of the laws of learning. Psychol. Rev. 77, 406–418.

Siviy, S.M., Panksepp, J., 1987. Sensory modulation of juvenile play in rats. Dev. Psychobiol. 20, 39–55.

Smith, J.W., 1979. The study of ultrasonic communication. Am. Zool. 18, 531–538.

Thomas, D.A., Barfield, R.J., 1985. Ultrasonic vocalization of the female rat (*Rattus norvegicus*) during mating. Anim. Behav. 33, 720–725.

Thomas, D.A., Talalas, L., Barfield, R.J., 1981. Effects of devocalization of the male on mating behavior in rats. J. Comp. Physiol. Psychol. 95, 630–637.

Thomas, D.A., Howard, S.B., Barfield, R.J., 1982. Male-produced postejaculatory vocalizations and the mating behaviour of estrous female rats. Behav. Neural. Biol. 36, 403–410.

Thompson, B., Leonard, K.C., Brudzynski, S.B., 2006. Amphetamine-induced 50kHz calls from rat nucleus accumbens: a quantitative mapping study and acoustic analysis. Behav. Brain Res. 168, 64–73.

Vianna, D.M.L., Brandão, M.L., 2003. Anatomical connections of the periaqueductal gray: specific neural substrates for different kinds of fear. Braz. J. Med. Biol. Res. 36, 557–566.

Voits, M., Beckett, S.R.G., Marsden, C.A., Fink, H., 1999. Role of cholecystokinin type B receptors in ultrasound induced behavior in rats. Peptides 20, 383–386.

White, N.R., Barfield, R.J., 1987. Role of the ultrasonic vocalization of the female rat (*Rattus norvegicus*) in sexual behavior. J. Comp. Psychol. 101, 73–81.

White, N.R., Barfield, R.J., 1989. Playback of female rat ultrasonic vocalizations during sexual behavior. Physiol. Behav. 45, 229–233.

White, N.R., Barfield, R.J., 1990. Effects of male pre-ejaculatory vocalizations on female receptive behavior in the rat (*Rattus norvegicus*). J. Comp. Psychol. 104, 140–146.

White, N.R., Gonzales, R.N., Barfield, R.J., 1993. Do vocalizations of the male rat elicit calling from the female? Behav. Neural Biol. 59, 16–78.

Wöhr, M., Schwarting, R.K.W., 2007. Ultrasonic communication in rats: can playback of 50-kHz calls induce approach behavior? PLoS ONE 2, e1365.

Wöhr, M., Schwarting, R.K.W., 2008a. Maternal care, isolation-induced infant ultrasonic calling, and their relations to adult anxiety-related behavior in the rat. Behav. Neurosci. 122, 310–320.

Wöhr, M., Schwarting, R.K.W., 2008b. Ultrasonic calling during fear conditioning in the rat: no evidence for an audience effect. Anim. Behav. 76, 749–760.

Wöhr, M., Borta, A., Schwarting, R.K.W., 2005. Overt behavior and ultrasonic vocalization in a fear conditioning paradigm – a dose-response study in the rat. Neurobiol. Learn. Mem. 84, 228–240.

Wöhr, M., Houx, B., Schwarting, R.K.W., Spruijt, B., 2008. Effects of experience and context on 50-kHz vocalizations in rats. Physiol. Behav. 93, 766–776.

CHAPTER 4.3

Selective perception and recognition of vocal signals

Günter Ehret* and Simone Kurt

Institute of Neurobiology, University of Ulm, Ulm, Germany

Abstract: Perception and recognition of vocal signals in mammals is based on hearing abilities, which have a species-specific profile but share common mechanisms of sound processing in the auditory system. Common perceptual abilities in mammals refer to audiogram values, just noticeable differences in the perception of frequencies, intensities and duration of sounds, and to the mechanisms and limits of spectral and temporal resolution in hearing. After the discussion of these key features, we will show how perceptual limits and borders between perceptual classes in the auditory systems of mammals become effective in assessing the basic biological meaning of vocal messages. Finally, we will briefly address sound recognition as a consequence of the ability to learn subtle differences in vocalizations in order to identify the vocalizing animal as a member of a certain species and/or as an individual.

Keywords: audiogram; limits in perception; categorical perception; difference limens in audition; hearing abilities; perception of biological meaning; perceptual borders; spectral-temporal resolution

Abbreviations: CB: critical band; d: sound duration; Δd: difference in sound duration; Δf: difference in tone frequency; ΔL: difference in sound level; f: tone frequency; JMD: just meaningful difference; JND: just noticeable difference; L: sound level; VOT: voice onset time

I. Introduction

Mammals vocalize in many behavioral contexts, often starting in their lives with the first cries after birth. The vocalizations convey information about the sender. They may specify the sender's location, its membership of a species, and its characteristics as an individual such as age, sex, body size, health and experienced emotions and motivations. Other animals can take advantage of these vocalizations by perceiving them, decoding their informational content and adjusting their own behavior according to the recognized biological meaning (semiotic value). In this way, vocal communication is established. Since most social interactions that are accompanied by vocalizations happen among members of a given species, we can assume that the acoustical features of emitted sounds and the auditory mechanisms for perceiving the sounds are mutually adapted by natural selection in the evolution of that species.

By similarities in the morphology of the vocal tract of many mammalian species, communication sounds consist of basic elements such as tones, tonal complexes (a number of harmonically or non-harmonically related frequencies), or noise, all of constant or variable frequency, intensity and duration. In addition, the same or varying acoustic elements are often emitted in sequences of a certain temporal structure. These elements, with their acoustic parameters, determine the requirements of the ear and the auditory system for their perception and recognition. Thus, we can expect a basic set of mechanisms and abilities common to all mammals providing the "auditory tools" for communication sound perception. Based on the hypothesis of mutual adaptation of vocalizing and perceiving, however, we can also expect specializations in hearing and perception that are adapted to communication in special environments or in special

*Corresponding author. E-mail: guenter.ehret@uni-ulm.de

Stefan M. Brudzynski (Ed.)
Handbook of Mammalian Vocalization
ISBN 978-0-12-374593-4

DOI: 10.1016/B978-0-12-374593-4.00013-9

cases, such as dual communication (self and social) in bats or highly-developed semantic communication with speech in humans. In the following subsection, we will first discuss common perceptual abilities and related auditory mechanisms, and then provide an outlook on how these abilities might be used in general and in species-specific vocal communication.

II. Perceptual abilities and their physiological bases

II.A. Audiogram

The basis of auditory perception is sensation, i.e., sounds must be audible to the individual. By definition, sounds are audible if their frequency components are within the animal's audiogram, which is the curve illustrating the minimal sound pressure levels of just audible tones, as a function of the tone frequency (Fig. 1). A compilation of audiograms of many species can be found in Fay's psychophysics databook (Fay, 1988). Audiograms describe the frequency range of hearing of a species, together with frequency ranges of increased or reduced sensitivity. The species-specific shape of the audiogram is generated by the filter and amplifier characteristics of the outer, middle and inner (cochlea) ear (e.g., Ehret, 1989), i.e., the basis of hearing reflects very peripheral properties of processing in the auditory pathways.

Further, audiograms set the limits for sound communication in the frequency-intensity space, i.e., frequencies of communication sounds must be within the frequency range of the audiogram, and at or above the minimal audible sound pressure levels represented by the audiogram. For many mammals the

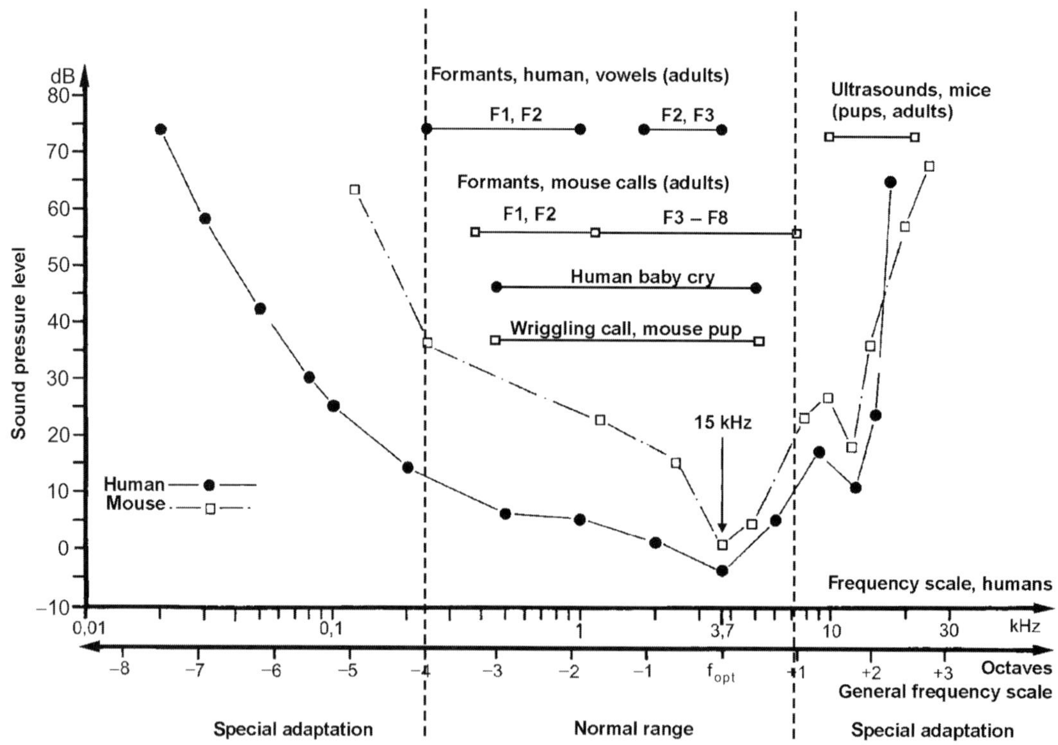

Fig. 1. Relationship between audiograms of humans and house mice and the frequency ranges of their vocalizations. The audiograms represent the auditory threshold curves, i.e., the minimum sound pressure levels in dB (y-axis) as a function of the frequency of a just audible tone (x-axis). Two frequency scales are shown. One applies to human hearing (in kHz); the other is expressed in octaves and calibrated to the frequency with the lowest hearing threshold in the audiogram. The octave scale is common to mammals. The audiogram of the mouse shows the frequency of best hearing (f_{opt}) at 15 kHz. The main frequency ranges of vowels of humans speech, of cries of human babies and of calls of adult mice all are in a frequency range around and about four octaves below f_{opt} (normal frequency range of vocalizations) which can be found also in many other mammals. Mice have special adaptations to hear and vocalize in the high ultrasonic range up to three octaves above f_{opt}. Other rodents and many bat species have such specializations. The audiograms are from Ehret (1974).

frequency range of the audiogram can be divided into three parts with regard to the frequency ranges of communications sounds (Fig. 1): (1) a central "normal" part where the main energy of most communication sounds is located, as illustrated for humans and mice; (2) a specialized part above the central part serving communication in the high ultrasonic ranges (e.g., rodents, bats; see Brudzynski and Fletcher, Chapter 3.3 in this volume); and (3) a specialized part below the central part for communicating over long distances with low-frequency sounds (e.g., elephants; see Garstang, Chapter 3.2 in this volume). Knowing the sound pressure levels of frequencies in communication sounds and their specific attenuation by the medium in which the sound spreads out, one can calculate the communication space of a sender (e.g., Haack et al., 1983). Since high frequencies, especially in the high ultrasonic range, are heavily damped in air, communication with ultrasounds is restricted to short distances around the sender.

Two other conditions have to be considered if one would like to take the audiogram of a species as a frame for estimating the audibility of communication sounds. First, audiograms usually describe the sensation of sounds of rather long duration (100 ms and longer, depending on the frequency spectrum). The perceptual thresholds of shorter sounds increase by about 10 dB for a decrease in sound duration by about one tenth, for example from 100 ms to 10 ms (Fay, 1988; Ehret, 1989). Thus, for optimal detection, communication sounds should either be long, or consist of a train of short duration pulses with a high repetition rate so that their energy is integrated over time to reach a low detection threshold.

Second, the shape of the audiogram changes with the age of the animal. Young animals start hearing in a restricted frequency range (often close to the best frequency range of hearing in adults) at rather high detection thresholds. With increasing age during development, the audible frequency range increases towards lower and higher frequencies, and the thresholds decrease towards those of adults (Ehret, 1983, 1988). Old animals may suffer from hearing loss, especially at high frequencies (Ehret, 1974). Thus, newborns of cats, dogs, many rodent and bat species, and marsupials start hearing only several days after birth and may reach adult auditory sensitivity within about 3–12 weeks; in humans not before the age of two years (Ehret, 1983, 1988). This means that very young mammals and also very old ones may be active in producing sounds, but may not be able to perceive sounds as young adults do. This requires special strategies for acoustic communication with young infants. If infants are the receivers of communication sounds, adult bats (Gould, 1971; Brown, 1976), cats (Haskins, 1977), wolves (Shalter et al., 1977) and humans (Stern et al., 1982) use sounds of a simple frequency and time structure, with frequency components in the optimal frequency range of the young and with rhythmic repetitions of elements.

II.B. Just noticeable differences in frequency, intensity and duration

Mammals can detect minimal frequency differences (Δf) between tones in a series ($f, f + \Delta f, f, f + \Delta f, \ldots$) in the order of 0.1–10%, except for very low frequencies where Δf is constant at a small value (Fay, 1988; Ehret, 1989). Similarly, just noticeable differences (JNDs) in the sound level (ΔL) between tones in a series ($L, L + \Delta L, L, L + \Delta L, \ldots$) are in the order of 0.5–4 dB (Fay, 1988; Ehret, 1989). These JNDs (Δf, ΔL) decrease with increasing sound pressure level (Δf up to a level of about 60 dB above the audiogram curve; Ehret, 1989). This means that communication sounds have to be vocalized loud enough to make small fluctuations in frequency and intensity perceptible. JNDs in frequency and intensity can be derived from cochlear excitation patterns and hair cell densities in the organ of Corti of the cochlea (Maiwald, 1967; Ehret, 1989), and thus are peripheral in origin. Since the cochleae of mammals, except for some bats with a cochlear fovea region (Bruns, 1976; Suga and Jen, 1977), seem to be variants derived from a common scale (Greenwood, 1990), the mentioned frequency and sound level dependencies of JNDs may apply to most mammals.

These JNDs and their properties refer only to sounds in series, to allow comparisons without the involvement of memory. If differences in frequency and/or intensity between sounds have to be detected on the basis of stored and remembered sounds, central auditory processing and learning with the additional variables of emotion, motivation, memory formation and recall, and the issue of "meaning" come into play, which will be discussed below.

JNDs in duration (Δd) in a series of sounds ($d, d + \Delta d, d, d + \Delta d, \ldots$) are in the order of 4–100% and depend very much on the species and the method of measurement (Fay, 1988; Klink and Klump, 2004). The variability of JNDs for duration between species is much larger compared to that for frequencies and intensities. The probable reason is that duration discrimination cannot be directly related to common peripheral (cochlear) properties. Specific coding for sound duration (duration sensitive neurons) seems to occur first at the level of the auditory midbrain in the central nucleus of the inferior colliculus (Covey and Casseday, 1999; Brand et al., 2000) so that species differences in auditory processing in the brainstem can contribute to enlarge the species-dependent variation of JNDs in duration.

II.C. Spectral resolution and the perception of the frequency composition in vocalizations

Due to the vibrations of the vocal folds, most vocalizations consist of a fundamental frequency and harmonically-related overtones, several of which may be amplified by resonances in the vocal tract. These resonance frequencies determine the frequency ranges of formants in vocalizations. The analysis of the spectral structure of sounds requires frequency filters and spectral integrators. The characteristics of such filters – bandwidth as a function of center frequency, sound pressure level and type of sound processed – have been determined in various auditory perception tests in humans and, to a lesser degree, in mammals and other animals (Fay, 1988; Ehret, 1995).

Depending on the method of measurement, the filters have different names, such as critical band (CB), critical ratio (CR), or equivalent rectangular bandwidth (ERB) (Scharf, 1970; Moore, 1989). Perceptually important characteristics of CBs, the increase of bandwidth with increasing center frequency of the filter (Fig. 2) and the independence of the filter bandwidth and the spectral integration within the filter from the sound level, are derived from their cochlear origin and further processing up to the level of the inferior colliculus. One CB covers a frequency range represented by about 0.7–1 mm length of the cochlear tonotopy (Greenwood, 1961, 1990; Ehret, 1977, 1989) which leads to an increase of the CB bandwidth by almost one octave for a one octave increase in center frequency in the frequency range at and above the best hearing range (Fig. 1). In this frequency range, the cochlear frequency representation follows a logarithmic function. A consequence

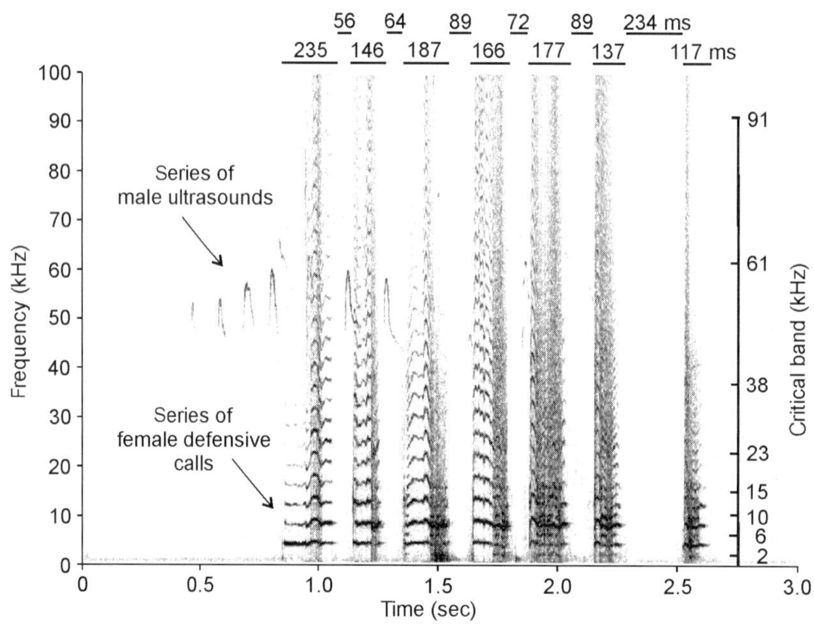

Fig. 2. Spectrograms of vocalizations of a male and a female mouse during sexual interaction. The sexually-interested male emits a series of ultrasounds when approaching the female. The non-receptive female emits a series of defensive calls while trying to keep the male off. The male's ultrasounds are pure tones with frequency modulations; the female's defensive calls consist of many harmonics and broadband noise. The durations of the defensive calls and of the inter-call intervals are indicated on top of the figure. The right-side y-axis scale shows the border frequencies of critical bands of the mouse as they are established to resolve the first two or three harmonics of the defensive calls and the male ultrasounds. Higher harmonics and the noisy parts of the defensive calls cannot be resolved. The spectrograms were kindly provided by Simone Gaub. The critical band scale is based on Ehret (1976).

is that mammals with a comparably short cochlea (about 6.5 mm length) and high-frequency hearing, such as the mouse, have a poor frequency resolution, i.e., large CBs of several kHz (compare Fig. 2), while humans with a long cochlea (about 32 mm) and low-frequency hearing have a high-frequency resolution, i.e., small CBs of less than 100 Hz up to several 100 Hz. Intensity-independent spectral filtering and integration is reached in neurons of the inferior colliculus (Ehret and Merzenich, 1985, 1988; Egorova and Ehret, 2008), which is a prerequisite for the experienced perceptual constancy of sounds in the spectral domain (e.g. timbre) over a wide range of intensities.

II.D. Pitch perception

Every harmonic in a complex sound, such as a vowel in human speech or a complex animal vocalization, leads, if heard in isolation, to a pitch percept according to its frequency. The harmonic complex, however, also gives rise to a global pitch percept, which corresponds to the fundamental frequency of the complex, even if it is physically absent (Terhardt, 1974a, 1978; Moore, 1989). Further, when a tone is sinusoidally amplitude-modulated with a frequency between about 50 Hz and several hundred Hz, a pitch (periodicity pitch) is heard, often corresponding to the amplitude modulation frequency (Langner, 1992). The perception of these forms of pitches may be a general property of hearing not only of humans, but also of other mammals (Heffner and Whitfield, 1976; Tomlinson and Schwarz, 1988; Deutscher et al., 2006) even if their main frequency range of hearing is in the high ultrasounds (Preisler and Schmidt, 1995).

Since pitch perception requires either harmonically-structured or continuously amplitude-modulated sounds, as found in many animal vocalizations, it can be useful for the discrimination of such sounds, and of animal sounds, with a rather regular structure in the frequency and/or time domain from other sounds with non-harmonically-related frequency components, often occurring as noise in the environment.

Pitch perception relies on the temporally precise coding of sound frequencies and amplitude modulations in the auditory nerve up to the inferior colliculus, from where it may be represented by a spike-rate code in maps or special areas in the auditory cortex (Schulze et al., 2002; Bendor and Wang, 2005; Kurt et al., 2008).

II.E. Perception of rhythms

When communication sounds are vocalized in series, the single elements of the series and the silent intervals between them may be rather regular in duration and, thus, create the percept of rhythm. Rhythms are heard if the intervals between short elements (clicks, tone or noise bursts) are of about 100–2,000 ms length, which equals a repetition rate of 10–0.5 Hz. If the silence intervals are longer than about 2,000 ms, every element is heard as a single event, if the intervals are shorter than about 100 ms, the rhythm percept changes to roughness, and if the intervals are shorter than about 20 ms, the percept is a pitch (Miller and Taylor, 1948; Besser, 1967; Terhardt, 1974b; Roederer, 1975; Krumbholz et al., 2000; Zanto et al., 2006). Thus, there is a continuum from the perception of single sounds to series of sounds in a rhythm, to roughness of a sound, which can be regarded as a percept of a temporally unresolved rhythm, to the pitch percept. The series of defensive calls of a non-receptive female mouse shown in Fig. 2 has inter-call intervals of mostly less than 100 ms, suggesting that this series produces some roughness percept. If one calculates the repetition rate of the calls, however, by adding the call durations and the durations of the following inter-call intervals, then an average repetition rate of about 3.6 Hz is the result, suggesting that this series (Fig. 2) is perceived as a rhythm.

Rhythm is encoded in the centers of the auditory system by the rhythmic occurrence of discharge peaks in response to sound elements. Although rhythm perception is possible without auditory cortex (Deutscher et al., 2006), neurons in the auditory cortex detect frequency and intensity changes in the sound elements and the interval lengths between the elements of a rhythmic series (Weinberger and McKenna, 1988; Ulanovsky et al., 2003) by increases of discharge rates so that changes in rhythmic vocalization series should be easily perceptible.

III. From perceptual abilities to the perception of biological meaning

III.A. General perceptual borders and categories in the time domain

In the above-mentioned perceptual shift from rhythm to roughness to pitch, a border is near a 50 Hz repetition rate or a 20 ms silence interval between sound elements. This 20 ms border in the perception of a temporal pattern seems to exist as a general time-domain border for mammals. It separates, for example, as the shortest phonetic border on the voice-onset-time (VOT) continuum, the perception of the stop consonants /ba/ and /pa/ in human speech (perception of /ba/ at VOT <25 ms, /pa/ at VOT >25 ms; Pisoni and Lazarus, 1974). The same border has been found for /ba/−/pa/ discrimination by VOT in the chinchilla (Kuhl and Miller, 1978), for the perception of temporal order (Hirsh, 1959) or gaps in sounds (Stevens and Klatt, 1974; Penner, 1975) by humans, for the categorization of mouse pup ultrasounds into biologically relevant and irrelevant ones by their mothers (Ehret, 1992), and for separating pursuable objects (potential prey) from non-pursuable objects in the evaluation of echo delays in the auditory cortex of moustached bats (Suga and O'Neill, 1979; Suga et al., 1983). In addition, the spectral integration of frequency components for the perception of series of auditory objects in a single stream by humans or mice is possible only if the frequency components start and end together within about 20–30 ms (Pisoni, 1977; Darwin and Sutherland, 1984; Geissler and Ehret, 2002).

There are two further perceptual borders in the time domain which may be general to mammals. One border is near the 100 ms inter-sound interval. In humans, a series of tones of alternating sound frequency are heard as a single stream of tones with alternating frequency if the inter-tone intervals are longer than about 100 ms. At shorter inter-tone intervals, the tones are grouped in two streams, one stream consisting of the tones of higher frequency and the other of tones of the lower frequency (Anstis and Saida, 1985). Similarly, the time window for integrating sounds picked up by one or the other ear into one single perceptual stream is close to 100 ms (Culling and Summerfield, 1998). The other boundary is near 400 ms inter-sound interval. Loudness summation of click sounds across silent intervals occurs only if the intervals are shorter than about 400 ms (Zwislocki et al., 1962; Scharf, 1978), and masking of the perception of a sound burst by a previous one loses its effectiveness if the interval between both is longer than about 400 ms (Zwicker and Feldtkeller, 1967). These perceptual borders near 100 ms and 400 ms interval durations between sound bursts are also found in the perception of a series of mouse pup wriggling calls by their mothers (Gaub and Ehret, 2005). Both perceptual borders (100 ms, 400 ms) become evident in event-related potentials (Budd and Michie, 1994; Yabe et al., 1997) measured in humans and in neuronal responses recorded from animals (Calford and Semple, 1995; Finlayson, 1999; Fishman et al., 2001).

The general perceptual borders in the time domain near 20 ms, 100 ms and 400 ms seem to represent inherent constants of the mammalian auditory system which can be, and actually are, exploited for categorizing communication sounds (vocalizations). In ethological terms, we may speak of an "inborn releasing mechanism," which is equivalent to a filter for separating communication sounds into different classes of biological significance or biological meaning.

III.B. General perceptual borders and categories in the spectral domain

Above we introduced critical bands (CBs) or equivalent psychoacoustical measures as important "tools" of the auditory system for the analysis of sounds in the spectral domain. One can go a step further and state that whatever we perceive in the spectral domain is the direct consequence of CB processing of sounds (Scharf, 1970; Moore, 1989). Since most animal vocalizations, including vowels of human speech, are spectrally rich, CB processing is the basis for perception in the spectral domain. In order to perceive a spectrally-complex type of sound against background noise and independently of the individual voice characteristics of the sender, the spectral peaks (main harmonics and/or formants) have to be resolved by CB filters. In this way, the fundamental or first formant and other harmonics/formants can be detected, so that the sound type may often be derived from the frequency ratios of the first three spectral peaks. This happens in the analysis and categorization of vowels of human speech independent of the pitch and intensity of the speaker's voice (Flanagan, 1972; Kuhl, 1979; Grieser and Kuhl, 1989), and in the perception of wriggling calls of mouse pups by their mothers (Ehret and Riecke, 2002). Also, ultrasounds of mouse pups are perceived categorically as sound energy in one single CB with a center frequency above about 35 kHz (Ehret and Haack, 1982). Fig. 2 shows that ultrasounds of a male mouse, emitted immediately before trying to mount a female, are analyzed in only one single CB with a bandwidth of about 23 kHz. Each of the first three harmonics of defensive calls of a non-receptive female mouse can also be analyzed in one CB, while higher harmonics and the noisy parts of the calls cannot be resolved (Fig. 2). Similarly, spectrally rich monkey calls may be perceived after a CB analysis in their auditory system (Fishman et al., 2000). As for the perceptual borders in the time domain, the CB mechanism serves the general function of an "inborn releasing mechanism" for the separation of biological meaning (semiotic value) in the spectral domain.

III.C. The perception of three basic biological meanings

Morton (1977) developed, and August and Anderson (1987) complemented, motivation–structural rules describing common acoustic properties in animal vocalizations with their relationship to the sender's emotional/motivational state at the moment of vocalizing (derived from 78 mammalian species). In summary, they divided the vocalizations into three groups expressing different emotions/motivations: (1) high-frequency sounds of rather tonal character express fear; (2) harsh, noisy, loud, broadband or low-frequency sounds express aggressiveness or hostility; (3) soft, rhythmic, low-frequency sounds express friendliness. Assuming a co-evolution of vocalizing and perception, the three categories of vocalizations suggest the perception of three categories of biological meaning. On the basis of the acoustic properties of vocalizations, hearing abilities, and the behavioral responses to species-specific calls of house mice and other mammals, Ehret (2006) proposed six rules for perception of communication sounds. Rule four deals with the perception of meaning (biological significance), which is expressed by the receiver's innate response behavior: (1) the relatively high-frequency tonal sounds expressing appeasing or fearful emotions/motivations are attractive and thus are perceived as "attraction;" (2) the soft, low-frequency rhythmic sounds expressing "friendly" emotions are often accompanied by peaceful interactions of animals in a group and thus are perceived as "cohesion" (staying together in a common context); (3) the harsh, loud, noisy and broadband sounds expressing aggressiveness may cause avoidance behavior, thus indicating "aversion."

The theory is (Ehret, 2006) that acoustic properties in the vocalizations determine the three biological meanings which, in turn, release an adequate response behavior in the receivers after being integrated with information from other sensory modalities and processed in the context of the perceiver's own state of arousal and emotion/motivation. The above-mentioned general perceptual mechanisms in the temporal and spectral domain are very well-suited for sorting out the three basic biological meanings of communication sounds. Non-resolved frequencies and temporal variations in a loud sound over a broad frequency range of the audiogram (see defensive calls, Fig. 2) lead to the percept of roughness and noisiness being associated with aversion. Well-resolved frequency components in sounds with major energy at or above the best frequency range of hearing (see male ultrasounds, Fig. 2) lead to the percept of a high pitch being associated with attraction. Series of rather soft low-frequency sounds lead to the percept of a rhythm being associated with cohesion.

In this way, all types of sounds varying in many acoustic parameters are perceptually classified for basic behavioral consequences, as they are approaching, avoiding, or staying with the sound source, which have to be compatible, however, with other sensory information and the perceiver's state of arousal and emotion/motivation. It is important to note that the described predetermined perception of the semiotic value/biological meaning does not build on JNDs in sounds and, therefore, is not working at the perceptual limits of the auditory system. The perception of the three basic biological meanings of vocalizations signaling basic affective states, and also other meanings, relies on just meaningful differences, JMDs, allowing even acoustically-distinct call types, which can easily be discriminated via JNDs, to be classified in the same category of meaning/semiotic category (Ehret, 2006; rule two). A characteristic feature of classifying sounds for the perception of meaning is to ignore perceivable variability in some acoustic parameters of sounds.

IV. Recognition of vocal signals and conclusions

We used the term "perception" of basic meanings/semiotic values, although the neural processing behind this perception is cognitive and emotive in the way that auditory information is integrated with the emotional state, motivational tendencies and memory traces in the brain of the receiver, before a certain response can occur. This integration, which leads to a modulation of the responsiveness, may also cause modulation of the semiotic value/meaning of a vocalization. It becomes very obvious when the responses of animals to vocalizations of others change with changing hormonal states, for example during the reproductive cycle or estrous cycle. The following experiment may illustrate such a change: the perception of wriggling calls of mouse pups was modulated during the estrous cycle of pup naïve females (Ehret and Schmid, 2009). During diestrous and proestrous females discriminated the acoustic quality of the sounds, during estrous they were highly responsive and did not discriminate, during metestrous their responsiveness was low and they did not discriminate. This example shows that only on a certain emotional/motivational background acoustic properties of sounds become relevant in perception. Considering this, it is conceivable that the non-receptive female (Fig. 2) was not attracted by the ultrasounds of the male, and defended herself against the mating intentions of the male, while the sexually highly-motivated male was not repelled by the defensive calls of the female.

Differing from such situations of modulation of the biological meaning of vocalizations by changes in the state of the organism, learning certain acoustic properties of vocalizations may become important for adjusting behavioral responses. This type of vocal recognition opens and expands the perception of basic biological meanings to the recognition of individuals and/or kin, sex and age in many species, especially those living in structured groups, such as vervet monkeys (Cheney and Seyfarth, 1982), rhesus monkeys (Rendall et al., 1996), spotted hyenas (Holekamp et al., 1999), elephants (McComb et al., 2000) and pigs (Illmann et al., 2002). In these situations of learning to associate sound properties with a certain biological meaning, which is derived from the consequences of interaction of the vocalizing with the perceiving animal, JNDs may become important for assessing the caller's behavioral state and identity, and discriminate these from the identities of other individuals by perceiving small acoustic differences in the voices. By learning to pay attention to subtle differences in vocalizations, the perceptual abilities of the species' auditory systems may fully be exploited.

References

Anstis, S., Saida, S., 1985. Adaptation to auditory streaming of frequency-modulated tones. J. Exp. Psychol. Hum. Percept. Perform. 11, 257–271.

August, P.V., Anderson, J.G.T., 1987. Mammal sounds and motivation-structural rules: a test to the hypothesis. J. Mammal. 68, 1–9.

Bendor, D., Wang, X., 2005. The neuronal representation of pitch in primate auditory cortex. Nature 436, 1161–1165.

Besser, G.M., 1967. Some physiological characteristics of auditory flutter fusion in man. Nature 214, 17–19.

Brand, A., Urban, A., Grothe, B., 2000. Duration tuning in the mouse auditory midbrain. J. Neurophysiol. 84, 1790–1799.

Brown, P.E., 1976. Vocal communication in the pallid bat, *Antrozous pallidus*. Z. Tierpsychol. 41, 34–54.

Bruns, V., 1976. Peripheral auditory tuning for fine frequency analysis by the Cf-FM bat, *Rhinolophus ferrumequinum*. II. Frequency mapping in the cochlea. J. Comp. Physiol. 106, 87–97.

Budd, T.W., Michie, P.T., 1994. Facilitation of the N1 peak of the auditory ERP at short stimulus intervals. NeuroReport 5, 2513–2516.

Calford, M.B., Semple, M.N., 1995. Monaural inhibition in cat auditory cortex. J. Neurophysiol. 73, 1876–1891.

Cheney, D.L., Seyfarth, R.M., 1982. Recognition of individuals within and between groups of free-ranging vervet monkeys. Am. Zool. 22, 519–529.

Covey, E., Casseday, J.H., 1999. Timing in the auditory system of the bat. Annu. Rev. Physiol. 61, 457–476.

Culling, J.F., Summerfield, Q., 1998. Measurements of the binaural temporal window using a detection task. J. Acoust. Soc. Am. 103, 3540–3553.

Darwin, C.J., Sutherland, N.S., 1984. Grouping frequency components of vowels: when is a harmonic not a harmonic. Quart. J. Exp. Psychol. 36A, 193–208.

Deutscher, A., Kurt, S., Scheich, H., Schulze, H., 2006. Cortical and subcortical sides of auditory rhythms and pitches. NeuroReport 17, 853–856.

Egorova, M., Ehret, G., 2008. Tonotopy and inhibition in the midbrain inferior colliculus shape spectral resolution of sounds in neural critical bands. Eur. J. Neurosci. 28, 675–692.

Ehret, G., 1974. Age-dependent hearing loss in normal hearing mice. Naturwissenschaften 61, 506–507.

Ehret, G., 1976. Critical bands and filter characteristics in the ear of the house mouse (*Mus musculus*). Biol. Cybernetics, 24, 35–42.

Ehret, G., 1977. Comparative psychoacoustics: perspectives of peripheral sound analysis in mammals. Naturwissenschaften 64, 461–470.

Ehret, G., 1983. Development of hearing and response behavior to sound stimuli: behavioural studies. In: Romand, R. (Ed.), Development of Auditory and Vestibular Systems. Academic Press, New York, NY, pp. 211–237.

Ehret, G., 1988. Auditory development: psychophysical and behavioral acpects. In: Meisami, E., Timiras, P.S. (Eds.), Handbook of Human Growth and Developmental Biology. CRC Press, Boca Raton, FL, pp. 141–154.

Ehret, G., 1989. Hearing in the mouse. In: Dooling, R.J., Hulse, S.H. (Eds.), The Comparative Psychology of Audition: Perceiving Complex Sounds. Lawrence Erlbaum, Hillsdale, NJ, pp. 3–32.

Ehret, G., 1992. Categorical perception of mouse-pup ultrasounds in the temporal domain. Anim. Behav. 43, 409–416.

Ehret, G., 1995. Auditory frequency resolution in mammals: from neuronal representation to perception. In: Manley, G.A., Klump, G.M., Köppl, C., Fastl, H.,

Oeckinghaus, H. (Eds.), Advances in Hearing Research. World Scientific, Singapore, pp. 387–397.
Ehret, G., 2006. Common rules of communication sound perception. In: Kanwal, J., Ehret, G. (Eds.), Behavior and Neurodynamics for Auditory Communication. Cambridge University Press, Cambridge, UK, pp. 85–114.
Ehret, G., Haack, B., 1982. Ultrasound recognition in house mice: Key-stimulus configuration and recognition mechanisms. J. Comp. Physiol. A, 148, 245–251.
Ehret, G., Merzenich, M.M., 1985. Auditory midbrain responses parallel spectral integration phenomena. Science 227, 1245–1247.
Ehret, G., Merzenich, M.M., 1988. Complex sound analysis (frequency resolution, filtering and spectral integration) by single units of the inferior colliculus of the cat. Brain Res. Rev. 13, 139–163.
Ehret, G., Riecke, S., 2002. Mice and humans perceive multiharmonic communication sounds in the same way. Proc. Nat. Acad. Sci. USA, 99, 479–482.
Ehret, G., Schmid, C., 2009. Reproductive cycle-dependent plasticity of perception of acoustic meaning in mice. Physiol. Behav. 96, 428–433.
Fay, R.R., 1988. Hearing in Vertebrates: A Psychophysics Databook. Hill-Fay Associates, Winnetka, IL.
Finlayson, P.G., 1999. Post-stimulatory suppression, facilitation and tuning for delays shape responses of inferior colliculus neurons to sequential pure tones. Hear. Res. 131, 177–194.
Fishman, Y.I., Reser, D.H., Arezzo, J.C., Steinschneider, M., 2000. Complex tone processing in primary auditory cortex of the awake monkey. II. Pitch versus critical band representation. J. Acoust. Soc. Am. 108, 247–262.
Fishman, Y.I., Reser, D.H., Arezzo, J.C., Steinschneider, M., 2001. Neural correlates of auditory stream segregation in primary auditory cortex of the awake monkey. Hear. Res. 151, 167–187.
Flanagan, J.L., 1972. Speech Analysis, Synthesis, and Perception. Springer-Verlag, Berlin, Germany.
Gaub, S., Ehret, G., 2005. Grouping in auditory temporal perception and vocal production is mutually adapted: the case of wriggling calls of mice. J. Comp. Physiol. A. 191, 1131–1135.
Geissler, D.B., Ehret, G., 2002. Time-critical integration of formants for perception of communication calls in mice. Proc. Natl. Acad. Sci. USA 99, 9021–9025.
Gould, E., 1971. Studies of maternal-infant communication and development of vocalization in the bats *Myotis* and *Eptesicus*. Commun. Behav. Biol. 5, 262–313.
Greenwood, D.D., 1961. Auditory masking and the critical band. J. Acoust. Soc. Am. 33, 484–502.
Greenwood, D.D., 1990. A cochlear frequency-position function for several species – 29 years later. J. Acoust. Soc. Am. 87, 2592–2605.
Grieser, D.A., Kuhl, P.K., 1989. Categorization of speech by infants: support for speech-sound prototypes. Dev. Psychol. 25, 577–588.
Haack, B., Markl, H., Ehret, G., 1983. Sound communication between parents and offspring. In: Willott, J.F. (Ed.), The Auditory Psychobiology of the Mouse. Charles C. Thomas, Springfield, IL, pp. 57–97.
Haskins, R., 1977. Effect of kitten vocalizations on maternal behavior. J. Comp. Physiol. Psychol. 91, 830–838.
Heffner, H., Whitfield, I.C., 1976. Perception of the missing fundamental by cats. J. Acoust. Soc. Am. 59, 915–919.
Hirsh, I.J., 1959. Auditory perception of temporal order. J. Acoust. Soc. Am. 31, 759–767.
Holekamp, K.E., Boydston, E.E., Szykman, M., Graham, I., Nutt, K.J., Birch, S., Piskiel, A., Singh, M., 1999. Vocal recognition in the spotted hyena and its possible implications regarding the evolution of intelligence. Anim. Behav. 58, 383–395.
Illmann, G., Schrader, L., Špinka, M., Šustr, P., 2002. Acoustical mother–offspring recognition in pigs (*Sus scrofa domestica*). Behaviour 139, 487–505.
Klink, K.B., Klump, G.M., 2004. Duration discrimination in the mouse (*Mus musculus*). J. Comp. Physiol. A, 190, 1039–1046.
Krumbholz, K., Patterson, R.D., Pressnitzer, D., 2000. The lower limit of pitch as determined by rate discrimination. J. Acoust. Soc. Am. 108, 1170–1180.
Kuhl, P.K., 1979. Models and mechanisms in speech perception. Species comparisons provide further contributions. Brain Behav. Evol. 16, 374–408.
Kuhl, P.K., Miller, J.D., 1978. Speech perception by the chinchilla: identification function for synthetic VOT stimuli. J. Acoust. Soc. Am. 63, 905–917.
Kurt, S., Deutscher, A., Crook, J.M., Ohl, F.W., Budinger, E., Moeller, C.K., Scheich, H., Schulze, H., 2008. Auditory cortical contrast enhancing by global winner-take-all inhibitory interactions. PLoS ONE 3, e1735.
Langner, G., 1992. Periodicity coding in the auditory system. Hear. Res. 60, 115–142.
Maiwald, D., 1967. Ein Funktionsschema des Gehörs zur Beschreibung der Erkennbarkeit kleiner Frequenz- und Amplitudenänderungen. Acustica 18, 81–92.
McComb, K., Moss, C., Sayialel, S., Baker, L., 2000. Unusually extensive networks of vocal recognition in African elephants. Anim. Behav. 59, 1103–1109.
Miller, G., Taylor, W., 1948. The perception of repeated bursts of noise. J. Acoust. Soc. Am. 20, 171–182.
Moore, B.C.J., 1989. An Introduction to the Psychology of Hearing, 3rd edn. Academic Press, London, UK.
Morton, E.S., 1977. On the occurrence and significance of motivation-structural rules in some bird and mammal sounds. Amer. Nat. 111, 855–869.
Penner, M.J., 1975. Persistence and integration: two consequences of a sliding integrator. Percept. Psychophys. 18, 114–120.
Pisoni, D.B., 1977. Identification and discrimination of the relative onset time of two component tones: Implications for voicing perception in stops. J. Acoust. Soc. Am. 61, 1352–1361.
Pisoni, D.B., Lazarus, J.H., 1974. Categorical and noncategorical modes of speech perception along the voicing continuum. J. Acoust. Soc. Am. 55, 328–333.
Preisler, A., Schmidt, S., 1995. Virtual pitch formation in the ultrasonic range. Naturwissenschaften 82, 45–47.
Rendall, D., Rodman, P.S., Emond, R.E., 1996. Vocal recognition of individuals and kin in free-ranging rhesus monkeys. Anim. Behav. 51, 1007–1015.

Roederer, J.G., 1975. Introduction to the Physics and Psychophysics of Music. Springer, New York, NY.

Scharf, B., 1970. Critical band. In: Tobias, J.V. (Ed.), Foundations of Modern Auditory Theory, Vol. 1. Academic Press, New York, NY, pp. 159–202.

Scharf, B., 1978. Loudness. In: Carterette, E.C., Friedman, M.P. (Eds.), Handbook of Perception, Vol. IV. Hearing. Academic Press, New York, NY, pp. 187–242.

Schulze, H., Hess, A., Ohl, F.W., Scheich, H., 2002. Superposition of horseshoe-like periodicity and linear tonotopic maps in auditory cortex of the Mongolian gerbil. Eur. J. Neurosci. 15, 1077–1084.

Shalter, M.D., Fentress, J.C., Young, G.W., 1977. Determinants of response of wolf pups to auditory signals. Behaviour 60, 98–114.

Stern, D.N., Spieker, S., MacKain, K., 1982. Intonation contours as signals in maternal speech to prelinguistic infants. Dev. Psychol. 18, 727–735.

Stevens, K.N., Klatt, D.H., 1974. Role of formant transitions in the voiced-voiceless distinction for stops. J. Acoust. Soc. Am. 55, 653–659.

Suga, N., Jen, P.H.S., 1977. Further studies on the peripheral auditory system of "CF-FM" bats specialized for fine frequency analysis of Doppler-shifted echoes. J. Exp. Biol. 69, 207–232.

Suga, N., O'Neill, W.E., 1979. Neural axis representing target range in the auditory cortex of the mustache bat. Science 206, 351–353.

Suga, N., O'Neill, W.E., Kujirai, K., Manabe, T., 1983. Specialization of "combination-sensitive" neurons for processing of complex biosonar signals in the auditory cortex of the mustached bat. J. Neurophysiol. 49, 1573–1626.

Terhardt, E., 1974a. Pitch, consonance and harmony. J. Acoust. Soc. Am. 55, 1061–1069.

Terhardt, E., 1974b. On the perception of periodic sound fluctuations (roughness). Acustica 30, 201–213.

Terhardt, E., 1978. Psychoacoustic evaluation of musical sounds. Percept. Psychophys. 23, 483–492.

Tomlinson, R.W., Schwarz, D.W., 1988. Perception of the missing fundamental in nonhuman primates. J. Acoust. Soc. Am. 84, 560–565.

Ulanovsky, N., Las, L., Nelken, I., 2003. Processing of low-probability sounds by cortical neurons. Nature Neurosci. 6, 391–398.

Weinberger, N.M., McKenna, T.M., 1988. Sensitivity of single neurons in auditory cortex to contour: toward a neurophysiology of music perception. Music Percept. 5, 355–390.

Yabe, H., Tervaniemi, M., Näätänen, R., 1997. Temporal window of integration revealed by MMN to sound ommission. NeuroReport 8, 1971–1974.

Zanto, T.P., Snyder, J.S., Large, E.W., 2006. Neural correlates of rhythmic expectancy. Adv. Cog. Psychol. 2, 221–231.

Zwicker, E., Feldtkeller, R., 1967. Das Ohr als Nachrichtenempfänger. Hirzel, Stuttgart, Germany.

Zwislocki, J., Hellmann, R.P., Verillo, R.T., 1962. Threshold of audibility for short pulses. J. Acoust. Soc. Am. 34, 1648–1652.

CHAPTER 4.4

Cortical processing of vocal sounds in primates

Christopher I. Petkov[1,2*], Christoph Kayser[2] and Nikos K. Logothetis[2,3]

[1]*Institute of Neuroscience, Newcastle University, Framlington Place, Newcastle upon Tyne, UK*
[2]*Max-Planck Institute for Biological Cybernetics, Tübingen, Germany*
[3]*Imaging Science and Biomedical Engineering (ISBE), Stopford Building, University of Manchester, Manchester, UK*

Abstract: The recent work on speech and vocal sound processing by the human brain finds itself at a crossroads with the studies in non-human primates on the neurobiological basis of vocal communication. Speech is a recent evolutionary adaptation, so direct animal homologs of the neural systems supporting speech perception are not expected. However, vocal expressions are richly informative for many social animals. Thus, the interest in how the human brain is processing the speaker's identity and affective aspects of the human voice, including the stimulus-bound aspects of speech, may be approached from an evolutionary perspective. From the other side, comparative biologists have started to close the gap between animal and human data by using the same noninvasive imaging techniques as those applied to the human brain, to study the brains of animals. Relying on the same techniques facilitates cross-species comparisons, and provides links to invasive studies of the brain processes at the neuronal level. In this chapter, we consider how the brains of primates analyze the features in vocal sounds, focusing in particular on the correspondence between the auditory cortex processes in the brains of monkeys, apes and man.

Keywords: human; monkey; vocalizations; communication; imaging; electrophysiology; comparative; mammals; auditory cortex; auditory processing

I. Introduction

When the topic of vocal communication is discussed, we naturally think of speech and the linguistic aspects of spoken language, because of our familiarity with this means of communication. Yet it is easy to forget, regardless of its semantics, how a single spoken word is a vocal expression that not only identifies many aspects of the speaker, but it may predict what they are about to say. Especially in threatening situations, the first vocal expression can have a more immediate behavioral impact than the linguistic content, which takes a longer time to express. Yet we are only beginning to understand how the brain evaluates the information contained in vocal sounds, and whether the brains of other animals have related processes.

Humans are not alone in the ability to extract acoustic and semantic elements from vocal sounds. Comparative biologists have been aware of this ability for some time from behavioral observations of animals during vocal production, reception and social interaction, both in field studies and laboratory settings. The behavioral observations have established a number of evolutionary relationships that are now guiding studies on the neurobiological basis of vocal communication. Although there has recently been

*Corresponding author. E-mail: chris.petkov@ncl.ac.uk

ISBN 978-0-12-374593-4
DOI: 10.1016/B978-0-12-374593-4.00014-0

considerable progress in brain science, comparative studies on human and animal communication have faced some impediments. In the first part of the chapter we describe some practical and conceptual roadblocks, and how they have been bypassed. We then review the cortical processing of vocal sounds, focusing our review on recent findings from studies of humans, apes and monkeys on the neurophysiological basis of vocal reception.

II. Different techniques, different species

Part of the ongoing problem with the comparative approach has been that many of the cellular techniques used in animals cannot easily be related to data on human brain function. By comparison, behavioral or anatomical studies often use the same approaches or paradigms to evaluate different species (Fay, 1988; Brodmann and Garey, 1999). For example, pioneering comparative work of Brodmann revealed anatomical relationships between the brains of human and non-human primates (for a recent review see Van Essen, 2005). However, for the study of the functional aspect of the brain, the comparative approach has often had to rely on incompatible methods.

In humans, noninvasive imaging techniques are now in standard use, such as functional magnetic-resonance imaging (fMRI), positron emission tomography (PET), or electro- and magneto-encephalography (EEG and MEG). At this level, the imaging reveals neuronal activity at a mesoscopic, rather than microscopic, scale. Yet the measurements done by each technique – such as hemodynamic in fMRI, metabolic in PET, electrical in EEG and magnetic in MEG – require careful comparisons before we can know in what conditions it is possible to interpret the results from one approach in relation to another (for a recent review see Logothetis, 2008). Animal studies, on the other hand, often rely on the responses of small groups of cells, or individual neurons that are recorded with microelectrodes. There are now a number of examples documenting dissociations between the fMRI imaging signal and at least some types of neuronal responses (Logothetis et al., 2001; Maier et al., 2008; Rauch et al., 2008). Thus, caution is needed when comparing the data obtained from invasive studies in animals and the mesoscopic imaging techniques in humans.

These difficulties can be eliminated to a large extent by using the same approach in evaluating the brain function of different species. The functional imaging techniques presently used in humans (Fox et al., 1986; Ogawa et al., 1992) have now been well-developed to image the brains of such animals as rodents (Ogawa et al., 1990), apes (Taglialatela et al., 2008, 2009), monkeys (Disbrow et al., 1999; Logothetis et al., 1999) and birds (Van der Linden et al., 2002; Van Meir et al., 2005). Direct cross-species comparisons can now clarify functional: (1) homology, i.e., evolutionarily conserved capabilities obtained from a common ancestor; (2) analogy, i.e., independent specializations that have solved a similar problem; (3) a loss of function; or (4) some combination of situations mentioned in points (1) to (3).

While the animal imaging of the visual and somatosensory systems was developed first (for instance, in monkeys see Disbrow et al., 1999; Logothetis et al., 1999), the animal auditory imaging has been gaining ground (Poremba et al., 2004; Kayser et al., 2005; Van Meir et al., 2005; Petkov et al., 2006; Taglialatela et al., 2009). Since these developments are recent, a question arises as to what extent the animal imaging results would be comparable to the copious imaging data available from humans? It is important to emphasize, at this point, that the animal imaging studies are not a surrogate for the invasive neuroanatomical and neurophysiological methods. The imaging provides a link to the rich invasively-obtained neuroanatomical and neurophysiological data from animals, and can initiate or guide additional invasive approaches. Considering the ongoing need of medical science for animal species that can model human brain function at the neuronal level, no current technique can both bridge the gap to the human imaging work, and adequately describe the neuronal processes at the same time.

III. Comparative neurobiology based on conserved behaviors

Some impediments for human–non-human animal comparisons are conceptual rather than experimental. For instance, human language is recognized as a unique adaptation that appeared recently in evolutionary history. Thus, many functional subsystems that support human verbal communication may not be evident even in our closest primate relatives, the great apes. The emphasis on the unique aspects of human communication (such as the semantic and higher-order grammatical aspects of speech and language) has led to a considerable debate about the origin(s) of language, but it has generated few empirical advances.

Tangible links between human and animal communication were needed, and these are now available from several sources. For example, strong counterarguments against the uniqueness of the many aspects of human language have been made, which serve to rekindle interest in the comparative approach. Hauser, Chomsky and Fitch have argued that it is mainly the higher-order generative aspects of grammar that may be unique – such as our ability to generate nearly infinite combinations of meaningful expressions from a finite set of rules (Hauser et al., 2002). They have termed these capabilities "the language faculty in the narrow sense." However, other aspects of human language – the "language faculty in the broad sense" – have been evolutionarily selected for, evidence for which comes from observing how animals communicate.

Assisting in rekindling the interest in comparative methods has been the sheer explosion of human imaging studies (especially with fMRI) in the last three decades. This has bolstered cognitive science, and opened up venues of human research that can be more closely related to the animal work. A number of groups studying speech perception have recently made an effort to relate to the animal literature, especially to the guiding principles of brain function that were developed from direct studies of the animal nervous system (see Fig. 1), and to update models developed on the neuroanatomical basis of speech (Scott and Johnsrude, 2003; Scott, 2005; Hickok and Poeppel,

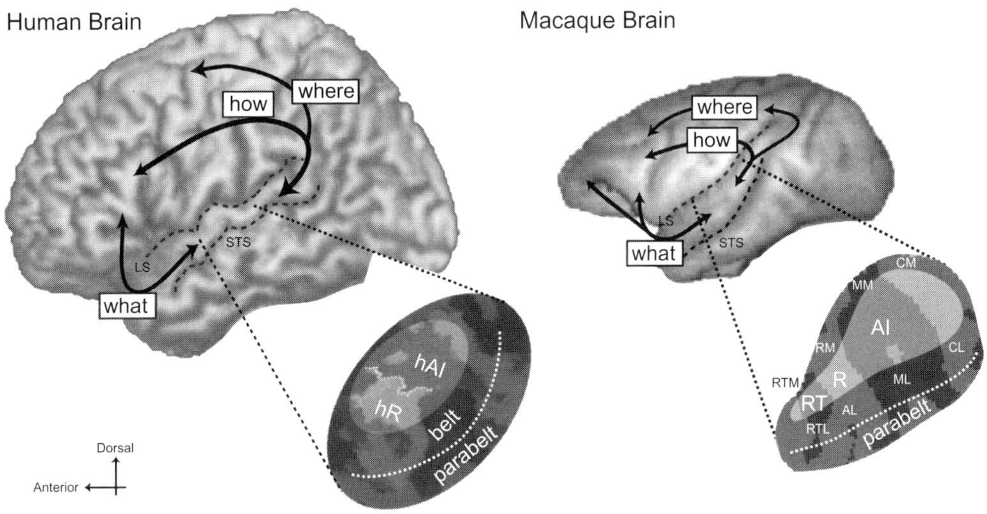

Fig. 1. Pathways supporting vocal communication in humans and monkeys. The pathways for parallel brain processing are labeled with black arrows. These are termed "what," "where" and "how," reflecting their respective contribution in evaluating what a sensory object is, where it is located in sensory space and how it was generated. Although the figure suggests a close correspondence between the human and monkey brain, it remains unclear how many pathways there are, their precise roles and the extent of the homologies between the species. The inserts illustrate the functional organization of the auditory cortex. Cortical processing begins with the primary auditory cortex (see the lighter region in the enlarged inserts, which includes field AI in monkeys and its presumed homolog hAI in humans). The colors of the insert reflect the direction of the tonotopic gradients of the auditory fields (dark blue: fields with high to low frequency preference toward the anterior direction; light green: fields with low to high frequency preference in the anterior direction). Data such as these can be used to functionally parcellate the auditory cortex and delineate borders between fields with mirror reversed tonotopic gradients. More detailed processing of sound follows in the hierarchically higher auditory "belt" and "parabelt" regions of auditory cortex. See the text for further details and references. This figure contains a rendered human brain image kindly contributed by J. Obleser and an example of the mapping of human auditory cortical fields contributed by E. Formisano. Abbreviations: AL: antero-lateral; CL: caudo-lateral; CM: caudo-medial; CPB: caudal parabelt; LS: lateral sulcus; ML: medio-lateral; MM: medio-medial; STS: superior temporal sulcus; RM: rostro-medial; RPB: rostral parabelt; RL: rostro-lateral; RT: rostro-temporal; RTL: rostro-temporal lateral; RTM: rostro-temporal medial (To see the full color version of this figure please refer to the color plate in the back of the book. Copies produced via our print on demand service do not contain color plates. If your copy does not have the color plate, please go to this website to view the figure in color http://www.elsevierdirect.com/companions/9780123745934).

2007). Others have been focusing on the acoustical or stimulus-bound aspects of sub-lexical speech processing (Dehaene-Lambertz et al., 2005; Liebenthal et al., 2005; Rimol et al., 2005; Obleser et al., 2006, 2007), or leaving linguistics entirely aside and describing how the auditory system processes the information contained in the human voice (Belin et al., 2000; Belin and Zatorre, 2003; von Kriegstein et al., 2003).

The human paralinguistic processing of information in the voice (Belin et al., 2004) makes more immediate links to the results of animal studies, since the voice processing regions constitute the substrate that gave evolutionary rise to human speech (Belin, 2006; Petkov et al., 2009 in press). Thus, some cross-species correspondence in how the brain evaluates vocal sounds and voice identity might be expected and, as we describe below, is in fact seen in human and monkey imaging data. We now consider the general principles of vocal sound processing, and how the components of vocalizations are processed by the cortex of primates.

IV. Principles of cortical organization and vocal sound processing

Since vocal sounds contain many acoustical features, we might reasonably expect different features to be processed in the receiver by different regions and networks in the brain. A first step in this direction is to identify brain regions that are responding to and processing vocal sounds or their parameters. A number of vocalization-sensitive sites in the brains of primates have been identified. In fact, the central auditory system of the primate brain is much more extensive and widely distributed than was initially thought (Poremba et al., 2003). In order to organize the available results we consider the processing of vocal sounds in the context of two general principles of brain function (see Fig. 1): hierarchical and parallel processing.

IV.A. Hierarchical processing of vocal sounds

The idea of hierarchical processing is one of the guiding principles of brain function. The peripheral sensory organs transduce the physical parameters of environmental stimuli into neural signals (action potentials), and these signals propagate along subcortical regions of the central nervous system through a number of processing stages until they reach cortex. At each stage, different parameters of the sensory information are analyzed (for detail, see Syka, Chapter 4.1 in this volume).

The first stage of auditory processing in the neocortex is the primary auditory cortex. Within the primary auditory cortex (PAC), field AI is one of the better known regions, although at least one or two other "primary-like" fields also reside in the PAC of primates (see the lighter central region in the insert of Fig. 1). The PAC, also called the auditory "core," receives strong input from the ventral division of the medial geniculate body, the auditory thalamus (Kaas and Hackett, 1998). Following subcortical processing, the PAC conducts additional analysis of the spectral and temporal features of sounds, including the features of species-specific vocalizations (Funkenstein and Winter, 1973; Newman and Wollberg, 1973; Wang et al., 1995; Nagarajan et al., 2002). The output can then be fed forward to the next stage of the auditory cortical processing, the auditory "belt," which contains seven or eight discrete subfields in monkeys (Hackett et al., 1998).

Field AI has received the most attention, and there is a considerable amount of detailed information obtained from AI about the neuronal processing of sound in primates. These results originate mainly from imaging studies of humans and electrophysiological recordings in monkeys. AI neurons are highly selective for different spectral and temporal features of sounds, hence they will respond to vocalizations (Funkenstein and Winter, 1973; Newman and Wollberg, 1973; Wang et al., 1995; Nagarajan et al., 2002). However, the AI neuronal responses do not seem to be responding to monkey vocalizations differently than to other natural sound categories or sound features, noises, spectral ripples, etc. There are, however, clusters of AI neurons in the squirrel monkey which strongly respond to vocalizations and show some preference for species-specific vocalizations versus those of other mammalian species, such as cats (Wang and Kadia, 2001). Field AI seems to be well-suited to extract the spectral and temporal components of vocalizations, including the envelope of the sounds (Nagarajan et al., 2002), although, at least in cats, other natural sounds may be analyzed in the same way (Nelken et al., 1999).

Some authors have used human speech sounds (syllables) as stimuli for monkey AI field, in order to probe the general processing capabilities of this region in primates (Steinschneider et al., 1990). How this information is relayed or used by the subsequent processing stages in the monkey auditory cortex is not clear.

There is increasing selectivity for complex sounds along the auditory hierarchy away from field AI, which finds general agreement with observations from imaging studies of the human auditory cortex using different stimulation paradigms (Binder et al., 1994; Wessinger et al., 2001; Uppenkamp et al., 2006). For instance, more selective responses to some sound features and to vocalizations are seen at the next stage of auditory processing in the anatomically defined auditory "belt" (Hackett et al., 1998). Results of electrophysiological studies have shown that neurons in the lateral belt of the monkey primary auditory cortex respond less to single frequency "pure" tones, preferring instead sounds with a more complex frequency structure, such as bandpassed noise or other complex sounds as vocalizations, some of which can resemble bandpassed noise (Rauschecker et al., 1995; Rauschecker and Tian, 2004; Tian and Rauschecker, 2004). Although it has been known that primary sensory field AI has an organized topographical representation of preferred sound frequencies (also called tonotopy or cochleotopy), a similar approach of using tones at different frequencies to map the tonotopic organization didn't work for much of the auditory belt, leaving its functional organization unclear. It was first hypothesized and then shown for the lateral belt fields that by using a bandpassed noise centered at different frequencies (rather than pure tones), several fields in the lateral belt can be revealed with mirror-reversed topographical organization (Rauschecker et al., 1995).

The preference of the lateral belt for complex sounds in combination with observations that lesions to AI disrupted the responses of belt neurons (Rauschecker et al., 1997) were taken as evidence for the hierarchical organization of the primate auditory cortex. Comparison of neuronal responses to bandpassed noises and to pure tones revealed numerous subfields in the core and belt regions of the monkey auditory cortex. A recent macaque monkey fMRI study (Petkov et al., 2006), obtained evidence for 11 discrete subfields in the core and belt regions (3 within primary auditory cortex and 8 in the medial and lateral belt, see Fig. 1), consistent with one of the anatomical models that had been proposed for macaques (Kaas and Hackett, 2000). Although for humans there are also a number of anatomical models of discrete fields in the auditory cortex, the strongest evidence for separate auditory cortical fields comes from a high-resolution human functional imaging study that revealed two fields in the auditory core that have a mirror-reversed organization, including the border in between them (Formisano et al., 2003). Namely, these are hAI and hR, as presumed homologs of the monkey fields AI and R (see Fig. 1). Regionally, there are anatomical and functional data that the human auditory cortex, as the monkey cortex, contains core, belt and parabelt regions (Hackett et al., 2001; Wessinger et al., 2001; Sweet et al., 2005) However, homologies between the organization of monkey and human auditory cortices cannot be assumed, even at the level of the primary auditory cortex (see the blowout in Fig. 1) (Fullerton and Pandya, 2007). There is insufficient information available and agreement as to the number, specific function, location and orientation of the human auditory cortical subfields for specific comparisons to be made with the data from other species.

The auditory belt in macaque monkeys also responds strongly to frequency modulated (FM) tones (Tian and Rauschecker, 2004). These sounds relate to the elements of vocal sounds because FM components are present in formant transitions and frequency contours in many vocal sounds. Japanese macaques are sensitive to the frequency inflections in the tonal calls of their conspecifics (Zoloth et al., 1979) and the tonal "coo" calls of conspecifics can have different referential meanings depending on where the frequency contours occur, either early or late in the vocalization (Hauser, 1991; Hauser and Marler, 1993). These observations are by no means restricted to primates and, in fact, the studies on primates were likely inspired by observations of how the bat auditory cortex evaluates the steady and FM components in their vocalizations during echolocation. For example, there are neurons in the bat auditory cortex that respond to combinations of vocalization components, including the constant frequency, FM or Doppler shifted components of the emitted biosonar signal (Suga et al., 1978). More recently, electrophysiological studies from the monkey auditory cortex have shown that some neurons in the lateral belt respond stronger to complete vocalizations rather than to filtered versions of the vocalizations, such as their low- or highpass filtered components (Rauschecker et al., 1995). This finding suggests that some selectivity for monkey vocalizations is present at the second processing stage in the monkey auditory cortex. However, a recent comparison of how the belt analyzes reversed vocalizations (an acoustical control) reported a lack of clear selectivity in the auditory core and belt regions for species-specific vocalizations, suggesting that higher level processing centers may be more selective for vocalizations (Recanzone, 2008).

After passing the core and belt regions, further processing is known to occur in the cortex of the temporal, parietal and frontal lobes (including interactions with, and contributions from, the subcortical regions). For instance, there is now evidence from functional imaging that selective responses to species-specific vocalizations can be seen several stages away from the PAC, anterior on the temporal lobe (Poremba et al., 2004; Petkov et al., 2008). However, as the signals leave the core and belt regions of the monkey cortex, the correspondence to the results from human studies becomes less clear. This is mainly because of a paucity of data, even though several research groups have been actively studying cortical responses to vocalizations outside the first two analysis stages of the auditory cortical hierarchy. Some of these results revealed that neuronal responses from regions such as the superior temporal gyrus code for different vocalizations (Russ et al., 2008) and the ventral prefrontal cortex also shows some representation of this coding (Romanski et al., 2005; Cohen et al., 2007). Furthermore, auditory sensitive parts of the primate insular cortex have been described (Bieser, 1998; Remedios et al., 2009), suggesting that a broad network of regions in the primate brain analyzes the different aspects of vocalizations.

IV.B. Parallel-processing pathways

Different aspects of sensory stimuli are processed in the cortex in parallel pathways (see Fig. 1). Work on the monkey visual system obtained evidence for dual parallel-processing pathways (Ungerleider and Mishkin, 1982), where it was noted that a dorsal brain pathway, composed of several visual fields, evaluates the spatial features of visual stimuli, a so-called "where" pathway. Another, ventral, pathway was more involved in evaluating the features of visual objects, often referred to as a "what" pathway.

The hypothesized dual-pathway model for the auditory cortex found support from studies of anatomical connectivity and electrophysiological recordings in monkeys (Rauschecker, 1998; Romanski et al., 1999; Tian et al., 2001). The dual pathway has now also gained support from human imaging studies (Alain et al., 2001) and work in other mammals, such as cats (Lomber and Malhotra, 2008). Electrophysiological recordings in monkeys showed that the anterior belt regions tended to better encode the non-spatial features that distinguish different monkey vocalizations (Tian et al., 2001). This was interpreted as evidence for a ventral or anterior auditory "what" pathway. The caudal parts of the lateral belt, on the other hand, tended to better encode the spatial features of the sounds, or "where" the sounds were located in auditory space. The spatial processing capabilities of neurons in the caudal belt and posterior parietal cortex agree with this model (Recanzone et al., 2000; Gifford and Cohen, 2005). However, parts of the anterior lateral belt and ventral prefrontal cortex (presumably a part of the "what" pathway) are also spatially sensitive (Cohen et al., 2004) or the posterior parietal cortex can encode non-spatial attributes of sounds (Gifford and Cohen, 2005), suggesting that the processing pathways are not completely segregated.

Unfortunately, we do not clearly understand, even in individual species, how many pathways there are, what their precise function is and to what extent the different pathways interact. Fig. 1 is based on recent reviews attempting to integrate the dual "what" and "where" auditory pathways (Scott and Johnsrude, 2003; Scott, 2005; Petkov et al., 2009, in press) with the idea that a third "how" or "how-to" pathway exists for evaluating how a sensory object may have been generated (Belin and Zatorre, 2000). Fig. 1 illustrates these ideas. Although the figure suggests a close correspondence between the human and monkey brain, the extent of the homology of the processing pathways is unclear. For example, recent comparative anatomical study of humans and monkeys has already revealed considerable cross-species differences in the structural organization of auditory cortex, even at the level of the primary auditory cortex (Fullerton and Pandya, 2007). Related to this, and as we noted in the previous section, a careful comparison of the primary auditory cortex in the inserts of Fig. 1 shows that the detailed mapping of cortical fields for humans and monkeys may correspond to a certain extent (Formisano et al., 2003; Petkov et al., 2006), mostly in relation to two fields in the primary auditory cortex. Also under active pursuit is the correspondence of the dorsal pathways in monkeys, humans and chimpanzees, as studied with MRI-based imaging of connectivity. Current impressions are that the connectivity of the dorsal pathways may differ considerably between humans and monkeys (Rilling et al., 2008), and needs to be re-evaluated in both of the species (Schmahmann et al., 2007).

The insufficient functional data from non-human primates is part of the problem, and as we noted above, direct comparisons across the species are facilitated by using the same techniques. We will now consider the recent evidence from the imaging of non-human primate brain functions related to vocal sound processing, and how they may correspond to the imaging data from the human brain.

V. Functional imaging of cortical areas sensitive to vocalizations in monkeys, chimpanzees and humans

Results of imaging studies (PET or fMRI) on chimpanzees and monkeys, which have used species-specific vocalizations as stimuli, are summarized in Fig. 2 (Poremba et al., 2004; Gil-da-Costa et al., 2006; Petkov et al., 2008; Taglialatela et al., 2009). For comparison with humans (although it is difficult to know what would be “speech-like” processing in non-human primates) we have summarized the results of several human studies on the stimulus-bound aspects of speech, such as the pre-lexical processing of speech sound components, see the darker purple circles in Fig. 2 (Dehaene-Lambertz et al., 2005; Liebenthal et al., 2005; Rimol et al., 2005; Obleser et al., 2006, 2007). We have also summarized results of human studies that specifically looked at non-linguistic voice processing, see the lighter orange and yellow circles in Fig. 2 (Belin et al., 2000; Belin and Zatorre, 2003; von Kriegstein et al., 2003).

From a more general perspective (regardless of the color of the circle summaries in Fig. 2), the results of the studies across the species find common ground that the processing of species-specific vocalizations is distributed along both anterior and posterior parts of the superior temporal lobe. These regions are likely to involve different brain pathways, regardless of what our understanding of those pathways may be at this point. This observation for the human brain is in line with updated models of human speech processing and perception, but it contrasts with traditional models of speech and language, which propose that speech processing occurs mainly in the posterior temporal lobe and in the left hemisphere (for a review see Scott and Johnsrude, 2003; Hickok and Poeppel, 2007; Petkov et al., 2009 in press). We also observed that similar recruitment of different parts of the temporal lobe appears to occur for the processing of species-specific vocalizations in chimpanzees and macaques (see Fig. 2).

Results of studies on chimpanzees suggested hemispheric lateralization, and showed a right hemisphere preference for the processing of the vocalizations of conspecifics (Taglialatela et al., 2009). In humans and macaques, as summarized in Fig. 2, the results showed that both hemispheres appear nearly equally involved in the general processing of speech sounds in humans and the conspecific vocalizations in monkeys. The level of lateralization in the human or non-human primate brains for the processing of vocal communications needs to be more systematically evaluated with direct comparisons across the species.

If we evaluate the sensitivity in the human brain to human voices (labeled in orange), including voice identity (in yellow), we see that the regions sensitive to the human voice are either neighboring or overlapping with areas evaluating the acoustical aspects in speech (compare for humans the location of the purple circles to the orange and yellow ones in Fig. 2). This observation suggests that speech and voice identity processing might share some neuronal resources, and a recent study has now explicitly demonstrated this phenomenon (Formisano et al., 2008). Thus, even if the human speech regions have no direct homologs in any of the extant primates, non-human primates may still possess precursory networks on which human speech may have evolved. Interestingly, there is now evidence from monkey fMRI for a voice-selective region in the anterior temporal lobe (Petkov et al., 2008) which shows a close correspondence to a region in the human brain for processing human voices (Belin et al., 2000; Belin and Zatorre, 2003; von Kriegstein et al., 2003).

VI. Voice identity coding in the anterior temporal lobe of monkey and man

The selectivity for human voices that has been seen in the responses from human anterior temporal lobe (Belin et al., 2000; Belin and Zatorre, 2003; von Kriegstein et al., 2003) shows an interesting correspondence with responses from a monkey voice processing region suggested by a recent fMRI study on rhesus macaques (Petkov et al., 2008). In the macaque monkey imaging study, first several regions were revealed in the auditory cortex which showed preferential responses to presentation of macaque vocalizations over vocalizations from other animals, natural sounds, or other acoustical control stimuli (see the orange summaries in Fig. 2 for the macaque

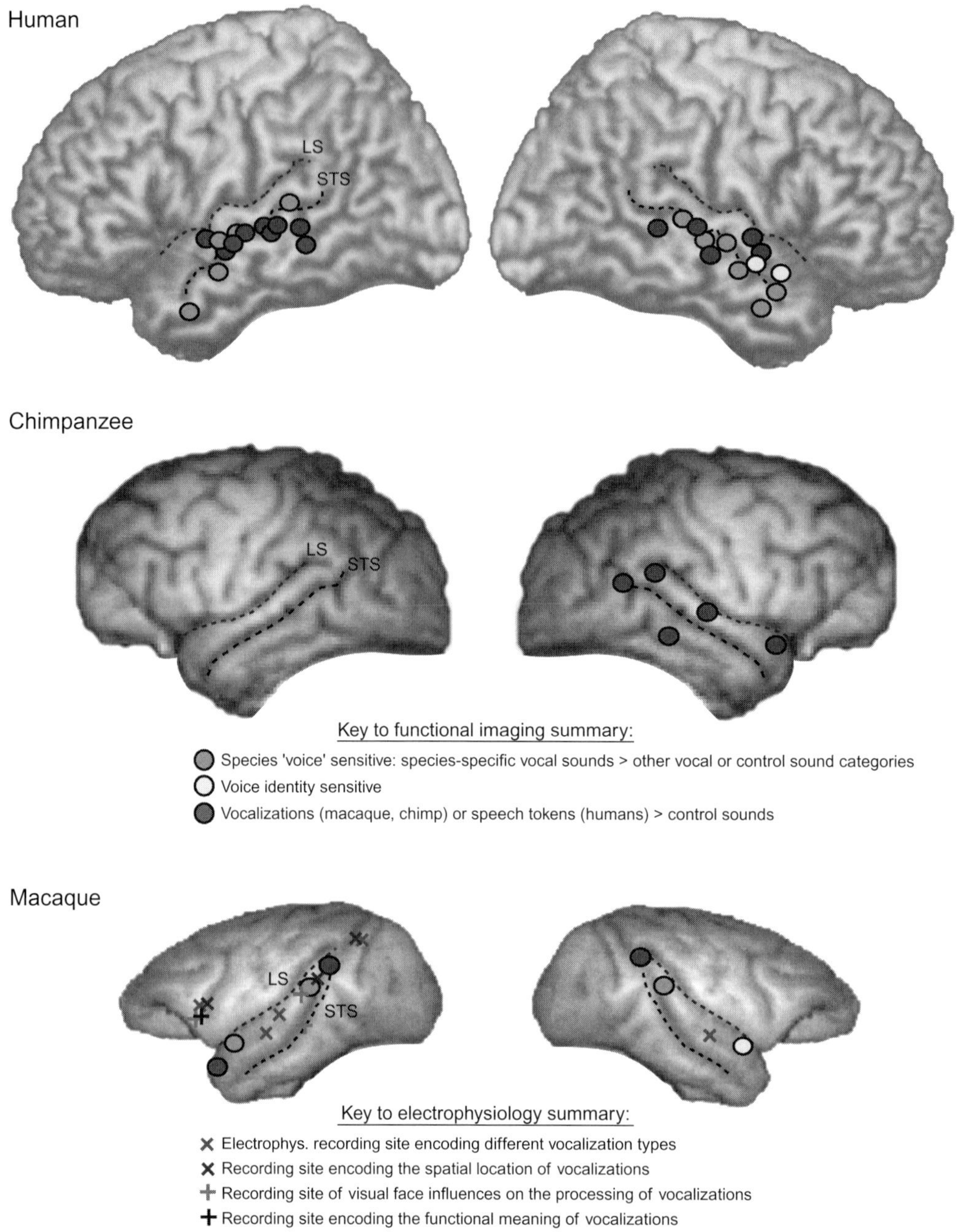

Fig. 2. Comparative summary of human, chimpanzee and macaque processing of species-specific communication sounds. Colored circles summarize several functional imaging results (see key) focusing on the stimulus-bound processing of vocal signals in the temporal lobe. For humans, we summarize the peaks of activity reported in studies of voice sensitivity (Belin et al., 2000; Belin and Zatorre, 2003; von Kriegstein et al., 2003), voice identity (Belin and Zatorre, 2003; von Kriegstein et al., 2003) and the sublexical or stimulus-bound aspects of speech (Dehaene-Lambertz et al., 2005; Liebenthal et al., 2005; Rimol et al., 2005; Obleser et al., 2006, 2007); for the exact coordinates of the summaries in humans and monkeys see Petkov et al., 2009 (in press). For chimpanzees we summarize a recent study evaluating chimp vocal sound processing in these great apes (Taglialatela et al., 2009). For the macaque brain we show the sensitivity to macaque vocalizations from both PET (Poremba et al., 2004; Gil-da-Costa et al., 2006) and fMRI (Petkov et al., 2008) studies. The monkey voice sensitive regions (orange circles) and voice-identity selective regions (yellow circle) identified in the monkey fMRI study (Petkov et al., 2008) can be directly compared with the human studies on human voice sensitivity and selectivity (compare the orange and yellow circles in the human and macaque). For macaques, we also identify electrophysiological recording sites (see the lower key), from the temporal (Rauschecker et al., 1995; Tian et al., 2001; Ghazanfar et al., 2005, 2008;

(*Continued*)

image). This result was interpreted as evidence of sensitivity for the acoustic features that distinguish the vocalizations of members of the species from other animal vocalizations and natural sounds. A separate experiment showed that only a small cluster of cells in the anterior temporal lobe was also sensitive to the vocal identity of different familiar conspecifics (yellow circle in Fig. 2 for the macaque image).

This anterior temporal lobe region in monkeys closely corresponds to the anterior temporal lobe region in humans that is known to be sensitive to human vocal sounds, including voice identity (Belin et al., 2000; Belin and Zatorre, 2003; von Kriegstein et al., 2003). This region is also involved in active voice recognition in humans (von Kriegstein et al., 2003). The human and monkey imaging data from these voice regions support the conclusion of a high-level correspondence in the processing of species-specific vocal features, and the functional homology between the human and macaque vocal processing that occurs in the anterior temporal regions in these primates (compare the yellow circle summaries in Fig. 2 for humans and monkeys).

What features of vocal sounds are critical in activating these vocal regions? At this point, one can only speculate; however, a reasonable hypothesis can be postulated on the basis of the behavioral literature. For instance, it has now been shown that macaque monkeys are sensitive to the high-frequency formant structure in vocal sounds (Fitch and Fritz, 2006), and can use these sounds to judge the size of the vocalizing individual (Ghazanfar et al., 2007). Thus, part of the sensitivity of the voice regions may be derived from the analysis of the formant frequencies, although these are not the only components in vocal sounds that can identify individuals.

A recent human imaging study showed that mainly the posterior superior temporal gyrus was sensitive to the formant or resonant structures in human vocal sounds (von Kriegstein et al., 2007). Other regions, including the anterior temporal lobe and the intraparietal sulcus, were sensitive not only to the resonances in the human vocal sounds, but also to the resonances in animal sounds and in sounds obtained from musical instruments. This result is interesting, because it suggests that some of the vocal sensitivity of the brain may be generally suited to evaluate the length and size of the filter not only of vocal sounds, but also of musical instruments. Comparable work from animals is needed to elucidate whether the regions subserving voice recognition in non-human primates may relate to, or can elucidate, the evolutionary origins of speech and music processing in the human brain.

VII. Emotion and meaning

If voice information is processed modularly by the brain, then some pathways and networks would code identity, others the emotional content and still others the meaning of the vocal expression (Belin et al., 2004). The coding of emotional content is often thought to involve the amygdala, since emotional vocalizations and faces elicit strong responses from the central nucleus neurons in the amygdala (Kuraoka and Nakamura, 2007; see also Kuraoka and Nakamura, Chapter 5.3 in this volume). However, in humans the amygdala belongs to a broad cortical network, including the superior temporal sulcus, which evaluates the intentions of other individuals during social communication. Thus, the relationship between the human and monkey social and affective networks could be better understood (Allison et al., 2000).

How the brains of primates evaluate the meaning of vocal communication has come from the detailed work on the speech perception network in the human brain. An interesting recent study with macaques suggests that the ventral prefrontal cortex is sensitive to the meaning of conspecific vocalizations, rather than acoustical differences in general (see Fig. 2) (Gifford et al., 2005). The authors showed that neurons in this part of the brain encoded transitions between vocalizations that are known to hold different meanings for the animals, but the neurons were rather insensitive to vocalizations that shared the same general meaning yet differed acoustically.

VIII. Cross-modal influences: voices and faces

Human language does not need to be verbal, and the ability to hear and recognize vocal communications

Fig. 2. (Continued) Russ et al., 2008), parietal (Gifford and Cohen, 2005) and prefrontal cortices (Romanski and Goldman-Rakic, 2002; Cohen et al., 2004; Gifford et al., 2005; Sugihara et al., 2006). This figure contains a rendered chimpanzee brain image kindly contributed by J. Taglialatela. For abbreviations, see Fig. 1 (To see the full color version of this figure please refer to the color plate in the back of the book. Copies produced via our print on demand service do not contain color plates. If your copy does not have the color plate, please go to this website to view the figure in color http://www.elsevierdirect.com/companions/9780123745934).

could benefit from multisensory information, especially in noisy situations. Thus, it makes sense to consider the cross-modal influences on cortical voice processing and to discuss possible connections for such interactions.

Human imaging studies have shown that different parts of the temporal lobe, including regions compatible with the primary auditory cortex, are influenced by visual stimuli (for a review see Ghazanfar and Schroeder, 2006). A recent fMRI study in the monkey brain localized the visual influences in auditory cortex, and showed them to occur mainly in the caudal portions of the auditory cortex (Kayser et al., 2007). Results of experiments that have used voice and face stimuli have shown that, when human subjects match the identity of cross-modal stimuli, interactions are seen between the face- and voice-sensitive areas in the human temporal lobe (von Kriegstein et al., 2005). However, the correspondence between the human and monkeys studies is still fairly general, in part because of some conflicting results from the animal studies. Electrophysiological studies in macaques reported cross-modal influences on neuronal activity in the lateral belt and ventral prefrontal regions that may be specific to the faces and voices of conspecifics (see red crosses in Fig. 2) (Ghazanfar et al., 2005; Sugihara et al., 2006). However, others have reported that the visual influences in the core and belt regions of the macaque auditory cortex do not appear to be restricted to voices and faces, but are also seen to a large extent with other natural audio-visual stimuli (Bizley et al., 2007; Kayser et al., 2008). This discrepancy requires further study.

IX. Conclusions and a look ahead

Just as in other research fields, understanding of how the brain encodes communication signals requires an integrative approach that takes into account homologies between humans and other species, as well as a combination of noninvasive and invasive methodologies. If the communication system in humans is treated as a special and unique one, it is difficult to hypothesize from results obtained just from human studies what the evolutionarily conserved mechanisms of the brain are, and what they were designed to achieve. This situation would raise a question as to whether even closely-related primate species can sufficiently model human vocal communication. However, exciting developments on different research fronts are providing more tangible links between the studies of the auditory cortex in humans and other species. Evaluation of brain structure and function across the species, using the same or comparable methods, revealed novel agreement and insights in how vocal sounds are processed by the different cortical regions of humans, monkeys and apes.

Acknowledgments

We are grateful to E. Formisano and J. Obleser for providing images of the human brain which were used to create some of the figures. We also thank J. Taglialatela for the chimpanzee brain anatomical. This work was supported by a grant from Newcastle University (Institute of Neuroscience) to C.P., the German Research Society (DFG) to C.K. and the Max-Planck Society to N.K.L. and C.K.

References

Alain, C., Arnott, S.R., Hevenor, S., Graham, S., Grady, C.L., 2001. "What" and "where" in the human auditory system. Proc. Natl. Acad. Sci. USA, 98, 12301–12306.

Allison, T., Puce, A., McCarthy, G., 2000. Social perception from visual cues: role of the STS region. Trends Cogn. Sci. 4, 267–278.

Belin, P., 2006. Voice processing in human and non-human primates. Philos. Trans. R. Soc. Lond. B Biol. Sci. 361, 2091–2107.

Belin, P., Zatorre, R.J., 2000. "What," "where" and "how" in auditory cortex. Nat. Neurosci. 3, 965–966.

Belin, P., Zatorre, R.J., 2003. Adaptation to speaker's voice in right anterior temporal lobe. Neuroreport 14, 2105–2109.

Belin, P., Fecteau, S., Bedard, C., 2004. Thinking the voice: neural correlates of voice perception. Trends Cogn. Sci. 8, 129–135.

Belin, P., Zatorre, R.J., Lafaille, P., Ahad, P., Pike, B., 2000. Voice-selective areas in human auditory cortex. Nature 403, 309–312.

Bieser, A., 1998. Processing of twitter-call fundamental frequencies in insula and auditory cortex of squirrel monkeys. Exp. Brain. Res. 122, 139–148.

Binder, J.R., Rao, S.M., Hammeke, T.A., Yetkin, F.Z., Jesmanowicz, A., Bandettini, P.A., Wong, E.C., Estkowski, L.D., Goldstein, M.D., Haughton, V.M. et al., 1994. Functional magnetic resonance imaging of human auditory cortex. Ann. Neurol. 35, 662–672.

Bizley, J.K., Nodal, F.R., Bajo, V.M., Nelken, I., King, A.J., 2007. Physiological and anatomical evidence for multisensory interactions in auditory cortex. Cereb. Cortex 17, 2172–2189.

Brodmann, K., Garey, L.J., 1999. Brodmann's Localisation in the Cerebral Cortex. Imperial College Press, London, UK.

Cohen, Y.E., Russ, B.E., Gifford, 3rd., G.W., Kiringoda, R., MacLean, K.A., 2004. Selectivity for the spatial and non-spatial attributes of auditory stimuli in the ventrolateral prefrontal cortex. J. Neurosci. 24, 11307–11316.

Cohen, Y.E., Theunissen, F., Russ, B.E., Gill, P., 2007. Acoustic features of rhesus vocalizations and their representation in the ventrolateral prefrontal cortex. J. Neurophysiol. 97, 1470–1484.

Dehaene-Lambertz, G., Pallier, C., Serniclaes, W., Sprenger-Charolles, L., Jobert, A., Dehaene, S., 2005. Neural correlates of switching from auditory to speech perception. Neuroimage 24, 21–33.

Disbrow, E., Roberts, T.P., Slutsky, D., Krubitzer, L., 1999. The use of fMRI for determining the topographic organization of cortical fields in human and non-human primates. Brain Res. 829, 167–173.

Fay, R.F., 1988. Hearing in Vertebrates: A Psychophysics Databook. Heffernan Press, Worcester, MA.

Fitch, W.T., Fritz, J.B., 2006. Rhesus macaques spontaneously perceive formants in conspecific vocalizations. J. Acoust. Soc. Am. 120, 2132–2141.

Formisano, E., De Martino, F., Bonte, M., Goebel, R., 2008. "Who" is saying "what"? Brain-based decoding of human voice and speech. Science 322, 970–973.

Formisano, E., Kim, D.S., Di Salle, F., van de Moortele, P.F., Ugurbil, K., Goebel, R., 2003. Mirror-symmetric tonotopic maps in human primary auditory cortex. Neuron 40, 859–869.

Fox, P.T., Mintun, M.A., Raichle, M.E., Miezin, F.M., Allman, J.M., Van Essen, D.C., 1986. Mapping human visual cortex with positron emission tomography. Nature 323, 806–809.

Fullerton, B.C., Pandya, D.N., 2007. Architectonic analysis of the auditory-related areas of the superior temporal region in human brain. J. Comp. Neurol. 504, 470–498.

Funkenstein, H.H., Winter, P., 1973. Responses to acoustic stimuli of units in the auditory cortex of awake squirrel monkeys. Exp. Brain. Res. 18, 464–488.

Ghazanfar, A.A., Schroeder, C.E., 2006. Is neocortex essentially multisensory? Trends Cogn. Sci. 10, 278–285.

Ghazanfar, A.A., Chandrasekaran, C., Logothetis, N.K., 2008. Interactions between the superior temporal sulcus and auditory cortex mediate dynamic face/voice integration in rhesus monkeys. J. Neurosci. 28, 4457–4469.

Ghazanfar, A.A., Maier, J.X., Hoffman, K.L., Logothetis, N.K., 2005. Multisensory integration of dynamic faces and voices in rhesus monkey auditory cortex. J. Neurosci. 25, 5004–5012.

Ghazanfar, A.A., Turesson, H.K., Maier, J.X., van Dinther, R., Patterson, R.D., Logothetis, N.K., 2007. Vocal-tract resonances as indexical cues in rhesus monkeys. Curr. Biol. 17, 425–430.

Gifford 3rd., G.W., Cohen, Y.E., 2005. Spatial and non-spatial auditory processing in the lateral intraparietal area. Exp. Brain Res. 162, 509–512.

Gifford 3rd., G.W., MacLean, K.A., Hauser, M.D., Cohen, Y.E., 2005. The neurophysiology of functionally meaningful categories: macaque ventrolateral prefrontal cortex plays a critical role in spontaneous categorization of species-specific vocalizations. J. Cogn. Neurosci. 17, 1471–1482.

Gil-da-Costa, R., Martin, A., Lopes, M.A., Munoz, M., Fritz, J.B., Braun, A.R., 2006. Species-specific calls activate homologs of Broca's and Wernicke's areas in the macaque. Nat. Neurosci. 9, 1064–1070.

Hackett, T.A., Stepniewska, I., Kaas, J.H., 1998. Subdivisions of auditory cortex and ipsilateral cortical connections of the parabelt auditory cortex in macaque monkeys. J. Comp. Neurol. 394, 475–495.

Hackett, T.A., Preuss, T.M., Kaas, J.H., 2001. Architectonic identification of the core region in auditory cortex of macaques, chimpanzees, and humans. J. Comp. Neurol. 441, 197–222.

Hauser, M.D., 1991. Sources of acoustic variation in rhesus macaque (*Macaca mulatta*) vocalizations. Ethology 89, 29–46.

Hauser, M.D., Marler, P., 1993. Food-associated calls in rhesus macaques (*Macaca mulatta*): I. Socioecological factors. Behav. Ecol. 4, 194–205.

Hauser, M.D., Chomsky, N., Fitch, W.T., 2002. The faculty of language: what is it, who has it, and how did it evolve? Science 298, 1569–1579.

Hickok, G., Poeppel, D., 2007. The cortical organization of speech processing. Nat. Rev. Neurosci. 8, 393–402.

Kaas, J.H., Hackett, T.A., 1998. Subdivisions of auditory cortex and levels of processing in primates. Audiol. Neurootol. 3, 73–85.

Kaas, J.H., Hackett, T.A., 2000. Subdivisions of auditory cortex and processing streams in primates. Proc. Natl. Acad. Sci. USA 97, 11793–11799.

Kayser, C., Petkov, C.I., Augath, M., Logothetis, N.K., 2005. Integration of touch and sound in auditory cortex. Neuron 48, 373–384.

Kayser, C., Petkov, C.I., Augath, M., Logothetis, N.K., 2007. Functional imaging reveals visual modulation of specific fields in auditory cortex. J. Neurosci. 27, 1824–1835.

Kayser, C., Petkov, C.I., Logothetis, N.K., 2008. Visual modulation of neurons in auditory cortex. Cereb. Cortex. 18, 1560–1574.

Kuraoka, K., Nakamura, K., 2007. Responses of single neurons in monkey amygdala to facial and vocal emotions. J. Neurophysiol. 97, 1379–1387.

Liebenthal, E., Binder, J.R., Spitzer, S.M., Possing, E.T., Medler, D.A., 2005. Neural substrates of phonemic perception. Cereb. Cortex 15, 1621–1631.

Logothetis, N.K., 2008. What we can do and what we cannot do with fMRI. Nature 453, 869–878.

Logothetis, N.K., Guggenberger, H., Peled, S., Pauls, J., 1999. Functional imaging of the monkey brain. Nat. Neurosci. 2, 555–562.

Logothetis, N.K., Pauls, J., Augath, M., Trinath, T., Oeltermann, A., 2001. Neurophysiological investigation of the basis of the fMRI signal. Nature 412, 150–157.

Lomber, S.G., Malhotra, S., 2008. Double dissociation of "what" and "where" processing in auditory cortex. Nat. Neurosci. 11, 609–616.

Maier, A., Wilke, M., Aura, C., Zhu, C., Ye, F.Q., Leopold, D.A., 2008. Divergence of fMRI and neural signals in V1 during perceptual suppression in the awake monkey. Nat. Neurosci. 11, 1193–1200.

Nagarajan, S.S., Cheung, S.W., Bedenbaugh, P., Beitel, R.E., Schreiner, C.E., Merzenich, M.M., 2002. Representation of spectral and temporal envelope of twitter vocalizations in common marmoset primary auditory cortex. J. Neurophysiol. 87, 1723–1737.

Nelken, I., Rotman, Y., Bar Yosef, O., 1999. Responses of auditory-cortex neurons to structural features of natural sounds. Nature 397, 154–157.

Newman, J.D., Wollberg, Z., 1973. Multiple coding of species-specific vocalizations in the auditory cortex of squirrel monkeys. Brain Res. 54, 287–304.

Obleser, J., Boecker, H., Drzezga, A., Haslinger, B., Hennenlotter, A., Roettinger, M., Eulitz, C., Rauschecker, J.P., 2006. Vowel sound extraction in anterior superior temporal cortex. Hum. Brain Map. 27, 562–571.

Obleser, J., Zimmermann, J., Van Meter, J., Rauschecker, J.P., 2007. Multiple stages of auditory speech perception reflected in event-related FMRI. Cereb. Cortex 17, 2251–2257.

Ogawa, S., Lee, T.M., Kay, A.R., Tank, D.W., 1990. Brain magnetic resonance imaging with contrast dependent on blood oxygenation. Proc. Natl. Acad. Sci. USA 87, 9868–9872.

Ogawa, S., Tank, D.W., Menon, R., Ellermann, J.M., Kim, S.G., Merkle, H., Ugurbil, K., 1992. Intrinsic signal changes accompanying sensory stimulation: functional brain mapping with magnetic resonance imaging. Proc. Natl. Acad. Sci. USA 89, 5951–5955.

Petkov, C.I., Kayser, C., Augath, M., Logothetis, N.K., 2006. Functional imaging reveals numerous fields in the monkey auditory cortex. PLoS Biol. 4, e215.

Petkov, C.I., Kayser, C., Steudel, T., Whittingstall, K., Augath, M., Logothetis, N.K., 2008. A voice region in the monkey brain. Nat. Neurosci. 11, 367–374.

Petkov, C.I., Logothetis, N.K., Obleser, J., 2009. Where are the human speech and voice regions and do other animals have anything like them? Neuroscientist 2 (in press).

Poremba, A., Malloy, M., Saunders, R.C., Carson, R.E., Herscovitch, P., Mishkin, M., 2004. Species-specific calls evoke asymmetric activity in the monkey's temporal poles. Nature 427, 448–451.

Poremba, A., Saunders, R.C., Crane, A.M., Cook, M., Sokoloff, L., Mishkin, M., 2003. Functional mapping of the primate auditory system. Science 299, 568–572.

Rauch, A., Rainer, G., Logothetis, N.K., 2008. The effect of a serotonin-induced dissociation between spiking and perisynaptic activity on BOLD functional MRI. Proc. Natl. Acad. Sci. USA 105, 6759–6764.

Rauschecker, J.P., 1998. Parallel processing in the auditory cortex of primates. Audiol. Neurootol. 3, 86–103.

Rauschecker, J.P., Tian, B., 2004. Processing of band-passed noise in the lateral auditory belt cortex of the rhesus monkey. J. Neurophysiol. 91, 2578–2589.

Rauschecker, J.P., Tian, B., Hauser, M., 1995. Processing of complex sounds in the macaque nonprimary auditory cortex. Science 268, 111–114.

Rauschecker, J.P., Tian, B., Pons, T., Mishkin, M., 1997. Serial and parallel processing in rhesus monkey auditory cortex. J. Comp. Neurol. 382, 89–103.

Recanzone, G.H., 2008. Representation of con-specific vocalizations in the core and belt areas of the auditory cortex in the alert macaque monkey. J. Neurosci. 28, 13184–13193.

Recanzone, G.H., Guard, D.C., Phan, M.L., Su, T.K., 2000. Correlation between the activity of single auditory cortical neurons and sound-localization behavior in the macaque monkey. J. Neurophysiol. 83, 2723–2739.

Remedios, R., Logothetis, N.K., Kayser, C., 2009. An auditory region in the primate insular cortex responding preferentially to vocal communication sounds. J. Neurosci. 29, 1034–1045.

Rilling, J.K., Glasser, M.F., Preuss, T.M., Ma, X., Zhao, T., Hu, X., Behrens, T.E., 2008. The evolution of the arcuate fasciculus revealed with comparative DTI. Nat. Neurosci. 11, 426–428.

Rimol, L.M., Specht, K., Weis, S., Savoy, R., Hugdahl, K., 2005. Processing of sub-syllabic speech units in the posterior temporal lobe: an fMRI study. Neuroimage 26, 1059–1067.

Romanski, L.M., Goldman-Rakic, P.S., 2002. An auditory domain in primate prefrontal cortex. Nat. Neurosci. 5, 15–16.

Romanski, L.M., Averbeck, B.B., Diltz, M., 2005. Neural representation of vocalizations in the primate ventrolateral prefrontal cortex. J. Neurophysiol. 93, 734–747.

Romanski, L.M., Tian, B., Fritz, J., Mishkin, M., Goldman-Rakic, P.S., Rauschecker, J.P., 1999. Dual streams of auditory afferents target multiple domains in the primate prefrontal cortex. Nat. Neurosci. 2, 1131–1136.

Russ, B.E., Ackelson, A.L., Baker, A.E., Cohen, Y.E., 2008. Coding of auditory-stimulus identity in the auditory non-spatial processing stream. J. Neurophysiol. 99, 87–95.

Schmahmann, J.D., Pandya, D.N., Wang, R., Dai, G., D'Arceuil, H.E., de Crespigny, A.J., Wedeen, V.J., 2007. Association fibre pathways of the brain: parallel observations from diffusion spectrum imaging and autoradiography. Brain 130, 630–653.

Scott, S.K., 2005. Auditory processing – speech, space and auditory objects. Curr. Opin. Neurobiol. 15, 197–201.

Scott, S.K., Johnsrude, I.S., 2003. The neuroanatomical and functional organization of speech perception. Trends Neurosci. 26, 100–107.

Steinschneider, M., Arezzo, J.C., Vaughan, Jr., H.G., 1990. Tonotopic features of speech-evoked activity in primate auditory cortex. Brain Res. 519, 158–168.

Suga, N., O'Neill, W.E., Manabe, T., 1978. Cortical neurons sensitive to combinations of information-bearing elements of biosonar signals in the mustache bat. Science 200, 778–781.

Sugihara, T., Diltz, M.D., Averbeck, B.B., Romanski, L.M., 2006. Integration of auditory and visual communication information in the primate ventrolateral prefrontal cortex. J. Neurosci. 26, 11138–11147.

Sweet, R.A., Dorph-Petersen, K.A., Lewis, D.A., 2005. Mapping auditory core, lateral belt, and parabelt cortices in the human superior temporal gyrus. J. Comp. Neurol. 491, 270–289.

Taglialatela, J.P., Russell, J.L., Schaeffer, J.A., Hopkins, W.D., 2008. Communicative signaling activates "Broca's" homolog in chimpanzees. Curr. Biol. 18, 343–348.

Taglialatela, J.P., Russell, J.L., Schaeffer, J.A., Hopkins, W.D., 2009. Visualizing vocal perception in the chimpanzee brain. Cereb. Cortex 19, 1151–1157.

Tian, B., Rauschecker, J.P., 2004. Processing of frequency-modulated sounds in the lateral auditory belt cortex of the rhesus monkey. J. Neurophysiol. 92, 2993–3013.
Tian, B., Reser, D., Durham, A., Kustov, A., Rauschecker, J.P., 2001. Functional specialization in rhesus monkey auditory cortex. Science 292, 290–293.
Ungerleider, L.G., Mishkin, M., 1982. Two cortical visual systems. In: Ingle, D.J., Goodale, M.A., Mansfield, R. J.W. (Eds.), Analysis of Visual Behavior. MIT Press, Cambridge, MA, pp. 549–580.
Uppenkamp, S., Johnsrude, I.S., Norris, D., Marslen-Wilson, W., Patterson, R.D., 2006. Locating the initial stages of speech-sound processing in human temporal cortex. Neuroimage 31, 1284–1296.
Van der Linden, A., Verhoye, M., Van Meir, V., Tindemans, I., Eens, M., Absil, P., Balthazart, J., 2002. *In vivo* manganese-enhanced magnetic resonance imaging reveals connections and functional properties of the songbird vocal control system. Neuroscience 112, 467–474.
Van Essen, D.C., 2005. Surface-based comparisons of macaque and human cortical organization. In: Dahaene, S., Duhamel, J., Hauser, M.D., Rizzolatti, G. (Eds.), From Monkey Brain to Human Brain. MIT Press, Cambridge, MA, pp. 1–19.
Van Meir, V., Boumans, T., De Groof, G., Van Audekerke, J., Smolders, A., Scheunders, P., Sijbers, J., Verhoye, M., Balthazart, J., Van der Linden, A., 2005. Spatiotemporal properties of the BOLD response in the songbirds' auditory circuit during a variety of listening tasks. Neuroimage 25, 1242–1255.
von Kriegstein, K., Eger, E., Kleinschmidt, A., Giraud, A.L., 2003. Modulation of neural responses to speech by directing attention to voices or verbal content. Brain Res. Cogn. Brain Res. 17, 48–55.
von Kriegstein, K., Kleinschmidt, A., Sterzer, P., Giraud, A.L., 2005. Interaction of face and voice areas during speaker recognition. J. Cogn. Neurosci. 17, 367–376.
von Kriegstein, K., Smith, D.R., Patterson, R.D., Ives, D.T., Griffiths, T.D., 2007. Neural representation of auditory size in the human voice and in sounds from other resonant sources. Curr. Biol. 17, 1123–1128.
Wang, X., Kadia, S.C., 2001. Differential representation of species-specific primate vocalizations in the auditory cortices of marmoset and cat. J. Neurophysiol. 86, 2616–2620.
Wang, X., Merzenich, M.M., Beitel, R., Schreiner, C.E., 1995. Representation of a species-specific vocalization in the primary auditory cortex of the common marmoset: temporal and spectral characteristics. J. Neurophysiol. 74, 2685–2706.
Wessinger, C.M., VanMeter, J., Tian, B., Van Lare, J., Pekar, J., Rauschecker, J.P., 2001. Hierarchical organization of the human auditory cortex revealed by functional magnetic resonance imaging. J. Cogn. Neurosci. 13, 1–7.
Zoloth, S.R., Petersen, M.R., Beecher, M.D., Green, S., Marler, P., Moody, D.B., Stebbins, W., 1979. Species-specific perceptual processing of vocal sounds by monkeys. Science 204, 870–873.

SECTION 5

Effects of Vocalization on the Organism's State and Behavior: Brain as an Amplifier of Vocal Signals

Reception of species-specific vocalizations has activating properties on the central nervous system and behavior. Vocalization can trigger defensive responses, maternal responses and other affective responses in the recipients. Species-specific vocalization plays a role of an interindividual stimulus regulating brain activity of the recipients. Thus, vocalizations may be used by the callers to influence the affective system of recipients and in this way influence their behavior.

CHAPTER 5.1

Vocalization as a social signal in defensive behavior

Yoav Litvin[1*], D. Caroline Blanchard[2] and Robert J. Blanchard[1]

[1]*Department of Psychology, University of Hawaii at Manoa, Honolulu, HI, USA*
[2]*Pacific Biosciences Research Center, University of Hawaii at Manoa, Department of Genetics and Molecular Biology, John A. Burns School of Medicine, Honolulu, HI, USA*

Abstract: Impending dangers elicit alarm vocalizations aimed at conspecifics, and these calls in turn elicit defensive responses in recipients. The notion that these vocalizations represent adaptive responses to threat, enhancing the survival and reproduction of potentially-related conspecifics, is supported by findings from laboratory and field studies. These observations show that vocalizations occur predominantly when the vocalizing animal is in the presence of conspecifics (colony); that higher rates may occur in females; that animals may reduce their own risk of negative consequences by calling preferentially from a place of relative safety; and that dominant males, which tend to sire differentially numbers of offspring within a group, may show higher calling rates, reflecting their status-dependent activities within the group and territory. Both laboratory findings in rats and field observations of various social species suggest that alarm vocalizations include emotional and referential or descriptive information with reference to the source of threat and its characteristics. Research into common mechanisms may shed light on the underpinnings of a variety of human psychiatric conditions associated with fear and anxiety.

Keywords: defense; conspecific; VBS (visible burrow system); predator; alarm vocalizations; defensive threat; anxiety; fear

I. Defense

Defensive behaviors constitute a diverse range of responses to immediate and potential threats in the environment (Blanchard and Blanchard, 2008). A defensive repertoire is constructed of such behaviors, each of which has proved successful in response to particular types of threats in particular situations. Natural threats may include dangerous features of the environment, predators and associated cues, and conspecific attack (Endler, 1986). Inanimate environmental threat sources, such as floods or fires, elicit relatively few and simple defenses, whereas interactions with animate threats, such as predators and conspecifics, may involve an intricate arms race that ultimately facilitates the development of elaborate patterns of species-specific defensive behaviors (Endler, 1997; Dawkins and Krebs, 1979; Blanchard and Blanchard, 2008).

In rodents, an extensive repertoire of innate defensive behaviors can be seen in predator–prey interactions in natural and semi-natural (laboratory) situations. To predator threat, rodents of small prey species exhibit a general cessation of ongoing non-defensive activities (e.g., grooming, playing, foraging, feeding and sexual behavior) in both adults (Blanchard and Blanchard, 1989) and preweanling pups (Takahashi, 1992), along with enhancement of avoidance, flight, freezing, risk assessment, defensive threat and defensive attack behaviors (Blanchard and Blanchard, 2008). Rodents will also readily bury novel, aversive, or potentially dangerous objects (Treit et al., 1981). Manifestation of specific behaviors is dependent on a number of factors: (1) *context* – an

*Corresponding author. E-mail: litvin@hawaii.edu

ISBN 978-0-12-374593-4
DOI: 10.1016/B978-0-12-374593-4.00015-2

animal will typically flee a threat within an environment in which escape is possible, yet when trapped it will freeze; (2) *stimulus ambiguity* – whereas ambiguous stimuli, such as predator odors, elicit risk assessment behaviors (e.g., stretch attend, stretch approach, olfactory investigation) and are often associated with a state of anxiety, discrete, present threats elicit flight, avoidance, defensive threat and attack, and are associated with a state of fear; (3) *defensive distance* – the distance between predator and prey shifts defensive coping strategies from avoidance to escape, with short distances to the threat and unavoidable contact culminating in defensive threat and attack postures.

In addition to environmental influences on the evolution of defense, organismic factors influence the relative importance of each behavior within a complete repertoire. For example, while practically all vertebrates and many invertebrates show some form of behavioral defense, the metabolic restrictions of cold-blooded reptiles may have promoted a shift toward passive defensive strategies, such as avoidance and hiding, at least in colder climates, whereas the emergence of warm-blooded mammals enabled rapid evolution of active strategies, such as escape and defensive threat and attack (Edmunds, 1974; see also Porges and Lewis, Chapter 7.2 in this volume). Within mammalia, long legs and flight as a predominant defense are interactive features, each promoting the success of the other, particularly for animals living in open plain habitats. In members of gregarious, but not solitary, species it is highly adaptive for animals to develop the ability to both emit and detect signals that may serve as warnings against potential or present dangers.

II. Defensive vocalizations

The evolution of intricate social networks in vertebrate animals has been paralleled by the development of communication skills that are also integrated into the defensive repertoires of some avian and many mammalian species, ranging from rodents to primates (Bradbury and Vehrencamp, 1998). Vocal communications of threat are manifested as "alarm vocalizations" (also "alarm calls," or "alarm cries") that serve to communicate information about imminent or potential threats in the environment. Alarm vocalizations may carry information about the threat source, such as its intensity and potential attack vector (e.g., terrestrial, aerial, fossorial), as well as information about the sender such as size, sex, age, or even social status. Mammals integrate defensive vocalizations, which span sonic to ultrasonic ranges, as a response to a variety of adverse circumstances; e.g., when presented with unpleasant or painful stimuli (Kaltwasser, 1991), as a defensive threat (Darwin, 1872), when in danger (Greene and Meagher, 1998; Manser et al., 2002) and immediately following a successful retreat into safety (Blanchard et al., 1991; Blumstein and Armitage, 1997). These different vocalizations usually have different functions when directed at the threat source or at (nonattacking) conspecifics as the target of the communication. Consequently, some ambiguity has arisen as to the precise definition and usage of the term "alarm vocalization" or "alarm cry" (Litvin et al., 2007), either as a warning signal of the presence of danger, targeted toward conspecifics (Maynard Smith, 1965; Sherman, 1977) or as a defensive threat vocalization intended directly to deter an oncoming threat, such as a predator (Fitzgibbon and Fanshawe, 1988; Hasson, 1991; Blumstein, 1999; Swaisgood et al., 1999), or indeed both (Shelley and Blumstein, 2004). These distinctions signify very different causality and functions, but field studies and laboratory observations have demonstrated both types of vocalizations in a variety of animals. An analysis of the differences between alarm vocalizations and defensive threats has been presented elsewhere (Litvin et al., 2007). Here, we refer to "alarm vocalizations" in the context of social signals directed at conspecifics.

II.A. Laboratory studies

There are three types of ultrasonic vocalizations (USV) which have been identified and described in laboratory rats (*Rattus norvegicus*) (Blanchard et al., 1986; Brudzynski et al., 1993). When isolated from the dam, rat pups emit circa 40 kHz short vocalizations; an infant distress call or isolation vocalization (Brudzynski et al., 1999). Juvenile and adult rats produce two kinds of USV; a high-pitched and short, circa 50 kHz USV, and a low-pitched and longer, circa 22 kHz USV. 50 kHz USV are emitted in non-aggressive conspecific social interactions (Blanchard et al., 1993), during play (Knutson et al., 1998), and during male ejaculation (McIntosh and Barfield, 1980). Such calls have also been reported during fighting in male conspecifics (Sales and Pye, 1974; Burgdorf et al., 2008). Furthermore, rats will work for playback of these calls, suggesting that they produce a positively valenced appetitive state.

Adult rats emit 22 kHz USV in a variety of situations (for review see Sales and Pye, 1974; Brudzynski, 2005): (1) during the post-ejaculatory period (Barfield and Geyer, 1972; Barfield and Thomas, 1986); (2) in intraspecific defensive/submissive postures in the context of intermale social interactions (Lore et al., 1976; Thomas et al., 1983; Portavella et al., 1993); (3) when in the presence of a predator (Blanchard et al., 1991, 1992); (4) when startled during opiate and cocaine withdrawal (Vivian and Miczek, 1991; Mutschler and Miczek, 1998; Covington and Miczek, 2003); and (5) when subjected to chronic pain (Calvino et al., 1996). In contrast to readily "self-administering" playback of 50 kHz recordings, rats avoid playback of 22 kHz vocalizations, further suggesting that the latter are aversive (Burgdorf et al., 2008). Collectively, these findings indicate that 22 kHz USV are elicited in stressful situations and can be used as indices of aversive affective states (Bell, 1974; Brudzynski, 2001; Knutson et al., 2002).

Careful observation and analysis of alarm vocalizations within a laboratory setting has enabled their use as indices of affective states in rats and mice (Bell, 1974; Brudzynski, 2001; Knutson et al., 2002; see also Brudzynski, Chapter 7.3 in this volume) in a variety of both conditioned and unconditioned models of anxiety (Vivian et al., 1994; Brudzynski and Chiu, 1995; Miczek et al., 1995; Sanchez, 2003; Koo et al., 2004; Nunes Mamede Rosa et al., 2005). This laboratory has observed and analyzed alarm vocalizations in the semi-natural context of a visible burrow system (VBS) consisting of an open surface area connected to tunnels and chambers and containing small mixed-sex groups of adult rats and sometimes also pups born in the habitat. Following a week of habituation to the context, a cat was introduced into the surface area for 15 minutes and USV were recorded for an additional 175 minutes after removal of the cat. A control/sham procedure 2–3 days later presented a toy cat in the VBS, followed by a second (real) cat exposure after an additional interval of the same length.

The resident rats displayed a consistent and highly reliable sequence of behaviors in response to cat exposure. On the first exposure, rats quickly fled into the burrows, subsequently emitting 22 kHz USV (actually 18–24 kHz USV) that lasted for about 30 minutes after the cat was removed and gradually subsided thereafter. Although the dominant male was typically the only rat on the surface and actually in visual contact with the cat, a single initial vocalization often recruited additional vocalizations from other members of the group (Blanchard et al., 1991). During cat exposure and for several hours afterward, rats reduced or ceased non-defensive activities such as eating, copulation, or conspecific social interactions while avoiding the surface of the VBS. When presented with the toy cat control, rats also quickly fled into the burrows and emitted 22 kHz USV, yet these behaviors did not persist; they quickly resumed normal non-defensive activities. A second cat exposure produced identical behaviors to the first exposure, including flight, risk assessment and 22 kHz USV, indicating little habituation to cat presentation in the VBS. However, when rats were exposed individually to a cat either in a VBS or an open field cage that provided no escape, they did not produce USV (Blanchard et al., 1991), congruent with the notion that USV are indeed alarm vocalizations oriented towards conspecifics (compare with results obtained in single cages; see Wöhr and Schwarting, Chapter 4.2 in this volume).

Playback of 22 kHz USV induced a noticeable reduction in the behavior of the listening rats, further strengthening the notion that these are conspecific alarm calls (Brudzynski and Chiu, 1995). In addition, playback of 22 kHz USV appeared to induce vocalizations from conspecifics as young as 10 days of age when in a social setting (Blanchard, unpublished observations). Rat pups on postnatal day 10 responded to the playback by emitting USV of approximately 27 kHz. These calls were markedly different from the typical 40 kHz isolation calls of young pups separated from the dam. In contrast, separated rat pups on postnatal day 14 inhibit the typical 40 kHz infant distress calls (isolation vocalizations) on encountering a threatening stimulus (adult male conspecific) (Takahashi, 1992). Collectively, these findings suggest that 27 kHz vocalizations serve a different function than 40 kHz pup vocalizations and may be related to the classic 22 kHz alarm vocalizations of adult rats.

It is useful to analyze the timing and duration of 22 kHz USV within the complete defensive repertoire displayed in the context of the VBS. On introduction of the cat into the VBS, rats reliably exhibited several types of defenses, with flight closely followed by emission of USV. During the most intense periods of USV emission, rats were most often freezing, but this immobile period invariably involved orientation toward the tunnel leading to the surface area of the VBS, constituting a rather passive form of risk assessment. After the USV died down, more active forms of risk assessment gradually appeared, including approach to the openings of the surface area. Rats might extend their heads closer and closer to these openings over time, orienting their eyes and ears to the surface, but without actually entering it (Blanchard and Blanchard, 1989). These observations fit well with an interpretation that circa 22 kHz USV in rats serve to warn colony members of the presence of danger, whereas risk assessment activities enable the subject to assess the likelihood of this danger, initially from a relatively safe location (the tunnel or burrow) and later, in the absence of any further indication of danger, with forays onto the surface itself. Thus, these behaviors, both emission of USV and risk assessment activities, serve as components of a defensive decision-making sequence, with USV promoting defensiveness and risk assessment reducing it (given that, as in these studies, the cat is no longer present). This process provides an indication of the essential conservatism of the defensive process in that the interval between the gradual phasing out of alarm USV in these colonies and the first entry of any rat onto the surface area was about five hours. Better safe than sorry.

However, these behaviors did not end the defensive sequence. When the first rat (always the dominant) re-entered the surface area, it would quickly sprint across to the near corner of the open space, exiting one tunnel and entering another in a few seconds or less. Such "corner runs" may be an efficient way to assess the potential danger of a situation further, by soliciting an attack from the no-longer-visible threat source, when such attack is unlikely to be successful because the rat's period of vulnerability is so short. As such, it may represent an additional form of risk assessment.

Sex differences in emission of USV to a live cat included differences in the frequency and duration of emission, as well as in the sonographic characteristics of these calls, with females reliably producing more frequent and longer USV than males. While cat exposure produced high levels of USV in both sexes, perhaps indicative of a ceiling effect (maximal response), potential threats (cat odor alone) produced marked sex differences. Sonographic analyses showed consistent differences in the patterns of USV emitted by the two sexes. Females showed a mean frequency of 22.2 kHz with a narrow frequency distribution, while males showed a mean frequency of 20.1 kHz with a wider distribution, congruent with the expected relationship between body size and vocalization frequency (see Fletcher, Chapter 3.1 in this volume). Six major subtypes of vocalizations were described based on the sonographic structure: horizontal; linear ascending; linear descending; U-shaped; negatively accelerated ascending; and negatively descending pulses. Females showed higher levels of horizontal, linear descending, U-shaped and negatively accelerated ascending USV, while males largely emitted negatively accelerated descending pulses. These sex differences are in accordance with the notion that in addition to their emotional content, 22 kHz USV carry significant informative value; it may be adaptive for females to produce more USV due to their care of pups, a particularly vulnerable group of conspecifics that may be closely related to the caller. Gender variation in the sonographic characteristics of alarm vocalizations may also provide a valid means of recognition of the caller's gender or other features. The wide range of USV characteristics that show variation may also contribute to the identification of unique signals through which individual rats may be recognized. Recent studies of responses of rats to olfactory stimuli from cats indicated their ability to discriminate between individuals (Staples et al., 2008). These findings suggest that the complexity of 22 kHz vocalizations may provide a parallel in terms of individual identification of conspecifics.

II.B. Field studies

Recent studies of animal communication question the classical distinction between the semantic communication of human language and the emotion-related calls of animals. Field observations show that threat-linked animal vocalizations may be evaluated in terms of both their informative value, i.e., the reliability with which a call signifies the presence of a threat, and their referential specificity, i.e., the precision with which particular stimuli elicit a specific call type (Seyfarth and Cheney, 2003).

The field literature contains many examples of vocalizations oriented toward conspecifics that include referential or descriptive information with reference to the source of threat and its characteristics. African meerkat sentinels (*Suricata suricatta*) (meerkat groups usually consist of 3 to 30 individuals) cease foraging and climb to high ground or to the tops of shrubs to assess better the danger of possible predator threats (Clutton-Brock et al., 1999). When such a threat is detected, meerkats emit alarm vocalizations that have been shown to convey information regarding both predator type and the urgency of danger to its group (Manser et al., 2002). Vocalizations of black-capped chickadees (*Poecile atricapilla*) convey predator size to conspecifics, subsequently modulating their defensive reactions (Templeton et al., 2005). Diana monkeys (*Cercopithecus diana*) produce differential alarm calls for aerial and terrestrial predators, as well as for predators, nonpredators and other general disturbances (Zuberbühler et al., 1997). Belding's ground squirrels (*Spermophilus beldingi*) emit alarm calls that have been shown to be nepotistic warnings preferentially directed at relatives (Sherman, 1977); young squirrels learn to respond more quickly to calls signifying fast-moving predators than to those indicative of slow-moving ones, thus demonstrating both their referential value and their importance for survival (Mateo, 1996). Gunnison's prairie dogs (*Cynomys gunnisoni*) produce four different types of alarm calls depending on the type of predator, which result in two different responses by conspecifics (Kiriazis and Slobodchikoff, 2006). Several species of marmots (*Marmota flaviventris*, *M. olympus*, *M. caligata*, *M. monax*) emit alarm calls directed towards pups in order to warn them of danger (Blumstein and Armitage, 1997, 1998; Daniel and Blumstein, 1998). The alarm calls of a number of species are modified by the immediacy of a threat, with distant threats eliciting one type of call and immediate threats a distinctly different call (Klump and Shalter, 1984; Macedonia and Evans, 1993). Similarly, defensive responses of Richardson's ground squirrels (*Spermophilus richardsonii*) vary in accordance with the proximity of the conspecific caller, with neighbor squirrel calls eliciting higher levels of vigilance than non-neighbor calls, indirectly indicative of the threat urgency (Hare, 1998).

A number of phenomena observed in rat VBS studies are supported by field work with other mammals. The dependence of vocalizations on the presence of conspecifics is in line with observations of similar calls in other rodent species (Blumstein and Armitage, 1997, 1998). Female Belding's ground squirrels (Sherman, 1977) and marmots (Blumstein and Armitage, 1997, 1998) showed significantly higher alarm vocalizations, akin to findings showing enhanced levels of 22 kHz USV in female rats (Blanchard et al., 1992), observations suggesting greater adaptive value for female alarm calls as their young are more likely to be in close proximity. Meerkats and marmots emit alarm vocalizations from a place of relative safety. Similarly, rats in the VBS are more likely to emit USV from within the burrows. Field studies in meerkats (Clutton-Brock et al., 1999) and marmots (Blumstein and Armitage, 1997) showed a link between dominance and risk assessment, suggesting that dominant animals are also more likely to produce alarm vocalizations. In the VBS, dominant males are more likely than subordinates to occupy the surface area, and as a result are more likely to first encounter threat-related stimuli, and to initiate alarm vocalizations.

III. Conclusions

The study of defensive behaviors and their biology may be relevant to analysis of a number of human emotion-related behaviors. In this context, alarm vocalizations are particularly interesting as they may shed light on the social and communicatory aspects of anxiety and other threat-related emotions. Laboratory and field studies suggest that alarm vocalizations in a variety of species serve to convey both emotion and information related to the caller and the threat source. Their conservation over mammalian species, and in particular their strong association with species showing gregarious social systems, provide substantial support for a view that human alarm or warning vocalizations may provide a strong functional parallel to those of non-human mammals.

References

Barfield, R.J., Geyer, L.A., 1972. Sexual behavior: ultrasonic postejaculatory song of the male rat. Science 176, 1349–1350.

Barfield, R.J., Thomas, D.A., 1986. The role of ultrasonic vocalizations in the regulation of reproduction in rats. Ann. N.Y. Acad. Sci. 474, 33–43.

Bell, R.W., 1974. Ultrasounds in small rodents: arousal-produced and arousal producing. Dev. Psychobiol. 7, 39–42.

Blanchard, D.C., Blanchard, R.J., 2008. Defensive behaviors, fear and anxiety. In: Blanchard, R.J., Blanchard, D.C., Griebel, G., Nutt, D.J. (Eds.) Handbook of Anxiety and Fear. Elsevier Academic Press, Amsterdam, The Netherlands.

Blanchard, R.J., Blanchard, D.C., 1989. Antipredator defensive behaviors in a visible burrow system. J. Comp. Psychol. 103, 70–82.

Blanchard, R.J., Flannelly, K.J., Blanchard, D.C., 1986. Defensive behavior of laboratory and wild *Rattus norvegicus*. J. Comp. Psychol. 100, 101–107.

Blanchard, R.J., Blanchard, D.C., Agullana, R., Weiss, S.M., 1991. Twenty-two kHz alarm cries to presentation of a predator, by laboratory rats living in visible burrow systems. Physiol. Behav. 50, 967–972.

Blanchard, R.J., Yudko, E.B., Blanchard, D.C., Taukulis, H.K., 1993. High-frequency (35–70 kHz) ultrasonic vocalizations in rats confronted with anesthetized conspecifics: effects of gepirone, ethanol, and diazepam. Pharmacol. Biochem. Behav. 44, 313–319.

Blanchard, R.J., Agullana, R., McGee, L., Weiss, S., Blanchard, D.C., 1992. Sex differences in the incidence and sonographic characteristics of antipredator ultrasonic cries in the laboratory rat (*Rattus norvegicus*). J. Comp. Psychol. 106, 270–277.

Blumstein, D.T., 1999. The evolution of functionally referential alarm communication: multiple adaptations, multiple constraints. Evol. Comm. 3, 135–147.

Blumstein, D.T., Armitage, K.B., 1997. Alarm calling in yellow-bellied marmots: 1. The meaning of situationally variable alarm calls. Anim. Behav. 53, 143–171.

Blumstein, D.T., Armitage, K.B., 1998. Why do yellow-bellied marmots call? Anim. Behav. 56, 1053–1055.

Bradbury, J.W., Vehrencamp, S.L., 1998. Principles of Animal Communication. Sinauer Associates, Inc., Sunderland, MA.

Brudzynski, S.M., 2001. Pharmacological and behavioral characteristics of 22 kHz alarm calls in rats. Neurosci. Biobehav. Rev. 25, 611–617.

Brudzynski, S.M., 2005. Principles of rat communication: quantitative parameters of ultrasonic calls in rats. Behav. Genet. 35, 85–92.

Brudzynski, S.M., Chiu, E.M., 1995. Behavioural responses of laboratory rats to playback of 22 kHz ultrasonic calls. Physiol. Behav. 57, 1039–1044.

Brudzynski, S.M., Kehoe, P., Callahan, M., 1999. Sonographic structure of isolation-induced ultrasonic calls of rat pups. Dev. Psychobiol. 34, 195–204.

Brudzynski, S.M., Bihari, F., Ociepa, D., Fu, X.W., 1993. Analysis of 22 kHz ultrasonic vocalization in laboratory rats: long and short calls. Physiol. Behav. 54, 215–221.

Burgdorf, J., Kroes, R.A., Moskal, J.R., Pfaus, J.G., Brudzynski, S.M., Panksepp, J., 2008. Ultrasonic vocalizations of rats (*Rattus norvegicus*) during mating, play, and aggression: behavioral concomitants, relationship to reward, and self-administration of playback. J. Comp. Psychol. 122, 357–367.

Calvino, B., Besson, J.M., Boehrer, A., Depaulis, A., 1996. Ultrasonic vocalization (22–28 kHz) in a model of chronic pain, the arthritic rat: effects of analgesic drugs. Neuroreport 7, 581–584.

Clutton-Brock, T.H., O'Riain, M.J., Brotherton, P.N., Gaynor, D., Kansky, R., Griffin, A.S., Manser, M., 1999. Selfish sentinels in cooperative mammals. Science 284, 1640–1644.

Covington 3rd., H.E., Miczek, K.A., 2003. Vocalizations during withdrawal from opiates and cocaine: possible expressions of affective distress. Eur. J. Pharmacol. 467, 1–13.

Daniel, J.C., Blumstein, D.T., 1998. A test of the acoustic adaptation hypothesis in four species of marmots. Anim. Behav. 56, 1517–1528.

Darwin, C., 1872. The Expression of the Emotions in Animals and Man. Appleton & Co., New York, NY.

Dawkins, R., Krebs, J.R., 1979. Arms races between and within species. Proc. R. Soc. Lond. B. Biol. Sci. 205, 489–511.

Edmunds, M., 1974. Defence in Animals. A Survey of Anti-predator Defences. Longman Group, New York, NY.

Endler, J.A., 1986. Defense Against Predators. University of Chicago Press, Chicago, IL.

Endler, J.A., 1997. Interactions between predators and prey. In: Krebs, J.R., Davies, N.B. (Eds.), Behavioural Ecology: An Evolutionary Approach. Blackwell Publishing, Inc., London, UK.

Fitzgibbon, C.D., Fanshawe, J.H., 1988. Stotting in Thomson gazelles: an honest signal of condition. Behav. Ecol. Sociobiol. 23, 69–74.

Greene, E., Meagher, T., 1998. Red squirrels, *Tamiasciurus hudsonicus*, produce predator-class specific alarm calls. Anim. Behav. 55, 511–518.

Hare, J.F., 1998. Juvenile Richardson's ground squirrels, *Spermophilus richardsonii*, discriminate among individual alarm callers. Anim. Behav. 55, 451–460.

Hasson, O., 1991. Pursuit deterrent signals: the communication between prey and predator. Trends. Ecol. Evol. 6, 325–329.

Kaltwasser, M.T., 1991. Acoustic startle-induced ultrasonic vocalization in the rat: a novel animal model of anxiety? Behav. Brain Res. 43, 133–137.

Kiriazis, J., Slobodchikoff, C.N., 2006. Perceptual specificity in the alarm calls of Gunnison' prairie dogs. Behav. Process. 73, 29–35.

Klump, G.M., Shalter, M.D., 1984. Acoustic behaviour of birds and mammals in the predator context. I. Factors affecting the structure of alarm signals. II. The functional significance and evolution of alarm signals. Z. Tierpsych. 66, 189–226.

Knutson, B., Burgdorf, J., Panksepp, J., 1998. Anticipation of play elicits high-frequency ultrasonic vocalizations in young rats. J. Comp. Psychol. 112, 65–73.

Knutson, B., Burgdorf, J., Panksepp, J., 2002. Ultrasonic vocalizations as indices of affective states in rats. Psychol. Bull. 128, 961–977.

Koo, J.W., Han, J.S., Kim, J.J., 2004. Selective neurotoxic lesions of basolateral and central nuclei of the amygdala produce differential effects on fear conditioning. J. Neurosci. 24, 7654–7662.

Litvin, Y., Blanchard, D.C., Blanchard, R.J., 2007. Rat 22 kHz ultrasonic vocalizations as alarm cries. Behav. Brain. Res. 182, 166–172.

Lore, R., Flannelly, K., Farina, P., 1976. Ultrasounds produced by rats accompany intraspecific fighting. Aggress. Behav. 2, 175–181.

Macedonia, J.M., Evans, C.S., 1993. Variation among mammalian alarm call systems and the problem of meaning in animal signals. Ethology 93, 177–197.

Manser, M.B., Seyfarth, R.M., Cheney, D.L., 2002. Suricate alarm calls signal predator class and urgency. Trends Cogn. Sci. 6, 55–57.

Mateo, J.M., 1996. The development of alarm-call response behaviour in free-living juvenile Belding's ground squirrels. Anim. Behav. 52, 489–505.

Maynard Smith, J., 1965. The evolution of alarm calls. Naturalist 99, 59–63.
McIntosh, T.K., Barfield, R.J., 1980. The temporal patterning of 40–60 kHz ultrasonic vocalizations copulation in the rat. Behav. Neural Biol. 29, 349–358.
Miczek, K.A., Weerts, E.M., Vivian, J.A., Barros, H.M., 1995. Aggression, anxiety and vocalizations in animals: GABAA and 5-HT anxiolytics. Psychopharmacology (Berl.) 121, 38–56.
Mutschler, N.H., Miczek, K.A., 1998. Withdrawal from i.v. cocaine "binges" in rats: ultrasonic distress calls and startle. Psychopharmacology (Berl) 135, 161–168.
Nunes Mamede Rosa, M.L., Nobre, M.J., Ribeiro Oliveira, A., Brandao, M.L., 2005. Isolation-induced changes in ultrasonic vocalization, fear-potentiated startle and prepulse inhibition in rats. Neuropsychobiology 51, 248–255.
Portavella, M., Depaulis, A., Vergnes, M., 1993. 22–28 kHz ultrasonic vocalizations associated with defensive reactions in male rats do not result from fear or aversion. Psychopharmacology (Berl) 111, 190–194.
Sales, G., Pye, D., 1974. Ultrasonic Communication by Animals. John Wiley and Sons, New York, NY.
Sanchez, C., 2003. Stress-induced vocalisation in adult animals. A valid model of anxiety? Eur. J. Pharmacol. 463, 133–143.
Seyfarth, R.M., Cheney, D.L., 2003. Meaning and emotion in animal vocalizations. Ann. N.Y. Acad. Sci. 1000, 32–55.
Shelley, E.L., Blumstein, D.T., 2004. The evolution of vocal alarm communication in rodents. Behav. Ecol. 16, 169–177.
Sherman, P.W., 1977. Nepotism and the evolution of alarm calls. Science 197, 1246–1253.
Staples, L.G., Hunt, G.E., van Nieuwenhuijzen, P.S., McGregor, I.S., 2008. Rats discriminate individual cats by their odor: possible involvement of the accessory olfactory system. Neurosci. Biobehav. Rev. 32, 1209–1217.
Swaisgood, R.R., Owings, D.H., Rowe, M.P., 1999. Conflict and assessment in a predator-prey system: ground squirrels versus rattlesnakes. Anim. Behav. 57, 1033–1044.
Takahashi, L.K., 1992. Ontogeny of behavioral inhibition induced by unfamiliar adult male conspecifics in preweanling rats. Physiol. Behav. 52, 493–498.
Templeton, C.N., Greene, E., Davis, K., 2005. Allometry of alarm calls: black-capped chickadees encode information about predator size. Science 308, 1934–1937.
Thomas, D.A., Takahashi, L.K., Barfield, R.J., 1983. Analysis of ultrasonic vocalizations emitted by intruders during aggressive encounters among rats (*Rattus norvegicus*). J. Comp. Psychol. 97, 201–206.
Treit, D., Pinel, J.P., Fibiger, H.C., 1981. Conditioned defensive burying: a new paradigm for the study of anxiolytic agents. Pharmacol. Biochem. Behav. 15, 619–626.
Vivian, J.A., Miczek, K.A., 1991. Ultrasounds during morphine withdrawal in rats. Psychopharmacology (Berl) 104, 187–193.
Vivian, J.A., Farrell, W.J., Sapperstein, S.B., Miczek, K. A., 1994. Diazepam withdrawal: effects of diazepam and gepirone on acoustic startle-induced 22 kHz ultrasonic vocalizations. Psychopharmacology (Berl) 114, 101–108.
Zuberbühler, K., Noe, R., Seyfarth, R.M., 1997. Diana monkey long distance calls: messages for conspecifics and predators. Anim. Behav. 53, 589–604.

CHAPTER 5.2

Effect of altricial pup ultrasonic vocalization on maternal behavior

Markus Wöhr[1], Diego Oddi[2] and Francesca R. D'Amato[2*]

[1]*Experimental and Physiological Psychology, Philipps-University of Marburg, Marburg, Germany*
[2]*CNR Institute of Neuroscience, Rome, Italy*

Abstract: Ultrasonic calls are emitted by altricial rodent pups during separation from the mother and serve to orient and facilitate maternal retrieval to the nest. The relevance of associated olfactory and visual cues from the pup has been demonstrated and ultrasonic vocalizations alone have been proved to promote maternal retrieval. Mothers differ from pup-naive females in showing a greater behavioral and physiological responsiveness to pups' ultrasonic calls. Apart from retrieval behavior, no clear evidence exists that these vocalizations directly elicit nursing or grooming of the pups. Rather, grooming pups and rough contact stimulate pup calling. The relevance of these calls during mother–offspring interaction is discussed and it is suggested that they play a regulatory role in modulating infant responsiveness to stress.

Keywords: rodents; rats; mice; pups; ultrasonic calls; isolation calls; retrieval behavior; maternal behaviors; acoustic features of pup calls; mother–infant interaction

I. Introduction

Altricial pups of different rodent species are entirely dependent on their mother for nutrition, body temperature control, protection from intra- and inter-specific attacks and environmental insults. The pups have the capability to produce a variety of sounds from birth, the majority of which are pure ultrasounds (ultrasonic vocalizations; USVs). These calls are emitted in concomitance with some well-identified environmental conditions that could affect infant survival. Cold, absence of contact with conspecifics, unfamiliar olfactory stimuli and rough manipulation are among the most common stimuli that elicit pup USVs (Costantini and D'Amato, 2006). Blumberg and Sokoloff (2001) argued that these calls represent an acoustic by-product of laryngeal braking caused by abdominal compression in response to external stimulation. These authors stated that USVs simply reflect a state of arousal, and not a motivation to socially interact. However, several studies in the literature postulated that such calls represent the basis on which natural selection operated to confer them intraspecific communicative value (Ehret, 2005), as suggested by the continuation of the capability to emit and hear USVs into adulthood, mainly in response to threatening situations and in social affiliative contexts (Costantini and D'Amato, 2006).

The physical characteristics of these vocalizations present at least two advantages: they are not perceived by some predators and these high-frequency sounds are attenuated with distance more than lower frequency sounds, emphasizing the role that these calls play in close social interactions during infancy, as well as in adult animals (Branchi et al., 1998; Brudzynski and Fletcher, Chapter 3.3 in this volume). The cost:benefit ratio favors vocalizing pups, even if there is still the risk of being localized and attacked by unfamiliar infanticidal conspecifics (Elwood et al., 1990).

There is a great variability among pups in both the structure and emission rate of USVs. The first studies have simply explored the rate of emission of pup USVs in different experimental conditions and

*Corresponding author. E-mail: francesca.damato@cnr.it

ISBN 978-0-12-374593-4

DOI: 10.1016/B978-0-12-374593-4.00016-4

they found age and strain differences (Noirot, 1968; Oswalt and Meyer, 1975). The use of more complex equipment added other parameters to be considered such as calling rate, frequency range, presence of harmonics, bout structure of call sequences, sound amplitude and frequency modulation. Many factors may contribute to the great variability of calls particularly observed in pups. Some of them depend on individual characteristics (age, sex, strain, nutritional state); other factors may depend on stimuli the pups are exposed to (temperature, handling, odor, contact). The interactions between internal and external factors contribute to the complexity of each USV sequence.

The question is: do USVs simply express the individual internal state or do they have a communicative value? In other words, does the mother simply orient towards a source of noise, or can the pup communicate with the mother and modify her behavior?

II. Mother's behavioral response to ultrasonic vocalizations emitted by isolated pups

In 1956 Zippelius and Schleidt discovered that infant mice emit USVs during separation from mother and litter with frequencies between 70 and 80kHz. Already in this very first study, they have suggested: (1) that such calls reflect a negative affective state and named them – following Konrad Lorenz – "Pfeifen des Verlassenseins" ("whistles of loneliness"); and (2) that such calls serve communicative purposes, since they have observed that mothers leave the nest to retrieve pups which were scattered outside the nest. Furthermore, Zippelius and Schleidt (1956) showed that mothers are selectively retrieving active pups, which are able to vocalize, but not anesthetized or sacrificed pups, demonstrating that pup USVs are important stimuli for maternal retrieval behavior to occur.

Thus, the first study conducted to explore the role of pup USVs showed that they oriented and elicited the mother's exploratory behavior, facilitating retrieval of isolated pups. Later, Sewell (1970) demonstrated that the crucial stimuli for the induction of maternal search behavior are indeed pup USVs and not visual or olfactory cues associated with them, by performing a playback experiment. In order to do so, tape recordings were made of USVs from a five-day-old litter of wood mice (*Apodemus sylvaticus*). Recordings were presented to lactating mothers through a loudspeaker, which was placed on one or the other side of a T-partition in the end of the cage away from the nest. Mothers left the nest and displayed exploratory behavior during 48 out of 56 presentations of USVs (86%) and 38 out of these 48 responses were correct, i.e., the mother entered the partition where USVs were presented (79%). The mothers generally reached the loudspeaker within 5–30 seconds of the onset of playback. Importantly, mothers did not respond to background noise or artificial 45kHz pulses. Overall, Sewell (1970) showed for the first time that visual and olfactory cues are not necessary, since a vocalizing pup can be replaced by a loudspeaker playing back natural calls. These results clearly demonstrate that pup USVs have communicative value, supporting the early findings of Zippelius and Schleidt (1956). Since this pioneering study, many experiments have shown that maternal search behavior can be induced by presenting solely pup USVs (Smith, 1976; Ehret and Haack, 1981, 1982, 1984; Ehret, 1987; Koch and Ehret, 1989; Uematsu et al., 2007; Wöhr et al., 2008) (for review see Ehret, 2005).

As in mice, rat pup USVs are also associated with maternal search and retrieval behavior (Allin and Banks, 1972; Smotherman et al., 1974; Brunelli et al., 1994; Hashimoto et al., 2001; Farrell and Alberts, 2002a,b; Wöhr and Schwarting, 2008). Shortly after the first playback experiment in mice (Sewell, 1970), Allin and Banks (1972) also used this approach to test the communicative value of rat pup USVs. They found that rat mothers leave their nest in response to playback and that they localize the stimulus source very well. In 80% of all cases, responding lactating females searched for their pups in one out of six areas where pup USVs were presented. However, response rates were in general very low, i.e., only half of the mothers responded to the playback during at least one out of six presentations. Later, Smotherman et al. (1974) found that lactating females do not prefer the arm of a Y-maze where pup USVs were presented compared to the arm without tone presentation. Since they observed stimulus-directed search behavior only when the USVs were accompanied by relevant olfactory cues, they postulated that olfactory cues from an isolated pup are "a necessary condition for retrieval" (Smotherman et al., 1974, p. 61). Recently, the assumption that pup odors are crucial for the maternal response to vocalizing pups was supported by Farrell and Alberts (2002b), who have shown that peripherally-induced anosmia attenuated maternal proximal orientation towards a vocalizing pup. However, there is at least one clear demonstration that USVs alone are sufficient cues to direct the mother to a calling pup (Wöhr and Schwarting, 2008).

Apart from maternal retrieval behavior, pup USVs were suggested to induce nest building in mice (Noirot, 1969) and rats (Hashimoto et al., 2001), to shorten the latency to transport litters from an endangered nest in rats (Brewster and Leon, 1980), as well as to induce anogenital licking in rats (Brouette-Lahlou et al., 1992) but not in mice (Noirot, 1969).

III. Mother's physiological response to ultrasonic vocalizations emitted by isolated pups

It is well-known that some behaviors of pups, for example suckling stimulation, modulate the expression of maternal behavior by promoting arched-back nursing postures (Stern and Johnson, 1990; Stern et al., 2002). The role of USVs in promoting maternal activation is less well-known. Terkel et al. (1979) reported that prolactin levels in mothers increased in response to USVs and, as is now well-known, several maternal behaviors such as retrieving, nest building and lactation position are promoted by prolactin (Mann and Bridges, 2001). However, Voloschin and Tramezzani (1984) demonstrated that, where nipple attachment was prevented, vocalizations of pups failed to increase prolactin release. More recently, Hashimoto et al. (2001), using a cannula inserted in the jugular vein, showed significant increase in prolactin in lactating rat during exposure to USVs, in concomitance with increases in maternal search, retrieving and nest building.

IV. Does responsiveness to ultrasonic vocalizations depend on maternal state?

Mothers, but not virgins, recognize the communicative significance of pup USVs. In two alternative-choice experiments, mouse mothers preferentially approached pup-like ultrasounds compared with a neutral, non-communicative sound, while pup-naive virgin females do not (Ehret et al., 1987). As for rats, Farrell and Alberts (2002a,b) demonstrated that mothers approached and maintained proximal orientation to a vocalizing pup to a greater extent than did virgin females, and they showed that estrogen plus progesterone administration facilitated proximal orientation toward vocalizing pups in the latter. Immunohistochemical studies of c-Fos expression indicated differences between mothers and pup-naive females in the auditory cortical fields after exposure to repetitive pup calls (Fichtel and Ehret, 1999). Moreover, Liu et al. (2006) presented the first electrophysiological evidence that, in lactating female mice, this behavioral preference is correlated with changes in the population activity of the auditory cortical neurons to the acoustic structure of these vocalizations.

V. Call features: which call features are important to induce maternal behavior?

Although lactating female mice respond to a large variety of ultrasounds, it has been demonstrated that they prefer USVs and synthesized call models with certain acoustic properties (frequency and time domain, minimum duration, critical bands) typical of pups of their own species (for review see Ehret, 2005). Uematsu et al. (2007) used a nanocrystalline silicon thermo-acoustic emitter to investigate mothers' responses to digitally-reproduced pup USVs in mice. When synthesized ultrasonic sounds were transformed from actual pup USVs altering duration or frequency, mothers did not approach the speaker. This is particularly remarkable, since mice discriminate pup calls from different USVs of the mouse repertoire produced by adult males, on the basis of the repetition rate of calls (Liu et al., 2003). Furthermore, there is some evidence that in competitive situations, i.e., in situations where more than one pup is calling for the mother, call amplitude is important (Wöhr et al., 2008). By means of a pup discrimination task, Wöhr et al. (2008) have shown that mothers spent more time in contact with pups emitting calls with high amplitudes. Remarkably, number of calls was not related to the maternal preference.

Little is known about the communicative value of various call parameters in rats. Based on developmental changes in call characteristics and accompanying changes in maternal retrieval behavior, Brudzynski and his colleagues have suggested that specific call features, such as the extent of frequency modulation, are likely to be critical for communication between pups and their mother, particularly in inducing retrieval behavior (Brudzynski et al., 1999). It was argued that alternating frequency sweeps can be more easily detected than a steady sound. This hypothesis was supported in a recent study, where Wöhr and Schwarting (2008) have shown that mothers responded to natural pup calls by approaching the loudspeaker, but did not respond to an artificial 40 kHz sine wave tone. The fact that a 40 kHz sine wave tone was insufficient to induce search behavior indicates that frequency modulation or bout structure contains important communicative value.

VI. Relationship between maternal behavior and pups' ultrasonic vocalizations: who is affecting whom?

Based on the observations that USVs are able to induce maternal care, Bell (1979) hypothesized that changed levels of pup calling behavior are responsible for so-called handling effects (Levine, 1957). He argued that "the altered maternal behavior induced by differences in rate of ultrasonic calling by pups appears to be the major mediating factor in the large body of early experience research" (Bell, 1979, p. 416). In fact, pups were found to emit more ultrasound during the handling protocol than controls (Bell et al., 1971; Hennessy et al., 1980; D'Amato and Cabib, 1987). Handling of pups (15 minutes of daily separation from the mother during the first 10–15 days of life) is usually followed by an increase in caregiving behavior towards pups, particularly maternal licking (Francis et al., 1999), which affects the biological basis of emotional behavior, thus making pups less anxious (Caldji et al., 1998; Francis et al., 1999; Menard et al., 2004; Zhang et al., 2005; Menard and Hakvoort, 2007). Consistently, when pups are repeatedly separated from the mother, a decrease in the number of USVs during the separation session was reported (Zimmerberg et al., 2003b; D'Amato et al., 2005). For instance, a recent study showed that pups isolated for several hours received a higher level of maternal care at the end of the separation than pups isolated for a shorter period of time, although they vocalized less than the latter (Zimmerberg et al., 2003a). Other studies observed no relationship between pup calling and maternal care (Marino et al., 2002) or a negative one (Wöhr and Schwarting, 2008). Specifically, Wöhr and Schwarting (2008) demonstrated that rat pups whose mothers displayed more pup licking behavior during the first six days of life emitted fewer USVs and showed reduced calling time during isolation on postnatal day 11 than pups with mothers showing low levels of licking. These results suggested that maternal licking may affect pup calling behavior rather than vice versa.

Another approach to evaluate the relationship between maternal behavior and USVs emitted by pups during isolation is the comparison between data from different mouse strains. It is known that mouse and rat strains differ both in maternal care (e.g., Anisman et al., 1998; Wood et al., 2001) and USV emission (Sales and Smith, 1978; Hennessy et al., 1980; Cohen-Salmon et al., 1985; D'Amato et al., 2005; Wöhr et al., 2008). In rats, animals with high rates of calling, as recorded in maternally separated pups (Brunelli, 2005) or in adolescent tickled rats (Burgdorf et al., 2005), were mated to obtain lines of high/low emitters (Harmon et al., 2008). Unfortunately, most of these studies failed to analyze maternal behavior in parallel, not clarifying whether the differences in USVs emitted during social isolation could rather be related to differences in maternal behavior among these groups. The role of the mother in affecting pup calls was demonstrated by D'Amato and Populin (1987), showing that normal hearing pups reduced the amount of USVs when reared by deaf mothers, and by Weller et al. (2003), showing that the number of USVs in genetically-identical pups was affected by the genotype of the mother. Also, Wöhr et al. (2008) demonstrated, by embryo-transfer procedure, that the amount of USVs emitted in isolation is dependent on the dyadic interaction between mother and pup.

Taking advantage of the existence of spontaneous differences in maternal caregiving behavior between C57 and BALB mouse dams (Anisman et al., 1998), D'Amato et al. (2005) obtained evidence that a high level of maternal care resulted in a reduction of pup USVs. Specifically, C57 mothers showed higher maternal care than BALB dams, with C57 pups emitting fewer calls than BALB ones. Moreover, both brief handling and separation during the first postnatal days and adoption procedure were related to an increase in maternal care, associated with a decrease in USVs in isolated pups (Darnaudéry et al., 2004; D'Amato et al., 2005).

These data stress the existence of an inverse relation between maternal care and USVs, but no studies have clarified if the pup's calling is modulated by maternal behavior, or rather if the latter is shaped by USVs. If the second hypothesis is true, we should expect mother–offspring communication also to take place during in-nest episodes.

Branchi et al. (1998) demonstrated that tactile stimulation also elicited USVs in isolated mouse pups and the sound characteristics of these calls differed from those elicited by other stimuli (e.g., nest odor, low temperature, etc.). Unfortunately, the behavior of the mothers in response to these different calls has not been analyzed. To test whether this is the case, a brief preliminary experiment was conducted using

NMRI albino mice. A mother and its offspring (four males and four females) were continuously monitored in the home cage for the last four hours of the light phase (from 15:00 to 19:00 p.m.) of postnatal day 8. An ultrasound microphone was suspended over the nest, and the vocalizations of the pups recorded and stored for subsequent spectrographic analysis (Avisoft Bioacoustics, Berlin, Germany). As shown in Fig. 1a, this interval of time was sufficient to observe several maternal behaviors, both inside and outside the nest. Since the expression of the different behaviors varied greatly, the number of vocalizations was corrected as a ratio of the time spent by the dam in each behavior (Fig. 1b). Arched back and blanket posture scores (see Shoji and Kato, 2006, for the description of these maternal behaviors) were collapsed into a "passive nursing" measure, while "active maternal care" comprised grooming pups and nest building behaviors. The label "out" (Fig. 1b) was a composite measure including various activities out of the nest (i.e., eating, drinking, moving, sniffing and self-grooming). Results showed that pups vocalized not only in the ultrasonic range, but also and primarily (63% of the calls) in the audible spectrum. These "wriggling calls" (WC) were virtually absent when the mother was outside the nest, consistent with the notion that such signals occur during mother–infant interaction, for example in response to rough contact (Ehret, 1975; Ehret and Bernecker, 1986). Noticeably, although the absolute number of USVs was predictably higher during periods of separation (52% of the calls against 24% emitted during lactation and 23% during bouts of active maternal care), the corrected values showed a different situation. The majority of the vocalizations were produced when the mother and pups remained in close contact (Fig. 1b). Although these brief separations from the mother are possibly perceived as less threatening compared with the usual separation paradigms, in which the pup is isolated from both the dam and littermates, it is intriguing that USVs occurred even when the dam was actively nursing the offspring. Spectrographic analysis also showed that vocalizations produced when the mother was grooming pups or building the nest were more modulated than in other situations, a characteristic that Guo and Holy (2007) linked to a higher activation in adult male mice. In addition, internal frequency shifts are thought to be perceived as siren-like signals, thus being more likely noticed by the mother (Brudzynski et al., 1999). These preliminary data suggest the possibility that ultrasounds not only facilitate maternal proximity, but also contribute to shape the dam's behavior even during close interactions. It should be noted that an analogous function has been assigned to wriggling calls, whose playback has been shown to elicit changes in the maternal nursing position (Ehret and Bernecker, 1986). It is conceivable that both wriggling calls and

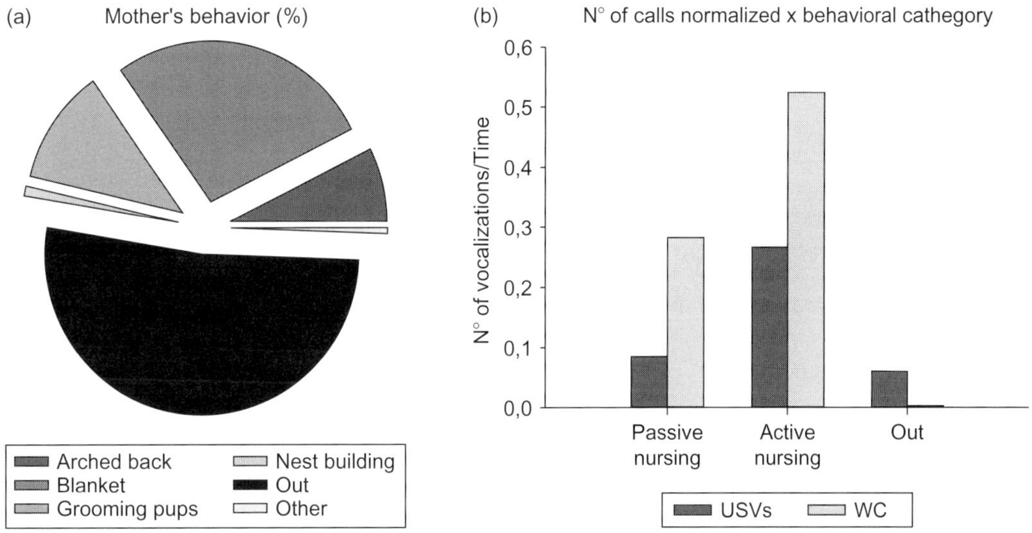

Fig. 1. (a) Maternal behaviors observed in an NMRI dam during a four-hour continuous sampling session on postnatal day 8; (b) number of ultrasonic vocalizations (USVs) and wriggling calls (WC) recorded during the four-hour continuous recording, normalized according to the maternal behavior observed in concomitance with the vocalizations. Passive nursing: arched back and blanket postures; Active nursing: grooming pups and nest building behaviors; Out: any behavior performed by the mother not in contact with pups.

USVs share a similar function inside the nest, acting in synergy. On the other hand, during separation, the USVs retain the survival advantages of attracting the mother while reducing the chance of predatory attacks.

VII. Conclusions

Meaney and his research group, as well as others, have demonstrated that maternal behavior, in particular licking and grooming of the pups, affects the behavioral and physiological response to stress of the offspring at adulthood (Liu et al., 1997; Caldji et al., 1998; Francis et al., 1999; Menard et al., 2004; Zhang et al., 2005; Menard and Hakvoort, 2007), and this effect on emotionality can be detected in very young pups by measuring their USVs during isolation (Wöhr and Schwarting, 2008). It has already been suggested that USVs may represent a precocious index of pup emotionality (Dichter et al., 1996; Zimmerberg et al., 2005) and they induce maternal searching and pup retrieval. We suggest that USVs also regulate mother–offspring interaction and that the amount of care received by pups in the undisturbed conditions is inversely related to the amount of calls the pups emit during isolation. The more arched-back nursing and grooming the pup received, the fewer the calls it will emit when out of the nest. If we consider USVs as a measure of attachment of pups to the mother/nest environment, the higher levels of maternal care can be predictive of a reduced distress response during brief separation.

Acknowledgments

This research was supported by Telethon grant (GGP05220). D.O. was supported by a FILAS, regione Lazio Ph. D. grant, Italy. M.W. was supported by the Gustav-Adolf-Lienert-Stiftung, Germany.

References

Allin, J.T., Banks, E.M., 1972. Functional aspects of ultrasound production by infant albino rats (*Rattus norvegicus*). Anim. Behav. 20, 175–185.

Anisman, H., Zaharia, M.D., Meaney, M.J., Meraly, Z., 1998. Do early-life events permanently alter behavioral and hormonal responses to stressors? Int. J. Dev. Neurosci. 16, 149–164.

Bell, R.W., 1979. Ultrasonic control of maternal behavior: developmental implications. Am. Zool. 19, 413–418.

Bell, R.W., Nitschke, W., Gorry, T.H., Zachman, T.A., 1971. Infantile stimulation and ultrasonic signaling: a possible mediator of early handling phenomena. Dev. Psychobiol. 4, 181–191.

Blumberg, M.S., Sokoloff, G., 2001. Do infant rats cry? Psychol. Rev. 108, 83–95.

Branchi, I., Santucci, D., Vitale, A., Alleva, E., 1998. Ultrasonic vocalizations by infant laboratory mice: a preliminary spectrographic characterization under different conditions. Dev. Psychobiol. 33, 249–256.

Brewster, J.A., Leon, M., 1980. Relocation of the site of mother–young contact: maternal transport behavior in Norway rats. J. Comp. Physiol. Psychol. 94, 69–79.

Brouette-Lahlou, I., Vernet-Maury, E., Vigouroux, M., 1992. Role of pups' ultrasonic calls in a particular maternal behavior in Wistar rat: pups' anogenital licking. Behav. Brain Res. 50, 147–154.

Brunelli, S.A., 2005. Selective breeding for an infant phenotype: rat pup ultrasonic vocalization (USV). Behav. Genet. 35, 53–65.

Brunelli, S.A., Shair, H.N., Hofer, M.A., 1994. Hypothermic vocalizations of rat pups (*Rattus norvegicus*) elicit and direct maternal search behavior. J. Comp. Psychol. 108, 298–303.

Brudzynski, S.M., Kehoe, P., Callahan, M., 1999. Sonographic structure of isolation-induced ultrasonic calls of rat pups. Dev. Psychobiol. 34, 195–204.

Burgdorf, J., Panksepp, J., Brudzynski, S.M., Kroes, R., Moskal, J.R., 2005. Breeding for 50-kHz positive affective vocalization in rats. Behav. Genet. 35, 67–72.

Caldji, C., Tannenbaum, B., Sharma, S., Francio, D., Plotsky, P.M., Meaney, M.J., 1998. Maternal care during infancy regulates the development of neural systems mediating the expression of fearfulness in the rat. Proc. Natl. Acad. Sci. USA 95, 5335–5340.

Cohen-Salmon, C., Carlier, M., Roubertoux, P., Jouhaneau, J., Semal, C., Paillette, M., 1985. Differences in patterns of pup care in mice. V. Pup ultrasonic emissions and pup care behavior. Physiol. Behav. 35, 167–174.

Costantini, F., D'Amato, F.R., 2006. Ultrasonic vocalizations in mice and rats: social contexts and functions. Acta Zool. Sinica 52, 619–633.

D'Amato, F.R., Cabib, S., 1987. Chronic exposure to a novel odor increases pups' vocalizations, maternal care, and alters dopaminergic functioning in developing mice. Behav. Neural Biol. 48, 197–205.

D'Amato, F.R., Populin, R., 1987. Mother–offspring interaction and pup development in genetically deaf mice. Behav. Genet. 17, 465–475.

D'Amato, F.R., Scalera, E., Sarli, C., Moles, A., 2005. Pups call, mothers rush: does maternal responsiveness affect the amount of ultrasonic vocalizations in mouse pups? Behav. Genet. 35, 103–112.

Darnaudéry, M., Koehl, M., Barbazanges, A., Cabib, S., Le Moal, M., Maccari, S., 2004. Early and later adoptions differently modify mother–pup interactions. Behav. Neurosci. 118, 590–596.

Dichter, G.S., Brunelli, S.A., Hofer, M.A., 1996. Elevated plus-maze behavior in adult offspring of selectively bred rats. Physiol. Behav. 60, 299–304.

Ehret, G., 1975. Schallsignale der Hausmaus (*Mus musculus*) [Sound signals in the house mouse (*Mus musculus*)]. Behaviour 52, 38–56.

Ehret, G., 1987. Left hemisphere advantage in the mouse brain for recognizing ultrasonic communication calls. Nature 325, 249–251.

Ehret, G., 2005. Infant rodent ultrasounds – a gate to the understanding of sound communication. Behav. Genet. 35, 19–29.

Ehret, G., Haack, B., 1981. Categorial perception of mouse pup ultrasound by lactating females. Naturwissenschaften 68, 208.

Ehret, G., Haack, B., 1982. Ultrasound recognition in house mice: key-stimulus configuration and recognition mechanisms. J. Comp. Physiol. 148, 245–251.

Ehret, G., Haack, B., 1984. Motivation and arousal influence sound-induced maternal pup-retrieving behavior in lactating house mice. Z. Tierpsychol. 65, 25–39.

Ehret, G., Bernecker, C., 1986. Low-frequency sound communication by mouse pups (*Mus musculus*): wriggling calls release maternal behaviour. Anim. Behav. 34, 821–830.

Ehret, G., Koch, M., Haack, B., Markl, H., 1987. Sex and parental experience determine the onset of an instinctive behavior in mice. Naturwissenschaften 74, 47.

Elwood, R.W., Kennedy, H.F., Blakely, H.M., 1990. Responses of infant mice to odors of urine from infanticidal, noninfanticidal, and paternal male mice. Dev. Psychobiol. 23, 309–317.

Farrell, W.J., Alberts, J.R., 2002a. Maternal responsiveness to infant Norway rat (*Rattus norvegicus*) ultrasonic vocalizations during the maternal behavior cycle and after steroid and experiential induction regimens. J. Comp. Psychol. 116, 286–296.

Farrell, W.J., Alberts, J.R., 2002b. Stimulus control of maternal responsiveness to Norway rat (*Rattus norvegicus*) pup ultrasonic vocalizations. J. Comp. Psychol. 116, 297–307.

Fichtel, I., Ehret, G., 1999. Perception and recognition discriminated in the mouse auditory cortex by c-Fos labeling. Neuroreport 10, 2341–2345.

Francis, D.D., Champagne, F.A., Liu, D., Meaney, M.J., 1999. Maternal care, gene expression, and the development of individual differences in stress reactivity. Ann. N.Y. Acad. Sci. 896, 66–84.

Guo, Z., Holy, T.E., 2007. Sex selectivity of mouse ultrasonic songs. Chem. Senses 32, 463–473.

Harmon, K.M., Cromwell, H.C., Burgdorf, J., Moskal, J.R., Brudzynski, S.M., Kroes, R.A., Panksepp, J., 2008. Rats selectively bred for low levels of 50 kHz ultrasonic vocalizations exhibit alterations in early social motivation. Dev. Psychobiol. 50, 322–331.

Hashimoto, H., Saito, T.R., Furudate, S., Takahashi, K.W., 2001. Prolactin levels and maternal behavior induced by ultrasonic vocalizations of the rat pup. Exp. Anim. 50, 307–312.

Hennessy, M.B., Li, J., Lowe, E.L., Levine, S., 1980. Maternal behavior, pup vocalizations, and pup temperature changes following handling in mice of 2 inbred strains. Dev. Psychobiol. 13, 573–584.

Koch, M., Ehret, G., 1989. Estradiol and parental experience but not prolactin are necessary for ultrasound recognition and pup-retrieving in the mouse. Physiol. Behav. 45, 771–776.

Levine, S., 1957. Infantile experience and resistance to physiological stress. Science 126, 405.

Liu, D., Diorio, J., Tannenbaum, B., Caldji, C., Francis, D., Freedman, A., Sharma, S., Pearson, D., Plotsky, P.M., Meaney, M.J., 1997. Maternal care, hippocampal glucocorticoid receptors, and hypothalamic-pituitary-adrenal responses to stress. Science 277, 1659–1662.

Liu, R.C., Linden, J.F., Schreiner, C.E., 2006. Improved cortical entrainment to infant communication calls in mothers compared with virgin mice. Eur. J. Neurosci. 23, 3087–3097.

Liu, R.C., Miller, K.D., Merzenich, M.M., Schreiner, C.E., 2003. Acoustic variability and distinguishability among mouse ultrasound vocalizations. J. Acoust. Soc. Am. 114, 3412–3422.

Mann, P.E., Bridges, R.S., 2001. Lactogenic hormone regulation of maternal behavior. Prog. Brain Res. 133, 251–262.

Marino, M.D., Cronise, K., Lugo, Jr., J.N., Kelly, S.J., 2002. Ultrasonic vocalizations and maternal–infant interactions in a rat model of fetal alcohol syndrome. Dev. Psychobiol. 41, 341–351.

Menard, J.L., Hakvoort, R.M., 2007. Variations of maternal care alter offspring levels of behavioural defensiveness in adulthood: evidence for a threshold model. Behav. Brain Res. 176, 302–313.

Menard, J.L., Champagne, D.L., Meaney, M.J., 2004. Variations of maternal care differentially influence "fear" reactivity and regional patterns of cFos immunoreactivity in response to the shock-probe burying test. Neuroscience 129, 297–308.

Noirot, E., 1968. Ultrasounds in young rodents. II. Changes with age in albino rats. Anim. Behav. 16, 129–134.

Noirot, E., 1969. Selective priming of maternal responses by auditory and olfactory cues from mouse pups. Dev. Psychobiol. 2, 273–276.

Oswalt, G.L., Meyer, G.W., 1975. Olfactory, thermal, and tactual influences on infantile ultrasonic vocalization in rats. Dev. Psychobiol. 8, 129–135.

Sales, G.D., Smith, J.C., 1978. Comparative studies of the ultrasonic calls of infant murid rodents. Dev. Psychobiol. 11, 595–619.

Sewell, G.D., 1970. Ultrasonic communication in rodents. Nature 227, 410.

Shoji, H., Kato, K., 2006. Maternal behavior of primiparous females in inbred strains of mice: a detailed descriptive analysis. Physiol. Behav. 89, 320–328.

Smith, J.C., 1976. Responses of adult mice to models of infant calls. J. Comp. Physiol. Psychol. 90, 1105–1115.

Smotherman, W.P., Bell, R.W., Starzec, J., Elias, J., Zachman, T.A., 1974. Maternal responses to infant vocalizations and olfactory cues in rats and mice. Behav. Biol. 12, 55–66.

Stern, J.M., Johnson, S.K., 1990. Ventral somatosensory determinants of nursing behaviour in Norway rats. I. Effects of variations in the quality and quantity of pup stimuli. Physiol. Behav. 47, 993–1011.

Stern, J.M., Yu, Y.L., Crockett, D.P., 2002. Dorsolateral columns of the spinal cord are necessary for both suckling-induced neuroendocrine reflexes and the kyphotic nursing posture in lactating rats. Brain Res. 947, 110–121.

Terkel, J., Damassa, D.A., Sawyer, C.H., 1979. Ultrasonic cries from infant rats stimulate prolactin release in lactating mothers. Horm. Behav. 12, 95–102.

Uematsu, A., Kikusui, T., Kihara, T., Harada, T., Kato, M., Nakano, K., Murakami, O., Koshida, N., Takeuchi, Y., Mori, Y., 2007. Maternal approaches to pup ultrasonic vocalizations produced by a nanocrystalline silicon thermo-acoustic emitter. Brain Res. 1163, 91–99.

Voloschin, L.M., Tramezzani, J.H., 1984. Relationship of prolactin release in lactating rats to milk ejection, sleep state, and ultrasonic vocalization by the pups. Endocrinology 114, 618–623.

Weller, A., Leguisamo, A.C., Towns, L., Ramboz, S., Bagiella, E., Hofer, M., Hen, R., Brunner, D., 2003. Maternal effects in infant and adult phenotypes of 5HT1A and 5HT1B receptor knockout mice. Dev. Psychobiol. 42, 194–205.

Wöhr, M., Schwarting, R.K., 2008. Maternal care, isolation-induced infant ultrasonic calling, and their relations to adult anxiety related behavior in the rat. Behav. Neurosci. 122, 310–330.

Wöhr, M., Dahlhoff, M., Wolf, E., Holsboer, F., Schwarting, R.K., Wotjak, C.T., 2008. Effects of genetic background, gender, and early environmental factors on isolation-induced ultrasonic calling in mouse pups: an embryo-transfer study. Behav. Genet. 38, 579–595.

Wood, G.K., Marcotte, E.R., Quirion, R., Srivastava, L.K., 2001. Strain differences in the behavioural outcome of neonatal ventral hippocampal lesions are determined by the postnatal environment and not genetic factors. Eur. J. Neurosci. 14, 1030–1034.

Zhang, T.Y., Chrétien, P., Meaney, M.J., Gratton, A., 2005. Influence of naturally occurring variations in maternal care on prepulse inhibition of acoustic startle and the medial prefrontal cortical dopamine response to stress in adult rats. J. Neurosci. 25, 1493–1502.

Zimmerberg, B., Rosenthal, A.J., Stark, A.C., 2003b. Neonatal social isolation alters both maternal and pup behaviors in rats. Dev. Psychobiol. 42, 52–63.

Zimmerberg, B., Kim, J.H., Davidson, A.N., Rosenthal, A.J., 2003a. Early deprivation alters the vocalization behavior of neonates directing maternal attention in a rat model of child neglect. Ann. N.Y. Acad. Sci. 1008, 308–313.

Zimmerberg, B., Brunelli, S.A., Fluty, A.J., Frye, C.A., 2005. Differences in affective behaviors and hippocampal allopregnanolone levels in adult rats of lines selectively bred for infantile vocalizations. Behav. Brain Res. 159, 301–311.

Zippelius, H.M., Schleidt, W.M., 1956. Ultraschall-Laute bei jungen Mäusen [Ultrasonic vocalizations of young mice]. Naturwissenschaften 43, 502.

CHAPTER 5.3

Vocalization as a specific trigger of emotional responses

Koji Kuraoka[1,2] and Katsuki Nakamura[2,3,4*]

[1]*Japan Society for the Promotion of Science, Chiyoda, Tokyo, Japan*
[2]*Primate Research Institute, Kyoto University, Inuyama, Aichi, Japan*
[3]*CREST, JST, Kawaguchi, Saitama, Japan*
[4]*National Institute of Neuroscience, National Center of Neurology and Psychiatry, Kodaira, Tokyo, Japan*

Abstract: For several decades the functions of mammalian vocalization have been discussed, mainly from the vantage of the caller. In this article, however, the functions of mammalian vocalization are discussed from the position of the listener. Mammalian vocalization can induce changes in the internal state of the receivers, including their emotional state. We review data of autonomic and other physiological responses to mammalian vocalization, as well as neuronal responses in the subcortical brain structures, including the amygdala, of the rhesus monkey. The discussion leads to the conclusion that animal vocalization can trigger an emotional response in the listener, allowing it to avert a life-threatening situation.

Keywords: rhesus monkey; threat vocalization; affect-inducing vocalization; amygdala; autonomic response; skin temperature; skin conductance; neuronal activity; single neuron activity; facial expression

I. Introduction

Since Darwin described the association between animal vocalization and emotion in his book *The Expression of the Emotions in Animal and Man* (Darwin, 1872/1965), researchers have been interested in animal vocalization, as well as facial expressions or body gestures as communication conveying information about the signaler's emotional state. Myers (1976) wrote that vocalization in non-human primates, such as rhesus monkeys and chimpanzees, functions as a signaling system pertaining to social and emotional behavior. Moreover, Morton (1977) discussed the evolution of harsh, low-frequency sounds emitted by hostile animals and more pure tone-like, high-frequency sounds emitted by fearful or appeasing animals. After the finding that vervet monkeys produce acoustically distinct alarm calls in response to predators, such as leopards, eagles and snakes (Seyfarth et al., 1980a,b), referential vocalization has also been reported in various other mammals, such as diana monkeys and suricates. Diana monkeys produce long-distance calls in response to such dangers as predators or falling trees (Zuberbühler et al., 1997), while the alarm calls of suricates encode information about both the type of predator and the level of urgency (Manser, 2001; Manser, Chapter 11.2 in this volume). These signals represent threat or danger to the mammals receiving the calls, and indicate the situations that should be avoided for survival. Seyfarth and Cheney (2003a), who have provided evidence that animal vocalization can be referential, stated that "the emotional nature of animal vocalization has never been in doubt" (p. 34). It is widely believed that animal vocalization related to external threats has emotional function.

Many previous studies examined functions of animal vocalization from the caller's point of view. On the other hand, vocalization is addressed to the receivers and it can affect the internal physiological or cognitive state of the listener. For example, high notes such as a scream uttered by a young animal make a strong impression on listeners; conversely, listeners may be

*Corresponding author. E-mail: katsuki@pri.kyoto-u.ac.jp

Stefan M. Brudzynski (Ed.)
Handbook of Mammalian Vocalization
ISBN 978-0-12-374593-4

DOI: 10.1016/B978-0-12-374593-4.00017-6

calmed by sounds such as birdsong and the chirps of tree-frogs (Darwin, 1872/1965, p. 91). Owren and Rendall investigated functions of mammalian vocalization from the listener's point of view, and suggested that the primary function of animal vocalization is not to convey information about the emotional state of the caller, but rather to influence the emotional state of the receiver. This hypothesis is termed the "affect–induction" model of animal vocalization (Owren and Rendall, 1997, 2001). Owren and Rendall proposed this model on the basis of two possible effects of vocalization: direct and indirect effects. Some types of vocalization have a primary effect on the emotional state of the listener; this is a direct effect. They mentioned squeaks, shrieks and screams of monkeys as examples that have a direct effect on the listener's arousal. In contrast, some other types of vocalization can induce a learned emotional response as a result of the listener's previous experience of hearing a similar sound; this is an indirect effect. The threats, coos and grunts of monkeys are examples that have an indirect effect on the listener's emotional state.

The measurement of autonomic responses is an effective index of emotional state in mammals. Although several articles have already reviewed emotional behaviors in response to species-specific vocalization (Manser et al., 2002; Seyfarth and Cheney, 2003b), few studies have reviewed the autonomic responses to such vocalization. In this chapter, we review the autonomic responses to species-specific vocalization and the neuronal systems in the relevant brain structures that subserve the processing of the received vocalization. This will contribute to better understanding of the behavioral and emotional states induced in listeners when they perceive species-specific vocalization, and understanding of the relationship between mammalian vocalization and the emotional state of the listeners.

II. Autonomic response to species-specific vocalization

Different types of emotional states can induce different patterns of autonomic responses, such as those involving changes in the heart rate, skin temperature and skin conductance response (SCR) (Ekman et al., 1983; Levenson, 1992). Therefore, autonomic responses were regarded as reliable indices of the emotional states of animals and have been investigated in response to species-specific vocalization in non-human primates.

Berntson and Boysen (1989) measured cardiac responses to conspecific vocalization in apes. They presented three types of chimpanzee vocalization (threat bark, stress scream and alarm call) and noise-control stimuli to young chimpanzees and orangutans, and then measured their heart rate responses. Although they could not find any differences in the heart rate response in orangutans after exposure to different types of vocalization, different heart rate responses were observed in chimpanzees. In the chimpanzees, heart rate decreased after being presented with stress screams and alarm calls, whereas it increased in response to threat barks. Based on frequency-domain analyses of heart period variability, the authors also found that the increase in heart rate in response to the threat barks resulted from sympathetic activations. They argued that the pattern of cardiac acceleration to threat vocalization reflected a defensive and aversive response. The same group of researchers also compared changes in the cardiac rhythm in young chimpanzees in response to conspecific screams and laughter (Berntson et al., 1989). The results showed a cardiac deceleration in response to screams and a cardiac acceleration in response to laughter. Analysis of respiratory sinus arrhythmia (Grossman and Svebak, 1987) clarified that the former was attributed to an increase in parasympathetic activity, whereas the latter was to an increase in sympathetic activity. The deceleration of the cardiac rhythm in response to conspecific screams suggests the elicitation of an orienting response. Although the authors acknowledged that the acceleration of cardiac rhythm in response to conspecific laughter was debatable, they proposed the possibility that this effect was related to a defensive response, considering the phylogenetic origin of laughter.

Heart rate changes in response to conspecific vocalization were also recorded in mammals other than non-human primates. In susliks and reindeer, conspecific vocalization altered the heart rates of receivers if they were visually isolated (Movchan, 1996). In addition, 20-day-old Belding's ground squirrels exhibited an increase or a decrease in heart rate depending on the type of the conspecific alarm calls they heard (Mateo, 1996). These findings suggest that different alarm calls may elicit different emotional responses, at least in ground squirrels, even shortly after birth. The observations of heart rate changes in response to conspecific vocalizations are not confined to terrestrial mammals. Miksis et al. (2001) reported that conspecific vocalization elicited heart rate acceleration in aquatic mammals. They recorded cardiac responses of captive bottle dolphins in a noninvasive way by taking broadband acoustic measurements from a hydrophone. The dolphins were trained to stay still at a depth of three meters for 90–120 seconds. Conspecific vocalizations such as a jaw-clap threat or a familiar signature whistle were then presented. An acceleration of cardiac rhythm was observed after presentation of the vocalization. The results showed that, like terrestrial mammals, aquatic mammals can demonstrate defensive behavior in response to a threat vocalization, and this behavior is reflected in altered autonomic activity. However, in contrast to terrestrial mammals, dolphins showed the defensive response even to signature whistles, which are otherwise non-threatening vocalizations. Clearly, further studies are needed to clarify these findings.

Emotional responses in the receivers of conspecific vocalization can be demonstrated by other parameters. We measured facial expressions and autonomic responses other than heart rate to replay of conspecific vocalizations in rhesus monkeys. We measured SCRs and changes in skin temperature around the nasal region as physiological indices for assessing negative emotional states. Facial skin temperature is known to reflect psychological state (Levine et al., 2001). In the previous study, we demonstrated that the nasal skin temperature of rhesus monkeys decreased when they faced a threatening person, which induced an aversive emotional state in the monkey (Nakayama et al., 2005). In the currently conducted study, we investigated whether a decrease in the nasal temperature would be observed when macaque monkeys heard species-specific threat calls, or were presented with the relevant threatening facial expressions.

The subjects were three male rhesus monkeys (4–5 years old). Although each of these monkeys had been housed in an individual cage, they had facial and vocal communications with other monkeys in a colony; they could see other individuals and hear their voices, but could not make body contact with each other. Measurements were conducted in a temperature-controlled experimental room, where each monkey sat in a primate chair with its head fixed to the chair. Stimuli were two-second digitized sounds of aggressive threat calls or two-second video clips of threatening facial expressions obtained from the stimulus animals. All three stimulus monkeys were unfamiliar to the subject monkeys. Aggressive threat calls are often expressed by a dominant individual motivated to attack, thus, these calls are thought to elicit a frightening emotional state in the listeners. The threat calls were presented through a speaker at about 70 dB sound pressure level measured at the position of the ears. Each monkey was tested with only one vocalization or facial expression per day to avoid habituation to this procedure. We measured the facial skin temperature before and after the presentation of threat calls or threat facial expressions by using an infrared thermographic (IRT) system (TH9100MLN; NEC-Sanei, Tokyo) placed 35 cm from the monkey's face and examined the changes in the skin temperature of the nasal region. To quantify the change in skin temperature in response to the vocal and facial stimuli, we selected the nasal region as a target area. The IRT system can measure heat radiating from the skin at 30 Hz sampling rate. We recorded facial temperature every second. We also recorded SCRs to presentation of the stimuli at 250 Hz by two Ag/AgCl electrodes filled with electrolyte paste attached to the middle phalanx of the index and middle fingers.

Fig. 1 shows examples of thermograms of the nasal region of one monkey before and after the presentation of a vocal stimulus. The thermogram of the nasal region was superimposed on the photographic image of the monkey. In this thermogram, the lighter and darker shade of gray indicates higher and lower temperature, respectively. As shown in Fig. 1, the facial skin temperature around the upper nasal region decreased in response to the species-specific threat call. First, we compared the average nasal skin temperature between the pre-stimulus and six post-stimulus periods. During the 10 second pre-stimulus period, immediately preceding the stimulus presentation, the average nasal skin temperatures were 35.6 ± 0.3°C and 35.8 ± 0.3°C (mean ± SEM) for the threat calls and facial expressions, respectively. On the other hand, during the 10 second post-stimulus period, beginning 50 seconds after the stimulus presentation, the average temperatures were 35.4 ± 0.4°C and 35.5 ± 0.3°C for the threat calls and facial expressions, respectively.

Fig. 2 shows the temporal changes in the nasal skin temperature during the 10 seconds pre-stimulus and 60 seconds post-stimulus periods every 10 seconds. The results were analyzed by a two-way analysis of variance (ANOVA), where the two main effects were the type of stimulus (i.e., call or facial expression) and the stimulus period (seven 10 second bins, as shown in Fig. 2). The result showed that the main effect of stimulus period was significant ($p = 0.04$), whereas the main effect of the type

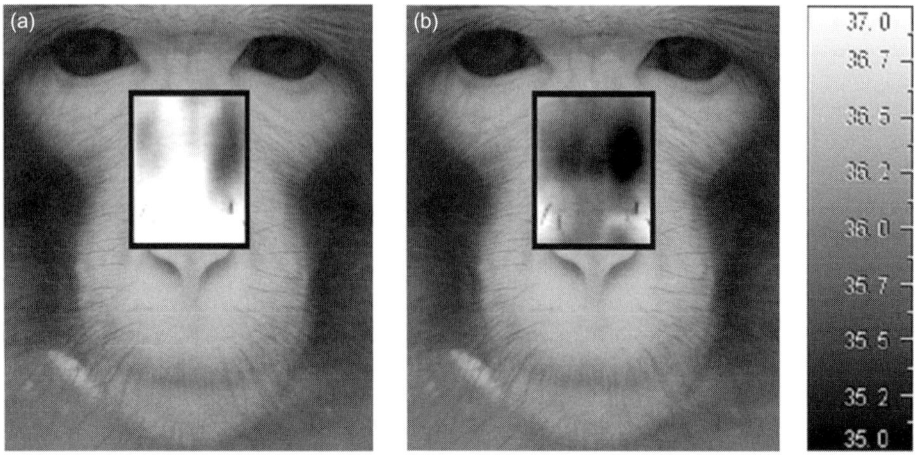

Fig. 1. An example of a decrease in the facial skin temperature in a rhesus monkey. The images around the nasal region (within the box), obtained by the IRT system, are superimposed on the photographic image of the monkey. Lighter and darker gray shades indicate higher and lower temperature, respectively. Temperature is given in degrees celsius on the right margin. (a) Facial skin temperature before presentation of a vocal stimulus. (b) Facial skin temperature after presentation of a vocal stimulus.

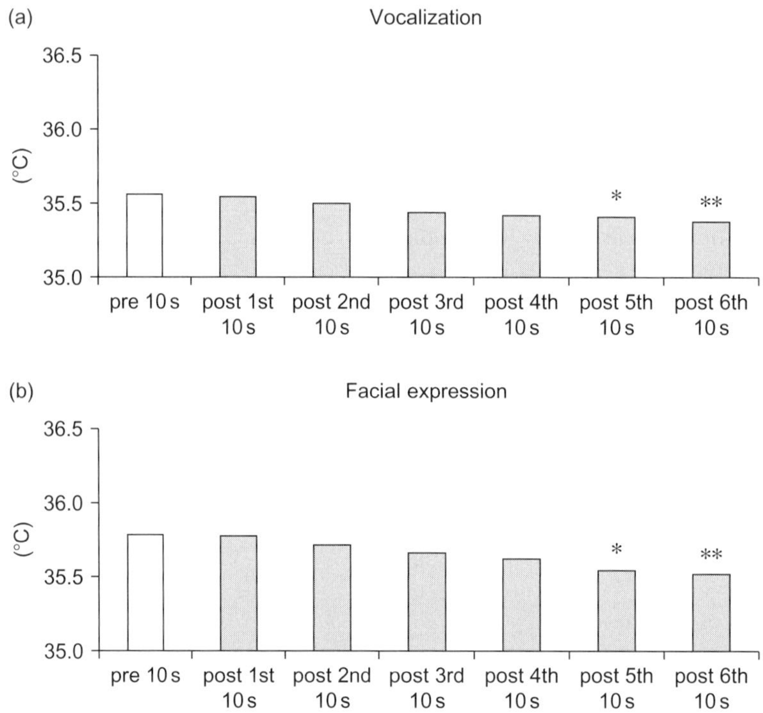

Fig. 2. Temporal changes in nasal skin temperature 10 seconds before and 60 seconds after the stimuus. Average nasal skin temperature in response to (a) replay of vocalization; and (b) presentation of facial expressions (Dunnett's *t*-test, *: $p < 0.05$, **: $p < 0.01$, compared to pre-10 seconds).

of stimulus and the interaction were not significant ($p > 0.05$ for both). Thus, both species-specific threat calls and threat facial expressions could significantly change nasal skin temperature. To elucidate temporal change in nasal skin temperature in response to the stimuli, we carried out a multiple comparison test between the 10 second pre-stimulus and the 10 second post-stimulus periods. The results showed that the stimulus-induced decrease in nasal skin temperature appeared within 40 seconds of presenting the calls or facial expressions.

We also determined the SCRs to both threat calls and facial expressions. The average SCRs to threat calls was 0.18 ± 0.05 microsiemens (μS) and that to threat facial expressions was 0.12 ± 0.02 μS. Although the responses were slightly higher to calls than to facial expressions, we could not find any significant difference in the SCRs between the two types of stimuli.

In summary, when the rhesus monkeys heard species-specific threat calls or saw threatening facial expressions, the facial skin temperature around the upper nasal region decreased and SCRs were induced. Both the decrease in the skin temperature and induction of SCRs reflect the activation of the sympathetic nervous system (Boucsein, 1992; Kistler et al., 1998). Thus, as the heart rate increases in response to threat calls in chimpanzees or to alarm calls in ground squirrels, these autonomic responses in macaque monkeys may also suggest defensive or aversive responses to aggressive threat calls or facial expressions.

III. Neuronal systems involved in processing of the perceived species-specific vocalization

Mammalian vocalization is known to activate both cortical and subcortical brain structures. Gil-da-Costa et al. (2004) examined the brain activity of macaque monkeys in response to species-specific vocalization by performing positron emission tomography (PET). They presented two conspecific calls (i.e., coo and scream) and a non-biological sound to three rhesus monkeys. A coo call is often expressed in response to food presented, and by monkeys separated from their group members or from mothers. This call represents a response expressing the need for body contact or the opportunity to feed. In contrast, a scream call is often expressed by a subordinate attacked or threatened by a dominant individual. Thus, unlike coo calls, scream calls convey more negative (threatening) emotional information. Compared to the non-biological sound, the conspecific calls strongly activated the visual cortical regions, such as V2, V3 and V4 in the occipital lobe, MT and MST in the parietal lobe, TEO, TE and STS in the temporal lobe. When the activations in response to the two types of calls (coos and screams) were compared, the screams were found to elicit stronger activation in the ventral medial prefrontal cortex (area 32), in the hippocampus, and in the amygdala. All these brain regions are associated with emotional functions. Gil-da-Costa et al.'s PET data suggest that emotion-related brain structures can be activated to a greater degree by scream calls than coo calls.

The influences of species-specific vocalization on brain function have also been investigated in rodents. In rats, ultrasonic vocalizations have been reported to have an effect on listeners' emotionality. The playback of a 22 kHz vocalization (alarm calls) induced defensive behaviors compatible with a negative state (Brudzynski and Chiu, 1995), whereas a 50 kHz social vocalization induced approach behavior compatible with a positive state (Wöhr and Schwarting, 2007). These two types of ultrasonic vocalizations induced c-Fos expression in various brain regions. C-Fos expression is known to reflect neural activation in the brain (Zhu et al., 1996). C-Fos expression in response to a 22 kHz vocalization was found in the perirhinal cortex, the amygdala and the periaqueductal gray, while response to a 50 kHz vocalization was found sparsely in the ventral striatum, the ventral pallidum, the medial forebrain bundle and the parafascicular thalamic nucleus (Sadananda et al., 2008; see also Wöhr and Schwarting, Chapter 4.2 in this volume).

In another rodent, *Octodon degus* (the common degu, native to Chile), maternal vocalization showed various influences on the brains of infants. By using a (^{14}C)-2-fluoro-deoxyglucose autoradiography method, Braun and Poeggel (2001) found metabolic activations in the precentral medial cortex, the anterior cingulate cortex and the lateral thalamus in response to emotionally-relevant maternal nursing calls; however, they found no activation in the amygdala or the hippocampus at their developmental stage. Nonetheless, this group has demonstrated later the effect of maternal vocalization of *Octodon degus* on the deregulation of the monoamine receptor in the amygdala and in the hippocampus of infants separated from the parents (Ziabreva et al., 2003). When infants of *O. degus* were separated from their parents shortly after birth, dopaminergic D_1 receptor density increased both in the amygdala and the hippocampus, serotonergic 5-$HT1_A$ receptor density increased only in the hippocampus, and $GABA_A$ receptor density decreased both in the amygdala and the hippocampus. The authors found that the presentation of maternal vocalization to infants separated from their parents had an effect on the studied receptor densities.

In many of the studies mentioned above, species-specific vocalization was found to activate the amygdala in various mammals – especially in negative (threatening and stressful) situations. The amygdala has been implicated in emotional processing of the incoming stimuli (Rolls, 2000) and thus it is reasonable to assume that neurons in the amygdala will respond to species-specific vocalization. Although several studies have examined the responses of single neurons in the cortex, such as those in the auditory and the ventrolateral prefrontal cortex, to species-specific vocalization (Wang et al., 1995; Tian et al., 2001; Romanski and Goldman-Rakic, 2002), few studies have reported the activity of such neurons in the amygdala.

In our experiments, the activity of single neurons was recorded in the amygdala of macaque monkeys during the presentation of both species-specific vocalization and species-specific facial expressions (Kuraoka and Nakamura, 2007). We prepared video clips of nine species-specific emotional expressions (three stimulus monkeys, A, B and C, showing three emotional expressions: aggressive threat; coo; and scream). We trained three female rhesus monkeys to fixate on a central spot, and we then extracellularly recorded the activity of amygdala neurons during the stimulus presentation. To compare the responses of amygdala neurons to vocalization with that to facial expressions, we presented separately either the auditory stimulus (vocalization) or the visual stimulus (motion picture of facial expression). We found that approximately half of the responsive neurons in the central nucleus of the monkey amygdala responded to species-specific vocalization, while all the recorded neurons responded to facial expressions. An example of the activity of these responding neurons is shown in Fig. 3. Half of the recorded neurons responded to two threat calls and one scream call (Fig. 3a). Interestingly, the same neurons also responded to the relevant threat or scream facial expressions, although the responses were weaker (Fig. 3b). The remaining half of the recorded neurons responded to facial expressions, but not to vocalization. In contrast, we found no neurons showing responses to vocalization, but not to the facial expressions. These results indicate that although vocalizations have a strong effect on neuronal activity, facial expressions have a more extensive effect on neuronal activity in the amygdala.

This predominant effect of neuronal responses to visual threat over responses to auditory stimuli might reflect the priority of the visual system over other sensory modalities in non-human primates (Redican, 1975). Rhesus monkeys use vocal signals in social interaction only 5% of the time (Altmann, 1967); thus, amygdala neuronal activity might reflect the visually-dominated characteristics of monkey behavior. Nevertheless, it is interesting to note that species-specific vocalization alone elicited neuronal activity in the central nucleus of the monkey amygdala. The central nucleus of the amygdala sends its output to other brain areas that are closely related to the production of emotional responses, such as the thalamus, hypothalamus and brainstem (Amaral et al., 1992; Pitkänen et al., 1997). Because monkeys need to respond quickly to highly emotional and socially tense situations signaled by threat or scream calls, the emotionally-charged information would communicate directly with other subcortical structures via the central nucleus of the amygdala. Thus, the neuronal responses to species-specific vocalization in the central nucleus of the amygdala would be part of self-protection and survival mechanisms, rapidly producing appropriate emotional responses in the receivers.

IV. Vocalization triggers emotional state in listeners

If mammalian vocalization is produced as an emotional expression of a caller, as is the case with, e.g., fearful vocalizations of hares, cattle and horses mentioned by Darwin (Darwin, 1872/1965, p. 83), these signals would inform the listeners about potential dangerous events. In addition to the referential aspect of vocalization itself, reported for some species, much of the call information is often about the emotion-inducing nature of objects or events. For example, the alarm calls of vervet monkeys, diana monkeys and suricates announce to listeners the presence and/or type of predator (Seyfarth et al., 1980a,b; Zuberbühler et al., 1997; Manser, 2001). These references represent threat or danger to the listeners and thus the vocalization is capable of inducing an emotional state in the receivers.

As outlined in the previous two sections, when mammals perceive species-specific aversive vocalization, they show various autonomic and somatic responses and the emotion-related brain structures, such as the amygdala, are activated. This strongly suggests that mammals can develop emotion-related brain response to species-specific vocalization. The amygdala rapidly processes information about dangerous external events by bypassing the cortical route,

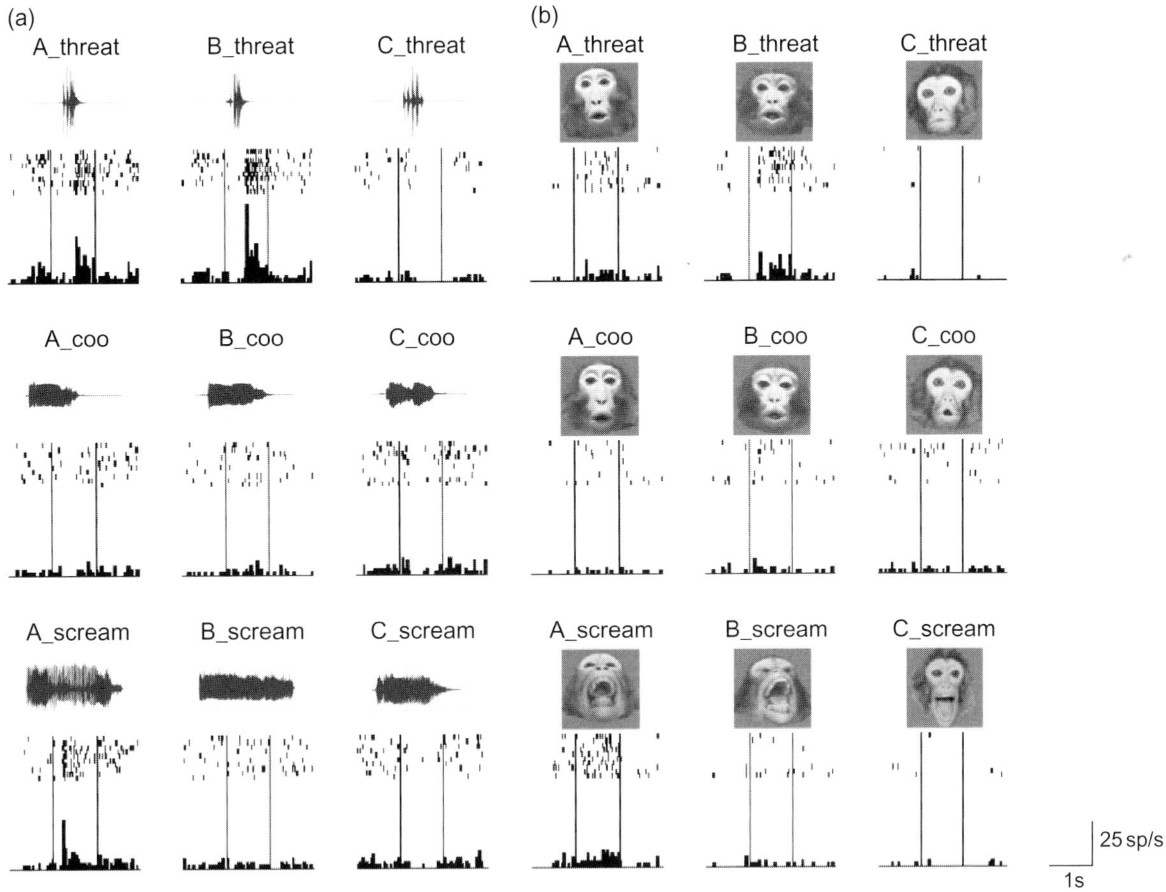

Fig. 3. Example of neuronal response in the central nucleus of the rhesus monkey amygdala. (a) Responses to vocalization; and (b) responses to facial expressions. Neuronal responses are shown as histograms with scatter plots above them, and with the sonographic image of the call or photographic image of the monkey face in the top portion. The two vertical lines in each diagram indicate the onset and offset of the stimulus. The first character (A, B, C) of each stimulus-monkey label corresponds to monkey identity and is followed by the label of the stimulus: aggressive threat, coo vocalization, and scream (for example, A_threat in (a) indicates the aggressive threat vocalization of monkey A, while A_threat in (b) indicates the aggressive threat face of monkey A, etc.).

and the information causes changes in autonomic and other physiological responses (LeDoux, 1994). The mammalian amygdala directly receives projections from the acoustic thalamus (LeDoux et al., 1990), and the activity of the amygdala is known to correlate with autonomic responses to emotionally-charged stimuli, such as changes in the heart rate (Critchley et al., 2005) and SCR (Williams et al., 2005), which are thought to create the animal's readiness for an upcoming emotional, aversive event endangering its survival. This wiring enables the production of emotional responses, which can be triggered by species-specific vocalization in order to adapt quickly to environmental challenges.

V. Summary

When mammals hear conspecifics' aversive vocalizations, the subcortical brain structures, including the amygdala, are activated, and emotional responses occur in preparation for a "fight-or-flight" response. In this sense, it can be inferred that mammalian vocalization can trigger emotional responses in listeners.

Acknowledgments

This work was supported by the Japan Society for the Promotion of Science (JSPS) Research Fellowships for Young Scientists and by the Core Research for Evolutional Science and Technology (CREST) from the Japan Science and Technology Agency (JST).

References

Altmann, S.A., 1967. The structure of primate social communication. In: Altmann, S.A. (Ed.), Social Communication Among Primates. University of Chicago Press, Chicago, IL, pp. 325–362.

Amaral, D.G., Price, J.L., Pitkänen, A., Carmichael, S.T., 1992. Anatomical organization of the primate amygdaloid complex. In: Aggleton, J.P. (Ed.), The Amygdala: Neurobiological Aspects of Emotion, Memory, and Mental Dysfunction. Wiley-Liss, New York, NY, pp. 1–66.

Berntson, G.G., Boysen, S.T., 1989. Specificity of the cardiac response to conspecific vocalizations in chimpanzees. Behav. Neurosci. 103, 235–245.

Berntson, G.G., Boysen, S.T., Bauer, H.R., Torello, M.S., 1989. Conspecific screams and laughter: cardiac and behavioral reactions of infant chimpanzees. Dev. Psychobiol. 22, 771–787.

Boucsein, W., 1992. Electrodermal Activity. Plenum Press, New York, NY and London, UK.

Braun, K., Poeggel, G., 2001. Recognition of mother's voice evokes metabolic activation in the medial prefrontal cortex and lateral thalamus of *Octodon degus* pups. Neuroscience 103, 861–864.

Brudzynski, S.M., Chiu, E.M., 1995. Behavioural responses of laboratory rats to playback of 22 kHz ultrasonic calls. Physiol. Behav. 57, 1039–1044.

Critchley, H.D., Rotshtein, P., Nagai, Y., O'Doherty, J., Mathias, C.J., Dolan, R.J., 2005. Activity in the human brain predicting differential heart rate responses to emotional facial expressions. NeuroImage 24, 751–762.

Darwin, C., 1872/1965. The Expression of the Emotions in Man and Animals. University of Chicago Press, Chicago, IL.

Ekman, P., Levenson, R.W., Friesen, W.V., 1983. Autonomic nervous system activity distinguishes among emotions. Science 221, 1208–1210.

Gil-da-Costa, R., Braun, A., Lopes, M., Hauser, M.D., Carson, R.E., Herscovitch, P., Martin, A., 2004. Toward an evolutionary perspective on conceptual representation: species-specific calls activate visual and affective processing systems in the macaque. Proc. Natl. Acad. Sci. USA 101, 17516–17521.

Grossman, P., Svebak, S., 1987. Respiratory sinus arrhythmia as an index of parasympathetic cardiac control during active coping. Psychophysiol. 24, 228–235.

Kistler, A., Mariauzouls, C., von Berlepsch, K., 1998. Fingertip temperature as an indicator for sympathetic responses. Int. J. Psychophysiol. 29, 35–41.

Kuraoka, K., Nakamura, K., 2007. Responses of single neurons in monkey amygdala to facial and vocal emotions. J. Neurophysiol. 97, 1379–1387.

LeDoux, J.E., 1994. Emotion, memory and the brain. Sci. Am. 270, 32–39.

LeDoux, J.E., Farb, C., Ruggiero, D.A., 1990. Topographic organization of neurons in the acoustic thalamus that project to the amygdala. J. Neurosci. 10, 1043–1054.

Levenson, R.W., 1992. Autonomic nervous system differences among emotions. Psychol. Sci. 3, 23–27.

Levine, J.A., Pavlidis, I., Cooper, M., 2001. The face of fear. Lancet 357, 1757.

Manser, M.B., 2001. The acoustic structure of suricates' alarm calls varies with predator type and the level of response urgency. Proc. R. Soc. London B. Biol. Sci. 268, 2315–2324.

Manser, M.B., Seyfarth, R.M., Cheney, D.L., 2002. Suricate alarm calls signal predator class and urgency. Trends Cogn. Sci. 6, 55–57.

Mateo, J.M., 1996. Early auditory experience and the ontogeny of alarm-call discrimination in Belding's ground squirrels (*Spermophilus beldingi*). J. Comp. Psychol. 110, 115–124.

Miksis, J.L., Grund, M.D., Nowacek, D.P., Solow, A.R., Connor, R.C., Tyack, P.L., 2001. Cardiac responses to acoustic playback experiments in the captive bottlenose dolphin (*Tursiops truncatus*). J. Comp. Psychol. 115, 227–232.

Morton, E.S., 1977. On the occurrence and significance of motivation–structural rules in some bird and mammal sounds. Am. Nat. 111, 855–869.

Movchan, V.N., 1996. Influence of various types of species-specific acoustic signals on the heart rate of mammals. Neurosci. Behav. Physiol. 26, 295–300.

Myers, R.E., 1976. Comparative neurology of vocalization and speech: proof of a dichotomy. Ann. N.Y. Acad. Sci. 280, 745–757.

Nakayama, K., Goto, S., Kuraoka, K., Nakamura, K., 2005. Decrease in nasal temperature of rhesus monkeys (*Macaca mulatta*) in negative emotional state. Physiol. Behav. 84, 783–790.

Owren, M.J., Rendall, D., 1997. An affect-conditioning model of non-human primate vocal signaling. In: Owings, D.H., Beecher, M.D., Thompson, N.S. (Eds.), Perspectives in Ethology, 12. Plenum Press, New York, NY, pp. 299–346.

Owren, M.J., Rendall, D., 2001. Sound on the rebound: bringing form and function back to the forefront in understanding non-human primate vocal signaling. Evol. Anthropol. 10, 58–71.

Pitkänen, A., Savander, V., LeDoux, J.E., 1997. Organization of intra-amygdaloid circuitries in the rat: an emerging framework for understanding functions of the amygdala. Trends Neurosci. 20, 517–523.

Redican, W.K., 1975. Facial expressions in non-human primates. In: Rosenblum, L.A. (Ed.), Primate Behaviour: Developments in Field and Laboratory Research, vol. 4. Academic Press, New York, NY and London, UK, pp. 103–194.

Rolls, E.T., 2000. Neurophysiology and functions of the primate amygdala, and the neural basis of emotion. In: Aggleton, J.P. (Ed.), The Amygdala: A Functional

Analysis. Oxford University Press, New York, NY, pp. 447–478.
Romanski, L.M., Goldman-Rakic, P.S., 2002. An auditory domain in primate prefrontal cortex. Nat. Neurosci. 5, 15–16.
Sadananda, M., Wöhr, M., Schwarting, R.K.W., 2008. Playback of 22-kHz and 50-kHz ultrasonic vocalizations induces differential c-fos expression in rat brain. Neurosci. Lett. 435, 17–23.
Seyfarth, R.M., Cheney, D.L., 2003a. Meaning and emotion in animal vocalizations. Ann. N.Y. Acad. Sci. 1000, 32–55.
Seyfarth, R.M., Cheney, D.L., 2003b. Signalers and receivers in animal communication. Annu. Rev. Psychol. 54, 145–173.
Seyfarth, R.M., Cheney, D.L., Marler, P., 1980a. Monkey responses to three different alarm calls: evidence of predator classification and semantic communication. Science 210, 801–803.
Seyfarth, R.M., Cheney, D.L., Marler, P., 1980b. Vervet monkey alarm calls: semantic communication in a free-ranging primate. Anim. Behav. 28, 1070–1094.
Tian, B., Reser, D., Durham, A., Kustov, A., Rauschecker, J.P., 2001. Functional specialization in rhesus monkey auditory cortex. Science 292, 290–293.
Wang, X., Merzenich, M.M., Beitel, R., Schreiner, C.E., 1995. Representation of a species-specific vocalization in the primary auditory cortex of the common marmoset: temporal and spectral characteristics. J. Neurophysiol. 74, 2685–2706.
Williams, L.M., Barton, M.J., Kemp, A.H., Liddell, B.J., Peduto, A., Gordon, E., Bryant, R.A., 2005. Distinct amygdala-autonomic arousal profiles in response to fear signals in healthy males and females. NeuroImage 28, 618–626.
Wöhr, M., Schwarting, R.K.W., 2007. Ultrasonic communication in rats: can play-back of 50-kHz calls induce approach behaviour? PloS ONE 2, e1365.
Zhu, X.O., McCabe, B.J., Aggleton, J.P., Brown, M.W., 1996. Mapping visual recognition memory through expression of the immediate early gene c-fos. Neuroreport 7, 1871–1875.
Ziabreva, I., Poeggel, G., Schnabel, R., Braun, K., 2003. Separation-induced receptor changes in the hippocampus and amygdala of *Octodon degus*: influence of maternal vocalizations. J. Neurosci. 23, 5329–5336.
Zuberbühler, K., Noe, R., Seyfarth, R.M., 1997. Diana monkey long-distance calls: messages for conspecifics and predators. Anim. Behav. 53, 589–604.

CHAPTER 5.4

Vocalizations as tools for influencing the affect and behavior of others

Drew Rendall[1*] and Michael J. Owren[2]

[1]*Behaviour and Evolution Research Group, Department of Psychology, University of Lethbridge, Lethbridge, Alberta, Canada*
[2]*Department of Psychology, Georgia State University, Atlanta, GA, USA*

Abstract: Emotional systems are central to the behavior of many animals, either as part of the impetus to behavioral action, or as an important accompaniment to it. In this chapter, we consider the role of the emotions, or affective systems construed more broadly, in vocal communication. We focus not on the role that affective systems might play in motivating calling, nor on the related issue of how vocal signals might then serve as indicators of caller emotional state. Instead, we focus on how vocal signals influence the affective systems of listeners in ways that steer and impel behavioral responding in them. We emphasize: (1) that affective systems can be important functional targets of signaling in animals, including humans; (2) that they include phylogenetically widespread neuro-affective sensitivity to some sounds; (3) that such sensitivity can impel responses from listeners in quite direct fashion, or combine with general processes of conditioning and learning to steer listener behavior more indirectly; and (4) that the effect that signals have on core affective processes and behavior in listeners can also serve to scaffold more complex communicative processes and outcomes.

Keywords: vocal communication; affective communication; emotion; cognition; affect; primates; humans; language

I. Affect-induction

It is probably intuitive, even axiomatic, that sounds can exert considerable affective influence on listeners. After all, most of us have direct experience of the way music can bring forth a variety of powerful emotions, and how our emotional responses sometimes can be quite difficult to control or resist, even when we try. We probably also realize that this affective influence is not at all accidental, but rather that a great deal of music is designed specifically to have these *affective effects*. It should be no less intuitive that other animals might show similar affective sensitivity to sound and, furthermore, that many of the vocal signals they commonly produce might also be designed to have such effects. Systematic research in this area is still in its infancy and so the range of potential effects remains an open question. However, the principles are well-illustrated by one extensively documented example, namely the acoustic startle reflex (ASR).

The ASR is an involuntary response demonstrated in a wide range of animal taxa (in mammals, e.g., from rats and cats to monkeys, apes and humans), and is thought to occur in every hearing species. It is particularly triggered by sounds that are loud and have abrupt onsets (i.e., short signal rise times) that give them a harsh and plosive quality. When experienced at close range, such sounds induce a cascade of physiological and behavioral changes in listeners that include the immediate cessation of ongoing activity and abrupt shifts in attention toward, and in movement away from, the sound stimulus, and a host of reflexive autonomic changes, such as heart rate and blood pressure changes, increased muscle tonus and modulation of overall brain activity and glucose metabolism (reviewed in Eaton, 1984; Koch, 1999).

*Corresponding author. E-mail: d.rendall@uleth.ca

Stefan M. Brudzynski (Ed.)
Handbook of Mammalian Vocalization
ISBN 978-0-12-374593-4

DOI: 10.1016/B978-0-12-374593-4.00018-8

Altogether, these changes describe a very broad and dramatic systemic response to sound that prepares the organism for a "fight-or-flight" response.

Notably, this suite of effects is induced with latencies on the order of 10 ms and requires no substantive cortical mediation. Instead, it is induced by a very short, direct circuit connecting the auditory nerve to brainstem regions controlling whole body arousal and activation: axons in the auditory nerve project to cochlear root neurons in the brainstem, which project to giant neurons in the nucleus pontis caudalis of the reticular formation, from which projections then radiate to a large number of motoneurons in the brainstem and spinal cord. This simple circuit has been studied extensively in rats and cats and is thought to be the same in humans, attesting to its very deep and broadly conserved nature (reviewed in Koch, 1999; Davis, 2006).

At the same time, related research on brainstem regulation of emotional states shows that the initiation and maintenance of broadly negative or positive affect and accompanying organismal states are centrally determined by the ascending cholingergic and dopaminergic systems, respectively, and that these also originate in closely associated areas of the reticular formation (reviewed in Brudzynski, 2007; see also Brudzynski, Chapter 7.3 in this volume). It is also the case that the midbrain acoustic thalamus that receives projections directly from cochlear root neurons in turn projects directly to the amygdala, a well-established center for integrating emotional processing and responses in mammals (LeDoux et al., 1990).

In other words, sounds of a particular sort are only a few short synapses away from brainstem, midbrain and limbic system regions that regulate major aspects of organismal autonomic function and whole body arousal and activation. As a result, it would be very surprising indeed if many vocal signals did not have the capacity to induce a range of affective effects in listeners.

II. Familiar examples of affective influence

In fact, there are a number of other natural examples of important affective effects of sounds on listener behavior. One widespread example that might actually capitalize on the ASR involves the loud distress calls produced by many prey species when captured by a predator, what are sometimes referred to as "death screams." These calls are commonly described as "sudden," "powerful," "high-intensity" and "piercing," and they are proposed to function as a last-ditch effort by a captured individual to startle its captor with a powerful signal that causes the predator to hesitate momentarily and release its grip, or accidently drop its victim, allowing some opportunity for escape (e.g., Driver and Humphries, 1969; Wise et al., 1999).

There are additional familiar examples that involve less dramatic, but no less functional, effects. For example, pastoral herders and domestic animal handlers have long capitalized on the impact of sounds to manage the behavior and activity of their animal charges (McConnell, 1991). To capture attention and increase motor activity, they typically use rapidly repeated pulses of signals with abrupt onsets (e.g., tongue clicks and lip smacks) or signals with dramatic frequency upsweeps (e.g., whistles). In contrast, to decrease motor activity, or to soothe excited animals, they use signals with smooth onsets and gradually descending pitch (e.g., whistles or hums). Humans are also responsive to the same sound patterns (Fernald, 1992; Kaplan and Owren, 1994). We use whistles with a rapid frequency upsweep to capture a companion's attention, and similar dramatic frequency variations are characteristic of the speech directed to young infants where it serves to focus and maintain attention and modulate arousal. Additional familiar examples in humans include fingernails scraping on chalk-boards, infant crying and contagious laughter, all of which have quite direct affective effects on listeners, either negative or positive.

III. Primate vocal communication

Non-human primates are an especially interesting group in which to consider the potential affective influence of vocalizations on listeners. Although well-studied as a taxon, such effects have become a focus of primate research only recently (e.g., Berntson et al., 1989; Owren and Rendall, 1997, 2001; Gil-da-Costa et al., 2004; Kuraoka and Nakamura, Chapter 5.3 in this volume). Instead, research has tended to focus on the potential "higher-level" cognitive processes that organize communication behavior in primates, because of their large brains and phylogenetic proximity to humans. However, the neurophysiological substrates for affective influence are clearly very broadly conserved and, indeed, humans who have even larger brains and (presumably) more sophisticated cognitive capacities than any of the non-human primates show considerable susceptibility to affective influence. Therefore, it is likely that affective influence is also an important part of the vocal signals of non-human primates.

III.A. Affective effects of alarm calls

Indeed, many of the vocalizations of non-human primates appear to bear the mark of design for direct effects on listener affect and behavior. For example, the alarm vocalizations that are produced during encounters with predators are structurally similar across a range of primate species, and they preserve acoustic features that are well-designed for capturing and manipulating attention and arousal in listeners (Owren and Rendall, 2001). Alarm calls tend to be short, abrupt-onset, broadband calls. These punctuate designs of alarm calls make them stand out against background noise and make them easy to localize. Accordingly, listeners' responses involve immediate orienting in the direction of the calls, coupled with reflexive movements preparatory to flight. In fact, the same basic alarm call structure is evident in a range of other mammals (Owings and Morton, 1998) and also some birds (Marler, 1955), suggesting that the direct acoustic effects that calls with this structure have on listener attention, affect and behavior may by highly conserved in evolution. They likely stem from the development of functional sensitivity to punctuate sounds in ancestral vertebrates as an aid in identifying and localizing predators, and for capturing prey. Importantly, developmental studies in primates have shown that such generalized startle responses to species' alarm calls are induced even in very young infants in the absence of significant experience with either the calls or predators (e.g., Herzog and Hopf, 1984; Seyfarth and Cheney, 1986), as would be expected from the operation of widely conserved and low-level brainstem and subcortical processes associated with sound localization, orienting and autonomic responding.

III.B. Affective effects of "squeaks, shrieks and screams"

Such evolved sensitivity to certain kinds of sound naturally creates additional opportunites for signalers to use vocalizations to engage others by influencing their attention, arousal and concomitant behaviors in many contexts, sometimes even overriding their ability to resist such influence. Among non-human primates, one entire class of vocalizations that has been labeled "squeaks, shrieks and screams" appears to capitalize on this potential. These sounds are numerous and diverse, and are produced by every primate species studied, as well as by many other mammals, birds and amphibians. This broad class of vocalizations is characterized by sharp signal onsets, dramatic frequency and amplitude fluctuations, and chaotic spectral structures, which are exactly the acoustic features that have direct impact on animal perception (Gil-da-Costa et al., 2004; Rendall et al., 2009). Vocalizations of this class are produced especially frequently by infants and juveniles (Owren et al., 1993), and this pattern is not simply a reflection of an immature vocal production system in young animals. Instead, squeaks, shrieks and screams are likely to be especially functional to youngsters who otherwise have relatively little ability to influence the behavior of older and larger individuals in their groups.

For example, a young weanling who has been repeatedly rejected from the nipple by its mother cannot physically force its mother to relent and allow nursing or close physical contact. However, it is not entirely helpless, because it is capable of producing loud protracted bouts of harsh and variable vocalizations that effectively influence the mother's attention mechanisms, increase her arousal state, and with repetition become quite aversive. Vocal protests like this in the context of weaning are ubiquitous across species, including humans, and they share a set of common acoustic features that include rapidly varying combinations of loud, noisy screams and piercing high-frequency tonal cries, with dramatic amplitude and frequency modulations (Owren and Rendall, 2001; Owings and Zeifman, 2004). Everyday experience and systematic perceptual studies on humans confirm that such screaming is extremely grating and aversive to listeners, and very effective in catalyzing responses from them (Lester and Boukydis, 1985; Bradley and Lang, 2000; Owings and Zeifman, 2004).

The functional value of such aversive screaming is not limited to immature animals, however, because adult animals can be equally impotent, particularly in highly social species with developed dominance hierarchies. In such species, daily activities involve regular interaction with more dominant individuals, who often antagonize and attack subordinate group members. Like immature animals, such low-ranking victims of aggression cannot offer serious physical resistance to their dominant counterparts. However, they can make themselves unappealing targets of further attack by screaming vociferously. When experienced at close range, the aversive qualities of such screaming may be extremely aversive and quite difficult to habituate to, thereby testing a dominant's commitment to further antagonization (Rendall et al., 2009).

III.C. Affective effects of "sonants and gruffs"

The vocal repertoire of most primate species contains an additional class of sounds referred to as "sonants and gruffs." This class of sounds is, in many ways, the structural antithesis of squeaks, shrieks and screams. Wheareas the latter call types tend to be relatively unpatterned and chaotic, sonants and gruffs tend either to be tonal, harmonically-rich calls, or to be characterized by a more diffuse broadband spectral structure that is nevertheless regularly patterned. Therefore, while the chaotic features of squeaks, shrieks and screams are well-suited to having direct impact on listener arousal and affect, the more patterned nature of sonants and gruffs gives them less inherent affective force, either negative or positive. However, the richly-structured spectra of these calls provide an excellent medium for revealing clear cues to caller identity through individual idiosyncracies. Such idiosyncracies routinely impart individually distinctive voice cues in acoustic features of these calls that are associated either with the pattern of dynamic action of the vocal folds or through resonance properties of the vocal tract cavities (Rendall et al., 1998; Ghazanfar et al., 2007).

Somewhat paradoxically then, many of these calls are used in face-to-face social interactions where such vocal cues to identity would seem to be wholly redundant. However, in the social groups of many primate species, one's influence on other group members hinges on individual identity and social status, and therefore simply announcing one's identity vocally can also influence the affect and arousal of others. Furthermore, such identity cues provide additional explicit opportunities for influential individuals to leverage the social behavior of others by controlling the behavioral sequelae that follow from vocal exchanges, providing myriad opportunities for behavioral shaping through processes of conditioning and learning (see next section).

IV. Mechanisms of affective influence

Because research in this area is still in its infancy, we have a very incomplete understanding of the range of possible mechanisms through which vocal signals might exert functional affective influence on listeners. However, at least three broad classes of mechanism are likely. One of these includes the quite direct influence that vocal signals can have on listener affect through stimulation of autonomic systems organizing and impelling basic behavioral action, as already illustrated by the example of the acoustic startle reflex, as well as by some of the examples from the vocal repertoires of primates. Two additional broad classes of mechanism are possible.

IV.A. Affective influence through learning and conditioning

Vocal signals might also influence listener affect and behavior more indirectly through general processes of conditioning or learning.

IV.A.1. Affective learning

In natural environments, an individual's emotional response to events is a good heuristic to what aspects of the environment are important. As a result, vocal signals that induce powerful autonomic responses in listeners can effectively serve also to highlight or tag salient events in the world, and thereby support additional learning about them. A potential example involves the alarm call system of vervet monkeys. These monkeys produce a small number of different alarm vocalizations that are specific to the different major classes of predator that prey on them, and each predator requires the monkeys to engage a functionally different escape response. The alarm calls alone elicit the different escape responses from adult listeners, as though the calls themselves encoded referentially-specific, or semantic, information about the type of predator encountered, similar to the way human words function (Seyfarth et al., 1980; Manser, Chapter 11.2 in this volume). In contrast, predator-naïve infant vervets do not respond in adult-like fashion with differentiated escape responses to the different alarm calls. The different alarm calls nevertheless preserve a common set of affect-inducing acoustic features, and infants do respond strongly to all of the alarm calls with a generalized startle response (Seyfarth and Cheney, 1986). Over time, infants' responses begin to differentiate into the more adult-like repertoire of escape options. Although the details of this process remain undocumented, it is well-established that the up-regulation of brain activation, glucose metabolism and stress hormones that accompany emotion-inducing experiences facilitates learning and memory (McGaugh, 2003). Therefore, the powerful effects that the alarm calls of vervets have from the very start on attentional and affective systems likely serve to tag the significance of these events for naïve infants and promote additional learning about the different predators involved and the specific behavioral responses that follow and are appropriate to them.

Learning effects like this that are scaffolded on a foundation of affect induction might be critically functional not only in non-intentional species including primates, but also in many other species. Such species lack the social cognitive abilities that would motivate adult members to instruct naïve infants about predators and other dangers (or other important aspects of the local environment, e.g., appropriate food items), because those adults do not appreciate to begin with that infants are naïve about these dangers (Cheney and Seyfarth, 2005). Yet, infants who learn about them early in life must certainly benefit. These converging constraints describe an informational bottleneck to infant surival – infants lack survival information and adults cannot appreciate and remediate infants' deficits. In this situation, alarm vocalizations that by themselves induce powerful affective responses in infants offer a functional, evolutionary "work-around" to the problem. Such vocalizations promote additional learning about the environment without either the adults or the infants being aware of this fact.

IV.A.2. Learned affect

If vocalizations can induce affective responses in listeners that establish the conditions for learning about the environment, then it follows that signalers can also condition listeners to the affective consequences of vocalizations if they pair calls of a particular kind with salient, emotion-inducing acts. For example, in primates (and many other social species), more dominant group members routinely antagonize their subordinates. During such aggression, the dominant typically produces distinctive threat vocalizations while biting and chasing the subordinate. As a result, the dominant's threat calls predict the associated, aggression-induced affect. Thereafter, the dominant can elicit similar negative affect in previous victims by use of the calls alone. Conversely, the same dominant individual may produce an acoustically distinctive affiliative sound before approaching a subordinate with peaceable intent. This interaction occurs commonly in attempts by dominant females to inspect and interact with the infants of other females in the group. In many cases, these interactions include grooming by the dominant of the approached female or her infant. In this situation, then, the subordinate might come to associate the calls with a different set of emotional responses, like those experienced during the positively-toned grooming episode that often follows approach and calling.

In both negative and positive outcomes, a socially influential animal has the opportunity to use a listener's own learning processes to create vocal "leverage" over its affective states and behavior. As in the case of more direct autonomic effects, subcortical structures are critically important with, for instance, learned fear of auditory stimuli having been shown to be mediated by nuclei in the thalamus and amygdala (LeDoux, 2000).

The ability to induce differentiated affective outcomes in this way requires the use of at least subtly different call types. It also hinges on the individual distinctiveness of these calls because, during the course of any given day, every group member will regularly hear threat and affiliative vocalizations produced by many other group members, but most of these will be directed at, and involve the chasing or grooming of, others. Furthermore, the threat calls of subordinate animals will generally be of little affective consequence for dominants. Hence, the capacity to induce learned affect in others depends on the identity of the caller *vis à vis* the recipient. At the same time, this functional requirement might help to explain why the variety of calls used in such face-to-face contexts in different primate species have regularly proven to contain clear cues to caller identity when such cues have otherwise seemed entirely redundant in these contexts.

IV.B. Affective and behavioral resonance

An additional possible mechanism through which vocal signals might exert affective influence is through a process of affective and behavioral resonance. This possibility emerges from the increasing realization that the neurophysiological organization of behavior depends on reciprocal influence between systems guiding the production of behavior and systems involved in perceiving, interpreting and responding to the behavior of others. The landmark finding on this front was the discovery of mirror and canonical neuron systems in primate brains which are activated both by seeing an object, or seeing an action performed by another individual, and by acting on that same object, or performing the same action oneself (Gallese et al., 1996; Scott et al., Chapter 5.5 in this volume).

This perceptuo-motor integration generates an unconscious behavioral resonance between individuals via incipient "motor sympathy" for one another's actions. The effects have been shown to include visuo-motor sympathy for certain communicative gestures in primates (Ferrari et al., 2003) and for facial expressions of emotions in humans (Carr et al., 2003; Hennenlotter et al., 2005). They have also been shown to extend beyond the visuo-motor system. For example, auditory-motor mirror neurons that integrate the sound of an act with the behavior required to generate it have been reported in non-human primates (Kohler et al., 2002; Keysers et al., 2003). Auditory-motor sympathy has also been shown in humans for non-verbal vocalizations, where the sound of non-verbal exclamations of positive affect potentiates in listeners activity in motor areas involved in facial expressions associated with producing the same positively-toned exclamations (Warren et al., 2006). In addition, Gallese et al. (2004) have proposed viscero-motor mirror circuits as a foundation for direct emotional resonance via simulation of at least some of the felt emotions of others.

Although the neuroscientific evidence for affective and behavioral resonance is still nascent, the implications for vocal-affective influence are clear. Given the reciprocally integrated neurophysiological networks for perception and action, and in so far as vocal signals are reliably driven by emotional states in signalers (see other chapters in Section 6 in this volume; see also Fichtel et al., 2000), then the process by which listeners perceive emotion-laden signals includes the capacity for experiencing some similar emotions and thus also potentiating behavioral responses that might naturally flow from that shared experience. Although not yet confirmed empirically, this type of process offers a concrete neurophysiological account for some previously well-known, but poorly understood, examples of affective and behavioral resonance in humans such as contagious laughter, and contagious crying and comfort-seeking. It also invites consideration of the broader possiblities for unconscious vocal-affective influence in other species and in other contexts.

V. Affective bootstrapping of complex communication

Additional possibilities for vocal-affective influence involve its bootstrapping of more complex communicative outcomes. For example, among non-human primates, it seems likely that the predator-specific, referential quality of alarm calls in some species, such as vervet monkeys, is bootstrapped developmentally from the more direct effects that these calls have on the attentional and affective systems of predator-naïve infants, as reviewed above. Similar effects appear to play a role in bootstrapping some of the even more complex properties of human language. Here, the traditional wisdom has long been that children's progression into full-blown, adult language must be governed by some innate coding of its deep organizational properties in a special language module in the brain (Chomsky, 1957), because otherwise the obstacles infants face in distilling the complex semantics and syntax of language merely through their exposure to adult speech have seemed insurmountable. In recent years, many findings have served to undermine this view, but importantly for present purposes they include how speech sounds, as physical signals, influence attentional and affective systems of listeners in ways that might promote language acquisition.

V.A. Affective learning of language structure

In many cultures, the speech that adults use when talking to infants is quite different from the speech they use when talking to other adults. Infant-directed speech is often simplified compared to adult-directed speech, but it also often involves exaggerated prosodic features, such as wider excursions of voice pitch; more variable amplitude, tempo and delivery; and more varied patterns of word stress. These modifications to infant-directed speech mean that, as a physical acoustic signal, it is inherently both more salient and more variable than adult-directed speech, and these properties also make it more functional in capturing and focusing infant attention and modulating core arousal. The resulting modulatory effects have been shown to facilitate semantic learning and to highlight additional organizational properties of language. For example, exaggerated pitch excursions and word stress in conjunction with manual gesturing and manipulation of concrete objects facilitates semantic labeling. Variable tempo and pausing help to highlight phonetic boundaries, clausal boundaries and higher syntactical units (reviewed in Kuhl, 2000).

V.B. Bio-phonetics

It is also possible that some of the relatively rich semantics of adult language is scaffolded on simpler sound–meaning relationships. For example, the first natural sound–meaning relationships for infants are those marking the caregiver's identity. Young infants of most species, including humans, show an early preference for mothers' voice whose unique features they quickly associate with the comfort, support and accompanying positive affect that she represents. Over time, infants learn additional social discriminations that have other functional behavioral consequences. These include discriminating the age, sex and body size of social companions, because these basic distinctions herald important differences in the social status, behavioral dominance and implicit threat or challenge that others represent. Many of these social distinctions are also signaled by salient differences in the acoustic features of the vocalizations these individuals produce, such as in voice pitch (fundamental frequency) and voice resonances (or formants) that vary predictably among age–sex classes and among individuals of varying body size (Titze, 1989; Fitch and Giedd, 1999; Rendall et al., 2005). Hence, many of the earliest sound–meaning relationships that young infants acquire are those that represent meaningful social distinctions among group members, and the affective and behavioral consequences they predict.

It is, therefore, possible that the semantics of human languages exploit some of these biologically, preprepared sound–meaning relationships which would then offer infants a natural aid in semantic learning. One illustrative example involves the phenomenon of vowel-specific marking of semantic diminution and augmentation; words that convey "smallness" are disproportionately characterized by high-front vowels whose spectral density is biased toward higher frequencies, and words that convey "largeness" are characterized by low-back vowels whose spectral density is biased toward lower frequencies (Hinton et al., 1995). This phenomenon occurs cross-linguistically and represents the semantic extension in languages of the natural sound-symbolic relationship that exists in the wider world as noted above. Hence, a young infant's affective familiarity with these basic sound-symbolic relationships could subsequently facilitate semantic learning, at least of words that obey the cross-cultural pattern of semantic diminution and augmentation. And these linguistic patterns in turn are likely to have arisen naturally but unintentionally through historical processes of cultural selection, which favored the use and survival of word forms that convey their meaning more "naturally" in the sense that they are easier to learn, recall and deploy, precisely because they exploit biologically preprepared sound–meaning relationships.

V.C. Affective semantics

Accepting this possibility suggests even wider scope for semantic constructs to be grounded in the perceptuo-affective "impacts" of sound structure – what might be termed *affective semantics.* For example, Köhler (1947) famously reported a bias for human subjects to match particular nonsense words, such as *naluma* and *takete*, with unfamiliar objects whose form was *rounded* or *jagged*, respectively. This bias has been confirmed in other populations and in young children, and it has been extended to include other objects and other nonsense words, such as *bouba* and *kiki* (e.g., Westbury, 2005; Maurer et al., 2006). One explanation for this implicit semantic bias is that it reflects reciprocal linkage between the visuo-sensory processing of a rounded object and activation of motor areas responsible for coordinating the articulation of the round-mouth vowels both in *bouba* and *naluma*, but not *kiki* or *takete* (Ramachandran and Hubbard, 2001).

A related alternative is that the semantic bias reflects the differential affective quality of sounds with different spectral density and signal onsets, as reviewed earlier for animal vocalizations. In this case, the consonants /k/ and /t/ are unvoiced and have relatively plosive onsets and noisy spectra. Therefore, the words *kiki* and *takete* might more naturally conjure "harsh/fractured" and similar semantic constructs, and so be matched to jagged objects preferentially. In contrast, the consonants /b/, /n/, /l/ and /m/ are all either voiced, or nearly so, and therefore have smoother onsets and more patterned spectral structures. Hence, the words *bouba* and *naluma* might more naturally conjure "smooth/connected" and similar semantic constructs, and so be matched to rounded objects. Recent experiments using words that both replicate and cross the original consonant and vowel contexts (i.e., bouba-kiki; kouka-bibi) provide some support for this account (D. Rendall and A. Nielsen, unpublished data). By extension, a vastly larger set of affective semantic effects might await future discovery and might ultimately be shown to account for the form of at least some real words.

VI. Conclusions

Vocalizations can exert direct and indirect influences on listener affect and behavior. Some of the effects are taxonomically widespread, evolutionarily conserved and very difficult for listeners to control or resist. Of course, as functional as such affective effects of vocalizations can be, they do not undercut the role of cognition, nor do they preclude the possibility of more complex communicative processes and outcomes in many species. On the contrary, given the complementary and integrated nature of affective and cognitive systems generally, many of the core affective effects of vocal signals probably serve, in part, to scaffold increasing communicative complexity. This point is underscored by the possibility that at least some of the semantic complexity of human language might capitalize on affectively-based, sound–meaning relationships. Such preprepared, or early acquired, sound–sense relationships represent a form of intrinsic (i.e., original) meaning that provides a natural foundation from which to construct increasingly complex semantic systems, both developmentally and evolutionarily. The corollary is that the communicative importance of the affective influence of vocal signals do not disappear when brains get larger and the potential for cognitive, evaluative control of behavior increases. Rather, complex communicative processes might often specifically exploit and build on the phylogenetically-ancient and widespread affective effects of vocal signals.

Acknowledgments

For continued generous research support we thank the Natural Sciences and Engineering Research Council of Canada (NSERC), the National Institutes of Mental Health (NIH), the L.S.B. Leakey Foundation, the University of Lethbridge and Georgia State University.

References

Berntson, G.G., Boysen, S.T., Bauer, H.R., Torello, M.S., 1989. Conspecific screams and laughter: cardiac and behavioral reactions of infant chimpanzees. Dev. Psychobiol. 22, 771–787.

Brudzynski, S.M., 2007. Ultrasonic calls of rats as indicator variables of negative or positive states: acetylcholine–dopamine interaction and acoustic coding. Behav. Brain Res. 182, 261–273.

Carr, L., Iacoboni, M., Dubeau, M.C., Mazziotta, J.C., Lenzi, G.L., 2003. Neural mechanisms of empathy in humans: a relay from neural systems for imitation to areas. Proc. Natl. Acad. Sci. USA 100, 5497–5502.

Cheney, D.L., Seyfarth, R.M., 2005. Constraints and preadaptations in the earliest stages of language evolution. Linguistic Rev. 22, 135–159.

Chomsky., 1957. Syntactic Structures. Mouton, The Hague, Netherlands.

Davis, M., 2006. Neural systems involved in fear and anxiety measured with fear-potentiated startle. Amer. Psychol. 6, 741–756.

Driver, P.M., Humphries, D.A., 1969. The significance of the high-intensity alarm call in captured passerines. Ibis 111, 243–244.

Eaton, R.C. (Ed.), 1984. Neural Mechanisms of Startle Behavior. Plenum Press, New York, NY.

Fernald, A., 1992. Human maternal vocalizations to infants as biologically relevant signals: an evolutionary perspective. In: Barkow, J.H., Cosmides, L., Tooby, J. (Eds.), The Adapted Mind. Oxford University Press, Oxford, UK.

Ferrari, P.F., Gallese, V., Rizzolatti, G., Fogassi, L., 2003. Mirror neuron responding to the observation of ingestive and communicative mouth actions in the monkey ventral premotor cortex. Eur. J. Neurosci. 17, 1703–1714.

Fitch, W.T., Giedd, J., 1999. Morphology and development of the human vocal tract. A study using magnetic resonance imaging. J. Acoust. Soc. Am. 106, 1511–1522.

Gallese, V., Keysers, C., Rizzolatti, G., 2004. A unifying view of the basics of social cognition. Trends Cog. Sci. 8, 396–403.

Gallese, V., Fadiga, L., Fogassi, L., Rizzolatti, G., 1996. Action recognition in the premotor cortex. Brain 119, 593–609.

Ghazanfar, A.A., Turesson, H.K., Maier, J.X., van Dinther, R., Patterson, R.D., Logothetis, N.K., 2007. Vocal-tract resonances as indexical cues in rhesus monkeys. Curr. Biol. 17, 425–430.

Gil-da-Costa, R., Braun, A., Lopes, M., Hauser, M.D., Carson, R.E., Herscovitch, P., Martin, A., 2004. Toward an evolutionary perspective on conceptual representation: Species-specific calls activate visual and affective processing systems in the macaque. Proc. Natl. Acad. Sci. USA 101, 17516–17521.

Hennenlotter, A., Schroeder, U., Erhard, P., Catrop, F., Haslinger, B., Stoecker, D., Lange, K.W., Ceballos-Baumannn, A.O., 2005. A common neural basis for receptive and expressive communication of pleasant facial affect. NeuroImage 26, 581–591.

Herzog, M., Hopf, S., 1984. Behavioral responses to species-specific warning calls in infant squirrel monkeys reared in social isolation. Amer. J. Primatol. 7, 99–106.

Hinton, L., Nichols, J., Ohala, J. (Eds.), 1995. Sound Symbolism. Cambridge University Press, Cambridge, UK.

Kaplan, P.S., Owren, M.J., 1994. Dishabituation of visual attention in 4-month-olds by infant-directed frequency sweeps. Inf. Behav. Develop. 17, 347–358.

Keysers, C., Kohler, E., Umilta, M.A., Nanetti, L., Fogassi, L., Gallese, V., 2003. Audiovisual mirror neurons and action recognition. Exp. Brain Res. 153, 628–636.

Koch, M., 1999. The neurobiology of startle. Progress in Neurobiol. 59, 107–128.

Kohler, E., Keysers, C., Umilta, M.A., Fogassi, L., Gallese, V., Rizzolatti, G., 2002. Hearing sounds, understanding actions: action representation in mirror neurons. Science 297, 846–848.

Köhler, W., 1947. Gestalt Psychology. Liveright Publishing, New York, NY.

Kuhl, P., 2000. A new view of language acquisition. Proc. Nat. Acad. Sci. USA 97, 11850–11857.

LeDoux, J.E., 2000. Emotion circuits in the brain. Ann. Rev. Neurosci. 23, 155–184.

LeDoux, J.E., Farb, C., Ruggiero, D.A., 1990. Topographic organization of neurons in the acoustic thalamus that project to the amygdala. J. Neurosci. 10, 1043–1054.

Lester, B.M., Boukydis, C.F.Z. (Eds.), 1985. Infant Crying: Theoretical and Research Perspectives. Plenum Press, New York, NY.

Marler, P., 1955. Characteristics of some animal calls. Nature 176, 6–8.

Maurer, D., Pathman, T., Mondloch, C.J., 2006. The shape of boubas: sound-shape correspondences in toddlers and adults. Dev. Sci. 9, 316–322.

McConnell, P.B., 1991. Lessons from animal trainers: the effect of acoustic structure on an animal's response. In: Bateson, P., Klopfer, P. (Eds.), Perspectives in Ethology. Human Understanding and Animal Awareness, vol. 9. Plenum Press, New York, NY.

McGaugh, J.L., 2003. Memory and Emotion: The Making of Lasting Memory. Weidenfeld & Nicolson, London, UK.

Owings, D.H., Morton, E.S., 1998. Animal Vocal Communication: A New Approach. Cambridge University Press, Cambridge, UK.

Owings, D., Zeifman, D.M., 2004. Human infant crying as an animal communication system: insights from an assessment-management approach. In: Oller, D.K., Griebel, U. (Eds.), Evolution of Communication Systems: A Comparative Approach. MIT Press, Cambridge, MA, pp. 151–170.

Owren, M.J., Rendall, D., 1997. An affect-conditioning model of non-human primate vocal signaling. In: Owings, D.H., Beecher, M.D., Thompson, N.S. (Eds.), Perspectives in Ethology. Communication, vol. 12. Plenum Press, New York, NY, pp. 299–346.

Owren, M.J., Rendall, D., 2001. Sound on the rebound: returning form-function relations to the forefront of acoustic primatology. Evol. Anthropol. 10, 58–71.

Owren, M.J., Dieter, J.A., Seyfarth, R.M., Cheney, D.L., 1993. Vocalizations of rhesus (*Macaca mulatta*) and Japanese (*M. fuscata*) macaques cross-fostered between species show evidence of only limited modification. Develop. Psychobiol. 26, 389–406.

Ramachandran, V., Hubbard, E., 2001. Synaesthesia – a window into perception, thought and language. J. Consciousness Studies, 8, 3–34.

Rendall, D., Owren, M.J., Rodman, P.S., 1998. The role of vocal tract filtering in identity cueing in rhesus monkey (*Macaca mulatta*) vocalizations. J. Acoust. Soc. Amer. 103, 602–614.

Rendall, D., Notman, H., Owren, M.J., 2009. Asymmetries in the individual distinctiveness and maternal recognition of infant contact calls and distress screams in baboons. J. Acoust. Soc. Amer. 125 (3), 1792–1805.

Rendall, D., Kollias, S., Ney, C., Lloyd, P., 2005. Pitch (F_0) and formant profiles of human vowels and vowel-like baboon grunts: the role of vocalizer body size and voice-acoustic allometry. J. Acoust. Soc. Amer. 117, 944–955.

Seyfarth, R.M., Cheney, D.L., 1986. Vocal development in vervet monkeys. Anim. Behav. 34, 1640–1658.

Seyfarth, R.M., Cheney, D.L., Marler, P., 1980. Monkey responses to three different alarm calls: evidence of predator classification and semantic communication. Science 210, 801–803.

Titze, I.R., 1989. Physiologic and acoustic differences between male and female voices. J. Acoust. Soc. Am. 85, 1699–1707.

Westbury, C., 2005. Implicit sound symbolism in lexical access: evidence from an interference task. Brain Lang. 93, 10–19.

Wise, K.K., Connover, M.R., Knowlton, F.F., 1999. Response of coyote to avian distress calls: testing the startle-predator and predator-attraction hypotheses. Behav. 136, 935–949.

CHAPTER 5.5

Brain mechanisms for processing perceived emotional vocalizations in humans

Sophie K. Scott[1*], Disa Sauter[2] and Carolyn McGettigan[1]

[1]*Institute of Cognitive Neuroscience, University College, London, UK*
[2]*Max Planck Institute for Psycholinguistics, Nijmegen, The Netherlands*

Abstract: Humans express emotional information in their facial expressions and body movements, as well as in their voice. In this chapter we consider the neural processing of a specific kind of vocal expressions, non-verbal emotional vocalizations e.g., laughs and sobs. We outline evidence, from patient studies and functional imaging studies, for both emotion-specific and more general processing of emotional information in the voice. We relate these findings to evidence for both basic and dimensional accounts of the representations of emotion. We describe in detail an fMRI study of positive and negative non-verbal expressions of emotion, which revealed that prefrontal areas involved in the control of orofacial movements were also sensitive to different kinds of vocal emotional information.

Keywords: voice; emotion; non-verbal expressions of emotion; functional magnetic resonance imaging (fMRI); superior temporal sulcus; premotor cortex; supplementary motor area; insula; amygdala

I. What are emotional vocalizations?

Humans are the talking apes, and our use of spoken language is considered to be one of our unique skills. Indeed, there is arguably no other sound in nature which has the complex spectrotemporal structure seen in speech. However, our voices also simultaneously express a multitude of other cues, about our mood, age, sex, health, social class and origins (Karpf, 2006). In this chapter we will outline the neural systems recruited to perceive a subset of human vocalizations: non-verbal vocal expressions of emotion. By non-verbal expressions of emotion, we specifically mean non-linguistic vocalizations, such as laughter, screams and sobbing. While speech itself is obviously frequently inflected with emotional information (e.g., we can hear whether a speaker is smiling or not (Tartter and Braun, 1994)), these non-verbal tokens are ideal for investigating the expression of emotional information when there is no concurrent verbal information,[1] and they thus closely mirror the kind of emotional information available via other channels, such as facial emotional expressions. In this chapter, we will address only some aspects of emotional speech, but will mainly emphasize non-verbal emotional vocalizations.

Speech production requires extremely fast and accurate movements of the articulators (the tongue, jaw, lips and soft palate), which shape the sounds produced at the larynx. Some of the complexity in speech can be seen in Fig. 1, which shows the acoustic structure of emotional speech and of non-verbal emotional vocalizations. Compared to the articulations of speech, non-verbal emotional expressions typically use less precise supraglottal articulations. The sound produced at the larynx is shaped by roughly positioned pharyngeal/oral/labial constrictions, which may in turn correlate with aspects of the associated facial expression.

*Corresponding author. E-mail: sophie.scott@ucl.ac.uk

[1]These kinds of sounds have been previously designated as "affect bursts" (Scherer, 1994): we avoid this term since the term "burst" connotes several acoustic features (rapid onsets, brief durations) which are not necessarily seen in the sounds themselves, which can start slowly and be relatively lengthy.

Stefan M. Brudzynski (Ed.)
Handbook of Mammalian Vocalization
ISBN 978-0-12-374593-4

DOI: 10.1016/B978-0-12-374593-4.00019-X

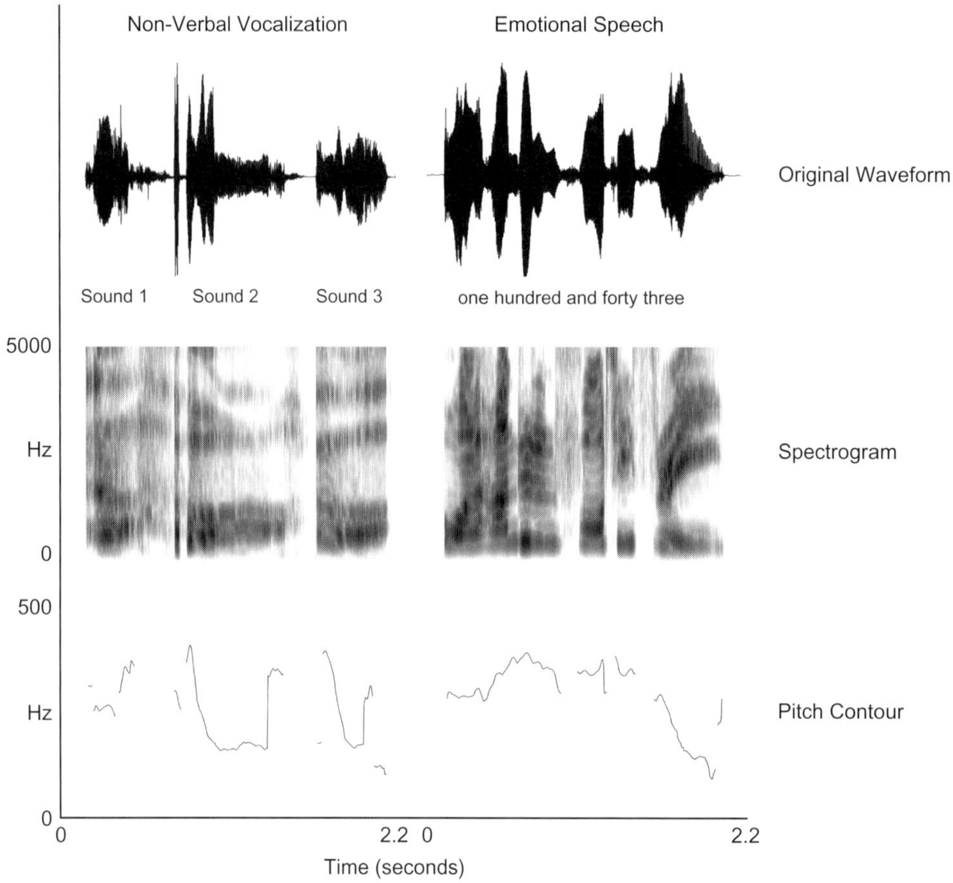

Fig. 1. A comparison of emotional speech and human emotional vocalizations. The top panel shows oscillograms of three examples of non-verbal disgust vocalizations on the left, and an oscillogram of a disgusted emotional speech example from the same (female) speaker. The middle and lower panel show the spectrograms and pitch profiles of the exemplars, respectively. The non-verbal vocalizations show less detailed spectral change than the speech, reflecting the cruder, relatively static positioning of the articulators in these emotional vocalizations. In contrast, the pitch profiles of the non-verbal vocalizations show big pitch changes throughout each utterance, while the pitch changes in the emotional speech are confined to the end of the utterance.

The spectral detail and changes in speech are much coarser in non-verbal vocalizations. The quality and control of the sound produced at the larynx has a big affect on the sounds found in emotional vocalizations. Thus, a sound like laughter, for example, has been described as more like a way of modified breathing than it is like a way of speaking (Abercrombie, 1967; cited in Kohler, 2008). These differences in the production of non-verbal emotional vocalizations are reflected in their acoustic structure. Unlike speech, which can be understood with only a certain degree of coarse spectrotemporal structure (Shannon et al., 1995), the recognition of emotional vocalizations is dependent on the preservation of information such as pitch and fine spectral structure, as well as cues such as those from the amplitude envelope (Sauter, 2006). This observation suggests that, unlike meaning in spoken language, the recognition of emotion in non-verbal emotional expressions is dominated by effects created at the glottis and postural properties (e.g., tension) in the articulators. These non-verbal emotional expressions are arguably more like animal vocalizations than they are like spoken language, since they do not require the dynamic spectral shaping caused by the rapid movements of the articulators (Jürgens, 1986).

Voice quality and pitch, as well as loudness and rate, are also important in the emotional inflection of spoken language. The recognition of emotional states in speech is strongly influenced by the pitch and pitch changes in the speech (Banse and Scherer, 1996;

Bänziger and Scherer, 2005) and by fine spectral properties (Murray and Arnott, 1993). However, the recognition rates for emotional information in speech are somewhat lower (around 50%) (Banse and Scherer, 1996) than they are for non-verbal, emotional, vocalizations (around 80%) (Sauter, 2006). While these differences may be due to the quality of the emotional stimuli (poor exemplars may not be representative of the potential emotional information in speech), this difference between emotional speech and non-verbal emotional vocalizations may also reflect the fact that emotion in speech is overlaid on the speech signal, and is thus somewhat more constrained in its expression than it is in non-verbal emotional vocalizations. Thus, there could be conflicts between the prosodic cues in sentence-level speech which denote the emotional information and those that cue linguistic information, e.g., the rising pitch of a question, the falling pitch of normally produced words, or the changes in pitch used to express linguistic emphasis and stress. Fig. 1 shows the pitch contours in both emotional speech and non-verbal emotional vocalizations. The pitch changes in the latter occur over most of the duration of the sounds, in contrast to that of the speech, where there are big changes in pitch only at the end of the utterance. These conflicts need not have a long-term effect on the communicative sense of the uttered speech, but could be enough to create difficulty in specifically making stimuli that unambiguously express a single emotion. In contrast, non-verbal emotional expressions are not produced in tandem with any other constraints other than those that affect the social and cultural ease with which these can be produced.

II. Two theories of emotion

There are two broad approaches to the study of emotion: those which identify different emotions as separate categories, which exist without reference to each other, and those in which different emotions are represented as points in emotional space, a space which is defined along two or three general dimensions (e.g., arousal, valence, approach–avoid). A dominant approach in the former class of theories is the theory of basic emotions. According to the theory of basic emotions, there is a subset of emotions that are recognized across different cultures (i.e., are universal), that are innate, that have distinctly different expressions, that can be processed in a rapid and automatic way, and that are implemented in different neural systems (Ekman et al., 1969; Ekman, 1992a,b). In contrast, a dominant dimensional account of emotions is the circumplex model (Russell, 1980). In this model, emotions are points in emotional space along the dimensions of arousal and valence.

III. Positive and negative emotions in the voice

Emotional vocalizations have been identified that mirror the categories of facial expressions of the basic emotions: fear; sadness; disgust; anger; happiness; and surprise (Scott et al., 1997). Notably, the majority of these six basic emotions – fear, sadness, anger and disgust – are negative in their valence. It has been suggested that further basic emotions, reflecting more positive moods, might be preferentially expressed with the voice, while being generally expressed facially with a smile (Ekman, 1992b). We specifically investigated this claim, developing a set of non-verbal emotional expressions of the hypothesized positive emotions of amusement, relief, sensual pleasure, contentment and triumph. We found good recognition rates for these stimuli (Sauter and Scott, 2007), with the exception of contentment, which seemed to be processed as a weaker version of sensual pleasure. While we have not tested whether it would be possible for people to distinguish between facial expressions of these emotions, this is good evidence that positive non-verbal expressions of emotion are well-recognized for the categories of amusement, triumph, relief and sensual pleasure, which goes some way to support Ekman's original argument. There is now evidence that, across cultures, people can infer pride from postural and facial cues (Tracy and Robins, 2008), which suggests that at least one more positive emotion can be read in the face and body. It is also, however, possible that faces and voices vary in how well they convey different kinds of emotion.

IV. Emotional vocalizations: general neural systems

Functional imaging (fMRI and PET) has been used to identify the cortical systems important in the perceptual processing of human vocalizations. The bilateral superior temporal sulcus (STS) has been designated a voice processing area (Belin et al., 2000), since it shows a robust response to human vocalizations across a wide range, relative to an acoustic control

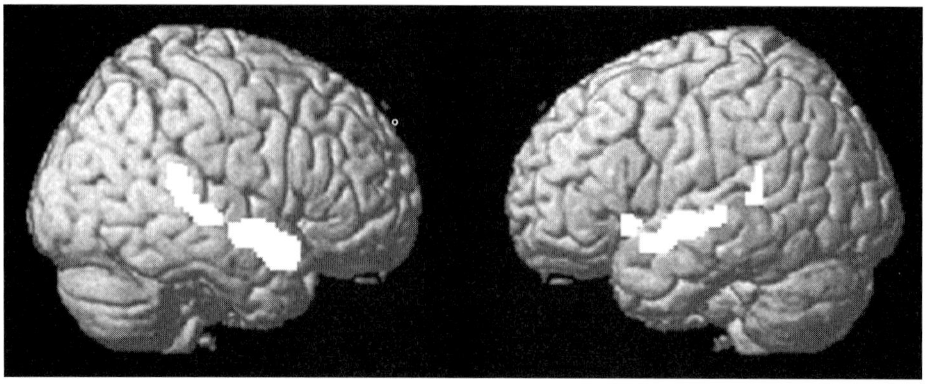

Fig. 2. Projection fields (white) of the contrast of emotional vocalizations over an acoustic control condition (data from Warren et al., 2006b) (data thresholded at false discovery rate = 0.05). The activity runs along the left and right superior temporal sulcus.

condition.[2] This STS system can be fractionated to some degree. The left mid-STS shows an enhanced response to sublexical phonetic structure in speech (Liebenthal et al., 2005), and the left anterior STS has been linked to the processing of intelligible speech (Scott et al., 2000; 2006) and to the formation of auditory word forms (Cohen et al., 2004). In contrast, the right STS has been argued to be important for processing pitch variation, since it shows a sensitivity to the dynamic pitch information in speech (Scott et al., 2000) and music (Patterson et al., 2002). In an adaptation study, the right anterior STS showed a specific response to changes in speakers, but not to changes in the words the different speakers said, suggesting a role in the processing of vocal identity (Belin et al., 2003). The right anterior STS has also been shown to be sensitive to signals with sufficient spectral detail for speaker identity to be available, over an acoustic control condition (Scott et al., 2006; Warren et al., 2006a).

The STS is also very sensitive to non-verbal emotional expressions; Fig. 2 is from an fMRI study comparing non-verbal expressions of emotion to a well-matched acoustic control condition (Warren et al., 2006b), and shows the comparison of all emotional vocalizations over spectrally rotated equivalents (the control condition). This shows extensive bilateral STS activation, extending further in the anterior STS on the right than on the left side of the brain. This result suggests that, as for speech and speaker information, the STS is an important stage in the perception and representation of non-verbal emotional stimuli, following acoustic processing in the superior temporal gyrus. Within the STS there may be a left/right distinction, with the left anterior STS fields being more sensitive to linguistic information, and the right anterior STS fields being more sensitive to talker identity (Belin and Zatorre, 2003) and emotion (Grandjean et al., 2005; Schirmer and Kotz, 2006) information in the voice. It has been suggested that, in face processing, visual speech is processed very distinctly from visual identity and emotional expression information, which are seen to share many properties at perceptual levels (Calder and Young, 2005). This broad left/right distinction between speech information and talker/emotion in the voice may thus mirror functional aspects of the organization of facial processing.

V. Emotional vocalizations: specific neural systems

In the theory of basic emotions (Ekman et al., 1969; Ekman, 1992a,b), it is hypothesized that there are distinct neural systems for different basic emotion categories. Neuropsychological studies of emotion recognition impairments following brain damage, and functional imaging studies of the neural systems recruited by different emotions in participants with undamaged brains provide good ways of testing this hypothesis, although typically small numbers of patients are studied in the former case. These techniques allow us to identify different neural systems

[2]Control conditions are important in functional imaging studies, since the observed activation is always determined relative to some other condition. In studies with communication sounds, such as speech or other vocalizations, this allows us to distinguish between neural activation associated with low-level acoustic stimulation, and activation associated with the specific properties of the sounds of interest.

that are recruited in the perception of different emotions, and that are distinct from general auditory processing in the superior temporal gyrus (STG) and STS. The "basic" emotions, comprising anger, disgust, fear, sadness, surprise and happiness, have been investigated mainly using facial expressions of emotion; however, there are some neuropsychological and neuroscientific investigations using non-verbal vocal expressions of emotion.

VI. Neuropsychological studies of emotion recognition

VI.A. Fear and anger

Patients with amygdala damage have difficulty in processing identity from faces, and also in processing facial signals communicating certain emotions, particularly fear and anger. Work with a patient who had bilateral amygdala resections (patient DR) found that she also showed a constellation of problems in dealing with vocal information. She showed deficits in perceiving non-emotional prosodic information in speech (e.g., the difference between a question and a command), while performing normally on tests of speech perception, sentence comprehension and recognition of environmental noises (Scott et al., 1997). Consistent with her problems in identifying people from photographs, she performed poorly on voice recognition tasks, both matching unfamiliar voices and recognizing familiar voices. In emotional processing, she was impaired on the recognition of sadness, fear and anger when tested with emotionally inflected single words: she did not perform well on the other emotions tested (happiness and disgust); however, her performance was not significantly different from that of age-matched control participants (suggesting that the emotional speech stimuli were not necessarily well-recognized). When tested with non-verbal emotional expressions, however, the results were clearer: DR was impaired on the recognition of fear and anger, and performed at ceiling for the recognition of the other emotions tested (disgust, sadness, surprise and happiness), while the control subjects recognized all the different emotional noises well. Thus, as in processing facial emotional information, DR had a specific problem with the processing of anger and fear from the voice, and this problem was clearer when tested with non-verbal emotional vocalizations. This finding has been replicated in other patients, and extended to emotional bodily postures of fear (Sprenglemeyer et al., 1999). These results have been taken as evidence that the amygdala (while clearly also involved in a variety of processes) has an important role in the processing of threat. Thus, the amygdala has been hypothesized to process angry cues from others because the angry person could be a threat, and to process fearful cues from others because whatever is threatening them could also threaten you. Some studies have failed to replicate this effect with vocal expressions of fear (e.g., Anderson and Phelps, 1998). However, in this case, the patient did show poorer recognition of the non-verbal recognition of fearful sounds than of any other non-verbal emotional expression; the control subjects also performed very variably on the recognition of these fear sounds, suggesting that the fear stimuli used in this study were generally not well-recognized.

VI.B. Disgust

There is some evidence that disorders affecting the basal ganglia can lead to specific deficits in processing emotions. Thus, patients with Huntingdon's chorea, a heritable dementia whose effects are initially mainly in the basal ganglia, have been shown to have deficits in the perception of disgust from the face and in emotional speech (Sprengelmeyer et al., 1996). A case study of a patient with specific damage to the putamen and the anterior insula (Calder et al., 2000) also showed deficits in the recognition of disgust from emotional facial expressions, and from both verbal and non-verbal vocal emotional expressions.

VI.C. Sadness and anger

Studies of patients with frontal lesions, grouped by lesion location (Hornak et al., 2003), have revealed that specific problems in processing non-verbal emotional vocalizations of sadness can be seen following damage to the ventromedial frontal lobe. In a subset of these patients, problems with vocal expressions of anger were also reported.

VI.D. Happiness

The presupplementary motor area (pSMA) has been implicated in the production of laughter (Fried et al., 1998). A depth electrode investigation of an epileptic patient showed that stimulation of the pSMA resulted in first smiling, then laughter and a sensation of mirth (Krolak-Salmon et al., 2006). Increasing the stimulating current led to signs of distress, such as moaning and tears (Krolak-Salmon et al., 2006). Recording from this area also showed an enhanced response to facial expressions of happiness (i.e., smiling faces) (Krolak-Salmon et al., 2006). The authors suggested that these effects might reflect an involvement of a brain mirror system in the recognition, experience and production of happiness, since the pSMA was involved in the production of smiles and laughter, in the experience of mirth and the perception of happiness in the face (Krolak-Salmon et al., 2006).

VII. Separate neural systems for different vocal expressions of emotion?

There is thus some evidence from patient studies that damage to different brain areas can lead to specific deficits in the recognition of different vocal expressions of emotion. Amygdala damage can lead to deficits in the recognition of fear and anger, insula and putamen damage can detrimentally affect disgust recognition, orbito-frontal lesions can affect sadness, anger and disgust recognition, and the expression of happiness (laughter) recognition is associated with stimulation of the pSMA. In many cases, these recognition deficits are seen across different modalities (e.g., face as well as voice), which would be congruent with the notion of basic expressions of emotion (Ekman et al., 1969; Ekman, 1992a,b). As might be expected, however, there is no clear evidence of simple one-to-one mapping between brain regions and recognition of specific emotions – the amygdala does not exclusively support fear recognition and likewise single emotions can be affected by a variety of brain regions – for example, disgust deficits can be seen following damage to the insula and putamen (Calder et al., 2000), or orbitofrontal lesions (Hornak et al., 2003). This pattern is borne out by functional imaging studies, which have tended to show more distributed patterns of activation to emotional vocalizations.

VIII. Functional imaging studies of emotion recognition

In line with some of the patient literature, some functional imaging studies of non-verbal emotional vocalizations have found a specific amygdala response to fearful vocalizations (e.g., Phillips et al., 1998). In this study, a range of different brain areas was also activated to the vocal expressions of fear, in addition to the amygdala activation. This study showed insula responses to facial expressions of disgust, but not to vocal expressions of disgust. In other studies, amygdala activation has been seen to a range of emotional vocalizations, both positive and negative (Sander et al., 2001, 2003; Fecteau et al., 2007). It is known from functional imaging studies that brain regions other than the amygdala are activated by fearful sounds, and not only fearful sounds activate the amygdala. This variation may reflect, in part, the range of emotional vocalizations used in the study (e.g., Phillips et al., 1998 only investigated fear and disgust). Functional imaging methods see all the brain areas recruited by a task or perceptual processes, and it can be difficult to distinguish between brain areas that are somewhat peripheral to the task or process of interest from those that are critical. Insula activation has been seen to both positive and negative non-verbal emotional expressions, as well as to disgust, e.g., laughter and crying vocalizations (Morris et al., 1999; Sander et al., 2003), which may reflect its role in specific emotions, or may reflect the more general role of the anterior insula in the representation of vocalizations (Wise et al., 1999).

IX. Dimensional accounts of emotion and functional imaging studies

As mentioned in the introduction, in contrast to the basic emotions approach to the nature of emotional processing, there are theories which posit that emotional states are represented as points in emotional space, with two (or more) coordinates. Thus, in the circumplex model of emotion (Russell, 1980), different emotions are plotted along the dimensions of valence and arousal. In vocal emotion processing, it has been suggested that, rather than specific emotions, "speech acoustics provide an external cue to the level of non-specific arousal associated with emotional processes" (Bachorowski, 1999, p. 55). Consistent with this, patients with Parkinson's disease, a degenerative disease of the basal ganglia, have been reported to show a general problem with prosody (Lloyd, 1999). Recent work has shown, however, that the Parkinson's patients remain sensitive to the intensity in emotional speech, even though they have deficits in recognizing the emotion conveyed, or in rating the valence of the emotional speech (Dara et al., 2007). This finding suggests that a single dimension in Russell's circumplex model, arousal, might be preserved while another (valence) could be selectively damaged. An alternative interpretation could be that quite simple properties of emotional speech, such as intensity, map onto the perceived emotional intensity – louder sounds are more aroused, across all emotion categories (Banse and Scherer, 1996). This interpretation might mean that it is easier to establish arousal than valence from a simple appraisal of the stimuli, in the presence of brain damage, allowing the patients to make accurate judgments of emotional intensity.

In a recent functional imaging study of non-verbal vocal expressions of emotion, we used fMRI to present people with non-verbal vocal expressions of two negative emotion categories (disgust and fear) and two positive emotion categories (amusement and triumph) (Warren et al., 2006b). These four emotional categories were chosen since they were all highly recognizable and also mapped clearly along the dimension of valence: the amusement and triumph sounds were rated as highly positive, and the fear and disgust sounds as highly negative (Sauter, 2006). We also included a condition in which participants were prompted to make orofacial movements, by instructing them to smile. This permitted an analysis of brain areas involved in both the perception of emotional vocalizations and the brain areas involved in orofacial movement. We also included a control condition of spectrally-rotated versions of the emotional stimuli, which we used to control for the brain response to low-level acoustic properties of the emotional sounds.

Comparing all the emotional categories with the acoustic control condition revealed activity in the left and right STS, the left insula and pSMA (Sauter, 2006). Comparing individual emotion categories with the control condition revealed activation in the bilateral anterior STS for every emotion. In addition to this, amusement showed activation in the inferior frontal gyrus, fear showed extensive activation of the bilateral anterior insula, and triumph showed left insula activation, pSMA and left and right premotor activation.

Further to the simple contrast of all the different emotion categories with the acoustic baseline described above, a contrast of brain areas that varied significantly across emotion conditions (relative to the acoustic control condition) was run, to identify any specific cortical responses to the kinds of emotional information in the sounds. This contrast identified significant patterns in activation across the superior and inferior temporal lobes, the precentral gyrus and prefrontal cortex, limbic and mesial temporal cortex, including the amygdala, hippocampus and basal ganglia. This contrast was used to mask a contrast of the areas which were activated by orofacial movement (smiling to command), revealing activation in the precentral gyrus, pSMA and insula, and in the temporal and occipital lobes, which was affected both by the emotional sounds subjects heard and when they (silently) smiled.

Within these areas co-activated by production and perception of emotional vocalizations, we identified regions that correlated with the valence and arousal ratings for the stimuli (shown in Fig. 3) which had been collected with a different set of subjects in a previous behavioral experiment. In terms of valence, the triumph sounds were rated most positive, followed by amusement, then fear and disgust (Sauter, 2006). The correlation with valence alone was seen in the right inferior frontal gyrus. In terms of arousal, the categories of triumph and fear were rated as the most aroused, followed by amusement and disgust, and this

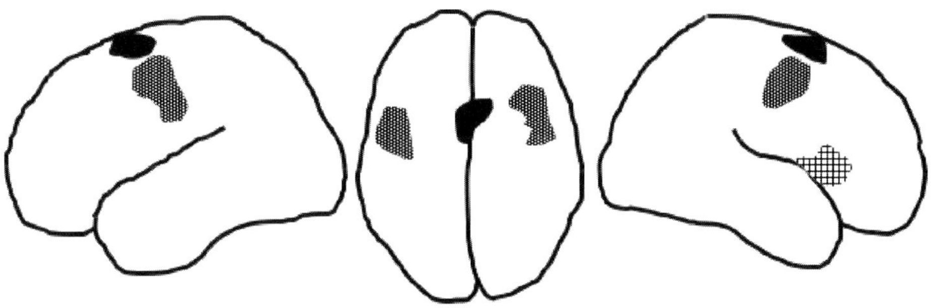

Fig. 3. Brain regions co-activated by orofacial movements and by emotional vocalizations (adapted from Warren et al., 2006b). Key: pre-supplementary motor areas (solid black); lateral premotor (black dots); inferior frontal gyrus (hatching). Activity in regions pre-supplementary motor areas correlate with the perceived arousal of the non-verbal emotional vocalizations: Activity in right inferior frontal gyrus correlates with perceived valence, and activity in bilateral premotor cortex correlates with both valence and arousal scores.

led to activation of the pSMA. This result provides evidence that emotional vocalizations of high arousal therefore engage a region involved in higher-order aspects of complex motor control, which may link to a specific preparation for responsive action.

The left and right precentral gyri were co-activated by increases in both valence and arousal. The precentral gyri are somatotopically organized, and the peaks of activation were greatest in the face motor area (Buccino et al., 2001; Carr et al., 2003; Leslie et al., 2004). The activation also extended into more ventral regions associated with motor control of the articulators (Murphy et al., 1997; Blank et al., 2002; Wilson et al., 2004). This activation extended into primary motor regions bilaterally, even though the subjects did not actually move their faces during the scanning session.

With the proviso that only four different emotional categories were tested, this finding is a striking demonstration that motor regions associated with producing facial expressions are activated by hearing emotional vocalizations, and that different regions within this system correlate with the arousal and valence of the stimuli. The perception of vocal expressions of positive or arousing emotions automatically engages preparation for responsive orofacial gestures. We tend to express emotions like cheering and laughing in groups, and laughter has been shown to be a very contagious behavior (Provine, 2000) – when we hear someone laughing, we are primed to join in, or at very least to smile. The finding of increased premotor cortex activation to the sounds of laughter (and cheering) would provide a mechanism for this link.

The finding that the dimensions of valence and arousal have specific neural bases within frontal areas associated with the perception of emotional vocalizations and the production of orofacial movements may provide evidence in support of the circumplex model of emotion (Russell, 1980). The circumplex model of emotion posits that different emotional states are processed and represented as points in an emotional space, along the dimensions of valence and arousal. The current data do not provide full support for this model, however, since no negative correlations were found. There were no increases in cortical activity with lower ratings of valence, despite the fact that both the fear and disgust stimuli were rated as highly negative in valence. Likewise, there were no increases in cortical activity with lower ratings of arousal. It seems highly unlikely that the unpleasantness of disgust and fear are encoded cortically simply as a lack of activation. Further work with a wider palate of emotional states may be the way to address this point and identify whether convincing correlations with more negative valence or decreased arousal can be determined, as would be predicted by dimensional accounts of emotion representations. Different imaging techniques will also facilitate the development of this work – e.g., Sauter and Eimer (in press) have shown that the timing of electrophysiological responses to non-verbal emotional vocalizations varies with the perceived arousal of the expressions.

X. Role of mirror systems in perception of emotional vocalizations

The finding that brain areas associated with motor control are activated by both emotion perception and orofacial gestures has some congruency with theories about the involvement of mirror systems in human communication (Warren et al., 2006b). Several recent studies have suggested that the premotor cortex is co-activated by produced and perceived speech (Wilson et al., 2004), and the results of Warren et al. (2006a,b) extend this into non-verbal vocalizations and orofacial gestures. Indeed, these results go somewhat beyond the results from speech perception studies, since the current investigation revealed activation relative to complex acoustic control conditions, unlike the speech studies (Scott et al., 2009).

Does the involvement of the "mirror" system in the perception of emotional vocalizations and production of orofacial movements suggest that the recognition of emotional vocalizations entails the imitation of the emotions in question? We would argue not, for several reasons. First, the mirror system activations seen in Warren and colleagues study (Warren et al., 2006b) are not likely to reflect the *emotional* contagion of the stimuli *per se*, since the disgust vocalizations led to the weakest activations, despite disgust being an exceedingly contagious emotion (Nemeroff and Rozin, 1994). The results are also unlikely to reflect a mirror system involvement in the recognition of the sounds, since the fear and disgust sounds were very highly recognizable (Sauter, 2006), yet led to lower activation in "mirror" motor areas. Instead of recognition of emotions, or emotional recognition, we argued that these results are related to behavioral responses to the stimuli – either preparation for general action to respond in pSMA (such as readiness to run away for the fear sounds) or preparation for responsive emotional orofacial expressions or vocalizations (such as smiling or laughing) in right IFG and bilateral precentral gyri. It is striking that behavioral mirroring in human communication tends to be in positive contexts – if we like someone, we tend to adopt the same position and gestures when talking to them, and will do so more if we are trying to express affiliation towards them (e.g., Chartrand and Bargh, 1999). A recent review (Scott et al., 2009) has linked motor cortex activation during speech perception to the employment of motor areas in the timing and coordination of communicative cooperative actions, such as turn taking in conversation. In this context, the lateral premotor system responses to more positive and more arousing emotional stimuli reflect the more social nature of the emotions used in this experiment. Laughter and cheering tend to be vocal expressions we produce in groups, and which play an important part in reinforcing positive social bonds between people (Provine, 2000).

Motor cortex in humans has been shown to respond to a variety of different sounds – including music, animal vocalizations and environmental sounds. The evidence for consistent activation of motor cortex by speech (relative to an acoustic control condition) is somewhat weaker (Scott et al., 2009). Thus, it may suggest that unlike the STS, motor areas are less sensitive to the particular acoustic structure of speech – i.e., its reliance on the processing of dynamic, broad spectral structure. In this context, the sensitivity of motor cortex to different emotional vocalizations may reflect the fact that they, like other non-speech sound, rely on the use of envelope, pitch and fine spectral cues for recognition (e.g., environmental noises, Gygi et al., 2004; Sauter, 2006).

XI. Conclusions

A wide range of different neural areas are recruited when people process non-verbal emotional vocalizations. Some are very generally involved in processing social communication, thus hearing emotional sounds activates the bilateral STS, with an emphasis on anterior regions in the right hemisphere. Some other brain regions appear to be specifically involved in the perception of certain emotions across different perceptual channels; damage to brain regions, such as the amygdala and insula, result in deficits in the recognition of some emotions from the voice (e.g., fear and disgust), mirroring deficits in the perception of the same emotions in the face. Functional imaging studies tend to reveal that a network of activity is seen to different vocal emotional expressions, and that some areas (such as the amygdala) are also activated by a range of different vocally-expressed emotions, in addition to those to which they have been linked by the neuropsychological literature. Some brain regions appear to be processing some of the dimensions hypothesized to be important in the conceptual representation of emotions; in frontal areas, brain regions can be identified that are correlated in activity with the perceived arousal (pSMA) or valence (right IFG) of emotional vocalizations, or be sensitive to both arousal and valence (right and left premotor cortex) (Warren et al., 2006a,b). This frontal activity seems to be a specific function of the behavioral correlates of the emotions used in this study, with the positive sounds of laughter and cheering being those commonly performed in groups of other people, and which have a specific role of reinforcing bonds between members of a group. We argue that, generally, mirroring behavior has strong positive affect in human communication – if we like someone, we copy their gestures and movements in conversations with them, and even start to use the same words. The motor system may thus underlie the coordination of behaviors such as conversation and contagious laughter, which are central to the affiliative aspects of human communication.

References

Abercrombie, D., 1967. Elements of General Phonetics. Edinburgh University Press, Edinburgh, UK.

Anderson, A.K., Phelps, E.A., 1998. Intact recognition of vocal expressions of fear following bilateral lesions of the human amygdala. Neuroreport 9 (16), 3607–3613.

Bachorowski, J.A., 1999. Vocal expression and perception of emotion. Curr. Dir. Psychol. Sci. 8, 53–57.

Banse, R., Scherer, K.R., 1996. Acoustic profiles in vocal emotion expression. J. Pers. Soc. Psychol. 70, 614–636.

Bänziger, T., Scherer, K.R., 2005. The role of intonation in emotional expressions. Speech Commun. 46, 252–267.

Belin, P., Zatorre, R.J., 2003. Adaptation to speaker's voice in right anterior temporal lobe. Neuroreport 14 (16), 2105–2109.

Belin, P., Zatorre, R.J., Lafaille, P., Ahad, P., Pike, B., 2000. Voice-selective areas in human auditory cortex. Nature 403 (6767), 309–312.

Blank, S.C., Scott, S.K., Murphy, K.M., Warburton, E., Wise, R.J.S., 2002. Speech production: Wernicke, Broca and beyond. Brain 125, 1829–1838.

Buccino, G., Binkofski, F., Fink, G.R., Fadiga, L., Fogassi, L., Gallese, V., Seitz, R.J., Zilles, K., Rizzolatti, G., Freund, H.J., 2001. Action observation activates premotor and parietal areas in a somatotopic manner: an fMRI study. Eur. J. Neurosci. 13, 400–404.

Calder, A.J., Young, A.W., 2005. Understanding the recognition of facial identity and facial expression. Nat. Rev. Neurosci. 6 (8), 641–651.

Calder, A.J., Keane, J., Manes, F., Antoun, N., Young, A.W., 2000. Impaired recognition and experience of disgust following brain injury. Nat. Neurosci. 3 (11), 1077–1078.

Carr, L., Iacoboni, M., Dubeau, M.C., Mazziotta, J.C., Lenzi, G.L., 2003. Neural mechanisms of empathy in humans: a relay from neural systems for imitation to limbic areas. Proc. Natl. Acad. Sci. USA, 100, 5497–5502.

Chartrand, T.L., Bargh, J.A., 1999. The chameleon effect: the perception-behavior link and social interaction. J. Pers. Soc. Psychol. 76, 893–910.

Cohen, L., Jobert, A., Le Bihan, D., Dehaene, S., 2004. Distinct unimodal and multimodal regions for word processing in the left temporal cortex. Neuroimage 23 (4), 1256–1270.

Dara, C., Monetta, L., Pell, M.D., 2007. Vocal emotion processing in Parkinson's disease: reduced sensitivity to negative emotions. Brain Res. 1188, 100–111.

Ekman, P., 1992a. Are there basic emotions? Psychol. Rev. 99, 550–553.

Ekman, P., 1992b. An argument for basic emotions. Cogn. Emot. 6, 169–200.

Ekman, P., Sorenson, E.R., Friesen, W.V., 1969. Pan-cultural elements in facial displays of emotion. Science 164, 86–88.

Fecteau, S., Belin, P., Joanette, Y., Armony, J.L., 2007. Amygdala responses to nonlinguistic emotional vocalizations. Neuroimage 36 (2), 480–487.

Fried, I., Wilson, C.L., MacDonald, K.A., Behnke, E.J., 1998. Electric current stimulates laughter. Nature 391, 650.

Grandjean, D., Sander, D., Pourtois, G., Schwartz, S., Seghier, M.L., Scherer, K.R., et al., 2005. The voices of wrath: brain responses to angry prosody in meaningless speech. Nature Neurosci. 8, 145–146.

Gygi, B., Kidd, G.R., Watson, C.S., 2004. Spectral-temporal factors in the identification of environmental sounds. J. Acoust. Soc. Am. 115 (3), 1252–1265.

Hornak, J., Bramham, J., Rolls, E.T., Morris, R.G., O'Doherty, J., Bullock, P.R., Polkey, C.E., 2003. Changes in emotion after circumscribed surgical lesions of the orbitofrontal and cingulate cortices. Brain 126 (7), 1691–1712.

Jürgens, U., 1986. The squirrel monkey as an experimental model in the study of cerebral organization of emotional vocal utterances. Eur. Arch. Psychiat. Neurol. Sci. 236 (1), 40–43.

Karpf, A., 2006. The Human Voice. Bloomsbury, London, UK.

Kohler, K.J., 2008. "Speech-smile," "speech-laugh," "laughter" and their sequencing in dialogic interaction. Phonetica 65, 1–18.

Krolak-Salmon, P., Hénaff, M-A., Vighetto, A., Bauchet, F., Bertrand, O., Mauguière, F., et al., 2006. Experiencing and detecting happiness in humans: the role of the supplementary motor area. Ann. Neurol. 59, 196–199.

Leslie, K.R., Johnson-Frey, S.H., Grafton, S.T., 2004. Functional imaging of face and hand imitation: towards a motor theory of empathy. Neuroimage 21, 601–607.

Liebenthal, E., Binder, J.R., Spitzer, S.M., Possing, E.T., Medler, D.A., 2005. Neural substrates of phonemic perception. Cereb. Cortex 15 (10), 1621–1631.

Lloyd, A.J., 1999. Comprehension of prosody in Parkinson's disease. Cortex 35 (3), 389–402.

Morris, J.S., Scott, S.K., Dolan, R.J., 1999. Saying it with feeling: neural responses to emotional vocalizations. Neuropsychologia 37 (10), 1155–1163.

Murphy, K., Corfield, D.R., Guz, A., Fink, G.R., Wise, R.J., Harrison, J., Adams, L., 1997. Cerebral areas associated with motor control of speech in humans. J. Appl. Physiol. 83, 1438–1447.

Murray, I.R., Arnott, J.L., 1993. Toward the simulation of emotion in synthetic speech: a review of the literature on human vocal emotion. J. Acoust. Soc. Am. 93, 1097–1108.

Nemeroff, C., Rozin, P., 1994. The contagion concept in adult thinking in the United States: transmission of germs and of interpersonal influence. Ethos 22, 158–186.

Patterson, R.D., Uppenkamp, S., Johnsrude, I.S., Griffiths, T.D., 2002. The processing of temporal pitch and melody information in auditory cortex. Neuron 36 (4), 767–776.

Phillips, M.L., Young, A.W., Scott, S.K., Calder, A.J., Andrew, C., Giampietro, V., Williams, S.C.R., Bullmore, E.T., Brammer, M., Gray, J.A., 1998. Neural responses to facial and vocal expressions of fear and disgust. Proc. Royal Soc. B. 265 (1408), 1809–1817.

Provine, R.R., 2000. Laughter: A Scientific Investigation. Faber and Faber, London, UK.

Russell, J.A., 1980. A circumplex model of affect. J. Person. Soc. Psychol. 39, 1161–1178.

Sander, K., Brechmann, A., Scheich, H., 2003. Audition of laughing and crying leads to right amygdala activation in a low-noise fMRI setting. Brain Res. Brain Res. Protoc. 11 (2), 81–91.

Sauter, D., 2006. An investigation into vocal expressions of emotions: the roles of valence, culture, and acoustic factors. PhD thesis, University of London, UK.

Sauter, D.A. and Eimer, M. Rapid detection of emotion from human vocalizations, J. Cogn. Neurosci. (in press).

Sauter, D., Scott, S.K., 2007. More than one kind of happiness: can we recognize vocal expressions of different positive states? Motiv. Emotion 31, 192–199.

Scherer, K.R., 1994. Affect bursts. In: van Goozen, S., van de Poll, N.E., Sergeant, J.A. (Eds.), Emotions: Essays on Emotion Theory. Erlbaum, Hillsdale, NJ, pp. 161–196.

Schirmer, A., Kotz, S.A., 2006. Beyond the right hemisphere: brain mechanisms mediating vocal emotional processing. Trends Cogn. Sci. 10, 24–30.

Scott, S.K., Young, A.W., Calder, A.J., Hellawell, D.J., Aggleton, J.P., Johnson, M., 1997. Impaired auditory recognition of fear and anger following bilateral amygdala lesions. Nature 385, 254–257.

Scott, S.K., Blank, S.C., Rosen, S., Wise, R.J.S., 2000. Identification of a pathway for intelligible speech in the left temporal lobe. Brain 123, 2400–2406.

Scott, S.K., Rosen, S., Lang, H., Wise, R.J.S., 2006. Neural correlates of intelligibility in speech investigated with noise vocoded speech – a Positron Emission Tomography study. J. Acoust. Soc. Am. 120 (2), 1075–1083.

Scott, S.K., McGettigan, C., Eisner, F., 2009. A little more conversation, a little less action: candidate roles for motor cortex in speech perception. Nature Rev. Neurosci. 10 (4), 295–302.

Shannon, R.V., Zeng, F.G., Kamath, V., Wygonski, J., Ekelid, M., 1995. Speech recognition with primarily temporal cues. Science 270, 303–304.

Sprengelmeyer, R., Young, A.W., Calder, A.J., Karnat, A., Lange, H., Hömberg, V., Perrett, D.I., Rowland, D.,

1996. Loss of disgust. Perception of faces and emotions in Huntington's disease. Brain 119 (5), 1647–1665.
Tartter, V.C., Braun, D., 1994. Hearing smiles and frowns in normal and whisper registers. J. Acoust. Soc. Am. 96 (4), 2101–2107.
Tracy, J.L., Robins, R.W., 2008. The nonverbal expression of pride: evidence for cross-cultural recognition. J. Pers. Soc. Psychol. 94, 516–530.
Warren, J.D., Scott, S.K., Price, C.J., Griffiths, T.D., 2006a. Human brain mechanisms for the early analysis of voices. Neuroimage 31 (3), 1389–1397.
Warren, J.E., Sauter, D.A., Eisner, F., Wiland, J., Dresner, M.A., Wise, R.J., Rosen, S., Scott, S.K., 2006b. Positive emotions preferentially engage an auditory-motor "mirror" system. J. Neurosci. 26 (50), 13067–13075.
Wilson, S.M., Saygin, A.P., Sereno, M.I., Iacoboni, M., 2004. Listening to speech activates motor areas involved in speech production. Nat. Neurosci. 7, 701–702.
Wise, R.J.S., Greene, J., Büchel, C., Scott, S.K., 1999. Brain systems for word perception and articulation. The Lancet 353 (9158), 1057–1061.

SECTION 6

Limbic Generation of Vocalization: Vocalization as an Index of Behavioral State

Mammalian vocalizations are generated by affective states of callers as a result of social-affective interactions and generated by limbic circuitries. The vocalizations have not only the capacity of influencing behavior of the recipients, but they can also serve as an index of a relevant affective state of the vocalizing animal. Evidence for this can be found in many groups, ranging from rodents to primates, and also in infants.

CHAPTER 6.1

Emotional causes and consequences of social-affective vocalization

Jaak Panksepp*

Department of Veterinary and Comparative Anatomy, Pharmacology and Physiology, College of Veterinary Medicine, Washington State University, Pullman, Washington, USA

Abstract: Instinctual (inborn, genetically-controlled) emotional vocalizations can be used to index internal affective states. Vocal learning adds layers of cognitive complexity to this intrinsic vocal repertoire of animals. This chapter focuses on: (1) the separation calls (cries) of young animals as indicators of emotional distress that have ancestral relationships to human grief and sadness, emotional responses which appear to be related to the genesis of depression; and (2) the positive affective vocalizations made by juvenile rats during vigorous playful interactions, which reflect positive social-joy processes of the brain, that may counteract sadness and depression. Such emotional vocalizations provide new gateways for understanding human psychiatric disorders. Further, the way in which basic emotional sounds may be related to language acquisition and human cultural inventions such as music are considered.

Keywords: emotional vocalizations; separation calls/cries; play vocalizations; brain mechanisms of affect; neurochemical mechanisms; evolution of music; evolution of language; comparative perspectives on vocalizations; guinea pigs; rats; domestic chickens

I. Introduction

Brain research suggests that the evolution of species-typical vocal capacities was intimately linked to the neural evolution of brain emotional networks and the emergence of internally-experienced affective states that signaled key survival issues. Experienced affects may be internal reward/punishment codes by which reinforcement-based learning occurs. If so, externally evident behaviors that can index such central states are of great value in advancing our understanding of basic, evolutionarily-provided psychological abilities and provide ways of probing brain mechanisms that link instinctual (inborn, genetically-controlled) adaptive behaviors to learning (Panksepp, 2005a; Zinken et al., 2008).

Brain stimulation-induced instinctual emotional vocalization has been the basis for mapping some of the basic emotional circuits of the brain (Panksepp, 1998, 2005b). Such brain circuits have been shown to mediate primal affective states, as measured by unconditional and conditioned approach and avoidance patterns. Thus, emotional vocalizations are one major gateway to the hidden nature of affective states – the internal value codes that tell animals where they are situated on the gradients of survival.

The basic emotional networks of the brain reflect ancestral memories – brain functions of such general importance that they were built into the genetically-dictated subcortical infrastructure of the brain. In addition to the diverse emotional action patterns of animals and humans that Charles Darwin (1872/1965) first described in some cross-species detail, emotional vocalizations emerged early in mammalian evolution as a way for animals to communicate their emotional states to others. It is likely that the very first vocalization on the face of the earth was the screech of pain, which served as a platform for evolutionary development of a variety of vocal adaptations to alert others to danger – for instance via separation-distress calls to help mothers find lost

*Corresponding author. E-mail: jpanksepp@vetmed.wsu.edu

Stefan M. Brudzynski (Ed.)
Handbook of Mammalian Vocalization
ISBN 978-0-12-374593-4

DOI: 10.1016/B978-0-12-374593-4.00020-6

infants (Panksepp, 1998), and various alarm calls to alert community members to the presence of shared dangers. This evolutionary continuity is indexed not only by the proximity of pain and separation-distress vocalization circuitry in the medial periaqueductal gray regions of the midbrain, but also by their shared neurochemical mechanisms such as those mediated by endogenous opioids (Panksepp et al., 1988).

It is likely that separation calls were the first emotional vocalizations that evolved from the more ancient neural networks for pain. Such ancient emotional vocalizations provided a basis for the shaping of a variety of more complex alarm calls that signaled danger (Litvin, et al., 2007; Furrer and Manser, 2009). On the positive emotional side, animals needed to coordinate various species-sustaining activities, and among the first calls to have evolved may have been signals of socio-sexual readiness for interaction, which may have then branched off to mother–infant proximity calls that in some species, such as dolphins, may become molded into individual characteristic "signature whistles" (Fripp and Tyack, 2008). Among such calls the best studied are the 50 kHz ultrasonic calls (chirps) that rats emit during various positive social activities, especially during play and sexual solicitations (Knutson et al., 2002; Panksepp, 2007; Burgdorf et al., 2008).

Indeed, 50 kHz calls communicate readiness for social interactions, from the playful engagements of youth to sexual readiness in adults, and this positive emotional call may have been an evolutionary preadaptation for the emergence of infantile social laughter in the human species. As we will see, the relationship of this call to a laughter-type vocalization is now well-supported by animal studies (Panksepp and Burgdorf, 2003; Panksepp, 2007). Human neuroanatomical studies of pathological crying and laughter (Wild et al., 2003) affirm that in the human species such primal vocal control networks are situated in subcortical brainstem regions that we share in a homologous way with other animals (Newman, 1988).

II. Emotional vocalizations and affective experiences

Although probably most, if not all, animal instinctual vocalizations are driven by emotional mechanisms, I will focus on the above two categories of vocalizations in this chapter. However, first, I will provide a few reflections about the empirical gateways to primordial mental functions – ancient affective feelings – that such emotional vocalizations provide. A general principle concerning the valenced feelings of emotions that has emerged during the past few decades is the likelihood that raw emotional affects – internal, subjectively experienced brain processes that cannot be directly studied – arise from the same ancient neurodynamic processes (e.g., large-scale nonlinear network attractor landscapes) that control objective, externally observable emotional action patterns (Panksepp, 2005a,b). For instance, brain stimulation within subcortical networks that evoke the fixed action patterns of RAGE, FEAR and PANIC/Separation Distress, are states of the nervous systems which animals develop and "seek" to terminate and avoid (via "intentionality in action"). On the other hand, brain stimulation that evokes displays of "primary-process" of SEEKING, LUST, CARE and PLAY generate central states that animals develop and "seek" to approach and sustain when given the opportunity (Panksepp, 1998). The capitalization of these systems is to highlight that a specific emotional network of the brain is the referent.

It is important to re-emphasize that current evidence indicates that the measurable instinctual emotional behaviors and the hidden affective processes that can only be studied indirectly, share the same neural substrates. From this dual-aspect "monism" point of view, one can use objectively observed behaviors as proxies for primary-process subjective states of the brain/mind. This strategy has opened up empirical approaches to the study of the neural infrastructure of emotional states in animal models (Panksepp, 1998, 2008; Brudzynski, 2007). One of the earliest demonstrations of this principle in primates was the vocal control mapping studies published by Uwe Jürgens, who first demonstrated that stimulation of electrode sites that reflected positive emotions sustained self-stimulation behaviors. Further, stimulation that evoked aversive emotional-type vocalizations was avoided by animals (Jürgens, 1976). Such findings help us understand the neural infrastructure of the primary-process emotional states of organisms.

III. From primary to tertiary emotional vocalizations

Please note that the designators for "primary-process" emotions listed above are capitalized. This convention is used to provide distinct scientific names for genetically prescribed circuits for a series of inherited emotional capacities. First, "primary-" and "secondary-processes" are not used here in the way psychoanalysis deployed these concepts ever since Sigmund Freud introduced them in his *Interpretation of Dreams* (1900, Chapter VII). Psychoanalysis uses such terms for primitive affect-governed modes of thinking as contrasted to more controlled, logical thoughts. In the current usage: (1) primary-process emotions arise from evolutionarily-provided subcortical operating systems (Panksepp, 1998, 2005b); while (2) secondary-process emotions reflect basic emotional learning and memory processes as reflected, for example, in classical and operant conditioning; and (3) tertiary-processes are the higher emotional functions that include thoughtful deliberations commonly based on episodic "autobiographical" memories and capacities for symbolic thought and communication.

In this chapter, we will only be concerned with primary-process social–emotional vocal activities. Because of differences in biophysical variables, ranging from body size and posture to respiratory rhythms, the primary-process categories of emotional sounds can exhibit considerable acoustic variability, but it is unlikely that the study of those details will add much to our understanding of the functionality of such vocalizations. For instance, it seems unlikely that the great variations in infant rat isolation calls (Brudzynski et al., 1999; Scattoni and Branchi, Chapter 3.5 in this volume) reflect distinct affective nuances, as opposed to simply the existence of pure acoustic variability depending on body posture, respiratory function and vocal fold dynamics. If so, a detailed analysis of minor acoustic differences is unlikely to provide any additional information about the underlying affective states. In any event, those issues need to be evaluated through functional studies, such as differential behavioral responses in playback experiments (see Burgdorf et al., 2008; Wöhr and Schwarting, Chapter 4.2 in this volume).

In this context, it is noteworthy that there are acoustic differences that are functionally meaningful, as with the "flat" and "frequency modulated (FM)" types of 50 kHz ultrasonic calls. These two types of calls are often conflated but clearly index different affective processes, and have different neurochemical sensitivities (Burgdorf et al., 2007; Burgdorf and Moskal, Chapter 6.2 in this volume). The FM type highlights an attractive sound, while the flat 50 kHz call does not have the same emotional-attractive impact as monitored by the desire of animals to listen to such calls (Burgdorf et al., 2008).

Also, it is worth noting that the intrinsic communicative intent – the "intentionality in action" – that is integral to primary-process social emotional vocalizations may have provided a platform for ever-more complex social communications and group activities as a function of secondary- and tertiary-process elaboration of primary processes. A great deal of vocal variety emerges from social learning in humans (Zinken, et al., 2008), and this is also open territory for further analysis of distinct communicative meanings of emotional vocalizations in animals (Cheney and Seyfarth, 1990; Fripp and Tyack, 2008; Furrer and Manser, 2009). Indeed, it is worth considering how emotional vocalizations may have prepared the way for propositional speech in brain evolution, especially by providing a motivation for social communication (Panksepp, 2009) – an urge that seems to be lost after anterior cingulate damage that results in akinetic mutism (Devinsky et al., 1995; Mega and Cohenour, 1997).

IV. Emotional vocalizations and the evolution of language

The developmental and evolutionary progressions toward increased emotional complexities within the audible range of frequencies may have ultimately prepared the way for the emergence of tertiary-process linguistic competence within the highest neocortical regions of the human brain. Evidence for such a scenario can be found in the way human children apparently learn language, with the intermediary of musical prosody, which may have been the initial tertiary-process mode of communication in the human line that set the stage for the development of the present human proposition communications (for more complete discussion, see Panksepp and Trevarthen, 2008; Panksepp, 2009). Consider these facts:

1. Some kind of "proto-musical" ability precedes language in human development (Fernald, 1989).
2. Mothers seem to draw out communicative intent from infants through the use of the melodic "motherese" – a high-pitched prosodic convention used across cultures when mothers engage their infants in baby talk (Trevarthen and Daniel, 2005).
3. Music is the most sophisticated human emotional communication. Since the affective power of music is largely conveyed by arousal in subcortical brain regions that integrate primary-process emotional arousals (e.g., Blood and Zatorre, 2001; Menon

and Levitin, 2005), subcortical emotional vocal control systems may have constituted a social-communicative preadaptation for the emergence of language.

4. Linguistic and musical capacities are tightly coupled in the brain (Levitin and Menon, 2003; Schwartz, et al., 2003; Callan et al., 2006). Indeed, recovery from language deficits following left hemisphere strokes can be promoted by recruiting the musical abilities of the right hemisphere (Racette et al., 2006).
5. Thus, perhaps human language differentiated in evolution from our early musical abilities (Mithen, 2005). If so, it is not surprising that ontogentically language is still linked to the musical interactions between mother and child (Trehub, 2003), and evolutionarily to emotional vocalizations.
6. Provisional conclusion: vocal music ultimately evolved from the primal social–emotional vocalizations we still share with other animals. The ability to communicate first affectively and then musically may have established a necessary, but not sufficient, neuropsychological groundwork for semantic communication abilities in human evolution.

A corollary of this type of analysis may be that the complex social–communicative abilities that are being found increasingly in other animals (Cheney and Seyfarth, 1990; Fripp and Tyack, 2008; Furrer and Manser, 2009) may be premised on the solid platform of emotional vocalizations that all mammals share. The emergence of complex secondary- and tertiary-process vocal learning capacities have long been demonstrated in a variety of birds and mammals (Jarvis, 2004), with additional animals from elephants (Poole, et al., 2005) to mice (Holy and Guo, 2005) emerging as species with potentially some type of learned emotional proto-language capacities.

V. Relationship between mammalian separation calls and human sadness and grief

The first social emotional communicative signals that are evident in most mammals and precocious birds, which can get lost soon after birth, are separation calls that are made when young animals lose contact with parents. The neural representations of these calls were first mapped in guinea pigs (Herman, 1979), and found to be well-represented in a neural network continuum starting in the dorsal periaqueductal gray, projecting to dorsomedial thalamus, with abundant representation in basal forebrain nuclei such as the dorsal preoptic area, ventral septal area and especially concentrated in the bed nucleus of the strial terminalis. A few vocalization sites were found in the amygdala, but they showed habituation and often disappeared across repeated testing trials, and a few sites were evident in the anterior cingulate region. A very similar trajectory of the separation distress system was evident in young domestic chickens (Bishop, 1984).

The first neurochemical systems found to inhibit separation cries robustly were endogenous opioids that stimulated mu-receptors (Panksepp et al., 1980, 1988), providing the initial neurochemical theory for social bonding, namely that it is an addictive process that resembles opiate addiction (Panksepp, 1981, 1998). Opioids were demonstrated to be natural endogenous modulators of vocalization within the separation distress circuitry of guinea pigs (Herman and Panksepp, 1981). These opioid effects had direct implications for understanding the mechanisms of social attachment, but also the neural sources of human loneliness and sadness and psychiatric disorders such as depression (Watt and Panksepp, 2009).

It was eventually demonstrated that human sadness corresponded to the separation-distress vocalization circuitry mapped in other animals (Damasio et al., 2000; Panksepp, 2003), and that human sadness was accompanied by low opioid activity in this circuitry (Zubieta et al., 2003), as was depression (Kennedy and Zubieta, 2004). It is also known that very low doses of opiates, such as buprenorphine, can yield robust antidepressant effects among individuals who have had little benefit from other pharmaco-therapies (Bodkin et al., 1995). These results highlight the profound extent to which a study of animal emotional vocalizations can promote understanding of corresponding affective processes in the human brain (see Burgdorf and Moskal, Chapter 6.2 in this volume).

From hundreds of pharmacological and neuropeptide agents evaluated across the years, tested mainly in the efficient avian model where neuropeptides could be administered freehand directly into the ventricular system, only a few had comparably robust effects. Among the neuropeptides, the items that stood out were the oxytocin/vasotocin and prolactin systems (Panksepp, 1998). Among psychopharmaceuticals, the efficacy of clonidine was quite high (Harris and Newman,

1987), but contrary to expectations this effect was due to postsynaptic as opposed to presynaptic norepinephrine receptors (Rossi et al., 1983). There were many other neuroactive drugs that had comparatively modest effects, and although their effects reached statistical significance, they came nowhere close to the robust effects of the above agents.

Among agents that facilitated separation calls, the neuropeptide corticotrophin releasing factor (CRF) clearly facilitated distress calls in strict dose-response and time-dependent ways (Panksepp et al., 1998), and the vocalization was also promoted by glutamatergic drive, as indicated by the ability of glutamate agonists to increase distress calls and NMDA receptor antagonists to inhibit calls (Panksepp, 1998). Across the years, work on separation distress calls has shifted to convenient rodent models, but one must wonder whether these extremely altricial species exhibit true separation calls or only short-term "I'm-out-of-the-nest-and-cold" calls. It is important to remember that altricial rodents only show these calls before they are motorically competent, and before their eyes and ears are fully open. At these times, it is difficult for young rats to get lost on their own, but quite easy for the mother to drag them out of the nest when they are firmly latched to nipples. We must recall that there is currently little data that rat or mouse pups are actually socially attached to their own mothers in the initial days of their life. The ease of cross-fostering may suggest that they are not.

As discussed in detail elsewhere (Panksepp, 2003), there are many peculiarities in rodent separation calls, such as very short, few minute periods of calling following separation, as opposed to the hours of crying seen in truly bonded species. Their pharmacological sensitivities are also unusual, with the most peculiar being the massive elevation of vocalizations after clonidine (Kehoe and Harris, 1989) when this agent markedly reduced vocalizations in other species. In contrast, other altricial mammals such as feline and canine species exhibit clear maternal bonding, and once they become motorically competent, they exhibit very robust and sustained distress vocalizations when socially separated. Mice and rats do not.

VI. A primal positive affective vocalization: mammalian play (laughter?) vocalizations

Rodent ultrasonic calls are well-summarized elsewhere in this volume, and there is good evidence that the low 22 kHz long calls can be used to index negative affective states, while the 55 kHz short calls (chirps), especially the frequency modulated ones, can be used to index positive affect (Knutson et al., 2002; Burgdorf et al., 2007, 2008). Here I would just briefly discuss the remarkably high presence of 50 kHz chirps during juvenile play. They appear most abundantly in one of the earlier maturing childhood emotional systems of the mammalian brain, namely of social play. This primary-process affective state provides optimal clarity for thinking about the nature of these vocalizations. It is a joyous social engagement response that percolates into all positively-motivated adult activities. Because of their high association with playfulness, we raised the possibility that they index an emotion of joy that is akin to the prominent laughter humans exhibit during social engagements of happy play activities (Scott and Panksepp, 2003), namely it may be a primal form of laughter. This led to a long series of studies deploying human hand play (also known as "tickling") as the optimal facilitator of 55 kHz frequency modulated calls.

Why should we seriously consider that this chirpy rodent call has any ancestral relationship to human laughter? The list of fulfilled criteria for such a conclusion is quite large, as summarized in detail in the study by Panksepp and Burgdorf (2003):

1. 50 kHz chirping call is evoked most robustly during rat play and tickling (heterospecific hand play).
2. Tickling is most effective on the dorsolateral region of the body where rats direct most of their play solicitations.
3. There are age-related declines in tickling responsivity in humans and rats.
4. All negatively valenced stimuli reduce the tickle response.
5. There are positive relationships between levels of playing and tickle responsivity.
6. Tickling is rewarding for rats and related to the degree to which they exhibit high levels of 50 kHz chirps.
7. Prior social isolation is an important ingredient for evocation of the tickling response. It is hard to tickle animals taken directly from their normal social environments.
8. Human, as well as rat, tickle-induced vocalizations exhibit rapid classical conditioning, so that simply waving fingers can easily evoke laughter-type responses.

9. Tickle-induced approach behavior is strongly correlated with rates of 50 kHz chirps.
10. Rats that have been petted by one experimenter and tickled by another dramatically prefer the company of the individual that has tickled them.
11. Low or high levels of tickle-induced chirping can be readily amplified by selective breeding programs.
12. Animals readily learn instrumental responses to getting tickled, and they prefer the company of animals that chirp a lot over those that do not.

Additional data for laughter-like responses in various higher mammals is gradually emerging. For example, the happy-type breathy sounds made by chimpanzees during play and tickling outwardly resemble laughter (Matsusaka, 2004). Play vocalizations have been noted during the rough-and-tumble play of squirrel monkeys (Biben and Symmes, 1986). Finally, there has also been preliminary work on play vocalizations in dogs (Simonet et al., 2001). In addition, it is also worth considering that human children begin to exhibit their social joy with laughter at a very young age, laughter is evident in blind and severely mentally impaired children (Reddy et al., 2001) and the fact that the neural foundations for human laughter are situated in very ancient regions of the brain (Mobbs et al., 2003). There is no special evolutionary problem with other species having homologous functional networks for laughter-type responses within their brains. The key question, of course, is whether this represents, at its base, a homology or only convergent evolution. Because of neuroanatomical correspondences, so far the evidence suggests a homology.

As already noted, the mapping of the separation-distress response has already led to human brain imaging studies that corroborate the conclusion that the general trajectory of animal separation-distress systems corresponds to brain mechanisms that generate sadness and depression in humans. The recent mapping of the 50 kHz call response in rodents (Burgdorf et al., 2007) provides a roadmap for thinking about the potential sources of human laughter and joy, if there are fundamental primal psycho-neurological homologies here. Recent brain imaging work, showing that laughter activates nucleus accumbens, a major terminal region for mesolimbic dopaminergic fibers (Mobbs et al., 2003), suggests some kind of evolutionary correspondence between the rodent and human primal laughter circuitry in humans.

VII. Conclusions

In sum, there is abundant evidence for several homologous types of vocalizations in "lower" mammals that have counterparts in humans, which allows animal work to clarify some human emotional issues that probably could not ever be studied in detail in humans. The most prominent among these are pain vocalizations, separation-distress vocalizations and social joy-play, laughter-type vocalizations. We know little about these processes in the human brain. In contrast, we know a great deal about the neural and biochemical specifics underlying many emotional expressions, from separation-distress calls to social play patterns and laughter-type vocalizations, in animals (Panksepp, 1998, 2007).

The evolutionary function of emotional vocalizations, as of affective feelings, is to provide critically important information for behavior choices humans and other animals must make to maximize the many types of affective "comfort zones" and to minimize "discomfort zones" needed for survival. Emotional vocalizations have an intrinsic prosody, with a capacity to evoke emotional resonances between animals that even humans can appreciate. Our own, much richer, emotional sounds, coupled with expansive cerebral hemispheres with new cognitive capacities, could have provided the critical impetus for additional cultural discoveries with sound, resulting in the highly repetitive (rhythmic) prosodic flow of sound that we now call music. The larynx may be regarded as the original biological instrument of music, rooted integrally in the preadaptation of prosodic emotional sounds.

To the best of our current knowledge, various social affective states are engendered by the basic social emotional circuits – LUST-eroticism, CARE-nurture, PLAY-joy, PANIC-sadness. These emotional systems were already well-encoded in the primitive emotional vocalizations of our mammalian ancestors (Newman, 1988, 2007; Jürgens and Hage, 2007). These calls also help reinforce social bonds, a vocal effect that can be demonstrated even in the "lower" mammals (Ziabreva et al., 2003).

These emotional preadaptations presumably led to primitive forms of singing in our own species (Panksepp and Trevarthen, 2008), and such sounds may have served as preadaptations for an essential bridgework for the emergence of human language. For instance, emotional vocalizations provide a motivational system by which mothers encourage human infant singing-babbling and prelinguistic pitch patterning, which sets the stage for an interest in sounds that eventually promotes the emergence of propositional language. These ancient emotional vocalizations represent an empirical strategy whereby we can decode the psychological primitives that are built into mammalian brain by evolution, as cross-mammalian shared tools for living (Panksepp, 1998, 2005b).

References

Biben, M., Symmes, D., 1986. Play vocalizations of squirrel monkeys *Saimiri sciureus*. Folia Primatol. 46, 173–182.

Bishop, P. (1984) Brain and Opiate Modulation of Avian Affective Vocalizations. Unpublished Ph.D. dissertation. Bowling Green State University, Bowling Green, OH.

Blood, A.J., Zatorre, R.J., 2001. Intensely pleasurable responses to music correlate with activity in brain regions implicated in reward and emotion. Proc. Nat. Acad. Sci., USA, 98, 11818–11823.

Bodkin, J.A., Zornberg, G.L., Lukas, S.E., Cole, J.O., 1995. Buprenorphine treatment of refractory depression. J. Clin. Psychopharmacol. 15, 49–57.

Brudzynski, S.M., 2007. Ultrasonic calls of rats as indicator variables of negative or positive states: acetylcholine-dopamine interactions and acoustic coding. Behav. Brain Res. 182, 261–273.

Brudzynski, S.M., Kehoe, P., Callahan, M., 1999. Sonographic structure of isolation-calls of rat pups. Dev. Psychobiol. 34, 195–204.

Burgdorf, J., Wood, P.L., Kroes, R.A., Moskal, J.R., Panksepp, J., 2007. Neurobiology of 50-kHz ultrasonic vocalizations in rats: electrode mapping, lesion, and pharmacological studies. Behav. Brain Res. 182, 274–283.

Burgdorf, J., Kroes, R.A., Moskal, J.R., Pfaus, J.G., Brudzynski, S.M., Panksepp, J., 2008. Ultrasonic vocalizations of rats (*Rattus norvegicus*) during mating, play, and aggression: behavioral concomitants, relationship to reward and self-administration of playback. J. Comp. Psychol. 122, 357–367.

Callan, D.E., Tsytsarev, V., Hanakawa, T., Callan, A.M., Katsuhara, M., Fukuyama, H., Turner, R., 2006. Song and speech: brain regions involved with perception and covert production. Neuroimage 31, 1327–1342.

Cheney, D.L., Seyfarth, R.M., 1990. How Monkeys See the World: Inside the Mind of Another Species. University of Chicago Press, Chicago, IL.

Damasio, A.R., Grabowski, T.J., Bechara, A., Damasio, H., Ponto, L.L.B., Parvizi, J., Hichwa, R.D., 2000. Subcortical and cortical brain activity during the feeling of self-generated emotions. Nature Neurosci. 3, 1049–1056.

Darwin, C., 1872/1965. The Expression of the Emotions in Man and Animals. University of Chicago Press, Chicago, IL.

Devinsky, O., Morell, M.J., Vogt, B.A., 1995. Contributions of anterior cingulate cortex to behavior. Brain 118, 279–306.

Fernald, A., 1989. Intonation and communicative intent in mothers' speech to infants: is the melody the message? Child Devel. 60, 1497–1510.

Freud, S., 1900. Die Traumdeutung [The Interpretation of Dreams]. Franz Deuticke, Leipzig and Wien, Germany.

Fripp, D., Tyack, P., 2008. Pospartum whistle production in bottlenose dolphins. Marine Mammal Sci. 24, 479–502.

Furrer, R.D., Manser, M.B., 2009. The evolution of urgency-based and functionally referential alarm calls in ground-dwelling species. Am. Nat. (in press).

Harris, J.C., Newman, J.D., 1987. Mediation of separation distress by alpha 2-adrenergic mechanisms in a non-human primate. Brain Res. 410, 353–356.

Herman, B.H., 1979. An Exploration of Brain Social Attachment Substrates in Guinea Pigs. Unpublished Ph.D. dissertation. Bowling Green State University, Bowling Green, OH.

Herman, B.H., Panksepp, J., 1981. Ascending endorphinergic inhibition of distress vocalization. Science 211, 1060–1062.

Holy, T.E., Guo, Z., 2005. Ultrasonic songs of male mice. PLoS Biol. 3, e386.

Jarvis, E., 2004. Learned birdsong and the neurobiology of human language. Ann. N.Y. Acad. Sci. 1016, 749–777.

Jürgens, U., 1976. Reinforcing concomitants of electrically elicited vocalizations. Exp. Brain Res. 26, 203–214.

Jürgens, U., Hage, S.R., 2007. On the role of the reticular formation in vocal pattern generation. Behav. Brain Res. 182, 308–314.

Kehoe, P., Harris, J.D., 1989. Ontogeny of noradrenergic effects on ultrasonic vocalizations in rat pups. Behav. Neurosci. 103, 1099–1107.

Kennedy, S.E., Zubieta, J.K., 2004. Neuroreceptor imaging of stress and mood disorders. CNS Spectrums 9, 292–301.

Knutson, B., Burgdorf, J., Panksepp, J., 2002. Ultrasonic vocalizations as indices of affective states in rats. Psych. Bull. 128, 961–977.

Levitin, D.J., Menon, V., 2003. Musical structure is processed in "language" areas of the brain: a possible role for Brodmann Area 47 in temporal coherence. Neuroimage 20, 2142–2152.

Litvin, Y., Blanchard, D.C., Blanchard, R.J., 2007. Rat 22kHz ultrasonic vocalizations as alarm cries. Behav. Brain. Res. 182, 166–172.

Matsusaka, T., 2004. When does play panting occur during social play in wild chimpanzees? Primates 24, 221–229.

Mobbs, D., Greicius, M.D., Abdel-Azim, E., Menon, V., Reiss, A.L., 2003. Humor modulates the mesolimbic reward centers. Neuron 40, 1041–1048.

Mega, M.S., Cohenour, R.C., 1997. Akinetic mutism: disconnection of frontal-subcortical circuits. Neuropsychiat. Neuropsychol. Behav. Neurol. 10, 254–259.

Menon, V., Levitin, D.J., 2005. The rewards of music listening: response and physiological connectivity of the mesolimbic system. Neuroimage 28, 175–184.

Mithen, S., 2005. The Singing Neanderthals: The Origins of Music, Language, Mind and Body. Weidenfeld and Nicolson, London, UK.

Newman, J.D. (Ed.), 1988. The Physiological Control of Mammalian Vocalization. Plenum Press, New York, NY.

Newman, J.D., 2007. Neural circuits underlying crying and cry responding in mammals. Behav. Brain Res. 182, 155–165.

Panksepp, J., 1981. Brain opioids: a neurochemical substrate for narcotic and social dependence. In: Cooper, S. (Ed.), Progress in Theory in Psychopharmacology. Academic Press, London, UK, pp. 149–175.

Panksepp, J., 1998. Affective Neuroscience: The Foundations of Human and Animal Emotions. Oxford University Press, New York, NY.

Panksepp, J., 2003. Can anthropomorphic analyses of "separation cries" in other animals inform us about the emotional nature of social loss in humans? Psych. Rev. 110, 376–388.

Panksepp, J., 2005a. On the embodied neural nature of core emotional affects. J. Consc. Studies 12, 161–187.

Panksepp, J., 2005b. Affective consciousness: core emotional feelings in animals and humans. Consc. Cognit. 14, 19–69.

Panksepp, J., 2007. Neuroevolutionary sources of laughter and social joy: modeling primal human laughter in laboratory rats. Behav. Brain Res. 182, 231–244.

Panksepp, J., 2008. The affective brain and core-consciousness: how does neural activity generate emotional feelings? In: Lewis, M., Haviland, J.M., Barrett, L.F. (Eds.), Handbook of Emotions. Guilford, New York, NY, pp. 47–67.

Panksepp, J., 2009b. The emotional antecedents to the evolution of music and language. Musicae Scientiae (in press).

Panksepp, J., Burgdorf, J., 2003. "Laughing" rats and the evolutionary antecedents of human joy? Physiol. Behav. 79, 533–547.

Panksepp, J., Trevarthen, C., 2008. Motive impulse and emotion in acts of musicality and in sympathetic emotional response to music. In: Maloch, S., Trevarthen, C. (Eds.), Communicative Musicality. Oxford University Press, Oxford, UK (in press).

Panksepp, J., Herman, B.H., Vilberg, T., Bishop, P., DeEskinazi, F.G., 1980. Endogenous opioids and social behavior. Neurosci. Biobehav. Rev. 4, 473–487.

Panksepp, J., Normansell, L.A., Herman, B., Bishop, P., Crepeau, L., 1988. Neural and neurochemical control of the separation distress call. In: Newman, J.D. (Ed.), The Physiological Control of Mammalian Vocalizations. Plenum Press, New York, NY, pp. 263–300.

Poole, J.H., Tyack, P.L., Stoeger-Horwath, A.S., Watwood, S., 2005. Elephants are capable of vocal learning. Nature 434, 455–456.

Racette, A., Bard, C., Peretz, I., 2006. Making non-fluent aphasics speak: sing along! Brain 129, 22571–22584.

Reddy, V., Williams, E., Vaughan, A., 2001. Sharing laughter: the humor of pre-school children with Down Syndrome. Downs Synd. Res. Pract. 7, 125–128.

Rossi III., J., Sahley, T.L., Panksepp, J., 1983. The role of brain norepinephrine in clonidine suppression of isolation-induced distress in the domestic chick. Psychopharmacol. 79, 338–342.

Schwartz, D.A., Howe, C.Q., Purves, D., 2003. The statistical structure of human speech sounds predicts musical universals. J. Neurosci. 23, 7160–7168.

Scott, E., Panksepp, J., 2003. Rough-and-tumble play in human children. Aggress. Behav. 29, 539–551.

Simonet, P., Murphy, M., Lance, A., 2001. Laughing Dog: Vocalizations of Domestic Dogs during Play Encounters. Animal Behavior Society Conference, Cornvallis, OR.

Trehub, S.E., 2003. Musical predispositions in infancy: an update. In: Peretz, I., Zatorre, R. (Eds.), The Cognitive Neuroscience of Music. Oxford University Press, New York, NY, pp. 3–20.

Trevarthen, C., Daniel, S., 2005. Rhythm and synchrony in early development, and signs of autism and Rett syndrome in infancy. Brain Dev. 27 (Suppl. 1), S25–S34.

Watt, D.F., Panksepp, J., 2009. Depression: an evolutionarily conserved mechanism to terminate separation distress? A review of aminergic, peptidergic, and neural network perspectives. Neuropsychoanalysis (in press).

Wild, B., Rodden, F.A., Ruch, W., 2003. Neural correlates of laughter and humor. Brain 126, 2121–2138.

Ziabreva, I., Poeggel, G., Schnabel, R., Braun, K., 2003. Mother's voice "buffers" separation-induced receptor changes in the prefrontal cortex of *Octodon degus*. Neuroscience 119, 433–441.

Zinken, J., Knoll, M., Panksepp, J., 2008. Universality and diversity in the vocalization of emotions. In: Izdebski, K. (Ed.), Emotions in the Human Voice, vol. 1. Foundations. Plural Publishing Inc., San Diego, CA, pp. 185–202.

Zubieta, J.K., Ketter, T.A., Bueller, J.A., Xu, Y., Kilbourn, M.R., Young, E.A., Koeppe, R.A., 2003. Regulation of human affective responses by anterior cingulate and limbic mu-opioid neurotransmission. Arch. Gen. Psychiat. 60, 1145–1153.

CHAPTER 6.2

Frequency modulated 50 kHz ultrasonic vocalizations reflect a positive emotional state in the rat: neural substrates and therapeutic implications

Jeffrey Burgdorf* and Joseph R. Moskal

The Falk Center for Molecular Therapeutics, Department of Biomedical Engineering, The McCormick School of Engineering and Applied Sciences, Northwestern University, Evanston, IL, USA

Abstract: The evidence that frequency modulated (FM) 50 kHz ultrasonic vocalizations (USVs) reflect a positive emotional state in rats is reviewed. Positive emotional states in humans are measured by facial/vocal displays (e.g., Duchenne smiling and laughter), approach behavior and subjective self-reporting of feeling states. In laboratory animals, only facial–vocal displays, along with approach behavior, can be measured. FM 50 kHz USVs are uniquely elevated by hedonic stimuli and suppressed by aversive stimuli. Rates of FM 50 kHz USVs are positively correlated to the rewarding value of the eliciting stimulus. Playbacks of these vocalizations are also rewarding. The neural and pharmacological substrates of 50 kHz USVs are consistent with those of human positive affective states. By experimentally eliciting FM 50 kHz USVs, the novel molecular underpinning of positive affect can be elucidated and may be similar to those in humans. In humans, positive emotional states confer resilience to depression and anxiety, as well as promote overall health. Therefore, novel antidepressants that promote positive affect-induced resilience to depression may emerge from this research.

Keywords: ultrasonic vocalizations; emotion; rat; human; frequency modulation; dopamine; depression; nucleus accumbens; 50 kHz calls

I. Positive affective states in humans

I.A. Measuring positive affective states in humans

Subjective well-being appears to be a unitary concept in humans, with self-reported well-being having a high correlation with independent third-party rating (spouse or friend) and the objective physiological measures of Duchenne smiling and EEG lateralization (Ekman et al., 1990; Rosenkranz et al., 2003; Lyubomirsky et al., 2005). The most consistent personality traits associated with subjective well-being are a positive correlation with extroversion and a negative correlation with neuroticism (Diener et al., 2003). In both adults and adolescents, the activity eliciting the most positive affective state is socializing with friends or romantic partners (Csikszentmihalyi and Hunter, 2003; Kahneman and Krueger, 2006; Stone et al., 2006). It is important to note that not all socializing is hedonic. The same studies showed that socializing with friends or one's romantic partner elicit positive affective states, while interacting with supervisors and other family members is not consistently hedonic. Therefore, positive affective states are primarily related to positive pro-social interactions.

Experimental studies that elicit positive affective states generally use social positive affective stimuli (i.e., positive feedback, giving a small gift, or watching a video tape eliciting positive affective state). Positive affective states that are elicited in an experimental

*Corresponding author. E-mail: j-burgdorf@northwestern.edu

Stefan M. Brudzynski (Ed.)
Handbook of Mammalian Vocalization
ISBN 978-0-12-374593-4

DOI: 10.1016/B978-0-12-374593-4.0021-8

setting by these social stimuli have been shown to increase gregariousness, optimism and openness to new experiences (Lyubomirsky et al., 2005). These effects of positive affective state inductions have been referred to as "broaden and build" (Fredrickson et al., 2004).

A functional distinction can be made between positive pro-social affective states primarily associated with subjective well-being and consummatory pleasures. Experimental study of nonsocial hedonic stimuli (i.e., food or thermal regulation) showed that these pleasures function primarily to maintain homeostasis. For example, a warm stimulus would be experienced as pleasurable by a cold individual, with the magnitude of the pleasure being proportional to the ability of the stimulus to return the body to homeostatic conditions (Cabanac, 1971). This emotionally-driven change in sensation associated with a return to homeostasis is referred to as sensory alliesthesia (Cabanac, 1971, 1992).

Positive affective states, as studied longitudinally in humans, confer resilience to depression and anxiety, and lead to an increase in overall health and a decrease in mortality from all causes (Lyubomirsky et al., 2005). The psychological and physical health benefits of positive affective states appear to be mediated through increased resilience, defined as continued global functioning despite the presence of stressors. For example, following a major life stressor individuals exhibiting greater resilience are less likely to develop psychological disorders, such as anxiety or depression (Fredrickson et al., 2003). Longitudinal studies also showed that positive affective states precede the health benefit effect of positive affect (Lyubomirsky et al., 2005). Therefore, positive affect is not simply a secondary consequence of overall good health. Positive affective states are stable across the lifespan, with major positive and negative life events having little long-term effect on these states (Lykken and Tellegen, 1996). Conversely, individuals who have low levels of positive affective states are at greater risk of developing anxiety disorders, depression and global health problems (Lyubomirsky et al., 2005). Interventions that increase positive affective states have been shown to reduce levels of depression and anxiety (Duckworth et al., 2005).

I.B. Neurobiology of positive affective states in the human brain

The primary neuroanatomical underpinnings of positive emotional states are associated with the ascending mesolimbic dopamine system, and have relied primarily on correlational brain imaging studies (i.e., functional magnetic resonance imaging or positron emission tomography), as well as on the direct elicitation of positive affective states through drug administration or electrical brain stimulation. Brain imaging studies using recall of positive affective memories (Damasio et al., 2000), listening to positive music (Blood and Zatorre, 2001), male orgasm (Holstege et al., 2003) and positive anticipation of monetary reward (Knutson et al., 2001) have all been shown to activate aspects of the ascending mesolimbic dopamine system that includes the ventral tegmental area, nucleus accumbens, medial prefrontal and orbital frontal cortices (Burgdorf and Panksepp, 2006). The euphoric effects of intravenous amphetamine have been shown to be directly related to dopamine activity in the nucleus accumbens (Drevets et al., 2001; Oswald et al., 2005). Direct electrical brain stimulation of the accumbens has been shown to elicit Duchenne laughter and self-report of positive affect (Okun et al., 2004). Patients given the opportunity to self-administer electrical stimulation to the nucleus accumbens (then called the nucleus accumbens septi as a ventral extension of the lateral septum) or to an area at or near the ventral tegmental area, repeatedly self-administered this stimulation and reported that the stimulation elicited a positive affective state (Heath, 1960, 1972).

I.C. Molecular underpinnings of positive affective states

The molecular mechanisms that are involved in the regulation of positive affective states are largely unknown. In order to establish a causal link between a molecular mechanism associated with positive affect, the following conditions must be fulfilled: (1) concentrations of key molecules associated with the mechanism under investigation should be significantly altered in critical brain regions following positive affective stimuli; (2) these molecular changes should change in the opposite direction or not change significantly at all, following presentation of negative affective stimuli; (3) direct injection of the target molecules or agonists should produce a positive affective state; and (4) pharmacological agonism/antagonism of the key molecules should increase/decrease positive affective states. Thus far, no molecular mechanism reported to be associated with positive affective states that has been characterized could meet all four of these criteria.

Endogenous opiates acting on μ-receptors (endomorphins, met-enkephalin, and β-endorphin) and dopamine have been the most extensively examined (reviewed in Burgdorf and Panksepp, 2006). Mu(μ)-opiate and dopamine levels in the mesolimbic positive affect circuit have been found to be positively correlated with the euphoric effect of exercise and amphetamine respectively (Drevets et al., 2001; Boecker et al., 2008). Intravenous administration of μ-opiate and dopamine agonists produced positive affective states in humans (Zacny et al., 1994; Drevets et al., 2001). Mu(μ)-opiate antagonists have been shown to blunt the positive affective state elicited by exercise and alcohol (Janal et al., 1984; Davidson et al., 1999), and dopamine antagonists decreased positive affective states associated with psychostimulants (Jönsson et al., 1971; Romach et al., 1999; Newton et al., 2001) and could produce a state of dysphoria (Voruganti et al., 2001). However, aversive stimuli also increase μ-opiate and dopamine levels in the nucleus accumbens (Tidey and Miczek, 1996; Marinelli et al., 2004). Therefore, the μ-opiate and dopamine systems are not completely specific to positive emotions.

II. Measuring positive affective states in laboratory animals

In order to establish that an animal behavior reflects a positive affective state, several criteria must be met. In humans, positive affective states are measured primarily via subjective self-report and behaviorally by facial/vocal displays such as felt- or Duchenne-smiling (Ekman et al., 1990). Therefore, in laboratory animal experiments, where we can rely only on observations, a positive affective state should be expressed as facial or vocal displays with the predicted changes in approach/avoidance behavior. In humans, positive affective states are elicited primarily by rewarding social interaction, food and exercise, and are decreased by negative affective stimuli (Csikszentmihalyi and Hunter, 2003; Kahneman and Krueger, 2006; Stone et al., 2006). Therefore, in laboratory animals, the same categories of positive affective (appetitive) stimuli should increase the facial/vocal displays and aversive stimuli should decrease them. Finally, what is known about the neurobiological mechanisms of the facial/vocal displays in animals should be consistent with the neurobiological mechanisms of human positive affective states. To date, only two such animal behaviors meet all of these criteria; emission of ultrasonic vocalizations (USVs) that are discussed below, and hedonic taste reactivity (reviewed in Berridge et al., 2008).

II.A. 50 kHz social vocalizations in rats

Fifty kHz ultrasonic vocalizations (50 kHz USVs) have been shown to reflect a positive affective state in rats. Rewarding social interactions (i.e., mating and rough-and-tumble play in juveniles), anticipation of food and action of euphorigenic drugs of abuse increased the number of emitted 50 kHz USVs (Burgdorf et al., 2000, 2001a, 2007, 2008; Panksepp and Burgdorf, 2000) (see example in Fig. 1), whereas aversive stimuli such as social defeat, frustrative non-rewarding situations, sickness-inducing doses of lithium chloride and foot-shock all decreased the number of 50 kHz USVs (Burgdorf et al., 2000, 2001b, 2008). The rewarding value of the stimuli eliciting positive affective states was positively correlated with the rates of 50 kHz USVs elicited by positive social, drug and electrical brain stimulation rewards (Burgdorf et al., 2007, 2008). Mu(μ)-opiate and dopamine agonists, as well as electrical brain stimulation of the mesolimbic dopamine system, also increased rates of 50 kHz USVs in rats (Burgdorf et al., 2000, 2007).

Also, alternative non-hedonic interpretations of the emission of 50 kHz USVs (e.g., non-positively valenced arousal, non-positively valenced seeking behavior, or

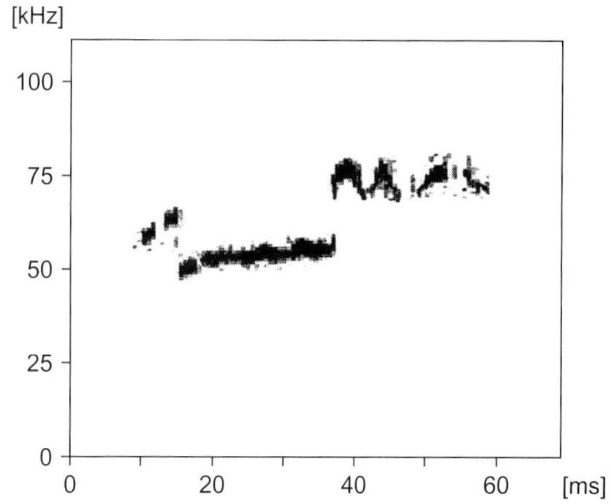

Fig. 1. An exemplary sonogram of a developed frequency modulated (FM) 50 kHz call of an adult rat induced by amphetamine. It contains two forms of frequency modulation: a step (seen at 18 ms); and a trill (seen from 40 to 60 ms). Total duration of the call is 50 ms. Courtesy of Melanie Komadoski.

Table 1. Non-affective hypotheses of 50 kHz USVs emission with their rebuttal

I. 50 kHz USVs are an artifact of locomotor activity-induced thoratic compressions (Blumberg, 1992). Only 10% of 50 kHz were coincident with thoratic compressions, and could be dissociated from locomotion (Panksepp and Burgdorf, 2003).

II. 50 kHz USVs are a non-affective contact call (Schwarting et al., 2007). Flat 50 kHz calls appear to be a contact call, occurring at the highest rates during non-positive affective social interactions. However, FM 50 kHz calls appeared to be selective for positive affective social interactions (Burgdorf et al., 2008).

III. 50 kHz calls are evident during aggression (Berridge, 2003). 50 kHz calls occur primarily before the onset of aggression, and the vast majority of the 50 kHz calls were of the non-affective flat variety (Panksepp and Burgdorf, 2003; Burgdorf et al., 2008).

IV. 50 kHz calls reflect a non-positive affective "wanting" state (Schwarting et al., 2007). 50 kHz USVs were increased in the anticipation of delivered reward, which in humans has been shown to elicit a positive affective state (Knutson et al., 2001). However, during extinction bursts or "frustrative non-reward" such appetitive behavior decreased rates of 50 kHz calls and increased rates of aversive 22 kHz calls (Burgdorf et al., 2000).

V. Adult and infant rat ultrasonic calls reflect a state of high arousal that is not specific to positive affective states (Bell, 1974). Highly arousing aversive stimuli such as predatory odor, foot-shock and bright light, decrease rates of 50 kHz calls, whereas rewarding stimuli increase rates of 50 kHz calls (Knutson et al., 2002).

non-affective social contacts) are not supported by the available experimental data (for details, see Table 1).

II.B. 22 kHz aversive vocalizations in adult rats and isolation calls (35–40 kHz) in infants

Adult 22 kHz USVs and infant isolation calls (35–40 kHz) may represent a negative emotional state associated with human anxiety and/or depressive states (e.g., aversive facial expressions such as crying, and behavioral inhibition) (Knutson et al., 2002). Despite significant sonographic differences between these adult and infant vocalizations (see Panksepp, Chapter 6.1 in this volume), both of them share similar characteristics to the aversive and dangerous situations which elicit them. In humans, these affective states are often elicited by social loss and anticipation of perceived threats (Knutson et al., 2002). In rats, infant isolation 35–40 kHz USVs are best elicited by separating the pup from the mother. Twenty-two kHz USVs are best elected by social defeat and the presence of a predator (Blanchard and Blanchard, 1989; Brunelli and Hofer, 2007; Panksepp et al., 2007). Emission of 22 kHz USVs calls is strongly related to avoidance behavior and freezing during social defeat testing (Panksepp et al., 2007). Environments paired with drugs causing aversive states also elicit 22 kHz USVs, and rates of emitted calls are positively correlated with drug-induced conditioned place avoidance (Burgdorf et al., 2001b). Anxiolytic benzodiazepines and antidepressants reduce rates of 22 kHz calls and 35–40 kHz USVs (Carden and Hofer, 1990; Covington and Miczek, 2003).

Using social defeat as a method to elicit negative emotional states associated with 22 kHz USVs, we conducted a transcriptomic analysis of gene expression in the periaqueductal gray, one of the regions found to be critical for the generation of negative affect and 22 kHz USVs in rats (Kroes et al., 2007). These studies revealed that mRNA expression of genes associated with acetylcholine metabolism and receptor function was altered in the PAG following social defeat. This finding is consistent with the previously reported role of the tegmental cholinergic system in the generation of 22 kHz USVs (Brudzynski, 2001). Carbachol has been shown to be the best elicitor of these vocalizations in both rats, cats and squirrel monkeys (Lu and Jürgens, 1993; Brudzynski, 2007; see also Brudzynski, Chapter 7.3 in this volume). Recent studies in humans demonstrated that depressed patients have alteration in cholinergic transmission (Wang et al., 2004), and scopolamine has been shown to be a potent rapid antidepressant (Furey and Drevets, 2006).

II.C. Selective breeding for differential rates of 50 kHz and 22 kHz ultrasonic vocalizations

In order to further elucidate the molecular mechanisms that are involved in the regulation of positive and negative emotional states, rats were selectively bred for differential rates of hedonic 50 kHz USVs (Burgdorf et al., 2005, 2008). Animals selectively bred for low rates of 50 kHz USVs (low line) had a concomitant increase in 22 kHz USVs and showed elevated levels of anxiety in the open field, in the social contact test and in infant distress vocalization tests, as compared to randomly bred animals (Burgdorf et al., 2008). Conversely, animals selectively bred for high rates of 50 kHz USVs (high line) had a concomitant decrease in the 22 kHz USVs, and showed lower levels of anxiety in the open field test, decreased rates of aggression and increased sensitivity to sucrose reward, compared to randomly breed animals (Burgdorf et al., 2008). These animals have been selectively bred for 18 generations to date and have displayed stable differences in USVs from adolescence through adulthood (three months).

Studies on the molecular mechanisms associated with the USV patterns of the high line and low line animals to date are consistent with a depressant-resilient and a depressant-prone phenotype, as discussed above. For example, high line animals exhibited higher levels of the μ-opiate acting Met-enkephalin-like immunoreactivity in the hypothalamus and other related limbic structures (Burgdorf et al., 2008). Injections of the μ-opiate agonist DAMGO into the ventral tegmental area (a region included in the hypothalamus dissection) increased rates of 50 kHz USVs and was rewarding to the animals (Burgdorf et al., 2007). Low line animals exhibited higher levels of cholecystokinin-like immunoreactivity in the posterior neocortex. Cholecystokinin (CCK) content in the posterior cortex was elevated by social defeat and was correlated with 22 kHz USVs rate of the defeated animal (Panksepp et al., 2004). It has also been shown that social defeat, which elevates levels of 22 kHz USVs, increased CCK-like immunoreactivity in cortical microdialysates (Becker et al., 2001) and CCK administration promoted social defeat-induced behaviors, including 22 kHz USVs (Becker et al., 2007).

III. Conclusions

Fifty kHz ultrasonic vocalizations (50 kHz USVs) have been shown to reflect a positive affective state in rats. By studying positive pro-social emotional states in rats, the neuroantomical basis and molecular mechanisms of this form of positive affect can now be elucidated. These studies should lead to a deeper understanding of the brain mechanisms of positive affect in humans, and should lead to the development of novel therapeutics for the treatment of depression and other affective disorders.

References

Bell, R.W., 1974. Ultrasounds in small rodents: arousal-produced and arousal-producing. Dev. Psychobiol. 7, 39–42.

Berridge, K.C., 2003. Pleasures of the brain. Brain Cogn. 52, 106–128.

Blanchard, R.J., Blanchard, C.D., 1989. Antipredator defensive behaviors in a visible burrow system. J. Comp. Psychol. 103, 70–82.

Blood, A.J., Zatorre, R.J., 2001. Intensely pleasurable responses to music correlate with activity in brain regions implicated in reward and emotion. Proc. Natl. Acad. Sci. USA 98, 11818–11823.

Blumberg, M.S., 1992. Rodent ultrasonic short calls: locomotion, biomechanics, and communication J. Comp. Psychol. 106, 360–365.

Boecker, H., Sprenger, T., Spilker, M.E., Henriksen, G., Koppenhoefer, M., Wagner, K.J., Valet, M., Berthele, A., Tolle, T.R., 2008. The runner's high: opioidergic mechanisms in the human brain. Cereb. Cortex 18, 2523–2531.

Brudzynski, S.M., 2001. Pharmacological and behavioral characteristics of 22 kHz alarm calls in rats. Neurosci. Biobehav. Rev. 25, 611–617.

Burgdorf, J., Panksepp, J., 2006. The neurobiology of positive emotions. Neurosci. Biobehav. Rev. 30, 173–187.

Burgdorf, J., Knutson, B., Panksepp, J., 2000. Anticipation of rewarding electrical brain stimulation evokes ultrasonic vocalization in rats. Behav. Neurosci. 114, 320–327.

Burgdorf, J., Knutson, B., Panksepp, J., Ikemoto, S., 2001a. Nucleus accumbens amphetamine microinjections unconditionally elicit 50-kHz ultrasonic vocalizations in rats. Behav. Neurosci. 115, 940–944.

Burgdorf, J., Knutson, B., Panksepp, J., Shippenberg, T.S., 2001b. Evaluation of rat ultrasonic vocalizations as predictors of the conditioned aversive effects of drugs. Psychopharmacology (Berl.) 155, 35–42.

Burgdorf, J., Wood, P.L., Kroes, R.A., Moskal, J.R., Panksepp, J., 2007. Neurobiology of 50-kHz ultrasonic vocalizations in rats: electrode mapping, lesion, and pharmacology studies. Behav. Brain Res. 182, 274–283.

Burgdorf, J., Kroes, R.A., Moskal, J.R., Pfaus, J.G., Brudzynski, S.M., Panksepp, J., 2008. Ultrasonic vocalizations of rats (*Rattus norvegicus*) during mating, play, and aggression: behavioral concomitants, relationship to reward, and self-administration of playback. J. Comp. Psychol. 122, 357–367.

Cabanac, M., 1971. Physiological role of pleasure. Science 173, 1103–1107.

Cabanac, M., 1992. Pleasure: the common currency. J. Theor. Biol. 155, 173–200.

Carden, S.E., Hofer, M.A., 1990. Independence of benzodiazepine and opiate action in the suppression of isolation distress in rat pups. Behav. Neurosci. 104, 160–166.

Covington, H.E., Miczek, K.A., 2003. Vocalizations during withdrawal from opiates and cocaine: possible expressions of affective distress. Eur. J. Pharmacol. 467 (1–3), 1–13.

Csikszentmihalyi, M., Hunter, J., 2003. Happiness in everyday life: the uses of experience sampling. J. Happiness Stud. 4, 1–15.

Damasio, A.R., Grabowski, T.J., Bechara, A., Damasio, H., Ponto, L.L., Parvizi, J., Hichwa, R.D., 2000. Subcortical and cortical brain activity during the feeling of self-generated emotions. Nat. Neurosci. 3, 1049–1056.

Davidson, D., Palfai, T., Bird, C., Swift, R., 1999. Effects of naltrexone on alcohol self-administration in heavy drinkers. Alcohol Clin. Exp. Res. 23, 195–203.

Diener, E., Oishi, S., Lucas, R., 2003. Personality, culture, and subjective well-being: emotional and cognitive evaluations of life. Annu. Rev. Psychol. 54, 403–425.

Drevets, W.C., Gautier, C., Price, J.C., Kupfer, D.J., Kinahan, P.E., Grace, A.A., Price, J.L., Mathis, C.A., 2001. Amphetamine-induced dopamine release in human ventral striatum correlates with euphoria. Biol. Psychiatry 49, 81–96.

Duckworth, A.L., Steen, T.A., Seligman, M.E.P., 2005. Positive psychology in clinical practice. Annu. Rev. Clin. Psychol. 1, 629–651.

Ekman, P., Davidson, R.J., Friesen, W.V., 1990. The Duchenne smile: emotional expression and brain physiology. II. J. Pers. Soc. Psychol. 58, 342–353.

Fredrickson, B.L., Tugade, M.M., Waugh, C.E., Larkin, G.R., 2003. What good are positive emotions in crises? A prospective study of resilience and emotions following the terrorist attacks on the United States on September 11th, 2001. J. Pers. Soc. Psychol. 84, 365–376.

Heath, R.G., 1972. Pleasure and brain activity in man. Deep and surface electroencephalograms during orgasm. J. Nerv. Ment. Dis. 154, 3–18.

Holstege, G., Georgiadis, J.R., Paans, A.M., Meiners, L.C., van der Graaf, F.H., Reinders, A.A., 2003. Brain activation during human male ejaculation. J. Neurosci. 23, 9185–9193.

Janal, M.N., Colt, E.W., Clark, W.C., Glusman, M., 1984. Pain sensitivity, mood and plasma endocrine levels in man following long-distance running: effects of naloxone. Pain 19, 13–25.

Jönsson, L.E., Anggård, E., Gunne, L.M., 1971. Blockade of intravenous amphetamine euphoria in man. Clin. Pharmacol. Ther. 12, 889–896.

Knutson, B., Adams, C.M., Fong, G.W., Hommer, D., 2002. Anticipation of increasing monetary reward selectively recruits nucleus accumbens. J. Neurosci. 22, 3303–3305.

Kroes, R.A., Panksepp, J., Burgdorf, J., Otto, N.J., Moskal, J.R., 2006. Modeling depression: social dominance-submission gene expression patterns in rat neocortex. Neuroscience 137, 37–49.

Kroes, R.A., Burgdorf, J., Otto, N.J., Panksepp, J., Moskal, J.R., 2007. Social defeat, a paradigm of depression in rats, preferentially activates the cholinergic signaling pathway in the periaqueductal gray and generation of 22-kHz calls. Behav. Brain Res. 182, 290–300.

Lykken, D., Tellegen, A., 1996. Happiness is a stochastic phenomenon. Psychol. Sci. 7, 186–189.

Lyubomirsky, S., King, L., Diener, E., 2005. The benefits of frequent positive affect: does happiness lead to success? Psychol. Bull. 131, 803–855.

Marinelli, P.W., Quirion, R., Gianoulakis, C., 2004. An *in vivo* profile of beta-endorphin release in the arcuate nucleus and nucleus accumbens following exposure to stress or alcohol. Neuroscience 127, 777–784.

Newton, T.F., Ling, W., Kalechstein, A.D., Uslaner, J., Tervo, K., 2001. Risperidone pre-treatment reduces the euphoric effects of experimentally administered cocaine. Psychiatry Res. 102, 227–233.

Okun, M.S., Bowers, D., Springer, U., Shapira, N.A., Malone, D., Rezai, A.R., Nuttin, B., Heilman, K.M., Morecraft, R.J., Rasmussen, S.A., Greenberg, B.D., Foote, K.D., Goodman, W.K., 2004. What's in a "smile?" Intra-operative observations of contralateral smiles induced by deep brain stimulation. Neurocase 10, 271–279.

Oswald, L.M., Wong, D.F., McCaul, M., Zhou, Y., Kuwabara, H., Choi, L., Brasic, J., Wand, G.S., 2005. Relationships among ventral striatal dopamine release, cortisol secretion, and subjective responses to amphetamine. Neuropsychopharmacology 30, 821–832.

Panksepp, J., Burgdorf, J., 2000. 50-kHz chirping (laughter?) in response to conditioned and unconditioned tickle-induced reward in rats: effects of social housing and genetic variables. Behav. Brain Res. 115, 25–38.

Panksepp, J., Burgdorf, J., 2003. "Laughing" rats and the evolutionary antecedents of human joy? Physiol. Behav. 79, 533–547.

Panksepp, J., Burgdorf, J., Beinfeld, M.C., Kroes, R.A., Moskal, J.R., 2004. Regional brain cholecystokinin changes as a function of friendly and aggressive social interactions in rats. Brain Res. 1025, 75–84.

Panksepp, J., Burgdorf, J., Beinfeld, M.C., Kroes, R.A., Moskal, J.R., 2007. Brain regional neuropeptide changes resulting from social defeat. Behav. Neurosci. 121, 1364–1371.

Romach, M.K., Glue, P., Kampman, K., Kaplan, H.L., Somer, G.R., Poole, S., Clarke, L., Coffin, V., Cornish, J., O'Brien, C.P., Sellers, E.M., 1999. Attenuation of the euphoric effects of cocaine by the dopamine D1/D5 antagonist ecopipam (SCH 39166). Arch. Gen. Psychiatry 56, 1107–1108.

Stone, A.A., Schwartz, J.E., Schkade, D., Schwarz, N., Krueger, A., Kahneman, D., 2006. A population approach to the study of emotion: diurnal rhythms of a working day examined with the Day Reconstruction Method. Emotion 6, 139–149.

Schwarting, R.K., Jegan, N., Wöhr, M., 2007. Situational factors, conditions and individual variables which can determine ultrasonic vocalizations in male adult Wistar rats. Behav. Brain Res. 182, 208–222.

Tidey, J.W., Miczek, K.A., 1996. Social defeat stress selectively alters mesocorticolimbic dopamine release: an *in vivo* microdialysis study. Brain Res. 721, 140–149.

Voruganti, L., Slomka, P., Zabel, P., Costa, G., So, A., Mattar, A., Awad, A.G., 2001. Subjective effects of AMPT-induced dopamine depletion in schizophrenia: correlation between dysphoric responses and striatal D(2) binding ratios on SPECT imaging. Neuropsychopharmacology 25, 642–650.

Zacny, J.P., Lichtor, J.L., Thapar, P., Coalson, D.W., Flemming, D., Thompson, W.K., 1994. Comparing the subjective, psychomotor and physiological effects of intravenous butorphanol and morphine in healthy volunteers. J. Pharmacol. Exp. Ther. 270, 579–588.

CHAPTER 6.3

Vocal expression of emotion in a nocturnal prosimian primate group, mouse lemurs

Elke Zimmermann*

Institute of Zoology and Center of Systems Neuroscience, University of Veterinary Medicine, Hannover, Germany

Abstract: This chapter reviews our current knowledge on vocal expression of emotional states in a prosimian primate group, the mouse lemurs, by applying a cross-taxa concept of emotion. In this concept, emotions are defined based on their adaptive significance. Using standardized ethological protocols, we were able to predictably induce changes in vocal expression of mouse lemurs by manipulating defined stimuli or situations. Whereas infants produced context-specific vocalizations linked to the external environment, adults additionally linked vocal features to the respective interaction partner. In a courtship paradigm, mouse lemur males showed dynamic changes in vocal expression paralleling the intensity of arousal evoked by the condition of females and their behavior. In contrast to rodents, neither vomeronasalectomy nor bulbectomy affected vocal production. Castration led to a decrease in vocal rate of a specific vocal expression. The emission of the latter, however, was not related to plasma testosterone in intact males. Altogether, the findings suggested that vocal control was more linked to individual social experience than to olfactory input or gonadal androgens. A comparison of three context-specific acoustically distinct vocal expressions across different mouse lemur species indicated that comparable vocal features were used across species to convey a particular individualized context, and thereby a specific appetitive or aversive state.

The findings support the hypothesis that vocalizations of mouse lemurs express the emotional state of a signaler reliably when linked to the respective individualized context.

Keywords: vocalization; communication; aversive state; appetitive state; individualized context-related concept of emotion; context-specific vocalizations; cross-taxa comparison; evolution; mouse lemur; primate

I. Introduction

Since the publication of *The Expressions of Emotions in Man and Animals* (Darwin, 1872), researchers have postulated that the expression of emotions in human speech evolved from affective features in mammalian vocalizations (e.g., Andrew, 1963; Jürgens et al., 1967; van Lawick-Goodall, 1968). Morton (1977) was one of the first researchers who applied such an evolutionary concept systematically to vocal communication. August and Anderson (1987), as well as Peters (2002), refined Morton's rules to separate friendly (appetitive) from fearful (aversive) contexts by using additional acoustic parameters.

Whereas the refined Morton rules provided a powerful tool to predict the acoustic pattern of vocal expressions in non-human mammals and humans (for humans: e.g., Scherer et al., 2001; for non-human primates: Hauser, 1996; Gouzoules and Gouzoules, 2000; Fichtel et al., 2001; Snowdon, 2003; Hammerschmidt and Jürgens, 2007; for non-primate mammals: Compton et al., 2001; Robbins and McCreery, 2003; Schehka et al., 2007; Bastian and Schmidt, 2008), they fail to explain the often highly diverse sound patterns emitted in superficially similar interactions of the same sender living in a complex society with elaborated inter-individual relationships (Altenmüller et al., 2009, submitted). Thus, rhesus macaques and chimpanzees emit screams during agonistic interactions, which differ considerably in frequency contour (e.g., Gouzoules et al., 1984; Slocombe and Zuberbühler, 2005). The

*Corresponding author. E-mail: elke.zimmermann@tiho-hannover.de

ISBN 978-0-12-374593-4
DOI: 10.1016/B978-0-12-374593-4.00022-X

most likely call structure in such a social scenario can only be predicted when the prior experience and the social environment of the sender (together termed the individualized context) are considered. To explain the vocal behavior in such situations and to provide the basis for cross-taxa comparisons of vocal expressions, Altenmüller et al. (2009, submitted) developed an individualized context-related concept of emotion (ICCE), which allows an empirical and quantitative comparison of communication sounds produced by a sender in a well-defined state in a particular social scenario. In this concept, emotions are defined based on their adaptive significance, but independent of subjective self-reports restricted to humans. Emotions in the ICCE are defined as brain decision mechanisms, which originally evolved for governing fitness-relevant behavioral and physiological responses to a particular stimulus or situation. They can be measured on the behavioral level by the type of behavioral response and by changes in the intensity of this response to a specific stimulus or situation. Comparative research exploring this concept is just beginning to emerge and, until now, it is restricted to non-primate mammals (e.g., tree shrews: Schehka et al., 2007; bats: Bastian and Schmidt, 2008).

This chapter aims to review our current knowledge on vocal expression of emotions in a prosimian primate group, the mouse lemurs, by applying the ICCE. First, I will introduce why mouse lemurs represent an excellent model for such an approach. I will present standardized ethological protocols simulating natural situations, by which we were able to systematically manipulate stimuli and/or situations and thereby vocal expressions. I will then demonstrate how vocal production in males is affected by manipulations of olfactory input and gonadal androgens, respectively. Finally, I will compare 3–4 cryptic mouse lemur species, to show how the vocal expression of emotion varies across species of the same genus.

II. Mouse lemurs as a model group for primate affective communication in the acoustic domain

The Malagasy mouse lemurs (*Microcebus* ssp.) represent an excellent model group to gain insight into early evolutionary roots of primate affective communication, since mouse lemurs are suggested to reflect the most ancestral primate condition (Martin, 1972). They are nocturnal and small-bodied, about the size of a gerbil, but inhabit the fine branch niche of dense parts of forests. Their highly mobile, bat-like ears and their broad auditory sensitivity (Niaussat and Petter, 1980) are ideal prerequisites for vocal communication. Mouse lemurs live in dispersed multi-male/multi-female societies, in which home ranges within and between sexes overlap (Radespiel, 2006). Individuals forage solitarily during the night, but related females form temporarily stable sleeping groups during the day (Lutermann et al., 2006). Groups occasionally change their sleeping sites, mainly tree holes and leaf nests; however, the composition of sleeping groups remains stable over time. Group-specific calls are used by females for coordination of group movement and for guiding group reunion (Braune et al., 2005). Vocal activity in mates is enhanced during the breeding season when males actively search for estrous females (Zimmermann and Lerch, 1993; Craul et al., 2004). Females give birth to one to four infants which may be nursed cooperatively by all females of the same sleeping group; however, females transport their own infants individually (Eberle and Kappeler, 2006). Mouse lemurs evolved an infant-parking system (Ross, 2001), where mothers park their infants temporarily in nests of dense vegetation during foraging. Since mouse lemurs live in such a dispersed and individualized network of complex social relations, in which acoustic signals are ideally suited to coordinate social interactions, they represent a promising group to study the vocal expression of emotions. Furthermore, at present, 16 cryptic species are known (e.g., Olivieri et al., 2007; Radespiel et al., 2008) providing a possibility of exploring the variation of vocal expression in comparable social situations across different species.

III. Relationship between specific stimuli/situations and vocal expressions in the gray mouse lemur

The gray mouse lemur (*Microcebus murinus*) represents the species for which most information about vocal communication is available. Field and laboratory studies focusing on spontaneously-occurring social or predator-induced interactions demonstrated that mouse lemurs use an elaborate set of vocalizations (Fig. 1) with individual-, population-, or species-specific signatures (Zimmermann and Lerch, 1993; Zimmermann, 1995; Buesching et al., 1998; Hafen et al., 1998; Zietemann, 2000; Zimmermann et al., 2000; Zimmermann and Hafen, 2001). Different call

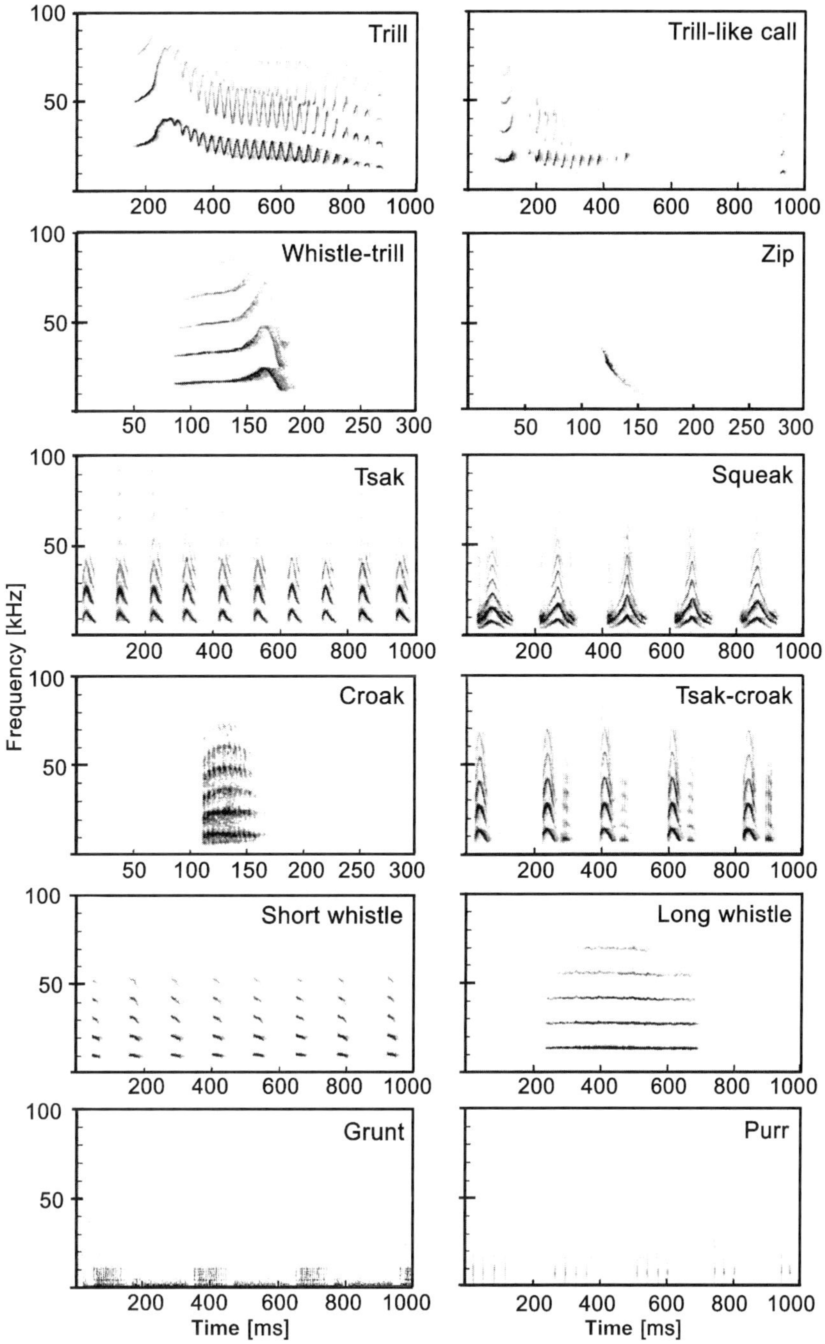

Fig. 1. A set of sonograms showing the vocal repertoire of the adult gray mouse lemur (*Microcebus murinus*) displaying major call types. The names of individual call types are labeled on the sonograms. Note that the timescale may be different between call types.

types were identified on the basis of their acoustic structure, and these call types could be combined into higher order structures (Zimmermann et al., 2007). Until recently, however, it was not clear to what extent a high intra-individual variation in vocal expression may be explained by specific environmental or social situations.

We have established standardized ethological protocols based on situations occurring in the natural (including the social) environment to explore the effect of context variation on vocal expression. We assumed that mouse lemurs use vocalizations to express their emotional state if there is a predictable link between a given external stimulus or situation and specific features in the vocal expression.

In order to get information about infant vocalizations, we exposed sucklings in a separation paradigm to three distinct contexts to evoke calls, and to perform an audio-videographic analysis of the sound-correlated infant behavior and a simultaneous multiparametric sound analysis (Scheumann et al., 2007b). In the isolation context, an infant was removed from its siblings and mother, placed on a wooden bar in an experimental cage and left alone. In the threat-induced context, the infant was additionally manipulated by the experimenter, e.g., by grasping it or nudging it with the finger. In the isolation context, infants tried to change their situation either by moving around in the cage, scanning the environment and searching for the mother. In the threat-induced context, the infants were showing defensive threat displays, withdrawal, or attacking the experimenter. The two dissimilar types of behavioral responses suggested that infants perceived the two contexts as differing in aversiveness. Indeed, we found that the emission of a specific call type, named the infant whistle (Fig. 2), was reliably predicted by the isolation context. The threat-induced context was primarily associated with emission of another call type, termed the infant tsak (Fig. 2).

In the third context, the grooming context, an infant was placed on the warm hand of the experimenter and groomed by her. Infants were habituated to this procedure since they were measured and weighed twice a week. In contrast to the former contexts, infants tried to remain on the hand or prolong this situation, e.g., by licking the experimenters' hand, by stretching the arms or lifting the head so that they could be groomed. In this behavioral context, the animals emitted the third type of calls, the infant purr (Fig. 2). Considering this behavioral response, we assumed this context to be pleasant or appetitive for the infant. All these findings confirmed our hypothesis that infants may express different emotional states acoustically in response to different contexts. These calls may convey specific needs to their mother. Peak frequency was the best predictor for the context. This context-specificity of vocal expression in infants may be based on phylogenetically old neuronal networks, guiding inborn adaptive responses.

In adults, vocal production is additionally linked to the social interaction partner. Using a courtship

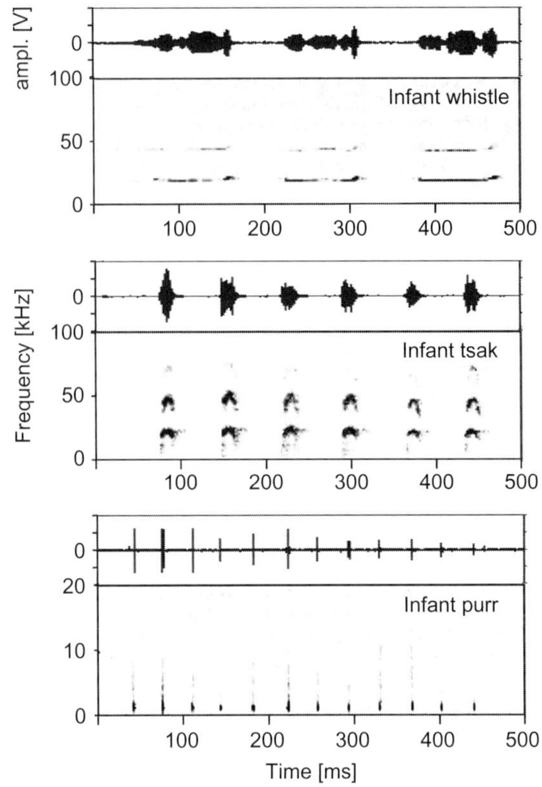

Fig. 2. Context-specific vocalizations produced by sucklings of the gray mouse lemur. Waveforms (upper traces) and sonagrams (below) of infant whistle calls, infant tsak calls and infant purr calls.

paradigm and applying the same recording and analyzing techniques as for infants, we revealed that mouse lemurs showed dynamic changes in vocal expression which paralleled arousal evoked by the interaction partner (Dietz and Zimmermann, 2005). In this paradigm, a male and a female were placed in two separate experimental cages with a connecting door which was opened after a specified amount of time. Experiments were conducted during the breeding season. The male entered the cage of the female and tried to court her. During the interaction of the partners, males emitted a series of trill calls (Fig. 3a), which were often associated with tail-flicking. Tail-flicking could be associated with the condition and behavior of the female, and might signal higher emotionality and potentiated the male's sexual arousal. We assumed that the intensity of the emotional state (here the sexual arousal state) of males can be measured by the behavioral display of tail-flicking. Encounters with tail-flicking were thus considered to be of high emotion intensity (high arousal state), while

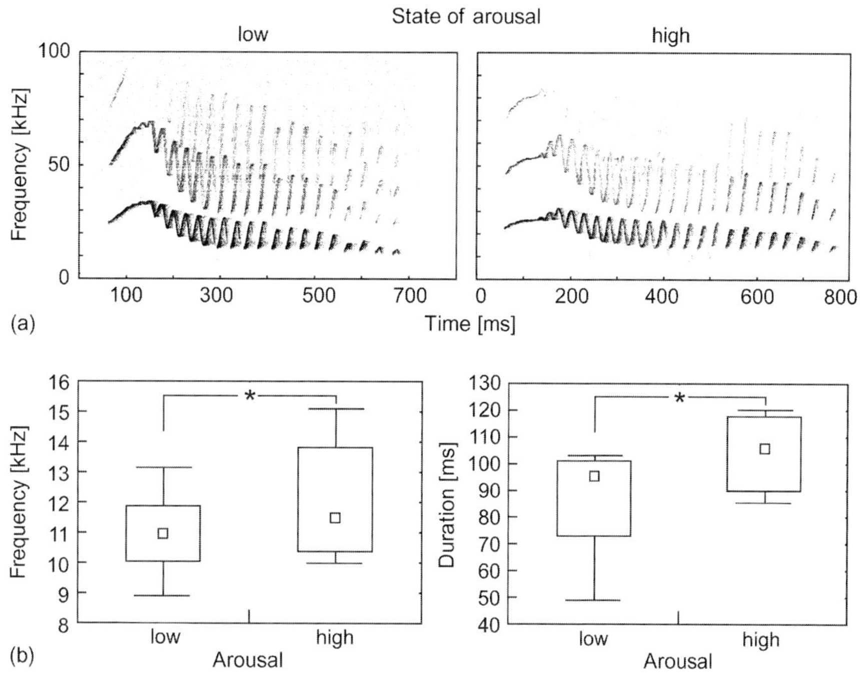

Fig. 3. Sonagrams (a) displaying the variation of trill call structure in the same male gray mouse lemur during two arousal states. (b) A multiparametric acoustic analysis with seven spectral and four temporal parameters revealed significant differences between the arousal states in the frequency contour (frequency: fundamental frequency at the end of the call; duration: duration of whistle part (onset to maximum fundamental frequency of first part of the call), $N = 6$ males, Wilcoxon matched pairs test; $T = 1$, $Z = 1.99$, $p < 0.046$).

encounters without tail-flicking were of low emotion intensity (low arousal state). Males always showed appetitive behavior towards females in the breeding season and tried to establish contact with them, irrespective of the reproductive state and behavior of the female. Based on this behavioral response of males, we considered the respective situation as pleasant for them.

In order to explore to what extent the arousal state of males may be expressed acoustically, we manipulated sexual attractiveness of females and thereby arousal of males. In one condition, the female was in an anestrous state, in the other she was in an estrous state. Encounters with tail-flicking were restricted to the estrous condition in all studied males ($N = 6$). We compared trill calls of males emitted during low arousal (no tail-flicking) with those of the same males during high arousal (tail-flicking). Vocal expression reflected the intensity of emotionality and the concomitant arousal state reliably. We found a significant link between tail-flicking and vocal features related to the frequency contour of trill calls (Fig. 3a,b). These findings confirm the hypothesis that mouse lemurs may express dynamic changes in arousal state acoustically. Listeners are provided with vocal cues about the intensity of emotionality of the vocalizer.

Olfactory cues relating to the reproductive state of the female or gonadal androgens represent sufficient predictors of the male's vocal behavior in mice and rats (e.g., Lumia et al., 1987; Vandenbergh, 1988; Simerly, 1990; Holy and Guo, 2005; Nyby, Chapter 7.6 in this volume), but not in mouse lemurs (Zimmermann, 1996). We paired males of different gonadal status (males castrated as adults and intact males), or of varying olfactory input (males vomeronasalectomized or bulbectomized as adults) with females in the courtship paradigm. Vocal rate of trill calls declined in castrated males compared to control males. There was, however, no significant relationship between calling rate and plasma testosterone levels in gonadally intact males. Furthermore, trill calling in males was rank-dependent (Zimmermann and Lerch, 1993). However, it increased rapidly in subdominant males as soon as the dominant male was removed (Lindemann and Zimmermann, personal observations). Neither vomeronasalectomy nor bulbectomy affected trill calling rate. Besides, estrous female urine, presented to males on cotton tips, did not evoke trill calling behavior in males of the Goodman's mouse lemur (Borchert et al., 2008). These results imply that testicular hormones have a stronger impact on trill calling motivation than olfactory information from females. Furthermore, trill calling rate seems to be dependent not only on gonadal status, but also on social experience and thereby on an individualized behavioral context.

IV. Variation of vocal expression in comparable situations across different mouse lemur species

To date, information on the presence and extent of cross-taxa commonalities in vocal features coding for different emotions is lacking, at least for primates. Such a systematic comparative approach, however, is important to disentangle common supraspecies from species-specific adaptations in emotional expression and the respective processing networks in the brain.

We have performed a cross-taxa comparison of calls in three different scenarios for three to four cryptic mouse lemur species (Zimmermann, 2009, submitted) to get first insight into potential commonalities in vocal features. We used a startle and a courtship paradigm to elicit and compare vocalizations under standardized conditions across species.

In the startle paradigm, a caged mouse lemur was sitting in a nest box and was startled by the experimenter, e.g., by grasping with the hand inside the box. Under this condition and similar conditions in the field (stranger or predator trying to enter a sleeping site), a mouse lemur gets startled, then tries to change its situation by an open-mouth display, often accompanied by loud, plosive and noisy grunts (Fig. 4a), and may finally attack the intruder. Uniformity in the acoustic pattern of this display, reflecting aversion across species, may be explained by similar physiological constraints (e.g., Fitch and Hauser, 2003) and similar predation pressures (Scheumann et al., 2007a).

In the courtship paradigm, females rejecting approaches of males express their unwillingness to mate by a similar open-mouth display associated, however, with a series of short, sonic calls with chevron-shaped frequency contour, the tsak calls (Fig. 4b). If the male continues approaching the female, she attacks him and tries to chase him away. Aversion against close-contact approaches is conveyed across the species by a uniform chevron-shaped frequency contour in a sonic frequency range. Quantitative differences between species were found in the maximum fundamental frequency and in the bandwidth. They could be attributed neither to body size, nor to body mass (Zietemann, 2000). Since species differ in the degree of their arousal to disturbances at the sleeping sites (Radespiel et al., 2003), we assume that inter-specific differences in general arousal may provide the explanation. Stronger tension of the vocal cords due to sympathetic arousal is known to enhance fundamental frequency (e.g., Schehka et al., 2007).

Males of all studied species emit a series of loud, ultrasonic and broadband calls (Fig. 4c; trill) while courting for females. These calls have the longest duration of all mouse lemur calls and exhibited the most complex frequency modulated acoustic contour. Request for social contact, and thereby appetitiveness, is coded across all these species by ultrasonic vocalizations with similar absolute frequency ranges and bandwidths, despite remarkable individual-, population- and species-specific differences in vocal contours (e.g., Braune et al., 2008). Broadband, frequency modulated ultrasonic vocalizations, found across all the species, provide advantages for sound localization. Uniformity in the frequency range across the species may be explained by similar morphological constraints and predation pressures.

Our comparative approach provided evidence that different species use common vocal features to convey similar emotions when studied in individualized contexts. Thus, two different types of aversion were coded either by plosive, noisy and broadband calls in a low frequency, or by tonal chevron-shaped calls in a medium frequency range. In contrast to that, one type of the appetitive state was conveyed by tonal ultrasonic calls with a complex acoustic contour.

V. Synopsis and outlook

In this review, I have shown that we were able to induce predictable changes in vocal expression of mouse lemurs by manipulating environmental or social situations using standardized ethological protocols. Across species, specific acoustic features in vocalizations were reliably linked to individualized contexts and states of the vocalizer.

Aversion against external unexpected stimuli was encoded in adult mouse lemurs across species by series of loud, plosive and low-frequency noisy grunts covering the same range. Vocalizations with comparable acoustic features given under the same circumstances

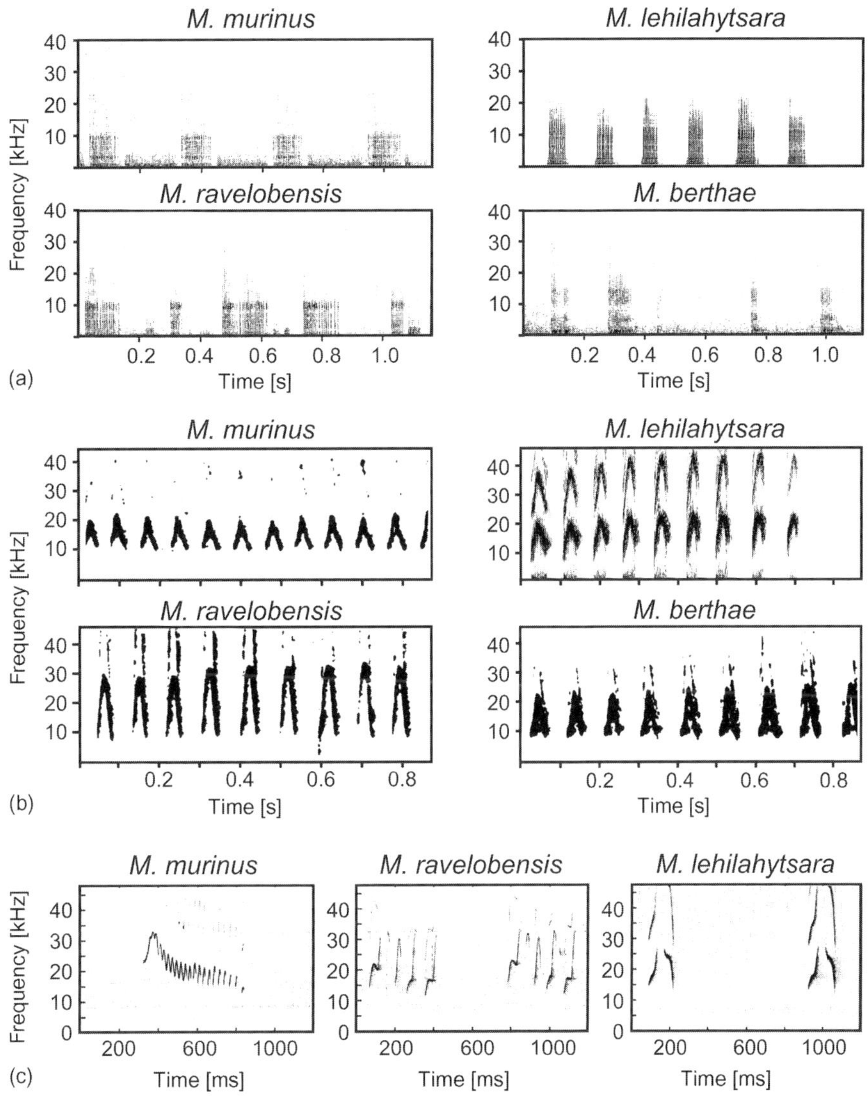

Fig. 4. Sonagrams (a) displaying inter-specific acoustic variation of grunt calls, produced in the startle paradigm. Grunts show a uniform acoustic pattern across species. Sonagrams (b) displaying inter-specific acoustic variation of tsak calls, produced by females, unwilling to mate, in the courtship paradigm. Tsaks show statistical differences in maximum fundamental frequency and bandwidth across species (Zietemann, 2000). Sonagrams (c) displaying inter-specific variation of trill calls, produced by courting males in the courtship paradigm. Despite considerable species-, population- and individual-specific differences, trills cover the same ultrasonic frequency range and bandwidth across species.

are also found in other nest- or hole-sleeping prosimians (all small nocturnal lemurs, and slow or slender loris and bushbabies: Zimmermann, 1985a,b; Scheumann et al., 2000a), as well as in burrow-using mammals (e.g., tree shrews: Binz and Zimmermann, 1989). Even predators of these small mammals, such as carnivores, emit similar low-frequency and noisy growls or grunts when cornered, e.g., during defensive threat displays (e.g., in cats: Leyhausen, 1979; Brudzynski, 2007; dogs: Feddersen-Petersen, 2000). Comparable selective forces in such a scenario may have fixed these uniform acoustic responses during evolution, most likely because of their effectiveness in evoking reflexive escape responses in strangers and predators (e.g., Owren and Rendall, 2001; Scheumann et al., 2007a; Rendal and Owren, Chapter 5.4 in this volume). In cats, neurophysiological and neuropharmacological studies have revealed that growling vocalizations may be induced by activation of specific components of the ascending cholinergic system of the brain (Brudzynski, 2007). A similar neurochemical system was reported to evoke alarm calls in rats (Brudzynski, 2007). Further research is necessary to explore to what extent vocal responses expressing comparable negative states across species and genera are caused by similar neuronal processing networks.

A different type of aversion, aversion directed towards approaches of interaction partners, is conveyed in adult mouse lemurs across all species by a uniform chevron-shaped frequency contour in a sonic medium frequency range, the tsak. Infant mouse lemurs also utter similar threat-induced calls towards the experimenter. Distress calls of many mammalian species (e.g., house mouse: Ehret, 2006; tree shrews: Schehka et al., 2007; macaques: Gouzoules and Gouzoules, 2000; Maestripieri et al., 2000; Jovanovic and Gouzoules, 2001; chimpanzees: Slocombe and Zuberbühler, 2005) share chevron-shaped frequency contours and broad bandwidth in a medium frequency range compared to the communication range of the respective species.

Infants of the gray mouse lemur express another aversive state, namely isolation from mother and/or peers, by an acoustically different vocalization, the whistle. This call with its high frequency and tonal structure is acoustically comparable with infant isolation calls of other mammalian groups, such as Carnivora, Cetacea, Artiodactyla, Rodentia, Chiroptera and other primates, including humans (see Newman, 2004, 2007 for review). Based on neurochemical studies it is assumed that it is the opiate system in the brain which is most clearly associated with isolation calls (Panksepp et al., 1980; Newman, 2007).

Different types of appetitiveness are conveyed in mouse lemurs by at least two remarkably different vocalizations, purrs and trills. Infants of the gray mouse lemur use a series of purr calls to maintain grooming, as do adult mouse lemurs (Zimmermann, 1995) and members of various other mammalian groups (e.g., Carnivora, Scandentia, Artiodactyla, Rodentia, Chiroptera: Peters, 2002). Purrs are always soft, low-frequency noisy vocalizations with extremely short duration and high repetition rate, which are restricted to positive tactile interactions in close proximity. In contrast, across species, adult mouse lemurs direct ultrasonic trill calls towards conspecifics during contact approaches at some distance, advertising their motivation for social bonding. Ultrasonic trill calls are tonal high-frequency vocalizations with the most complex frequency contour in the mouse lemur vocal repertoire. Comparable calls were found not only in male or pre-estrous or estrous females (Büesching et al., 1998) during the breeding season, but also in captive females searching for contact with isolated infants (Scheumann and Zimmermann, 2007), or in wild-living females during reunions and movements of social groups (Braune et al., 2005). Tonal high-frequency calls with complex frequency contours were observed during courtship or joint territorial displays of pair partners, and were also reported for other mammalian groups, such as bats (Leippert 1974; Behr and von Helversen, 2004), tree shrews (Binz and Zimmermann, 1987), rats (Brudzynski, 2007), mice (e.g., Holy and Guo, 2005; Ehret, 2006) and pair-living primates (e.g., sportive lemurs: Méndez-Cárdenas and Zimmermann, 2009), marmosets (Snowdon, 2003) and gibbons (e.g., Tenaza, 1976; Mitani, 1988; Geissmann, 2002), and were suggested to strengthen social bonds. Whereas appetitive ultrasonic calls in rodent males can be elicited by olfactory cues of females and varied by plasma testosterone level (e.g., Lumia et al., 1987; Vandenbergh, 1988; Simerly, 1990; Holy and Guo, 2005), these vocalizations are less dependent on hormones and olfactory input, but specifically linked to social partners in primates. In rats, the activation of the ascending dopaminergic system was postulated to induce a positive state with tonal high-frequency vocalizations, called 50 kHz calls (Brudzynski, 2007; Burgdorf and Moskal, Chapter 6.2 in this volume). Parkinson's patients in humans with deficient dopamine expression do also show deficiencies in expressing positive emotions acoustically (Möbes et al., 2008). These findings support the notion that some components of the brain system evoking vocal expression of positive emotions in humans may be based on the same neurochemical network as that in the brain of non-human mammals.

Altogether, this review provides support for the hypothesis that mouse lemurs use acoustically distinct vocal expressions as indicators of individualized contexts and related aversive or appetitive states. Ongoing psychoacoustical studies will explore to what extent listeners will assess and categorize these vocalizations.

Acknowledgments

I would like to thank Stefan Brudzynski and an anonymous reviewer for clarifying comments, Sabine Schmidt and Marina Scheumann for critical commenting on the first draft of this manuscript,

Elisabeth Engelke for her help in editing and Rüdiger Brüning for technical support with the figures. Research described in this review was supported by various grants of the German Research Council, the Volkswagen Foundation and the German Academic Exchange Council. The paper benefited from stimulating discussions with members of the research group DFG FOR 499.

References

Altenmüller, E., Schmidt, S., Zimmermann, E., 2009. A cross-taxa concept of emotion in acoustic communication: an ethological perspective (submitted).

Andrew, R.J., 1963. The origin and evolution of the calls and facial expressions of the primates. Behav. 20, 1–109.

August, P.V., Anderson, J.G.T., 1987. Mammal sounds and motivation–structural rules: a test of the hypothesis. J. Mammal. 68, 1–9.

Bastian, A., Schmidt, S., 2008. Affect cues in vocalizations of the bat, *Megaderma lyra*, during agonistic interactions. J. Acoust. Soc. Am. 124, 598–608.

Behr, O., von Helversen, O., 2004. Bat serenades – complex courtship songs of the sac-winged bat (*Saccopteryx bilineata*). Behav. Ecol. Sociobiol. 56, 106–115.

Binz, H., Zimmermann, E., 1989. The vocal repertoire of adult tree shrews (*Tupaia belangeri*). Behaviour 109, 142–162.

Borchert, K., Scheumann, M., Zimmermann, E., 2008. The role of olfaction in social communication of the Goodman's mouse lemur. 22nd Congress of the International Primatological Society, Edinburgh, 3.8.2008–8.8.2008. Primate Eye 96, 282.

Braune, P., Schmidt, S., Zimmermann, E., 2005. Spacing and group coordination in a nocturnal primate, the golden brown mouse lemur (*Microcebus ravelobensis*): the role of olfactory and acoustic signals. Behav. Ecol. Sociobiol. 58, 587–596.

Braune, P., Schmidt, S., Zimmermann, E., 2008. Acoustic divergence in the communication of cryptic species of nocturnal primates (*Microcebus ssp.*). BMC Biol. 6, 19.

Brudzynski, S.M., 2007. Ultrasonic calls of rats as indicator variables of negative or positive states: acetylcholine-dopamine interaction and acoustic coding. Behav. Brain Res. 182, 261–273.

Buesching, C.D., Heistermann, M., Hodges, J.K., Zimmermann, E., 1998. Multimodal oestrus advertisement in a small nocturnal prosimian, *Microcebus murinus*. Folia Primatol. 69 (Suppl. 1), 295–308.

Compton, L.A., Clarke, J.A., Seidensticker, J., Ingrisano, D.R., 2001. Acoustic characteristics of white-nosed coati vocalizations: a test of motivation-structural rules. J. Mammal. 82, 1054–1058.

Craul, M., Zimmermann, E., Radespiel, U., 2004. First experimental evidence for female mate choice in a nocturnal primate. Primates 45, 271–274.

Darwin, C., 1872. The Expressions of Emotions in Man and Animals. D. Appleton and Company, New York, NY.

Dietz, M., Zimmermann, E., 2005. The effect of arousal on a social call of a nocturnal primate (*Microcebus murinus*). 6th Meeting of the German Neuroscience Society, Göttingen, 17.2.2005–20.2.2005. Supplement to Neuroforum, 11: 52A.

Eberle, M., Kappeler, P.M., 2006. Family insurance: kin selection and cooperative breeding in a solitary primate (*Microcebus murinus*). Behav. Ecol. Sociobiol. 60, 582–588.

Ehret, G., 2006. Common rules of communication sound perception. In: Kanwal, J.S., Ehret, G. (Eds.), Behavior and Neurodynamics for Auditory Communication. Cambridge University Press, Cambridge, UK, pp. 85–114.

Feddersen-Petersen, D.U., 2000. Vocalization of European wolves (*Canis lupus lupus* L.) and various dog breeds (*Canis lupus f. fam.*). Arch. Tierz. 43, 387–397.

Fichtel, C., Hammerschmidt, K., Jürgens, U., 2001. On the vocal expression of emotion. A multi-parametric analysis of different states of aversion in the squirrel monkey. Behaviour 138, 97–116.

Fitch, W.T., Hauser, M.D., 2003. Unpacking "honesty:" vertebrate vocal production and the evolution of acoustic signals. In: Simmons, A.M., Popper, A.N., Fay, R.R. (Eds.), Acoustic Communication. Springer, New York, NY, pp. 65–137.

Geissmann, T., 2002. Duet-splitting and the evolution of gibbons songs. Biol. Rev. 77, 57–76.

Gouzoules, H., Gouzoules, S., 2000. Agonistic screams differ among four species of macaques: the significance of motivation–structural rules. Anim. Behav. 59, 501–512.

Gouzoules, S., Gouzoules, H., Marler, P., 1984. Rhesus monkey (*Macaca mulatta*) screams: representational signalling in the recruitment of agonistic aid. Anim. Behav. 32, 182–193.

Hafen, T., Neveu, H., Rumpler, Y., Wilden, I., Zimmermann, E., 1998. Acoustically dimorphic advertisement calls separate morphologically and genetically homogenous populations of the grey mouse lemur (*Microcebus murinus*). Folia Primatol. 69 (Suppl. 1), 342–356.

Hammerschmidt, K., Jürgens, U., 2007. Acoustical correlates of affective prosody. J. Voice 21, 531–540.

Hauser, M.D., 1996. The Evolution of Communication. The MIT Press, Cambridge, MA.

Holy, T.E., Guo, Z., 2005. Ultrasonic songs of male mice. PLoS Biol. 3, 2177–2186.

Jovanovic, T., Gouzoules, H., 2001. Effects of nonmaternal restraint on the vocalizations of infant rhesus monkeys (*Macaca mulatta*). Am. J. Primatol. 53, 33–45.

Jürgens, U., Maurus, M., Ploog, D., Winter, P., 1967. Vocalization in the squirrel monkey (*Saimiri sciureus*) elicited by brain stimulation. Exp. Brain Res. 4, 114–117.

Leyhausen, P., 1979. Cat Behavior: The Predatory and Social Behavior of Domestic and Wild Cats. Garland STPM Press, New York, NY.

Lumia, A.R., Zebrowski, A.F., McGinnis, M.Y., 1987. Olfactory bulb removal decreases androgen receptor binding in amygdala and hypothalamus and disrupts masculine sexual behavior. Brain Res. 404, 121–126.

Lutermann, H., Schmelting, B., Radespiel, U., Ehresmann, P., Zimmermann, E., 2006. The role of survival for the evolution of female philopatry in a solitary forager, the grey mouse lemur (*Microcebus murinus*). Proc. Biol. Sci. 273, 2527–2533.

Maestripieri, D., Jovanovic, T., Gouzoules, H., 2000. Crying and infant abuse in rhesus monkeys. Child Dev. 71, 301–309.

Martin, R.D., 1972. Adaptive radiation and behaviour of the malagasy lemurs. Philos. Trans. R. Soc. Lond. B. Biol. Sci. 264, 295–352.

Méndez-Cárdenas, M.G., Zimmermann, E., 2009. Duetting – a mechanism to strengthen pair bonds in a dispersed pair-living primate (*Lepilemur edwardsii*)? Am. J. Phys. Anthropol. 139 (4), 523–532.

Mitani, J.C., 1988. Male gibbon (*Hylobates agilis*) singing behavior: natural history, song variations and function. Ethology 79, 177–194.

Möbes, J., Joppich, G., Stiebritz, F., Dengler, R., Schröder, C., 2008. Emotional speech in Parkinson's disease. Mov. Disord. 23, 824–829.

Morton, E.S., 1977. On the occurrence and significance of motivation–structural rules in some bird and mammal sounds. Am. Nat. 111, 855–869.

Newman, J.D., 2004. The primate isolation call: a comparison with precocial birds and non-primate mammals. In: Rogers, L.J., Kaplan, G. (Eds.), Comparative Vertebrate Cognition: Are Primates Superior to Non-Primates? Kluwer Academic/Plenum Publishers, New York, NY, pp. 171–187.

Newman, J.D., 2007. Neural circuits underlying crying and cry responding in mammals. Behav. Brain Res. 182, 155–165.

Niaussat, M.M., Petter, J.J., 1980. Etude de la sensibilité auditive d'un lémurien malgache: *Microcebus murinus* (J.-F. Miller, 1777). Mammalia 44, 553–558.

Olivieri, G., Zimmermann, E., Randrianambinina, B., Rasoloharijaona, S., Rakotondravony, D., Guschanski, K., Radespiel, U., 2007. The ever-increasing diversity in mouse lemurs: three new species in north and northwestern Madagascar. Mol. Phylogenet. Evol. 43, 309–327.

Owren, M.J., Rendall, D., 2001. Sound on the rebound: bringing form and function back to the forefront in understanding non-human primate vocal signaling. Evol. Anthropol. 10, 58–71.

Panksepp, J., Herman, B.H., Vilberg, T., Bishop, P., Deeskinazi, F.G., 1980. Endogenous opioids and social behavior. Neurosci. Biobehav. Rev. 4, 473–487.

Peters, G., 2002. Purring and similar vocalizations in mammals. Mammal Rev. 32, 245–271.

Radespiel, U., 2006. Ecological diversity and seasonal adaptations of mouse lemurs (*Microcebus spp.*). In: Gould, L., Sauther, M.L. (Eds.), Lemurs: Ecology and Adaptation. Springer, New York, NY, pp. 211–234.

Radespiel, U., Ehresmann, P., Zimmermann, E., 2003. Species-specific usage of sleeping sites in two sympatric mouse lemur species (*Microcebus murinus* and *M. ravelobensis*) in northwestern Madagascar. Am. J. Primatol. 59, 139–151.

Radespiel, U., Olivieri, G., Rasolofoson, D.W., Rakotondratsimba, G., Rakotonirainy, O., Rasoloharijaona, S., Randrianambinina, B., Ratsimbazafy, J.H., Ratelolahy, F., Randriamboavonjy, T., Rasolofoharivelo, T., Craul, M., Rakotozafy, L., Randrianarison, R.M., 2008. Exceptional diversity of mouse lemurs (*Microcebus spp.*) in the Makira Region with the description of one new species. Am. J. Primatol. 70, 1033–1046.

Robbins, R.L., McCreery, E.K., 2003. African wild dog pup vocalizations with special reference to Morton's model. Behav. 140, 333–351.

Ross, C., 2001. Park or ride? Evolution of infant carrying in primates. Int. J. Primatol. 22, 749–771.

Schehka, S., Esser, K.H., Zimmermann, E., 2007. Acoustical expression of arousal in conflict situations in tree shrews (*Tupaia belangeri*). J. Comp. Physiol. A Neuroethol. Sens. Neural Behav. Physiol. 193, 845–852.

Scherer, K.R., Banse, H.G., Wallbot, H.G., 2001. Emotion inferences from vocal expression correlate across languages and cultures. J. Cross Cult. Psychol. 32, 76–92.

Scheumann, M., Zimmermann, E., 2007. Mother–offspring communication in gray mouse lemurs. 30th International Ethological Conference, Halifax, Canada, 15.8.2007–23.8.2007, p. 129.

Scheumann, M., Rabesandratana, A., Zimmermann, E., 2007a. Predation, communication, and cognition in lemurs. In: Gursky, S.L., Nekaris, K.A.I. (Eds.), Primate Anti-Predator Strategies. Springer, New York, NY, pp. 100–126.

Scheumann, M., Zimmermann, E., Deichsel, G., 2007b. Context-specific calls signal infants' needs in a strepsirrhine primate, the gray mouse lemur (*Microcebus murinus*). Dev. Psychobiol. 49, 708–718.

Simerly, R.B., 1990. Hormonal control of neuropeptide gene expression in sexually dimorphic olfactory pathways. Trends Neurosci. 13, 104–110.

Slocombe, K.E., Zuberbühler, K., 2005. Agonistic screams in wild chimpanzees (*Pan troglodytes schweinfurthii*) vary as a function of social role. J. Comp. Psychol. 119, 67–77.

Snowdon, C.T., 2003. Expression of emotion in non-human animals. In: Davidson, R.J., Scherer, K.R., Goldsmith, H.H. (Eds.), Handbook of Affective Sciences. Oxford University Press, Oxford, UK, pp. 457–480.

Tenaza, R.R., 1976. Songs, choruses and countersinging of Kloss' gibbons (*Hylobates klossii*) in Siberut Island, Indonesia. Z. Tierpsychol. 40, 37–52.

Vandenbergh, J.G., 1988. Pheromones and mammalian reproduction. In: Knobil, E., Neill, J. (Eds.), The Physiology of Reproduction. Raven Press, New York, NY, pp. 1679–1696.

Van Lawick-Goodall, J., 1968. A preliminary report on expressive movements and communication in the Gombe Stream chimpanzees. In: Jay, P.C. (Ed.), Primates: Studies in Adaptation and Variability. Holt, Rinehart and Winston, New York, NY, pp. 313–374.

Zietemann, V. (2000). Artendiversität bei Mausmakis: die Bedeutung der akustischen Kommunikation. Dissertation, Universität Hannover, Hannover, Germany.

Zimmermann, E., 1985a. The vocal repertoire of the adult Senegal bushbaby (*Galago senegalensis senegalensis*). Behaviour 94, 212–233.

Zimmermann, E., 1985b. Vocalizations and associated behaviours in adult slow loris (*Nycticebus coucang*). Folia Primatol. 44, 52–64.

Zimmermann, E., 1995. Acoustic communication in nocturnal prosimians. In: Alterman, L., Doyle, G.A., Izard, M.K. (Eds.), Creatures of the Dark: The Nocturnal Prosimians. Plenum Press, New York, NY, pp. 311–330.

Zimmermann, E., 1996. Castration affects the emission of an ultrasonic vocalization in a nocturnal primate, the grey mouse lemur (*Microcebus murinus*). Physiol. Behav. 60, 693–697.

Zimmermann, E., 2009. Primate serenades – communication call variation and implications for species diversity and adaptation in nocturnal prosimians. In: Masters, J., Gamba, M. (Eds.), Leaping Ahead – Advances in Prosimian Biology. Springer, New York, NY (submitted).

Zimmermann, E., Lerch, C., 1993. The complex acoustic design of an advertisement call in male mouse lemurs (*Microcebus murinus*, Prosimii, Primates) and sources of its variation. Ethology 93, 211–224.

Zimmermann, E., Hafen, T.G., 2001. Colony specificity in a social call of mouse lemurs (*Microcebus ssp.*). Am. J. Primatol. 54, 129–141.

Zimmermann, E., Braune, P., Dietz, M., 2007. Ultrasonic songs of a nocturnal primate: specific variations in a complex communication sound convey changes in arousal of male mouse lemurs. 30th International Ethological Conference, Halifax, Canada, 15.8.2007–23.8.2007, p. 141.

Zimmermann, E., Vorobieva, E., Wrogemann, D., Hafen, T., 2000. Use of vocal fingerprinting for specific discrimination of gray (*Microcebus murinus*) and rufous mouse lemurs (*Microcebus rufus*). Int. J. Primatol. 21, 837–852.

CHAPTER 6.4

Rat infant isolation vocalizations and their modulation by social cues as a model of expression of infantile emotionality

Jeff Muller*, Susan Brunelli and Harry Shair

Department of Developmental Neuroscience, New York State Psychiatric Institute and Department of Psychiatry, College of Physicians and Surgeons, Columbia University, New York, NY, USA

Abstract: Social bonds are an important element of mammalian lives. Here we review the pharmacology and functional neuroanatomy of structures controlling infant rat isolation-induced vocalizations. There is strong evidence that these vocalizations reflect an emotional process, because emotion-linked brain regions participate in the infant vocalizations, and anxiolytics reduce the vocalizations. Selective breeding for rate of infant isolation vocalization produces high and low vocalizing lines that also show differences in adult emotional behavior. Modulation of isolation vocalization rate by social cues offers another way to probe the emotional state of the infant. Presentation of littermates and dams in a novel testing chamber reduces a pup's isolation vocalizations. The "quieting" induced by dams versus littermates is differentially dependent on dopamine. In another social modulation paradigm, recent contact with potential caregivers increases the subsequent rate of isolation vocalizations. This "potentiation" of isolation vocalization is also modulated by dopamine, critically in the ventral striatum, which is a region important for reward-related phenomena.

Keywords: infant isolation vocalization; dopamine; D1, D2 receptors; ventral striatum; nucleus accumbens; pharmacology; behavior; rat pups; emotional expression

I. Introduction

We have directed our research efforts to the understanding of an important element of the infant's repertoire in developing social bonds, the cry. Hearing a human child's cry, we can appreciate both powerful feelings evoked in ourselves and sense the emotional state of the child. The cry, as a communication, as a trigger for interaction, and as an act reflecting motivation and initiative, plays a critical role in laying a foundation for healthy social relatedness in humans and other mammals (see Hofer, Chapter 2.3 and Wöhr et al., Chapter 5.2 in this volume). Our approach is to study behavior, physiological processes, and their neural basis in a rodent model (see Fig. 1) to identify basic building blocks from which complex human emotional bonds arise.

We elicit infant cries by isolating a rat pup in unfamiliar surroundings. Acute social isolation often produces crying in human infants. Most other mammalian infants also vocalize when socially isolated. Hearing the cries (isolation vocalizations), rat dams process the calls differently in the brain than other sounds (Ehret, 1987), have physiological responses to the calls (Terkel et al., 1979) and will go out from the nest to retrieve isolated pups (Allin and Banks, 1972; Smotherman et al., 1974; Smith and Sales, 1980; Brunelli et al., 1994).

*Corresponding author.
Dr Jeff Muller, Department of Developmental Neuroscience NYSPI, Unit 40, 1051 Riverside Drive, New York, NY 10032, USA

ISBN 978-0-12-374593-4
DOI: 10.1016/B978-0-12-374593-4.00023-1

Fig. 1. A rat dam and two week old pup as a model of mother–infant interactions in mammals.

Are the isolation-induced calls of rat pups and other infant mammals an expression of emotion? The answer is unknown, if not unknowable. The interpretation of the isolation vocalizations as an affective "distress" response has existed for decades, although it has more recently been critiqued (Fredericson, 1952; Blumberg and Sokoloff, 2001). It is possible to consider the cry as a simple reflexive response to specific cues. Physiological needs do indeed influence vocalization rate. For example, gastric filling reduces isolation-induced vocalization (Nelson and Alberts, 2002). Ambient temperature is an important factor, warmth reduces and cold increases the vocalizations (Allin and Banks, 1971; Oswalt and Meier, 1975; Blumberg et al., 1992). Vocalization does occur in response to isolation in a rat pup without an intact forebrain, although its onset is delayed for many minutes (Middlemis-Brown et al., 2005). Hence, it is possible to regard these data as support for the view that the cry behavior is not associated with an emotional state, but a stimulus response reflex. There are, however, reasons to believe these hypotheses are too simplistic (Newman, 2007). While it would be impractical to rule out every possible reflex, none of the proposed physiological paths has withstood rigorous testing (Hofer and Shair, 1992; Hofer et al., 1993; Shair et al., 2003). Furthermore, these same physiological factors can also be hypothesized to induce complex affective neural states of hunger and cold, possibly dependent on the activity of forebrain and diencephalic structures.

There is also compelling evidence that isolation vocalizations are suppressed by fear. Cues associated with predators suppress isolation vocalizations (Wiedenmayer, 2009). For example, pups reared without exposure to their sire or other adult males will not vocalize in an isolation that is preceded by contact with an adult male (Brunelli et al., 1998). This suppression is mediated by neuroanatomical areas known to underlie fear, such as the amygdala and the periaquecductal gray of the brainstem (Wiedenmayer et al., 2000; Chen et al., 2006). Although suppression of isolation vocalizations by an emotional event is not evidence that these vocalizations themselves are an expression of emotional state, there are other situations in which emotional behaviors compete. For example, the induction of fear by the presentation of cues previously associated with a painful stimulus suppressed bar pressing for food in a hungry rodent (Bouton and Bolles, 1980).

The most compelling arguments that the vocalizations arise from an affective state are first, the prima facia resemblance between human crying with its subjective experience of distress; second, the broad range of animals that emit analogous vocalizations under analogous circumstances; and third, that anxiolytics reduce and anxiogenics increase isolation vocalizations (Insel et al., 1986). Moreover, these vocalizations are associated with increases in other behaviors also associated with separation distress, such as increased activity, urination and self-grooming (Bowlby, 1973; Hofer, 1975; Hofer and Shair, 1987). We will also present in this chapter evidence that rat pup cries are modulated by social factors that are transduced in a manner not dependent on simple, single sensory modality reflexes, consistent with the view that the isolation vocalizations are complex affective phenomena. Furthermore, the demonstration that selection for low or high rates of infant vocalization during isolation also alters adolescent and adult affective behavior supports the idea that the selected genes underlie both infant and adult affective behavior (Brunelli and Hofer, 2007).

Finally, if the vocalizations are an affective phenomenon, then areas of the brain involved in emotion should be necessary for the vocalizations. Several brain areas implicated in emotion are involved in isolation vocalization. Our recent findings show the nucleus accumbens has a critical role in infant isolation-induced vocalization and its social modulation. As will be discussed below, the anterior cingulate and the amygdala are also involved in infant isolation vocalizations.

Certainly, the isolation vocalizations are a complex behavior. Understanding their neural basis will undoubtedly provide a foundation for understanding human affective states, including social bonds. Here we review pharmacological and functional anatomical evidence of the neural basis of isolation-induced vocalizations and their modulation by social factors.

II. Pharmacology of infant rat isolation vocalization

Systemic pharmacological experiments have demonstrated that a number of neurotransmitter systems contribute to infant rat isolation vocalization (for a review see Hofer, 1996). The primary excitatory neurotransmitter glutamate has a role in the production of vocalizations. Systemic administration of an NMDA-antagonist reduced isolation vocalization (Winslow et al., 1990). Also, the primary inhibitory neurotransmitter GABA, and other transmitters that act on GABA's receptor complexes, are involved. Benzodiazepines reduced vocalizations (Insel et al., 1986; Olivier et al., 1998), as did allopregnanolone (Zimmerberg et al., 1994).

Several of the classic modulatory transmitters have also been implicated in isolation-induced infant vocalization. As for norepinephrine, stimulation of α_1 receptors increased vocalization (Hård et al., 1988) as did desipramine, a somewhat selective antagonist of the norepinephrine transporter (Winslow and Insel, 1990). Serotonin (5-HT) also affected isolation vocalizations. 8-OH-DPAT, a relatively specific 5-HT 1A agonist, reduced vocalization (Hård and Engel, 1988) as early as three days postnatally (Joyce and Carden, 1999). Clomipramine, a catecholamine transporter blocker with preferential affinity for the serotonin (5-HT) transporter, decreased vocalizations (Winslow and Insel, 1990). Agonists of 5-HT 1C and 2 receptors also decreased, and a 5-HT 1C and 2 receptor antagonist increased, isolation vocalizations (Winslow and Insel, 1991b). Both dopamine D1 and D2 receptor family agonists reduced isolation vocalizations (Dastur et al., 1999).

Peptidergic transmitters have also been implicated in neural basis of isolation-induced vocalization. Opioid agonists reduced and antagonists increased vocalization (Kehoe and Blass, 1986). However, another report has found no effect of opioid antagonists on vocalization (Winslow and Insel, 1991a). Mu and delta receptor agonists increased, whereas a kappa receptor agonist decreased, vocalization (Carden et al., 1991). Administration of vasopressin into the central nervous system of infant rats decreased vocalization (Winslow and Insel, 1993). Intracerebral ventricular (ICV) administration of oxytocin reduced vocalization (Insel and Winslow, 1991). Further, vocalization was reduced in mice lacking a functional oxytocin gene (Winslow and Insel, 2002). ICV administration of corticotropin releasing hormone (CRH) reduced vocalization and ICV administration of a CRH antagonist increased vocalization (Insel and Harbaugh, 1989). Choleocystokinin (CCK) administration reduced isolation vocalization (Weller and Blass, 1988). A CCK antagonist blocked the ability of intra-oral administration of milk or fat to reduce isolation-induced vocalization (Blass and Shide, 1993). Finally, an antagonist for Substance P's receptor NK1 reduced isolation vocalization in infant guinea pigs (Steinberg et al., 2002). These findings will need to be accounted for in a model of the neural basis of isolation-induced vocalization. Certainly, GABA, benzodiazepines and serotonin, as well as CRH, CCK and Substance P, are all transmitters or hormones that have been implicated in anxiety. Understanding the neural circuits underlying infant isolation vocalizations will provide important information to localize the sources and mechanisms of these pharmacological results.

III. Neuroanatomical substrate of infant isolation vocalization

Any vocalization is a complex behavior recruiting much of the central nervous system. As studied primarily in adults, vocalization requires motor and sensory neurons of the spinal cord and brainstem, as well as nuclei of the brainstem that coordinate activity of the two (Jürgens, 2009); see also Hage, Chapter 8.3 in this volume.

There are relatively few neuroanatomical studies of mammalian infant isolation vocalizations, but most implicate areas known to be involved in emotional behavior. Lesions of the periaqueductal gray, specifically its ventrolateral part, interfere with infant isolation vocalization as with calls in adults (Wiedenmayer et al., 2000). Although infant isolation vocalizations can be produced in rats without intact forebrains, the rate of vocalization is dramatically reduced in the first several minutes of the isolation (Middlemis-Brown et al., 2005). The involvement of the amygdala has been shown in infant guinea pigs where local infusion of a substance P antagonist into the basolateral nucleus attenuated isolation vocalizations (Boyce et al., 2001). We have shown that inactivation of the nucleus accumbens and neighboring ventral striatum with muscimol, a selective $GABA_A$-receptor agonist, completely blocked acute isolation-induced vocalization, both at room temperature and in a 10°C cold environment (Muller, Shair and Moore, submitted manuscript). Associations between vocalization rates and neurotransmitter levels have been studied in lines of rats selectively bred for infant isolation vocalization rates (Brunelli and Hofer, 2007). The details of these follow.

IV. Association of selective breeding for infant isolation vocalization with emotional behavior later in life

To examine the role of heritable factors in infant isolation vocalization rate, rat pups from the N:NIH strain, an outbred mix of eight commonly used laboratory rat strains, were identified at postnatal day 10 for vocalization rate during isolation, and selectively bred in adulthood. The breeding quickly demonstrated that isolation-induced vocalization was a heritable trait as the lines began to diverge in the third generation (Brunelli et al., 1997; Brunelli and Hofer, 2007). Clearly differentiated lines of "low vocalizers," "high vocalizers," and an unselected control line have been maintained for over 30 generations. See Fig. 2 for data up to the twentieth generation.

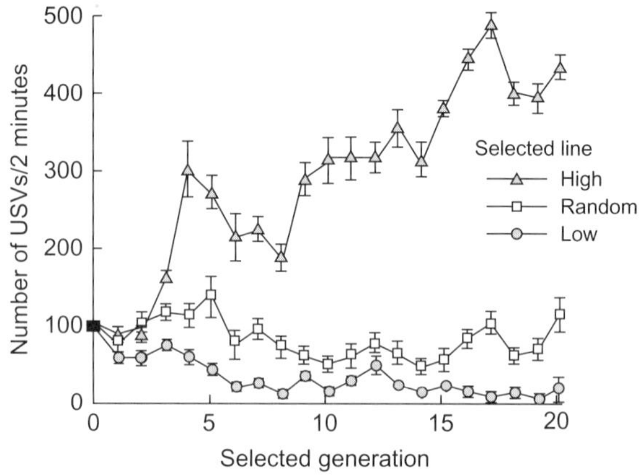

Fig. 2. Selective breeding for infant isolation vocalization. Vocalization rate of rat pups from the N:NIH strain, an outbred mix of eight commonly used laboratory rat strains were identified at postnatal day 10 for vocalization rate during isolation, and selectively bred in adulthood. The lines began to diverge in the third generation. Clearly differentiated lines of "low vocalizers," "high vocalizers" and an unselected-control line have been maintained for over 30 generations. Shown here are mean vocalization rate in two minutes of isolation for each of the first 20 generations of 3 lines: unselected ("Random:" squares), selected for low vocalization rate ("Low:" circles), selected for high vocalization rate ("High:" triangles). Vertical lines indicate SEM. Reprinted from Brunelli (2005).

One goal of this selective breeding was to investigate the behavioral and neurobiological factors linking early and later life affective behaviors. As the first step, we showed that the selected lines differed in later-life affective behaviors. In the open field test, adult males and females from the high line took longer to emerge from a start enclosure, and spent less time in the center of the field than low line males and females, suggesting more anxiety in the high than low line. In the forced swim test, both male and females of the high line spent more time than the low line immobile. Hence, the high line's behavior is also consistent with a depression-like phenotype. As adolescents, both high and low lines made fewer pins and vocalized less during a play session than the unselected group. The low line showed even lower levels of pinning and vocalization than the high line, suggesting that the low line may have a more pronounced deficit in social interactions. Adult male low line animals showed a greater likelihood of fighting than unselected controls, consistent with increased aggression. Among adult female subjects in proestrous, the high line showed more incidents and higher degree of lordosis behavior than did the low line (Frye et al., 2006). High line females, compared to low line, also showed other behaviors associated with successful mating including more exits and longer return latencies following mating. In summary, associations have been found between selection for high levels of infant isolation vocalization and adult anxiety and depression-like behavior. Selection for low levels of infant isolation vocalization are associated with reduced adolescent play behavior, increased adult aggression and reduced sexual performance.

It is tempting to speculate that the lines may be a model for human temperaments that create a predisposition toward externalizing versus internalizing behavior (Weiss et al., 1998; Krueger, 1999). The anxious and "depressed" high line animals might model a predisposition for internalizing. The low line animals, showing aggression and poor social interaction, might model an externalizing predisposition. Much remains to be understood about these associations, but results are consistent with the possibility that selection for genes involved in emotional behavior was successful, which altered both infant and adult measures of emotion.

Neurobiological studies of the selectively bred lines at postnatal day 10 have revealed associations between isolation vocalization rates and alterations of neurotransmitter activity in brain structures mediating emotional behavior (Brunelli and Kehoe, 2005). Low line pups had higher levels of dopamine metabolites and higher turnover rate of dopamine in the anterior cingulate and periaqueductal gray than did unselected controls and the high line, indicating greater dopamine transmission in these areas. In the bed nucleus of the stria terminalis, a region that may be involved in anxiety (Davis and Shi, 1999), serotonin levels were higher in the high and low lines than in the unselected controls. In the striatum, with its links to frontal brain structures involved in affective motor behavior, serotonin metabolites were higher in the high line, hence turnover was also greater.

In adulthood, measurements of allopregnanolone, an endogenous metabolite of progesterone that acts to enhance $GABA_A$ receptor effects, showed that levels of this metabolite varied with behavior and by brain region. In the amygdala, levels were lower in the high line than the low line (Zimmerberg et al., 2005). In the study showing reduced sexual behavior in the low line (Frye et al., 2006), midbrain progesterone and allopregnanolone levels were higher in high line than low line. In hippocampus and amygdala the higher levels of progesterone and its metabolites in the high line were associated with reduced responsiveness to stress; in the midbrain higher levels were associated with more successful sexual behavior exhibited by high line adult females. Together, these infant and adult studies suggest that selection for high and low rates of infant isolation-induced vocalizations have produced lifelong alterations in neurochemical systems and functions of brain structures involved in affective regulation.

V. Contact quieting or comfort response

Thus far in the review, we have considered the neurobiological mechanisms of isolation-induced infant vocalizations. However, an important focus of our research is the study of the modulation of isolation vocalization by recent or ongoing exposure to social cues. Understanding the neural basis of how social cues influence isolation vocalization may illuminate important social and affective processes. Further, as we shall see, the social specificity of some of the behaviors suggests a means of investigating the building blocks of social bonds.

Contact quieting is defined as the reduction in the rate of isolation-induced vocalizations elicited by the presence of a (usually familiar) stimulus, a phenomenon that was demonstrated for odors and textures over 30 years ago (Oswalt and Meier, 1975). The introduction of littermates or the dam into a novel isolation chamber produces an even larger reduction of an infant's isolation calls (Hofer and Shair, 1978). Familiar cues appear to work in an additive fashion: the more cues, the more quieting (Hofer and Shair, 1980). Contact quieting is considered a comfort response, but must be carefully distinguished from fear-related suppression of vocalization. Contact quieting is associated with the presence of appetitive responses (nosing companion) and the absence of fear responses (no freezing).

VI. Neural basis of contact quieting

Perhaps because familiar cues have an additive effect on contact quieting, researchers including us have tended to assume that the reduction in vocalization is regulated by a single set of neural mechanisms, which may be activated to a greater or lesser extent by differing comforting stimuli. We have recently demonstrated this assumption to be false (Shair et al., 2009).

Endogenous dopamine plays a critical role in dam-induced contact quieting. The D1 receptor must be activated within a narrow range of stimulation for the occurrence of contact quieting to the presence of the dam. That is, both D1-family receptor antagonism and agonism can prevent dam-induced reduction in vocalization (see Fig. 3). D2-family receptor antagonism also blocks contact quieting to the dam (see Fig. 4). Unlike D1 receptors, increased activation of D2 receptors with an agonist enhanced contact quieting to the dam (see Fig. 4). Thus, the comfort response of the isolated rat pup to its dam appears to require endogenous activation of both D1 and D2 receptors. Littermate-induced contact quieting, on the other hand, was not affected by injection of the D1 antagonist (Fig. 3) and only partially antagonized by administration of D2 antagonist (Fig. 4). As yet, the necessary locus of dopamine action to allow dam-induced contact quieting is unknown. Unlike the case for maternal potentiation (see below), the nucleus accumbens is not critical. Infusion of a D2 antagonist in the ventral striatum did not prevent contact quieting, nor did it block the effect of a systemic D2 agonist on vocalization (Shair et al., 2009).

The role of other neuroactive substances has also been investigated. A benzodiazepine antagonist does not block contact quieting to either dam or littermate (Carden and Hofer, 1990a,b), nor does the

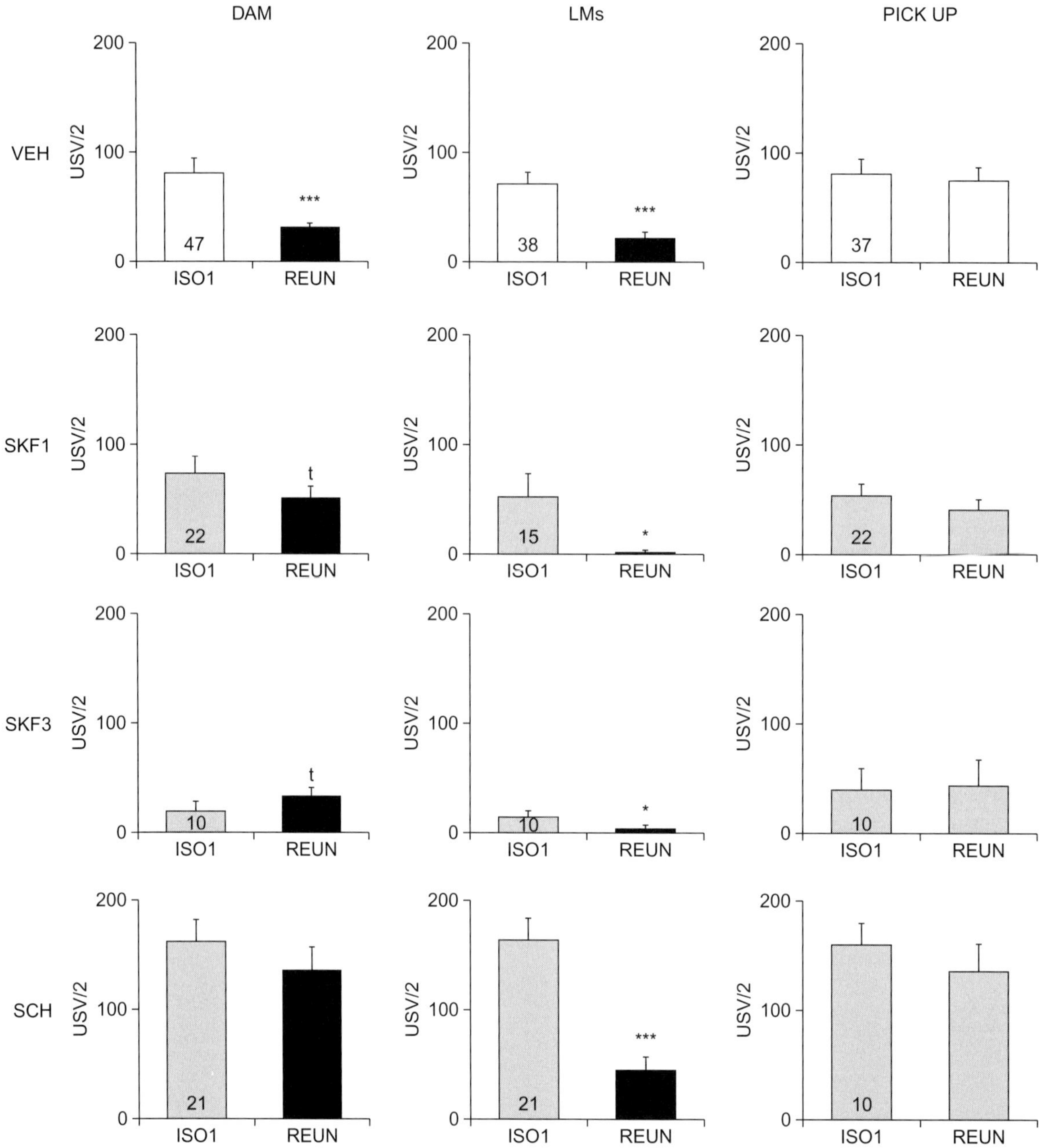

Fig. 3. Effect on contact quieting of DA/D1-selective ligands: mean (±SEM) number of vocalizations by isolated rat pups during a two minute initial isolation (USV/2) in a novel cage (ISO1) and a two minute second period (REUN). Reunions were with the anesthetized dam (DAM) or three anesthetized littermates (LMs). PICK UP pups were handled and replaced in the novel cage for a second isolation as controls. Drug conditions: normal saline (VEH), the D1 agonist SKF81297 (SKF) at one of two doses (SKF1 or SKF3, i.e., 1.0 or 3.0 mg/kg), or the D1 antagonist SCH23390 (SCH, .3 mg/kg). Asterisks indicate a significant change in the number of vocalizations from ISO1 to REUN by t-test; $*p < 0.05$, $***p < 0.001$. The lack of contact quieting is indicated by non-significant changes from ISO1 to REUN. The number inside the ISO1 bar is the number of animals tested in that condition. Reprinted from Shair et al. (2009).

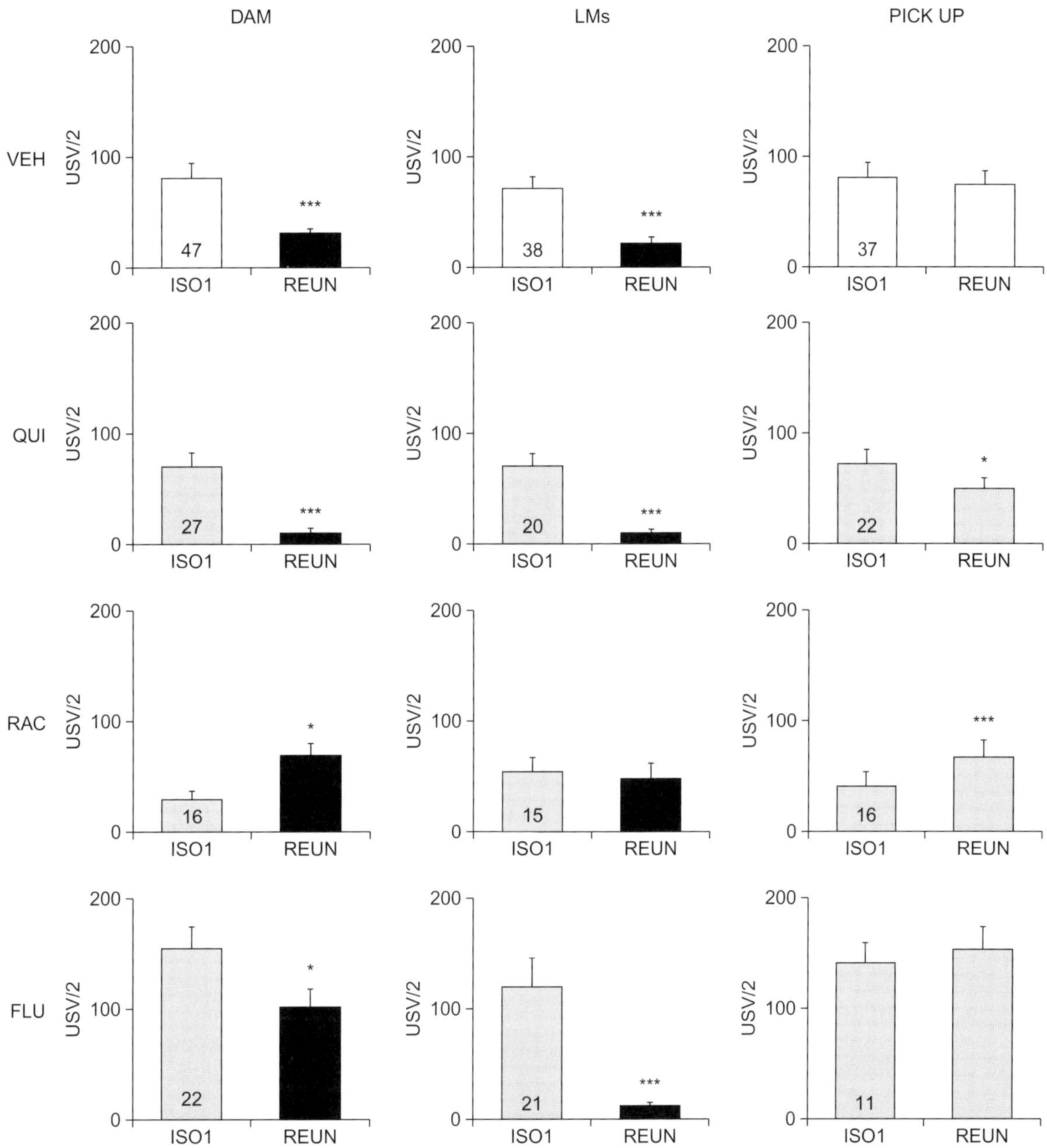

Fig. 4. Effect on contact quieting of DA D2-selective ligands and a dual D1/D2 antagonist: mean (±SEM) number of vocalizations by isolated rat pups during a two minute initial isolation (USV/2) in a novel cage (ISO1) and a two minute second period (REUN). Reunions were with the anesthetized dam (DAM) or three anesthetized littermates (LMs). PICK UP pups were handled and replaced in the novel cage for a second isolation as controls. Drug conditions: normal saline (VEH, data are the same as in Fig. 1), the D2 agonist quinpirole (QUI, 1.0 mg/kg), the D2 antagonist raclopride (RAC, 1.0 mg/kg), or the dual antagonist alpha flupenthixol (FLU, 0.33 mg/kg). Asterisks indicate a significant change in the number of vocalizations from ISO1 to REUN by t-test; $*p < 0.05$, $***p < 0.001$. "t" indicates $p < 0.10$. The number inside the ISO1 bar is the number of animals tested in that condition. The lack of contact quieting is indicated by non-significant changes or significant increases from ISO1 to REUN. Reprinted from Shair et al. (2009).

cholescystikinen antagonist devazepide block contact quieting to the dam (not tested with littermates) (Weller and Gispan, 2000). Various hypotheses have been advanced about the role of endogenous opioids in contact quieting. It may be that the distress of isolation recruits compensatory coping mechanisms, one of which involves the reduction of distress by opioid release (Winslow and Insel, 1991a). Consistent with this idea, there is, in fact, a reduction in rate of vocalization over time in isolation. Another (not mutually exclusive) possibility is that the contact itself elicits opioid release which is responsible for the reduction in vocalization (Nelson and Panksepp, 1998). There is general agreement that systemically-administered opioid agonists reduce isolation vocalization, but the evidence for a role of endogenous opioids in contact quieting is less clear. An early report that naltrexone reduced the quieting effect of the dam (Carden and Hofer, 1990a) has not been supported in other work (Blass et al., 1990; Shair et al., 2005). There is more support for a role of opioid release in contact quieting to littermates (Carden and Hofer, 1990a; Carden et al., 1996), but see also the study by Winslow and Insel (1991a).

VII. Maternal and paternal potentiation

One behavioral phenomenon that most directly links isolation-induced vocalization to social and emotional behavior is that the rate of isolation-induced vocalization can be enhanced by recent contact with the dam, but not by other social stimuli that elicit contact quieting, e.g., littermates or home cage shavings (Hofer et al., 1994). After a short exposure to the parent, the number of emitted isolation calls by a pup dramatically increases compared to the number of vocalizations in an isolation not preceded by such contact. This paradigm, known as maternal potentiation, may reflect the modulation of an affective state by a caregiver's social cues. In fact, the enhancement of vocalization is found to be elicited not only by the dam, but by other adult females from the colony, consistent with the species-specific observation that pups can be cared for by females other than their dam (Brunelli et al., 1998). Maternal potentiation is not present at birth, and emerges only after the first week of life, which is consistent with the idea that potentiation is experience-dependent (Hofer et al., 1998). Furthermore, pups reared only by the dam showed a fear response on exposure to an adult male during the second and third weeks of life (Takahashi, 1994). If, on the other hand, a pup had experienced the presence of the sire in its first weeks of life, the sire elicited the enhancement of vocalization rate, i.e., paternal potentiation (Brunelli et al., 1998). We presume that some testosterone-related odor is an important stimulus, as the presence of a castrated adult male during early life does not lead to potentiation of isolation vocalization after contact with intact adult males (Brunelli et al., 1998). Given the greater chance of survival provided by the dam, it is not surprising that the pup can distinguish between its dam and littermates. So, too, other adults offer protection and care that littermates cannot. Because of the limited number of social stimuli that elicit potentiation and the role of early experience in altering those stimuli, we and others have hypothesized that potentiation of isolation-induced vocalization is a marker for an early attachment-like social bond.

Our usual experimental procedure is to isolate a rat pup briefly, give it a brief interaction with a stimulus animal, and then a brief reisolation. Times of each epoch usually range from 1 to 5 minutes. We have shown that the first isolation period is not necessary for potentiation to occur. Thus, it is not the mere repeated testing that enhances vocalization. Similarly, we have shown that temperature of the dam is not critical. A similarly warm litterpile does not elicit potentiation, but an anesthetized, cold (10°C) dam does (Hofer et al., 1993). Anosmic pups can show potentiation to an awake behaving dam, but not an anesthetized dam (Shair et al., 1999), suggesting that both somatosensory and olfactory information about the dam's presence can play a role in the pup's response. Finally, although we define potentiation by an increase in the rate of vocalizations, potentiated vocalizations are louder, tend to be longer, and have a different bout structure than vocalizations in the first isolation (Myers et al., 2004). A recent review provides a more complete description of what is known of the formation and expression of the potentiation (Shair, 2007). As considered in an earlier chapter of this volume by Hofer (Chapter 2.3), potentiation of vocalization may represent an evolutionarily-selected balance between vocalizing to promote survival and suppression of vocalization to avoid predator detection. Selection pressure probably promotes vocalization when circumstances make it most likely to elicit retrieval, i.e., when information exists that the dam has recently been present. Suppression of vocalization, in contrast, is associated with stimulus cues associated with predators, such as unrelated adult male rats and other predatory species (Wiedenmayer and Barr, 2001).

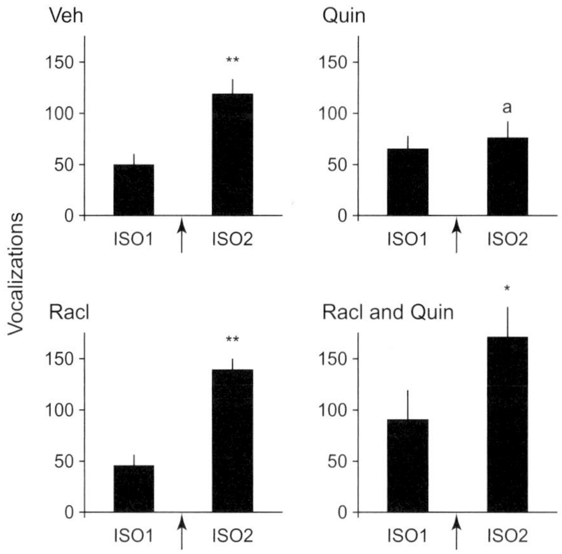

Fig. 5. Effect on maternal potentiation of DA D2 selective ligands. Mean (±SEM) number of vocalizations during a two minute initial isolation (ISO1) and a two minute reisolation (ISO2). Arrows indicate a two minute reunion with the anesthetized dam between isolations. In the group that received a systemic injection of vehicle (Veh), the pups called more in the second isolation than in the first, a phenomenon called "maternal potentiation." Injection with the D2 agonist quinpirole (Quin; 1 mg/kg) disrupted potentiation without preventing isolation-induced vocalization in either isolation. Injection of the D2 antagonist raclopride (Racl; 1 mg/kg) did not prevent potentiation. Raclopride (1 mg/kg) restored potentiation when injected before quinpirole (1 mg/kg) (Racl and Quin). An "a" indicates that the Quin group's difference from ISO1 to ISO2 was significantly less than the vehicle group's ($p = 0.003$). Asterisks indicate a significant increase in number of vocalizations from ISO1 to ISO2. *$p < 0.01$. **$p < 0.001$. Reprinted from Muller et al. (2005).

VIII. Neural basis of maternal potentiation

It has been known that dopamine agonists decrease isolation-induced vocalizations (Dastur et al., 1999), but any more specific role in vocalizations was unknown. Our work has now demonstrated that the D2 agonist quinpirole, systemically administered, selectively disrupted maternal potentiation at doses that do not influence vocalization in the first isolation (Muller et al., 2005) (see Fig. 5). A systemically-administered D1 agonist SKF81297 also significantly reduced potentiation, but at doses that also dramatically reduced the underlying isolation vocalizations (Muller et al., 2009) (see Fig. 6). In other words, activation of D2 receptors appears to produce a far more selective effect on disruption of maternal potentiation than does activation of D1 receptors. The role of endogenous dopamine remains to be clarified. Systemic administration of a dual D1 and D2 antagonist alpha flupenthixol antagonized maternal potentiation, but at doses that significantly reduced locomotion, raising the possibility of non-specific effects (Muller et al., 2009) (see Fig. 7).

Mu-opioid receptors appear to have no selective role in the expression of maternal potentiation (Shair et al., 2005), as seen by pharmacological antagonism. However, using mu-opioid knockout mice (in which the mu-opioid gene has been replaced by a nucleotide sequence that does not produce functional protein), it has been shown that the absence of such receptors prevents the development of maternal potentiation (Moles et al., 2004).

Progress identifying the neural basis of maternal potentiation supports the view of the affective basis of the phenomenon. The nucleus accumbens is implicated in the rewarding aspect of feeding and drugs of abuse (Berridge, 1996; Robbins and Everitt, 1996; Ikemoto and Panksepp, 1999; Salamone, 2002; Kelley et al., 2005) and its orthologous structure in birds is implicated in song production (Doupe et al., 2005). We have shown that the nucleus accumbens is implicated in maternal potentiation. Local injection of the D2 agonist quinpirole in the accumbens disrupted maternal potentiation (Muller et al., 2008), although quinpirole injection into the dorsal striatum had no such effect (see Fig. 8).

Taken together, the pharmacological and local injection results suggest an important role for D2 receptors, particularly in the ventral striatum, in maternal potentiation. D2 receptor activity in the nucleus accumbens must be at a low level for the expression of potentiation. What role endogenous dopamine has in the phenomenon has not been clarified by results from the application of the systemic D1/D2 dual antagonist because, although potentiation was disrupted, it may have been due to non-specific effects on arousal. It is worth noting that in spite of the possibly sedating effects of the systemic dual antagonist, the isolation-induced vocalization rate increased above vehicle levels, demonstrating that at least some isolation vocalization behavior is independent of the sedating effects of dopamine antagonists.

One explanation for the role of dopamine in maternal potentiation may be supported by the recent finding that transient changes in dopaminergic firing are linked to the reward value of stimuli (Schultz, 2007). Cues predictive of reward elicited a transient increase in dopaminergic cell firing. The absence of expected

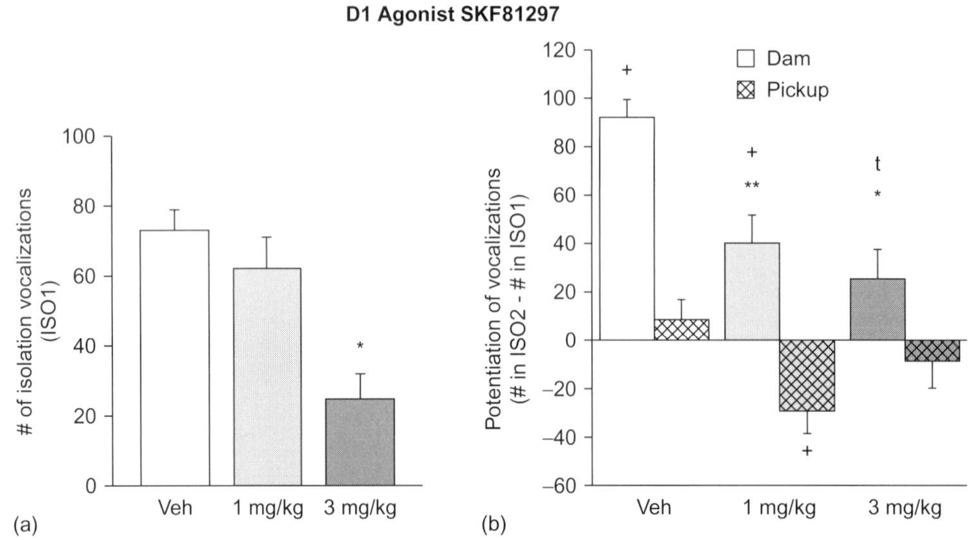

Fig. 6. Effect of a D1-family agonist on isolation vocalizations and maternal potentiation. (a) Mean (±SEM) number of vocalizations during a two minute initial isolation (ISO1). Injection with the D1 agonist SKF81297 reduced vocalization in a dose dependent manner. The 3 mg/kg dose was significantly lower than vehicle (*$p < 0.05$). (b) Mean (±SEM) difference in number of vocalizations between an initial isolation (ISO1) and a reisolation (ISO2) that followed a two minute period either with the anesthetized dam (Dam) or without (Pickup control). Compared to the appropriate vehicle control group (Veh), the D1 receptor agonist lowered the vocalization change score in most dose and reunion conditions. The "+" symbol indicates a significant difference from 0 for that group, at least $p < 0.05$; "t" (trend) indicates $p < 0.1$. Asterisks indicate that a dam-reunion drug-injected group is significantly less than the dam-reunion vehicle group by Dunnett's *post hoc* test. *$p < 0.05$, **$p < 0.01$. Reprinted from Muller et al. (2009).

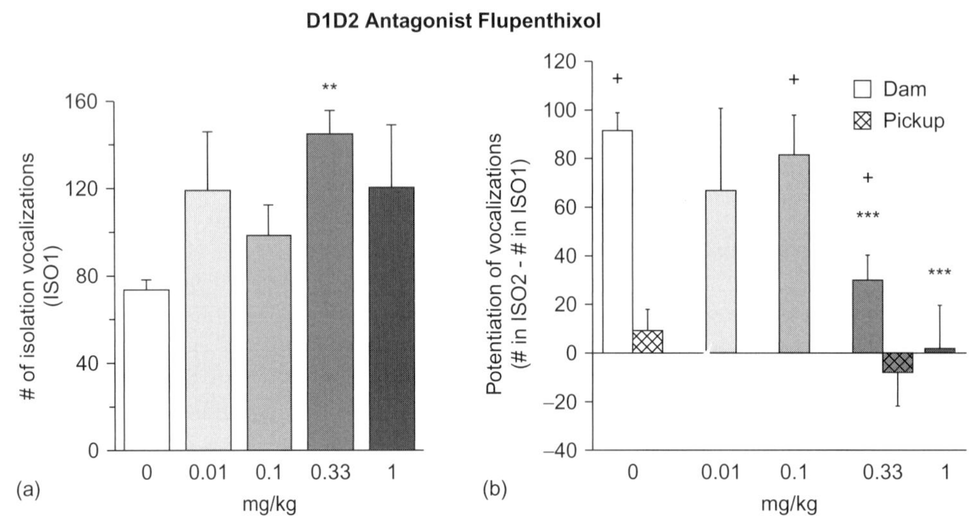

Fig. 7. Effect of a combined D1 and D2 antagonist on isolation vocalization and maternal potentiation. (a) Mean (±SEM) number of vocalizations during a two minute initial isolation (ISO1). Injection with the dual antagonist, flupenthixol (doses 0.01–1.0 mg/kg) increased vocalization (**$p < 0.01$). (b) Mean (±SEM) difference in number of vocalizations between an initial isolation and a reisolation that followed two minute reunion phase either with the anesthetized dam (Dam) or a pickup condition (Pickup-pup handling as in the dam-contact reunion, but without the dam) (ISO2–ISO1). Flupenthixol reduced potentiation at the 0.33 mg/kg dose and eliminated it at the 1 mg/kg dose. Asterisks indicate that a drug group is significantly different the dam-reunion vehicle group, ***$p < 0.001$. "+" indicates ISO2–ISO1 is significantly different than 0. Reprinted from Muller et al. (2009).

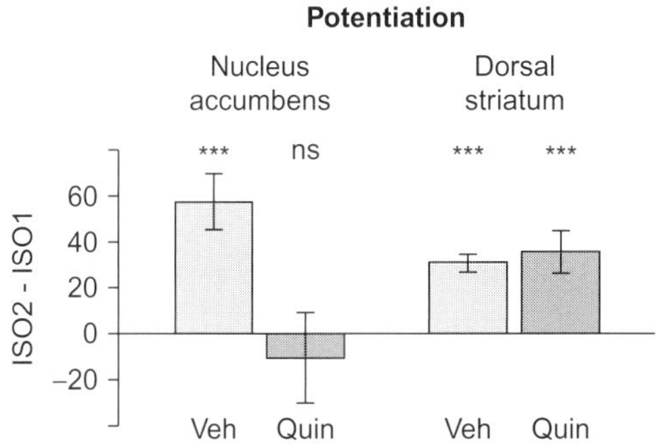

Fig. 8. Effect of a D2 agonist in the ventral striatum (nucleus accumbens) on maternal potentiation. Isolation-induced vocalizations are counted in two isolations, one preceded by contact with littermates (ISO1) and a second preceded by contact with the dam (ISO2). Difference in potentiation scores is displayed on the y-axis, derived by subtracting ISO1 from ISO2. Unlike vehicle injections (Veh) and quinpirole injections (Quin) into the dorsal striatum, quinpirole injections in the ventral striatum (nucleus accumbens) abolished potentiation. The interaction of change score by drug presence by injection location was significant $F(1, 32) = 9.4$, $p = 0.0044$: quinpirole injected into the ventral striatum disrupted potentiation (paired t-test, $t(18) = 0.55$, $p = 0.59$); but in all other conditions potentiation was present (saline in the ventral striatum paired t-test, $t(18) = 4.92$, $p < 0.001$; saline in dorsal striatum paired t-test, $t(14) = 7.79$, $p < 0.001$; quinpirole in dorsal striatum paired t-test, $t(14) = 3.66$, $p < 0.003$). ns, not significant; ***$p < 0.001$. There were 19 pups in the ventral striatum group, 15 pups in the dorsal striatum group. Reprinted from Muller et al. (2008).

reward elicited a decrease in dopaminergic cell firing below the baseline rate. Based on this finding, we hypothesize that the dam's presence could trigger a transient increase in dopamine release, and her unexpected removal could trigger a decrease in dopamine level. It is known that the dam can serve as a reward for pup's operant behavior (Amsel et al., 1977), and dopamine is implicated in the pup's learning of this task (McDougall et al., 1991). We do not yet have evidence of the role of the dam's reward value in maternal potentiation. Whether this putative decrease in dopamine release is permissive for another process which is responsible for potentiation, or whether the drop in dopamine itself is responsible for the increased vocalization also remains to be clarified. This latter hypothesis is possible because, in spite of the finding that neither a D1 nor a D2 antagonist by itself disrupts potentiation, the possibility remains that antagonism of both receptor families simultaneously is needed to disrupt potentiation. In fact, as described above, we found that a dual antagonist of both D1 and D2 receptor families does block potentiation (Muller et al., 2009). The effects of the dual antagonism need further investigation, however, because the doses that prevent maternal potentiation also drastically reduce locomotor activity, suggesting a non-specific sedating effect may be involved.

IX. Conclusions

We have presented evidence consistent with the idea that isolation vocalizations are an emotional phenomenon, representing a critical building block of human social bonds. Systemic administration of anxiolytics reduces and administration of anxiogenics increases the vocalizations. Selective breeding for low and high levels of isolation vocalization have produced a high line with anxious and depression-like behavior, and a low line with heightened aggression and social deficits. Brain regions implicated in emotion are critical to the isolation vocalizations, including the periaqueductal gray, the amygdala, the anterior cingulate and the ventral striatum. The affective behavioral differences in the selected lines are associated with differences in levels of allopregnanalone, which acts on the GABA-receptor complex, in the midbrain, amygdala and hippocampus. Serotonin turnover is altered in the bed nucleus of the stria terminalis and the striatum. Dopamine turnover differs in the anterior cingulated and periaqueductal gray. Other recent work has demonstrated that the reduction of isolation vocalizations by the presentation of littermates and dams are two distinct underlying processes differentially dependent on dopamine. Finally, the potentiation of isolation vocalizations by potential caregivers is also modulated by dopamine, critically in the ventral striatum.

Many questions continue to interest us. To what extent are the cues that trigger isolation vocalization hardwired? To what degree does experience play a role in regulating what will evoke or ameliorate isolation distress? To what degree do different social companions differentially evoke both the states of security and distress? Is this infant state a marker for temperamental difference linked to lifelong traits such as anxiety, depression, sociability, sexuality and aggression? What are the neurobiological correlates underlying these state–trait associations? What role do gene–environment interactions play in their development? Is sadness a meaningful description of the state induced by infant social isolation? As such, could it be useful in understanding the neurobiology of depression in which prolonged, unrelenting sadness is a key symptom? For our purposes, these questions are best considered heuristic ones that help us consider the overall direction and goals of our work. We plan to test the hypothesis that the reward value of the dam is responsible for its enhancement of isolation vocalization by manipulating the reward value of the dam and of otherwise neutral stimuli. We expect that the knowledge gained will provide insight into human social bonds and their role in human health and pathology.

References

Allin, J.T., Banks, E.M., 1972. Functional aspects of ultrasound production by infant albino rats (*Rattus norvegicus*). Anim. Behav. 20 (1), 175–185.

Amsel, A., Radek, C.C., Graham, M., Letz, R., 1977. Ultrasound emission in infant rats as an indicant of arousal during appetitive learning and extinction. Science 197 (4305), 786–788.

Berridge, K.C., 1996. Food reward: brain substrates of wanting and liking. Neurosci. Biobehav. Rev. 20 (1), 1–25.

Blass, E.M., Shide, D.J., 1993. Endogenous cholecystokinin reduces vocalization in isolated 10-day-old rats. Behav. Neurosci. 107 (3), 488–492.

Blass, E.M., Fillion, T.J., Weller, A., Brunson, L., 1990. Separation of opioid from nonopioid mediation of affect in neonatal rats: nonopioid mechanisms mediate maternal contact influences. Behav. Neurosci. 104 (4), 625–636.

Bouton, M.E., Bolles, R.C., 1980. Conditioned fear assessed by freezing and by the suppression of three different baselines. Anim. Learn. Behav. 8, 429–434.

Bowlby, J., 1973. Attachment and Loss, Vol. 2: Separation, Anxiety and Anger. Basic Books, New York, NY.

Boyce, S., Smith, D., Carlson, E., Hewson, L., Rigby, M., O'Donnell, R., Harrison, T., Rupniak, N.M., 2001. Intraamygdala injection of the substance P [NK(1) receptor] antagonist L-760735 inhibits neonatal vocalisations in guinea-pigs. Neuropharmacology 41 (1), 130–137.

Brunelli, S.A., 2005. Selective breeding for an infant phenotype: rat pup ultrasonic vocalization (USV). Behav. Genet. 35 (1), 53–65.

Brunelli, S.A., Kehoe, P., 2005. Differences in catecholamine utilization at postnatal day 10 in rats. Paper presented at the 38th Annual Meeting of the International Society for Developmental Psychobiology, Washington, DC.

Brunelli, S.A., Hofer, M.A., 2007. Selective breeding for infant rat separation-induced ultrasonic vocalizations: developmental precursors of passive and active coping styles. Behav. Brain Res. 182 (2), 193–207.

Brunelli, S.A., Shair, H.N., Hofer, M.A., 1994. Hypothermic vocalizations of rat pups (*Rattus norvegicus*) elicit and direct maternal search behavior. J. Comp. Psychol. 108 (3), 298–303.

Brunelli, S.A., Masmela, J.R., Shair, H.N., Hofer, M.A., 1998. Effects of biparental rearing on ultrasonic vocalization (USV) responses of rat pups (*Rattus norvegicus*). J. Comp. Psychol. 112 (4), 331–343.

Carden, S.E., Barr, G.A., Hofer, M.A., 1991. Differential effects of specific opioid receptor agonists on rat pup isolation calls. Brain Res. Dev. Brain Res. 62 (1), 17–22.

Davis, M., Shi, C., 1999. The extended amygdala: are the central nucleus of the amygdala and the bed nucleus of the stria terminalis differentially involved in fear versus anxiety? Ann. N.Y. Acad. Sci. 877, 281–291.

Doupe, A.J., Perkel, D.J., Reiner, A., Stern, E.A., 2005. Birdbrains could teach basal ganglia research a new song. Trends Neurosci. 28 (7), 353–363.

Ehret, G., 1987. Left hemisphere advantage in the mouse brain for recognizing ultrasonic communication calls. Nature 325 (6101), 249–251.

Hård, E., Engel, J., 1988. Effects of 8-OH-DPAT on ultrasonic vocalization and audiogenic immobility reaction in pre-weanling rats. Neuropharmacol. 27 (10), 981–986.

Hård, E., Engel, J., Lindh, A.S., 1988. Effect of clonidine on ultrasonic vocalization in preweaning rats. J. Neural. Transm. 73 (3), 217–237.

Hofer, M.A., 1975. Studies on how early maternal separation produces behavioral change in young rats. Psychosom. Med. 37 (3), 245–264.

Hofer, M.A., 1996. Multiple regulators of ultrasonic vocalization in the infant rat. Psychoneuroendocrinology 21 (2), 203–217.

Hofer, M.A., Shair, H.N., 1978. Ultrasonic vocalization during social interaction and isolation in 2-week-old rats. Dev. Psychobiol. 11 (5), 495–504.

Hofer, M.A., Shair, H.N., 1980. Sensory processes in the control of isolation-induced ultrasonic vocalization by 2-week-old rats. J. Comp. Physiol. Psychol. 94 (2), 271–279.

Hofer, M.A., Shair, H.N., 1987. Isolation distress in two-week-old rats: influence of home cage, social companions, and prior experience with littermates. Dev. Psychobiol. 20 (4), 465–476.

Hofer, M.A., Brunelli, S.A., Shair, H.N., 1993. Ultrasonic vocalization responses of rat pups to acute separation and contact comfort do not depend on maternal thermal cues. Dev. Psychobiol. 26 (2), 81–95.

Hofer, M.A., Brunelli, S.A., Shair, H.N., 1994. Potentiation of isolation-induced vocalization by brief exposure of rat pups to maternal cues. Dev. Psychobiol. 27 (8), 503–517.

Hofer, M.A., Masmela, J.R., Brunelli, S.A., Shair, H.N., 1998. The ontogeny of maternal potentiation of the infant rats' isolation call. Dev. Psychobiol. 33 (3), 189–201.

Ikemoto, S., Panksepp, J., 1999. The role of nucleus accumbens dopamine in motivated behavior: a unifying interpretation with special reference to reward-seeking. Brain Res. Rev. 31, 6–41.

Insel, T.R., Harbaugh, C.R., 1989. Central administration of corticotropin releasing factor alters rat pup isolation calls. Pharmacol. Biochem. Behav. 32 (1), 197–201.

Insel, T.R., Winslow, J.T., 1991. Central administration of oxytocin modulates the infant rat's response to social isolation. Eur. J. Pharmacol. 203 (1), 149–152.

Insel, T.R., Hill, J.L., Mayor, R.B., 1986. Rat pup ultrasonic isolation calls: possible mediation by the benzodiazepine receptor complex. Pharmacol. Biochem. Behav. 24 (5), 1263–1267.

Joyce, M.P., Carden, S.E., 1999. The effects of 8-OH-DPAT and (+/−)-pindolol on isolation-induced ultrasonic vocalizations in 3-, 10-, and 14-day-old rats. Dev. Psychobiol. 34 (2), 109–117.

Jürgens, U., 2009. The neural control of vocalization in mammals: a review. J. Voice 23 (1), 1–10.

Kehoe, P., Blass, E.M., 1986. Opioid-mediation of separation distress in 10-day-old rats: reversal of stress with maternal stimuli. Dev. Psychobiol. 19 (4), 385–398.

Kelley, A.E., Baldo, B.A., Pratt, W.E., Will, M.J., 2005. Corticostriatal-hypothalamic circuitry and food motivation: integration of energy, action and reward. Physiol. Behav. 86 (5), 773–795.

McDougall, S.A., Nonneman, A.J., Crawford, C.A., 1991. Effects of SCH 23390 and sulpiride on the reinforced responding of the young rat. Behav. Neurosci. 105 (5), 744–754.

Middlemis-Brown, J.E., Johnson, ED., Blumberg, M.S., 2005. Separable brainstem and forebrain contributions to ultrasonic vocalizations in infant rats. Behav. Neurosci. 119 (4), 1111–1117.

Moles, A., Kieffer, B.L., D'Amato, F.R., 2004. Deficit in attachment behavior in mice lacking the mu-opioid receptor gene. Science 304 (5679), 1983–1986.

Muller, J.M., Moore, H., Myers, M.M., Shair, H.N., 2008. Ventral striatum dopamine D2 receptor activity inhibits rat pups' vocalization response to loss of maternal contact. Behav. Neurosci. 122 (1), 119–128.

Muller, J.M., Moore, H., Myers, M.M., Shair, H.N., 2009. Dopamine's role in social modulation of infant isolation-induced vocalization: II. Maternally modulated infant separation responses are regulated by D1- and D2-family dopamine receptors. Dev. Psychobiol. 51 (2), 158–172.

Muller, J.M., Brunelli, S.A., Moore, H., Myers, M.M., Shair, H.N., 2005. Maternally modulated infant separation responses are regulated by D2-family dopamine receptors. Behav. Neurosci. 119 (5), 1384–1388.

Myers, M.M., Ali, N., Weller, A., Brunelli, S.A., Tu, A. Y., Hofer, M.A., et al., 2004. Brief maternal interaction increases number, amplitude, and bout size of isolation-induced ultrasonic vocalizations in infant rats (*Rattus norvegicus*). J. Comp. Psychol. 118 (1), 95–102.

Nelson, E.E., Panksepp, J., 1998. Brain substrates of infant–mother attachment: contributions of opioids, oxytocin, and norepinephrine. Neurosci. Biobehav. Rev. 22 (3), 437–452.

Nelson, E.E., Alberts, J.R., 2002. Gastric saline infusion reduces ultrasonic vocalizations and brown fat activity in suckling rat pups. Dev. Psychobiol. 40 (2), 160–167.

Newman, J.D., 2007. Neural circuits underlying crying and cry responding in mammals. Behav. Brain Res. 182 (2), 155–165.

Oswalt, G.L., Meier, G.W., 1975. Olfactory, thermal, and tactual influences on infantile ultrasonic vocalization in rats. Dev. Psychobiol. 8 (2), 129–135.

Robbins, T.W., Everitt, B.J., 1996. Neurobehavioural mechanisms of reward and motivation. Curr. Opin. Neurobiol. 6 (2), 228–236.

Salamone, J.D., 2002. Functional significance of nucleus accumbens dopamine: behavior, pharmacology and neurochemistry. Behav. Brain Res. 137 (1–2), 1.

Schultz, W., 2007. Behavioral dopamine signals. Trends Neurosci. 30 (5), 203–210.

Shair, H.N., 2007. Acquisition and expression of a socially mediated separation response. Behav. Brain Res. 182 (2), 180–192.

Shair, H.N., Masmela, J.R., Hofer, M.A., 1999. The influence of olfaction on potentiation and inhibition of ultrasonic vocalization of rat pups. Physiol. Behav. 65 (4–5), 769–772.

Shair, H.N., Brunelli, S.A., Hofer, M.A., 2005. Lack of evidence for mu-opioid regulation of a socially mediated separation response. Physiol. Behav. 83 (5), 767–777.

Shair, H.N., Muller, J.M., Moore, H., 2009. Dopamine's role in social modulation of infant isolation-induced vocalization: I. Reunion responses to the dam, but not littermates, are dopamine dependent. Dev. Psychobiol. 51 (2), 131–146.

Smith, J.C., Sales, D.G., 1980. Ultrasonic behavior and mother-infant interactions in rodents. In: Bell, R.W., Smotherman, W.P. (Eds.), Maternal Influences and Early Behavior. Spectrum Publications, New York, NY, pp. 105–134.

Smotherman, W.P., Bell, R.W., Starzec, J., Elias, J., Zachman, T.A., 1974. Maternal responses to infant vocalizations and olfactory cues in rats and mice. Behav. Biol. 12 (1), 55–66.

Takahashi, L.K., 1994. Stimulus control of behavioral inhibition in the preweanling rat. Physiol. Behav. 55 (4), 717–721.

Terkel, J., Damassa, D.A., Sawyer, C.H., 1979. Ultrasonic cries from infant rats stimulate prolactin release in lactating mothers. Horm. Behav. 12 (1), 95–102.

Weller, A., Blass, E.M., 1988. Behavioral evidence for cholecystokinin-opiate interactions in neonatal rats. Am. J. Physiol. 255 (6 Pt 2), R901–R907.

Weller, A., Gispan, I.H., 2000. A cholecystokinin receptor antagonist blocks milk-induced but not maternal-contact-induced decrease of ultrasonic vocalization in rat pups. Dev. Psychobiol. 37 (1), 35–43.

Wiedenmayer, C.P., 2009. Plasticity of defensive behavior and fear in early development. Neurosci. Biobehav. Rev. 33 (3), 432–441.

Wiedenmayer, C.P., Barr, G.A., 2001. Developmental changes in responsivity to threat are stimulus-specific in rats. Dev. Psychobiol. 39 (1), 1–7.

Wiedenmayer, C.P., Goodwin, G.A., Barr, G.A., 2000. The effect of periaqueductal gray lesions on responses to age-specific threats in infant rats. Brain Res. Dev. Brain Res. 120 (2), 191–198.

Winslow, J.T., Insel, T.R., 1990. Serotonergic and catecholaminergic reuptake inhibitors have opposite effects on the ultrasonic isolation calls of rat pups. Neuropsychopharmacology 3 (1), 51–59.

Winslow, J.T., Insel, T.R., 1991a. Endogenous opioids: do they modulate the rat pup's response to social isolation? Behav. Neurosci. 105 (2), 253–263.

Winslow, J.T., Insel, T.R., 1991b. Serotonergic modulation of the rat pup ultrasonic isolation call: studies with 5HT1 and 5HT2 subtype-selective agonists and antagonists. Psychopharmacology (Berl.) 105 (4), 513–520.

Winslow, J.T., Insel, T.R., 1993. Effects of central vasopressin administration to infant rats. Eur. J. Pharmacol. 233 (1), 101–107.

Winslow, J.T., Insel, T.R., 2002. The social deficits of the oxytocin knockout mouse. Neuropeptides 36 (2–3), 221–229.

Winslow, J.T., Insel, T.R., Trullas, R., Skolnick, P., 1990. Rat pup isolation calls are reduced by functional antagonists of the NMDA receptor complex. Eur. J. Pharmacol. 190 (1–2), 11–21.

Zimmerberg, B., Brunelli, S.A., Hofer, M.A., 1994. Reduction of rat pup ultrasonic vocalizations by the neuroactive steroid allopregnanolone. Pharmacol. Biochem. Behav. 47 (3), 735–738.

Zimmerberg, B., Brunelli, S.A., Fluty, A.J., Frye, C.A., 2005. Differences in affective behaviors and hippocampal allopregnanolone levels in adult rats of lines selectively bred for infantile vocalizations. Behav. Brain Res. 159 (2), 301–311.

SECTION 7

Hypothalamic/Limbic Integrative Function for Vocal/Behavioral Outcome

The hypothalamic and related diencephalic and mesencephalic limbic regions are the primary structures responsible for generation of vocalizations and concomitant affective states. This system includes somatic, autonomic and endocrine mechanisms for an integrated behavioral outcome in affective situations, including pain. The hypothalamic and limbic structures are controlled by the ascending reticular activating system, which is capable of changing the state of the organism and initiating affective states with concomitant vocalization. This system is also sensitive to hormonal and pheromonal regulation.

CHAPTER 7.1

Limbic, hypothalamic and periaqueductal gray circuitry and mechanisms controlling rage and vocalization in the cat

Allan Siegel*, Suresh Bhatt, Rekha Bhatt and Steven S. Zalcman

Departments of Neurology & Neuroscience and Psychiatry, UMDNJ-New Jersey Medical School, Newark, NJ, USA

Abstract: There are several forms of vocalization present in the cat which include meowing, howling, growling and hissing. The present chapter focuses on one type of vocalization, namely hissing, that occurs in response to a real or perceived threat as part of an overall behavioral pattern referred to as "defensive rage behavior." Specific attention is given to analysis of the neural substrates and mechanisms by which this response is mediated. A distinction is made between two classes of structures: (1) those associated with the integration and expression of defensive rage behavior; and (2) those associated with the modulation of those structures from which this behavior is elicited. The first class includes the medial hypothalamus and midbrain periaqueductal gray (PAG), and the second class includes limbic and related structures. The major focus considers the principal underlying neuroanatomical circuitry and neurochemical mechanisms associated with defensive rage. Integration of defensive rage takes place in the medial hypothalamus and a similar response mechanism is expressed in the dorsal PAG, which receives direct inputs from the medial hypothalamus. The vocalization component of the defensive rage response is mediated through descending fibers that supply the motor nuclei of cranial nerves V and VII. Potentiation or suppression of this behavior is modulated through limbic nuclei, which include the amygdala, septal area and prefrontal cortex that supply the medial hypothalamus and/or PAG. Neurotransmitters that facilitate defensive rage by acting within the medial hypothalamus or PAG include glutamate, substance P, catecholamines, serotonin (acting through 5-HT_2 receptors), acetylcholine and cholecystokinin. Neurotransmitters that suppress this behavior include enkephalins, GABA and serotonin (acting through 5-HT_{1A} receptors). In addition, recent evidence has shown that cytokines present in the medial hypothalamus and PAG, which include interleukin-1 and interleukin-2, act through classical neurotransmitters to powerfully modulate this response.

Keywords: affective defense; amygdala; cytokines; defensive rage; enkephalins; glutamate; hissing; hypothalamus; limbic system; midbrain periaqueductal gray (PAG); monoamines; prefrontal cortex; septal area; substance P; vocalization

I. Introduction

There are a variety of types of vocalization expressed by felines. These include meowing, howling, growling and hissing. This chapter concerns the neural substrates and mechanisms that govern the expression of hissing. Hissing occurs in response to the presence of a real or perceived threat as part of a constellation of responses collectively referred to as "defensive rage" or "affective defense" behavior. The components of the response include hissing, growling, pupillary dilatation, arching of the back, retraction of the ears, marked increases in blood pressure and heart rate, piloerection and striking at a moving object, such as a conspecific, within the visual field of the cat (Leyhausen, 1979). It should be pointed out that hissing and growling reflect different ways by which the cat responds to an

*Corresponding author. E-mail: Siegel@UMDNJ.edu

Stefan M. Brudzynski (Ed.)
Handbook of Mammalian Vocalization
ISBN 978-0-12-374593-4

DOI: 10.1016/B978-0-12-374593-4.00024-3

environmental threat and thus represent integral components of the overall defensive rage response. These forms of threat display can be effective in removing the presence of a fearful environmental stimulus. The discussion below considers the anatomical substrates and neural mechanisms that underlie the expression of defensive rage, including the hissing component of this behavior. Included in this chapter is a brief review of the historical development of this area of investigation which provided important clues for the directions taken by later studies.

II. Historical background

There have been many investigations that were designed to identify the neural bases of threat and rage behavior over the last 80 years. Because of the paucity of techniques available, the earliest studies were limited in the approaches that were applied for the study of the neural bases of threat and rage behavior. The most common approach utilized in studies conducted during the period of time extending from the 1920s through the mid-1950s involved ablations of brain regions believed to be important for the occurrence of this behavior. Accordingly, a brief review of the history of the development of this field of research is provided in order to provide insight into how the direction of later research was sculptured.

II.A. Regional ablation

The first recorded study of the neurobiology of rage and threat behavior was conducted on dogs (Goltz, 1892). After removal of the cortex, dogs displayed stereotyped barking, growling and biting. Similar findings were observed several decades later in a decorticate preparation in which the dogs would elicit growling and snarling following mild scratching of their backs (Rothmann, 1923). Likewise, parallel observations were observed in cats following decortications in which innocuous stimulation evoked growling, hissing, spitting and piloerection (Dusser De Barenne, 1920). The conclusion reached from these studies was that rage and threat behavior, including associated vocalization, was not dependent on an intact cerebral cortex and therefore the expression of this behavior must be localized in subcortical structures.

The next generation of studies attempted to identify the regional levels at which rage and threat are localized. After ablations of the cortex and thalamus cats were still able to display components of rage behavior, indicating that the key regions associated with integration of rage behavior were located in regions caudal to the thalamus (Cannon, 1929). A systematic examination of the regions involving different levels of the diencephalon were carried out in a series of studies (Bard, 1928, 1934; Bard and McK.Rioch, 1937; McK.Rioch, 1938). These investigators showed that rage responses could be induced as long as the posterior hypothalamus was intact. Further evidence of the importance of this region was provided by a series of studies in which lesions of the posterior hypothalamus in cat and monkey produced marked reductions in rage and related responses (Ranson and Ingram, 1932; Ranson, 1939).

II.B. More precise brain lesions and electrical stimulation

Following on these studies, other experimenters attempted to identify the sites and regions of the brain directly associated with the expression of rage and threat behavior more precisely by the placement of localized lesions or by electrical stimulation of specific sites in the brain. The importance of the posterior hypothalamus was underscored by several studies that showed that decorticate cats whose thalami had been destroyed still elicited rage, as long as the posterior hypothalamus was intact (Wang and Akert, 1962; Emmers et al., 1965). In addition, lesions of the posterior hypothalamus could also block rage in response to noxious stimulation (Carli et al., 1966). The importance of the hypothalamus in rage behavior was also reflected by classical electrical stimulation studies conducted independently by Hess and Brugger (Hess and Brugger, 1943), Magoun et al. (1937) and Ranson and Magoun (1939). Each of these studies demonstrated that electrical stimulation of the hypothalamus could induce rage responses in the cat and provided the template for many studies that were to follow in later generations. A later study by Wasman and Flynn (1962) further defined the rostrocaudal continuum of the medial hypothalamus from which electrical stimulation could elicit defensive rage behavior.

II.C. Other regions of the brain

Two other regions of the brain have been implicated in expression and control of rage behavior. These include the midbrain PAG and several limbic structures. Lesions of the PAG could eliminate the faciovocal component of the rage response induced by nociceptive stimulation (Kelly et al., 1946) or rage induced by electrical stimulation of the hypothalamus (Hunsperger, 1956, 1959). These studies thus provided the first evidence that the PAG plays an important role in the elicitation of rage and threat behavior.

Several now classic studies utilized ablations of the amygdala and adjoining cortex to identify a role for this region of temporal lobe in modulating rage behavior. The first, conducted in monkeys, resulted in abnormal placidity, hypersexuality, increased oral tendencies and lack of recognition of objects (Kluver and Bucy, 1939). The second, conducted in cats, resulted in marked increases in rage behavior (Bard and Mountcastle, 1948). The dramatic opposing effects of these two studies conducted on the same region of temporal lobe were certainly paradoxical, but provided the impetus for many studies that were to follow on the role of the amygdala in aggression and rage behavior. Studies described below in this chapter have been able to account for the differences in the effects reported by these two investigations.

III. Regions of the brain from which rage, threat and vocalization can be expressed and their anatomical relationships

III.A. The hypothalamus

III.A.1. Sites associated with defensive rage

Identification of the sites in the hypothalamus has been achieved most effectively through the use of electrical or chemical brain stimulation. As noted above, the original observations utilizing electrical stimulation first established that rage behavior is associated with the medial hypothalamus (Magoun et al., 1937; Hess and Brugger, 1943). It remained for later studies to provide a more thorough mapping of this region of the forebrain. For example, later studies conducted by Wasman and Flynn (1962) and Chi and Flynn (1971) and then by Siegel and colleagues (see Siegel et al., 1999; Siegel, 2005) further clarified the sites within the medial hypothalamus from which defensive rage could be elicited, as did a series of experiments utilizing chemical stimulation of the hypothalamus (Romaniuk et al., 1973; Romaniuk et al., 1974; Brudzynski, 1981a,b; Brudzynski and Eckersdorf, 1988; Brudzynski et al., 1995). A more detailed discussion of the effects of cholinergic stimulation and its role in threat display is considered in a separate chapter in this volume by Brudzynski. The results of these studies may be summarized as follows; as shown in Fig. 1, the regions from which defensive rage behavior can be elicited include the rostrocaudal extent of the medial hypothalamus, beginning at the border of the preoptic region and extending up to and continuous with the midbrain PAG (see discussion below concerning the sites from which defensive rage can be elicited from the PAG). Within the medial hypothalamus, the most effective region for elicitation of defensive rage includes its rostral third, extending from the level of the ventromedial nucleus rostrally up to the border with the preoptic region.

III.A.2. Anatomical relationships: afferent connections and their functional relationships

There are three major sources of input to the medial hypothalamus that relate to the regulation and control of defensive rage behavior and associated vocalization. These include: limbic nuclei; brainstem monoamine neurons; and ascending fibers from the PAG.

Limbic structures possess unique properties, in that they receive secondary or tertiary sensory inputs as well as inputs from reticular formation monoamine neurons and they project their axons directly or indirectly to the hypothalamus or PAG. Because of this arrangement, activation of a limbic structure from its afferent source could then provide an output signal targeting the medial hypothalamus or PAG, resulting in significant modulation of the mechanism regulating defensive rage.

Amygdala. As noted above, the amygdala has been shown to generate opposing effects on defensive rage behavior. Experiments detailing the specific effects of electrical stimulation of specific sites within different nuclei of amygdala have characterized the role of different nuclei of amygdala on defensive rage. Activation of the medial nucleus and adjoining regions of the cortical and basomedial nuclei, as well as pyriform cortex, potentiates defensive rage behavior, while more lateral regions, including the lateral and central nuclei, suppress defensive rage (Stoddard-Apter and MacDonnell, 1980; Brutus et al., 1986; Shaikh et al., 1991b, 1993; Shaikh and Siegel, 1993). Two efferent pathways of the amygdala subserve modulation of defensive rage. The potentiating effects of the medial amygdala are mediated via the stria terminalis, which terminates directly in the rostral half of the medial

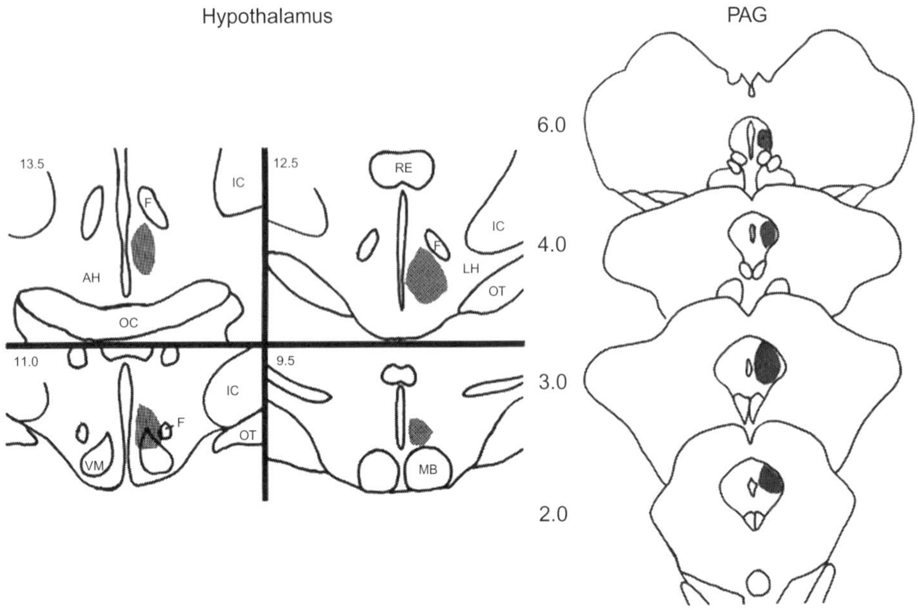

Fig. 1. Maps of the regions (enclosed dark areas) within the medial hypothalamus and dorsolateral midbrain periaqueductal gray (PAG) from which defensive rage can be elicited by electrical stimulation. Numbers in upper left margin for hypothalamus and left side for the PAG indicate the anterior-posterior levels in stereotaxic coordinates (Jasper and Ajmone-Marsan, 1954). Abbreviations: AH: anterior hypothalamus; F: fornix; IC: internal capsule; LH: lateral hypothalamus; MB: mammillary bodies; OC: optic chiasm; OT: optic tract; RM: nucleus reuniens; VM: ventromedial nucleus.

hypothalamus (Krettek and Price, 1978). The principal neurotransmitter is substance P, which acts through neurokinin 1 (NK_1) receptors in the medial hypothalamus (Shaikh et al., 1993). The second pathway, which mediates suppression of defensive rage, arises from the central, basolateral and lateral nuclei of amygdala and appears to supply mainly the dorsal PAG rather than the hypothalamus. Suppression is quite powerful and is mediated through opioid μ receptors in the PAG (Shaikh et al., 1991a).

Septal area. A study by Stoddard-Apter and MacDonnell (Stoddard-Apter and MacDonnell, 1980; Stoddard-Apter et al., 1983) identified the pathway from the septal area to the hypothalamus associated with modulation of defensive rage behavior in the cat. Initially, these investigators demonstrated that stimulation of sites in the region of the medial septal area facilitated the occurrence of defensive rage. Then the pathway from the site of modulation in the septal area was identified by the use of autoradiographic tracing methods. The principal projection was shown to project directly to the medial hypothalamus. Thus, the region of the medial septal area and adjoining neurons in the medial aspect of the lateral septal nucleus, via a monosynaptic projection, can facilitate defensive rage. Moreover, since the septal area receives input from both the dorsal and ventral regions of hippocampal formation (Siegel and Tassoni, 1971; Edinger et al., 1973; Meibach and Siegel, 1977), the septal area could also be viewed as a relay nucleus of the hippocampal formation by which it could modulate aggression and rage behavior associated with the hypothalamus.

Prefrontal cortex. The prefrontal cortex powerfully suppresses different kinds of aggressive behavior, including defensive rage and predatory attack behavior (Siegel et al., 1974; de Bruin et al., 1983; de Bruin, 1990). A basic question concerns how such modulation on defensive rage behavior may become manifest. A unique feature of the prefrontal cortex is that it receives inputs from all regions of cerebral cortex, mediodorsal thalamic nucleus and lateral amygdala (Bacon, 1973; Divac et al., 1978). This feature endows this region with the possibility of providing the neural basis by which conditioning of any one or more of these signals can be achieved. Such conditioning may have a component that reaches consciousness and which could then allow for the voluntary suppression of the rage mechanism. Suppression of defensive rage could be achieved via its output pathways to the hypothalamus and PAG. There are several routes by which impulses from the prefrontal cortex could reach the hypothalamus. The first is an indirect route through the mediodorsal nucleus, which then projects to the hypothalamus via a series of interneurons passing rostrally within midline thalamic nuclei, and ultimately reaching the anterior hypothalamus through the nucleus reuniens (Siegel et al., 1972, 1977). A second route is also indirect, involving a synapse in the amygdala (Rosenkranz and Grace, 2002). Projections from the amygdala can then reach either the hypothalamus or PAG. The third is a direct route from the prefrontal cortex to the hypothalamus (Ongur et al., 1998). A fourth route is also possible, namely, direct projections from the prefrontal cortex to the dorsal aspect of the PAG, where suppression of defensive rage could also be achieved (An et al., 1998). Although the neurotransmitters from the prefrontal cortex to their target regions have not been systematically examined, it is most likely that glutamate is involved since it is the primary neurotransmitter released from output neurons of the cerebral cortex. Since glutamate is an excitatory neurotransmitter and the prefrontal cortex suppresses defensive rage, it is suggested that suppression is mediated through relays such as those present in the lateral and central amygdala, which are known to powerfully inhibit this form of aggression.

Monoamine inputs. Monoamines provide significant input to both the hypothalamus and PAG. Noradrenergic neurons, arising mainly from the nucleus locus ceruleus, significantly modulate defensive rage behavior. The primary action of norepinephrine in the medial hypothalamus is to facilitate defensive rage behavior by acting through noradrenergic α_2-receptors (Barrett et al., 1990). Likewise, there are well-known dopaminergic inputs to the hypothalamus from the ventral tegmental area. An early study showed that peripheral blockade of dopamine receptors resulted in suppression of hypothalamically-elicited defensive rage in the cat (Maeda, 1976). Also consistent with these findings, a later study demonstrated that activation of dopamine D_2 receptors in the medial hypothalamus facilitates defensive rage (Sweidan et al., 1991). Serotonin, arising from raphe neurons in the pons and midbrain, abundantly supply the hypothalamus and PAG. It is of interest to note that serotonin acting through 5-HT_{1A} receptors in the medial hypothalamus or PAG has an effect on defensive rage that is opposite to its effects when acting through 5-HT_2 receptors in these regions (Shaikh et al., 1997; Hassanain et al., 2005). Specifically, activation of 5-HT_{1A} receptors suppress, while 5-HT_2 receptors facilitate, defensive rage in each of these regions.

Ascending fibers from the PAG. One additional input of importance should be noted, namely ascending fibers from the PAG (see Gruber-Dujardin, Chapter 8.1 in this volume). These ascending fibers pass mainly from the dorsal PAG to the posterior half of the medial hypothalamus. It is likely that these fibers constitute a limb of a reciprocal pathway linking the medial hypothalamus with the PAG, the ascending component of which serves to provide positive feedback, thus allowing for the defensive rage response to be of relatively long duration. The neurotransmitter associated with ascending fibers from the PAG has not been identified, but it is reasonable to believe that it may be glutamate.

III.A.3. Anatomical relationships: efferent connections

The first attempt at tracing the pathways associated with predatory attack and defensive rage behavior was carried out by Chi and Flynn (1971). In a portion of the study relating to the pathway associated with defensive rage, a lesion was placed at the tip of the electrode in the medial hypothalamus from which defensive rage was elicited, and the degenerating axons were traced by the use of a silver degeneration technique from the lesion site caudally into the midbrain PAG. This finding was the first to establish a functional relationship between the medial hypothalamus and PAG with respect to the descending pathway mediating defensive rage behavior (see Hage, Chapter 8.3 in this volume). This initial finding was replicated and extended in a series of studies utilizing autoradiographic, 2-deoxyglucose and retrograde tracing techniques (Fuchs et al., 1985a,b; Schubert et al., 1996). These studies revealed the following anatomical relationship between the medial hypothalamus and PAG with respect to defensive rage – namely that while rage behavior can be elicited from wide regions along the rostrocaudal axis of the medial hypothalamus, the origins of the projections from the medial hypothalamus to the PAG are limited to the rostral third of this structure. The region of the ventromedial nucleus contributes to the expression of defensive rage by virtue of rostrally directed fibers that synapse in the anterior medial hypothalamus. The descending projections from the medial hypothalamus supply principally the rostral half of the dorsal aspect of the PAG, although some fibers are distributed to more caudal levels as well. This pathway is the principal descending pathway from the hypothalamus for the expression of defensive rage and its functions are mediated through glutamate acting on NMDA receptors in the dorsal PAG (Lu et al., 1992; Schubert et al., 1996). Small quantities of descending projections pass through the brainstem to more caudal levels of reticular formation of the pons, while ascending fibers from the medial hypothalamus project back to limbic nuclei, including the amygdala and septal area, where they presumably serve as a feedback to these structures, which provide significant modulation of the defensive rage response at the levels of both the hypothalamus and PAG.

III.A.4. Local interneurons and their functions

In addition to the afferent fibers to the hypothalamus that significantly modulate defensive rage behavior, at least one other class of neurons plays an important role in the regulation of this form of aggression and threat behavior. These include small-sized neurons that are present in both the lateral and medial hypothalamus. Both neurons contain GABA and form reciprocal inhibitory connections between these two regions of hypothalamus that act through $GABA_A$ receptors (Han et al., 1996; Cheu and Siegel, 1998). The significance of these neurons can be understood in the following way. While the medial hypothalamus mediates defensive rage and threat behavior, the lateral hypothalamus mediates predatory attack behavior. Since neither of these forms of aggression can function effectively in the presence of the other (because they reflect competing responses), they are said to be mutually incompatible. For example, when predation is about to occur, hissing and related forms of vocalization would alarm the prey object and allow it to escape. Likewise, when a defense mechanism is needed for survival, any tendency towards establishing a predatory mode that involves attack in the absence of vocalization would disrupt the vocalization essential for the successful occurrence of defensive behavior. Thus, when one form of aggression is present, the other is suppressed by the GABA-ergic mechanism.

III.B. The PAG

III.B.1. Sites associated with defensive rage

The primary sites in the PAG where howling and hissing can be elicited are limited to the rostral half of this neuropil (Shaikh et al., 1987). The sites are principally localized to the lateral and dorsal aspects of this level of the PAG.

III.B.2. Anatomical relationships

Several of the key anatomical relationships involving afferent and efferent connections of the PAG have been indicated earlier in this chapter. Therefore, this section briefly summarizes these connections.

The following afferent fibers that supply the PAG are associated with either the expression of defensive rage or with its modulation. Afferent fibers arising from the rostral third of the medial hypothalamus mediate defensive rage, and its functions are mediated by glutamate acting through NMDA receptors in the rostral half of the dorsolateral PAG. In this manner, axons from limbic structures that underlie modulation of defensive rage include the following: fibers from the central nucleus of amygdala which mediate suppression via opioid μ receptors in the PAG; fibers from the prefrontal cortex that also presumably mediate suppression of defensive rage and whose neurotransmitter is presumed to be glutamate; and monoaminergic fibers from the brainstem. The effects of serotonin receptors have been tested in the PAG (Shaikh et al., 1997) and their effects mimic those observed in the hypothalamus, in which $5\text{-}HT_{1A}$ receptors suppress while $5\text{-}HT_2$ receptors facilitate defensive rage (Hassanain et al., 2003a).

Concerning the efferent connections of the PAG, this structure constitutes the most caudal region of the neuraxis of the brain where integration of defensive rage takes place. The outputs integrate two aspects of the defensive rage response – autonomic and somatomotor components. The autonomic component is mediated through descending fibers that either directly project to the intermediolateral cell column of the thoracic or lumbar spinal cord, or indirectly through synapses in the solitary nucleus or ventrolateral medulla (Bandler and Tork, 1987; Shaikh et al., 1987; Bandler, 1988; Carrive et al., 1989; Bandler et al., 1991; Lovick, 1993). By either or both routes, excitation of neurons in the dorsal PAG will induce significant increases in blood pressure, heart rate and pupillary dilatation which are principal autonomic features of defensive rage behavior. The somatomotor components of the behavior are mediated through descending pathways from the PAG that directly synapse on motor neurons of cranial nerves V and VII, as well as the nucleus ambiguus, which relate to jaw opening and the vocalization component of the defensive rage response (Shaikh et al., 1987; Holstege, 1989). Other downstream fibers project directly or indirectly to spinal cord motor nuclei that control movements of the upper limbs essential for striking a threatening object in the environment (Holstege, 1988, 1991).

As noted above, ascending fibers of the PAG supply the caudal half of the medial hypothalamus and presumably serve as a positive feedback mechanism for extending the duration of the attack response (Shaikh et al., 1987). The circuitry governing the expression and modulation of defensive rage behavior is depicted in Fig. 2.

III.B.3. Neurochemical relationships

While the effects of various neurotransmitters and receptors in the PAG have been noted earlier in the chapter, a more inclusive general summary is included at this time. Neurotransmitters and their associated receptors that have been shown to facilitate defensive rage behavior include glutamate, released from hypothalamic-PAG fibers, which acts through NMDA receptors (Lu et al., 1992; Schubert et al., 1996); substance P, acting through NK_1 receptors (Gregg and Siegel, 2003); cholecystokinin (CCK), acting through CCK_B receptors (Luo et al., 1998); and serotonin, acting through 5-HT_2 receptors (Shaikh et al., 1997). Neurotransmitter-receptors that suppress defensive rage include $GABA_A$ receptors (Shaikh and Siegel, 1990); opioid-μ receptors, which receive enkephalinergic input from the central nucleus of amygdala (Shaikh et al., 1991b); and 5-HT_{1A} receptors (Shaikh et al., 1997). It is not known whether the fiber systems

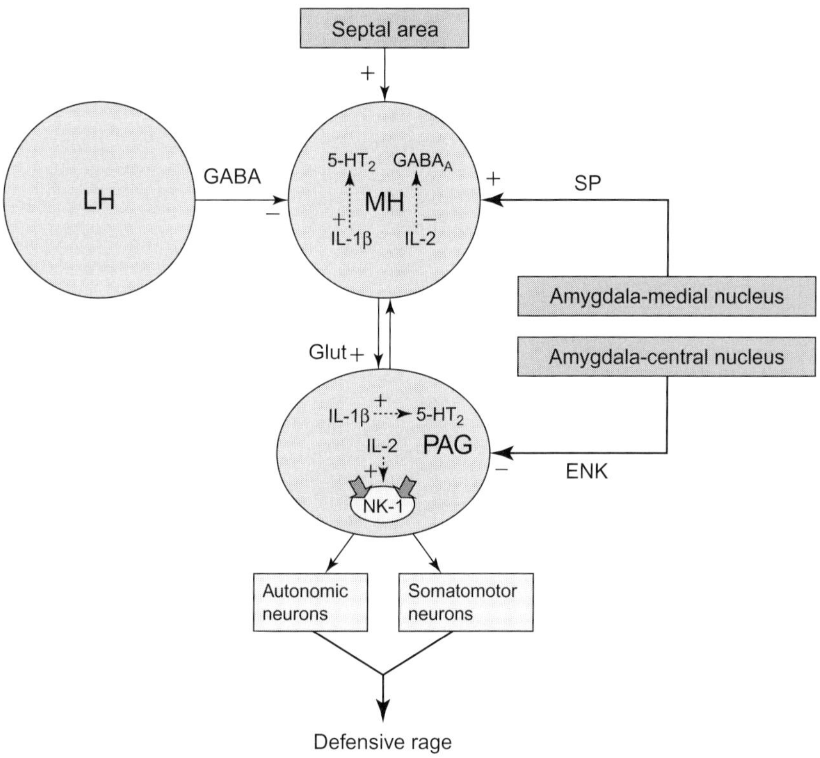

Fig. 2. Schematic diagram depicting principal neural relationships and mechanisms involving the septal area and amygdala as components of the limbic system, hypothalamus, and midbrain periaqueductal gray (PAG). Changes in excitability within the septal area and amygdala alter excitability levels and functions of the medial hypothalamus and PAG, thus affecting defensive rage behavior. Defensive rage behavior is mediated through neurons in the medial hypothalamus and a descending glutamatergic pathway to the midbrain PAG. The PAG constitutes the lowest region of the neuraxis of the brain at which integration of the autonomic and somatomotor components of defensive rage behavior occurs. GABAergic neurons from the lateral hypothalamus provide a major source of inhibition on defensive rage at the level of the medial hypothalamus. In addition, this figure also indicates the role of several proinflammatory cytokines, IL-1 and IL-2. IL-1 in the medial hypothalamus and PAG facilitate defensive rage and act through 5-HT_2 receptors in both regions. IL-2 in the medial hypothalamus suppresses defensive rage and its functions are mediated through $GABA_A$ receptors. In contrast, IL-2 in the PAG facilitates defensive rage behavior by acting through substance P NK_1 receptors in the PAG. Abbreviations: 5-HT: serotonin; GABA: γ-aminobutyric acid; IL-interleukin; LH: lateral hypothalamus; MH: medial hypothalamus; NK_1: neurokinin-1.

Table 1. Defensive rage behavior: role of neurotransmitter-receptors and cytokines

Compound	Anatomical location	Receptor	Defensive rage
Small molecule neurotransmitters			
Monoamines:			
Dopamine	Medial hypothalamus	Dopamine D_2	↑
Norepinephrine	Medial hypothalamus	Noradrenergic α-2	↑
Serotonin	Medial hypothalamus, PAG	$5\text{-}HT_{1A}$	↓
Serotonin	Medial hypothalamus, PAG	$5\text{-}HT_2$	↑
Other small molecules:			
Acetylcholine	Medial hypothalamus	Muscarinic	↑
Glutamate	PAG	NMDA	↑
GABA	Hypothalamus, PAG	$GABA_A$	↓
Neuropeptides:			
Substance P	Medial hypothalamus, PAG	Substance P NK_1	↑
Cholecystokinin (CCK)	PAG	CCK_B	↑
Enkephalin	PAG	Opioid-μ	↓
Pro-inflammatory cytokines:			
Interleukin 1	Hypothalamus, PAG	IL-1 + $5\text{-}HT_2$	↑
Interleukin 2	Hypothalamus	IL-2 + $GABA_A$	↓
Interleukin 2	PAG	IL-2 + NK_1	↑

Key: ↑: facilitation; ↓: suppression

that mediate their functions through NK_1, CCK_B and $GABA_A$ receptors are extrinsic to the PAG or whether they constitute local interneurons within the neuropil. These findings are also summarized in Table 1.

IV. Cytokines

Cytokines are pleiotropic cell-to-cell signaling proteins or glycoproteins that regulate immune function and modulate neurotransmitter activity. They are produced by different cells types in periphery and brain. Recent studies, summarized below and shown in Fig. 2, have demonstrated that they modulate defensive rage behavior in the cat.

In one series of studies, activation of interleukin I (IL-1β) receptors in either the medial hypothalamus or in the rostral half of the dorsal PAG, at sites from which defensive rage could be elicited, significantly facilitates defensive rage behavior (Hassanain et al., 2003b; Hassanain et al., 2005; Bhatt et al., 2008). Further analysis by these investigators revealed that the potentiating effects of IL-1β are mediated through serotonin $5\text{-}HT_2$ receptors in both the medial hypothalamus and PAG.

A similar analysis was also conducted with respect to IL-2. However, from this analysis, it was revealed that while activation of IL-2 receptors in the medial hypothalamus suppressed defensive rage behavior, activation of IL-2 receptors in the PAG facilitated defensive rage (Bhatt et al., 2005; Bhatt and Siegel, 2006). These investigators further determined that the suppressive effects in the medial hypothalamus were mediated through $GABA_A$ postsynaptic receptors, while the facilitating effects in the PAG were mediated through substance P NK_1 receptors. Additional analysis revealed that the epsilon (ε) subunit of the $GABA_A$ receptor is absent in the medial hypothalamus, but present in the PAG. It has been reported previously that in order for GABA to be effective as an inhibitory neurotransmitter, the ε subunit must be absent (Sergeeva et al., 2005). The conclusions thus far from these studies are that: (1) activation of cytokine receptors within the hypothalamus and PAG significantly modulate defensive rage behavior; (2) the modulating effects are mediated through classical neurotransmitter-receptor mechanisms; and (3) the effects of cytokines on rage behavior reflect site specificity. Considerably more research will be required to further elucidate the relationships of these and other cytokines on defensive rage behavior.

V. Conclusions

Brain stimulation has been utilized to induce defensive rage, which involves threat display and marked vocalization, including hissing and growling. This approach has provided the basis by which investigators have been able to identify: (1) the regions of the brain associated with the expression and modulation of threat display; (2) the neuroanatomical circuitry; and (3) the neural mechanisms subserving this behavior. Principal regions that are associated with the expression of defensive rage and vocalization include the medial hypothalamus and dorsolateral PAG. Regions which modulate this behavior comprise much of the limbic system, the most significant of which include the amygdala, septal area and prefrontal cortex. A variety of neurotransmitters released upon neurons in the medial hypothalamus and PAG significantly regulate this behavior. These transmitters include acetylcholine, monoamines, glutamate and GABA, as well as a number of peptides which include substance P, enkephalin and cholecystokinin. In addition, several proinflammatory cytokines, including IL-1 and IL-2, have also recently been shown to powerfully modulate threat behavior and vocalization by acting through specific neurotransmitter systems.

Acknowledgments

Supported by NIH Grant NS 07941-36.

References

An, X., Bandler, R., Öngür, D., Price, J.L., 1998. Prefrontal cortical projections to longitudinal columns in the midbrain periaqueductal gray in macaque monkeys. J. Comp. Neurol. 401, 455–479.

Bacon, W.E., 1973. Aversive conditioning in neonatal kittens. J. Comp. Physiol. Psychol. 83, 306–313.

Bandler, R., 1988. Brain mechanisms of aggression as revealed by electrical and chemical stimulation: suggestion of a central role for the midbrain periaqueductal grey region. Prog. Psychobiol. Physiol. Psychol. 13, 67–154.

Bandler, R., Tork, I., 1987. Midbrain periaqueductal grey region in the cat has afferent and efferent connections with solitary tract nuclei. Neurosci. Lett. 74, 1–6.

Bandler, R., Carrive, P., Zhang, S.P., 1991. Integration of somatic and autonomic reactions within the midbrain periaqueductal grey: viscerotopic, somatotopic and functional organization. Prog. Brain Res. 87, 269–305.

Bard, P., 1928. A diencephalic mechanism for the expression of rage with special reference to the sympathetic nervous system. Am. J. Physiol. 84, 490–515.

Bard, P., 1934. On emotional expression after decortication with some remarks on certain theoretical views. Part 1. Psychol. Rev. 41, 309–329.

Bard, P., McK.Rioch, D., 1937. A study of four cats deprived of neocortex and additional portions of the forebrain. Bull. Johns Hopkins Hosp. LX 2, 73–124.

Bard, P., Mountcastle, V.B., 1948. Some forebrain mechanisms involved in expression of rage with special reference to suppression of angry behavior. Ass. Res. Nerv. Ment. Dis. 27, 362–399.

Barrett, J., Edinger, H., Siegel, A., 1990. Intrahypothalamic injections of norepinephrine facilitate feline affective aggression via alpha-2 adrenoceptors. Brain Res. 525, 285–293.

Bhatt, S., Siegel, A., 2006. Potentiating role of interleukin 2 (IL-2) receptors in the midbrain periaqueductal gray (PAG) upon defensive rage behavior in the cat: role of neurokinin NK1 receptors. Behav. Brain Res. 167, 251–260.

Bhatt, S., Zalcman, S., Hassanain, M., Siegel, A., 2005. Cytokine modulation of defensive rage behavior in the cat: role of GABA(A) and interleukin-2 receptors in the medial hypothalamus. Neuroscience 133, 17–28.

Bhatt, S., Bhatt, R., Zalcman, S.S., Siegel, A., 2008. Role of IL-1 beta and 5-HT2 receptors in midbrain periaqueductal gray (PAG) in potentiating defensive rage behavior in cat. Brain Behav. Immun. 22, 224–233.

Brudzynski, S.M., 1981a. Carbachol-induced agonistic behavior in cats: aggressive or defensive response? Acta Neurobiol. Exp. 41, 15–32.

Brudzynski, S.M., 1981b. Growling component of vocalization as a quantitative index of carbachol-induced emotional-defensive response in cats. Acta Neurobiol. Exp. 41, 33–51.

Brudzynski, S.M., Eckersdorf, B., 1988. Vocalization accompanying emotional-aversive response induced by carbachol in the cat. Neuropsychopharmacology 1, 311–320.

Brudzynski, S.M., Eckersdorf, B., Golebiewski, H., 1995. Regional specificity of the emotional-aversive response induced by carbachol in the cat brain: a quantitative mapping study. J. Psychiatry Neurosci. 20, 119–132.

Brutus, M., Shaikh, M.B., Siegel, A., Edinger, H., 1986. Effects of experimental temporal lobe seizures upon hypothalamically elicited aggressive behavior in the cat. Brain Res. 366, 53–63.

Cannon, W.B., 1929. Bodily Changes in Pain, Fear, Hunger and Rage. Appleton-Century-Crofts, New York, NY.

Carli, G., Malliani, A., Zanchetti, A., 1966. Distruzione della sostanza grigia periacquedottale e comportamento di falsa fabbia del gatto decorticato acuto. Boll. Soc. Ital. Biol. Speriment 42, 295–296.

Carrive, P., Bandler, R., Dampney, R.A.L., 1989. Viscerotopic control of regional vascular beds by discrete groups of neurons within the midbrain periaqueductal gray. Brain Res. 493, 385–390.

Cheu, J.W., Siegel, A., 1998. GABA receptor mediated suppression of defensive rage behavior elicited from the medial hypothalamus of the cat: role of the lateral hypothalamus. Brain Res. 783, 293–304.

Chi, C.C., Flynn, J.P., 1971. Neuroanatomic projections related to biting attack elicited from hypothalamus in cats. Brain Res. 35, 49–66.

de Bruin, J.P.C., 1990. Orbital prefrontal cortex, dopamine, and social agonistic behavior of male Long Evans rats. Aggress. Behav. 16, 231–248.

de Bruin, J.P.C., Van Oyen, H.G.M., Van De Poll, N., 1983. Behavioural changes following lesions of the orbital prefrontal cortex in male rats. Behav. Brain Res. 10, 209–232.

Divac, I., Kosmal, A., Bjorklund, A., Lindvall, O., 1978. Subcortical projections to the prefrontal cortex in the rat as revealed by the horseradish peroxidase technique. Neuroscience 3, 785–796.

Dusser De Barenne, J.G., 1920. Recherches experimentales sur les fonctions du systeme nerveux central, faites en particulier sur deux chats dont le neopallium avait ete enleve. Arch. Neer. Phys. del'Homme et des Animaux IV, 31–123.

Edinger, H., Siegel, A., Troiano, R., 1973. Single unit analysis of the hippocampal projections to the septum in the cat. Exp. Neurol. 41 (3), 569–583.

Emmers, R., Chun, R.W.M., Wang, G.H., 1965. Behavior and reflexes of chronic thalamic cats. Arch. Ital. Biol. 103, 178–193.

Fuchs, S.A.G., Edinger, H.M., Siegel, A., 1985a. The organization of the hypothalamic pathways mediating affective defense behavior in the cat. Brain Res. 330, 77–92.

Fuchs, S.A.G., Edinger, H.M., Siegel, A., 1985b. The role of the anterior hypothalamus in affective defense behavior elicited from the ventromedial hypothalamus of the cat. Brain Res. 330, 93–108.

Goltz, P.F., 1892. The dog without cerebrum. Pfluger's Archiv fur die gesamte Physilogie 51, 1–2.

Gregg, T.R., Siegel, A., 2003. Differential effects of NK1 receptors in the midbrain periaqueductal gray upon defensive rage and predatory attack in the cat. Brain Res. 994, 55–66.

Han, Y., Shaikh, M.B., Siegel, A., 1996. Medial amygdaloid suppression of predatory attack behavior in the cat: II. Role of a GABAergic pathway from the medial to the lateral hypothalamus. Brain Res. 716, 72–83.

Hassanain, M., Bhatt, S., Siegel, A., 2003a. Differential modulation of feline defensive rage behavior in the medial hypothalamus by 5-HT1A and 5-HT2 receptors. Brain Res. 981, 201–209.

Hassanain, M., Zalcman, S., Bhatt, S., Siegel, A., 2003b. Interleukin-1 beta in the hypothalamus potentiates feline defensive rage: role of serotonin-2 receptors. Neuroscience 120, 227–233.

Hassanain, M., Bhatt, S., Zalcman, S., Siegel, A., 2005. Potentiating role of interleukin-1 beta (IL-1 beta) and IL-1 beta type 1 receptors in the medial hypothalamus in defensive rage behavior in the cat. Brain Res. 1048, 1–11.

Hess, W.R., Brugger, M., 1943. Das subkortikale Zentrum der affektiven Abwehrreaktion. Helv. Physiol. Pharmacol. Acta 1, 33–52.

Holstege, G., 1988. Direct and indirect pathways to lamina I in the medulla oblongata and spinal cord of the cat. Prog. Brain Res. 77, 47–94.

Holstege, G., 1989. Anatomical study of the final common pathway for vocalization in the cat. J. Comp. Neurol. 284, 242–252.

Holstege, G., 1991. Descending motor pathways and the spinal motor system. Limbic and non-limbic components. Prog. Brain Res. 87, 307–421.

Hunsperger, R.W., 1956. Role of substantia grisea centralis mesencephali in electrically-induced rage reactions. In: Kappers, J. (Ed.), Progress in Neurobiology. Elsevier, New York, NY, pp. 289–294.

Hunsperger, R.W., 1959. Les representation centrales des reactions affectives dans le cerveau anterieur et dans le tronc cerebral. Neuro-Chirurgie 5 (2), 207–233.

Jasper, H.H., Ajmone-Marsan, C.A., 1954. Stereotaxic Atlas of the Diencephalon of the Cat. National Research Council of Canada, Ottawa, Canada.

Kelly, A.H., Beaton, L.E., Magoun, H.W., 1946. A midbrain mechanism for facio-vocal activity. J. Neurophysiol. 9, 181–189.

Kluver, H., Bucy, P.C., 1939. Preliminary analysis of functions of the temporal lobes in monkeys. Arch. Neurol. Psychiat. 42, 979–1000.

Krettek, J.E., Price, J.L., 1978. Amygdaloid projections to subcortical structures within the basal forebrain and brainstem in the rat and cat. J. Comp. Neurol. 178, 225–254.

Leyhausen, P., 1979. Cat Behavior. The Predatory and Social Behavior of Domestic and Wild Cats. Garland STPM Press, New York, NY.

Lovick, T.A., 1993. The periaqueductal gray–rostral medulla connection in the defence reaction: efferent pathways and descending control mechanisms. Behav. Brain Res. 58, 19–25.

Lu, C.-L., Shaikh, M.B., Siegel, A., 1992. Role of NMDA receptors in hypothalamic facilitation of feline defensive rage elicited from the midbrain periaqueductal gray. Brain Res. 581, 123–132.

Luo, B., Cheu, J.W., Siegel, A., 1998. Cholecystokinin B receptors in the periaqueductal gray potentiate defensive rage behavior elicited from the medial hypothalamus of the cat. Brain Res. 796, 27–37.

Maeda, H., 1976. Effects of psychotropic drugs upon the hypothalamic rage response in cats. Folia Psychiat. Neurol. Jpn. 30, 539–546.

Magoun, H.W., Atlas, D., Ingersoll, E.H., Ranson, S.W., 1937. Associated facial, vocal and respiratory components of emotional expression: an experimental study. J. Neurol. Psychopath. 17, 241–255.

McK.Rioch, D., 1938. Certain aspects of the behavior of decorticate cats. Psychiat. 1, 339–345.

Meibach, R., Siegel, A., 1977. Efferent connections of the hippocampal formation in the rat. Brain Res. 124, 197–224.

Ongur, D., An, X., Price, J.L., 1998. Prefrontal cortical projections to the hypothalamus in macaque monkeys. J. Comp. Neurol. 401, 480–505.

Ranson, S.W., 1939. Somnolence caused by hypothalamic lesions in the monkey. Arch. Neurol. Psychiat. 41 (1), 1–23.

Ranson, S.W., Ingram, W.R., 1932. Catelepsy caused by lesions between the mammillary bodies and third nerve in the cat. Am. J. Physiol. 101, 690.

Ranson, S.W., Magoun, H.W., 1939. The hypothalamus. Ergebn. Physiol. 41, 56–163.

Romaniuk, A., Brudzynski, S., Gronska, J., 1973. The effect of chemical blockade of hypothalamic cholinergic system on defensive reactions in cats. Acta Physiol. Pol. XXIV, 809–816.

Romaniuk, A., Brudzynski, S., Gronska, J., 1974. The effects of intrahypothalamic injections of cholinergic and adrenergic agents on defensive behavior in cats. Acta Physiol. Pol. XXV, 297–305.

Rosenkranz, J.A., Grace, A.A., 2002. Cellular mechanisms of infralimbic and prelimbic prefrontal cortical inhibition and dopaminergic modulation of basolateral amygdala neurons *in vivo*. J. Neurosci. 22, 324–337.

Rothmann, H., 1923. Zusammenfassender Bericht uber den Rothmannschen groshirnlosen Hund nach klinischer und anatomischer Untersuchung. Zeit. gesam. Neurol. Psychiatr. 87, 247–313.

Schubert, K., Shaikh, M.B., Siegel, A., 1996. NMDA receptors in the midbrain periaqueductal gray mediate hypothalamically evoked hissing behavior in the cat. Brain Res. 726, 80–90.

Sergeeva, O.A., Andreeva, N., Garret, M., Scherer, A., Haas, H.L., 2005. Pharmacological properties of GABAA receptors in rat hypothalamic neurons expressing the epsilon-subunit. J. Neurosci. 25, 88–95.

Shaikh, M.B., Siegel, A., 1990. GABA-mediated regulation of feline aggression elicited from midbrain periaqueductal gray upon affective defense behavior in the cat. Brain Res. 507, 51–56.

Shaikh, M.B., Siegel, A., 1993. Neural models of aggressive behavior. In: Cohen, H. (Ed.), Neuroscience for Rehabilitation. J.B. Lippincott and Co., Philadelphia, PA, pp. 318–332.

Shaikh, M.B., Barrett, J.A., Siegel, A., 1987. The pathways mediating affective defense and quiet biting attack behavior from the midbrain central gray of the cat: an autoradiographic study. Brain Res. 437, 9–25.

Shaikh, M.B., Lu, C.-L., Siegel, A., 1991a. Affective defense behavior elicited from the feline midbrain periaqueductal gray is regulated by mu- and delta-opioid receptors. Brain Res. 557, 344–348.

Shaikh, M.B., Lu, C.-L., Siegel, A., 1991b. An enkephalinergic mechanism involved in amygdaloid suppression of affective defense behavior elicited from the midbrain periaqueductal gray in the cat. Brain Res. 559, 109–117.

Shaikh, M.B., Steinberg, A., Siegel, A., 1993. Evidence that substance P is utilized in medial amygdaloid facilitation of defensive rage behavior in the cat. Brain Res. 625, 283–294.

Shaikh, M.B., De Lanerolle, N.C., Siegel, A., 1997. Serotonin 5-HT_{1A} and 5-$HT_{2/1C}$ receptors in the midbrain periaqueductal gray differentially modulate defensive rage behavior elicited from the medial hypothalamus of the cat. Brain Res. 765, 198–207.

Siegel, A., 2005. The Neurobiology of Aggression and Rage. CRC Press, Boca Raton, FL.

Siegel, A., Tassoni, J.P., 1971. Differential efferent projections from the ventral and dorsal hippocampus of the cat. Brain Behav. Evol. 4, 185–200.

Siegel, A., Chabora, J., Troiano, R., 1972. Effects of electrical stimulation of the pyriform cortex upon hypothalamically-elicited aggression in the cat. Brain Res. 47, 497–500.

Siegel, A., Edinger, H., Lowenthal, H., 1974. Effects of electrical stimulation of the medial aspect of the prefrontal cortex upon attack behavior in cats. Brain Res. 66, 467–479.

Siegel, A., Edinger, H., Koo, A., 1977. Suppression of attack behavior in the cat by the prefrontal cortex: role of the mediodorsal thalamic nucleus. Brain Res. 127, 185–190.

Siegel, A., Roeling, T.A.P., Gregg, T.R., Kruk, M.R., 1999. Neuropharmacology of brain-stimulation-evoked aggression. Neurosci. Biobehav. Rev. 23, 359–389.

Stoddard-Apter, S.L., MacDonnell, M.F., 1980. Septal and amygdalar efferents to the hypothalamus which facilitate hypothalamically elicited intraspecific aggression and associated hissing in the cat. An autoradiographic study. Brain Res. 193, 19–32.

Stoddard-Apter, S., Levin, B., Siegel, A., 1983. A sympathoadrenal and cardiovascular correlates of aggressive behavior in the awake cat. J. Auton. Nerv. Syst. 8, 343–360.

Sweidan, S., Edinger, H., Siegel, A., 1991. D2 dopamine receptor-mediated mechanisms in the medial preoptic-anterior hypothalamus regulate affective defense behavior in the cat. Brain Res. 549, 127–137.

Wang, G., Akert, K., 1962. Behavior and reflexes of chronic striatal cats. Extrait. Arch. Ital. Biol. 100, 48–85.

Wasman, M., Flynn, J.P., 1962. Directed attack elicited from hypothalamus. Arch. Neurol. 6, 220–227.

CHAPTER 7.2

The polyvagal hypothesis: common mechanisms mediating autonomic regulation, vocalizations and listening

Stephen W. Porges* and Gregory F. Lewis

Brain-Body Center, Department of Psychiatry, University of Illinois at Chicago, Chicago, IL, USA

Abstract: This chapter applies hypotheses and principles derived from the polyvagal theory (Porges, 1995, 2001, 2007) to interpret mammalian vocalizations. The theory emphasizes the parallel phylogenetic shift in both the neural regulation of the autonomic nervous system and the evolutionary emergence of an integrated social engagement system that includes features optimizing conspecific mammalian vocal communication. The chapter presents the polyvagal hypothesis as a new way to interpret adaptive functions and acoustic features of mammalian vocalizations. The chapter emphasizes both the neural mechanisms involved in social communication, including the reciprocal relationship between production and reception of vocalizations, and the perceptual advantage that mammals have by vocalizing within a frequency band optimized by the physics of the middle ear.

Keywords: autonomic nervous system; polyvagal theory; middle ear muscles; heart rate variability; vagus; social engagement system; perceptual advantage

I. Introduction

Vocalizations are an intricate component of the complex biobehavioral repertoire of mammals. Since vocalizations are involved in coordinating behaviors for mammals that are both prosocial and survival oriented, the predominance of research on mammalian vocalizations has focused on describing the adaptive functions of specific vocal features. In contrast, there is a paucity of research investigating the neurophysiological linkage between autonomic regulation and both the production of vocalizations and the processing of acoustic information. This chapter applies hypotheses and principles derived from the polyvagal theory (Porges, 1995, 2001, 2007) to interpret mammalian vocalizations. The theory emphasizes the parallel phylogenetic shift in both the neural regulation of the autonomic nervous system and the evolutionary emergence of an integrated social engagement system that includes features optimizing conspecific vocal communication. The chapter will present the polyvagal hypothesis as a new way to interpret the adaptive functions of mammalian vocalizations. The chapter will emphasize neural mechanisms involved in social communication, including the reciprocal relationship between production and reception of vocalizations.

*Corresponding author.
Stephen W. Porges, PhD, Brain-Body Center (MC 912), Department of Psychiatry, University of Illinois at Chicago, 1601 W. Taylor Street, Chicago, IL 60612, USA.

II. The polyvagal theory: three phylogenetic response systems

The polyvagal theory (Porges, 1995, 2001, 2007) links parallel phylogenetic changes in the neural regulation

Stefan M. Brudzynski (Ed.)
Handbook of Mammalian Vocalization
ISBN 978-0-12-374593-4

DOI: 10.1016/B978-0-12-374593-4.00025-5

of the autonomic nervous system with adaptive behaviors. The theory emphasizes the phylogenetic shift in the features of the vagus, the primary parasympathetic cranial nerve involved in regulating visceral state. The theory is named polyvagal, since only in mammals does the vagus contain two distinct efferent pathways. In addition to an unmyelinated pathway originating in the dorsal motor nucleus of the vagus shared with other vertebrates (i.e., reptilians, amphibians, teleosts and elasmobrachs), mammals have a myelinated pathway originating in the nucleus ambiguus and this circuit communicates in the brainstem with source nuclei regulating the striated muscles of the face and head that are involved in several of the features of social communication (e.g., facial expression, vocalizations, listening).

The polyvagal theory provides a physiological basis to link uniquely mammalian anatomical and physiological circuits involved in visceral state regulation to both the expressive and receptive domains of social communication. The development of these features reflects the phylogenetic distinction between reptiles and mammals, and includes a cluster of behaviors dependent on the function of structures that phylogenetically emerged with mammals. These changes include detached middle ear bones, the advent of a diaphragm, and a myelinated vagal system regulating supradiaphragmatic organs that is distinct from an unmyelinated vagal system regulating subdiaphragmatic organs. The classic definition of mammals focuses on the observation of mammary glands and hair. However, the fossil record, since it cannot be explored through these features, relies on the identification of detached middle ear bones as the defining feature of mammals. Coincident with the separation of the middle ear bones, other phylogenetic transitions resulted in brainstem areas regulating the vagus becoming intertwined with the areas regulating the striated muscles of the face and head. The result of this transition was a dynamic social engagement system with social communication features (e.g., facial expression, head movements, vocalizations and listening) interacting with visceral state regulation.

The polyvagal theory articulates how each of three phylogenetic stages of the development of the vertebrate autonomic nervous system is associated with a distinct autonomic subsystem that is retained and expressed in mammals. These autonomic subsystems are phylogenetically ordered and behaviorally linked to social communication (e.g., facial expression, vocalization, listening), mobilization (e.g., fight–flight behaviors), and immobilization (e.g., feigning death, vasovagal syncope and behavioral suppression). Social communication (i.e., social engagement system, see Section III) involves the myelinated vagus, which serves to foster calm behavioral states by inhibiting the sympathetic influences to the heart and dampening the hypothalamic–pituitary axis (HPA) (e.g., Bueno et al., 1989). The mobilization system is dependent on the functioning of the sympathetic nervous system. The most phylogenetically primitive component, the immobilization system, is dependent on the unmyelinated vagus, which is shared with most vertebrates. With increased neural complexity due to phylogenetic development, the organism's behavioral and affective repertoire is enriched. The three circuits can be conceptualized as dynamically adjusting to provide adaptive responses to safe, dangerous and life-threatening events and contexts.

Unlike reptiles, the mammalian nervous system did not evolve solely to survive in dangerous and life-threatening contexts, but it evolved to promote social interactions and social bonds in safe environments. To accomplish this adaptive flexibility, the mammalian nervous system evolved a new neural strategy for safe environments, while retaining two more primitive neural circuits to regulate defensive strategies (i.e., fight–flight and death-feigning behaviors). It is important to note that social behavior, social communication and visceral homeostasis are incompatible with the neurophysiological states and behaviors promoted by the two neural circuits that support defense strategies. Thus, via evolution, the mammalian nervous system retains three neural circuits, which are in a phylogenetically organized hierarchy. In this hierarchy of adaptive responses, the newest circuit associated with social communication is used first, and if that circuit fails to provide safety, the older survival-oriented circuits are recruited sequentially.

By investigating the phylogeny of the regulation of the vertebrate heart (Morris and Nilsson, 1994; Porges, 1995, 1997; Taylor et al., 1999), four principles can be extracted that provide a basis for speculation regarding emergent behaviors and social communication. These principles lead to testable hypotheses relating the specific neural mechanisms supporting social engagement, fight–flight and death-feigning behaviors:

1. There is a phylogenetic shift in the regulation of the heart from endocrine communication, to unmyelinated nerves and finally to myelinated nerves.

2. There is a development of opposing neural mechanisms of excitation and inhibition to provide rapid regulation of graded metabolic output.
3. A face–heart connection evolved as the brainstem source nuclei of vagal pathways shifted ventrally from the older dorsal motor nucleus to the nucleus ambiguus. This resulted in an anatomical and neurophysiological linkage between the neural regulation of the heart via the myelinated vagus and the special visceral efferent pathways that regulate the striated muscles of the face and head, forming an integrated social engagement system (see Fig. 1) (for more details see Porges, 2001, 2003, 2007).
4. With increased cortical development, the cortex exhibits greater control over the brainstem via direct (e.g., corticobulbar) and indirect (e.g., corticoreticular) neural pathways originating in motor cortex and terminating in the source nuclei of the myelinated motor nerves emerging from the brainstem (e.g., specific neural pathways embedded within cranial nerves V, VII, IX, X and XI), controlling visceromotor structures (i.e., heart, bronchi) and somatomotor structures (muscles of the face and head).

III. The social engagement system

The study of comparative anatomy, evolutionary biology, and embryology provide important hints regarding the functional relationship between the neural control of the striated muscles of the face and head and emergent behaviors, including facial expressions, head movements and vocalizations. The nerves that control the muscles of the face and head share several common features. Pathways from five cranial nerves control the muscles of the face and head. Collectively, these pathways are labeled as special visceral efferent. Special visceral efferent nerves innervate striated muscles, which regulate the structures derived during embryology from the ancient gill arches (Truex and Carpenter, 1969). The special visceral efferent pathways regulate the muscles of mastication (e.g., ingestion), muscles of the middle ear (e.g., listening to conspecific vocalizations), muscles of the face (e.g., emotional expression), muscles of larynx and pharynx (e.g., prosody and intonation), and muscles controlling head tilt and turning (e.g., gesture). In fact, the neural pathway that enables the eyelids to be closed in a graded manner (i.e., winking) in humans is also involved in tensing the stapedius muscle in

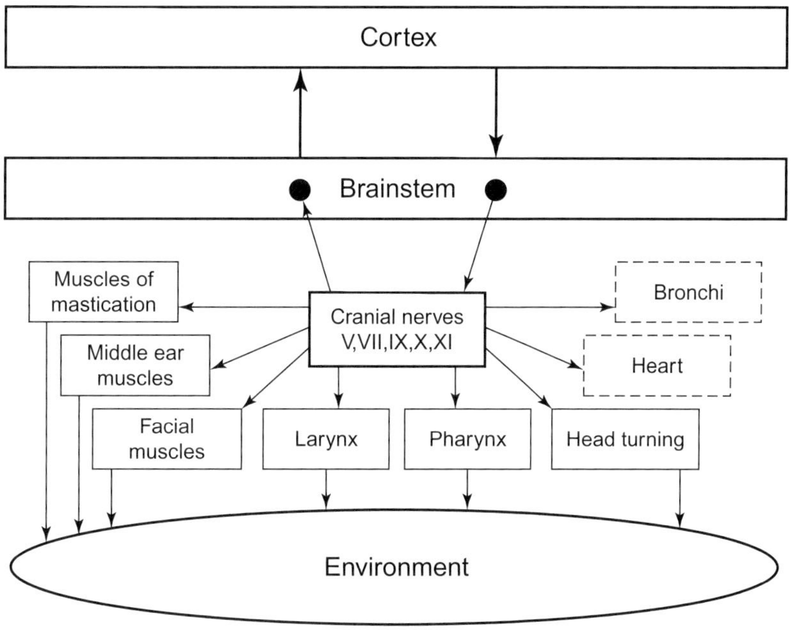

Fig. 1. The social engagement system. The social engagement system consists of a somatomotor component (i.e., special visceral efferent pathways that regulate the striated muscles of the face and head) and a visceromotor component (i.e., the myelinated vagus that regulates the heart and bronchi). Solid blocks indicate the somatomotor component. Dashed blocks indicate the visceromotor component.

the middle ear to facilitate hearing the human voice (Djupesland, 1976).

The source nuclei of the circuits regulating the striated muscles of the face and head interact in the brainstem with the source nucleus of the myelinated vagus, forming an integrated social engagement system. This system (as illustrated in Fig. 1) provides the neural structures involved in social and emotional behaviors. The social engagement system has a control component in the cortex (i.e., upper motor neurons) that regulates brainstem nuclei (i.e., lower motor neurons) to control eyelid opening (e.g., looking), facial muscles (e.g., emotional expression), middle ear muscles (e.g., extracting vocalizations from background noise), muscles of mastication (e.g., ingestion), laryngeal and pharyngeal muscles (e.g., vocalization) and head turning muscles (e.g., social gesture and orientation). Collectively, these muscles function as filters that limit social stimuli (e.g., observing facial features and listening to vocalizations) and determine engagement with the social environment. The neural control of these muscles determines social experiences by changing facial features (especially in humans and other primates), modulating laryngeal and pharyngeal muscles to regulate intonation of vocalizations, and coordinating both facial and vocal motor tone with respiratory actions (see also Smotherman et al., Chapter 9.2 in this volume). In addition, the frequency of breathing is encoded into the phrasing of vocalizations and may convey urgency by expressing short phrases associated with short exhalations (i.e., rapid breathing) or convey calmness by expressing long phrases associated with long exhalations (i.e., slow breathing).

The source nuclei (i.e., lower motor neurons) of the special visceral efferent pathways are located in the brainstem and communicate with the source nuclei of the mammalian myelinated vagus, which functions as a neural pathway inhibiting visceral functions (e.g., slowing heart rate, lowering blood pressure, etc.) to actively reduce arousal to promote calm states consistent with the metabolic demands of growth and restoration. Direct corticobulbar pathways reflect the influence of frontal areas of the cortex (i.e., upper motor neurons) on the regulation of this system. Moreover, afferent feedback, through the sensory vagus to medullary areas (e.g., nucleus of the solitary tract) influences forebrain areas and regulates states of arousal and alertness. In addition, the anatomical structures involved in the social engagement system have neurophysiological interactions with the HPA axis, the neuropeptides of oxytocin and vasopressin, and the immune system (for overview see Porges, 2001).

IV. Phylogeny of the middle ear

As vertebrates evolved from reptiles to mammals, the structures at the end of the mandible (i.e., jaw bone) that define components in the middle ear became detached (Rowe, 1996; Luo, 2007; Luo et al., 2001; Wang et al., 2001) and formed the small bones (auditory ossicles) of the middle ear. For humans and other mammals, sound in the environment impinges on the eardrum and is transduced from the eardrum to the inner ear via the ossicles in the middle ear. Bárány (1938) argued that the main purpose of the ossicular chain in land mammals was to reduce the effect of low-frequency sounds via bone conduction. In addition to the filtering imposed by bone separation, further attenuation can be achieved when the stapedius muscle (stabilizing stapes and innervated via a branch of the facial nerve) and the tensor tympani muscle (innervated via a branch of the trigeminal nerve) are contracted. Tension reduces the compliance of the ossicular chain and dampens the amplitude of the low-frequency acoustic stimulation from the environment reaching the inner ear. This process is similar to tightening the skin on a kettledrum. When the skin is tightened, the pitch of the drum is higher. When the ossicular chain is tightened, similar to the stretched skin, the movement of the eardrum is reduced and only higher frequencies bouncing against the eardrum are transmitted to the inner ear and to the auditory processing areas of the brain. The impact of these muscles on the perceived acoustic environment is to markedly attenuate low-frequency sounds, which facilitates the extraction of sounds in a higher frequency band associated with human voice and other mammalian vocalizations.

The detachment of the middle ear bones from the mandible was paralleled by two phylogenetic shifts: (1) the jawbone no longer restricted the cranium to expand and fostered the cortical development that characterizes modern mammals (Rowe, 1996); and (2) the mammalian middle ear enabled low-amplitude relatively high-frequency airborne sounds (i.e., sounds in the frequency of vocalizations) to be heard, even when the acoustic environment was dominated by low-frequency sounds.

The evolution of the mammalian middle ear enabled mammals to communicate in a frequency band that could not be detected by reptiles that, due to a dependence on bone conduction, were able to hear predominantly lower frequencies. This ability to hear low-amplitude high-frequency airborne sounds in an acoustic environment dominated by loud low-frequency sounds is accomplished when the middle ear muscles are tensed to create rigidity along the ossicular chain. This mechanism unmasks the high-frequency sounds associated with mammalian vocalizations from background sounds. Without stiffening the ossicular chain, mammals would lose this advantage and the soft airborne sounds of vocalizations would be easily lost in the acoustic background (see Borg and Counter, 1989). In fact, individuals who can voluntarily contract middle ear muscles exhibit an attenuation of approximately 30 dB at frequencies below 500 Hz, while there is no or minimal attenuation at frequencies above 1,000 Hz (see Kryter, 1985).

The earliest mammals were small and vocal communication outside the acoustic range of their predominant predators (i.e., reptiles) was crucial to survival. The physics of the middle ear resulted in a specific range of frequencies requiring a lower sound pressure level to be detected. In small mammals this frequency band was noticeably higher than the frequency band that large reptiles, due to a dependency on bone conduction, could easily detect. However, as mammals evolved, selective pressure resulted in larger mammals surviving. The size of the middle ear's structures increased with increased body size and the resonant frequency of the middle ear became lower. Thus, with large mammals (e.g., elephants and whales) the frequencies enhanced by middle ear structures foster infrasound communication with wavelengths that can travel over long distances and overlap with the lower frequencies that reptiles detect via bone conduction.

V. Impact of middle ear structures on sensitivity to conspecific vocalizations

The perception of sound is not equal at all frequencies. We hear sounds at low frequencies as if they were softer than their actual physical energy. In contrast, humans are relatively accurate in estimating the acoustic energy of frequencies associated with voice. This phenomenon initially reported as the Fletcher–Munson equal loudness contours (Fletcher and Munson, 1933), illustrated how human perception attenuated the "loudness" of low-frequency sounds. As measurement technologies improved, researchers refined the perceived loudness contours, and sound meters were modified to include a scale known as dB(A), which adjusted for the perceived differences in loudness as a function of frequency (i.e., the acoustic energy of lower frequencies had to be greatly increased to be perceived at the equivalent loudness of higher frequencies). This contrasts to sound pressure level, which describes the physical energy of the signal and does not apply any perceptually based weighting to the frequencies that constitute the acoustic stimulation.

The perceptual process of detecting conspecific vocalizations in background noise, illustrates the antimasking function of the middle ear muscles (attenuating the sounds at low frequencies). In addition to the antimasking function of the middle ear muscles, the middle ear structures act as a natural amplifier and contribute to the advantage in detecting conspecific vocalizations. Amplification occurs when the acoustic energy of vocalizations is characterized by a frequency band that overlaps with the resonance frequencies of the middle ear structures. Thus, due to the selective active antimasking by the middle ear muscles and the passive amplification by the middle ear structures, there are species-specific equal loudness contours. As a general rule, conspecific vocalizations occupy this frequency band of perceptual advantage.

Dammeijer and colleagues (2007) evaluated the effect of noise exposure on the stapedius muscle in the rat. Their data suggest that even in the absence of loud noise the stapedius is active, with contraction observed at sound pressure levels much lower than those needed to elicit the acoustic reflex threshold (Pilz et al., 1997). In addition, the data were consistent with the assumed purpose of the stapedius in unmasking high-frequency signals by attenuating low-frequency low-level constant noise in everyday life (Pang and Guinan, 1997). The middle ear muscles are composed primarily of relatively small fast-twitch fibers (de Jong et al., 1988). Since the muscles are characterized, at least in the rat, by relatively high anaerobic, glycolytic and aerobic oxydative enzyme activity, there is a natural resistance to fatigue. In addition, the numerous motor end plates, in combination with the numerous axon bundles, are indicative of small motor units, and support the assumption that the middle ear muscles are able to perform finely graded contractions. In our laboratory, we are demonstrating a similar intensity-tuned function of the middle ear muscles in humans.

The antimasking role of the middle ear muscles in auditory processing is especially relevant when considering the impact of low-frequency background sounds on cochlear mechanisms. Standing waves from pure tone stimuli occupy larger regions of the basilar membrane as the intensity increases, reducing the sensitivity of cochlear filtering for intense stimuli. This process is observed in the flattening of the equal loudness profile. Thus, by attenuating low-frequency sound waves, even below the level of the acoustic reflex threshold, the graded contractions of the middle ear muscles would improve the frequency sensitivity and selectivity mediated by the outer hair cells in the frequency band of mammalian vocalizations.

The physics of the middle ear structures impose other filter characteristics. Although the stiffening of the ossicular chain functions as a highpass filter by contracting the middle ear muscles and dampening the influence of low-frequency sounds on the inner ear, the physical characteristics of the ossicular chain also influence the acoustic energy reaching the inner ear. Ossicle inertia determines the highest frequencies that can pass through the middle ear (Hemila et al., 1995). The high-frequency limit is inversely proportional to the ossicular mass. Although, in general, larger mammals have greater ossicular mass, there are instances of larger mammals being able to detect higher frequencies due to adaptive shifts in ossicle mass to facilitate the detection of sounds associated with predator, prey and cohort. For example, cats have excellent sensitivity at very high frequencies and may benefit from hearing high-frequency sounds produced by small rodents (Rosenzweig and Amon, 1955; Forsman and Malmquist, 1988).

VI. The frequency band of perceptual advantage

In very small mammals, the middle ear and inner ear structures can convey acoustic information in a range well above the audible sounds that humans can reliably detect. The upper frequency limit for audible sounds in humans is approximately 20,000 Hz. Thus, 20,000 Hz is often used to delineate audible from ultrasound in the acoustic spectrum. However, the terminology is misleading, since acoustic stimuli characterized as ultrasonic are within the "audible" range of several mammalian species.

When audiograms of mammals are plotted on a logarithmic x-axis (for review see Fay, 1988), the lowest threshold is observed in a band of frequencies functionally defined by the high pass of the middle ear muscles and the low pass features of the combined effect of olivary–cochlear mechanisms and the inertia of the ossicular chain. This frequency band of perceptual advantage (i.e., lower thresholds to hear) is specific to each mammalian species with the smaller mammals, in general, having an advantage to hear higher frequencies. However, since low-frequency sounds dominate most acoustic environments this frequency band of perceptual advantage is optimized only when lower frequencies do not overwhelm the acoustic apparatus. Thus, the importance of contracting the middle ear muscles is to reduce acoustic energy in the lower frequencies. In general, it is within this frequency band of perceptual advantage that mammalian species produce most of their relevant conspecific vocalizations. For example, with the human, although the generally accepted frequencies for human hearing are between 20 and 20,000 Hz, the human frequency band of perceptual advantage includes a band of frequencies from approximately 500 Hz to about 4,000 Hz. Within these frequencies, the second and third formant in both male and female human speech always occur and in many cases so does the first formant. This select band, which conveys the information of human voice, is functionally amplified by the antimasking mechanisms of the middle ear muscles that dampen low-frequency activity and olivary–cochlear mechanisms that dampen high-frequency activity. Similar mechanisms in the rat provide a frequency band of perceptual advantage from approximately 5 kHz to 50 kHz (see Bjork et al., 1999).

Auditory information can be weighted within this frequency band to improve the extraction of human speech. Two such methods are known as the "index of articulation" (Kryter, 1962) and the more recent "speech intelligibility index" (American National Standards Institute, ANSI, 1997). These indices emphasize the relative importance of specific frequencies in conveying speech-related information embedded in human vocalizations. In the normal ear, acoustic energy within the primary frequencies of these indices is not attenuated, as it passes through the middle ear structures to the inner ear. The frequency band defining the index of articulation is similar to the frequency band that composers have historically selected to express melodies. It is also the frequency band that mothers have used to calm their infants by singing lullabies. Modulation of the acoustic energy within the frequencies of human voice that characterize music, similar to vocal prosody, will recruit and modulate the neural regulation of the middle ear muscles, functionally calm the behavioral and physiological state by increasing vagal regulation of the heart, and promote more spontaneous social engagement behaviors. Vocal music duplicates the effect of vocal prosody and triggers neural mechanisms that regulate the entire social engagement system with the resultant changes in facial affect and autonomic state. Basically, we start to look and feel better when we listen to melodies.

The auditory system is capable of compressing the range of acoustic stimuli. Much of this compression occurs at the periphery. For example, stapedius muscle contraction reduces sound transmission at low frequencies. Zwislocki (2002), said: "placement of stimulus compression in the auditory periphery must have been an important evolutionary adaptation through which the remaining system can operate within a biologically more easily achievable range" (p. 14,601). As Zwislocki suggests, by compressing the energy of loud low-frequency sounds, the inner ear and higher brain structures can process the higher frequency content of the acoustic signal in the range of vocalizations. From an engineering point of view, placing a mechanism to compress the nonlinearity at the periphery of a system reduces the dynamic range required of the remaining parts of the system. This compression mechanism would function as an automatic gain control to filter the acoustic energy at frequencies lower than the vocalizations and would allow higher brain structures to extract meaning and syntax by processing the acoustic energy in this narrower frequency range.

Although humans and other mammals can vocalize outside the frequency band of perceptual advantage, the within-species social communication is usually characterized by frequency modulated vocalizations within this frequency band. In contrast, danger and pain signals may be shrill cries (i.e., high pitch with diminished frequency modulation) at the upper edge of this frequency band. In addition, aggressive signaling may push vocalizations to lower frequencies outside this band (e.g., the roar of a lion). This preference to vocalize in a social context within the frequencies most easily detected by conspecifics has clear adaptive features, but it also creates challenges. In particular, the frequencies of these vocalizations are dependent on the processing of airborne acoustic energy and are above the frequencies easily conveyed through bone conduction. The higher frequencies of airborne mammalian vocalizations (audible and ultrasound) are characterized by very short wavelengths that dissipate rapidly with distance from the source. In contrast, low frequencies have long wavelengths that travel over long distances.

The short wavelength of mammalian vocalizations evolved with convergent mechanisms to aid in adaptive social engagement behaviors (see Porges, 2007). For many mammals, including humans, facial expressivity and behavioral gestures (e.g., use of the hands by primates) are coordinated with the shifts in prosody (intonation) to reduce ambiguity of the acoustic message (Corballis, 2003). Thus, the signals of distress and danger often require concordant facial cues and hand gestures (see also Eberl, Chapter 11.1 in this volume). Areas in the temporal cortex are sensitive to this cross-modal binding of auditory–visual inputs during vocalizations. Congruent speech-related visual input activates supra-additive multisensory neurons in the superior temporal cortex. In contrast, reduced activation in these areas has been reported during tasks requiring the integration of auditory and visual language inputs in schizophrenics, a disorder frequently associated with auditory hallucinations (Surguladze et al., 2001). Functionally, the simultaneous observation of facial and head movements, while listening to human vocalizations, improves speech intelligibility (e.g., McGurk and MacDonald, 1976; Munhall et al., 2004) and has been reported to increase the ability to extract speech from background sounds by approximately 10–20 dB (Sumby and Polack, 1954; Chen and Rao, 1998).

One of the consequences of depending on high frequencies for social communication is that infants cannot stray far from the protection of their mother. In many small mammalian species (e.g., rats, mice) the predominance of ultrasound vocalizations by infants further restricts the distance that the caregiver can move from the litter. In rats the frequencies used to communicate change developmentally. As rat pups develop and express exploratory behaviors, vocalizations shift from infantile ultrasound to adult type communication, which may also include audible sounds (Takahashi, 1992). As the rat pups mature, well-organized mobilization behaviors support exploration and the pups extend the distance they explore from the mother. Paralleling this shift towards adult vocalizations are increases in the neural regulation of the larynx and pharynx, structures involved in the production and articulation of vocalizations, with a parallel increase in the neural regulation of the heart via the myelinated vagus (Larson and Porges, 1982).

VII. Adaptive cost for active listening to vocalizations

There may be a cost for actively dampening the sensitivity to low-frequency sounds and engaging the neural mechanisms involved in listening to the frequency band of perceptual advantage. Listening to the frequency band of perceptual advantage requires the neural implementation of an active filter that reduces the acoustic information at low frequencies that can reach the brain. Since the sounds associated with predators, especially the movements of larger animals, are characterized by low-frequency sounds, engaging in this active "listening" process has maladaptive consequences by reducing the ability to detect predators. Thus, the advantage of listening to conspecific vocalization also comes at a cost. In the "wild" the potential cost of social communication is reduction of predator detection. The adaptive consequence of this vulnerability is to restrict or to limit listening to vocalizations within the frequency band of perceptual advantage (an important component of social engagement and social communication) primarily in safe environments such as nests and burrows.

VIII. The social engagement system and the polyvagal "vocalization" hypothesis

As proposed by the polyvagal theory, the functioning and development of the striated muscles of the face and head involved in listening and in production of vocalizations parallel the maturation of the myelinated vagus (see Larson and Porges, 1982). This developmental covergence, among several neural circuits that constitute an integrated functional social engagement system (see Fig. 1), results in facilitating several adaptive behaviors including: (1) an improved ability to regulate physiological state via myelinated vagal pathways (i.e., vagal brake) to both self-sooth and maintain calm states, as well as to mobilize by withdrawing the vagal brake to explore, forage and defend; (2) increased neural regulation of the larynx and pharnyx to promote conspecific vocalizations in the frequency band of perceptual advantage for the species to selectively signal peers and caregivers with a vocal mechanism that produces vocalizations with the highest adaptive value; and (3) improved thermoregulatory activity that reduces the need for the caregiver, as the autonomic nervous system matures.

The polyvagal theory emphasizes a phylogenetic parallel in the changing neural regulation of the autonomic nervous system and the neural regulation of the striated muscles of the face and head. This point is relevant to the study of mammalian vocalizations, since the striated muscles of the face and head are involved in both the detection of vocalizations (during listening) and in the production of these sounds through the coordination of the laryngeal and pharyngeal muscles with respiratory mechanisms.

The convergent phylogenetic changes in the neural regulation of the structures involved in the production and detection of mammalian vocalizations leads to the polyvagal hypothesis. Specifically, only mammals have a diaphragm to coordinate vocalizations with respiratory effort and volume. Consistent with the dependence on the mammalian diaphragm is the uniquely mammalian distinction between the two branches of the vagus: one dealing with supradiaphragmatic organs and the other dealing with subdiaphragmatic organs. The neural regulation of the subdiaphragmatic vagus is involved in abdominal breathing, while the neural regulation of supradiaphragmatic vagus is coordinated with the laryngeal and pharyngeal muscles that shape the acoustic features and provide facial expressions consistent with prosodic features of vocalizations. Moreover, slow exhalation, the respiratory process associated with expressive social vocalizations, enhances the impact of the myelinated vagus on the heart, promoting calm states.

The polyvagal hypothesis proposes that acoustic characteristics of vocalization not only serve to communicate to conspecifics relevant features in the environment, but also reflect the physiological state of the producer of the vocalizations. Mammals, but not reptiles or other phylogenetic antecedents to mammals, have a myelinated vagus, a diaphragm, detached middle ears, and neural circuits in the brainstem linking and coordinating the regulation of the myelinated vagus with the regulation of the striated muscles of the face and head. Specifically, engagement of this circuit conveys and expresses states of calmness and safety, and would be associated with greater vagal influences to the heart and lungs to promote calmer physiological states, increased neural tone to the middle ear muscles to optimize listening in the frequency band of perceptual advantage, and increased neural tone to the laryngeal and pharyngeal muscles to shift the acoustic energy of vocalizations to lower frequencies and increased frequency modulation within the band of perceptual advantage. In contrast, retraction of this circuit conveys and expresses states of danger and distress, and would be associated with faster heart rate and breathing and higher pitched vocalizations. Thus, in humans the features of vocal prosody are expressed during social interactions and reduced during both mental and physical illness. Similarly, the vocalizations of small mammals may have a parallel to human prosody. Rats, while playing or experiencing more positive affective states, modulate their ultrasonic vocalizations across a range of frequencies within the rat's band of perceptual advantage, while vocalizations communicating negative states such as danger are characterized by vocalizations at a relatively constant frequency through a different neural mechanism (e.g., Brudzynski, 2007).

Consistent with the polyvagal hypothesis, medically-compromised human infants have a high-pitched cry with little frequency modulation that is articulated in short bursts (Lester and Zeskin, 1982; Porter et al., 1988). Intonations of infant cries are regulated by neural tone, via source nuclei in the nucleus ambiguus, to the laryngeal muscles and the heart. In the physiologically stressed infant, decreased neural tone theoretically reduces the inhibitory effect on the heart and bronchi and the contraction of laryngeal muscles, thus producing dramatic increases in both heart rate and respiration rate and the fundamental frequency of the cry.

Porter and her collegues (Porter et al., 1988) reported a convergence between withdrawal of cardiac vagal tone (i.e., measured by respiratory sinus arrhythmia and mediated via the myelinated vagal pathways originating in the nucleus ambiguus) and the shift in fundamental frequency of newborn pain cries in response to circumcision. Cardiac vagal tone was significantly reduced during the severe stress of circumcision, and these reductions were paralleled by significant increases in the pitch of the infants' cries. These results document the important role that the social engagement system has in signaling and responding to "stress" and pain, and provide a demonstration that the vocalizations convey information regarding visceral and emotional state.

IX. Summary

The polyvagal theory emphasizes the phylogenetic shifts in the neural regulation of the autonomic nervous system and how this evolutionary shift in neural regulation converged with the regulation of the middle ear muscles to facilitate mammalian vocal communication. The theory emphasizes the different neural circuits that support defensive behaviors (i.e., fight–flight and freeze) and social interactions. According to the theory, during defensive states, when the middle ear muscles are not contracted, acoustic stimuli are prioritized by intensity and during safe social engagement states, acoustic stimuli are prioritized by frequency. During safe states, hearing of the frequencies associated with conspecific vocalizations is selectively being amplified, while other frequencies are attenuated. During the defensive states, the loud low-frequency sounds signaling a predator could be more easily detected and the soft higher frequencies of conspecific vocalizations are lost in background sounds. During social engagement behaviors, an integrated social engagement system regulates a shift in autonomic state to dampen sympathetic activity and to increase parasympathetic tone, while simultaneously increasing the neural tone to the striated muscles of the face and head (i.e., facial expressions, increased "emotional" cueing of the eyes associated with increased eye contact, increased prosody and enhanced listening by contracting the middle ear muscles). During social interactions, the stiffening of the ossicular chain actively changes the transfer function of the middle ear, and functionally dampens low-frequency sounds and improves the ability to extract conspecific vocalizations. However, the selectivity to listen to conspecific vocalizations comes at a cost, and the detection of lower acoustic frequencies generated by predators becomes more difficult. Thus, the identification and construction of safe contexts (e.g., burrows, nests, or houses) plays an important role in enabling the social engagement system to promote prosocial behavior.

References

Bárány, E., 1938. A contribution to the physiology of bone conduction. Acta Otolaryngol. Suppl. 26, 1–233.

Bjork, E., Nevalainen, T., Hakumaki, M., Voipio, H.-M., 1999. R-weighting provides better estimation for rat hearing sensitivity. Lab. Anim. 34, 136–144.

Borg, E., Counter, S.A., 1989. The middle-ear muscles. Sci. Am. 261, 74–80.

Brudzynski, S.M., 2007. Ultrasonic calls of rats as indicator variables of negative and positive states: acetylcholine-dopamine interaction and acoustic coding. Behav. Brain Res. 182, 261–273.

Bueno, L., Gue, M., Fargeas, M.J., Alvinerie, M., Junien, J.L., Fioramonti, J., 1989. Vagally mediated inhibition of acoustic stress-induced cortisol release by orally administered kappa-opioid substance in dogs. Endocrinology 124, 1788–1793.

Chen, T., Rao, R.R., 1998. Audio-visual integration in multimodal communication. Proc. IEEE 86 (5), 837–852.

Corballis, M., 2003. From mouth to hand: gesture, speech, and the evolution of right-handedness. Behav. Brain Sci. 26, 199–260.

Dammeijer, P., Dijk, P., Chenault, M., Manni, J., Mameren, H., 2007. Stapedius muscle fibre characterization in the noise exposed and auditory deprived rat. Hear. Res. 233, 54–66.

deJong, F., Kingma, H., Wirtz, P., Berge, H., Marres, E., 1988. Indications of a differentiated regulation of sound transmission by the middle ear muscles of the rat. Am. J. Otol. 9 (1), 70–75.

Djupesland, G., 1976. Nonacoustic reflex measurement: procedures, interpretations and variables. In: Feldman, A., Wilber, L. (Eds.), Acoustic Impedance and Admittance: The Measurement of Middle Ear Function. Williams and Wilkins Co., Baltimore, MD, pp. 217–235.

Fay, R.R., 1988. Comparative psychoacoustics. Hear. Res. 34, 295–306.

Fletcher, H., Munson, W.A., 1933. Loudness: its definition, measurement, and calculation. J. Acoust. Soc. Am. 5, 82–108.

Forsman, K.A., Malmquist, M.G., 1988. Evidence for echolocation in the common shrew, *Sorex araneus*. J. Zool. 216, 655–662.

Hemilä, S., Nummela, S., Reuter, T., 1995. What middle ear parameters tell about impedance matching and high frequency hearing. Hear. Res. 85, 31–44.

Kryter, K.D., 1962. Methods for the calculation and use of the articulation index. J. Acoust. Soc. Am. 34, 1689–1697.

Kryter, K.D., 1985. The Effects of Noise on Man. Academic Press, New York, NY.

Larson, S.K., Porges, S.W., 1982. The ontogeny of heart period patterning in the rat. Dev. Psychobiol. 15 (6), 519–528.

Lester, B.M., Zeskin, P.S., 1982. A biobehavioral perspective on crying in early infancy. In: Fizgerald, H., Lester, B., Yogman, M. (Eds.), Theory and Research in Behavioral Pediatrics. Plenum Press, New York, NY, pp. 133–180.

Luo, Z.X., 2007. Transformation and diversification in early mammal evolution. Nature 450 (7172), 1011–1019.

Luo, Z.X., Crompton, A.W., Sun, A.L., 2001. A new mammaliaform from the early Jurassic and evolution of mammalian characteristics. Science 292, 1535–1540.

McGurk, H., MacDonald, J., 1976. Hearing lips and seeing voices. Nature 264, 746–748.

Morris, J.L., Nilsson, S., 1994. The circulatory system. In: Nilsson, S., Holmgren, S. (Eds.), Comparative Physiology and Evolution of the Autonomic Nervous System. Harwood Academic Publishers, Chur, Switzerland, pp. 193–246.

Munhall, K.G., Jones, J.A., Callan, D.E., Kuratate, T., Vatikiotis-Bateson, E., 2004. Visual prosody and speech intelligibility: head movment improves auditory speech perception. Psychol. Sci. 15, 133–137.

Pang, X.D., Guinan, J.J., 1997. Effects of stapedius-muscle contractions on the masking of auditory-nerve responses. J. Acoust. Soc. Am. 102 (6), 3576–3586.

Pilz, P.K., Ostwald, J., Kreiter, A., Schnitzler, H.U., 1997. Effect of the middle ear reflex on sound transmission to the inner ear of rat. Hear. Res. 105, 171–182.

Porges, S.W., 1995. Orienting in a defensive world: mammalian modifications of our evolutionary heritage. A polyvagal theory. Psychophysiology 32, 301–318.

Porges, S.W., 1997. Emotion: an evolutionary by-product of the neural regulation of the autonomic nervous system. Ann. N.Y. Acad. Sci. 807, 62–77.

Porges, S.W., 2001. The polyvagal theory: phylogenetic substrates of a social nervous system. Int. J. Psychophysiol. 42, 123–146.

Porges, S.W., 2007. The polyvagal perspective. Biol. Psychol. 74, 116–143.

Porter, F.L., Porges, S.W., Marshall, R.E., 1988. Newborn pain cries and vagal tone: parallel changes in response to circumcision. Child Dev. 59, 495–505.

Rosenzweig, M.R., Amon, A.H., 1955. Binaural interaction in the medulla of the cat. Cell. Mol. Life Sci. 11 (12), 498–500.

Rowe, T., 1996. Coevolution of the mammalian middle ear and neocortex. Science 273, 651–654.

Sumby, W.H., Pollack, I., 1954. Visual contribution to speech intelligibility in noise. J. Acoust. Soc. Am. 26, 212–215.

Surguladze, S.A., Calvert, G.A., Brammer, M.J., Campbell, R., Bullmore, E.T., Giampietro, V., David, A.S., 2001. Audio-visual speech perception in schizophrenia: an fMRI study. Psychiatry Res. 106, 1–14.

Takahashi, L., 1992. Ontogeny of behavioral inhibition induced by unfamiliar adult male conspecifics in preweanling rats. Physiol. Behav. 52, 493–498.

Taylor, E.W., Jordan, D., Coote, J.H., 1999. Central control of the cardiovascular and respiratory systems and their interactions in vertebrates. Physiol. Rev. 79, 855–916.

Truex, R.C., Carpenter, M.B., 1969. Human Neuroanatomy. Williams & Wilkins, Baltimore, MD.

Wang, Y., Hu, Y., Meng, J., Li, C., 2001. An ossified Meckels cartilage in two Cretaceous mammals and origin of the mammalian middle ear. Science 294, 357–361.

Zwislocki, J.J., 2002. Auditory system: Peripheral nonlinearity and central additivity, as revealed in the human stapedius-muscle reflex. Proc. Natl. Acad. Sci. USA 99, 14601–14606.

CHAPTER 7.3

Medial cholinoceptive vocalization strip in the cat and rat brains: initiation of defensive vocalizations

Stefan M. Brudzynski*

Department of Psychology, Brock University, St. Catharines, Ontario, Canada

Abstract: Pharmacological studies performed on cat and rat brains are reviewed, which have allowed for identification of a widespread cholinoceptive system in the mammalian brain responsible for initiation of defensive vocalizations characteristic of aversive behavioral situations. Intracerebral injections of a predominantly muscarinic agent, carbachol, induced growling and hissing vocalizations in cats and 22 kHz ultrasonic alarm vocalizations in rats. Brain systems inducing these calls and their neurochemical organization in both the species show a very high degree of homology. This cholinoceptive substrate for these vocalizations, termed the medial cholinoceptive vocalization strip, is innervated by the ascending fibers from the brainstem cholinergic neurons located in the laterodorsal tegmental nucleus. This nucleus forms a cholinergic component of the ascending activating reticular system and its functions are discussed.

Keywords: medial cholinoceptive vocalization strip; pharmacological stimulation; reticular activating system; carbachol; acetylcholine; growling vocalization; 22 kHz vocalization; mapping; atropine; cat; rat

I. Introduction

Identification of the medial cholinoceptive vocalization strip has evolved slowly over nearly 40 years of research, conducted initially on cats and then on rats in a number of laboratories. The main evidence for the existence of a cholinoceptive substrate in the brain responsible for production and control of vocalization and its underlying emotional processes arose from behavioural–pharmacological studies in the early 1960s. Although chemicals were applied to the organism by different routes in these studies, it was the method of direct intracerebroventricular or intracerebral application of pharmacological agents into the restricted brain regions that allowed for bypassing the blood–brain barrier and that had given impetus to this line of study.

*Corresponding author. E-mail: sbrudzyn@brocku.ca

I.A. History of early behavioural–pharmacological studies in cats

Probably the first cholinergically-induced vocalization with concomitant emotional behavior was observed after injection of DFP (diisopropylfluorophosphonate) or large doses of acetylcholine into the lateral ventricle of a cat (Feldberg and Sherwood, 1954a,b). In these early studies, chemical agents were usually given systemically or intraventricularly in large doses and they affected almost the entire central nervous system, causing a complex behavioral outcome with numerous adverse or toxic side-effects. Thus, the observations were predominantly focused on autonomic manifestations (e.g., emesis) and pathological symptoms (e.g., motor dysfunctions, catatonia and epileptic seizures). Such outcomes created initial difficulty in identifying and localizing brain cholinoceptive regions responsible for vocalization and emotional arousal.

Stefan M. Brudzynski (Ed.)
Handbook of Mammalian Vocalization
ISBN 978-0-12-374593-4

DOI: 10.1016/B978-0-12-374593-4.00026-7

Early pharmacological studies provided substantial evidence that the cholinergic agents, both muscarinic and nicotinic, have a strong activating pharmacological effect on the central nervous system when given directly into the brain, i.e., bypassing the blood–brain barrier (Feldberg and Sherwood, 1954a,b; MacLean, 1957; Feldberg, 1963). Intracerebral injections of the cholinomimetic, pilocarpine, in cats induced a number of manifestations described as rage response with autonomic symptoms and growling vocalization (Borison, 1959; Zablocka and Esplin, 1964). Similar responses were reported after intraperitoneal or subcutaneous application of tremorine and were termed tremorine-induced rage response (Baker et al., 1960; Funcke et al., 1962; Koff and Langfitt, 1966). Tremorine is metabolized to oxotremorine, a potent muscarinic agonist (Cho et al., 1961; George et al., 1962), which was responsible for initiation of the tremorine-induced rage. Also, intravenous or intraventricular application of such cholinomimetics as oxotremorine, pilocarpine, arecoline, physostigmine (eserine), neostigmine, muscarine, McN-A-343 and carbachol were able to induce a similar rage response with accompanying manifestations (Leslie, 1965; Andjelkovič et al., 1971; Beleslin et al., 1973, 1974; Beleslin and Samardžič, 1977, 1979). Results of these early studies have also shown that the cholinergically-induced emotional manifestations and vocalization were antagonized by atropine, scopolamine and l-hyoscyamine, indicating muscarinic nature of the response (Baker et al., 1960; Leslie, 1965; Koff and Langfitt, 1966; Várszegi and Decsi, 1967; Romaniuk et al., 1973a; Beleslin et al., 1974).

Responses to these cholinergic muscarinic drugs were replicated by a direct intracerebral application of acetylcholine with physostigmine, physostigmine itself, or carbachol in a form of intracerebral crystalline microimplants in cats and rabbits (Nashold and Gills, 1960; Hernández-Peón et al., 1963). Chemostimulation by microimplants of crystals is disadvantageous from the pharmacological point of view. The crystals dissolve slowly in the interstitial fluid, creating local solutions of varying and probably locally very high concentrations. The induced symptoms often formed a combination of activation of physiologically opposite mechanisms (e.g., emotional activation and sleep atonia), activation of some receptors in a non-specific way at large doses, as well as deactivation of local circuits by an overstimulation effect (for that last effect, see Brudzynski and Eckersdorf, 1988). Nevertheless, it was possible to induce emotional responses (rage) in cats, with growling and hissing vocalizations, which were potentiated by provocation or approach of the experimenter, and that could even lead to a defensive attack. The most significant finding was that the brain presented a regional sensitivity to cholinergic stimulation and that cholinosensitive regions were stretched from rostral mesencephalon, through basal diencephalon to basal telencephalon (Nashold and Gills, 1960). The chemosensitive structures included the septum, anterior commissural region, preoptic area, medial hypothalamus, perifornical region and many other neighboring structures, as well as the periaqueductal gray (Hernández-Peón et al., 1963).

The question arose of how critical is the cholinoceptive region for induction of vocalization and emotional state. A series of localized electrolytic lesions placed in the cat limbic system were able to block the systemic tremorine-induced rage response and vocalization (Koff and Langfitt, 1966). The most effective lesions were found in many limbic structures including the posterior part of the hypothalamus (mammillary nuclei), septum, anterior fornical area, lateral amygdala and posterior hippocampus. It was also noticed that lesions placed in other subcortical and cortical regions did not have such a blocking effect (Koff and Langfitt, 1966). In another study, surgical isolation of the hypothalamus (surgical deafferentation) prevented development of the rage response after systemic oxotremorine (Gellén et al., 1972). Based on these results, the hypothalamus seemed to be the particular structure, or at least one of the critical structures, for expression of the cholinergically-initiated vocalization and underlying emotional response.

The series of studies that followed dealt with a direct intracerebral cholinergic stimulation of the hypothalamus and other limbic structures (e.g., septum) with carbachol or acetylcholine (Myers, 1964; Endröczi et al., 1964; Baxter, 1967; Várszegi and Decsi, 1967; MacPhail and Miller, 1968; Vahing and Allikmets, 1970; Romaniuk et al., 1973b; Allikmets, 1974; Decsi, 1974). All of these studies consistently reported a drug-induced emotional response, termed "rage response," "emotional-defensive response," or "emotional-aversive response" (Brudzynski, 1981b; Brudzynski and Eckersdorf, 1988) with prolonged growling vocalization and, to a lesser degree, hissing or spitting (a modified form of hiss). The response was accompanied by characteristic somatic and autonomic symptoms, such as back arching, retreat, striking with paws and sometimes attacking (when provoked), salivation, piloerection, mydriasis, hyperpnoea, tremor, etc. (Brudzynski, 1981a); all known from the studies with electrical stimulation of the relevant regions of cat hypothalamus and other brainstem structures (Hunsperger, 1956; Hunsperger and Bucher, 1967; Brown et al., 1969a,b; see also Siegel et al., Chapter 7.1 in this volume).

A similar emotional response was obtained by injection of acetylcholine to some other extrahypothalamic structures, such as the medial amygdala (bordering far lateral hypothalamus, Allikmets et al., 1969), or by injection of carbachol to the medial nucleus caudatus or ventral thalamic nucleus (Hull et al., 1967), as well as to the septum, intralaminar nuclei of thalamus and the red nucleus (Decsi, 1974; Decsi and Nagy, 1977). Carbachol could also induce the emotional-defensive response with growling vocalization from the periaqueductal gray matter, ventral tegmentum and the mesencephalic reticular formation (Baxter, 1968; Decsi, 1974; Karmos-Várszegi and Karmos, 1977a,b). In spite of the fact that the anteromedial hypothalamus and preoptic region seemed to be the main target for cholinomimetics in inducing the emotional-defensive response, there were other regions of the brain producing a similar response. The response, however, was structure-specific, and intracerebral injections of carbachol to the globus pallidus, putamen, dorsal and ventral hippocampus, all amygdaloid nuclei, or white matter were ineffective in inducing vocalizations or autonomic symptoms (Decsi, 1974). Similarly, a microinjection of carbachol close to the locus coeruleus induced sleep-like atonia and not an emotional response (van Dongen et al., 1978).

A further careful mapping of the brain was needed for the carbachol-induced response and to redefine the cholinoceptive regions associated with induction of vocalization and the emotional response in the cat. It was also important to answer the question "What behavioral response will be mapped?" i.e., to clarify whether the cholinergically-induced response is defensive (retreat) or offensive (attack) in nature.

I.B. Nature of the carbachol-induced behavioral response in cats

Results of early studies on cats have indicated that the cholinergically-induced emotional response evoked from a number of brain structures was structure-specific and somewhat different for different brain regions, and it might not represent the offensive type of aggression. Frequent attempts of cats to escape from the experimental cage during the muscarine- or carbachol-induced response led to the conclusion that the cholinergically-induced response with attacks represents a "fear and irritable kind of aggression" (Beleslin and Samardžić, 1977).

Hernández-Peón and his colleagues were the first researchers to observe that attacks evoked during the cholinergically-induced rage were dependent on the stimulated structure of the brain (Hernández-Peón et al., 1963). It was later reported that cats, which showed a consistent attack response against mice before injection, retreated from a mouse placed in front of them under the carbachol-induced response; they also ignored milk presented to them (Baxter, 1967; Hull et al., 1967). On the other hand, carbachol injected into the ventral tegmentum made cats kill mice violently, while this response did not occur from the anterior hypothalamic region, and cats injected with carbachol in the hypothalamic region still retreated from a mouse (Karmos-Várszegi and Karmos, 1977a,b). As a matter of fact, cats injected with carbachol in the anterior hypothalamus were retreating from any other animal or large object placed in the cage in front of them (unpublished observations). Further detailed studies have clarified that, during the carbachol-induced affective state, cats always retreated from the researcher if they were given space to do so, and never attacked. The response was not aggressive but defensive in nature, justifying the term "emotional-defensive response" (Brudzynski, 1981a). The response was also aversive in nature. An extended hand toward the animal, which represented an "indifferent" stimulus without any aversive response in the control conditions, acquired aversive properties after administration of carbachol. Cats always retreated from the approaching human hand, and the magnitude of their growling vocalization was proportional to the distance between the cat and the hand (Brudzynski et al., 1993b).

The conclusions about the defensive nature of the cholinergically-induced affective state with growling vocalization were consistent with ethological observations that growling vocalization in cats appears only in defensive situations, e.g., it would be emitted by the defending cat cornered by an aggressive opponent (Leyhausen, 1979).

II. Quantitative mapping studies of cholinergically-induced vocalization in cats

The question arose, "What features of the cholinergically induced response should be measured for the brain mapping purpose?" Results of several studies have shown that, although many of the carbachol-induced symptoms were dose-dependent (Várszegi and Decsi, 1967; Beleslin and Stefanović-Denić, 1986), the time of growling vocalization, and to a lesser degree the number of growls, were not only dose-dependent but they reflected in the best way the time-course and the dynamics of the pharmacological response (Decsi et al., 1969; Brudzynski, 1981b; Brudzynski and Eckersdorf, 1988). Time of growling was also a predictive measure of the magnitude of the carbachol-induced response. Cumulated time of growling was inversely proportional to the distance between the cat and the gloved human hand as a threat stimulus and reflected the defensive and aversive nature of the response (Brudzynski et al., 1993b). Thus, the cumulated time of growling vocalization appeared to be the best parameter for quantitative mapping of the emotional response in the cat's brain.

Although initially some qualitative mapping attempts were done in the cat's brain (Allikmets, 1974; Brudzynski et al., 1973), a comprehensive quantitative mapping of the carbachol-induced response was accomplished more than 20 years later, summarizing responses from 215 injection sites (Brudzynski, et al., 1995). The brain regions from which the response could be induced by a single unilateral dose of 10 μg of carbachol in unprovoked cats extended along two axes in the brain: (1) longitudinally along the neuraxis from tegmentum, through the hypothalamus to the basal forebrain including nucleus of the diagonal band; and (2) vertically along the fornix, from the mediobasal hypothalamus to the septum (Fig. 1). This extended strip of tissue showed anatomic specificity but varied response intensity. An intensive response was obtained from many hypothalamic–preoptic and septal areas. Weak responses were recorded from some regions (e.g., from ventromedial hypothalamus, ventral thalamus, supraoptic nucleus), or the response could not be induced from some other structures (e.g., from the ventral portion of the posterior hypothalamus, ventral tegmental area, amygdala, large fiber tracks and lateral ventricle) (Fig. 2) (Brudzynski et al., 1995).

It was noted during the mapping study that there was a significant linear negative correlation between the magnitude of the vocalization response and the distance from the injection site to the wall of the third ventricle in the hypothalamus. At the same time, intraventricular injections were ineffective; even those that were done in close proximity to the intraventricular ependyma both in the lateral and third ventricle (Brudzynski et al., 1995). It was evident that the periventricular tissue, also termed periventricular stratum (Sutin, 1966), stretched along the third ventricle, and a strip of tissue which continued medially up to the septum were the most sensitive to carbachol stimulation and induced the longest vocalizations.

The muscarinic nature of this response needed then to be reconfirmed locally in the anteromedial hypothalamic region, since there were reports that nicotinic antagonists, given in very large doses, could partially or totally antagonize the carbachol-induced response. Carbachol-induced vocalization was reduced by more than 50% by large intracerebral doses of such ganglionic blocking agents as hexamethonium, tetraethylammonium, or mecamylamine. Interestingly, very high doses of noradrenaline (50 μg) injected into the same brain site before carbachol entirely blocked the response as well (Decsi et al., 1969). In addition, d-tubocurarine, a nicotinic antagonist which could induce a high-pitched meowing in cats when injected alone (Decsi and Karmos-Várszegi, 1969), showed an antagonistic effect to the carbachol-induced response. The d-tubocurarine antagonism seemed to be non-competitive, because it occurred even if the d-tubocurarine was injected into the brain site contralateral to the carbachol injection (Decsi et al., 1969).

When low doses, close to equimolar amounts, of the muscarinic antagonists were given intracerebrally before carbachol, the muscarinic nature of the response was evident. Carbachol-induced growling vocalization was dose-dependent, could not be antagonized by equimolar mecamylamine, a nicotinic antagonist, but was almost totally antagonized by pretreatment with equimolar amounts of atropine. At the same time, local injection of physostigmine, which potentiates cholinergic transmission, induced a similar response to that after carbachol (Brudzynski et al., 1990, 1995).

Other studies using two salts of atropine, atropine sulfate, which penetrates the blood–brain barrier, and atropine methyl nitrate, which cannot penetrate that barrier, provided indirect evidence that the endogenous acetylcholine in the basal forebrain and diencephalon plays a role in naturally occurring emotional-aversive responses with growling vocalization in cats (Brudzynski et al., 1990). A comparable growling vocalization in cats appeared after presentation of a dog (but not a cat) or after intracerebral injection of carbachol or physostigmine. The carbachol-induced response was blocked by systemic atropine sulfate, but was not affected by systemic atropine methyl nitrate (Brudzynski et al., 1990).

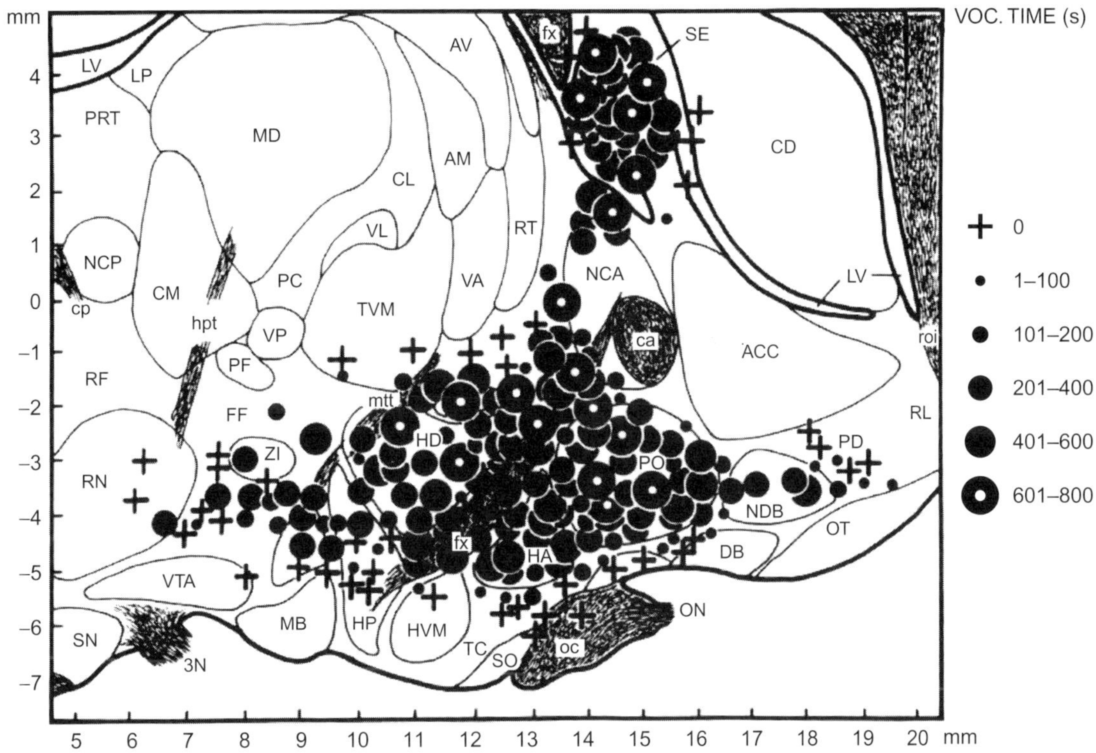

Fig. 1. Quantitative map of 215 unilateral injection sites for carbachol (10 μg) in the cat's brain shown on the parasagittal cross-section 2 mm lateral from the midsagittal plane. The injection sites are marked with symbols proportional to the magnitude of the growling vocalization (cumulative vocalization time over 30 minutes, VOC. TIME [s]). The symbols are listed on the right-hand side margin. The size of the circle corresponds to the magnitude of the vocalization time. The stereotaxic scales are in mm. Selected abbreviations: ACC: nucleus accumbens; ca: commissura anterior; CD: nucleus caudatus; CM: nucleus of centrum medianum; DB: diagonal band of Broca; FF: fields of Forel; fx: fornix; HA: anterior hypothalamic area; HD: dorsal hypothalamic area; HP: posterior hypothalamic area; HVM: ventromedial nucleus of hypothalamus; LV: lateral ventricle; MB: mammillary bodies; MD: thalamic mediodorsal nucleus; mtt: mammillothalamic tract; 3 N: oculomotor nerve; NCA: nucleus of anterior commissure; NDB: nucleus of the diagonal band; oc: optic chiasm; ON: optic nerve; OT: olfactory tubercle; PD: pre-diagonal area; PO: preoptic area; RF: reticular formation; RN: nucleus ruber; RT: reticular nucleus of thalamus; SE: septum; SN: substantia niagra; SO: supraoptic nucleus; TC: tuber cinereum; TVM: ventromedial nucleus of thalamus; VA: ventral anterior nucleus of thalamus; VTA: ventral tegmental area; ZI: zona incerta. Reprinted from Brudzynski et al. (1995) J. Psychiatr. Neurosci., 20: 119–132 with permission.

Thus, the central muscarinic cholinergic system was involved in production of growling vocalization and the underlying emotional-defensive response.

III. Quantitative mapping studies of cholinergically-induced vocalization in rats

Despite extensive studies of the effects of intracerebral carbachol in rats for at least 20 years, it was not possible to record vocalization in rats without the aid of a bat detector, which can lower sound frequency to an audible range and reveal the ultrasonic sounds. It was possible, however, to demonstrate the defensive nature of the response in rats and a number of autonomic changes. Grossman (1972) has found that neither cholinergic agents (carbachol, physostigmine) nor anticholinergic agents (atropine, scopolamine) given into the rat brain in crystalline form had any significant influence on shock-induced aggressive behavior or fighting for dominance, and suggested that aggression is not critically dependent on the activity of the cholinergic system. However, results of later studies have demonstrated that intrahypothalamic injections of carbachol or physostigmine in rats have increased the shock-induced defensive fighting, while injection of scopolamine reduced this fighting (Bell and Brown, 1980). Thus, the cholinergic agents caused potentiation of the

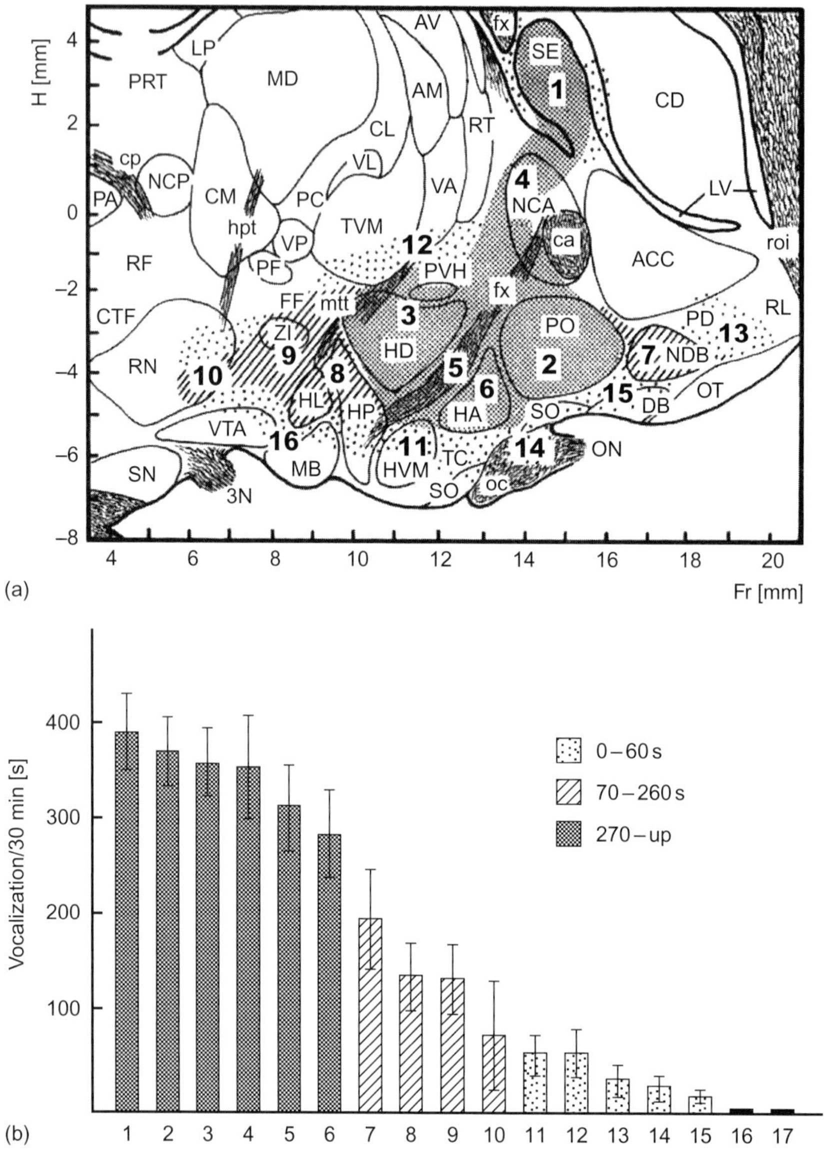

Fig. 2. A summary map of carbachol-induced responses shown on a parasagittal cross-section of the cat brain. (a) The responses are presented in cumulative vocalization as an average per anatomical structure and ranked according to their magnitude. (b) Numbers on the horizontal axis in (b) refer to the bold numbers in (a) and correspond to the following structures: 1: septal area (n = 27); 2: medial preoptic area (n = 29); 3: dorsal hypothalamic area and paraventricular nucleus (n = 26); 4: nucleus of commissure anterior (n = 8); 5: perifornical area (n = 19); 6: anterior hypothalamic area (n = 17); 7: nucleus of diagonal band (n = 8); 8: posterior and posterolateral hypothalamic area (n = 17); 9: fields of Forel and zona incerta (n = 16); 10: rostral nucleus ruber (n = 3); 11: ventromedial hypothalamic nucleus (n = 6); 12: ventral thalamic nuclei (n = 7); 13: prediagonal area (n = 7); 14: optic chiasm and supraoptic nucleus (n = 9); 15: diagonal band (n = 4); 16: mammillary bodies and ventral tegmental area (n = 2); 17: lateral ventricle (not marked in A, n = 4). Stereotaxic coordinates are in mm. Vertical lines in (b) represent SEMs. Reprinted from Brudzynski et al. (1995) J. Psychiatr. Neurosci., 20: 119–132 with permission.

defensive, but not offensive, behavior in rats, resembling the results obtained in cats. This interpretation is consistent with earlier findings that the centrally acting muscarinic antagonist, scopolamine, decreased defensive freezing in response to the presence of a cat and increased the number of approaches to the cat (Plotnik et al., 1974). Also, injections of cholinergic agents into the rat brain (intraventricularly or intrahypothalamically) induced autonomic manifestations, such as an increase in blood pressure, heart rate, muscular tremor, hyperthermia, and caused hyperglycemia and changes in the cortical EEG (George et al., 1966; Avery, 1970; Crawshaw, 1973; Carmona and Slangen, 1974; Hoffman and Phillips, 1976; Korner and Ramu, 1976), suggesting autonomic activation and general arousal.

Using a simple bat detector, Brudzynski and Bihari (1990) reported for the first time that unilateral injection of carbachol (1 μg) into the anteromedial hypothalamic area in the rat induced typical 22 kHz ultrasonic vocalizations (alarm calls) with duration of individual calls up to 1000 ms. Cholinergically-induced 22 kHz vocalizations from the anterior hypothalamic and preoptic areas did not differ substantially in any of the acoustic parameters from calls evoked naturally by hand-touch of naïve unhabituated rats or by foot-shock (Brudzynski et al., 1991b). Although the duration of single 22 kHz calls was 25% longer in the hand-touch and foot-shock conditions (up to 2,000 ms) than that in carbachol-induced calls (up to 1,500 ms), calls were of substantial duration in all studied situations. These results provided evidence that intracerebral carbachol induced typical alarm calls that were known to be emitted predominantly in anticipation of unavoidable aversive stimuli (van der Poel and Miczek, 1991; Blanchard et al., 1992; Brudzynski and Ociepa, 1992).

Results of these studies have strongly suggested that the rat central cholinergic system is involved in production of 22 kHz alarm calls. The magnitude of the vocal response was dose-dependent and could be antagonized by local pretreatment with equimolar concentrations of muscarinic antagonists (Brudzynski and Bihari, 1990; Brudzynski, 1994). Intracerebral pretreatment with an equimolar amount of atropine sulfate decreased the response by 85%. A quantitative functional mapping of the vocal response in the rat brain (at an average effective dose of 1 μg) revealed a medial cholinoceptive brain system stretched along the periventricular areas from tegmentum to the preoptic area and up to the septum (Brudzynski, 1994; Dencev et al., 1996) (Fig. 3). The strongest responses measured by a cumulated time of vocalization in seconds were obtained from the medial preoptic region, anterior hypothalamic area and lateral septal locations (Brudzynski, 1994; Dencev et al., 1996). Negligible or no responses were observed in the nucleus of the diagonal band, anterior commissure, substantia innominata, lateral and particularly far-lateral hypothalamus, ventromedial hypothalamic nucleus, ventral portion of the posterior hypothalamus, medial septal nuclei and ventricles. This cholinoceptive system showed remarkable homology with that in the cat brain.

Intracerebral carbachol also induced concomitant manifestations (other than vocalization) characteristic for a defensive behavioral response, including a crouched body posture, lowered head, freezing response and/or decrease in locomotor activity (Brudzynski and Mogenson, 1986). These manifestations were induced from the same brain regions as the 22 kHz ultrasonic vocalization. Brain mapping of carbachol-induced decrease in locomotor activity revealed a comparable brain system to that mapped for rat vocalization (Brudzynski et al., 1989).

IV. The medial cholinoceptive vocalization strip

The regions from which carbachol could induce 22 kHz vocalization in the rat appeared to be homologous to the system identified in the cat brain for cholinergic induction of growling vocalization (Brudzynski, 1998, 2001). This strip of tissue in both studied species was termed the medial cholinoceptive vocalization strip (Fig. 4, stippled area) (Brudzynski, 1998, 2001). It was postulated on the basis of the studies that in addition to anatomical homology, activation of this system is involved in homolog behavioral responses in these species. Both growling in cats and alarm calls in rats represent defensive and aversive types of vocalization, and both are accompanied by defensive behavioral patterns and autonomic activation. The precise communicative values of growling in cats and 22 kHz calls in rats might be different because of a substantial difference in social organization in these two species. Nevertheless, carbachol-induced vocalization could be induced from homolog brain structures and the released responses were associated with aversive contexts, threatening stimuli and/or unavoidable danger.

Both, cats' growling and rats' alarm calls are remarkably prolonged vocalizations. In cats, the developed single growl could last up to 11 seconds while in rats, single call duration reached over 3.9 seconds (Brudzynski et al., 1993a). Both cat and rat vocalizations were at the low end of the audible (to humans) frequency range for cats and at the low end of the ultrasonic frequency range for rats. It was also shown that both species responded to the respective vocalizations in their species-specific way, and consistently showed defensive behavioral patterns (Blanchard et al., 1991; Sales, 1991; Brudzynski and Chiu, 1995;

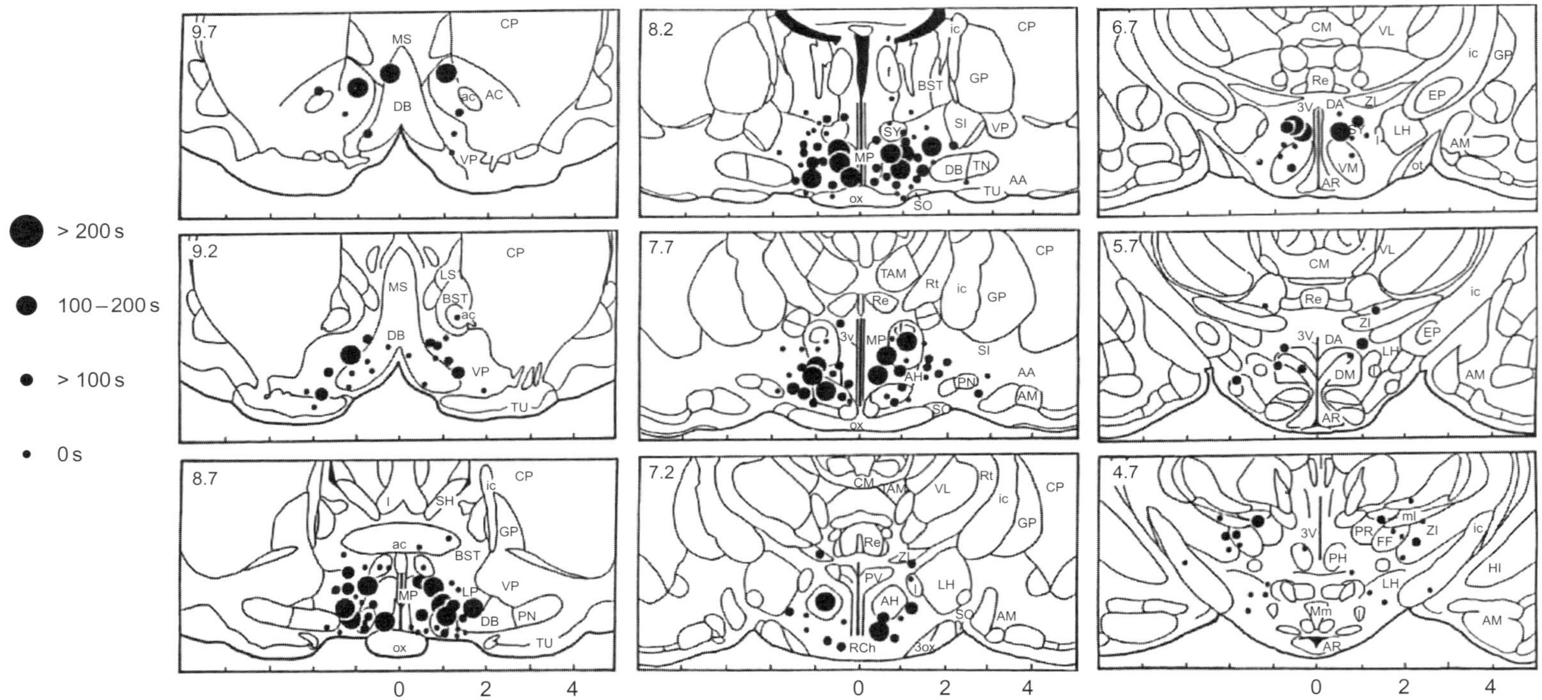

Fig. 3. Quantitative map of 208 unilateral injection sites for carbachol (1 μg) in the rat brain shown on the frontal sections between stereotaxic planes 5 and 10 mm anterior from the interaural zero plane. Successive stereotaxic planes are shown in the upper left-hand corner of each section. The injection sites are labeled with filled circles proportional in size to the magnitude of the induced 22 kHz ultrasonic vocalization. Vocalization was measured by cumulative vocalization time in seconds over 10 minutes of recording. The symbols are listed on the left-hand side margin. The size of the circle corresponds to the magnitude of the vocalization time. The stereotaxic scales are in mm. The septal injection sites are not shown on this diagram. Selected abbreviations: AA: anterior amygdaloid area; ac: anterior commissure; AC: nucleus accumbens; AH: anterior hypothalamic area; AM: amygdaloid complex; AR: arcuate hypothalamic nucleus; BST: bed nucleus of stria terminalis; CP: caudatoputamen; DA: dorsal hypothalamic area; DB: diagonal band; DM: dorsomedial hypothalamic nucleus; f: fornix; FF: fields of Forel; GP: globus pallidus; HI: hippocampal formation; LH: lateral hypothalamus; LP: lateral preoptic area; Mm: mammillary bodies; MP: medial preoptic area; MS: medial septal nucleus; PH: posterior hypothalamic area; PN: magnocellular preoptic nucleus; PR: pre-rubral field; PV: paraventricular hypothalamic nucleus; RCh: retrochiasmatic nucleus; Re: reuniens nucleus; Rt: reticular thalamic nucleus; SH: septohypothalamic nucleus; SI: substantia innominata; SO: supraoptic nucleus; TAM: anteromedial thalamic nucleus; TU: olfactory tubercle; SY: striohypothalamic nucleus; 3V: third ventricle; VM: ventromedial hypothalamic nucleus; VP: ventral pallidum; ZI: zona incerta. Reprinted from Brudzynski (1994) Behav. Brain Res., 63: 133–143 with permission.

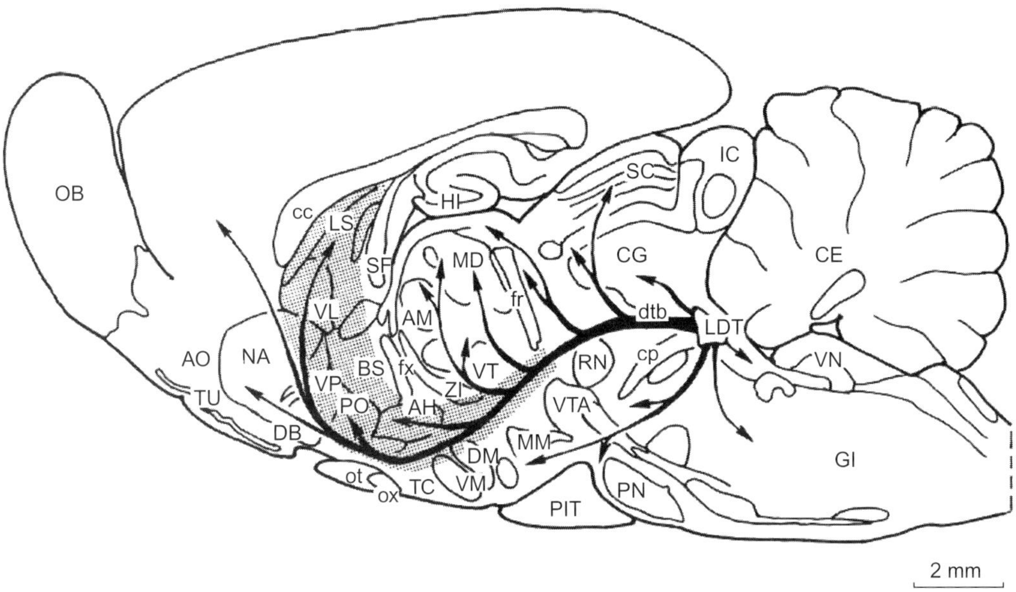

Fig. 4. Parasagittal cross-section of the rat brain 0.9 mm lateral from the midsagittal plane. The medial cholinoceptive vocalization strip is indicated by a stippled area. Ascending cholinergic projections from the laterodorsal tegmental nucleus (LDT) are superimposed on the diagram and shown by black arrows according to the study by Sato and Fibiger (1986). Selected abbreviations: AH: anterior hypothalamic area; AM: anteromedial thalamic nucleus; AO: anterior olfactory nucleus; BS: bed nucleus of stria terminalis; cc: corpus callosum; CG: central (periaqueductal) gray; cp: superior cerebellar peduncle; DB: horizontal limb of the diagonal band; DM: dorsomedial hypothalamic nucleus; dtb: dorsal tegmental bundle; fr: fasciculus retroflexus; fx: fornix; HI: hippocampal formation; LDT: laterodorsal tegmental nucleus; LS: dorsal portion of the lateral septal nucleus; MD: mediodorsal thalamic nucleus; MM: mammillary nuclei; NA: nucleus accumbens; ot: optic tract; ox: optic chiasm; PIT: pituitary gland; PN: pontine nuclei; PO: medial preoptic area; RN: red nucleus; SF: septofimbrial nucleus; TC: area of the tuber cinereum; TU: olfactory tubercle; VL: ventral portion of the lateral septal nucleus; VM: ventromedial nucleus of the hypothalamus; VP: ventral pallidum; VT: ventromedial thalamic nucleus; VTA: ventral tegmental area; ZI: zona incerta. Reprinted from Brudzynski (2001) Neurosci. Biobehav. Rev., 25: 611–617 with permission.

Brudzynski, 2001; for defensive patterns in cats, see Leyhausen, 1979; see also Wöhr and Schwarting, Chapter 4.2 in this volume).

It became apparent from the analysis of the location of the medial cholinoceptive vocalization strip that it represents a widespread terminal field of the ascending cholinergic pathways and the strip closely follows a portion of the ascending mesolimbic cholinergic pathways from the laterodorsal tegmental nucleus (Satoh and Fibiger, 1986; Cornwall et al., 1990). The terminal field is extensive and initiation of the emotional-defensive response with vocalization required stimulation of large brain regions (with a small dose) within the strip or more localized stimulation with a large dose of the drug (Brudzynski and Eckersdorf, 1988). However, activation of the source of the ascending cholinergic projection in the brainstem could induce this response from a very limited brain region. Such experiments were performed on both cats and rats.

Bilateral injection of a low dose of kainic acid into the region of the periaqueductal gray matter (with spread to the neighboring nuclei) in cats induced an emotional-defensive response with growling vocalization, while injection of the same dose into the anteromedial hypothalamic area did not induce such a response (Eckersdorf et al., 1996). Similar behavioral responses with growling, howling and hissing vocalization were induced in cats from a larger "defensive region" of the periaqueductal gray after unilateral injection of kainite and other excitatory amino acids, such as l-aspartate and d,l-homocysteate (Bandler and Carrive, 1988). Comparable microinjections of excitatory amino acids were performed into the periaqueductal gray matter of rats (Bandler et al., 1985; Bandler and Depaulis, 1988). Injections induced a number of behavioral defensive responses; however, the authors did not record ultrasonic vocalizations. Low doses of excitatory amino acids act as neuronal activator and can depolarize neurons within the area of the diffusion of the injected agent without affecting axons (Goodchild et al., 1982). These results have suggested that the excitatory amino acids, at least partially, could activate tegmental cholinergic neurons with the ascending pathways and induce defensive responses in this way.

The main group of cholinergic cell bodies is localized in the laterodorsal tegmental nucleus (Ch6 group), pedunculopontine nucleus (Ch5 group) and some neighboring structures, including the ventral periaqueductal gray (Honda and Semba, 1995; Motts et al., 2008; Wang and Morales, 2009). However, the main ascending projections to the diencephalic and forebrain regions originate in the laterodorsal tegmental nucleus (Satoh and Fibiger, 1986; Hallanger and Wainer, 1988; Cornwall et al., 1990; Woolf et al., 1990) (see Fig. 4). Direct injections of an excitatory amino acid, glutamate, directly into the laterodorsal tegmental nucleus in the rat induced the most complete and intensive defensive response with ultrasonic alarm calls (22 kHz vocalizations, Brudzynski and Barnabi, 1986; Bihari et al., 2003). The response was intensive, with a short latency of 15 seconds, which is 12-fold shorter than latency for carbachol-induced 22 kHz calls, and with a faster succession of emitted calls than that after carbachol.

This result was interpreted such that activation of cholinergic neurons of the laterodorsal tegmental nucleus, which have ascending projections to the medial cholinoceptive vocalization strip, caused a widespread release of acetylcholine from the terminals within all or most areas of the cholinoceptive strip, bringing about the fully-blown defensive response with alarm vocalization. In order to provide supportive evidence for this explanation, larger than usual volumes of atropine or scopolamine (muscarinic antagonists) were injected into the medial cholinoceptive strip to antagonize this response. After direct pretreatment of the preoptic/anterior hypothalamic areas with atropine or scopolamine, the glutamate-induced alarm vocalization from the laterodorsal tegmental nucleus was significantly reduced (Brudzynski and Barnabi, 1986). A similar experiment was repeated for the lateral septal region, from which carbachol can also release 22 kHz calls. Pretreatment of the lateral septum with scopolamine significantly reduced the defensive response, with alarm calls induced by direct injection of glutamate into the laterodorsal tegmental nucleus (Bihari et al., 2003).

In summary, the medial cholinoceptive vocalization strip represents a widespread terminal field of the ascending cholinergic projections from the laterodorsal tegmental nucleus. It is not known at present how large a portion of this strip has to be activated to initiate the behavioral response with vocalization. It could be a small number of neurons in a limited region, or larger pools of neurons spread over a considerable area. This question may be illustrated by a previous experiment on cats with mild kainite lesions. Doses of the kainic acid, which caused limited damage around the injection site, could not decrease the carbachol-induced emotional-defensive response to carbachol, which was subsequently injected into the kainite-treated brain site. However, similar treatment with kainite of the periaqueductal region in the vicinity of cholinergic cell bodies significantly reduced subsequent induction of the response with vocalization (Eckersdorf et al., 1987, 1996). This result indicated that, at the dose level studied, the hypothalamic neurons might be less sensitive to excitotoxic damage than tegmental neurons, and/or even small damage at the tegmental level has much larger functional consequences than comparable damage in the hypothalamus or other local part of the vocalization strip.

V. Function of the medial cholinoceptive vocalization strip

The medial cholinoceptive vocalization strip represents a portion of the ascending brainstem influences on the upper brain. The ascending cholinergic pathways from tegmentum to the diencephalic and forebrain regions form a fragment of brain system associated with the functions of the ascending reticular system. The ascending reticular system has been regarded for a long time as an important regulator of limbic functions (Moruzzi and Magoun, 1949; Shute and Lewis, 1967). The mesolimbic cholinergic component of the ascending reticular system (Lewis and Shute, 1967; Shute and Lewis, 1967) originates mainly from the laterodorsal tegmental nucleus. This nucleus shows many neurocytological features of the reticular core, with its neurons having long radiating dendrites with overlapping dendritic trees, typical of the reticular formation (Jones, 1995). The ascending cholinergic system appeared to be associated with many physiological states, such as wakefulness, paradoxical sleep, arousal and emotional arousal, including the states associated with emission of alarm vocalizations.

Since the ascending mesolimbic cholinergic system has diffuse character and innervates widespread areas of the brain, from the internal innervations of the reticular formation itself (Jones, 1995) to the medial cholinoceptive vocalization strip, function of this system has to be organized in a selective manner, i.e., this system must be able to selectively activate some structures while at the same time not activating other structures. The selective ascending influences may be organized anatomically by distinct pathways (e.g., dorsal and ventral pathway originating from the laterodorsal tegmental nucleus; Jones, 1995), or may depend on differences in the neurochemical substrate (different subtypes of postsynaptic cholinergic receptors on the target structures), or may depend on both these factors. The cholinergic component of the ascending reticular activating system traveling to the thalamus is mediated by nicotinic and muscarinic receptors (Curro et al., 1991), the reticular activating system controls the dopaminergic neurons by M5 muscarinic and nicotinic receptors (Forster et al., 2002; Yeomans et al., 2001; Wang et al., 2008), while effects of cholinergic input to the medial cholinoceptive vocalization strip could be mediated by M2 muscarinic receptors (Brudzynski et al., 1991a). The postsynaptic cholinergic receptors may also be critical for neuronal postsynaptic excitatory or inhibitory effects of the ascending cholinergic influences. Thus, the release of acetylcholine from these terminals in the medial cholinoceptive vocalization strip was shown to have widespread inhibitory effects on neuronal firing (Brudzynski et al., 1998), while the ascending cholinergic fibers reaching the lateral regions of the basal forebrain, supraoptic nucleus and pituitary region were found to be excitatory (Dreifuss and Kelly, 1970; Moss et al., 1972; Levine et al., 1986; Gribkoff et al., 1988; Lin et al., 1993; Jones, 2004).

Activation of the medial cholinoceptive vocalization strip not only initiates emission of vocalization, but is also associated with the initiation of a negative affective state observed by characteristic manifestations in animals (Brudzynski, 2007). In rats, this state was characterized as anxiety (Brudzynski and Holland, 2005) and production of 22 kHz alarm calls in rats may be treated as an index of a negative affective state (Knutson et al., 2002; Brudzynski, 2007). The ascending cholinergic reticular projections are capable of rapid changes in the state of the organism, including its affective component. The positive state was postulated to be initiated by the ascending dopaminergic system with the concomitant 50 kHz vocalization (Brudzynski, 2007; Burgdorf et al., 2008). The 50 kHz calls have been reflecting the appetitive state of the organism (see Burgdorf and Moskal, Chapter 6.2 in this volume). Thus, the ascending cholinergic and dopaminergic systems may work in a mutually exclusive way by a direct or indirect cholinergic–dopaminergic dialog (for details, see Brudzynski, 2007).

From the behavioral perspective, it is beneficial for social animals to signal their affective state to conspecifics. The best evidence for that comes from experiments on rats. Acoustic parameters of 22 and 50 kHz calls are generally non-overlapping (Brudzynski, 2007) and the recipients would have no difficulty in distinguishing between these signals. It was shown that rats recognize both types of vocalizations and show defensive responses to 22 kHz calls (Blanchard et al., 1991; Sales, 1991; Brudzynski and Chiu, 1995; Brudzynski, 2001) and approach responses to 50 kHz calls (Wöhr and Schwarting, 2007; Sadananda et al., 2008; see also Wöhr and Schwarting, Chapter 4.2 in this volume).

VI. Summary

The medial cholinoceptive vocalization strip represents a widespread target for the cholinergic portion of the ascending reticular activating system. The ascending cholinergic fibers originate from the laterodorsal tegmental nucleus and terminate in the vocalization strip localized in many medially-located hypothalamic/limbic structures. Activation of these ascending pathways and release of acetylcholine in the vocalization strip is postulated to be responsible for induction of a negative affective state with concomitant defensive and alarming vocalizations (growling in cats and 22 kHz alarm calls in rats). Thus emission of these vocalizations is indicative of the activity of this ascending cholinergic subsystem and the resulting negative affective state.

Acknowledgments

A portion of presented research results and preparation of this chapter were supported by the research grant from the Natural Sciences and Engineering Research Council of Canada.

References

Allikmets, L.H., 1974. Cholinergic mechanisms in aggressive behaviour. Med. Biol. 52, 19–30.

Allikmets, L.H., Vahing, V.A., Lapin, I.P., 1969. Dissimilar influences of imipramine, benactizine, and promazine of effects of micro-injections of noradrenaline, acetylcholine and serotonin in the amygdala in the cat. Psychopharmacologia (Berl.) 15, 392–403.

Andjelkovič, D., Beleslin, D.B., Vasič, B.V., 1971. Effect of eserine injected intraventricularly on behaviour and on activity of cholinesterase in some structures of the cerebral ventricles of the conscious cat. J. Pharm. Pharmacol. 23, 984–985.

Avery, D.D., 1970. Hyperthermia induced by direct injections of carbachol in the anterior hypothalamus. Neuropharm. 9, 175–178.

Baker, W.W., Hosko, M.J., Rutt, W.J., McGrath, J.R., 1960. Tremorine-induced rage and its antagonism by atropine. Proc. Soc. Exp. Biol. & Med. 104, 214–217.

Bandler, R., Carrive, P., 1988. Integrated defence reaction elicited by excitatory amino acid microinjection in the midbrain periaqueductal grey region of the unrestrained cat. Brain Res. 439, 95–106.

Bandler, R., Depaulis, A., 1988. Elicitation of intraspecific defence reactions in the rat from midbrain periaqueductal grey by microinjection of kainic acid, without neurotoxic effects. Neurosci. Lett. 88, 291–296.

Bandler, R., Depaulis, A., Vergnes, M., 1985. Identification of midbrain neurones mediating defensive behaviour in the rat by microinjections of excitatory amino acids. Behav. Brain Res. 15, 107–119.

Baxter, B.L., 1967. Comparison of the behavioral effects of electrical or chemical stimulation applied at the same brain loci. Exp. Neurol. 19, 412–432.

Baxter, B.L., 1968. Elicitation of emotional behavior by electrical or chemical stimulation applied to the same loci in the cat mesencephalon. Exp. Neurol. 21, 1–10.

Beleslin, D.B., Samardžić, R., 1977. Muscarine- and carbachol-induced aggressions: fear and irritable kinds of aggression. Psychopharmacology (Berl.) 55, 233–236.

Beleslin, D.B., Samardžić, R., 1979. Comparative study of aggressive behavior after injection of cholinomimetics, anticholinesterases, nicotinic, and muscarinic ganglionic stimulants into the cerebral ventricles of conscious cats: failure of nicotinic drugs to evoke aggression. Neuropsychopharmacology 60, 147–153.

Beleslin, D.B., Radmanovič, B.Ž., Grbovič, L., 1973. Adaptation to some central effects of carbachol after its intraventricular injection to conscious cats. Neuropharmacology 12, 495–496.

Beleslin, D.B., Grbovič, L., Radmanovič, B.Ž., 1974. The pharmacology of gross behavioural effects of cholinomimetic substances injected into the cerebral ventricles of unanaesthetized cats: evidence for central muscarinic mediation. Neuropharmacology 13, 1163–1169.

Bell, R., Brown, K., 1980. Shock-induced defensive fighting in the rat: evidence for cholinergic mediation in the lateral hypothalamus. Pharmacol. Biochem. Behav. 12, 487–491.

Bihari, A., Hrycyshyn, A.W., Brudzynski, S.M., 2003. Role of the mesolimbic cholinergic projection to the septum in the production of 22 kHz alarm calls in rats. Brain Res. Bull. 60, 263–274.

Blanchard, R.J., Blanchard, D.C., Agullana, R., Weiss, S.M., 1991. Twenty-two kHz alarm cries to presentation of a predator, by laboratory rats living in visible burrow system. Physiol. Behav. 50, 967–972.

Blanchard, R.J., Agullana, R., McGee, L., Weiss, S., Blanchard, D.C., 1992. Sex differences in the incidence and sonographic characteristics of antipredator ultrasonic cries in the laboratory rat (*Rattus norvegicus*). J. Comp. Psychol. 106, 270–277.

Borison, H.L., 1959. Effect of ablation of medullary emetic chemoreceptor trigger zone on vomiting responses to cerebral intraventricular injection of adrenaline, apomorphine, and pliocarpine in the cat. J. Physiol. (Lond.), 147, 172–177.

Brown, J.L., Hunsperger, R.W., Rosvold, H.E., 1969a. Defence, attack and flight elicited by electrical stimulation of the hypothalamus of the cat. Exp. Brain Res. 8, 113–129.

Brown, J.L., Hunsperger, R.W., Rosvold, H.E., 1969b. Interaction of defence and flight reactions produced by simultaneous stimulation at two points in the hypothalamus of the cat. Exp. Brain Res. 8, 130–149.

Brudzynski, S.M., 1981a. Carbachol-induced agonistic behavior in cats: aggressive or defensive response? Acta Neurobiol. Exper. 41, 15–32.

Brudzynski, S.M., 1981b. Growling component of vocalization as a quantitative index of carbachol-induced emotional-defensive response. Acta Neurobiol. Exper. 41, 33–51.

Brudzynski, S.M., 1994. Ultrasonic vocalization induced by intracerebral carbachol in rats: localization and a dose-response study. Behav. Brain Res. 63, 133–143.

Brudzynski, S.M. (1998). Role of the mesolimbic cholinergic pathways in the initiation of vocalization in cats and rats. Online presentation of the 5th Internet World Congress for Biomedical Sciences (INABIS '98) at McMaster University, Hamilton, ON Canada (http://www.mcmaster.ca/inabis98/brudzynski/brudzynski0219/index.htm), © Internet Association for Biomedical Sciences, 1998, pp. 1–8.

Brudzynski, S.M., 2001. Pharmacological and behavioral characteristics of 22 kHz alarm calls in rats. Neurosci. Biobehav. Rev. 25, 611–617.

Brudzynski, S.M., 2007. Ultrasonic calls of rats as indicator variables of negative or positive states. Acetylcholine-dopamine interaction and acoustic coding. Behav. Brain Res. 182, 261–273.

Brudzynski, S.M., Barnabi, F., 1986. Contribution of the ascending cholinergic pathways in the production of ultrasonic vocalization in the rat. Behav. Brain. Res. 80, 145–152.

Brudzynski, S.M., Bihari, F., 1990. Ultrasonic vocalization in rats produced by cholinergic stimulation of the brain. Neurosci. Lett. 109, 222–226.

Brudzynski, S.M., Chiu, E., 1995. Behavioural responses of laboratory rats to playback of 22 kHz ultrasonic calls. Physiol. Behav. 57, 1039–1044.

Brudzynski, S.M., Eckersdorf, B., 1988. Vocalization accompanying emotional-aversive response induced by carbachol in the cat: reproducibility and dose-response study. Neuropsychopharmacology 1, 311–320.

Brudzynski, S.M., Holland, G., 2005. Acoustic characteristics of air puff-induced 22-kHz alarm calls in direct recordings. Neurosci. Biobehav. Rev. 29, 1169–1180.

Brudzynski, S.M., Mogenson, G.J., 1986. Decrease in locomotor activity by injections of carbachol into the anterior hypothalamic/preoptic area of the rat. Brain Res. 376, 38–46.

Brudzynski, S.M., Ociepa, D., 1992. Ultrasonic vocalization of laboratory rats in response to handling and touch. Physiol. Behav. 52, 655–660.

Brudzynski, S.M., Gronska, J., Romaniuk, A., 1973. Studies on localization of defensive reactions in the hypothalamus of cats by the method of chemostimulation. Acta Physiol. Pol. 24, 631–641.

Brudzynski, S.M., McLachlan, R.S., Girvin, J.P., 1989. Cholinergically mediated reduction of locomotor activity from the basal forebrain of the rat. Exp. Neurol. 105, 197–205.

Brudzynski, S.M., Eckersdorf, B., Golebiewski, H., 1990. Evidence for involvement of endogenous acetylcholine in emotional-aversive response in the cat. Prog. Neuro-Psychopharmacol. & Biol. Psychiat. 14, 807–812.

Brudzynski, S.M., McLachlan, R.S., Girvin, J.P., 1991a. Involvement of M1 and M2 muscarinic receptors of the basal forebrain in cholinergically mediated changes in the rat locomotion. Prog. Neuro-Psychopharmacol. & Biol. Psychiat. 15, 279–284.

Brudzynski, S.M., Ociepa, D., Bihari, F., 1991b. Comparison between cholinergically and naturally induced ultrasonic vocalization in the rat. J. Psychiat. Neurosci. 16, 221–226.

Brudzynski, S.M., Bihari, F., Ociepa, D., Fu, X-W., 1993a. Analysis of 22 kHz ultrasonic vocalization in laboratory rats: long and short calls. Physiol. Behav. 54, 215–221.

Brudzynski, S.M., Eckersdorf, B., Golebiewski, H., 1993b. Emotional-aversive nature of the behavioral response induced by carbachol in cats. J. Psychiat. Neurosci. 18, 38–45.

Brudzynski, S.M., Eckersdorf, B., Golebiewski, H., 1995. Regional specificity of the emotional-aversive response induced by carbachol in the cat brain. A quantitative mapping study. J. Psychiatr. Neurosci. 20, 119–132.

Brudzynski, S.M., Kadishevitz, L., Fu, X.W., 1998. Mesolimbic component of the ascending cholinergic pathways: electrophysiological-pharmacological study. J. Neurophysiol. 79, 1675–1686.

Burgdorf, J., Kroes, R.A., Moskal, J.R., Pfaus, J.G., Brudzynski, S.M., Panksepp, J., 2008. Ultrasonic vocalizations of rats (*Rattus norvegicus*) during mating, play, and aggression: behavioral concomitants, relationship to reward, and self-administration of playback. J. Comp. Psychol. 122, 357–367.

Carmona, A., Slangen, J., 1974. EEG and heart rate changes elicited by chemical stimulation of the rat lateral hypothalamus. Pharm. Biochem. Behav. 2, 531–536.

Cho, A.K., Haslett, W.L., Jenden, D.J., 1961. The identification of an active metabolite of tremorine. Biochem. Biophys. Res. Commun. 5, 276–279.

Cornwall, J., Cooper, J.D., Phillipson, O.T., 1990. Afferent and efferent connections of the laterodorsal tegmental nucleus in the rat. Brain Res. Bull. 25, 271–284.

Crawshaw, L.I., 1973. Effects of intracerebral acetylcholine injection on thermoregulatory responses in the rat. J. Comp. Physiol. Psychol. 83, 32–35.

Decsi, L., 1974. Behavioral effect of intracerebrally injected carbachol on unrestrained cats. Pharmacol. Biochem. Behav. 2, 141–143.

Decsi, L., Karmos-Várszegi, M., 1969. Fear and escape reaction evoked by the intrahypothalamic injection of d-tubocurarine in unrestrained cats. Acta Physiol. Acad. Sci. Hung. 36, 95–104.

Decsi, L., Nagy, J., 1977. Adrenergic modulation of a cholinergic emotional reaction in the cat's thalamus. Psychopharmacology (Berl.) 54, 303–305.

Decsi, L., Várszegi, M.K., Méhes, J., 1969. Direct chemical stimulation of various subcortical brain areas in unrestrained cats. In: Lissak, K. (Ed.), Recent Development of Neurobiology in Hungary, Vol. 2. Académiai Kiadó, Budapest, Hungary, pp. 182–211.

Dencev, A., Hrycyshyn, A.W., Brudzynski, S.M., 1996. Cholinergic projection to the basal forebrain involved in the initiation of ultrasonic vocalization in the rat. Abstr. Int. Behav. Neurosci. Soc. 5, 60.

Dreifuss, J.J., Kelly, J.S., 1970. Excitation of identified supraoptic neurons by the iontophoretic application of acetylcholine. J. Physiol. (Lond.), 212, 68–69.

Eckersdorf, B., Głębiewski, H., Lewińska, M.K., 1987. Kainic acid lesions of the cat's antero-medial hypothalamus and emotional-defensive response evoked by carbachol injection to the same loci. Behav. Brain Res. 24, 161–166.

Eckersdorf, B., Głębiewski, H., Konopacki, J., 1996. Kainic acid versus carbachol induced emotional-defensive response in the cat. Behav. Brain Res. 77, 201–210.

Endröczi, E., Schreiberg, G., Lissák, K., 1964. The role of the central nervous activating and inhibitory structures in the control of pituitary-adrenocortical function. Effects of intracerebral cholinergic and adrenergic stimulation. Acta Physiol. Acad. Sci. Hung. 24, 211–221.

Feldberg, W., 1963. A Pharmacological Approach to the Brain from its Inner and Outer Surface. Evarts Graham Memorial Lecture, 1961. Edward Arnold (Publishers) Ltd, London, UK.

Feldberg, W., Sherwood, S.L., 1954a. Injections of drugs into the lateral ventricle of the cat. J. Physiol. 123, 148–167.

Feldberg, W., Sherwood, S.L., 1954b. Behaviour of cats after intraventricular injections of eserine and DFP. J. Physiol. 125, 488–500.

Forster, G.L., Yeomans, J.S., Takeuchi, J., Blaha, C.D., 2002. M5 muscarinic receptors are required for prolonged accumbal dopamine release after electrical stimulation of the pons in mice. J. Neurosci. 22, RC190.

Funcke, A.B.H., Bijlsma, U.G., Nauta, W.Th., 1962. Tremorine-induced rage and its antagonism by some antiparkinson drugs. Arch. Int. Pharmacodyn. 135, 447–453.

Gellén, B., György, L., Dóda, M., 1972. Influence of the surgical isolation of the hypothalamus on oxotremorine-induced rage reaction and sympathetic response in the cat. Acta Physiol. Acad. Sci. Hung. 42, 195–202.

George, R., Haslett, W.L., Jenden, D.J., 1962. The central action of a metabolite of tremorine. Life Sci. 1, 361–363.

George, R., Haslett, W.L., Jenden, D.J., 1966. The production of tremor by cholinergic drugs: central sites of action. Int. J. Neuropharmacol. 5, 27–34.

Goodchild, A.K., Dampney, R.A.L., Bandler, R., 1982. A method of evoking physiological response by stimulation of cell bodies, but not axons of passage, within

localized regions of the central nervous system. J. Neurosci. Methods, 6, 351–363.
Gribkoff, V.K., Christian, E.P., Robinson, J.H., Deadwyler, S.A., Dudek, F.E., 1988. Cholinergic excitation of supraoptic neurons in hypothalamic slices. Neuropharmacol. 27, 721–727.
Grossman, S.P., 1972. The ventromedial hypothalamus and aggression. Physiol. Behav. 9, 721–725.
Hallanger, A.E., Wainer, B.H., 1988. Ascending projections from the pedunculopontine tegmental nucleus and the adjacent mesopontine tegmentum in the rat. J. Comp. Neurol. 274, 483–515.
Hernández-Peón, R., Chávez-Ibarra, G., Morgane, P.J., 1963. Limbic cholinergic pathways involved in sleep and emotional behavior. Exp. Neurol. 8, 93–111.
Hoffman, W.E., Phillips, M.J., 1976. A pressor response to intraventricular injection of carbachol. Brain Res. 105, 157–162.
Honda, T., Semba, K., 1995. An ultrastructural study of cholinergic and non-cholinergic neurons in the laterodorsal and pedunculopontine tegmental nuclei in the rat. Neurosci. 68, 837–853.
Hull, C.D., Buchwald, N.A., Ling, G., 1967. Effects of direct cholinergic stimulation of forebrain structures. Brain Res. 6, 22–35.
Hunsperger, R.W., 1956. Affektreaktionen auf elektrische Reizung in Hirnstamm der Katze. Helv. Physiol. Pharmacol. Acta, 14, 70–92.
Hunsperger, R.W., Bucher, V.M., 1967. Affective behaviour produced by electric stimulation in the forebrain and brain stem of the cat. Prog. Brain Res. 27, 103–127.
Jones, B.E., 1995. Reticular formation: cytoarchitecture, transmitters, and projections. In: Paxinos, G. (Ed.), The Rat Nervous System, 2nd edn. Academic Press, San Diego, CA, pp. 155–171.
Jones, B.E., 2004. Activity, modulation and role of basal forebrain cholinergic neurons innervating the cerebral cortex. Prog. Brain Res. 145, 157–169.
Karmos-Várszegi, M., Karmos, G., 1977a. Comparative analysis of the emotional reactions elicited by direct cholinergic stimulation of the median hypothalamus and the ventral midbrain tegmentum. Proc. Hung. EEG Soc./ Electroenceph. Clin. Neurophysiol. 42, 853.
Karmos-Várszegi, M., Karmos, G., 1977b. A polygraphic study of the emotional reactions elicited by cholinergic stimulation of the ventral tegmentum and the anteromedial hypothalamus. Proc. Int. Union Physiol. Sci. 13, 369.
Knutson, B., Burgdorf, J., Panksepp, J., 2002. Ultrasonic vocalizations as indices of affective states in rats. Psychol. Bull. 128, 961–977.
Koff, G.Y., Langfitt, T.W., 1966. Tremorine-induced rage and the limbic system. Arch. Int. Pharmacodyn. 164, 272–285.
Korner, M., Ramu, A., 1976. Central hyperglycemic effect of carbachol in rats. Eur. J. Pharmacol. 35, 207–210.
Leyhausen, P., 1979. Cat Behaviour: The Predatory and Social Behaviour of Domestic and Wild Cats. Garland STMP Press, New York, NY.
Leslie, G.B., 1965. Central stimulant properties of compounds with peripheral muscarinic properties. Nature 208, 1291–1293.
Levine, M.S., Cepeda, C., Buchwald, N.A., 1986. Electrophysiological properties of basal forebrain neurons: responses to microphoretic application of putative neurotransmitters. Soc. Neurosci. Abstr. 12, 1468.
Lewis, P.R., Shute, C.C., 1967. The cholinergic limbic system: projections to hippocampal formation, medial cortex, nuclei of the ascending cholinergic reticular system, and the subfornical organ and supra-optic crest. Brain 90, 521–540.
Lin, J.Y., Li, C.S., Pan, J.T., 1993. Effects of various neuroactive substances on single-unit activities of hypothalamic arcuate neurons in brain slices. Brain Res. Bull. 31, 587–594.
MacPhail, E.M., Miller, N.E., 1968. Cholinergic brain stimulation in cats: failure to obtain sleep. J. Comp. Physiol. Psychol. 65, 499–503.
MacLean, P.D., 1957. Chemical and electrical stimulation of hippocampus in unrestrained animals. II. Behavioral findings. AMA Arch. Neurol. Psychiatry 78, 128–142.
Moruzzi, G., Magoun, H.W., 1949. Brain stem reticular formation and activation of the EEG. Electroenceph. Clin. Neurophysiol. 1, 455–473.
Moss, R.L., Urban, I., Cross, B.A., 1972. Microelectrophoresis of cholinergic and aminergic drugs on paraventricular neurons. Am. J. Physiol. 223, 310–318.
Motts, S.D., Slusarczyk, A.S., Sowick, C.S., Schofield, B.R., 2008. Distribution of cholinergic cells in guinea pig brainstem. Neurosci. 154, 186–195.
Myers, R.D., 1964. Emotional and autonomic responses following hypothalamic chemical stimulation. Canad. J. Physiol./Rev. Canad. Psychol. 18, 6–14.
Nashold Jr., B.S., Gills Jr., J.P., 1960. Chemical stimulation of telencephalon, diencephalon, and mesencephalon in unrestrained animals. J. Neuropathol. Exptl. Neurol. 19, 580–590.
Plotnik, R., Mollenauer, S., Snyder, E., 1974. Fear reduction in the rat following central cholinergic blockade. J. Comp. Physiol. Psychol. 88, 1074–1082.
Romaniuk, A., Brudzynski, S.M., Grońska, J., 1973a. The effect of chemical blockade of hypothalamic cholinergic system on defensive reactions in cats. Acta Physiol. Pol. 24, 809–816.
Romaniuk, A., Brudzynski, S.M., Grońska, J., 1973b. Rage reaction evoked by intrahypothalamic injection of carbachol in cats. Acta Physiol. Pol. 24, 623–630.
Sadananda, M., Wöhr, M., Schwarting, R.K., 2008. Playback of 22-kHz and 50-kHz ultrasonic vocalizations induces differential c-fos expression in rat brain. Neurosci. Lett. 435, 17–23.
Sales, G.D., 1991. The effect of 22 kHz calls and artificial 38 kHz signals on activity in rats. Behav. Processes 24, 83–93.
Satoh, K., Fibiger, H.C., 1986. Cholinergic neurons of the laterodorsal tegmental nucleus. Efferent and afferent connections. J. Comp. Neurol. 253, 277–302.
Shute, C.C., Lewis, P.R., 1967. The ascending cholinergic reticular system: neocortical, olfactory and subcortical projections. Brain 90, 497–520.
Sutin, J., 1966. The periventricular stratum of the hypothalamus. Int. Rev. Neurobiol. 9, 263–300.
Vahing, V.A., Allikmets, L.H., 1970. Behavioral and visceral reactions elicited by chemical stimulation of the

hypothalamus and septum in cats. Sechenov Physiol. J. 56, 38–47.

van der Poel, A.M., Miczek, K.A., 1991. Long ultrasonic calls in male rats following mating, defeat and aversive stimulation: frequency modulation and bout structure. Behaviour 119, 127–142.

van Dongen, P.A.M., Broekkamp, C.L.E., Cools, A.R., 1978. Atonia after carbachol microinjections near locus coeruleus in cats. Pharmacol. Biochem. Behav. 8, 527–532.

Várszegi, M.K., Decsi, L., 1967. Some characteristics of the rage reaction evoked by chemical stimulation of the hypothalamus. Acta Physiol. Acad. Sci. Hung. 32, 61–68.

Wang, H.L., Morales, M., 2009. Pedunculopontine and laterodorsal tegmental nuclei contain distinct populations of cholinergic, glutamatergic and GABA-ergic neurons in the rat. Eur. J. Neurosci, 29, 340–358.

Wang, H., Liang, S., Burgdorf, J., Wess, J., Yeomans, J., 2008. Ultrasonic vocalizations induced by sex and amphetamine in M2, M4, M5 muscarinic and D2 dopamine receptor knockout mice. PLoS ONE 3, e1893.

Wöhr, M., Schwarting, R.K., 2007. Ultrasonic communication in rats: can playback of 50-kHz calls induce approach behaviour? PLoS ONE 2, e1365.

Woolf, N.J., Harrison, J.B., Buchwald, J.S., 1990. Cholinergic neurons of the feline pontomesencephalon. II. Ascending anatomical projections. Brain Res. 520, 55–72.

Yeomans, J., Forster, G., Blaha, C., 2001. M5 muscarinic receptors are needed for activation of dopamine neurons and for rewarding brain stimulation. Life Sci. 68, 2449–2456.

Zablocka, B., Esplin, D.W., 1964. Central excitatory and neuromuscular paralyzant effects of pilocarpine in cats. Arch. Int. Pharmacodyn. 147, 490–496.

CHAPTER 7.4

Hypothalamic control of pain vocalization and affective dimension of pain signaling

George S. Borszcz* and Catherine A. Spuz

Behavioral and Cognitive Neuroscience Program, Department of Psychology, Wayne State University, Detroit, MI, USA

Abstract: Noxious tail-shock elicits vocalization afterdischarges (VADs) from rats that have distinct spectrographic characteristics and are a validated model of the affective response to pain. The ventromedial hypothalamus (VMH) is a core structure underlying the generation of affective behaviors to threats and receives nociceptive inputs. VAD-like vocalizations (vocalizations with the same spectral characteristics of VADs) are elicited by electrical or chemical stimulation of the dorsomedial-VMH. Pharmacological manipulation of $GABA_A$ receptors within the dorsomedial VMH altered the threshold for elicitation of VADs induced by this stimulation or by tail-shock, and altered the asymptotic level of fear conditioning supported by VMH stimulation or tail-shock. Partial kindling of the basolateral amygdala produced long-term increases in the amplitude of local field potentials recorded from the dorsomedial-VMH that correlated with long-term increases in VAD amplitude elicited by tail-shock. These findings demonstrate that the dorsomedial-VMH contributes to the processing of pain affect, and that the affective dimension of pain belongs to a broader class of sensory experience that represents threat to the individual.

Keywords: pain; affective pain; vocalization afterdischarges; nociception; ventromedial hypothalamus; periaqueductal gray; plasticity; defensiveness; fear; $GABA_A$; NMDA

Abbreviations: AP-5: 2-amino-5-phosphonopentanoic acid; BLA: basolateral amygdala; CNQX: 6-cyano-7-nitroquinoxaline-2,3-dione; CS: conditional stimulus; dlPAG: dorsolateral periaqueductal gray; dmVMH: dorsomedial portion of the ventromedial hypothalamus; GABA: gamma-aminobutyric acid; LFP: local field potential; LTP: long-term potentiation; NMDA: N-methyl-D-aspartic acid; SMR: spinal motor reflex; UR: unconditional response; US: unconditional stimulus; USV: ultrasonic vocalization; VAD: vocalization afterdischarge; VCR: vocalization–conditional response; VDS: vocalization during shock

I. Introduction

Mammalian vocalization is a vital component of defensive responding to environmental threats, among many social functions, in a variety of species (Darwin, 1872/1998). Ethological evaluation of the behavior of rodents, other mammals and humans revealed that individuals produce a series of defensive behaviors when confronted with a threat (Blanchard et al., 1986; Fanselow and Lester, 1988; Blanchard et al., 2001). Proximity of the individual to the threat (i.e., predatory imminence) and various environmental constraints (e.g., decrease in the availability of escape routes) govern the expression of particular behavioral patterns within a series of defensive responses. When contact with the predator is imminent and an escape

*Corresponding author. E-mail: borszcz@wayne.edu

ISBN 978-0-12-374593-4
DOI: 10.1016/B978-0-12-374593-4.00027-9

route is not available, rats, cats, monkeys and humans (as the most frequently studied organisms), but also other species engage in defensive aggression as exemplified by threat-attack behaviors that include the generation of vocalizations.

II. Vocalizations and the defensive aggression circuit

The neural circuit that controls the execution of defensive aggression is well-characterized (Siegel, 2005) and therefore provides a foundation to understand how imminent threats generate vocalizations (Fig. 1). The ventromedial hypothalamus (VMH), along with other interconnected medial hypothalamic nuclei (dorsal premammillary nucleus and anteriomedial hypothalamus–medial preoptic area), constitute a hypothalamic behavioral control system that governs the execution of innate defensive responses to environmental threats (Petrovich et al., 2001; Canteras, 2002). These hypothalamic nuclei exhibit c-Fos activation following exposure to either noxious or non-noxious threatening stimuli (Bullitt, 1990; Sandner et al., 1993; Beckett et al., 1997; Canteras et al., 1997; Liu et al., 1998; Rodella et al., 1998; Dielenberg et al., 2001), while inactivation or damage to these sites blocks naturally occurring defensive behaviors (Canteras et al., 1997; Cheu and Siegel, 1998; Markham et al., 2004). Stimulation of these medial hypothalamic nuclei elicits defensive responding accompanied by vocalizations in rats, cats and monkeys (Fernandez De Molina and Hunsperger, 1962; Lipp and Hunsperger, 1978; Milani and Graeff, 1987). In humans, stimulation of these sites generates reports of fear, anxiety and horror (Ervin et al., 1969; Iacono and Nashold, 1982; Tasker, 1982). Using positron emission tomography, activation of the medial hypothalamus in the human brain was observed during exposure to a traumatic painful stimulus that elicited an intense emotional experience (Hsieh et al., 1996). For all these species, vocalizations are part of their defensive responses to a potential or imminent threat (Fernandez De Molina and Hunsperger, 1962; Jürgens, 1979; Blanchard et al., 1986, 2001).

The dorsomedial portion of the VMH (dmVMH) sends glutaminergic projections to the dorsolateral column of the periaqueductal gray (dlPAG; Fig. 1) that interact with the postsynaptic NMDA receptors to generate defensive responding and vocalization (Beart et al., 1988; Schubert et al., 1996). These projections are activated by nociceptive input to the dmVMH.

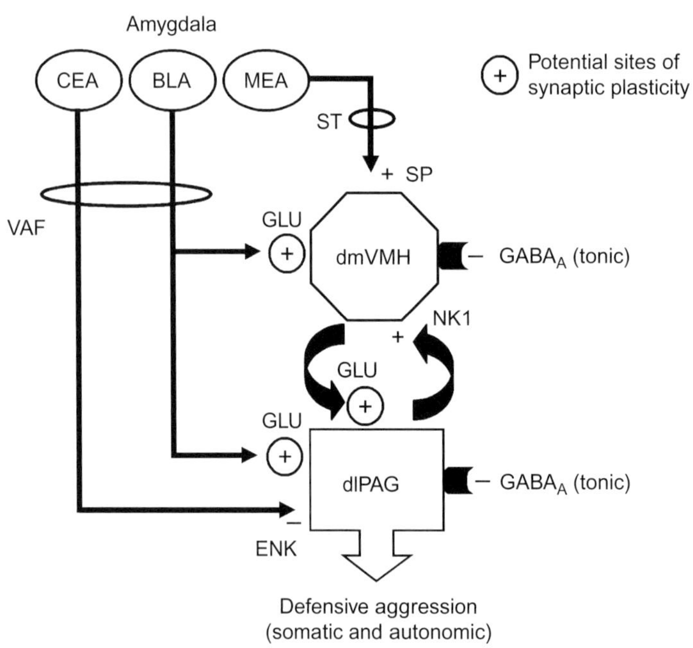

Fig. 1. Defensive aggression circuit. CEA: central amygdaloid nucleus; BLA: basolateral amygdaloid nucleus; MEA: medial amygdaloid nucleus; dmVMH: dorsomedial part of the ventromedial hypothalamus; dlPAG: dorsolateral and lateral periaqueductal gray; SP: substance P; GLU: glutamate; ENK: enkephalin; NK1: NKI receptors; VAF: ventral amygdalofugal pathway; ST: stria terminalis. Plus and minus signs indicated excitatory and inhibitory synapses, respectively.

Neurons within the dmVMH that exhibit c-Fos expression following presentation of a noxious cutaneous stimulus are double-labeled by administration of a retrograde tracer into the dlPAG (Parry et al., 2002). The dlPAG serves as the interface between the limbic forebrain sites that process threatening stimuli and execution of the innate defensive responses. Descending projections from the dlPAG to the brainstem coordinate the execution of the behavioral and autonomic responses that constitute defensive responding. Projections from the dlPAG to the rostral ventrolateral medulla initiate the autonomic components associated with defensive responding (Wang et al., 2002), whereas projections to the nucleus retroambiguus initiate activity in the laryngeal, articulatory and respiratory motor neurons that generate vocalizations (Jürgens and Pratt, 1979; Jürgens, 2002). In turn, neurons within the dlPAG send ascending excitatory projections to dmVMH sites from which defensive aggression is elicited (Bhatt et al., 2003). The reciprocal projections between dmVMH and dlPAG purportedly function as a positive feedback loop that sustains defensive responses during the course of a confrontation with an environmental threat.

The amygdala is the best characterized modulator of defensive responding generated from the medial hypothalamus and dlPAG (Fig. 1). Stimulation of the basolateral (BLA) and medial (MeA) amygdaloid nuclei generate defensive responses, and subthreshold stimulation potentiates defensive responses elicited by medial hypothalamic or dlPAG stimulation (Fernandez De Molina and Hunsperger, 1962; Egger and Flynn, 1967; Shaikh et al., 1994). Alternatively, the concurrent stimulation of central amygdaloid nucleus (CeA) with dmVMH or dlPAG results in inhibition of defensive responding elicited from these sites. The inhibition of dlPAG-elicited and dmVMH-elicited defensive aggression is mediated by enkephalinergic neurons in CeA that project to dlPAG and act on μ-opiate receptors (Shaikh et al., 1991; Shaikh and Siegel, 1994).

III. Nociceptive input to the defensive aggression circuit that generates vocalizations

As exposure to a noxious stimulus is the prototypical imminent threat to an individual, it is not surprising that noxious stimulation would engage neural circuits that govern execution of the innate defensive responses. Direct nociceptive projections to the dmVMH arise from cells in the superficial and deep layers of the spinal dorsal horn (Burstein et al., 1991; Newman et al., 1996; Zhang et al., 1999). The dmVMH also receives a dense projection from the superior lateral subnucleus of the mesencephalic division of the parabrachial area (Bester et al., 1997) that is innervated by interneurons of the spinal dorsal horn, which respond to noxious peripheral stimulation (Bester et al., 1995). Furthermore, the dmVMH is the only site of convergence of ascending projections from the dorsal horn interneurons innervated by peptide and non-peptide nociceptors (Todd, 2002).

In response to a brief noxious tail-shock, rats generate spinal motor reflexes (SMRs, i.e., tail-flick and hind limb movements) and vocalizations. Vocalization emitted during the shock (VDSs) is initiated by a mechanism organized within the medulla (Carroll and Lim, 1960; Borszcz et al., 1992), while vocalization afterdischarges (VADs), which occur immediately following the tail-shock, are integrated within the limbic forebrain and have distinct spectrographic characteristics (Fig. 2) compared to VDSs (Borszcz, 1995, 2006; Borszcz and Leaton, 2003). Research from this laboratory validated VADs as a rodent model of pain affect (Borszcz, 1993; Borszcz et al., 1994; Borszcz and Leaton, 2003). During these studies we also monitored defensive 22 kHz ultrasonic vocalization (USV). However, like other investigators, we found USV to be an unreliable measure of nociception (Jourdan et al., 2002).

IV. Stimulation of the hypothalamus generates affective vocalizations

Given the involvement of dmVMH in the production of defensive responses and the existence of nociceptive inputs to the dmVMH, we evaluated the contribution of the dmVMH to generation of pain affect as reflected by production of VADs (Borszcz, 2006). Electrical or chemical stimulation of dmVMH generated VAD-like vocalizations (vocalizations with the same spectrographic characteristics as VADs) (Fig. 3a,b). The amplitude, duration and 1/latency of VADs increased with the increase in the frequency of electrical stimulation of dmVMH (Fig. 3c). To assess whether VADs elicited from the dmVMH reflect activation of cell bodies rather than fibers of passage, the capacity of chemical stimulation of the dmVMH in eliciting VADs was evaluated. The dmVMH is under tonic $GABA_A$ inhibition, and injection of bicuculline ($GABA_A$ antagonist) into the dmVMH elicited vocalizations

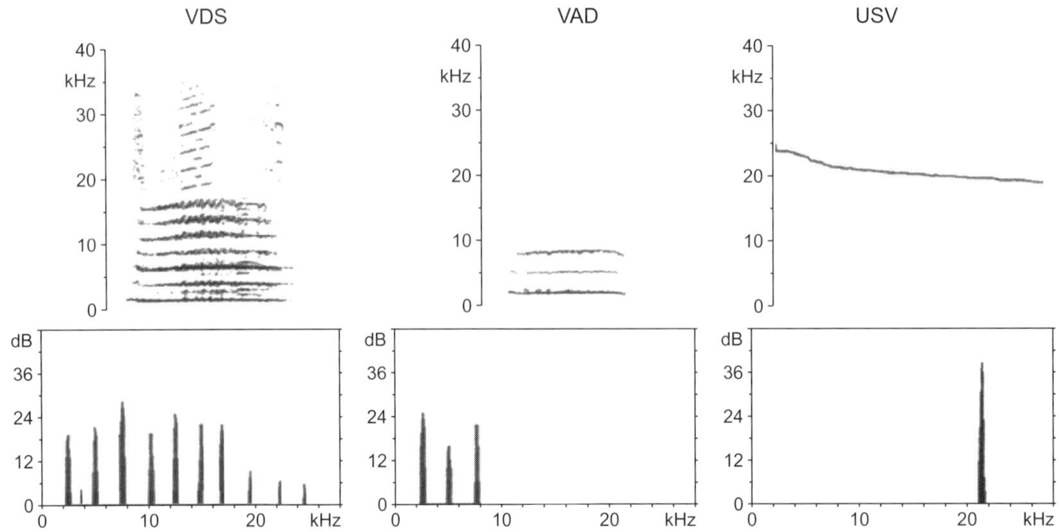

Fig. 2. Spectrograms (upper graphs) and power spectra (lower graphs) of vocalizations monitored during testing. VDS: vocalization during tail-shock; VAD: vocalization afterdischarge; USV: 22 kHz ultrasonic vocalization. Spectrograms show 2-second, 4-second and 500-millisecond bouts of time for VDS, VAD and USV, respectively.

in cats associated with defensive responses (Strzelczuk and Romaniuk, 1995). Microinjection of bicuculline into the dmVMH produced a dose-dependent increase in the VAD-like vocalizations. The number, peak amplitude and duration of VAD-like vocalizations recorded during each sampling epoch were directly related to the dose of bicuculline injected into dmVMH (Fig. 3d).

Although electrical and chemical stimulation of the dmVMH elicited VAD-like vocalizations, it had not been determined whether the dmVMH contributes to the generation of pain-elicited VADs. It is possible that VADs elicited by stimulation of the dmVMH or tail-shock are mediated by different integrative circuits within the limbic forebrain that converge upon a common set of brainstem neurons that coordinate the execution of the same vocal response (Jurgens, 2002). Therefore, we evaluated whether manipulation of $GABA_A$ receptors within the dmVMH, which modifies VADs elicited by electrical stimulation of the dmVMH, also modifies VADs elicited by tail-shock. VAD thresholds were recorded following administration of bicuculline in a subthreshold dose that did not elicit VADs (10 pmol), muscimol (a $GABA_A$ agonist; 100 pmol), or saline into dmVMH.

Drug treatment within the dmVMH produced corresponding changes in the thresholds of VADs elicited by dmVMH stimulation or tail-shock. Thresholds for VADs induced by electrical stimulation of the dmVMH or by tail-shock were raised by muscimol and lowered by bicuculline (Fig. 4a,b). The stimulus–response functions for VADs obtained by both these methods were shifted to the right by muscimol and shifted to the left by bicuculline. Although VAD thresholds were altered by drug treatments, performance of VADs was not affected, as indicated by the parallel stimulus–response functions and the equivalent maximum responding exhibited under the different drug treatments (Fig. 4a,b). Therefore, changes in VAD thresholds do not reflect the effects of drugs on the capacity of rats to vocalize. On the other hand, thresholds of VDS and SMR elicited by tail-shock were not altered by bicuculline or muscimol injections into the dmVMH.

V. Production of affective vocalizations predicts fear conditioning

Innate defensive responses elicited by noxious or threatening stimuli are accompanied by a negative affective state (i.e., unconditional fear). If a response to a noxious or threatening stimulus is a direct reflection of this affective state then its elicitation should predict fear conditioning. We provided evidence that the capacity of tail-shock to support fear conditioning correlates with its capacity to elicit VADs (Borszcz, 1993, 1995; Borszcz and Leaton, 2003). If the dmVMH contributes to processing the affective dimension of pain, then pharmacological manipulation

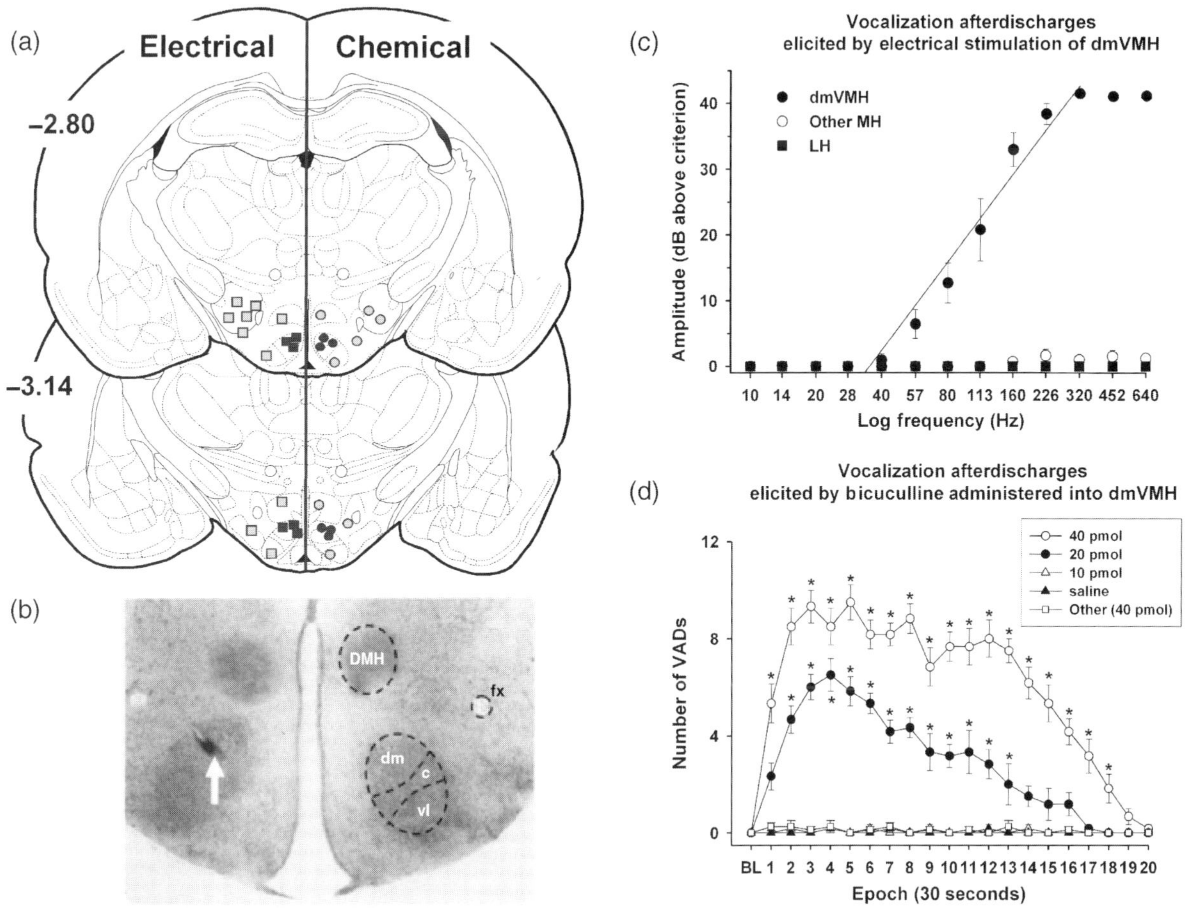

Fig. 3. (a) Histological reconstruction of sites that received unilateral electrical stimulation or chemical stimulation with bicuculline. For the sake of clarity all electrical stimulation sites are indicated on the left side of the diagrams and chemical stimulation sites on the right side of the diagrams. Black squares and circles indicate sites within the dorsomedial division of the ventromedial hypothalamus (dmVMH) where stimulation was effective in eliciting vocalization afterdischarges (VADs). Gray squares and circles indicate sites outside the dmVMH where stimulation was ineffective in eliciting VADs. Coordinates are in millimeters posterior to bregma. Diagrams are from the rat brain atlas of Paxinos and Watson (1998). (b) Nissl stained section of the medial hypothalamus showing a site (arrow) of electrical stimulation (electrodes were implanted at a 12 degree angle). Abbreviations: dm: dmVMH; c: core VMH; vl: ventrolateral VMH; DMH: dorsomedial hypothalamic nucleus; fx: fornix. (c) Mean psychophysical functions relating log frequency of brain stimulation to mean peak amplitude (±SEM) of VADs. Functions are from groups of rats that received stimulation of the dmVMH ($n = 6$), lateral hypothalamus (LH, $n = 5$), or other sites (Other, $n = 5$) in the vicinity of the dmVMH. (d) Mean (±SEM) number of VADs elicited during each 30 second epoch following administration of bicuculline into the dmVMH or sites in the vicinity of dmVMH (Other). Asterisks indicate significantly greater number of VADs when compared to the 30 second epoch immediately prior to bicuculline administration (BL = baseline, Dunnett's multiple comparison test, $p < 0.05$).

of the $GABA_A$ receptors in the dmVMH that altered thresholds of VADs elicited by tail-shock or dmVMH stimulation will also alter the capacity of the eliciting stimuli to support fear conditioning.

Fear conditioning was conducted using the Pavlovian conditional vocalization paradigm (Borszcz, 1995; Borszcz and Leaton, 2003), in which vocalizations (VCRs) are conditioned to a conditional stimulus paired with an aversive unconditional stimulus. The conditional stimulus (CS) was light provided by a 60 watt incandescent bulb in the otherwise dark isolation chamber. Intensity of the unconditional stimulus (US) was set as the mean frequency of dmVMH stimulation (112 Hz) or the mean current intensity of tail-shock (0.42 mA) that produced 2/3 maximal VAD responding under control conditions. Experimental groups received three days of training consisting of 15 trials of paired presentations of CS and US presented on a VI 2.5

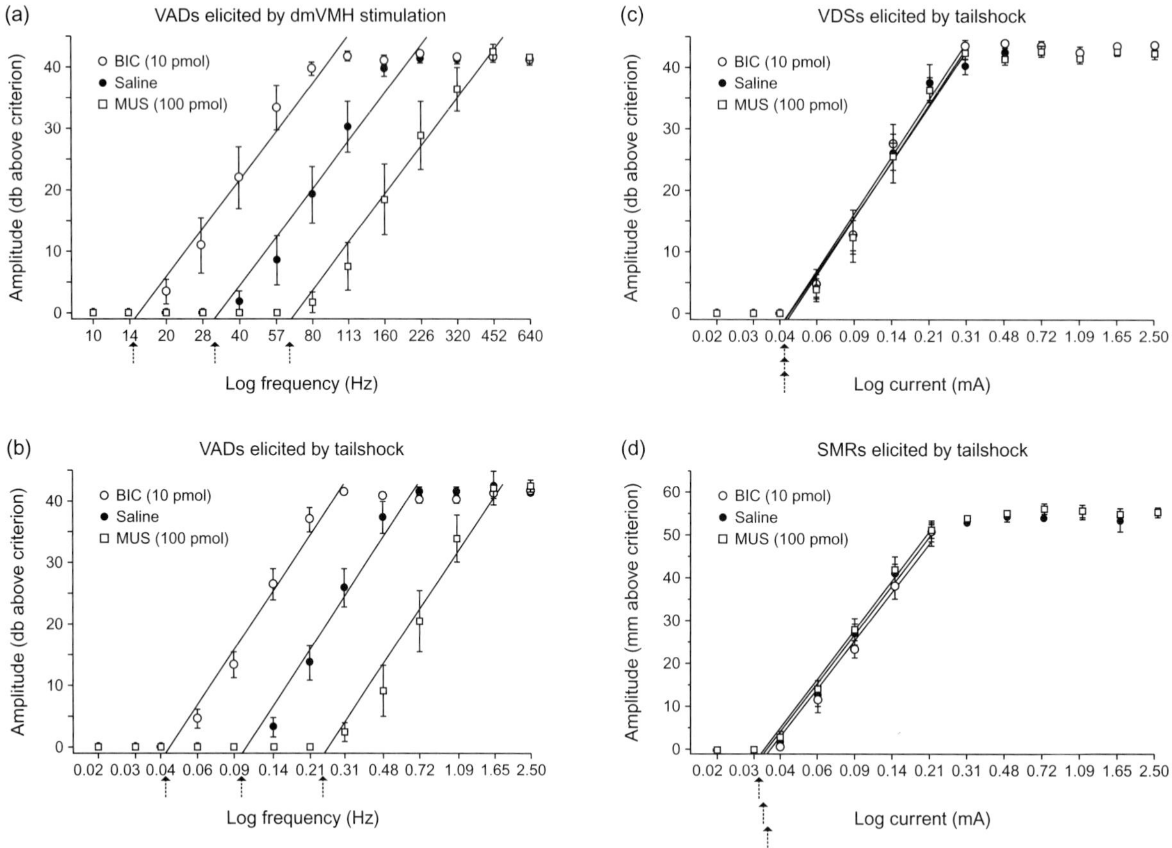

Fig. 4. (a) Mean psychophysical functions relating log frequency of dmVMH stimulation to mean (±SEM) peak amplitude of vocalization afterdischarges (VADs). (b, c, d) Mean psychophysical functions relating log current intensity of tail-shock to mean (±SEM) peak amplitude of vocalization afterdischarges (VADs), vocalizations during shock (VDSs) and spinal motor reflexes (SMRs). Responses were assessed following unilateral administration of saline, bicuculline (BIC) or muscimol (MUS) into the dorsomedial division of the ventromedial hypothalamus (dmVMH). Response thresholds (arrows) were calculated as the x-axis intercept.

minute schedule. A non-signaled presentation of the US (probe trial) was presented on the 16th trial of each session. Probe trials permitted assessment of unconditional responses (URs, i.e., VADs, VDSs and SMRs) uncontaminated by the occurrence of VCRs. Performance of URs was also recorded on the first training trial of Day 1. Forty-eight hours after the last training day, rats were exposed to the CS on three extinction trials and then given one probe trial. This session was not preceded by drug injections and permitted determination of whether differences in conditioning observed during training were attributable to the effects of drugs on the rats' ability to vocalize, or reflected differences in fear conditioning.

Both dmVMH stimulation and tail-shock were effective in supporting the development of fear conditioning in all groups that received paired presentations of CS and US. Consistent with our previous reports, VCRs and VADs shared spectral characteristics (Borszcz, 1995; Borszcz and Leaton, 2003), and administration of muscimol and bicuculline did not alter these spectral characteristics. However, manipulations of the $GABA_A$ receptors within the dmVMH modulated fear conditioning. The asymptotic level of fear conditioning was elevated by a subthreshold dose of bicuculline and reduced by muscimol administered into the dmVMH (Fig. 5a,b).

Performance of VADs during conditioning predicted the level of fear conditioning. Production of VADs on the first trial of training and on probe trials at the end of each conditioning session was enhanced in the group treated with bicuculline and suppressed in the group that received muscimol (Fig. 5c,d). This finding is consistent with our earlier reports that the

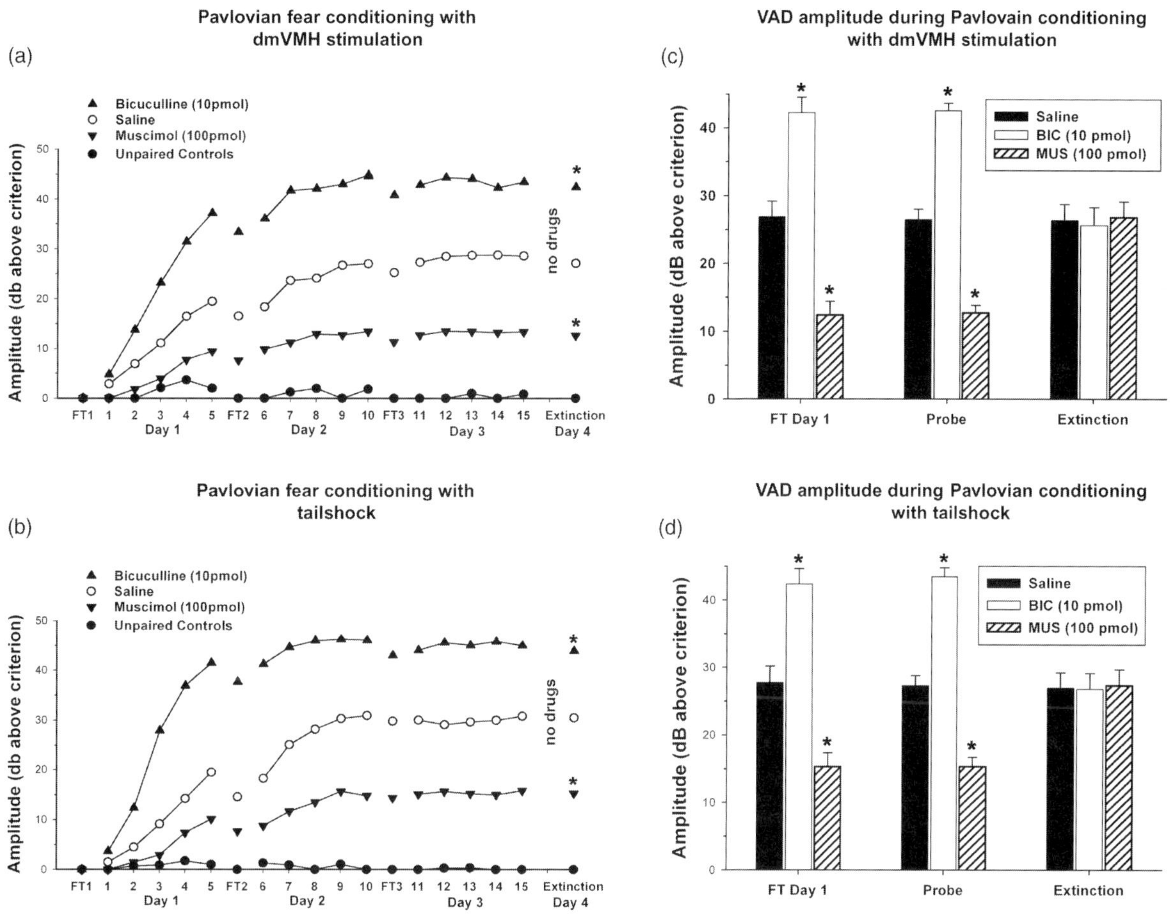

Fig. 5. Mean peak amplitude of conditional vocalizations (VCRs) during three days of training using either tail-shock (a) or the dorsomedial division of the ventromedial hypothalamus (dmVMH) stimulation (b) as the unconditional stimulus (US). For the sake of clarity data are presented as the means of blocks of three consecutive trials. Peak amplitude on the first trial (FT) of each training day is also represented. Prior to each day of training separate groups received injections of saline (A, n = 5; B, n = 6), bicuculline (A, n = 6; B, n = 7) or muscimol (A, n = 6; B, n = 6) into the dmVMH. Control groups received explicitly unpaired presentations of the CS and US (15 each) for three days. Separate control groups received injections of saline (A, n = 3; B, n = 3), bicuculline (A, n = 2; B, n = 2) or muscimol (A, n = 2; B, n = 2) into the dmVMH prior to each training day. As the data from control groups (within each US type) did not differ they were combined (Unpaired Controls). Prior to Extinction (Ext) training (Day 4) no drugs were administered and rats received three presentations of the CS alone. Asterisks indicate significant difference from groups that received saline prior to paired conditioning trials (Student's t-test for independent groups, $p < 0.05$). Mean (±SEM) peak amplitude of vocalization afterdischarges (VADs) during Pavlovian conditioning using either tail-shock (c) or dmVMH stimulation (d) as the unconditional stimulus (US). VADs were assessed on the first trial (FT) of training on Day 1 and on probe trials (US presented alone) that followed conditioning trials during the three days of training. VADs were also measured following the three trials of extinction training (Extinction). Separate groups were administered saline, bicuculline (BIC) or muscimol (MUS) prior to the three days of conditioning. No drugs were administered prior to extinction training. Asterisks indicate significant difference from saline treated groups (Student's t-test for independent groups, $p < 0.05$).

capacity of the tail-shock to support fear conditioning correlates with its generation of VADs (Borszcz, 1993, 1995; Borszcz and Leaton, 2003).

Differences in conditioning could not be attributed to the effects of drugs on expression of VCRs (ability of rats to vocalize). Expression of VCRs during the extinction session did not differ from that observed during the last day of training, despite the fact that drugs were not administered prior to this session. Evidence that drugs were not active during the extinction session was confirmed on the US probe trial that revealed a convergence of VAD performance across groups (Fig. 5c,d). Furthermore, no differences were observed in performance of VDSs during training among groups trained with tail-shock. Therefore, the observed differences in conditioned vocalizations reflect the contribution of the dmVMH in processing the negative affect associated with activation of the defensive aggression circuit either by direct stimulation or by exposure to a noxious stimulus.

VI. Enhancement of pain-induced vocalizations following changes in the synaptic plasticity in the hypothalamus

Several sites within the defensive aggression circuit are susceptible to induction of synaptic plasticity (Fig. 1). NMDA receptors are implicated in the induction of synaptic plasticity in a number of brain sites, and long-term changes in defensive responding are mediated by glutamatergic projections of the BLA that interact with NMDA receptors in the dlPAG and dmVMH. Partial kindling of the BLA leads to a long-term increase in defensive aggression in rats and cats to environmental threats, and intensifies defensive responding elicited by electrical stimulation of the dmVMH and dlPAG (Griffith et al., 1987; Hiyoshi et al., 1990). Partial kindling is generated with application of a few (3–4) brief high-frequency stimulations of the BLA and leads to an increase in the electric afterdischarges within the BLA (epileptic spiking in the EEG record that outlasts the stimulation) that only propagates to structures monosynaptically connected to the BLA. Behavioral seizures are not detectable, or are mild, following partial kindling (Rosen et al., 1996; Adamec and Young, 2000). Identical effects are observed following systemic administration of the anxiogenic drug FG-7142 (benzodiazepine inverse agonist) or exposure of rats to a predator (cat). All treatments resulted in NMDA-dependent long-term potentiation (LTP) in BLA projections to dlPAG and dmVMH. The effects of these treatments on defensive responding and LTP were blocked by prior systemic administration of an NMDA receptor antagonist, or by subsequent low-frequency stimulation (depotentiation) of the BLA (Adamec, 1998, 1999).

We evaluated the effects of partial kindling of the BLA on the threshold of tail-shock elicited VADs and induction of LTP in dmVMH. Rats implanted with stimulating electrodes in the right BLA and a recording electrode into the right dmVMH were first placed in an electrically grounded isolation chamber, and baseline local field potentials (LFPs) were recorded from the dmVMH in response to BLA stimulation (Fig. 6d, pre-kindling baseline). Each rat received 10 BLA stimulations (single 200 μs biphasic square wave pulses at 200 μA presented every 5 seconds) from which an averaged LFP was constructed and peak height and slope of the LFP were calculated.

Five minutes following completion of baseline LFP recording, rats received either their first session of partial kindling or sham kindling of the BLA. The

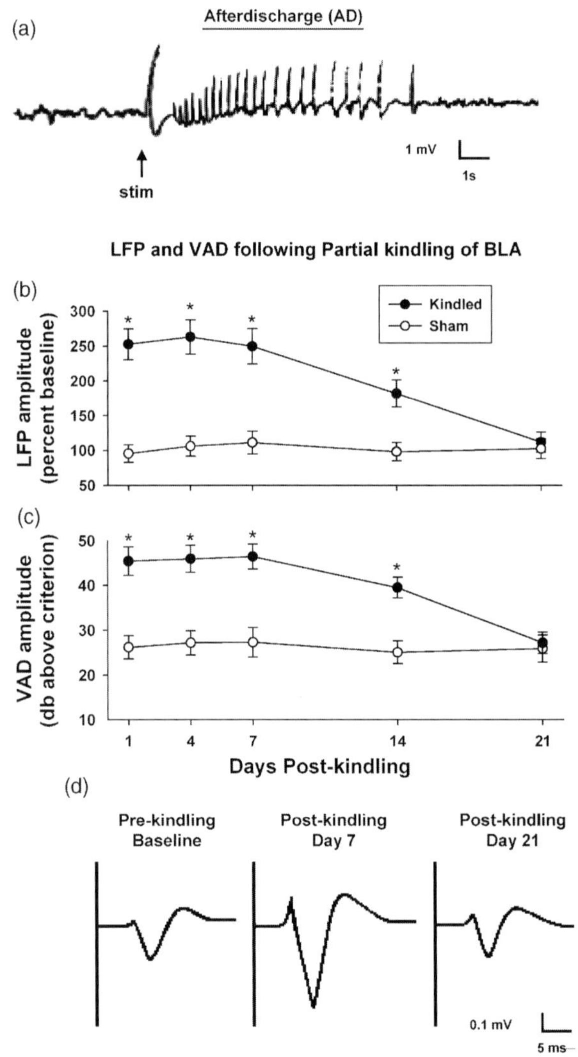

Fig. 6. (a) EEG recorded from the basolateral amydala (BLA) during the last session of partial kindling. Arrow indicates time of BLA stimulation. (b) Mean (±SEM) amplitude of local field potentials (LFP) recorded from the dorsomedial division of the ventromedial hypothalamus (dmVMH) in response to BLA stimulation following partial kindling or sham kindling. Asterisks indicate significantly elevated compared to sham kindled (Student's t-test for independent groups, $p < 0.05$). (c) Mean (±SEM) amplitude of VAD elicited by tail-shock following partial kindling or sham kindling. Asterisks indicate significantly elevated compared to sham kindled (Student's t-test for independent groups, $p < 0.05$). (d) LFPs recorded from the dmVMH in response to BLA stimulation.

kindling stimuli (400 μA of a 1 second, 100 Hz train of 1 ms biphasic square wave pulses) were delivered twice daily (inter-trial interval of 4 hours) for two consecutive days. Following each stimulation, the presence of electrical afterdischarges in the BLA and the behavioral seizure stage (if any, as scored according to the stages of Racine, 1972) were recorded.

Pain testing sessions occurred 1, 4, 7, 14 and 21 days following completion of partial kindling. During testing, rats were administered six tail-shocks at an intensity (0.42 mA) that elicited submaximal performance of VADs on all trials. Prior to each test session, LFPs in the dmVMH to BLA stimulation were recorded as described above. Also, EEG of the BLA and behavioral seizures were recorded for 10 minutes immediately preceding each test session.

Assessment of baseline LFPs did not generate electrical afterdischarges or behavioral seizures. However, all kindled rats exhibited electrical afterdischarges that increased in duration from day 1 to day 2 of kindling (Fig. 6a). BLA-kindled afterdischarge duration (mean ± SEM) for the first day was 11.6 ± 1.2 seconds, and for the second day 18.3 ± 1.1 seconds. On the last kindling session, no rat exhibited any stage of behavioral seizures. No kindled animal exhibited spontaneous electrical afterdischarges or behavioral seizures during the 10 minutes prior to pain testing on any day. The amplitude and slope of LFPs in dmVMH to BLA stimulation did not change in sham kindled rats over the 21 days of testing. However, kindled rats exhibited increases in the amplitude and slope of LFPs 1, 4, 7 and 14 days following the last kindling session (Fig. 6b,d). LFPs returned to baseline levels at 21 days after kindling. Kindling-induced changes in LFPs in dmVMH were positively correlated with changes in tail-shock elicited VADs. Kindled rats exhibited greater amplitude of VADs 1, 4, 7 and 14 days, but not at 21 days following the last kindling session (Fig. 6c). Rats that received sham kindling exhibited stable VAD performance over the course of repeated testing.

The contribution of NMDA and non-NMDA receptors in the dmVMH in increasing VADs by partial kindling of the BLA was also evaluated. The procedures were similar to those described above with the exception that NMDA (AP-5; 2-amino-5-phosphonopentanoic acid) or non-NMDA (CNQX; 6-cyano-7-nitroquinoxaline-2,3-dione) receptor antagonists were injected into the dmVMH via a chemitrode that permitted drug microinjection at the site of electrophysiological recording. Microinjection of AP-5 into dmVMH immediately prior to BLA kindling blocked, in a dose-dependent manner, the increase in VADs generated by partial kindling of the BLA. The increase in LFPs in dmVMH generated by partial kindling of the BLA was also reduced by pretreatment with AP-5. Treatment with CNQX did alter increases in VADs or LFPs generated by partial kindling of the BLA. We may conclude therefore that NMDA-induced synaptic plasticity in the dmVMH results in long-term increases in vocalizations that reflect the emotional response to painful stimuli.

VII. Conclusions

We have demonstrated that a particular form of vocalization (VADs) reflects an emotional response to pain in rats. VADs therefore provide an insight into the neural activity within the limbic–midbrain circuit responsible for processing the innate affective attributes of pain. This limbic–midbrain circuit appears to overlap with the circuit that generates innate defensive responses to imminent threats. The dmVMH is a core structure mediating defensive responses to noxious and non-noxious threatening stimuli and its contribution to production of the innate affective response to pain provides insight into the concept of pain affect. The primary affective dimension of pain belongs to a broader class of sensory experience that represents threat to the individual and engages neuronal circuits that govern the execution of innate defensive responses that enable the individual to cope with the threat immediately.

Induction of neuronal plasticity within the defensive aggression circuit produce long-term augmentation of pain affect as reflected in pain-induced vocalizations. This plasticity may also account for a long-term increase in defensiveness that accompanies pain. Alterations in the circuitry that controls defensive responding are implicated in such states as fear, anxiety, depression and anger (Dixon, 1998; Adamec and Young, 2000). These secondary emotional responses are components of the human pain experience and contribute to the suffering and disability associated with pain (Crombez et al., 1999; Ericsson et al., 2002). The pain-induced vocalization paradigm may therefore provide insight into the mechanisms through which emotional responses to pain produce enduring emotional disturbances in the organism.

Acknowledgments

Grant R01 NS045720 from the National Institute of Neurological Disorders and Stroke supported preparation of this manuscript.

References

Adamec, R.E., 1998. Evidence that NMDA-dependent limbic neural plasticity in the right hemisphere mediates pharmacological stressor (FG-7142)-induced lasting increases in anxiety-like behavior: study 3 – the effects on amygdala efferent physiology of block of NMDA receptors prior to injection of FG-7142 and its relationship to behavioral change. J. Psychopharmacol. 12, 227–238.

Adamec, R.E., 1999. Evidence that limbic neural plasticity in the right hemisphere mediates partial kindling induced lasting increases in anxiety-like behavior: effects of low-frequency stimulation (quenching?) on long-term potentiation of amygdala efferents and behavior following kindling. Brain Res. 839, 133–152.

Adamec, R., Young, B., 2000. Neuroplasticity in specific limbic system circuits may mediate specific kindling induced changes in animal affect-implications for understanding anxiety associated with epilepsy. Neuorosci. Biobehav. Rev. 24, 705–723.

Beart, P.M., Nicolopoulos, L.S., West, D.C., Headley, P.M., 1988. An excitatory amino acid projection from ventromedial hypothalamus to periaqueductal gray in the rat: autoradiographic and electrophysiological evidence. Neurosci. Lett. 85, 205–211.

Beckett, S.R., Duxon, M.S., Aspley, S., Marsden, C.A., 1997. Central c-fos expression following 20 kHz/ultrasound induced defence behaviour in the rat. Brain Res. Bull. 42, 421–426.

Bester, H., Besson, J.M., Bernard, J.F., 1997. Organization of efferent projections from the parabrachial area to the hypothalamus: a Phaseolus vulgaris-leucoagglutinin study in the rat. J. Comp. Neurol. 383, 245–281.

Bester, H., Menendez, L., Besson, J.M., Bernard, J.F., 1995. Spino (trigemino) parabrachiohypothalamic pathway: electrophysiological evidence for an involvement in pain processes. J. Neurophysiol. 73, 568–585.

Bhatt, S., Gregg, T.R., Siegel, A., 2003. NK1 receptors in the medial hypothalamus potentiate defensive rage behavior elicited from the midbrain periaqueductal gray of the cat. Brain Res. 966, 54–64.

Blanchard, R.J., Flannelly, K.J., Blanchard, D.C., 1986. Defensive behavior of laboratory and wild *Rattus norvegicus*. J. Comp. Psychol. 100, 101–117.

Blanchard, C.D., Hynd, A.L., Minke, K.A., Minemoto, T., Blanchard, R.J., 2001. Human defensive behaviors to threat scenarios show parallels to fear- and anxiety-related defense patterns of non-human mammals. Neurosci. Biobehav. Rev. 25, 761–770.

Borszcz, G.S., 1993. The capacity of motor reflex and vocalization thresholds to support avoidance conditioning in the rat. Behav. Neurosci. 107, 678–693.

Borszcz, G.S., 1995. Pavlovian conditional vocalizations of the rat: a model system for analyzing the fear of pain. Behav. Neurosci. 109, 648–662.

Borszcz, G.S., 2006. Contribution of the ventromedial hypothalamus to generation of the affective dimension of pain. Pain 123, 155–168.

Borszcz, G.S., Leaton, R.N., 2003. The effect of amygdala lesions on conditional and unconditional vocalizations in rats. Neurobiol. Learn. Mem. 79, 212–225.

Borszcz, G.S., Johnson, C.P., Anderson, M.E., Young, B.J., 1992. Characterization of tail-shock elicited withdrawal reflexes in intact and spinal rats. Physiol. Behav. 52, 1055–1062.

Borszcz, G.S., Johnson, C.P., Fahey, K.A., 1994. Comparison of motor reflex and vocalization thresholds following systemically administered morphine, fentanyl, and diazepam in the rat: assessment of sensory and performance variables. Pharmacol. Biochem. Behav. 49, 827–834.

Bullitt, E., 1990. Expression of c-fos-like protein as a marker for neuronal activity following noxious stimulation in the rat. J. Comp. Neurol. 296, 517–530.

Burstein, R., Dado, R.J., Cliffer, K.D., Giesler, Jr., G.J., 1991. Physiological characterization of spinohypothalamic tract neurons in the lumbar enlargement of rats. J. Neurophysiol. 66, 261–284.

Canteras, N.S., 2002. The medial hypothalamic defensive system: hodological organization and functional implications. Pharmacol. Biochem. Behav. 71, 481–491.

Canteras, N.S., Chiavegatto, S., Valle, L.E., Swanson, L.W., 1997. Severe reduction of rat defensive behavior to a predator by discrete hypothalamic chemical lesions. Brain Res. Bull. 44, 297–305.

Carroll, M.N., Lim, K.S., 1960. Observations on the neuropharmacology of morphine and morphinelike analgesia. Arch. Int. Pharmacodyn. Ther. 125, 383–403.

Cheu, J.W., Siegel, A., 1998. GABA receptor mediated suppression of defensive rage behavior elicited from the medial hypothalamus of the cat: role of the lateral hypothalamus. Brain Res. 783, 293–304.

Crombez, G., Vlaeyen, J.W., Heuts, P.H., Lysens, R., 1999. Pain-related fear is more disabling than pain itself: evidence on the role of pain-related fear in chronic back pain disability. Pain 80, 329–339.

Darwin, C., 1872/1998. The Expression of Emotions in Man and Animals, 3rd edn. Oxford University Press, New York, NY.

Dielenberg, R.A., Hunt, G.E., McGregor, I.S., 2001. "When a rat smells a cat:" the distribution of Fos immunoreactivity in rat brain following exposure to a predatory odor. Neurosci. 104, 1085–1097.

Dixon, A.K., 1998. Ethological strategies for defence in animals and humans: their role in some psychiatric disorders. Brit. J. Med. Psychol. 71, 417–445.

Egger, M.D., Flynn, J.P., 1967. Further studies on the effects of amygdaloid stimulation and ablation on hypothalamically elicited attack behavior in cats. Prog. Brain Res. 27, 165–182.

Ericsson, M., Poston, W.S., Linder, J., Taylor, J.E., Haddock, C.K., Foreyt, J.P., 2002. Depression predicts disability in long-term chronic pain patients. Disabil. Rehabil. 24, 334–340.

Ervin, F.R., Mark, V.H., Stevens, J., 1969. Behavioral and affective responses to brain stimulation in man. Proc. Ann. Meet. Am. Psychopathol. Assoc. 58, 54–65.

Fanselow, M.S., Lester, L.S., 1988. A functional behavoristic approach to aversively motivated behavior: predatory imminence as a determinant of the topography of defensive behavior. In: Bolles, R.C. (Ed.), Evolution and Learning. Erlbaum, Hillsdale, NJ.

Fernandez De Molina, A., Hunsperger, R.W., 1962. Organization of the subcortical system governing defence and flight reactions in the cat. J. Physiol. 60, 200–213.
Griffith, N., Engel, Jr., J., Bandler, R., 1987. Ictal and enduring interictal disturbances in emotional behaviour in an animal model of temporal lobe epilepsy. Brain Res. 400, 360–364.
Hiyoshi, T., Matsuda, M., Wada, J.A., 1990. Centrally induced feline emotional behavior and limbic kindling. Epilepsia 31, 259–269.
Hsieh, J.C., Stahle-Backdahl, M., Hagermark, O., Stone-Elander, S., Rosenquist, G., Ingvar, M., 1996. Traumatic nociceptive pain activates the hypothalamus and the periaqueductal gray: a positron emission tomography study. Pain 64, 303–314.
Iacono, R.P., Nashold, Jr., B.S., 1982. Mental and behavioral effects of brain stem and hypothalamic stimulation in man. Human Neurobiol. 1, 273–279.
Jourdan, D., Ardid, D., Eschalier, A., 2002. Analysis of ultrasonic vocalisation does not allow chronic pain to be evaluated in rats. Pain 95, 165–173.
Jürgens, U., 1979. Vocalization as an emotional indicator. A neuroethological study in the squirrel monkey. Behaviour 69, 88–117.
Jürgens, U., 2002. Neural pathways underlying vocal control. Neurosci. Biobehav. Rev. 26, 235–258.
Jürgens, U., Pratt, R., 1979. Role of the periaqueductal grey in vocal expression of emotion. Brain Res. 167, 367–378.
Lipp, H.P., Hunsperger, R.W., 1978. Threat, attack and flight elicited by electrical stimulation of the ventromedial hypothalamus of the marmoset monkey *Callithrix jacchus*. Brain Behav. Evol. 15, 260–293.
Liu, R.J., Qiang, M., Qiao, J.T., 1998. Nociceptive c-fos expression in supraspinal areas in avoidance of descending suppression at the spinal relay station. Neurosci. 85, 1073–1087.
Markham, C.M., Blanchard, D.C., Canteras, N.S., Cuyno, C.D., Blanchard, R.J., 2004. Modulation of predatory odor processing following lesions to the dorsal premammillary nucleus. Neurosci. Lett. 372, 22–26.
Milani, H., Graeff, F.G., 1987. GABA-benzodiazepine modulation of aversion in the medial hypothalamus of the rat. Pharmacol. Biochem. Behav. 28, 21–27.
Newman, H.M., Stevens, R.T., Apkarian, A.V., 1996. Direct spinal projections to limbic and striatal areas: anterograde transport studies from the upper cervical spinal cord and the cervical enlargement in squirrel monkey and rat. J. Comp. Neurol. 365, 640–658.
Parry, D.M., Semenenko, F.M., Conley, R.K., Lumb, B.M., 2002. Noxious somatic inputs to hypothalamic-midbrain projection neurones: a comparison of the columnar organisation of somatic and visceral inputs to the periaqueductal grey in the rat. Exp. Physiol. 87, 117–122.
Paxinos, G., Watson, C., 1998. The Rat Brain in Stereotaxic Coordinates. Academic Press, New York, NY.
Petrovich, G.D., Canteras, N.S., Swanson, L.W., 2001. Combinatorial amygdalar inputs to hippocampal domains and hypothalamic behavior systems. Brain Res. Brain Res. Rev. 38, 247–289.
Rodella, L., Rezzani, R., Gioia, M., Tredici, G., Bianchi, R., 1998. Expression of Fos immunoreactivity in the rat supraspinal regions following noxious visceral stimulation. Brain Res. Bull. 47, 357–366.
Rosen, J.B., Hamerman, E., Sitcoske, M., Glowa, J.R., Schulkin, J., 1996. Hyperexcitability: exaggerated fear-potentiated startle produced by partial amygdala kindling. Behav. Neurosci. 110, 43–50.
Sandner, G., Oberling, P., Silveira, M.C., Di Scala, G., Rocha, B., Bagri, A., Depoortere, R., 1993. What brain structures are active during emotions? Effects of brain stimulation elicited aversion on c-fos immunoreactivity and behavior. Behav. Brain Res. 58, 9–18.
Schubert, K., Shaikh, M.B., Siegel, A., 1996. NMDA receptors in the midbrain periaqueductal gray mediate hypothalamically evoked hissing behavior in the cat. Brain Res. 726, 80–90.
Shaikh, M.B., Siegel, A., 1994. Neuroanatomical and neurochemical mechanisms underlying amygdaloid control of defensive rage behavior in the cat. Braz. J. Med. Biol. Res. 27, 2759–2779.
Shaikh, M.B., Lu, C.L., Siegel, A., 1991. An enkephalinergic mechanism involved in amygdaloid suppression of affective defence behavior elicited from the midbrain periaqueductal gray in the cat. Brain Res. 559, 109–117.
Shaikh, M.B., Schubert, K., Siegel, A., 1994. Basal amygdaloid facilitation of midbrain periaqueductal gray elicited defensive rage behavior in the cat is mediated through NMDA receptors. Brain Res. 635, 187–195.
Siegel, A., 2005. The Neurobiology of Aggression and Rage. CRC Press, Boca Raton, FL.
Strzelczuk, M., Romaniuk, A., 1995. Fear induced by the blockade of GABAA-ergic transmission in the hypothalamus of the cat: behavioral and neurochemical study. Behav. Brain Res. 72, 63–71.
Tasker, R.R., 1982. Identification of pain processing systems by electrical stimulation of the brain. Human Neurobiol. 1, 261–272.
Todd, A.J., 2002. Anatomy of primary afferents and projection neurones in the rat spinal dorsal horn with particular emphasis on substance P and the neurokinin 1 receptor. Exp. Physiol. 87, 245–249.
Wang, M.R., Kuo, J.S., Chai, C.Y., 2002. Cardiovascular and vocalization reactions elicited by N-methyl-D-aspartate in the pretentorial periaqueductal grey of cats. Clin. Exp. Pharmacol. Physiol. 29, 759–771.
Zhang, X., Wenk, H.N., Gokin, A.P., Honda, C.N., Giesler, Jr., G.J., 1999. Physiological studies of spinohypothalamic tract neurons in the lumbar enlargement of monkeys. J. Neurophysiol. 82, 1054–1058.

CHAPTER 7.5

Responses of limbic, midbrain and brainstem structures to electrically-induced vocalizations

Francisco Gonzalez-Lima*

Departments of Psychology, Pharmacology and Toxicology, University of Texas at Austin, Austin, TX, USA

Abstract: This chapter describes the limbic, midbrain and brainstem regions that participate in the production of vocalizations elicited by midbrain reticular stimulation in the rat brain. Fluorodeoxyglucose autoradiography was used to map the functional activity of all brain regions simultaneously during the elicited vocalizations. This approach permitted us to propose a new model of functional pathways for innately programmed vocalization in the rat brain. The structures involved in species-typical vocalizations were grouped into five levels of hierarchical organization along the neuraxis: limbic; hypothalamic; midbrain; pontine; and medullar levels. These five hierarchical levels interact using two parallel sets of anatomical pathways: (1) a direct pathway, which corresponds to the well-known limbic–midbrain–medulla pathway for innate vocalization in mammals; and (2) a more elaborate, indirect pathway, which involves three specific regions at each of the five hierarchical levels that form a network comprising a total of 15 brain regions. This model is consistent with the results published on the neural control of vocalizations in mammals, while at the same time it significantly extends the scope of the general model of neural control of innate vocalizations in mammals.

Keywords: vocalization; neural pathways; fluorodeoxyglucose; midbrain reticular formation; prelimbic cortex; hypothalamus; periaqueductal gray; nucleus ambiguus; frontal cortex; amygdala

I. Introduction

The objective of this chapter is to describe the rat limbic, midbrain and brainstem regions that participate in production of vocalizations elicited by midbrain electrical stimulation, and to integrate these findings into the neuroanatomical organization of the central circuits for vocalization. The chapter is based on our previously reported findings using metabolic brain mapping and electrical brain stimulation (Gonzalez-Lima and Frysztak, 1991), as well as on unpublished findings with a similar approach. The proposed model of the functional pathways for vocalization in the rat brain represents an extension of Jürgens' general model of neural control of innate vocalizations in mammals (reviewed in Jürgens, 2009). It is hoped that this model may foster future research into the functional organization of pathways for vocalization in the mammalian brain, which have shown relatively little progress in the last 20 years. Two excellent reviews on this topic published by Jürgens and Ploog (1988) and Jürgens (2009), and separated by 21 years, have clearly shown limited advances in the understanding of the organization of pathways for vocalization in the mammalian brain.

The research approach, which was first introduced by Sokoloff and his co-workers, may help facilitate this objective (Sokoloff et al., 1977). Radiolabeled glucose analogs provided an opportunity to characterize patterns of neural activity using autoradiography in

*Corresponding author. E-mail: gonzalez-lima@mail.utexas.edu

Stefan M. Brudzynski (Ed.)
Handbook of Mammalian Vocalization
ISBN 978-0-12-374593-4

DOI: 10.1016/B978-0-12-374593-4.00028-0

non-human species (Kennedy et al., 1975; Gonzalez-Lima, 1987, 1992; Gonzalez-Lima and Scheich, 1984, 1986; Gonzalez-Lima et al., 1993) and positron emission tomography in humans (Phelps et al., 1979; Reivich et al., 1979; Huang et al., 1980). The major fuel for energy production in the brain is glucose, and glucose analogs such as fluorodeoxyglucose (FDG) and 2-deoxyglucose (2-DG) have been used to investigate regions of the brain involved in normal behaviors, including auditory and vocal responses studied in laboratory rodents (Martinez et al., 1982; Gonzalez-Lima and Scheich, 1986; Gonzalez-Lima, 1992; Gonzalez-Lima et al., 1993). The major advantage of this metabolic mapping approach is that it makes it possible visually to display regional activity throughout the entire brain of awake, behaving animals. Therefore, this approach can identify brain regions polysynaptically connected with a stimulated call site (de Lanerolle and Lang, 1988). However, regional mapping of glucose utilization is not designed to provide information about neurotransmitters or excitatory versus inhibitory postsynaptic potentials.

Autoradiographic mapping with glucose analogs has been successfully used in combination with electrical stimulation of the rat brain. For example, we have used electrical stimulation of the midbrain reticular formation (MRF) together with 2-DG or FDG to map functional responses in all the structures of the rat brain from the cerebrum (Gonzalez-Lima and Scheich, 1985) to the cerebellum (Gonzalez-Lima, 1987) and to the spinal cord (Gonzalez-Lima, 1986). Behavioral responses to MRF stimulation vary, depending on location of the electrode and stimulation parameters, from orienting reflexes, or arousing responses, to aversive and escape behavior. These behaviors in rats are often accompanied by species-typical vocalizations. Low-intensity MRF stimulation (300 μA, 0.5 s train of 40 Hz, 0.2 ms monophasic cathodal pulses) could often evoke these vocalizations without other behavioral effects (Schenk and Robinson, 1988; Gonzalez-Lima and Frysztak, 1991). We mapped the responses of different brain regions to MRF stimulation at sites that reliably evoked stereotyped vocalizations in the rat without any signs of aversion or heart rate changes, as described in the study by Gonzalez-Lima and Frysztak (1991).

Sonograms were generated to analyze the MRF-evoked vocalizations, and the brain results discussed here are from animals displaying calls that remained stable and stereotyped during the entire FDG session. Calls had a median fundamental component of 2–3 kHz FM downward sweep, and a median of four harmonics per call (including fundamental as the first harmonic). The duration of each call was 125–150 ms and call intensity was 60–70 dB SPL recorded from a 10–20 cm distance. This call type was similar to species-typical innate audible vocalizations displayed by rats during fighting or emitted as warning calls (Schenk and Robinson, 1988; Litvin et al., 2007), but vocalizing rats appeared otherwise undisturbed and showed no behavioral or electrocardiographic signs of arousal. The recorded vocalizations of each rat were played back to a yoked control rat during the FDG session, so that pairs of MRF-stimulated and non-stimulated control rats received comparable auditory input over the FDG sessions. After the FDG session, rats were decapitated and their brains removed and processed for autoradiography (Gonzalez-Lima et al., 1993). FDG uptake in the autoradiographs was quantified with an image processing system, from the frontal pole of the brain to the caudal medulla in each subject.

II. The magnitude of the activity in specific brain structures of vocalizing rats

The FDG uptake results are summarized in Fig. 1, which shows the percentage difference between the regional brain activity of vocalizing and yoked control rats. Brain regional differences of FDG uptake between electrically stimulated (vocalizing) and control (playback) groups were classified as large (20–90%), medium (15–20%) and small (5–15%) group differences. Structures with large activation effects in the vocalizing rats, listed in descending order of magnitude of the activity, included: the midbrain reticular formation (MRF); central gray-rostral part (CGr); area of nucleus ambiguus (NA); central gray-caudal part (CGc); ventromedial hypothalamic area (VmH); and periventricular hypothalamic nucleus (PeH). In the midbrain, electrical stimulation of the MRF call site activated a region of approximately one mm in diameter to a level of 90% greater than the level of activity of the corresponding MRF region in the control rats. The MRF region activated corresponded to the deep mesencephalic nucleus of the dorsal midbrain tegmentum, and it was centered at six mm posterior to bregma (bregma −6). The CG region activated in the vocalizing rats corresponded to the dorsolateral division throughout the extent of the CG, with more pronounced activation in the rostral (CGr) than the caudal (CGc) parts.

The column of cell bodies forming the nucleus ambiguus was activated throughout its rostrocaudal

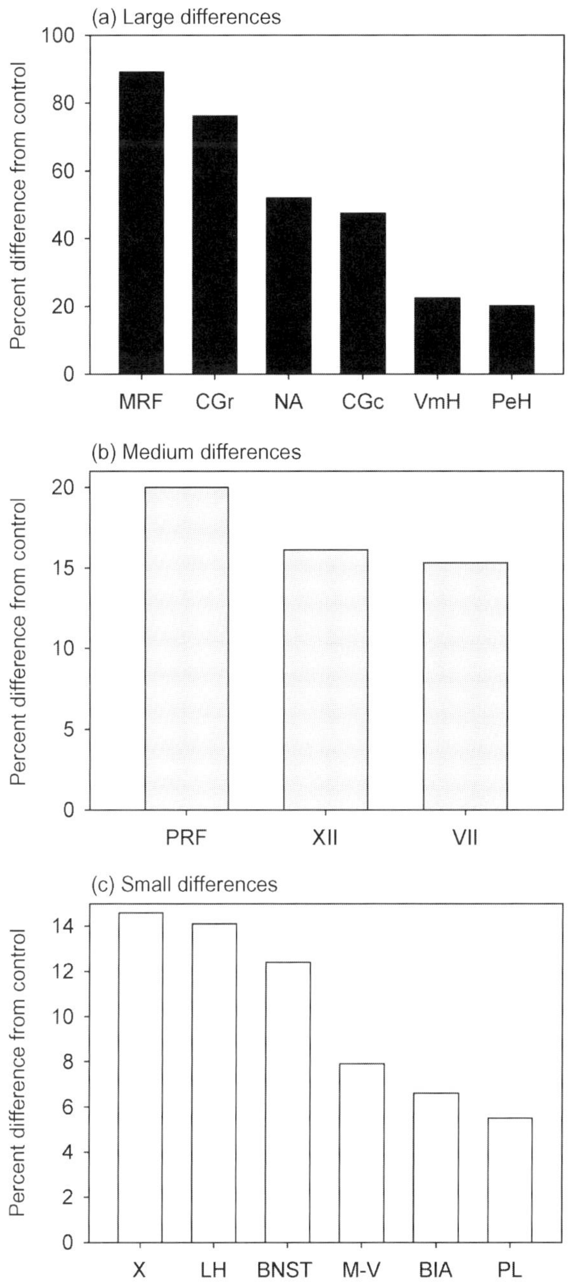

Fig. 1. Percent differences of fluorodeoxyglucose (FDG) uptake between electrically stimulated (vocalizing) and control (playback) groups, for brain regions showing large (a); medium (b); and small (c) group differences. Notice the decrement in the scales of percent differences in (a), (b) and (c). All values are expressed as increases in FDG uptake in the experimental stimulated group for the regions shown. Percentage differences were calculated with the formula: 100 × (Experimental − Control)/Control. Abbreviations: MRF: Midbrain reticular formation; CGr: Central gray, rostral; NA: Area of nucleus ambiguus; CGc: Central gray, caudal; VmH: Ventromedial hypothalamic region; PeH: Periventricular hypothalamic nucleus; PRF: Pontine reticular formation; XII: Cranial nerve XII nucleus; VII: Cranial nerve VII nucleus; X: Cranial nerve X nucleus; LH: Lateral hypothalamic region; BNST: Bed nucleus of stria terminalis; M–V: Cranial nerve V motor nucleus; BlA: Amygdala-basolateral nucleus; PL: Medial frontal cortex, prelimbic.

extent in the medulla, with increases in activity over 50% of that in the control brains in the caudal 2/3 portion of the nucleus, where laryngeal motoneurons are located in the rat, and with the peak activation centered at 13 mm posterior to bregma. However, the activated region in the NA was not limited to the location of cell bodies traditionally defined within the nucleus ambiguus. The NA-activated area included the immediately surrounding reticular formation, so that the diameter of the NA-activated area was about twice as big as the diameter of the nucleus ambiguus *per se*. Hence, our functionally-defined NA region includes both the nucleus ambiguus and its surrounding reticular formation, with both areas of activation of approximately equal size. This greater NA-activated area can be seen in Fig. 2.

Other regions with activations of over 20% above that in the control brains included the hypothalamic VmH and PeH areas located rostral to the activated midbrain regions (bregma −2). Regions with medium size effects were located caudal to the midbrain and included the pontine reticular formation (PRF) with the activation centered at bregma −8.8, the hypoglossal nucleus of cranial nerve XII (XII) activated at the same bregma location as the NA, and the facial nucleus of the cranial nerve VII (VII) with the activation centered at bregma −11. Smaller effects (<15%) were found in the dorsal vagal nucleus of the cranial nerve X (X), located dorsal to XII, and the trigeminal motor nucleus of the cranial nerve V (M–V). As in the case of the NA, the reticular formation immediately surrounding all the involved cranial nerve nuclei was also part of the area activated by MRF stimulation-induced vocalizations.

Small effects were also seen in higher brain regions associated with the limbic system, such as the lateral hypothalamic area (LH), bed nucleus of stria terminalis (BNST), basolateral nucleus of the amygdala (BlA), and prelimbic region of the medial frontal cortex (PL). The PL corresponds to the most anterior portion of the cingulate cortex in the rat, a region anterior to the corpus callosum and also called Brodmann area 32 or Cg3 cingulate cortex in rats (Zilles and Wree, 1985). The PL showed the lowest activation level (5–6%) and it was the most rostral brain region activated in the vocalizing rats at 3.2 mm anterior to bregma.

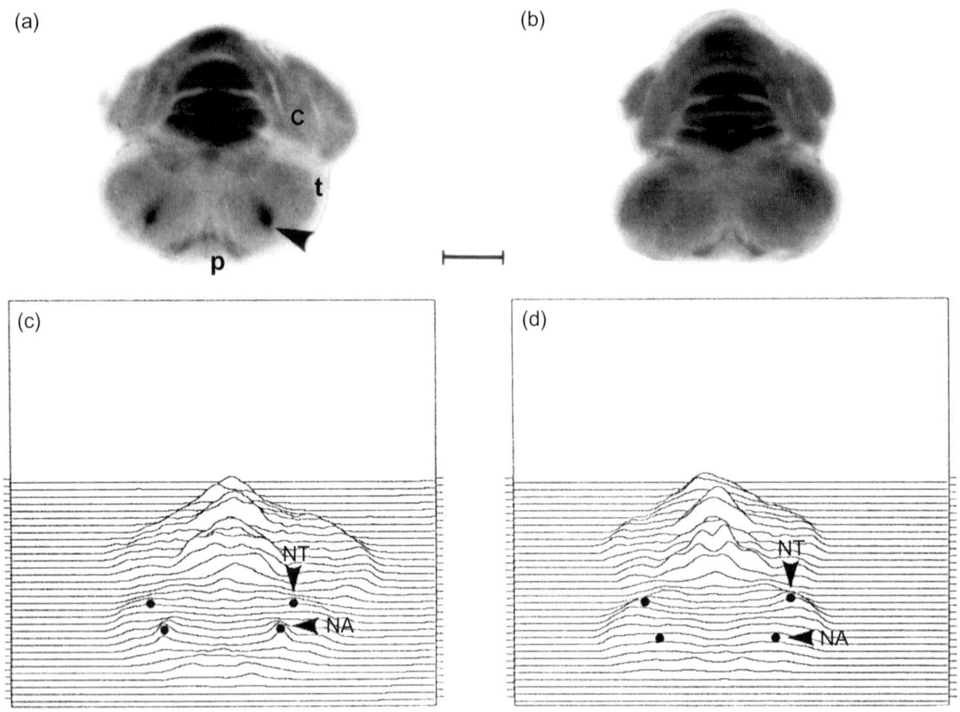

Fig. 2. The upper row shows fluorodeoxyglucose (FDG) autoradiographs at comparable levels of the medulla from vocalizing rats (a); and playback rats (b). The dark spots in (a) (arrowhead) correspond to the activation of nucleus ambiguus (NA) and surrounding reticular formation in the vocalizing rat. Note that activity in the corresponding NA region of (b) is at low background level. In contrast, the nucleus of the trigeminal spinal tract (NT) of vocalizing rats (region medial to "t" in (a)) showed a suppression of FDG uptake relative to controls (b). The lower row shows computer-generated densitometric profiles of the autoradiographs for vocalizing rats (c) and playback rats (d), with areas of higher FDG uptake shown as peaks. The same effects of NA activation and NT reduction during vocalization can be seen in the densitometric peaks labeled in (c) versus (d). Abbreviations: NA: area of nucleus ambiguus; NT: nucleus of t; t: trigeminal spinal tract; c: cerebellum; p: pyramidal tract. Scale bar = 2 mm.

III. Five-level hierarchical model

The brain regions showing increased metabolic activity during MRF-evoked vocalizations may be neuroanatomically grouped into five levels of hierarchical organization along the rat neuraxis: (1) limbic; (2) hypothalamic; (3) midbrain; (4) pontine; and (5) medullar levels, as summarized in Fig. 3. These five hierarchical levels interact via two parallel sets of anatomical pathways: a shorter more direct pathway comprising limbic–midbrain–medulla descending connections (path with shorter number of synaptic relays, connects levels 1–3–5, bypasses levels 2 and 4) (see Fig. 3), and a longer more indirect pathway (longer synaptic path, connects all five levels) where the hypothalamus interacts reciprocally with limbic and midbrain levels (levels 1–2–3) and the pons provides synaptic relays between the midbrain and medulla (levels 3–4–5). The direct pathway corresponds to the well-known limbic–midbrain–medulla pathway for innate vocalization in mammals, comprising descending connections between limbic cingulate cortex, periaqueductal gray and medullary reticular formation surrounding the nucleus ambiguus (Jürgens, 2009).

In the activated rat brain regions, this direct vocalization path will be represented by the PL–CGr–NA descending pathway. The additional indirect pathway identified by FDG mapping involves three specific regions at each of the five hierarchical levels, forming a network comprising 15 brain regions, as summarized in Fig. 4. Structures identified in these levels may contribute to production of motivated innate calls (limbic level 1), emotional defensive and threat calls (hypothalamic level 2), call gating (midbrain level 3), patterning motor coordination of calls by brainstem reticular neurons, and final neural output by phonatory motoneurons (pontomedullary levels 4 and 5).

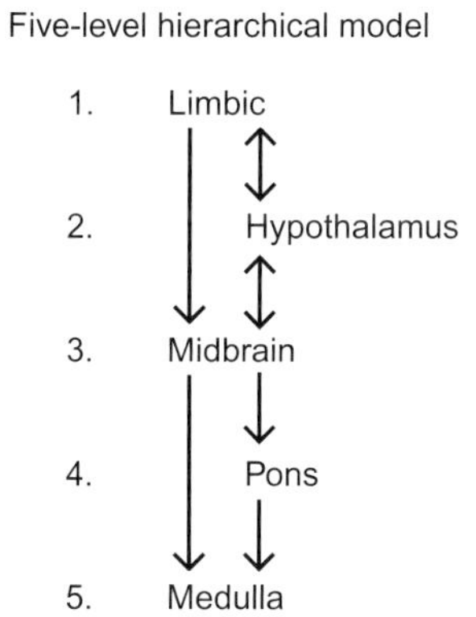

Fig. 3. Five-level hierarchical model showing the neuroanatomical levels with structures activated during production of innate vocalization as determined from FDG mapping. Two parallel sets of anatomical pathways are depicted in a simplified way: on the left-hand side there is a more *direct pathway* that connects levels 1–3–5, and bypasses levels 2 and 4; and on the right-hand side there is an *indirect pathway* that connects all the five levels, with the hypothalamus interacting reciprocally with limbic and midbrain levels (levels 1–2–3) and the pons as relay between the midbrain and medulla (levels 3–4–5).

III.A. Level 1: limbic

The structures activated by MRF stimulation at the limbic level included PL (Cg3 cingulate cortex), BNST and BlA (extended amygdala). No activation was found in the hippocampus. From PL, descending projections reach the CGr as part of the classic innate vocalization direct path (Müller-Preuss and Jürgens, 1976). PL also projects to the LH (Leichnetz and Astruc, 1976), which is part of an indirect pathway for MRF-evoked vocalization in the rat brain determined by FDG mapping. The medial portion of LH has anterior sites that elicit emotional-aversive vocalizations (Brudzynski, 1994). While the medial part of LH is associated with aversive calls, the lateral part is associated with appetitive calls, but both parts communicate with each other. Reciprocal connections also link the limbic and hypothalamic levels in parallel, one connecting BNST-VmH via the stria terminalis and another connecting BlA-LH via the ventral amygdalofugal pathway (Price, 2003). The circuit interactions between limbic–hypothalamic levels presumably process the motivational basis for the voluntary initiation of vocalizations (Jürgens and Ploog, 1988). For example, a severely reduced motivation to initiate a conversation was found in humans with bilateral damage of the anterior cingulate (Jürgens and Von Cramon, 1982).

III.B. Level 2: hypothalamus

The hypothalamic level comprised the LH, VmH and PeH activated by stimulation. These structures are part of the indirect path for vocalizations linking limbic with midbrain levels. LH is a key region for integrating motivated behavior with both artificial and natural reinforcing stimuli. For example, LH is the brain site that evokes the largest rates of electrical self-stimulation in rats, and LH shows the largest FDG uptake in motivated drinking behavior in thirsty versus satiated rats (Gonzalez-Lima et al., 1993). LH sends hypothalamoreticular fibers to MRF and CGc, and receives reticular afferents from MRF via the medial forebrain bundle (Lammers and Lohman, 1974).

Another major fiber bundle linking the hypothalamus with the midbrain is the dorsal longitudinal fasciculus, which provides reciprocal connections between PeH and CGr. In this pathway, the PeH may be regarded functionally as the hypothalamic extension of the CGr, comprising an interconnected zone of periventricular neurons extending from the third ventricle to the rostral periaqueductal gray. Hypothalamic regions are associated with emotional states and defensive behavior that commonly result in vocalizations. Pharmacological stimulation of anterior hypothalamic regions in rats resulted in 22 kHz ultrasonic vocalizations (Brudzynski, 1994). Injection of carbachol, an acetylcholine agonist, induced ultrasonic vocalization in rats and audible vocalization in cats, suggesting a common cholinergic contribution to the production of sonic and ultrasonic vocalization in mammals (Brudzynski and Bihari, 1990; Brudzynski, 2007; see Brudzynski, Chapter 7.3 in this volume).

III.C. Level 3: midbrain

The mesencephalic level comprised the CGr, CGc and MRF highlighted by stimulation. Mesencephalic structures are in a unique position for a gating function, because this level is a required anatomical relay of both the direct and indirect vocalization pathways. Whereas direct vocalization pathways may bypass the hypothalamus, no parallel vocalization pathways can bypass the midbrain level. In the direct pathway CGr serves as the major relay for descending PL projections and it, in turn, projects to NA. All three regions in this direct pathway showed large increases in FDG uptake in the vocalizing rats. The indirect pathways are mainly mediated by MRF projections to both CG

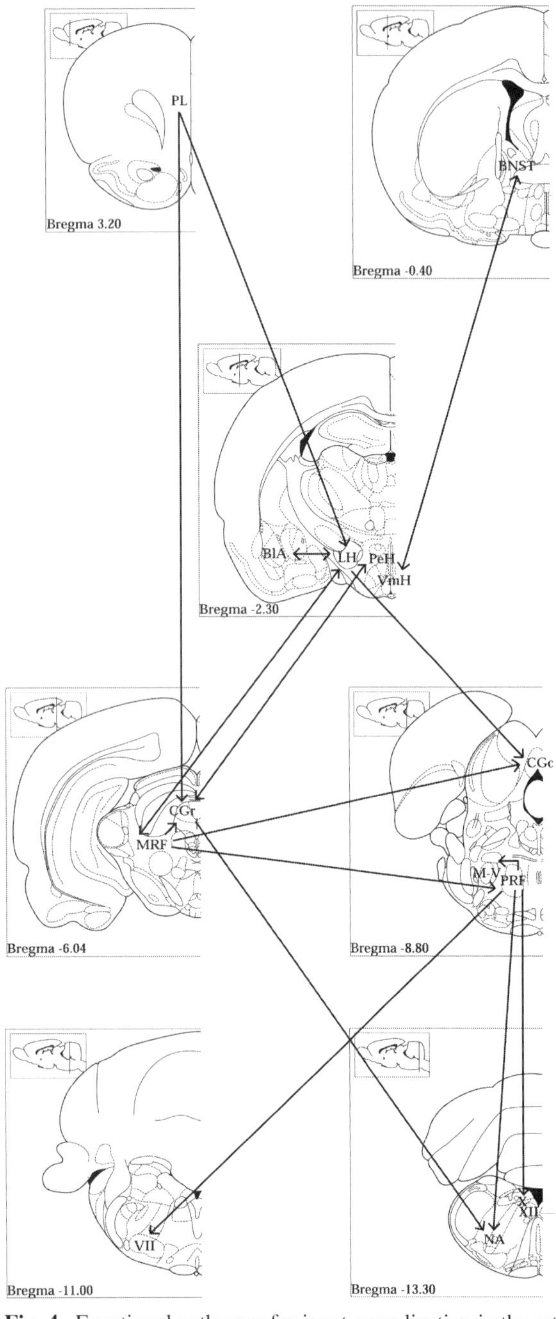

Fig. 4. Functional pathways for innate vocalization in the rat brain as determined from fluorodeoxyglucose (FDG) mapping. Specific structures activated by MRF-elicited vocalizations and their anatomical connections are illustrated in serial coronal sections of the rat brain. Bregma antero-posterior coordinates, at which the area of activation was centered for each specific region, are shown for all the regions activated. In the activated brain regions, the direct vocalization path is represented by the PL–CGr–NA descending pathway. The additional indirect pathway identified by FDG mapping involves three specific regions in each of the five levels, forming a network comprising 15 brain regions, as summarized in the text. Lines with one arrowhead represent descending projections, while lines with arrowheads at both ends represent reciprocal connections. Abbreviations are the same as in Fig. 1. Diagrams were adapted with permission from Paxinos, G. and Watson, C. (1997). The Rat Brain in Stereotaxic Coordinates, Compact disc, 3d edn. Academic Press, San Diego, CA.

and PRF. Stimulation in all the three structures elicits species-typical vocalizations in every mammalian species studied so far (Jürgens, 2009). Disconnection of the midbrain from lower levels resulted in mute animals and humans, suggesting a critical role of the midbrain gating mechanisms in vocal behavior (Jürgens and Pratt, 1979; Esposito et al., 1999). In an unrelated study (Gonzalez-Lima and Keene, 1982), we have observed that cats undergoing a pretrigeminal-postcollicular transection (cerveau isole cats) were unable to emit species-typical vocalizations (meowing, hissing, etc.), presumably due to interruption of the flow of vocal information between the midbrain and lower brainstem levels. Results of the studies with the combination of electrical stimulation and recordings are consistent with the interpretation that CG activity is important for vocalization intensity (global loudness, mean fundamental frequency), while activity of surrounding MRF and lower reticular formation sites are important for the more specific patterning of acoustic parameters of vocalizations (Jürgens, 2008; see also Hage, Chapters 8.2 and 8.3 in this volume).

III.D. Level 4: pons

The pontine level comprised three regions: PRF, M–V and VII, which are part of the indirect vocalization pathways. The PRF, which provides a major link between the MRF and the lower brainstem, showed the highest response level among these regions. PRF receives MRF inputs and, in turn, contributes to descending pathways to the lower motoneuron pools innervating vocalization muscles, as well as the reticular formation areas surrounding these motoneurons. For example, PRF projects to M–V and VII at the pontine level, and to NA, X and XII in the medulla. This provides a reticular formation network for motor coordination of all the various phonatory motoneuron pools, which are not directly interconnected between them (Thoms and Jürgens, 1987). Electrical and chemical stimulation of PRF produces vocalizations and stimulates pontine phonatory motoneurons that control articulatory muscles, such as M–V motoneurons that innervate muscles opening and closing the mandible via the mandibular nerve and VII motoneurons that drive lip muscles via the facial nerve (Cunningham and Sawchenko, 2000; see also Hage, Chapter 8.3 in this volume).

III.E. Level 5: medulla

In the medulla, there are three regions identified by MRF stimulation: NA; X; and XII, that form the final cranial output level for the control of vocalization. In particular, NA is the only region where the direct midbrain pathway from the CG and the indirect pontine pathways from the PRF converge anatomically as a final common relay for the control of phonatory motoneurons. Medullary reticular formation neurons surrounding the nucleus ambiguus are the recipient of both the CG and PRF afferents, which in turn project to the nucleus ambiguus motoneurons (see Hage, Chapter 8.2 in this volume). Cell bodies in the nucleus ambiguus send peripheral axons that innervate the intrinsic laryngeal muscles via the recurrent and external superior laryngeal nerves, and the levator veli palatine muscle via the pharyngeal plexus. Medullary reticular formation neurons surrounding X and XII and motoneurons in these cranial nerve nuclei constitute the lowest cranial component of the vocal motor output, especially XII motoneurons that innervate all tongue muscles via the hypoglossal nerve, as studied in the rat (Cunningham and Sawchenko, 2000).

In contrast to NA, the nearby nucleus of the trigeminal spinal tract (NT) showed a reduction of 15–20% in FDG uptake in the MRF-stimulated rats relative to controls (Fig. 2). Since the NT is involved in processing pain information, it is tempting to speculate that such suppression of NT activity during MRF-elicited vocalization may contribute a pain reduction (hypoalgesic) effect during emission of aversive vocalizations. This interpretation is consistent with the high activation in FDG uptake (>75%) seen in the CGr in the vocalizing rats because the CGr, when stimulated, can produce analgesia in rats (Schenk and Robinson, 1988; see also Borszcz and Spuz, Chapter 7.4 in this volume). Hence, species-typical vocalizations emitted during fighting may inhibit pain transmission (NT) and activate analgesia-controlling structures (CG), which would be ethologically advantageous.

IV. Conclusions

This study determined which specific limbic, midbrain and brainstem structures participate in the functional output of a vocalization response evoked by electrical MRF stimulation. Using FDG uptake, all brain regions were functionally mapped concurrently with the elicited vocalizations. This mapping of brain regions with FDG autoradiography during MRF-evoked vocalizations permitted us to propose a functional model of the activated pathways in the rat brain during emission of innate vocalization patterns. This model is consistent with the results reported in literature on the neural control of vocalization in mammals, while at the same time it significantly extends the scope of the present general model of the neural control of mammalian vocalization (Jürgens, 2009). Key features of the model include the following elements: (1) structures involved in species-typical vocalizations may be grouped into five levels of hierarchical organization along the rat neuraxis: limbic, hypothalamic, midbrain, pontine and medullar levels; (2) these five hierarchical levels interact via two parallel sets of anatomical pathways: a direct pathway, which corresponds to the well-known limbic–midbrain–medulla pathway for innate vocalization in mammals, comprising descending connections between PL-CGr-NA; and an indirect pathway, which involves three specific regions at each of the five hierarchical levels that form a network involving 15 brain regions; (3) the hypothalamus interacts reciprocally with limbic and midbrain levels and provides a parallel circuit for the motivational and emotional basis of vocalization; and (4) the reticular formation surrounding CG, the nucleus ambiguus and other activated cranial nerve motor nuclei, may provide an integrative network from midbrain to medullary levels for coordination of phonatory motoneurons and vocal pattern generation. It is hoped that this model will foster further research into the functional pathways for vocalization in the mammalian brain.

Acknowledgments

I thank Dr Henning Scheich who introduced me to FDG and the field of audio–vocal research and provided resources and advice in my initial studies, and Dr Douglas Barrett for his help with the figures. Supported in part by NIH grants R01 MH076847 and T32 MH65728.

References

Brudzynski, S.M., 1994. Ultrasonic vocalization induced by intracerebral carbachol in rats: localization and a dose-response study. Behav. Brain Res. 63, 133–143.

Brudzynski, S.M., 2007. Ultrasonic calls of rats as indicator variables of negative or positive states: acetylcholine-dopamine interaction and acoustic coding. Behav. Brain Res. 182, 261–273.

Brudzynski, S.M., Bihari, F., 1990. Ultrasonic vocalization in rats produced by cholinergic stimulation of the brain. Neurosci. Lett. 109, 222–226.

Cunningham, E.T., Sawchenko, P.E., 2000. Dorsal medullary pathways subserving oromotor reflexes in the rat: implications for the central neural control of swallowing. J. Comp. Neurol. 417, 448–466.

de Lanerolle, N.C., Lang, F.F., 1988. Functional neural pathways for vocalization in the domestic cat. In: Newman, J.D. (Ed.), The Physiological Control of Mammalian Vocalization. Plenum, New York, NY, pp. 7–19.

Esposito, A., Demeurisse, G., Alberti, B., Fabbro, F., 1999. Complete mutism after midbrain periaqueductal gray lesion. Neuroreport 10, 681–685.

Gonzalez-Lima, F., 1986. Activation of substantia gelatinosa by midbrain reticular stimulation demonstrated with 2-deoxyglucose in the rat spinal cord. Neurosci. Lett. 65, 326–330.

Gonzalez-Lima, F., 1987. Midbrain reticular stimulation produces patterns of metabolic activation and suppression in the cerebellum and vestibular nuclei: a 2-deoxyglucose study. Brain Res. 412, 275–284.

Gonzalez-Lima, F., 1992. Brain imaging of auditory learning functions in rats: studies with fluorodeoxyglucose autoradiography and cytochrome oxidase histochemistry. In: Gonzalez-Lima, F., Finkenstadt, T., Scheich, H. (Eds.), Advances in Metabolic Mapping Techniques for Brain Imaging of Behavioral and Learning Functions. Kluwer Academic Publishers, Boston, MA, pp. 39–109.

Gonzalez-Lima, F., Frysztak, R.J., 1991. Functional mapping of the rat brain during vocalizations: a 2-deoxyglucose study. Neurosci. Lett. 124, 74–78.

Gonzalez-Lima, F., Keene, J.J., 1982. Effect of stimulation of visual cortex on limbic and diencephalic neurons. Exp. Neurol. 77, 26–38.

Gonzalez-Lima, F., Scheich, H., 1984. Neural substrates for tone-conditioned bradycardia demonstrated with 2-deoxyglucose: I. Activation of auditory nuclei. Behav. Brain Res. 14, 213–233.

Gonzalez-Lima, F., Scheich, H., 1985. Ascending reticular activating system in the rat: a 2-deoxyglucose study. Brain Res. 344, 70–88.

Gonzalez-Lima, F., Scheich, H., 1986. Classical conditioning of tone-signaled bradycardia modifies 2-deoxyglucose uptake patterns in cortex, thalamus, habenula, caudate-putamen and hippocampal formation. Brain Res. 363, 239–256.

Gonzalez-Lima, F., Helmstetter, F.J., Agudo, J., 1993. Functional mapping of the rat brain during drinking behavior: a fluorodeoxyglucose study. Physiol. Behav. 54, 605–612.

Huang, S.C., Phelps, M.E., Hoffman, E.J., Sideris, K., Selin, C.E., Kuhl, D.E., 1980. Noninvasive determination of local cerebral metabolic rate of glucose in man. Am. J. Physiol. 238, E69–E82.

Jürgens, U., 2009. The neural control of vocalization in mammals: a review. J. Voice, 23, 1–10.

Jürgens, U., Pratt, R., 1979. Role of the periaqueductal gray in vocal expression of emotion. Brain Res. 167, 367–378.

Jürgens, U., Von Cramon, D., 1982. On the role of the anterior cingulate cortex in phonation: a case report. Brain Language 15, 234–248.

Jürgens, U., Ploog, D., 1988. On the motor coordination of monkey calls. In: Newman, J.D. (Ed.), The Physiological Control of Mammalian Vocalization. Plenum, New York, NY, pp. 7–19.

Kennedy, C., DesRosiers, M.H., Jehle, J.W., Reivich, M., Sharp, F.R., Sokoloff, L., 1975. Mapping of functional neural pathways by autoradiographic survey of local metabolic rate with 14-C-deoxyglucose. Science 187, 850–853.

Lammers, H.J., Lohman, A.H., 1974. Structure and fiber connections of the hypothalamus in mammals. Prog. Brain Res. 41, 61–78.

Leichnetz, G.R., Astruc, J., 1976. The efferent projections of the medial prefrontal cortex in the squirrel monkey (*Samiri sciureus*). Brain Res. 109, 455–472.

Litvin, Y., Blanchard, D.C., Blanchard, R.J., 2007. Rat 22 kHz ultrasonic vocalizations as alarm cries. Behav. Brain Res. 182, 166–172.

Martinez Jr., J.L., Petty, C., Messing, R.B., 1982. Regional brain uptake of 2-deoxy-D-glucose following training in a discriminated Y-maze avoidance task. J. Comp. Physiol. Psychol. 96, 721–724.

Müller-Preuss, P., Jürgens, U., 1976. Projections from the "cingular" vocalization area in the squirrel monkey. Brain Res. 103, 29–43.

Paxinos, G., Watson, C., 1997. The Rat Brain in Stereotaxic Coordinates, Compact disc, 3rd edn. Academic Press, San Diego, CA.

Phelps, M.E., Huang, S.C., Hoffman, E.J., Selin, C.E., Kuhl, D.E., 1979. Tomographic measurement of local cerebral glucose metabolic rate in man with (18F) fluorodeoxyglucose: validation of method. Ann. Neurol. 6, 371–388.

Price, J.L., 2003. Comparative aspects of amygdala connectivity. Ann. N. Y. Acad. Sci. 985, 50–58.

Reivich, M., Kuhl, D.E., Wolf, A., Greenberg, J.H., Phelps, M.E., Ido, T., Casella, V., Fowler, J., Hoffman, E.J., Alavi, A., Sokoloff, L., 1979. The (18F) fluorodeoxyglucose method for the measurement of local cerebral glucose utilization in man. Circulation Res. 44, 127–137.

Schenk, S., Robinson, B., 1988. Differentiation of the substrates for analgesia and vocalizations elicited by midbrain stimulation in rats: refractory period estimates. Behav. Brain Res. 31, 105–110.

Sokoloff, L., Reivich, M., Kennedy, C., Des Rosiers, M.H., Patlak, C.S., Pettigrew, K.D., Sakurada, O., Shinohara, M.,

1977. The [14C] deoxyglucose method for the measurement of local cerebral glucose utilization: theory, procedure, and normal values in the conscious and anesthetized albino rat. J. Neurochem. 28, 897–916.
Thoms, G., Jürgens, U., 1987. Common input of the cranial motor nuclei involved in phonation in the squirrel monkey. Exp. Neurol. 95, 85–99.
Zilles, K., Wree, A., 1985. Cortex: Areal and laminar structure. In: Paxinos, G. (Ed.), The Rat Nervous System, Volume 1: Forebrain and Midbrain. Academic Press, Orlando, FL, pp. 375–415.

CHAPTER 7.6

Adult house mouse (*Mus musculus*) ultrasonic calls: hormonal and pheromonal regulation

John G. Nyby*

Department of Biological Sciences, Lehigh University, Bethlehem, PA, USA

Abstract: An adult male house mouse typically begins calling at ultrasonic frequencies when presented with either a female mouse or her odors. Although direct experimental evidence has been difficult to obtain, ethological, genetic, endocrine, neurological, sound-spectrographic and perceptual findings all indirectly support male calling conveying important information to the female during sexual encounters. Since male calls correlate with the male's immediate sexual arousal and are also attractive to the female, these calls could help coordinate aspects of a mating bout. As might be expected of a reproductive behavior, male calling is hormonally and pheromonally regulated, similarly to other aspects of male reproductive physiology and behavior. Evidence is reviewed that androgen exerts a permissive activational effect during adulthood through its actions in the same brain area (the medial preoptic area) that regulates male-typical copulatory behavior. This activational effect also accounts for the sex difference in mouse calling. Evidence is also reviewed that male calling is pheromonally elicited by chemosignals in female urine.

Keywords: androgen; mouse; call; pheromone; reproduction; sex; ultrasonic vocalization; multisensory communication

I. Introduction

Infant and adult rodents of many species emit ultrasonic calls in a variety of contexts. Infant calls stimulate parental retrieval and inhibit rough parental handling. Adult ultrasonic calls, on the other hand, are most often associated with reproductive and agonistic behavior. Although it might make adaptive sense for nocturnal rodents, adult rodent calls do not appear to be used in echolocation (Anderson, 1954). Several reviews (Brown, 1976; Nyby and Whitney, 1978; Whitney and Nyby, 1983; Nyby, 2001) and a book (Sales and Pye, 1974) have previously described the mechanisms and functions of ultrasonic communication in adult rodents.

Compared to other adult rodents, house mice appear more selective about the situations in which they emit ultrasonic calls. For example, adult male Norway rats emit ultrasonic calls during reproductive and agonistic situations (Sales, 1972a; Sales and Pye, 1974; Brudzynski, 2007) and adolescent rats call during rough and tumble play and in response to human "tickling" (heterospecific interactions resembling rough and tumble play: Panksepp and Burgdorf, 2000; Burgdorf and Panksepp, 2001). Moreover, in golden hamsters (Floody and Pfaff, 1977b), Norway rats (White and Barfield, 1987; Sales, 1972b) and collared lemmings (Brooks and Banks, 1973), both sexes emit ultrasonic calls during sexual encounters. However, in adult house mice, ultrasonic calls are not normally emitted during agonistic or play situations, and in sexual encounters the male is the primary emitter, with the female emitting few or no calls (Whitney et al., 1973; Nunez et al., 1985; Warburton et al., 1989; White et al., 1998). Reproduction appears to be the only social context in which male mice readily emit ultrasonic calls (Sales, 1972a).

When isolated from males, female mice call quite readily to each other under certain conditions (Maggio and Whitney, 1985; D'Amato and Moles, 2001; Moles et al., 2007). However, such calling was inhibited following a brief interaction with a male

*Corresponding author. E-mail: jgn2@lehigh.edu

Stefan M. Brudzynski (Ed.)
Handbook of Mammalian Vocalization
ISBN 978-0-12-374593-4

DOI: 10.1016/B978-0-12-374593-4.00029-2

(Maggio et al., 1983). Since mice live in demes that typically include one or more adult males (Bronson, 1979), the occurrence and function of these female calls under naturalistic conditions is unclear. Because so little is known about female calling, this review will focus on the function of adult male calling and its hormonal and pheromonal regulation.

II. Function of adult male calls

Ethological, genetic, endocrine, neurological, sound-spectrographic and perceptual findings all point to male calls serving an important function in mouse sexual behavior.

Males of most domesticated laboratory strains typically begin emitting ultrasonic calls within a minute of encountering a non-cohabiting female or her odors, with call latency shortening even more after sexual experience. To my knowledge, males from every mouse strain studied emit these calls (e.g., Nyby and Whitney, 1978), reflecting the species-typical nature of this behavior. Although males of some strains are slower to begin, once they begin, they call similarly to males with shorter latencies (Nyby, 1983). Males emit at least two different types of calls in sexual encounters, a "70 kHz" call which occurs throughout precopulatory and copulatory periods and an interspersed "40 kHz" call which is seen most often in the later phases of a copulatory sequence (White et al., 1998). However, both show substantial frequency modulation and the 40 and 70 kHz designators simply represent where the calls were most readily detected with an ultrasound detector tuned to the relevant frequency.

One line of evidence bearing on functional significance is to observe what the male is doing while calling (Sales, 1972b; Nyby, 1983). In most cases the male is in close proximity to either a female or to an olfactory stimulus from a female (such as female urine). For example, during precopulatory investigation, most calls occur when the male is sniffing and following the female. During copulatory behavior, calls occur primarily when mounting or intromitting the female. However, interspersed among periods of precopulatory or copulatory behavior, relatively few calls occur when the male disengages from the female for short intervals and appears to have his attention elsewhere (Nyby, 1983).

Immediately after ejaculating, the male undergoes a refractory period of approximately 15–30 minutes during which he does not engage in copulatory behavior (Mosig and Dewsbury, 1976). For approximately the first 75% of the refractory period, when the male shows no active interest in the female, he does not call (Nyby, 1983). However, during the last 25% of the refractory period, the male resumes both olfactory investigation and calling (Nyby, 1983). Nyby (1983) hypothesized that the quiescent period demarcates the absolute refractory period with the resumption of calling demarcating the relative refractory period.

The refractory period is often followed by a second mating sequence, which can lead to a second ejaculation (Nyby, 1983). During the second copulatory sequence, ultrasonic calls are again associated primarily with the male's mounts and intromissions (Nyby, 1983). Thus, the calls accompany the male's immediate sexual attention to, and interest in, the female.

Behavioral genetics research is also consistent with male calls serving an important function (Whitney and Nyby, 1983). For example, breeding studies demonstrated that male calls possess low heritability and directional dominance in the direction of high levels of calling. This genetic architecture would be expected of an adaptive trait shaped by natural selection.

Sonographic analyses also point to male ultrasonic calls having communicatory significance. Male calls possess a richness and temporal complexity comparable to bird song, in which different syllable types occur in a highly nonrandom and repeated order (Holy and Guo, 2005). Thus, male calling also has the sonographic signature of a highly evolved communication system.

Female mice (with "normal" hearing) clearly hear the male's calls. Audiographic analyses indicated that mouse hearing has peak sensitivity near the frequencies of the male calls (Brown, 1971). Thus, female hearing appears selectively tuned to hear ultrasonic calls. Parental retrieval of offspring also is stimulated by infant ultrasonic calls which overlap adult male calls in frequency (e.g., Haack et al., 1983). Interestingly, infant calls are preferentially processed by the mother's left hemisphere, similar to the processing of both bird song and human speech (Ehret, 1987). Thus, not only do females detect male calls, they also appear to possess neural adaptations designed to process ultrasonic signaling. The female's ability to monitor the male's immediate sexual motivation, as reflected in his calling, would likely be adaptive in coordinating the reproductive process.

At the same time, some domesticated laboratory strains do not hear well in the ultrasonic range and, in some cases, may be deaf in adulthood (Ralls, 1967; Ehret, 1974). Inbreeding depression, and perhaps a reduced need for high-frequency hearing in domesticated mice, might account for this high-frequency hearing loss. Thus, domesticated laboratory mice should be used cautiously when studying the function of male calling.

The most direct evidence for male call functionality is that females find male calls attractive (Pomerantz et al., 1983). For example, female mice spent more time in a compartment with a tethered calling male (compared to a tethered devocalized male) or with a devocalized male accompanied by artificial ultrasounds (compared to one not accompanied by artificial ultrasounds). Therefore, one function of male calls may be to keep the female in close proximity.

However, other attempts to examine call function using devocalized males have not been very successful. For example, devocalized males readily mate, have normal mount latencies and do not appear reproductively impaired (Nunez et al., 1985; White et al., 1998). This shows male vocalizations are not required for successful reproduction in laboratory settings. At the same time, male calling could be more important for wild mice under naturalistic conditions.

Much of the mating in the wild occurs during the female's postpartum estrus, when the pups of the previous mating are still around and the female is highly aggressive towards strange males. Perhaps male calls serve to protect a male from aggression by lactating females and thereby enhance the male's likelihood of successful reproduction. In contrast to expectations, lactating females attacked vocalizing males even more rapidly than devocalized males (Bean et al., 1986a). However, since the lactating females of this experiment were probably not experiencing postpartum estrus at the time of testing, the hypothesis was not rigorously tested.

Other puzzling findings also need to be taken into account in formulating an explanation of why males call. While males do not call to other males, they call indiscriminately to females in all stages of estrous and even to ovariectomized females or hypophysectomized females (Nyby et al., 1979). One might expect males to be more selective. However, since the mouse estrous cycle is short (4 or 5 days), perhaps a communicatory bond would normally keep a nonreceptive female around until she becomes receptive. Alternatively, perhaps the lack of male selectivity is unique to domesticated mice. Relaxed natural selection along with artificial selection for high fecundity may have genetically altered aspects of this communication system during domestication. Ideally, it would be best to examine functional questions in wild mice in semi-natural settings; however, this approach has its own set of serious difficulties.

III. Androgenic regulation

As might be expected of a reproductively related behavior, adult male ultrasonic calls are androgenically regulated. Although both male and female infants emit ultrasonic distress calls, these calls cease by around 14–16 days of age as the pups begin to thermoregulate (Noirot, 1972). In males, calling to females emerges at puberty (Whitney et al., 1973; Warburton et al., 1989).

Traditional ablation/replacement studies demonstrated that castration reduces adult male calling (Dizinno and Whitney, 1977) and androgen replacement restores it (Nunez et al., 1978; Nunez and Tan, 1984; Warburton et al., 1989). To initiate calling in DBA/2J males, testosterone was aromatized to estradiol, which then activated calling by binding estrogen receptors (Nunez et al., 1978). However, subsequent work with Swiss Webster (Nunez and Tan, 1984) and C57BL/6J × AKR/J hybrid males (Bean et al., 1986b), found that either dihydrotestosterone (a nonaromatizable androgen) or estradiol restored calling in castrated males, suggesting that both androgen-receptor and estrogen-receptor pathways exist in these strains. Taken together, these findings indicated that house mice possess redundant neural receptor mechanisms for androgenic responsiveness, both of which are present in some strains, but not in others. One might expect that the hormonal regulation of calling would resemble that for copulatory behavior, where all rodent strains uniformly show estrogenic responsiveness but show substantial variation in androgen-receptor pathways (reviewed by Nyby and Simon, 1987).

Several lines of evidence support the sex difference in mouse reproductive calling being due to androgen's activational effects in adulthood, rather than to its organizational effects early in development. In one study (Nyby et al., 1977b), long-term gonadectomized males and females, who initially did not call to a stimulus female, received testosterone treatment over a period of several weeks. Over the course of treatment, both male and female calling to the stimulus female progressively increased in a similar fashion. The increase in calling was also accompanied in both sexes by an increase in mounting the stimulus female, suggesting that, in this context, female calling represented inappropriate androgen-stimulated male-typical behavior. Whether the slightly delayed female responsiveness in one of the two experiments reflected an organizational difference or whether females were simply more refractory because of lower initial androgen levels was not clear. However, if a sex difference existed in androgenic responsiveness, it was not profound (In contrast, female-to-female calling in females isolated from males, mentioned in the introduction, was not accompanied by mounting and thus did not appear to be inappropriate male-typical behavior (Maggio and Whitney, 1985)).

Another strategy to demonstrate organizational androgenic effects examines whether phenotypic differences between animals of the same sex can be accounted for by their intrauterine position. Specifically, individuals developing *in utero* between two males (2M mice) are exposed to more androgen early in development than individuals developing between two females (0M mice) (Clemens et al., 1978; vom Saal, 1981). As a result, 2M females should be more masculinized in adulthood than 0M females for traits organized by the prenatal effects of androgens. However, Jubilan and Nyby (1992) found intrauterine position had little effect on the ultrasonic vocalizations of either adult females or adult males to either a female mouse or her urine. Thus, these findings also did not support a prenatal organizational influence of androgens on ultrasonic vocalizations. In conclusion, females appear to have much the same potential to vocalize as males, and the adult sex difference observed in male/female pairings is likely due to adult sex differences in circulating androgen.

The medial preoptic area (MPOA) is an important brain area where androgens activate male ultrasonic calling, as evidenced by brain implants (Nyby et al., 1992; Matochik et al., 1994; Sipos and Nyby, 1996, 1998). Testosterone or testosterone propionate implants into the MPOA were highly effective at restoring ultrasonic calling in castrated C57Bl/6J × AKR males. However, testosterone implants into the anterior hypothalamus (Nyby et al., 1992), ventromedial hypothalamus (Nyby et al., 1992), septum (Nyby et al., 1992; Matochik et al., 1994), medial amygdala (Nyby et al., 1992; Matochik et al., 1994) and ventral tegmental area (Sipos and Nyby, 1996) were relatively ineffective. The MPOA is also the critical area for the androgenic activation of copulatory behavior in male mammals (Hart and Leedy, 1985), further contributing to the notion that male calling is a male-typical reproductive behavior.

As might be expected, if testosterone is a prohormone that must be aromatized for effectiveness, estradiol implants restored ultrasonic calling at more locations than testosterone, including the MPOA, ventromedial hypothalamus, anterior hypothalamus and lateral septum (Nyby et al., 1992). However, interpretation of the estrogenic effects is problematic as the effects could reflect the greater activity of estrogenic treatment, more diffusion by estradiol from the implant site, or a different CNS distribution of estrogen-sensitive versus androgen-sensitive neurons. More work is required to resolve these possibilities.

Many brain areas undoubtedly participate in regulating male ultrasonic calling. Newman (1999) hypothesized that all social behaviors including sexual, aggressive and parental behaviors, are regulated by a common neural network (the medial extended amygdala) including the corticomedial amygdala, lateral septum, ventromedial hypothalamus, midbrain central gray and tegmentum, anterior hypothalamus, as well as the medial preoptic area. In addition, the parts of the nervous system that control motor output to the larynx (where the calls are produced, Roberts (1975)) would also be involved. However, strategies other than androgen implants will likely be necessary to examine these areas (Nyby et al., 1992; Matochik et al., 1994; Sipos and Nyby, 1996, 1998).

Recent work (Wang et al., 2008) implicated acetylcholine signaling in the neural regulation of male calling. Significant reductions in both male mouse calling and reproductive behavior were seen in knockouts for the M2-muscarinic receptor. On the other hand, male M5-muscarinic receptor knockouts exhibited reduced calling, but sex behavior was undisturbed. However, the calling of male M4- and dopamine-D2 receptor knockouts appeared relatively undisturbed. Since M2 and M5 receptors occur on dopamine neurons (Wang et al., 2008), acetylcholine was suggested to regulate male calling through activating dopamine neurons. Whether these neurons are involved in the sensory perception, motivational regulation, or motor control is not clear.

IV. Pheromonal regulation

Male mouse ultrasonic calling depends heavily on pheromonal communication. As expected, if male calling is related to reproduction, many of the same exteroreceptive chemical cues that promote male reproductive physiology and behavior also elicit male calling. For example, not only are female mice good stimuli for eliciting male calling, female-associated odors are also quite effective. Effective odors include female-soiled cage shavings (Whitney et al., 1974), female urine (Nyby et al., 1977a), female saliva (Byatt and Nyby, 1986) and female vaginal fluids (Nyby et al., 1977a). Whether these stimuli all contain the same chemosignal or whether different naturally-occurring chemosignals elicit calling has not been determined.

Much of the research on the pheromonal elicitation of male calling has focused on female urine, since female urine also stimulates reflexive LH and testosterone release in males (Maruniak and Bronson, 1976), male copulatory behavior (Dixon and Mackintosh, 1971), while also reducing intermale aggression (Mugford and Nowell, 1971). In fact, in males encountering female urine, the presence of calling in the first minute was significantly correlated with (i.e., a good predictor of) whether the male exhibited reflexive testosterone release 30 minutes later (James et al., 2006). The many parallels between the different male-typical responses to female urine suggest that all may be elicited by the same urinary pheromones (Sipos et al., 1999).

At least two different ultrasound-eliciting chemosignals exist in female urine: a highly potent signal in fresh urine and a less potent one in aged urine. The chemosignal in freshly voided female urine fits the more rigorous definitions of a pheromone (Beauchamp et al., 1976) in which the male is biologically programmed to respond in a relatively invariant fashion. For example, freshly voided urine elicits calling from males regardless of whether they are sexually experienced (Sipos et al., 1992), responsiveness does not readily habituate with repeated exposures (Sipos et al., 1992), and fresh urine can serve as an unconditioned stimulus for causing neutral stimuli to acquire call-eliciting properties in a classical conditioning paradigm (Sipos et al., 1992). The activity of this potent chemosignal is destroyed by oxidation within 15–18 hours after voiding (Sipos et al., 1993), although pheromonal activity can be prolonged by adding an antioxidant (Sipos et al., 1994). Activity is not seen in urine collected overnight (12 hours) in a metabolic cage, as this method of collection likely hastens the oxidation that destroys the chemosignal (Sipos et al., 1992).

After the potent urinary pheromone has degraded, female urine continues to elicit male calling for at least 30 days (under certain circumstances), indicating the existence of a second chemosignal that elicits calling (Nyby and Zakeski, 1980). However, calling to this signal differs in significant ways from that to fresh urine. For example, responsiveness to this chemosignal is acquired as a result of adult sexual experience; sexually naïve males do not call to this urinary signal. Moreover, sexually experienced males rapidly habituate their calling after three or four exposures (Dizinno and Whitney, 1977). Furthermore, repeated pairing of aged urine with a neutral stimulus does not cause that stimulus to elicit calling (i.e., aged urine cannot serve as an unconditioned stimulus in a classical conditioning paradigm) (Sipos et al., 1992).

The relative concentrations of these two chemosignals in female urine changes quickly in the hours following urination, and thus female urine marks could provide important information to other mice about how recently a female has been in the area (Sipos et al., 1992). Whether these two chemosignals are single molecules or chemical mixtures is not known (although the latter seems most likely). We hypothesized that oxidation may turn the ephemeral, potent signal into the more stable, less-potent one (Sipos et al., 1992).

While males appear biologically programmed to call to the ephemeral urinary signal, they can "learn" to call to a variety of signals that normally do not elicit calling. For example, repeated sociosexual encounters with female mice odorized with the following stimuli caused males to call: urine from female rats (Kerchner et al., 1986); urine from hypophysectomized females (Maggio et al., 1983); a commercially available perfume (Nyby et al., 1978); clean cotton swabs (Sipos et al., 1995); and plastic bags (Nyby et al., 1979). In fact, almost any chemosensory stimulus that allows the male to learn to anticipate a female encounter appears to have the potential to serve as a conditioned stimulus for eliciting calling. This flexibility appears adaptive, as it might better allow the male to respond to females in different chemosensory environments. However, some biological constraints exist, since males could not be conditioned to vocalize to the urine of other male mice (unpublished observation).

One line of research (Nyby et al., 1978) examined whether "learning" to call to a neutral stimulus (perfume) would be influenced more by neonatal "imprinting" on the dam or by adult learning during sexual encounters. In this work, male mice encountered perfume on their dam from birth to weaning, and/or on adult females during adult sexual encounters. Very briefly, experience during adulthood with an odorized sex partner was far more important than infant experience, although infant exposure to the perfume may have enhanced the adult learning. However, perfume responsiveness, when it did occur, did not alter, or replace, calling to female urine (i.e., "learning" to call to a novel stimulus did not affect responsiveness to naturally-occurring pheromones).

Other work (Wysocki et al., 1982) examined the chemosensory mechanisms by which urinary pheromones are detected. Following olfactory bulbectomy, which deafferents all nasal chemosensory systems (including the main olfactory system and vomeronasal system), sexually naïve males did not call to either a female or to olfactory cues from the female (cage shavings or aged female urine). However, sexually experienced males, while similarly nonresponsive to chemosensory cues, continued to call to the female herself (presumably based on non-chemosensory cues). Thus, olfactory bulbectomy eliminated sensory detection of the call-eliciting chemosignals, but did not eliminate the male's ability or motivation to call to the female.

Further work (Wysocki et al., 1982) indicated that these calling deficits could be accounted for largely, but not completely, by vomeronasal organ removal. As was the case for olfactory bulbectomy, the reduction in male calling following vomeronasalectomy was mediated by sensory deafferentation rather than a reduction in sexual motivation. However, in contrast to the olfactory bulbectomy results, sexually experienced males exhibited low (but significant) levels of calling to aged female urine (i.e., urine collected overnight in a metabolic cage). This calling, mediated by non-vomeronasal chemoreception, required the male to have sexual experience before vomeronasalectomy. This finding suggested to us that while the vomeronasal system is the primary system for perceiving sexual pheromones, the olfactory system is capable of acquiring pheromone responsiveness as a result of sexual experience.

Subsequent work (Sipos et al., 1995) examined the roles of the vomeronasal and olfactory systems of sexually experienced males in perceiving the more potent but ephemeral female pheromone. Deafferenting neither the accessory olfactory system alone (by surgically removing the vomeronasal organ), nor the main olfactory system alone (by intranasal irrigation with $ZnSO_4$) had much effect on male calling to freshly voided urine. However, if both were simultaneously deafferented, calling to fresh urine was eliminated. Thus, in sexually experienced males, both chemosensory systems appeared equally important, and redundant, for detection of this potent but ephemeral pheromone. Whether sexual learning was involved in promoting olfactory system responsiveness to the potent ephemeral pheromone was not examined. However, whether innate or learned, it is not surprising that redundant perceptual systems would exist to detect such a biologically important pheromonal signal.

V. Concluding remarks

Male ultrasonic calling appears to be part of a multisensory communication system that helps coordinate reproductive physiology and behavior in male and female house mice. Male calling is hormonally and pheromonally regulated very similarly to other components of male-typical reproductive behavior. Calling by the male reflects his immediate sexual interest, and thus could provide important information to the female about the male's motivational state. Presumably this information would be important in helping time the occurrence of various aspects of mouse reproductive physiology and behavior. As would be expected of a reproductive behavior, male calling is androgenically regulated and elicited by female sex pheromones.

In fact, male calling may be an even "purer" operational measure of male sexual motivation than male copulatory behavior. In contrast to copulatory behavior, male calling can be elicited by chemosignals alone, and thus cannot be confounded by the degree of receptivity of the female sexual partner. This robust, reliable, easily measured behavior has much to recommend it for examining mechanistic questions relating to male reproduction and pheromonal communication.

And finally, while many lines of research implicate male calling as having communicatory significance during reproduction, establishing the precise nature of the female's response has been fraught with difficulties. Continued attempts, perhaps using wild mice under semi-natural conditions, should be a high priority for future research.

Acknowledgments

I thank Murray Itzkowitz and Stefan Brudzynski for reading and commenting on earlier versions of this manuscript. Parts of this manuscript were written while being supported by NIMH (R15 MH065956).

References

Anderson, J.W., 1954. The production of ultrasonic sounds by laboratory rats and other mammals. Science 119, 808–809.

Bean, N.J., Nunez, A.A., Wysocki, C.J., 1986a. 70-kHz vocalizations by male mice do not inhibit aggression in lactating mice. Behav. Neural Biol. 46, 46–53.

Bean, N.J., Nyby, J., Kerchner, M., Dahinden, Z., 1986b. Hormonal regulation of chemosignal-stimulated precopulatory behavior in male housemice (*Mus musculus*). Horm. Behav. 20, 390–404.

Beauchamp, G.K., Doty, R.L., Moulton, D.G., Mugford, R.A., 1976. The pheromonal concept in mammalian communication: a critique. In: Doty, R.L. (Ed.), Mammalian Olfaction, Reproductive Processes, and Behavior. Academic Press, New York, NY, pp. 144–157.

Bronson, F.H., 1979. The reproductive ecology of the house mouse. Q. Rev. Biol. 54, 265–299.

Brooks, R.J., Banks, E.M., 1973. Behavioral biology of the collared lemming (*Dicrostonyx groelandicus* (Traill)): an analysis of acoustic communication. Animal Behaviour Monographs 6, 1–82.

Brown, A.M., 1971. High frequency responsiveness in rodents at the level of the inferior colliculus. Nat. New Biol. 232, 223–224.

Brown, A.M., 1976. Ultrasound and communication in rodents. Comp. Biochem. Biophysiol. 53, 313–317.

Brudzynski, S.M., 2007. Ultrasonic calls of rats as indicator variables of negative or positive states: acetylcholine-dopamine interaction and acoustic coding. Behav. Brain Res. 182, 261–273.

Burgdorf, J., Panksepp, J., 2001. Tickling induces reward in adolescent rats. Physiol. Behav. 72, 167–173.

Byatt, S., Nyby, J., 1986. Hormonal regulation of chemosignals of female mice that elicit ultrasonic vocalizations from males. Horm. Behav. 20, 60–72.

Clemens, L.G., Gladue, B.A., Coniglio, L.P., 1978. Prenatal endogenous androgenic influences on masculine sexual behavior and genital morphology in male and female rats. Horm. Behav. 10, 40–53.

Dixon, A.K., Mackintosh, J.H., 1971. Effects of female urine upon the social behavior of adult male mice. Anim. Behav. 19, 138.

Dizinno, G., Whitney, G., 1977. Androgen influence on male mouse ultrasounds during courtship. Horm. Behav. 8, 188–192.

D'Amato, F.R., Moles, A., 2001. Ultrasonic vocalizations as an index of social memory in female mice. Behav. Neurosci. 115, 834–840.

Ehret, G., 1974. Age-dependent hearing loss in normal hearing mice. Naturwissenschaften 61, 506–507.

Ehret, G., 1987. Left hemisphere advantage in the mouse brain for recognizing ultrasonic communication calls. Nature 325, 249–251.

Floody, O.R., Pfaff, D.W., 1977b. Communication among hamsters by high-frequency acoustic signals: II Determinants of calling by females and males. J. Comp. Physiol. Psychol. 91, 807–819.

Haack, B., Markl, H., Ehret, G., 1983. Sound communication between parents and offspring. In: Willott, J.F. (Ed.), The Auditory Psychobiology of the Mouse. Charles C. Thomas, DeKalb, IL, pp. 57–97.

Hart, B.L., Leedy, M.G., 1985. Neurological basis of male sexual behavior. In: Adler, N., Pfaff, D., Goy, R.W. (Eds.), Handbook of Behavioral Neurobiology, Vol. 7, Reproduction. Plenum Press, New York, NY, pp. 373–422.

Holy, T.E., Guo, Z., 2005. Ultrasonic songs of male mice. PLoS Biol. 3, e386.

James, P.J., Nyby, J.G., Saviolakis, G.A., 2006. Sexually stimulated testosterone release in male mice (*Mus musculus*): roles of genotype and sexual arousal. Horm. Behav. 50, 424–431.

Jubilan, B.M., Nyby, J.G., 1992. The intrauterine position phenomenon and precopulatory behaviors of house mice. Physiol. Behav. 51, 857–872.

Kerchner, M., Vatza, E.J., Nyby, J.G., 1986. Ultrasonic vocalizations by male house mice (*Mus musculus*) to novel odors: roles of infant and adult experience. J. Comp. Psychol. 100, 253–261.

Maggio, J.C., Whitney, G., 1985. Ultrasonic vocalizing by adult female mice (*Mus musculus*). J. Comp. Psychol. 99, 420–436.

Maggio, J.C., Maggio, J.H., Whitney, G., 1983. Experience-based vocalization of male mice to female chemosignals. Physiol. Behav. 31, 269–272.

Maruniak, J., Bronson, F., 1976. Gonadotropic responses of male mice to female urine. Endocrinology 99, 963–969.

Matochik, J.A., Sipos, M.L., Nyby, J.G., Barfield, R.J., 1994. Intracranial androgenic activation of male-typical behaviors in house mice: motivation versus performance. Behav. Brain Res. 60, 141–149.

Moles, A., Costantini, F., Garbugino, L., Zanettini, C., D'Amato, F.R., 2007. Ultrasonic vocalizations emitted during dyadic interactions in female mice: a possible index of sociability? Behav. Brain Res. 182, 223–230.

Mosig, D.W., Dewsbury, D.A., 1976. Studies of copulatory behavior of house mice. Behav. Biol. 16, 463–473.

Mugford, R.A., Nowell, N.W., 1971. Endocrine control over production and activity of the anti-aggression pheromone from female mice. J. Endocrinol. 49, 225.

Noirot, E., 1972. Ultrasounds and maternal behavior in small rodents. Dev. Psychobiol. 5, 371–387.

Nunez, A.A., Tan, D.T., 1984. Courtship ultrasonic vocalizations in male Swiss-Webster mice: effects of hormones and sexual experience. Physiol. Behav. 32, 717–721.

Nunez, A.A., Nyby, J., Whitney, G., 1978. The effects of testosterone, estradiol, and dihydrotestosterone on male mouse (*Mus musculus*) ultrasonic vocalizations. Horm. Behav. 11, 264–272.

Nunez, A.A., Pomerantz, S.M., Bean, N.J., Youngstrom, T.G., 1985. Effects of laryngeal denervation on ultrasound production and male sexual behavior in rodents. Physiol. Behav. 34, 901–905.

Nyby, J.G., 1983. Ultrasonic vocalizations during sex behavior of male house mice (*Mus musculus*): a description. Behav. Neural Biol. 39, 128–134.

Nyby, J.G., 2001. Auditory communication among adults. In: Willott, J.F. (Ed.), Handbook of Mouse Auditory Research: From Behavior to Molecular Biology. CRC Press, Boca Raton, FL, London, New York, NY, Washington, DC, pp. 3–18.

Nyby, J.G., Simon, N.G., 1987. Nonaromatizable androgens may stimulate a male mouse reproductive behavior by binding estrogen receptors. Physiol. Behav. 39, 147–151.

Nyby, J.G., Whitney, G., 1978. Ultrasonic communication by adult myomorph rodents. Neurosci. Biobehav. Rev. 2, 1–14.

Nyby, J.G., Zakeski, D., 1980. Elicitation of male mouse ultrasounds: bladder urine and aged urine from females. Physiol. Behav. 24, 737–740.

Nyby, J.G., Wysocki, C.J., Whitney, G., Dizinno, G., 1977a. Pheromonal regulation of male mouse ultrasonic courtship (*Mus musculus*). Anim. Behav. 25, 333–341.

Nyby, J.G., Dizinno, G.A., Whitney, G., 1977b. Sexual dimorphism in ultrasonic vocalizations of mice (*Mus musculus*): gonadal hormone regulation. J. Comp. Physiol. Psychol. 91, 1424–1431.

Nyby, J.G., Whitney, G., Schmitz, S., Dizinno, G., 1978. Postpubertal experience establishes signal value of mammalian sex odor. Behav. Biol. 22, 545–552.

Nyby, J.G., Wysocki, C.J., Whitney, G., Dizinno, G., Schneider, J., 1979. Elicitation of male mouse (*Mus musculus*) ultrasonic vocalizations: I. Urinary cues. J. Comp. Physiol. Psychol. 93, 957–975.

Nyby, J.G., Matochik, J.A., Barfield, R.J., 1992. Intracranial androgenic and estrogenic stimulation of male-typical behaviors in house mice (*Mus domesticus*). Horm. Behav. 26, 24–45.

Panksepp, J., Burgdorf, J., 2000. 50-kHz chirping (laughter?) in response to conditioned and unconditioned tickle-induced reward in rats: effects of social housing and genetic variables. Behav. Brain Res. 115, 25–38.

Pomerantz, S.M., Nunez, A.A., Bean, N.J., 1983. Female behavior is affected by male ultrasonic vocalizations in house mice. Physiol. Behav. 31, 91–96.

Ralls, K., 1967. Auditory sensitivity in mice, *Peromyscus* and *Mus musculus*. Anim. Behav. 15, 123–128.

Roberts, L.H., 1975. The rodent ultrasound production mechanism. Ultrasonics 13, 83–88.

Sales, G.D., 1972a. Ultrasound and aggressive behaviour in rats and other small mammals. Anim. Behav. 20, 88–100.

Sales, G.D., 1972b. Ultrasound and mating behaviour in rodents with some observations on other behavioral situation. J. Zool., London, 168, 149–164.

Sales, G.D., Pye, D., 1974. Ultrasound in rodents. In: Ultrasonic Communication by Animals. Chapman and Hill, London, UK, pp. 149–201.

Sipos, M.L., Nyby, J.G., 1996. Concurrent androgenic stimulation of the ventral tegmental area and medial preoptic area: synergistic effects on male-typical behaviors in house mice. Brain Res. 729, 29–44.

Sipos, M.L., Nyby, J.G., 1998. Intracranial androgenic activation of male-typical behaviours in house mice: concurrent stimulation of the medial preoptic area and medial nucleus of the amygdala. J. Neuroendocrinol. 10, 577–586.

Sipos, M.L., Kerchner, M., Nyby, J.G., 1992. An ephemeral sex pheromone in the urine of female house mice (*Mus domesticus*). Behav. Neural Biol. 58, 138–143.

Sipos, M.L., Nyby, J.G., Serran, M.F., 1993. An ephemeral sex pheromone of female house mice (*Mus domesticus*): Pheromone fade-out time. Physiol. Behav. 54, 171–174.

Sipos, M.L., Alterman, L., Nyby, J.G., Perry, B., Vandenberg, J., 1994. An ephemeral pheromone of female house mice: degradation by oxidation. Anim. Behav. (in press).

Sipos, M.L., Wysocki, C.J., Nyby, J.G., Wysocki, L., Nemura, T.A., 1995. An ephemeral pheromone of female house mice: perception via the main and accessory olfactory systems. Physiol. Behav. 58, 529–534.

Sipos, M.L., Nyby, J.G., Snyder, R.W., Kerchner, M., 1999. Ultrasound-eliciting and aggression-reducing pheromones in female house mouse urine: are they the same? Physiol. Behav. (under review).

vom Saal, F.S., 1981. Variation in phenotype due to random intrauterine positioning of male and female fetuses in rodents. J. Reprod. Fertility 62, 633–650.

Wang, H., Liang, S., Burgdorf, J., Wess, J., Yeomans, J., 2008. Ultrasonic vocalizations induced by sex and amphetamine in M2, M4, M5 muscarinic and D2 dopamine receptor knockout mice. PLoS ONE 3, e1893.

Warburton, V.L., Sales, G.D., Milligan, S.R., 1989. The emission and elicitation of mouse ultrasonic vocalizations: the effects of age, sex and gonadal status. Physiol. Behav. 45, 41–47.

White, N.R., Barfield, R.J., 1987. Role of the ultrasonic vocalization of the female rat (*Rattus norvegicus*) in sexual behavior. J. Comp. Psychol. 101, 73–81.

White, N.R., Prasad, M., Barfield, R.J., Nyby, J.G., 1998. 40- and 70-kHz vocalizations of mice (*Mus musculus*) during copulation. Physiol. Behav. 63, 467–473.

Whitney, G., Nyby, J.G., 1983. Sound communication among adults. In: Willott, J.F. (Ed.), Auditory Psychobiology of the Mouse. Plenum, New York, NY, pp. 173–192.

Whitney, G., Coble, J.R., Stockton, M.D., Tilson, E.F., 1973. Ultrasonic emissions: do they facilitate courtship of mice? J. Comp. Physiol. Psychol. 84, 445–452.

Whitney, G., Alpern, M., Dizinno, G., Horowitz, G., 1974. Female odors evoke ultrasounds from male mice. Anim. Learn. Behav. 2, 13–18.

Wysocki, C.J., Nyby, J.G., Whitney, G., Beauchamp, G.K., Katz, Y., 1982. The vomeronasal organ: primary role in mouse chemosensory gender recognition. Physiol. Behav. 29, 315–327.

SECTION 8

Midbrain and Central Pattern Generators for Vocalization

The brainstem regions from the periaqueductal gray to reticular formation are involved in the final production of motor output for vocalization. The periaqueductal gray has a triggering and gating function for vocalization, while the network of dedicated vocal pattern generators is located in the pontine reticular formation. The network has direct access to the motoneuron pools controlling laryngeal, respiratory and articulatory components of vocalization.

CHAPTER 8.1

Role of the periaqueductal gray in expressing vocalization

Eva Gruber-Dujardin*

Department of Infectious Pathology, German Primate Center, Göttingen, Germany

Abstract: In this chapter, the current knowledge about the role of the midbrain periaqueductal gray (PAG) in the expression of vocalization is reviewed. Electrical and pharmacological stimulation of the PAG evokes species-specific calls with short latency and low habituation in many mammals. Lesion studies show that the destruction of the PAG and lateral bordering tegmentum can cause mutism without other motor impairments. Neuroanatomical studies revealed that vocalization-eliciting regions of the PAG receive a widespread input from all brain regions, reaching from the forebrain to lower brainstem areas. On the other hand, the vocalization-controlling PAG projects into intermediate and premotor neuron regions of the brainstem recently identified as the vocal pattern generator. Finally, neurophysiological data suggest that vocalization-correlated activity of PAG neurons that initiates specific call types, and sometimes even shows characteristics of an audio–vocal interface region, does not reflect the acoustical structure of a particular call type. It is proposed that the PAG is not involved in direct vocal motor coordination and pattern generation, but rather has a triggering/gating function. It seems to be responsible for the initiation and control of intensity of vocal responses by integrating various inputs, including cognitive, motivational, sensory, motor and arousal-controlling components, and transmitting that information to the vocal pattern generating areas within the lower brainstem.

Keywords: periaqueductal gray; vocalization; emotion; vocal relay station; brain-stimulation; vocalization-related input

Abbreviations: AC: nucleus accumbens; ALD: name of experimental animal; AP: preoptic area; APL: lateral preoptic area; APM: medial preoptic area; ARN: name of experimental animal; BC: brachium conjunctivum; BER: name of experimental animal; Cca 24: anterior cingulate cortex (area 24); CoI: colliculus inferior; CoS: colliculus superior; DG: dorsal tegmental nucleus of Gudden; dlPAG: dorsolateral PAG; Ep: epiphysis; FLI: name of experimental animal; fMRI: functional magnetic resonance imaging; FRPo: pontine reticular formation; FRTM: mesencephalic reticular formation; GABA: gamma amino-butyric acid; GM: medial geniculate body; Gsc 25: subcallosal gyrus (area 25); HP: posterior hypothalamus; III: third ventricle; KOL: name of experimental animal; LC: locus coeruleus; LLv: ventral nucleus of the lateral lemniscus; lPAG: lateral PAG; MUL: name of experimental animal; NIV: nucleus of the trochlear nerve; NST: bed nucleus of the stria terminalis; OS: superior olivary complex; P: posterior thalamic nucleus; PAG: periaqueductal gray; PM: medial preoptic nucleus; Py: pyramidal tract; vlPAG: ventrolateral PAG; WGA-HRP: wheatgerm-agglutinated horseradish-peroxidase

I. Introduction

Neuronal control of vocalization is generated by a complex interaction of numerous, widely-distributed brain areas, which are in parts intensively interconnected, reaching from motor neurons in the lower brainstem and spinal cord, directly controlling different phonatory muscle activities, towards higher, integrative, associative and motivation-controlling regions within the forebrain. By coordination and integration of activity of these brain regions, characteristic calls

*Corresponding author. E-mail: edujardin@dpz.eu

Stefan M. Brudzynski, (Ed.)
Handbook of Mammalian Vocalization
ISBN 978-0-12-374593-4

DOI: 10.1016/B978-0-12-374593-4.00030-9

within a species-specific repertoire can be produced. They can range from rather primitive, reflex-like and entirely genetically-determined vocal utterances (e.g., pain shriek), through vocalizations with predominantly innate acoustical pattern that could be volitionally initiated (e.g., vocal repertoire of different non-human primates, non-verbal human vocalizations or emotional intonation of human language), to vocal utterances with volitional control of the acoustical structure (e.g., during learning and imitation of vocal acts, such as in human language and bird or whale songs; for review, see Jürgens, 2008).

Depending on the complexity of the vocalization produced, different levels of neuronal processing are involved. At the lowest level of complexity, vocalization is emitted by an "innate releasing mechanism" in response to external stimuli and can already be elicited directly after birth (e.g., pain scream). Even deaf-born or anencephalic children (Monnier and Willis, 1953; Eibl-Eibesfeldt, 1973) or cats decerebrated at the forebrain–midbrain level (Bazett and Penfield, 1922; Magoun et al., 1937) are able to produce this type of vocalization. If decerebration is carried out between midbrain and pons, animals are mute (Bazett and Penfield, 1922). Thus, the entire forebrain seems to be dispensable for vocalization and the midbrain represents the highest essential level of vocal processing (Jürgens, 2002). More precisely, in the caudal half of the midbrain, a structure has been identified, which turned out to play a crucial role in the production of vocalization: the periaqueductal gray (PAG). The specific role of the PAG in vocal production is still under discussion, but there are several indications which suggest that the PAG is responsible for triggering a vocal response and its intensity, rather than being involved in direct motor coordination of vocal patterning. In this chapter the current knowledge about the role of the PAG in expressing vocalization is reviewed.

II. Subdivisions of the PAG

II.A. Anatomical–histological arrangement

The PAG is a comparatively large (in the squirrel monkey circa 5 mm long, 4 mm wide and 3 mm high), densely cellular, heterogenous midbrain structure, surrounding the mesencephalic aqueduct. Rostrally, it adjoins the periventricular gray substance of the thalamus and hypothalamus and caudally it continues into the dorsal pontine brainstem. Ventrally adjacent to the PAG are the occulomotor and dorsal raphe nuclei and dorsolaterally it is surrounded by the pretectal area, mesencephalic reticular formation, superior and inferior colliculi, respectively.

There are three different neuronal cell types described within the PAG (Depaulis and Bandler, 1991). The smallest Class I – neurons (circa 18 $\times$ 8.5 μm) are spindle-shaped and predominantly located medially, directly adjacent to the mesencephalic aqueduct (nucleus medialis). The slightly larger Class II – neurons (circa 12 μm in diameter) are fusiform to spherical with many dendritic arborizations, and are predominantly located dorsally in the nucleus dorsalis. The largest cells are Class III – neurons (circa 19 $\times$ 15 μm), which are spindle-shaped, round or polygonal with one prominent axon and one or two dendrites that are located at the lateral layers of the PAG, in the nucleus lateralis.

II.B. Functional divisions

Dividing the PAG according to its anatomical connections and functional representations, Bandler and colleagues distinguished four longitudinal columns extending in a rostrocaudal direction as shown in Fig. 1 (Bandler and Shipley, 1994; Bandler and Keay, 1996).

During the last decades, it has become more evident that the PAG is a crucial structure for the integration of different complex emotional behavioral patterns, such as flight and defense responses, sexual behavior, food intake, visceromotor responses and pain processing. It has turned out that the different longitudinal subunits of the PAG are therefore involved in different, sometimes even opposite, behavioral strategies (Bandler and Keay, 1996).

The dorsolateral column (dlPAG) seems to be important for the generation of defensive flight responses (Keay and Bandler, 2001). Electrical or pharmacological stimulation of the dlPAG in rats and rabbits evoked defensive flight responses (rat: Krieger and Graeff, 1985; Hilton and Redfern, 1986; rabbit: Paredes et al., 2000). It is also known from other studies that the dlPAG is essential for central fear processing (Fanselow, 1991; Behbehani, 1995). For example, stimulation of the dorsal PAG in humans produced sensations of fear (Iacono and Nashold, 1982) and injections of glycine antagonists into the dlPAG of rats had anxiolytic effects (Matheus et al., 1994).

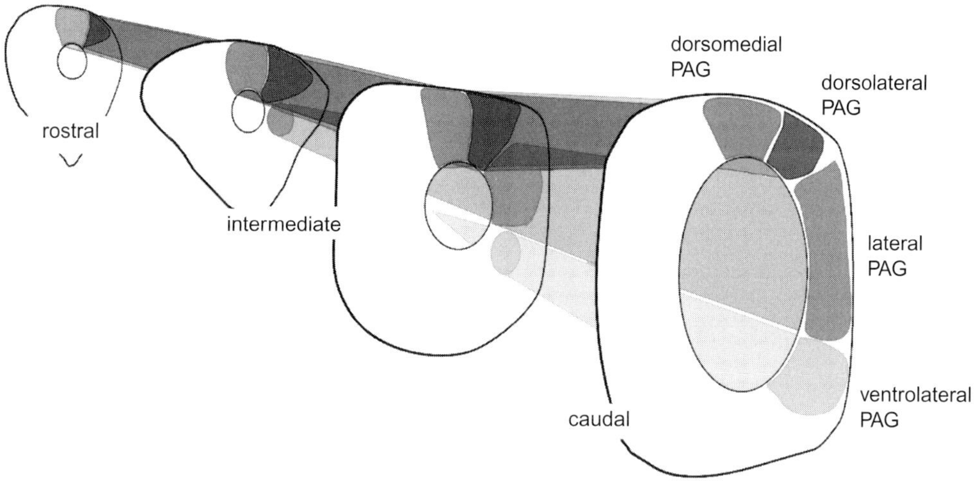

Fig. 1. Schematic drawing of rostrocaudally oriented longitudinal PAG columns. Modified from Bandler and Shipley (1994).

The lateral PAG (lPAG), or "defense region" (Bandler and Keay, 1996), is responsible for offensive aggression, which could be demonstrated by pharmacological stimulation with excitatory amino acids (EAA) in different species (cat: Bandler, 1982; Shaikh et al., 1987; Bandler and Carrive, 1988; rat: Depaulis et al., 1989). The animals show an aggressive response toward the threatening stimulus rather than trying to escape. Cats responded with flattening of the ears, arching of the back and piloerection, accompanied by hissing and growling vocalization (Shaikh et al., 1987). In rats, this behavior was characterized by an upright body posture and the emission of 22–28 kHz ultrasonic vocalizations (Bandler and Keay, 1996).

In the ventrolateral PAG (vlPAG), a different behavioral pattern can be evoked by electrical stimulation. It is characterized by a clear decrease in activity and quiescence of the animal, as well as decreased heart rate and blood pressure (Bandler and Keay, 1996; cat: Shaikh et al., 1987; Zhang et al., 1990; rat: Lovick, 1992; Depaulis et al., 1994; rabbit: Duan et al., 1997; Paredes et al., 2000). However, the functional aspect of this behavior is still being discussed. According to the studies by Duan et al. (1997), Paredes et al. (2000) and Shaikh et al. (1987), behavioral patterns elicited from the vlPAG are characterized by an increased state of vigilance. Other authors suggested that such behavior resembles a response to severe injury, chronic pain or repeated social defeat (Teskey and Kavaliers, 1991; Blanchard et al., 1993; Depaulis et al., 1994). According to their research, somatomotor activity and vigilance are reduced, and responsiveness to the environment decreases, sometimes resulting in complete freezing. It is thought to be a passive behavioral strategy to threatening and inescapable events, at the same time facilitating mechanisms of regeneration and healing. The fact that the vlPAG is involved in endogenous opioid-mediated analgesia supports this assumption (Morgan et al., 1991; Behbehani, 1995; Bandler and Keay, 1996; Keay and Bandler, 2001). Different types of vocalizations are represented within the longitudinal columns of the PAG, as part of a complex emotional/motivational behavioral control system which will be discussed in the following sections.

III. Functional approaches

III.A. Electrical and pharmacological periaqueductal gray stimulation

In 1915, brain stimulation in a chimpanzee revealed the first evidence of a possible role for the PAG in vocal control. Electrical stimulation of the rostral brainstem near the aqueduct elicited laughter-like vocalizations (Brown, 1915). About 20 years later, this effect has been reproduced and more systematically investigated in cats and rhesus monkeys by Magoun and co-workers. They explored the whole midbrain, pons and upper medulla by electrical stimulation and found numerous vocalization-eliciting sites, not only throughout the PAG, but also in the laterally adjacent tegmentum (Magoun et al., 1937). Vocalizations elicited from there were natural-sounding and species-specific and represented not just one but various different call types that could be achieved from different PAG sites (rhesus monkey: soft cries, screeches and barks; cat: spitting, screams and howls). Since then, the electrical elicitability of vocalization in the PAG has been confirmed in different species by many authors (squirrel monkey: Jürgens and Ploog, 1970; gibbon: Apfelbach, 1972; guinea pig: Martin, 1976; rat: Yajima et al., 1980; bat: Schuller and Radtke-Schuller, 1990). In humans, however, non-verbal vocalization elicited by electrical stimulation of the PAG has only been reported by one study, in which euphoric laughing was achieved (Sem-Jacobsen and Torkildsen, 1960), while many other authors were not able to elicit vocal responses from human PAG (Nashold et al., 1969; Hosobuchi et al., 1977; Richardson and Akil, 1977).

Evidence for PAG not only resembling an accumulation of vocalization-related bypassing fibers, but also being an essential synaptic relay station in the vocalization pathway, comes from neurochemical studies. Pharmacological stimulation of the PAG by injection of glutamate-agonists, cholinergic agonists, histamine or GABA-antagonists, elicited natural-sounding, species-specific vocalizations (Bandler and Carrive, 1988; Lu and Jürgens, 1993). It is suggested that the production of vocalization by the PAG depends predominantly on glutamatergic input from various sensory and limbic structures and vocalization-controlling neurons which are under a strong, tonic inhibitory control from GABA-ergic neurons (Jürgens, 1994).

However, vocalizations obtained by electrical or pharmacological stimulation of the PAG do not resemble secondary responses to stimulation-induced changes of the emotional/motivational state. As revealed by a self-stimulation study in the squirrel monkey, vocalizations elicited from the PAG are not accompanied by aversive or hedonistic side-effects (Jürgens, 1976).

Additional arguments for the PAG being involved in specific vocal control are given by characterization of the vocalization eliciting sites. There is not one specific vocalizations-related sub-area, but vocalizations can be elicited from all over the PAG. In different studies by Jürgens and co-workers, it has been shown that, depending on the exact site and intensity of stimulation or dose and concentration of the neurotransmitter used, different call types with different states of aversion could be elicited directly next to each other (Jürgens, 1976; Lu and Jürgens, 1993). Furthermore, the latencies of vocalization-eliciting sites within the PAG are very short and even multiple stimulations at the same site do not result in habituation (Jürgens and Ploog, 1970; Ortega et al., 1988).

A recent study in the squirrel monkey, in which the afferent input to different vocalization-eliciting PAG sites has been investigated (Dujardin and Jürgens, 2006), provided evidence for a vocalization-related subdivision of the PAG, which correlated with the different longitudinal columns described by Bandler and co-workers (see Fig. 2).

Fig. 2 shows sonagrams of PAG-elicited call types and the core zones of the corresponding injection site of a neuroanatomical tracer (wheatgerm-agglutinated horseradish-peroxidase, WGA-HRP) in the PAG. In the cases of ARN and BER (names of animals; see Fig. 2), electrical and pharmacological PAG stimulation yielded shriek calls (nomenclature of call types and concomitant aversive state follows Jürgens, 1979). Shriek calls are accompanied by highly aversive states, normally uttered during defensive threat and characterized by a noise-like (i.e., non-harmonic) acoustic structure. The WGA-HRP injection sites were located in the cranial left dlPAG, consistent with the area important in defensive flight responses (Keay and Bandler, 2001). Cases MUL and KOL (Fig. 2) received WGA-HRP injections into the left lPAG, also called the "defense region" (Bandler and Keay, 1996), at sites producing cackling calls. Cackling represents an offensive social mobbing call with moderate to high concomitant aversive states. Acoustically, it is characterized by a rhythmic frequency modulation. Cases ALD and FLI produced cluck calls, a short-distance contact call accompanied by low to neutral aversive states and characterized by a steep descending frequency sweep. In these cases, WGA-HRP injection sites were located in the right ventral PAG, a region involved in less aversive behavioral patterns, according to Bandler and Keay (1996).

Vocalization-eliciting sites are quite restricted and the different call types could be obtained only within small areas (circa 0.5 mm in diameter) that are functionally clearly separated from the surrounding PAG. Within one experimental animal, different vocalization-eliciting sites were located distant from each other (≥1 mm), but pooling the stimulation sites of all animals together revealed vocalization-eliciting areas of the same call type being located close to each other (Dujardin and Jürgens, 2006).

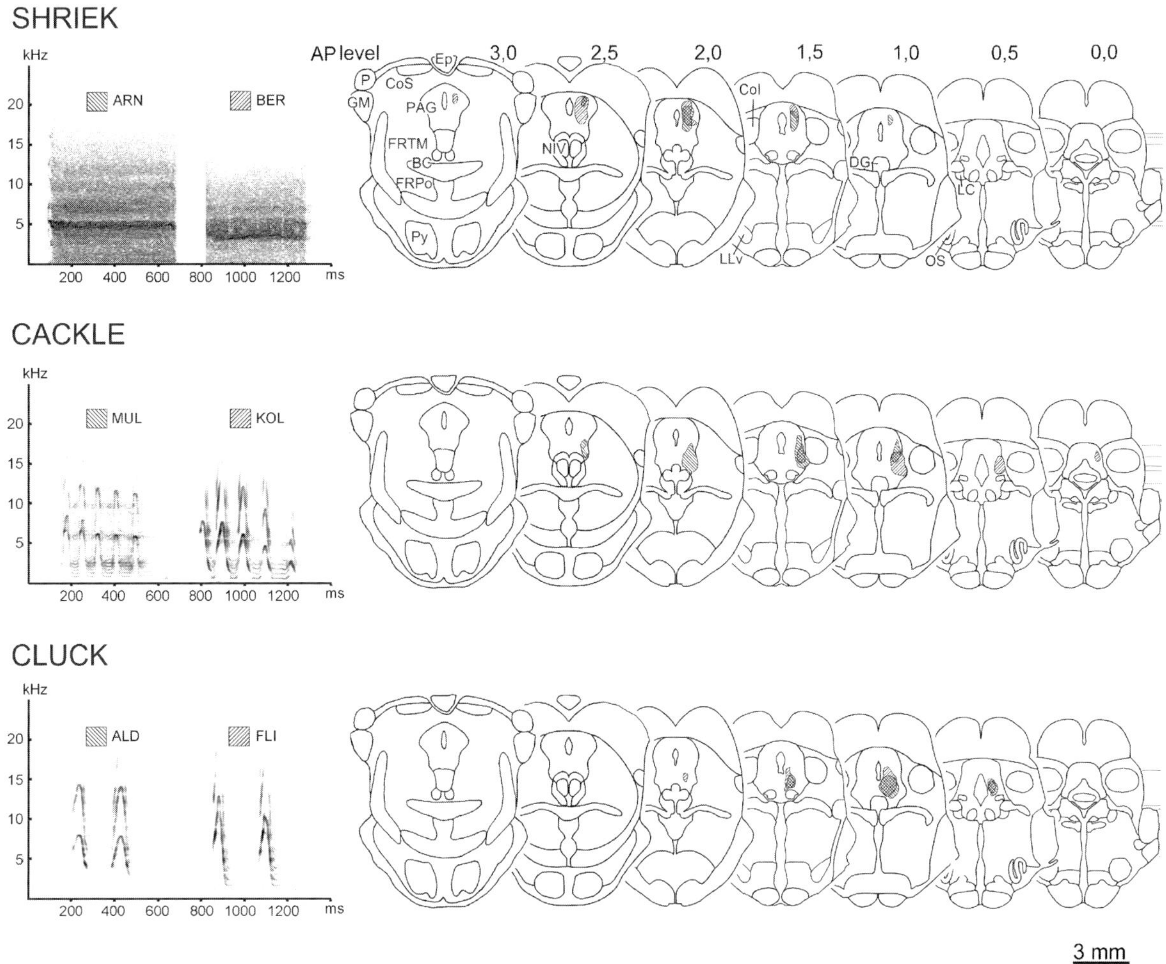

Fig. 2. Left: Sonagrams of call types elicited in the PAG by pharmacological stimulation with a glutamate agonist. Right: Series of brainstem diagrams illustrating the core zones of the periaqueductal injection sites of the neuroanatomical tracer WGA-HRP. Hatched zones represent the injection sites of two animals per call type group, superimposed upon each other. Crosshatched areas indicate the degree of overlap of these two injection sites. ARN, BER, MUL, KOL, ALD and FLI (on the sonograms) are names of the experimental animals. For names of the brain regions, see abbreviations. From Dujardin and Jürgens (2006).

III.B. Lesioning studies

Various lesioning studies also point to a crucial control of PAG in vocalization. Kelly and co-workers were the first to study the effects of large diencephalic and mesencephalic lesions on vocal behavior in the cat, and they found that lesions invading the PAG and laterally bordering tegmentum caused mutism without general akinesia (Kelly et al., 1946). This observation has been confirmed by many other authors for the same species (Adametz and O'Leary, 1959; Randall, 1964; Skultety, 1968), as well as for the rat (Chaurand et al., 1972), squirrel monkey (Jürgens and Pratt, 1979) and man (Botez and Carp, 1968; Esposito et al., 1999). Lesions do not need to destroy the PAG completely in order to be effective. Also, partial destruction led to loss of individual vocal responses, while others remained intact. In the squirrel monkey, for example, lesions restricted to the dorsomedial PAG canceled certain alarm calls to a visual stimulus, but retained vocal protest responses to tactile stimuli and non-vocal behavioral responses to visual stimuli (Jürgens and Pratt, 1979). Newman and MacLean (1982) were able to cancel long-distance contact calls in the squirrel monkey by destruction of the vlPAG and bordering tegmentum, while various other vocal responses remained unaffected. Also in cats and dogs, comparable observations have been made (Kelly et al., 1946; Skultety, 1962). Partial PAG lesions may eliminate some calls, even some primitive ones, while leaving others intact. One PAG-lesioned cat did not utter any vocal response to painful stimuli, but meowed spontaneously. This lack of vocal pain reaction was not due to a general analgesia, because the animal still showed non-vocal pain responses (Skultety, 1968).

It is of great importance to note that none of the above-mentioned studies reported any deterioration of the acoustic structure of vocalizations after PAG lesions, as would be the case after destruction of regions within the lower brainstem (Kirzinger and Jürgens, 1985). Furthermore, after complete inactivation of the PAG, natural-sounding species-specific vocalizations can still be evoked by electrical stimulation of the caudal midbrain, pons and medulla oblongata (Siebert and Jürgens, 2003). This observation, together with the above-mentioned results, suggests that the PAG serves triggering functions within vocal production rather than being involved in direct vocal pattern generation.

This assumption is also supported by the hierarchical position PAG has within the vocalization pathway (see review by Jürgens, 2008). As revealed by anatomical studies, the anterior cingulate cortex, a region important for volitional initiation of vocalization, directly projects into the PAG, which directly projects into the reticular formation of the lower brainstem (see also Hage, Chapters 8.2 and 8.3 in this volume). Electrical stimulation of all three structures induces species-specific vocalization. Destruction of PAG abolished vocalization elicited from the anterior cingulate cortex, but not from the reticular formation, whereas destruction of the anterior cingulate cortex left vocalization elicited from PAG and reticular formation unaffected (Jürgens and Pratt, 1979; Kirzinger and Jürgens, 1985; Siebert and Jürgens, 2003). Thus, these three structures form a hierarchical system, with the anterior cingulate representing the highest initiating level, the reticular formation the lowest, executive level, and the PAG situated at an intermediate level that fits best with the position of an integrative relay station. Furthermore, this conclusion makes it clear that pattern generation of natural-sounding, species-specific vocalizations does not depend on an intact PAG.

III.C. Neurophysiological data

Other indications showing that the PAG plays an important role in vocal control according to numerous single-unit electrophysiological recording studies. Vocalization-correlated single-unit activity has been found in the PAG of teleost fish (Kittelberger et al., 2006), cats (Adams, 1968), bats (Suga and Yajima, 1988), squirrel monkeys (Kirzinger and Jürgens, 1991; Düsterhoft et al., 2004) and macaques (Larson and Kistler, 1987). Most of the responsive neurons were located in the dorsal and lateral PAG including bordering tegmentum, but vocalization-responsive cells have also been described in the dorsomedial and vlPAG (Larson and Kistler, 1987). Different cell response patterns have been described, leading to controversial discussions concerning the role of the PAG in vocal production.

Larson and co-workers have investigated the vocalization-related activity of PAG neurons in the macaque. They have described four different response patterns: (1) "late burst" cells that became active shortly before or during vocalization; (2) "early burst" cells that started firing earlier than "late burst" cells and ceased before vocalization onset; (3) cells with "tonic increase" in firing rate that were tonically active and increased their activity just before vocalization onset; and (4) "tonic off" cells that were tonically active and decreased their discharge rate before or during vocalization (Larson, 1991). Further, they demonstrated that the neuronal activity of numerous "late burst" cells closely correlated with the electromyographic activity of specific laryngeal adductor and/or abdominal expiratory muscles, without being a simple auditory or proprioreceptive feedback mechanism (Larson and Kistler, 1986). Thus, the vocalization-responsive cells of the PAG seem to be involved in direct vocal motor control and, therefore, some authors concluded that the PAG functions as a vocal pattern generator.

In the squirrel monkey, detailed investigation of vocalization-correlated neuronal activity in the PAG has been carried out by Düsterhoft and co-workers (2004) using a telemetric single-unit recording technique. These authors described only a comparatively small fraction of PAG neurons (9.3%) showing vocalization-related activity. Four different neuronal response patterns could be distinguished, slightly deviating from those in the macaque. The majority of neurons showed an initiating activity burst immediately before vocalization onset, but not during vocalization, resembling the "early burst" cells described by Larson and co-workers. The second group of cells started right before vocalization onset and continued through part of the vocalization. Few neurons showed increased activity only during vocalization and not before. The common features of these three cell types were that they responsed selectively to different call types and none of them changed activity to perceived vocalization uttered by group mates. The fourth cell type behaved like an audio–vocal interface, which means it was active throughout the interval between a perceived vocalization of a group mate and the experimental animal's own vocal response, but it was not active when hearing a vocalization without responding to it. This indicates that auditory input is directly coupled to vocal output at the PAG level.

In a teleost fish (*Porichthys notatus*), it could be demonstrated that the activity pattern of PAG neurons does not only seem to predict the initiation, but also the duration of vocalization (Kittelberger et al., 2006). Furthermore, a correlation of neuronal activity has been found with the global loudness of a call (Larson, 1991) and it could be shown that by changing the electrical stimulation parameters, the intensity of the PAG-induced vocalization could be changed (Jürgens and Ploog, 1970). However, in contrast to the reticular formation, not a single neuron could be found in the PAG that fired in relation to changes in specific acoustical parameters during vocalization, such as frequency and/or amplitude modulation (Düsterhoft et al., 2004). Additionally, it is not possible to change the species-specific pattern of a PAG-elicited vocalization, for example, altering the rhythm of a call by changing the rhythm of the stimulation pulses (Jürgens and Pratt, 1979). These findings, together with the fact that no PAG neuron showed vocalization-correlated activity during all call types of a species-specific repertoire, and the fact that many neurons started firing before and not during vocalization, support the assumption that PAG processes vocalization in a general and initiating manner, determining the call type and global intensity rather than generating the modulation of the acoustical structure.

IV. Neuroanatomical studies

IV.A. Afferent projections of PAG

As revealed by numerous neuroanatomical studies, PAG receives input from various emotion- and motivation-controlling brain regions (see, for example, Mantyh, 1982a; Meller and Dennis, 1986; Chiba et al., 2001; Dujardin and Jürgens, 2005). It has been demonstrated that these limbic inputs are organized according to the different longitudinal PAG columns (Shipley et al., 1991; Bandler and Keay, 1996). Therefore, it is assumed that the single PAG subdivisions play different roles in coordination and processing of behavioral strategies related to different emotional situations (Bandler and Keay, 1996).

Besides the limbic input, PAG also receives afferent projections from sensory structures (e.g., Mantyh, 1982b; Carrive, 1993; Keay et al., 1997; Dujardin and Jürgens, 2005). Through these direct afferents, external stimuli can significantly influence processing of emotional behavior within the PAG.

In order to determine the input of functionally verified vocalization-controlling regions of the PAG, a neuroanatomical study was recently carried out in the squirrel monkey (Dujardin and Jürgens, 2005). The results showed that different vocalization-eliciting sites received their input from a largely identical set of structures, ranging from the rostral telencephalon to the caudal medulla oblongata. The heaviest projections came from the surrounding of the PAG, dorsomedial and ventromedial hypothalamus, medial preoptic region, substantia nigra, zona incerta and reticular formation of the mesencephalon, pons and medulla. The heaviest cortical input reached the vocalization-controlling PAG from the mediofrontal cortex. Moderate to weak cortical projections came from the insula, lateral prefrontal and premotor cortex, as well as from the superior and middle temporal cortex. Subcortical moderate to weak projections reached PAG from the central and medial amygdala, nucleus of the stria terminalis, septum, nucleus accumbens, lateral preoptic region, lateral and posterior hypothalamus, globus pallidus, pretectal area, deep layers of the superior colliculus, the pericentral inferior colliculus, mesencephalic trigeminal nucleus, locus coeruleus, raphe nuclei, vestibular nuclei, spinal trigeminal nucleus, solitary tract nucleus and nucleus gracilis (see Table 1).

In summary, vocalization-eliciting sites of the PAG receive afferent projections originating in structures all along the cerebral axis, including cognitive, limbic, sensory, motor and arousal-controlling systems. One of the heaviest projections thereby arose from the PAG itself. These intrinsic connections suggest that the different parts of the PAG are closely-related to each other functionally. Bandler and Shipley (1994) have shown that there is a particularly strong interaction along the

Table 1. The most important brain regions that project to all vocalization-eliciting areas of the PAG

	Limbic structures	Sensory structures	Motor structures	Associated/other structures
Forebrain	Anterior limbic cortex		Premotor cortex	Prefrontal cortex
	entorhinal cortex		PMA, SMS	insula
	septal region			temporal cortex
	nucleus accumbens			somatosensory association cortex
	ventral pallidum			
	nucleus striae terminalis		globus pallidus	
	amygdala			
	preoptic region		zona incerta	reticular thalamic nucleus
	hypothalamus			
	midline thalamus			
			substantia nigra	
Brainstem	PAG	**auditory**: inferior colliculi, pontine paralemniscal zone	occulomotor nuclei	ventrolateral reticular formation
		visual: superior colliculi	ambigual nucleus/ periambigual reticular formation	parabrachial nuclei
		somatosensory: parts of trigeminal nucleus and nucleus gracilis		locus coeruleus
		viscerosensory: nucleus of the solitary tract		raphe nuclei

The font size indicates the approximate intensity of vocalization-related input into the vocalization-controlling PAG from weak (smallest font) to heavy (largest font) (modified from Dujardin and Jürgens, 2005).

rostrocaudal axis of the PAG, but not along the dorsoventral or mediolateral axis, pointing to the columnar organization of this structure. Clearly, however, the studied projections were not restricted to a rostrocaudal orientation, but invaded all parts of the PAG.

The aim of a subsequent study was to find out whether there are quantitative differences in this vocalization-controlling input to PAG (Dujardin and Jürgens, 2006). In some regions, marked differences in the number of projecting neurons between three distinct call types (shriek, cackle and cluck) occurred (see Fig. 3). Such regions were the nucleus accumbens, preoptic area, posterior hypothalamus, anterior cingulate cortex, subcallosal gyrus and the nucleus stria terminalis. These quantitative differences were not uniform, but region-specific, showing regions in which the projection strength correlated with the aversive/hedonic quality of the PAG-elicited call type (states of aversion were evaluated according to Jürgens, 1979), or regions projecting more heavily to one PAG-elicited call type than another, irrespective of its aversive/hedonic quality.

Conspicuous aversion-related differences come from the posterior hypothalamus (HP) and preoptic area (AP). The posterior hypothalamus sent the strongest projections to PAG sites producing the most aversive call type (shriek), and the weakest projections to PAG sites producing the least aversive call type (cluck); in other words, the more aversive the PAG-elicited vocalization, the stronger the input from the posterior hypothalamus (Dujardin and Jürgens, 2006). Another neuroanatomical study that investigated the input into functionally verified vocalization-controlling PAG regions also demonstrated a specific input from the posterior hypothalamus into that part of the PAG producing "low whistle" calls in the guinea pig (Kyuhou and Gemba, 1998). "Low whistle" is a separation call, used by animals seeking

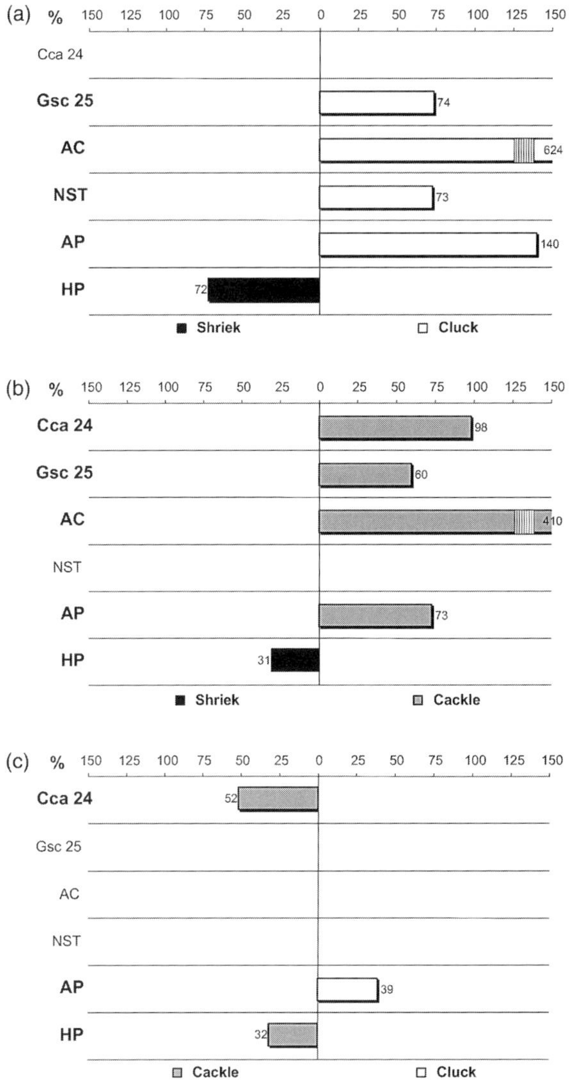

Fig. 3. (a)–(c) Diagrams showing relative differences of normalized numbers of retrogradely labeled neurons in regions with call type-specific differences. Pair-wise comparison between two call type groups, respectively. The numbers next to the bars give the percentage by which the call group with the heavier labeling differed from the call group with the weaker labeling. Bold face typed regions indicate call type-specific differences in that comparison. (a) Comparison between shriek and cluck calls; (b) Comparison between shriek and cackle calls; (c) Comparison between cackle and cluck calls. For names of the brain regions, see abbreviations. From Dujardin and Jürgens (2006).

social contact. The posterior hypothalamus of the guinea pig does not project to PAG sites producing non-aversive mating calls ("purr"). Additional evidence for a role of the posterior hypothalamus in the control of aversive behavior came from various experiments with electrical and pharmacological stimulation (see, for example Shekhar and Dimicco, 1987; Paredes et al., 2000).

Control of the aversive state accompanying vocalization also seems to be the function of the preoptic area (see Fig. 4), but with opposite polarity. Dujardin and Jürgens (2006) found a negative correlation between call type aversiveness and projection strength from that region. The higher the number of projecting neurons to PAG, the less aversive the PAG-elicited call type. Supporting evidence also came from the study by Kyuhou and Gemba (1998), which demonstrated a specific input from the medial preoptic area into that part of the PAG producing non-aversive, purr-like mating calls in the guinea pig (Kyuhou and Gemba, 1998). In general, the preoptic area is known to be involved in regulation of non-aversive emotions/motivations. Single-unit recording and c-Fos experiments have demonstrated an increased neuronal activity in the medial preoptic area during sexually-motivated behavior (Baum and Everitt, 1992; Struthers, 2001) and maternal care (Stack et al., 2002), while electrical brain stimulation experiments revealed self-stimulation behavior in this structure (Jürgens, 1976).

The most striking call type-specific but aversion-unrelated input came from the nucleus accumbens (AC). The AC showed the greatest differences in projection strengths of all structures tested in the study by Dujardin and Jürgens (2006), by projecting much more into chuck- and cluck-eliciting than into shriek-eliciting PAG sites. Between cackle and cluck calls, no clear differences could be found (Dujardin and Jürgens, 2006). It is known that the AC is part of the dopaminergic reward system, serving as an interface between limbic structures and motor systems (Groenewegen, 1996). Amphetamine microinjections into the AC of rats produced 50 kHz ultrasonic vocalizations (Burgdorf et al., 2001). Rats also emitted this call type during the presentation of cues that predict reward (e.g., access to a sexually receptive female: Barfield et al., 1979; play with conspecifics: Knutson et al., 1998; or food delivery: Burgdorf et al., 2001). In contrast, highly aversive 22 kHz ultrasonic vocalizations, displaying defensive alarm behavior and anxiety, could not be elicited by AC stimulation (Burgdorf et al., 2001). In squirrel monkeys, electrical stimulation of the AC elicited species-specific vocalizations which were all characterized by a typical frequency modulation (Jürgens and Richter, 1986) also found in cackle and cluck calls (Jürgens and Ploog, 1970). Similar to rats, the highly aversive shriek calls

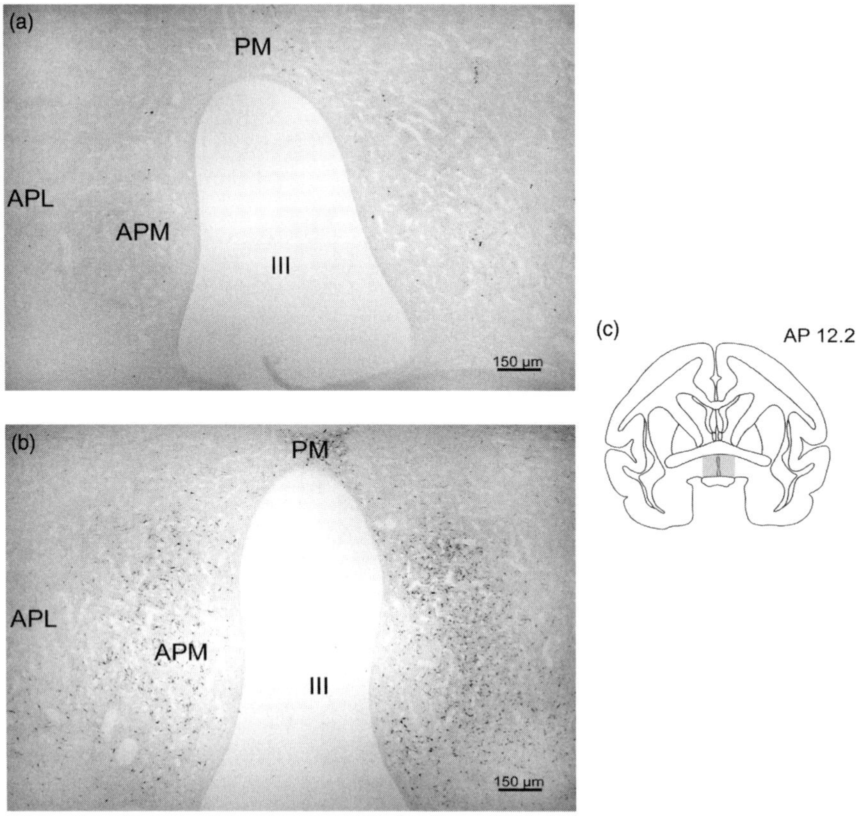

Fig. 4. (a)–(c) Photomicrographs showing: (a) sparse retrogradely labeled neurons in the preoptic region after injection of WGA-HRP in shriek-eliciting sites of the PAG compared to (b) robust labeling in the preoptic region after injection of WGA-HRP into cluck-eliciting sites; (c) schematic brain diagram with the gray square indicating the photographed regions. Brain regions, see abbreviations. Modified from Dujardin and Jürgens (2006).

were not elicited from the AC (Jürgens and Ploog, 1970). This is in harmony with the finding that PAG regions producing shriek calls received only little input from the AC, compared to cluck- and cackle-eliciting sites (Dujardin and Jürgens, 2006).

In the anterior cingulate cortex (CCA), input intensity was also call type-specific, without being related to the concomitant aversive state (Dujardin and Jürgens, 2006). The CCA sends most projections into PAG sites where cackle calls were elicited, a call type accompanied by intermediate states of aversion. Vocal operant conditioning experiments revealed the CCA to play a crucial role in volitional control over the initiation of vocalization in monkeys. Macaques whose CCA has been removed still responded to threatening stimuli with a warning call, but were not able to produce vocalization as an operant behavior to receive food reward or postpone an electric shock (Sutton et al., 1974; Aitken, 1981). Nevertheless, those macaques were still able to master an operant conditioning task with lever pressing, instead of vocalization, as the operant behavior (Sutton et al., 1974). In humans, bilateral lesions in the CCA affected voluntary control of affective prosody (Jürgens and Von Cramon, 1982) and the relationship between affective prosody and CCA has also been found in fMRI brain imaging studies (Barrett et al., 2004). Because human affective prosody is considered homologous to some parameters of monkey calls, it can be assumed that the initiation of cackling underlies a stronger voluntary control by the CCA than the initiation of shriek or cluck calls (Dujardin and Jürgens, 2006).

IV.B. Efferent projections from periaqueductal gray

Finally, all information relevant for vocal production converges in the PAG, where it is integrated and transferred to all motor neuron pools important for the production of vocalization. However, many alternatives have been discussed for this descending vocalization pathway in the past decades.

Direct connections between the PAG and motor nuclei involved in phonation are sparse and lacking completely in the case of expiratory motoneuron pools in the thoracic and upper lumbar spinal cord (Rye et al., 1988; Holstege, 1989). In the remaining motor nuclei, evidence for direct connections with the PAG is equivocal (see, for example, Mantyh, 1983; Panneton and Martin, 1983; Travers and Norgren, 1983; Holstege et al., 1984; Shaikh et al., 1987). However, a massive projection has been found from the PAG to the reticular formation around the nucleus retroambiguus, a region that first had been thought to be interconnected with all motor nuclei involved in phonation (Holstege, 1989). Together with the finding that extensive destruction of the retroambigual region blocks PAG-elicited vocalization (Shiba et al., 1995), it has been concluded that the nucleus retroambiguus represents an obligatory relay station of the descending vocalization pathway, controlling all phonatory motor neurons. More recent and more detailed neuroanatomical studies have made clear, however, that the nucleus retroambiguus does not have direct connections with the trigeminal motor, facial and hypoglossal nuclei (Fay and Norgren, 1997; Cunningham and Sawchenko, 2000; Travers et al., 2005) controlling jaw, lip and tongue movements, respectively. Most of the retroambigual neurons are directly connected with expiratory motor neurons in the thoracic and upper lumbar ventral horn (Billig et al., 2001; Boers et al., 2006), as well as with the nucleus ambiguus, controlling the vocal folds (Vanderhorst et al., 2001; Boers et al., 2002).

The PAG also sends descending efferent projections into the lateral parabrachial area, parvocellular reticular formation, just above the superior olive, lateral medullary reticular formation, and dorsal as well as ventral reticular nuclei, closely adjacent to the nucleus retroambiguus (Mantyh, 1983; Odeh and Antal, 2001). All these structures do have direct conjoint connections with all phonatory motor neuron pools, as revealed by many neuroanatomical studies (Thoms and Jürgens, 1987; Fay and Norgren, 1997; Bernard et al., 1999; Vanderhorst et al., 2001; Hannig and Jürgens, 2006). Moreover, recent neurophysiological investigations revealed that almost all of these structures play an important role in acoustical patterning of vocalization (Hage and Jürgens, 2006b), meeting all criteria that are necessary for a vocal pattern generator (for review see Jürgens and Hage, 2007; and see Hage, Chapter 8.2 in this volume). Furthermore, it is suggested that different vocal pattern generators might process different groups of call types (frequency modulated versus non-frequency modulated, for further detail, see Jürgens, 2000; Hage and Jürgens, 2006a,b). Neuroanatomical investigation of call type-specific differences in descending efferent PAG projections could verify this hypothesis.

V. Conclusions

In summary, the PAG can be seen as a crucial relay station within the vocalization pathway where different vocalization-inducing external and internal stimuli from widespread limbic, sensory, cognitive and arousal-controlling structures are integrated and linked with the vocal motor-coordinating/pattern generating areas in the lower brainstem. Neurophysiological data, as well as findings from lesioning and stimulation studies, suggest that different call types are represented in the PAG by distinct neuronal ensembles, rather than by different activity patterns of one ensemble. The fact that rather few neurons (9.3%) show vocalization-correlated activity within the PAG, together with the features of PAG vocalization-eliciting sites being functionally distinct and relatively widely scattered, indicates that vocalization control is just one of various different integrative functions of the midbrain PAG. However, it has been shown that the PAG is essential and indispensable for the production of animal vocalization and non-verbal human vocalizations, as well as the emotional intonation of speech. The PAG thereby can be seen as an interactive neuronal network. Regions controlling different functions (e.g., pain expression, defensive and fear behavior, autonomic/vegetative regulation, lordosis-reflex, vocalization) partly overlap in PAG and certainly many of these mechanisms occur together during complex emotional behavioral patterns with vocalization control representing an important part of the whole integrated behavioral system.

References

Adametz, J., O'Leary, J.L., 1959. Experimental mutism resulting from periaqueductal lesions in cats. Neurology 9, 636–642.

Adams, D.B., 1968. Cells related to fighting behaviour recorded from midbrain central grey neuropil of cat. Science 159, 894–896.
Aitken, P.G., 1981. Cortical control of conditioned and spontaneous vocal behavior in rhesus monkeys. Brain Lang. 13, 171–184.
Apfelbach, R., 1972. Electrically elicited vocalizations in the gibbon *Hylobates lar* (Hylobatidae), and their behavioral significance. Z. Tierpsychol. 30, 420–430.
Bandler, R., 1982. Induction of "rage" following microinjections of glutamate into midbrain but not hypothalamus of cats. Neurosci. Lett. 30, 183–188.
Bandler, R., Carrive, P., 1988. Integrated defence reaction elicited by excitatory amino acid microinjection in the midbrain periaqueductal grey region of the unrestrained cat. Brain Res. 439, 95–106.
Bandler, R., Keay, K.A., 1996. Columnar organization in the midbrain periaqueductal gray and the integration of emotional expression. In: Holstege, G., Bandler, R., Saper, C.B. (Eds.), The Emotional Motor System. Elsevier Science B.V., Amsterdam, The Netherlands, pp. 285–300.
Bandler, R., Shipley, M.T., 1994. Columnar organization in the midbrain periaqueductal gray: modules for emotional expression? Trends Neurosci. 17, 379–389.
Barfield, R.J., Auerbach, P., Geyer, L.A., Mcintosh, T.K., 1979. Ultrasonic vocalizations in rat sexual behavior. Am. Zool. 19, 469–480.
Barrett, J., Pike, G.B., Paus, T., 2004. The role of the anterior cingulate cortex in pitch variation during sad affect. Eur. J. Neurosci. 19, 458–464.
Baum, M.J., Everitt, B.J., 1992. Increased expression of c-fos in the medial preoptic area after mating in male rats: role of afferent inputs from the medial amygdala and midbrain central tegmental field. Neurosci. 50, 627–646.
Bazett, H.C., Penfield, W.G., 1922. A study of the Sherrington decerebrate animal in the chronic as well as the acute condition. Brain 45, 185–265.
Behbehani, M.M., 1995. Functional characteristics of the midbrain periaqueductal gray. Prog. Neurobiol. 46, 575–605.
Bernard, J.F., Villanueva, L., Carroue, J., le Bars, D., 1999. Efferent projections from the subnucleus reticularis dorsalis (SRD): a Phaseolus vulgaris leucoagglutinin studying the rat. Neurosci. Lett. 116, 257–262.
Billig, I., Hartge, K., Card, J.P., Yates, B.J., 2001. Transneuronal tracing of neural pathways controlling abdominal musculature in the ferret. Brain Res. 912, 24–32.
Blanchard, D.C., Sakai, R.R., McEwen, B., Weiss, S.M., Blanchard, R.J., 1993. Subordination stress: behavioral, brain, and neuroendocrine correlates. Behav. Brain Res. 58, 113–121.
Boers, J., Klop, E.M., Hulshoff, A.C., De Weerd, H., Holstege, G., 2002. Direct projections from the nucleus retroambiguus to cricothyroid motoneurons in the cat. Neurosci. Lett. 319, 5–8.
Boers, J., Kirkwood, P.A., De Weerd, H., Holstege, G., 2006. Ultrastructural evidence for direct excitatory retroambiguus projections to cutaneous trunci and abdominal external oblique muscle motoneurons in the cat. Brain Res. Bull. 249, 249–256.
Botez, M.I., Carp, N., 1968. Nouvelles données sur le problème du mécanisme de déclenchement de la parole. Revue Roum. Neurol. 5, 153–158.
Brown, T.G., 1915. Note on physiology of basal ganglia and midbrain of anthropoid ape especially in reference to act of laughter. J. Physiol. 49, 195–207.
Burgdorf, J., Knutson, B., Panksepp, J., Ikemoto, S., 2001. Nucleus accumbens amphetamine microinjections unconditionally elicit 50-kHz ultrasonic vocalizations in rats. Behav. Neurosci. 115, 940–944.
Carrive, P., 1993. The periaqueductal gray and defensive behavior: functional representation and neuronal organization. Behav. Brain Res. 58, 27–47.
Chaurand, J.P., Vergnes, M., Karli, P., 1972. Substance grise centrale du mésencéphale et comportement d'aggression interspécifique du rat. Physiol. Behav. 9, 475–481.
Chiba, T., Kayahara, T., Nakano, K., 2001. Efferent projections of infralimbic and prelimbic areas of the medial prefrontal cortex in the Japanese monkey, *Macaca fuscata*. Brain Res. 888, 83–101.
Cunningham, E.T., Sawchenko, P.E., 2000. Dorsal medullary pathways subserving oromotor reflexes in the rat: implications for the central neural control of swallowing. J. Comp. Neurol. 417, 448–466.
Depaulis, A., Bandler, R., 1991. The Midbrain Periaqueductal Gray Matter. Functional, Anatomical, and Neurochemical Organization. Plenum Press, New York, NY.
Depaulis, A., Bandler, R., Vergnes, M., 1989. Characterization of pretentorial periaqueductal gray matter neurons mediating intraspecific defensive behaviors in the rat by microinjections of kainic acid. Brain Res. 486, 121–132.
Depaulis, A., Keay, K.A., Bandler, R., 1994. Quiescence and hyporeactivity evoked by activation of cell bodies in the ventrolateral midbrain periaqueductal gray of the rat. Exp. Brain Res. 99, 75–83.
Duan, Y.F., Winters, R., McCabe, P.M., Green, E.J., Huang, Y., Schneiderman, N., 1997. Functional relationship between the hypothalamic vigilance area and PAG vigilance area. Physiol. Behav. 62, 675–679.
Dujardin, E., Jürgens, U., 2005. Afferents of vocalization-controlling periaqueductal regions in the squirrel monkey. Brain Res. 1034, 114–131.
Dujardin, E., Jürgens, U., 2006. Call type-specific differences in vocalization-related afferents to the periaqueductal gray of squirrel monkeys (*Saimiri sciureus*). Behav. Brain Res. 168, 23–36.
Düsterhoft, F., Häusler, U., Jürgens, U., 2004. Neuronal activity in the periaqueductal gray and bordering structures during vocal communication in the squirrel monkey. Neurosci. 123, 53–60.
Eibl-Eibesfeldt, I., 1973. The expressive behaviour of the deaf-and-blind-born. In: von Cranach, M., Vine, J. (Eds.), Social Communication and Movement. Academic Press, London, UK, pp. 163–194.
Esposito, A., Demeurisse, G., Alberti, B., Fabbro, F., 1999. Complete mutism after midbrain periaqueductal gray lesion. Neuroreport. 10, 681–685.
Fanselow, M., 1991. The midbrain periaqueductal gray as a coordinator of action in response to fear and aniety. In: Depaulis, A., Bandler, R. (Eds.), The Midbrain

Periaqueductal Gray Matter. Functional, Anatomical, and Neurochemical Organization. Plenum Press, New York, NY, pp. 151–173.
Fay, R.A., Norgren, R., 1997. Identification of rat brainstem multisynaptic connections to the oral motor nuclei using pseudorabies virus. Brain Res. Rev. 25, 255–275.
Groenewegen, H.J., 1996. The nucleus accumbens: gateway for limbic structures to reach the motor system? In: Holstege, G., Bandler, R., Saper, C.B. (Eds.), The Emotional Motor System. Elsevier, Amsterdam, The Netherlands, pp. 485–512.
Hage, S.R., Jürgens, U., 2006a. On the role of the pontine brainstem in vocal pattern generation: a telemetric single-unit recording study in the squirrel monkey. J. Neurosci. 26, 7105–7115.
Hage, S.R., Jürgens, U., 2006b. Localization of a vocal pattern generator in the pontine brainstem of the squirrel monkey. Eur. J. Neurosci. 23, 840–844.
Hannig, S., Jürgens, U., 2006. Projections of the ventrolateral pontine vocalization area in the squirrel monkey. Exp. Brain Res. 169, 92–105.
Hilton, S.M., Redfern, W.S., 1986. A search for brain stem cell groups integrating the defence reaction in the rat. J. Physiol. 378, 213–228.
Holstege, G., 1989. Anatomical study of the final common pathway for vocalization in the cat. J. Comp. Neurol. 284, 242–252.
Holstege, G., Tan, J., Van Ham, J., Bos, A., 1984. Mesencephalic projections to the facial nucleus in the cat. An autoradiographical tracing study. Brain Res. 311, 7–22.
Hosobuchi, Y., Adams, J.E., Linchitz, R., 1977. Pain relief by electrical stimulation of the central gray matter in humans and its reversal by naloxone. Science 197, 183–186.
Iacono R.P., Nashold, Jr., B.S., 1982. Mental and behavioral effects of brain stem and hypothalamic stimulation in man. Hum. Neurobiol. 1, 273–279.
Jürgens, U., 1976. Reinforcing concomitants of electrically elicited vocalizations. Exp. Brain. Res. 26, 203–214.
Jürgens, U., 1979. Vocalization as an emotional indicator. A neuroethological study in the squirrel monkey. Behaviour 69, 88–117.
Jürgens, U., 1994. The role of the periaqueductal grey in vocal behaviour. Behav. Brain Res. 62, 107–117.
Jürgens, U., 2000. Localization of a pontine vocalization-controlling area. J. Acoust. Soc. Am. 108, 1393–1396.
Jürgens, U., 2002. Neural pathways underlying vocal control. Neurosci. Biobehav. Rev. 26, 235–258.
Jürgens, U., 2008. The neural control of vocalization in mammals: a review. J. Voice. 23 (1), 1–10.
Jürgens, U., Hage, S.R., 2007. On the role of the reticular formation in vocal pattern generation. Behav. Brain. Res. 182, 308–314.
Jürgens, U., Ploog, D., 1970. Cerebral representation of vocalization in the squirrel monkey. Exp. Brain Res. 10, 532–554.
Jürgens, U., Pratt, R., 1979. Role of the periaqueductal grey in vocal expression of emotion. Brain Res. 167, 367–378.
Jürgens, U., Richter, K., 1986. Glutamate-induced vocalization in the squirrel monkey. Brain Res. 373, 349–358.
Jürgens, U., Kirzinger, A., Von Cramon, D., 1982. The effects of deep-reaching lesions in the cortical face area on phonation. A combined case report and experimental monkey study. Cortex 18, 125–139.
Keay, K.A., Bandler, R., 2001. Parallel circuits mediating distinct emotional coping reactions to different types of stress. Neurosci. Biobehav. Rev. 25, 669–678.
Keay, K.A., Feil, K., Gordon, B.D., Herbert, H., Bandler, R., 1997. Spinal afferents to functionally distinct periaqueductal gray columns in the rat: an anterograde and retrograde study. J. Comp. Neurol. 385, 207–229.
Kelly, A.H., Beaton, L.E., Magoun, H.W., 1946. A midbrain mechanism for faciovocal activity. J. Neurophysiol. 9, 181–189.
Kirzinger, A., Jürgens, U., 1985. The effects of brainstem lesions on vocalization in the squirrel monkey. Brain Res. 358, 150–162.
Kirzinger, A., Jürgens, U., 1991. Vocalization-correlated single-unit activity in the brain stem of the squirrel monkey. Exp. Brain Res. 84, 545–560.
Kittelberger, J.M., Land, B.R., Bass, A.H., 2006. Midbrain periaqueductal gray and vocal patterning in a teleost fish. J. Neurophysiol. 96, 71–85.
Knutson, B., Burgdorf, J., Panksepp, J., 1998. Anticipation of play elicits high-frequency ultrasonic vocalizations in young rats. J. Comp. Psychol. 112, 65–73.
Krieger, J.E., Graeff, F.G., 1985. Defensive behavior and hypertension induced by glutamate in the midbrain central gray of the rat. Braz. J. Med. Biol. Res. 18, 61–67.
Kyuhou, S., Gemba, H., 1998. Two vocalization-related subregions in the midbrain periaqueductal gray of the guinea pig. Neuroreport 9, 1607–1610.
Larson, C.R., 1991. On the relation of PAG neurons to laryngeal and respiratory muscles during vocalization in the monkey. Brain Res. 552, 77–86.
Larson, C.R., Kistler, M.K., 1986. The relationship of periaqueductal gray neurons to vocalization and laryngeal EMG in the behaving monkey. Exp. Brain Res. 63, 596–606.
Larson, C.R., Kistler, M.K., 1987. Brainstem neuronal activity associated with vocalization in the monkey. In: Baer, T., Sasaki, C., Harris, K.S. (Eds.), Laryngeal Function in Phonation and Respiration. College-Hill Publ, Boston, MA, pp. 154–167.
Lovick, T.A., 1992. Inhibitory modulation of the cardiovascular defence response by the ventrolateral periaqueductal grey matter in rats. Exp. Brain Res. 89, 133–139.
Lu, C.L., Jürgens, U., 1993. Effects of chemical stimulation in the periaqueductal gray on vocalization in the squirrel monkey. Brain Res. Bull. 32, 143–151.
Magoun, H.W., Atlas, D., Ingersoll, E.H., Ranson, S.W., 1937. Associated facial, vocal and respiratory components of emotional expression: an experimental study. J. Neurol. Psychopathol. 17, 241–255.
Mantyh, P.W., 1982a. Forebrain projections to the periaqueductal gray in the monkey, with observations in the cat and rat. J. Comp. Neurol. 206, 146–158.

Mantyh, P.W., 1982b. The ascending input to the midbrain periaqueductal gray of the primate. J. Comp. Neurol. 211, 50–64.
Mantyh, P.W., 1983. Connections of midbrain periaqueductal gray in the monkey. II. Descending efferent projections. J. Neurophysiol. 49, 582–594.
Martin, J.R., 1976. Motivated behaviors elicited from hypothalamus, midbrain, and pons of the guinea pig (*Cavia porcellus*). J. Comp. Physiol. Psychol. 90, 1011–1034.
Matheus, M.G., Nogueira, R.L., Carobez, A.P., Graeff, F.G., Guimarães, F.S., 1994. Anxiolytic effect of glycine antagonists microinjected into the dorsal periaqueductal grey. Psychopharmacology (Berl). 113, 565–569.
Meller, S.T., Dennis, B.J., 1986. Afferent projections to the periaqueductal grey in the rabbit. Neuroscience 19, 927–964.
Monnier, M., Willis, H., 1953. Die integrative Tätigkeit des Nervensystems beim meso-rhombo-spinalen Anencephalus (Mittelhirnwesen). Monatsschr. Psychiat. Neurol. 126, 239–273.
Morgan, M.M., Gold, M.S., Liebeskind, J.C., Stein, C., 1991. Periaqueductal gray stimulation produces a spinally mediated, opioid antinociception for the inflamed hindpaw of the rat. Brain Res. 545, 17–23.
Nashold, B.S., Wilson, W.P., Slaughter, D.G., 1969. Sensations evoked by stimulation in the midbrain of man. J. Neurosurg. 30, 14–24.
Newman, J.D., Maclean, P.D., 1982. Effects of tegmental lesions on the isolation call of squirrel monkeys. Brain Res. 232, 317–330.
Odeh, F., Antal, M., 2001. The projections of the midbrain periaqueductal grey to the pons and medulla oblongata in rats. Eur. J. Neurosci. 14, 1275–1286.
Ortega, J.D., De Rosier, E., Park, S., Larson, C.R., 1988. Brainstem mechanisms of laryngeal control as revealed by microstimulation studies. In: Fujimura, O. (Ed.), Vocal Physiology: Voice Production, Mechanisms and Functions. Raven Press, New York, NY, pp. 19–28.
Panneton, W.M., Martin, G.F., 1983. Brainstem projections to the facial nucleus of the opossum. A study using axonal transport techniques. Brain Res. 267, 19–33.
Paredes, J., Winters, R.W., Schneiderman, N., McCabe, P. M., 2000. Afferents to the central nucleus of the amygdala and functional subdivisions of the periaqueductal gray: neuroanatomical substrates for affective behavior. Brain Res. 887, 157–173.
Randall, W.L., 1964. The behaviour of cats (*Felis catus*) with lesions in the caudal midbrain region. Behaviour 23, 107–139.
Richardson, D.E., Akil, H., 1977. Pain reduction by electrical brain stimulation in man. Part 1. Acute administration in periaqueductal and periventricular sites. J. Neurosurg. 47, 178–183.
Rye, D.B., Lee, H.J., Saper, C.B., Wainer, B.H., 1988. Medullary and spinal efferents of the pedunculo-pontine tegmental nucleus and adjacent mesopontine tegmentum in the rat. J. Comp. Neurol. 269, 315–341.
Schuller, G., Radtke-Schuller, S., 1990. Neural control of vocalization in bats: mapping of brainstem areas with electrical microstimulation eliciting species-specific echolocation calls in the rufous horseshoe bat. Exp. Brain Res. 79, 192–206.
Sem-Jacobsen, C., Torkildsen, A., 1960. Depth recording and electrical stimulation in the human brain. In: Ramey, E.R., O'Doherty, D.S. (Eds.), Electrical Studies on the Unanesthetized Brain. Paul B. Hoeber, New York, NY, pp. 275–287.
Shaikh, M.B., Barrett, J.A., Siegel, A., 1987. The pathways mediating affective defense and quiet biting attack behavior from the midbrain central gray of the cat: an autoradiographic study. Brain Res. 437, 9–25.
Shekhar, A., Dimicco, J.A., 1987. Defense reaction elicited by injection of GABA antagonists and synthesis inhibitors into the posterior hypothalamus in rats. Neuropharmacol. 26, 407–417.
Shiba, K., Yoshida, K., Miura, T., 1995. Functional roles of the superior laryngeal nerve afferents in electrically induced vocalization in anesthetized cats. Neurosci. Res. 22, 23–30.
Shipley, M.T., Ennis, M., Rizvi, T.A., Behbehani, M.M., 1991. Topographical specificity of forebrain inputs to the midbrain periaqueductal gray: evidence for discrete longitudinally organized input columns. In: Depaulis, A., Bandler, R. (Eds.), The Midbrain Periaqueductal Gray Matter. Functional, Anatomical, and Neurochemical Organization. Plenum Press, New York, NY, pp. 417–449.
Siebert, S., Jürgens, U., 2003. Vocalization after periaqueductal grey inactivation with the GABA agonist muscimol in the squirrel monkey. Neurosci. Lett. 340, 111–114.
Skultety, F.M., 1962. Experimental mutism in dogs. Arch. Neurol. 6, 235–241.
Skultety, F.M., 1968. Clinical and experimental aspects of akinetic mutism. Report of a case. Arch. Neurol. 19, 1–14.
Stack, E.C., Balakrishnan, R., Numan, M.J., Numan, M., 2002. A functional neuroanatomical investigation of the role of the medial preoptic area in neural circuits regulating maternal behavior. Behav. Brain Res. 131, 17–36.
Struthers, W.M., 2001. Sex-induced fos in the medial preoptic area: projections to the midbrain. Neuroreport 12, 3065–3068.
Suga, N., Yajima, Y., 1988. Auditory-vocal integration in the midbrain of the mustached bat: periaqueductal gray and the reticular formation. In: Newman, J.D. (Ed.), The Physiological Control of Mammalian Vocalization. Plenum Press, New York, NY, pp. 87–107.
Sutton, D., Larson, C., Lindeman, R.C., 1974. Neocortical and limbic lesion effects on primate phonation. Brain Res. 71, 61–75.
Teskey, G.C., Kavaliers, M., 1991. Modifications of social conflict-induced analgesic and activity responses in male mice receiving chronic opioid agonist and antagonist treatments. Pharmacol. Biochem. Behav. 38, 485–493.
Thoms, G., Jürgens, U., 1987. Common input of the cranial motor nuclei involved in phonation in squirrel monkey. Exp. Neurol. 95, 85–99.
Travers, J.B., Norgren, R., 1983. Afferent projections to the oral motor nuclei in the rat. J. Comp. Neurol. 220, 280–298.
Travers, J.B., Yoo, J.-E., Chandran, R., Herman, K., Travers, S.P., 2005. Neurotransmitter phenotypes of intermediate

zone reticular formation projections to the motor trigeminal and hypoglossal nuclei in the rat. J. Comp. Neurol. 488, 28–47.

Vanderhorst, V.G., Terasawa, E., Ralston, H.J., 2001. Monosynaptic projections from the nucleus retroambiguus region to laryngeal motoneurons in the rhesus monkey. Neuroscience 107, 117–125.

Yajima, Y., Hayashi, Y., Yoshii, N., 1980. The midbrain central gray substance as a highly sensitive neural structure for the production of ultrasonic vocalization in the rat. Brain Res. 198, 446–452.

Zhang, S.P., Bandler, R., Carrive, P., 1990. Flight and immobility evoked by excitatory amino acid microinjection within distinct parts of the subtentorial midbrain periaqueductal gray of the cat. Brain Res. 520, 73–82.

CHAPTER 8.2

Localization of the central pattern generator for vocalization

Steffen R. Hage*

Department of Physiological Science, University of California, Los Angeles, CA, USA and Department of Animal Physiology, University of Tübingen, Tübingen, Germany

Abstract: The present chapter deals with the localization of the vocal pattern generator for species-specific vocalizations in the mammalian brain. First, a catalog is set, listing criteria which are necessary for a vocal pattern generator. According to this, a vocal pattern generator should show vocalization-correlated single unit activity, starting before vocal onset and reflecting specific acoustic features of the vocalization. Lesions should have an inhibitory or a deleterious effect on vocalization. Artificial electrical or pharmacological stimulation should yield artificial vocalization. In anatomical terms, a vocal pattern generator should have direct or, at least, oligosynaptic connections with all motoneuron pools involved in vocalization. Based on the current literature, only one structure in the lower brainstem meets all these criteria, indicating that vocal pattern generation takes place there. This structure reaches from the parvocellular pontine reticular formation, just above the superior olivary complex, through the lateral reticular formation around the facial nucleus and nucleus ambiguus down to the caudalmost medulla, including the dorsal and ventral reticular nuclei and the nucleus retroambiguus. This vocal pattern generating system seems to consist of a network of dedicated pattern generators, one for each call type, suited to modulation during development and through experience.

Keywords: mastication; nucleus ambiguus; pontine brainstem; reticular formation; single-unit recording; squirrel monkey; superior olivary complex; vocalization; vocal pattern generation

Abbreviations: BC: brachium conjunctivum; Cb: cerebellum; CT1 CT2 CTn: call type 1 2 n; GPo: griseum pontis; IC: inferior colliculus; PAG: periaqueductal gray; NA: nucleus ambiguus; NMV: motor trigeminal nucleus; NRA: nucleus retroambiguus; NSV: sensory trigeminal nucleus; NVII: facial nucleus; NXII: hypoglossal nucleus; SOC: superior olivary complex; VH: ventral horn of the spinal cord; VPG: vocal pattern generator

I. Introduction

Vocal communication in mammals essentially consists of two forms. One is characterized by vocal patterns learned by imitation or spontaneoulsy invented, for example human speech and songs of whales. The other is determined by genetically-preprogrammed vocal patterns (see also Chapter 8.3 in this volume). Almost all non-human mammalian vocalizations and the non-verbal vocal utterances of humans, such as laughing, crying or moaning, belong to this group (Hammerschmidt et al., 2001; Scheiner et al., 2004). These vocal utterances are innate, in the sense that they do not need to be heard initially in order to be produced. One example for this is that deaf-and-blind-born children are able to cry and laugh (Eibl-Eibesfeld, 1973). Other examples are deaf-born squirrel monkeys and, respectively, squirrel monkeys raised in isolation from their group mates; such animals show the same call structure in their vocal utterances as their normal group mates (Winter et al., 1973; Hammerschmidt et al., 2001). These findings point to a genetically-preformed pattern generation mechanism for vocal behavior in the mammalian brain.

*Corresponding author. E-mail: hage.steffen@gmail.com

ISBN 978-0-12-374593-4
DOI: 10.1016/B978-0-12-374593-4.00031-0

This chapter deals with the question of where in the mammalian brain the vocal pattern generator(s) is (are) localized. At first, necessary characteristics for a vocal pattern generator will be defined. Then, several brain structures of the vocal motor pathway, which were studied over the past several years in connection with vocal pattern generation, will be analyzed with respect to whether or not they meet these criteria.

II. Characteristics of a vocal pattern generator

To serve as, or be a part of, a vocal pattern generator, the following criteria need to be fulfilled. Most importantly, a vocal pattern generator has to contain neurons which fire in association with vocalization. More specifically, its firing patterns have to reflect specific acoustic features of the vocalizations, such as duration and/or pitch. This vocalization-correlated activity has to start before vocalization onset and last throughout vocalization, in order to be able to control the phonatory motoneuron pools during ongoing vocalization. Otherwise, a proper control of the phonatory motoneuron pools would not be possible. When a vocal pattern generator is electrically stimulated, it should produce vocalizations – but in an artificial manner, due to the artificial activity pattern forced upon the neurons by the electrical stimulation. (In contrast, it is expected that electrical stimulation of higher-order vocal control structures, projecting to the vocal pattern generator(s) will lead to species-specific vocalizations, indistinguishable from spontaneously uttered vocalizations.) Furthermore, vocalizations should be elicited not only by electrical, but also by pharmacological stimulation from a vocal pattern generator. Lesioning or pharmacological inactivation of a vocal pattern generator, or parts of it, should abolish vocalization or at least impair vocal production. Anatomically, a vocal pattern generator should control the phonatory motoneuron pools in a rather direct way, that is, mono- or oligosynaptically. Additionally, a vocal pattern generator has to receive afferent input from the relevant sensory structures. Only with this feedback information is it able to control and correct its vocal motor output in a precise way.

III. Localization of the vocal pattern generator according to its required characteristics

The location of the vocal pattern generator in the mammalian brain has been discussed not without controversy in the last decades, suggesting that there might not be a single structure, but a complex network of several brain structures involved in vocal pattern generation. Structures which were considered to play a significant role in vocal pattern generation are the periaqueductal gray of the midbrain, and several structures in the lower brainstem, such as the reticular formation and the nucleus retroambiguus (for more detail on the vocal motor network, see Hage, Chapter 8.3 and Shiba, Chapter 9.1 in this volume). In recent years, the question arose of whether or not the cerebral hemispheres might be involved in vocal pattern generation. In the following text, I will review all these structures as to their potential ability for being a part of the vocal pattern generating network. Finally, I will present a model of the mammalian vocal pattern generator based on the present literature.

III.A. Cerebral hemispheres

In recent years, behavioral studies on the ontogenetic development of monkey vocalization indicated specific changes in the fine vocal structure between infancy and adulthood (Egnor and Hauser, 2004; Pistorio et al., 2006). Differences in the fine vocal structure could also be detected in bats and monkeys between different groups of the same species, as so-called dialects, indicating learning and, therefore, an involvement of cortical structures in the vocal output (Esser and Schubert, 1998; Fischer et al., 1998). In fact, orofacial somatomotor responses have been found in the primate homolog of Broca's area, a motor cortical structure which is crucial for human speech (Petrides et al., 2005). Therefore, the debate was reopened into whether the cerebral hemispheres are involved in vocal pattern generation, or whether the vocal changes found in ontogenetic development result from maturation of the brainstem structures.

Early brain transection experiments in cats showed that the whole forebrain and the rostral half of the midbrain can be removed without interfering with the ability to produce species-specific vocalizations (Bazett and Penfield, 1922). Furthermore, human infants born without cortical hemispheres (anencephalia) are able to cry (Monnier and Willi, 1953). These and other experimental lesion studies in animals and neurological findings in brain-lesioned patients have made it clear that there are a number of brain structures which are necessary for the production of learned vocal patterns, but which are dispensable for the production of innate vocal patterns. Such structures are, for example, the primary motor cortex, putamen, ventrolateral thalamus and cerebellum (for more detail, see Barlow et al., Chapter 8.4 in this volume). Therefore, it can be ruled out that the cerebral cortex, the basal ganglia, the thalamus, and the cerebellum are involved in the pattern generation of species-specific vocalization. In the future, studies should focus on the role of these structures in modulating genetically-preprogrammed vocal behavior. Especially in primates, it would be extremely useful to learn more about the potential abilities of cortical structures to influence innate vocal behavior, which might help us to better understand the evolution of speech in humans.

III.B. Periaqueductal gray

One of the first brain structures in the brainstem which was assumed to play a specific role in vocal pattern generation was the periaqueductal gray (PAG). PAG lesions in rats, cats, dogs, squirrel monkeys and even human patients have been reported to cause mutism (for review, see Jürgens, 2002). Electrical stimulation of the PAG yielded species-specific vocalizations in several species. Pharmacological stimulation, that is, injection of glutamate agonists and GABA antagonists into the PAG, also led to initiation of vocalization (Bandler and Carrive, 1988; Lu and Jürgens, 1993; Jürgens, 2002). These results indicated that the PAG is not just a through-station, but rather a relay station of the vocal motor system, where processing of vocal information takes place. Finally, single-unit recording studies showed that PAG contains neurons with vocalization-correlated activity (Düsterhöft et al., 2004), in some cases even correlated with activity of laryngeal and respiratory muscles (Larson, 1991). From these observations, a number of authors concluded that the PAG is involved in vocal coordination and might serve as the vocal pattern generator for species-specific vocalization (Holstege, 1989; Larson, 1991; Davis et al., 1996).

In recent studies, it could be shown that inactivation of PAG, leaving the animal unable to vocalize to external stimuli and electrical stimulation of the forebrain, did not prevent the production of species-specific vocalization by electrical stimulation of the lower brainstem in squirrel monkeys and bats (Siebert and Jürgens, 2003; Siebert and Schuller, 2006). Furthermore, electrical stimulation of an intact PAG yields natural sounding species-specific vocalization, an effect which is expected of a brain structure projecting to the vocal pattern generator, rather than of the pattern generator itself. As mentioned before, electrical stimulation of the vocal pattern generator is expected to lead to artificial vocalizations. These observations speak against the PAG being a structure directly involved in vocal pattern generation.

Another fact pointing in the same direction comes from single-unit recordings in squirrel monkeys. Half of the vocalization-correlated neurons in the PAG showed an increase in neuronal activity exclusively before vocal onset. The other half of the neurons changed their activity before and during or only during vocalization; however, none of them showed frequency-correlated activity (Düsterhöft et al., 2004). In the earlier studies on macaque “coo” vocalizations, which lack marked frequency and amplitude modulations, some neurons showed correlation with fundamental frequency (Larson, 1991). A recent study in the squirrel monkey (Düsterhöft et al., 2004), which, apart from calls with a relatively constant frequency, also has calls with extensive frequency modulations, showed different results. In this study, when neuronal activity was analyzed during these frequency changes in a moment-to-moment way, correlation with frequency was not found. Düsterhöft and colleagues, on the other hand, found high call-type specificity for many of the PAG neurons. There were neurons firing rhythmically during frequency modulated “twitter” calls, but not during constant frequency “peep” calls, even though the fundamental frequency for both the call types was in the same frequency range. They also recorded neurons firing to high-pitched constant frequency calls (“peep”), but not to low-pitched constant frequency calls (“caw”) and *vice versa*. Some neurons fired to a combination of two calls (“growl-chuck”), but not to each call type alone. These observations suggest that the PAG codes vocalization in a more global manner, determining the call type and global intensity, rather than the fine structure of vocalization.

III.C. Lower brainstem

Up to this point, the data presented indicate that the vocal pattern generation mechanism is located in the lower brainstem. Vocalization depends on the coordinated activity of numerous muscles; these are the internal and external laryngeal muscles, abdominal and intercostal muscles and those articulatory muscles controlling lips, tongue, velum and jaw. The corresponding motoneurons are distributed over the nucleus ambiguus, the ventral horn of the thoracic and upper lumbar cord, the facial, hypoglossal and trigeminal motor nuclei (Jürgens, 2002). Therefore, the vocal pattern generation could theoretically be coordinated just by a direct interconnection between these motoneuron pools. So far, only a few neuroanatomical studies have focused on such interconnections. The results have shown, however, that there is an almost complete lack of interconnections between the motoneuron pools (Manaker et al., 1992; Fay and Norgren, 1997; Cunningham and Sawchenko, 2000; Travers et al., 2005). Therefore, based on present data, this possibility seems to be highly unlikely.

The other possibility is that there are one or several superordinate structures with direct connections to all phonatory motoneuron pools. In fact, there is an extensive region outside the phonatory motoneuron pools having direct connections with all the phonatory motoneuron pools conjointly. This region consists of the lateral parabrachial area, the parvocellular reticular formation just above the superior olive, the lateral medullary reticular formation and the dorsal and ventral reticular nuclei, together with the nucleus retroambiguus in the caudal-most part of the medulla (Thoms and Jürgens, 1987; Bernard et al., 1990; Fay and Norgren, 1997; VanderHorst et al., 2000; Hannig and Jürgens, 2006). Multiple-labeling studies have shown that even single cells in these structures project to more than one phonatory motoneuron pool (Li et al., 1993a,b; Cunningham and Sawchenko, 2000). Additionally, all structures with projections to all phonatory motoneuron pools receive direct input from the PAG (Mantyh, 1983; Odeh and Antal, 2001). Furthermore, all of these structures get direct input from the solitary tract nucleus (Beckstead et al., 1980; Gerrits and Holstege, 1996), providing these structures with proprioceptive information from lungs, larynx and oral cavity necessary for the production of normal vocalization (Thoms and Jürgens, 1981; Jürgens and Kirzinger, 1985; Shiba et al., 1995).

Further evidence that these brainstem structures are involved in vocal pattern generation comes from pharmacological stimulation studies. In contrast to glutamatergic stimulation of the PAG, which yields vocalizations with normal species-specific structure, vocalizations elicitable from structures in the caudal ventrolateral pons and medullary reticular formation have an abnormal structure (Jürgens and Richter, 1986). Furthermore, normal vocalizations elicited electrically from the PAG are deteriorated by simultaneous stimulation of the parvocellular reticular formation just above the superior olivary complex (Dressnandt and Jürgens, 1992). The artificial character of the stimulation-induced vocal output indicates that the stimulation interferes directly with the pattern generating process in this region.

Lesion studies give further evidence that pontine and reticular structures are involved directly in vocal pattern generation. Lesions in the caudal ventrolateral pons and medullary reticular formation including the nucleus retroambiguus block PAG-elicited vocalizations (Kirzinger and Jürgens, 1985; Zhang et al., 1995; Shiba et al., 1997). Blocking of PAG-elicited vocalization by pharmacological inactivation of lower brainstem structures has been reported in three studies. Inactivation of the nucleus retroambiguus and surrounding reticular formation with kainic acid lesion abolished PAG-elicited vocalization in cats (Shiba et al., 1997). In squirrel monkeys, PAG vocalization was blocked by injecting the glutamate antagonist kynurenic acid into the periolivary region in the ventrolateral pons (Jürgens, 2000). In this study, however, only a specific class of vocalizations could be blocked, namely vocalizations showing a characteristic frequency modulation (i.e., "trill," "cackle;" for representative examples of these classes) (see Fig. 1a). Other call types which lack such modulation (i.e., "peep," "caw," "shriek") remained unaffected. Similar findings were reported for the bat, where echolocation calls elicited in the PAG could be blocked by injecting kynurenic acid into the paralemniscal area (Fenzl and Schuller, 2005). PAG-induced social calls were still elicitable. Results of these studies indicate that the paralemniscal and the periolivary pontine regions play a crucial role in call production. The region around the nucleus retroambiguus seems to be involved in vocal production in a more general way.

The strongest support for the vocal pattern generating network being located in the lower brainstem is given by single-unit recording data. As reported, most neurons in the PAG fired just prior to vocal onset or prior to and before vocal onset with no clear correlation with call structure in detail. In contrast, neurons showing a correlation between their activity and specific acoustic parameters, such as fundamental frequency, amplitude or duration, have been recorded from the lower brainstem. They could be found in the phonatory motoneuron pools (nucleus ambiguus, facial nucleus), as well as in the parvocellular reticular formation, just above the superior olivary complex, the medullary reticular formation, dorsal and ventral reticular nuclei of the caudal medulla, and the nucleus retroambiguus (Yajima and Larson, 1993; Lüthe et al., 2000; Hage and Jürgens, 2006a). In the parvocellular reticular formation just above and caudal of the superior olivary complex, vocalization-correlated neurons also showed a call type-specific activity (Hage and Jürgens, 2006b).

While the neurons showed a call pattern-correlated activity to frequency modulated call types (i.e., "trill"), they were not active during non-frequency modulated call-types (i.e., "caw," see Fig. 1). In addition, this group of neurons has the potential to control the motoneuron pools temporally. With the exception of only one neuron, for each phonatory motoneuron, a reticular neuron could be found in this region firing prior to the motoneuron. This means that, in the phase of vocal initiation, the activity of the supraolivary reticular formation precedes that of phonatory motoneurons – a *conditio sine qua non* for a vocal pattern generator (Hage and Jürgens, 2006a) (see Fig. 2). In the other reticular structures with vocalization-correlated activity, comparisons were made only between the onset time of neuronal activity and that of the vocalization itself. In all these structures, neurons with prevocalization activity were found.

III.D. Location of the vocal pattern generator

The above-mentioned findings indicate that vocal pattern generation takes place in several structures of the lower brainstem: the parvocellular pontine reticular formation just above the superior olivary complex; the reticular formation around the facial nucleus and

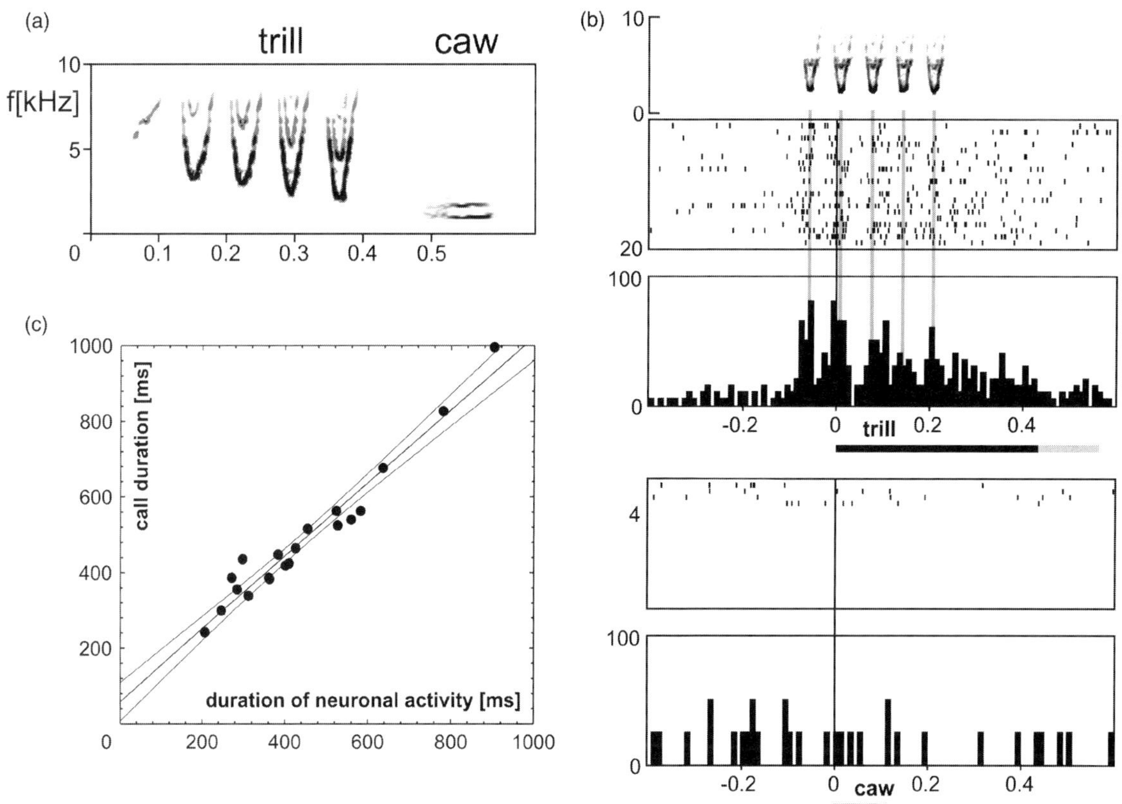

Fig. 1. Neuronal activity of a neuron of the vocal pattern generator for frequency modulated vocalizations during self-produced trill and caw calls (shown as sonagrams in (a)). The neuron is showing trill-syllable correlated activity (upper panel), but no activity during caw vocalizations (lower panel (b)). The neuronal activity is shown as raster (top panels) and peri-event time histogram (bottom panels). The black bars below the histograms indicate the onset and mean duration of vocalizations; the grey bars indicate standard deviation. Gray vertical lines relate the maxima of neuronal activity to the corresponding syllables of a representative trill call. The neuron is also showing a statistically significant correlation between duration of the neuronal activity and call duration (c). Pearson's correlation, $P < 0.001$, $r = 0.97$; each dot stands for one trill call. Different call durations are mainly due to different numbers of syllables in trill calls. Modified from Hage and Jürgens (2006b).

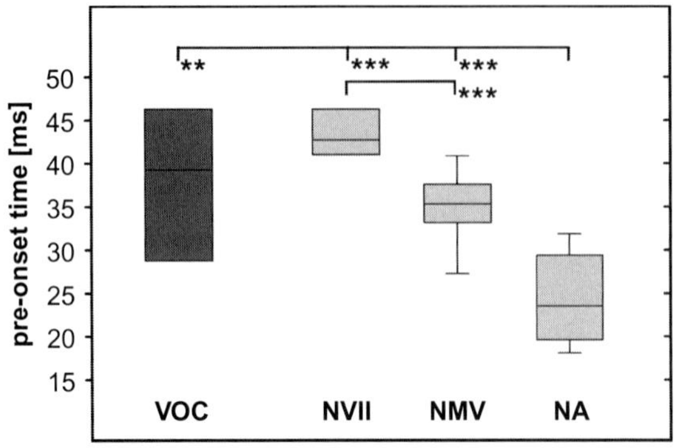

Fig. 2. Distribution of vocal pre-onset times in the vocal pattern generator for frequency modulated calls (VOC) and three motoneuron pools involved in vocalization (NVII facial nucleus; NMV motor trigeminal nucleus; NA nucleus ambiguus); see text for further explanation. Statistically significant differences between the groups are indicated by asterisks (*post hoc* Bonferroni's, $^{**}p < 0.01$, $^{***}p < 0.001$). Lines are medians, boxes surround one standard deviation and whiskers are 10th and 90th percentiles. Modified from Hage and Jürgens (2006a).

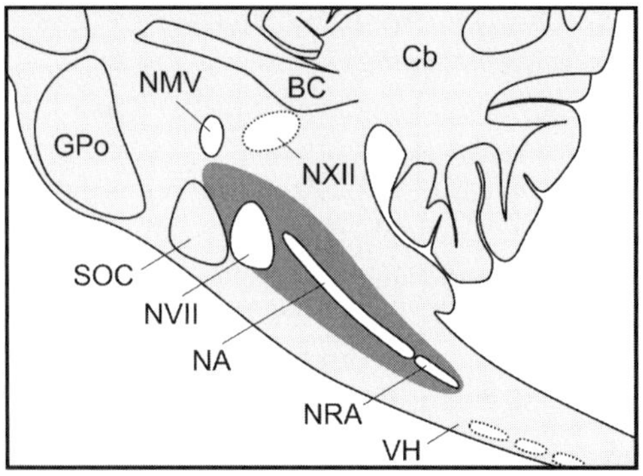

Fig. 3. The anatomical position of the vocal pattern generation network (gray area) and the motoneuron pools involved in vocalization (white structures) in the lateral pontine and medullary reticular formation of the squirrel monkey brainstem (sagittal section 3 mm lateral to midline), based on present data. Structures with dashed lines are positioned more medially from sagittal plane. For abbreviations, see list.

nucleus ambiguus; and the region around the nucleus retroambiguus including, probably, the nucleus retroambiguus itself (see Fig. 3).

The following section will discuss how these structures involved in vocal pattern generation might interact with each other to produce species-specific vocalization.

IV. The vocal pattern generator: dedicated, distributed or reorganizing?

Three basic forms of neural circuits that produce patterned output have been identified (Barlow and Estep, 2006). *Dedicated* circuits generate only one pattern. *Distributed* circuits consist of a single population of neurons which produce all kinds of patterns. *Reorganizing* circuits generate different patterns when the effectiveness of synaptic connections between different members of the total population of neurons changes. Based on present data, the vocal pattern generator seems to be a network of several dedicated pattern generators, one for each type or each group of vocalizations, respectively. It seems that the pattern generator for frequency modulated call types is located in the parvocellular pontine reticular formation just above the superior olivary complex and the reticular formation medial from the facial nucleus (Hage and Jürgens, 2006b).

However, the question remains whether each call type has its own pattern generator or whether more than one call type is produced by one pattern generator (see also Shiba, Chapter 9.1 in this volume). Neurons in the parvocellular pontine reticular formation fired during two call types, "trills" and "cackles." These call types are very similar in their acoustical structure. The only difference is that "cackle" calls have additional noisy components which "trill" calls lack. However, their behavioral context is quite different. While "trill" calls announce joyful events, "cackle" calls are typical for mobbing situations.

A behavioral study in squirrel monkeys pointed out that squirrel monkeys show an increase in their call intensity in response to an increase in the acoustical environment, the so-called Lombard reflex (Hage, 2008). This study showed that with an increase in environmental noise, more and more "trill" calls were replaced by "cackle" calls. Since the "cackle" calls were emitted in exactly the same behavioral context (food presentation) as the "trill" calls, a change in their emotional state could be excluded. It rather seemed that an increase in vocal intensity leads to noisy, overloaded "trill" calls, indistinguishable from "cackle" calls. This would mean that the pattern generator seems to produce the same acoustic structure for "trill" and "cackle" calls. An increase in intensity is produced by an increase in expiratory effort (Jürgens, 2002). For such a respiratory modulation, the nucleus retroambiguus would be a candidate structure, since it gets strong input from the PAG and has strong output to the respiratory motoneurons in the ventral horn of the spinal cord. Additionally, the nucleus retroambiguus is heavily interconnected with the structures involved in vocal pattern generation. A potential role of the nucleus retroambiguus in vocal intensity control in general might also explain why Larson (1991) found cells in the PAG correlated with global pitch. Even though vocal pitch is mainly controlled by the tension of the vocal folds, and therefore by the nucleus ambiguus, it also depends on the subglottic pressure, i.e., the expiratory effort (Alipour and Scherer, 2007).

These findings suggest that the nucleus retroambiguus is not directly involved in the vocal pattern-generating process of fine structure of species-specific vocalization, as suggested by earlier studies (Holstege, 1989; Shiba et al., 1997). Present data rather indicate that it might play a role in the modulation of the vocal pattern generator by increasing or decreasing expiratory effort. In Fig. 4, an overview of the suggested interconnections of structures involved in vocal pattern generation is depicted.

Dedicated pattern generators may be modulated during development and through experience (Barlow and Estep, 2006). This might explain developmental changes in mammalian vocalization, as well as "dialects" between different groups of the same species. From present data, we know that the structures of the vocal pattern generator get heavy input from the somatosensory structures, such as the solitary tract nucleus and the sensory trigeminal nucleus, as well as from several auditory structures, for example, the cochlear nucleus or the inferior colliculus. These sensory structures might modulate the vocal pattern generator in a specific way during development (maturation) and experience (learning).

V. Conclusions

In this review, the location of the mammalian vocal pattern generator was discussed. Based on currently available data, structures being involved in vocal pattern generation are the parvocellular pontine reticular formation just above the superior olivary complex, the reticular formation around facial nucleus and nucleus ambiguus and the region around the nucleus retroambiguus. The nucleus retroambiguus itself might play a role in vocal intensity control. These structures are interconnected in a complex neuronal network. It seems that the vocal pattern generator is a dedicated pattern generator. More studies, however, have to be carried out on the last issue.

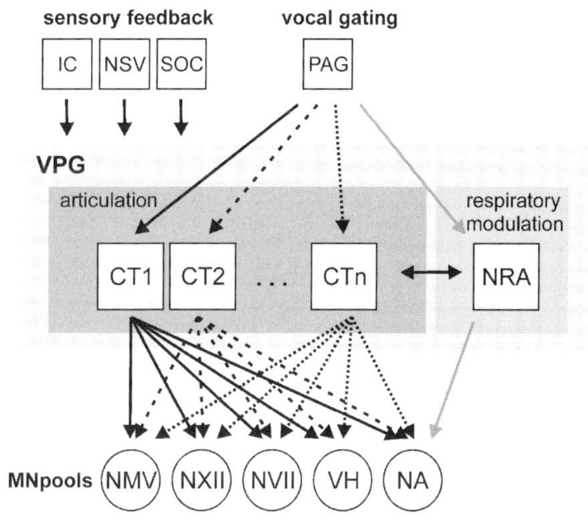

Fig. 4. Overview of input and output structures of the vocal pattern generating system (VPG), based on present anatomical and physiological data. Different call types are produced by different pattern generators (CT1, CT2, CTn), which are modulated during respiration by the nucleus retroambiguus (NRA). See text for further explanations. For abbreviations, see list.

Acknowledgments

This work was supported by the Deutsche Forschungsgemeinschaft (Ha 5400/1-1). I thank Uwe Jürgens for his support, fruitful discussions and helpful comments on the manuscript.

References

Alipour, F., Scherer, R.C., 2007. On pressure-frequency relations in the excised larynx. J. Acoust. Soc. Am. 122, 2296–2305.

Bandler, R., Carrive, P., 1988. Integrated defence reaction elicited by excitatory amino acid microinjection in the midbrain periaqueductal grey region of the unrestrained cat. Brain Res. 439, 95–106.

Barlow, S.M., Estep, M., 2006. Central pattern generation and the motor infrastructure for suck, respiration, and speech. J. Comm. Disord. 39, 366–380.

Bazett, H.C., Penfield, W.G., 1922. A study of the Sherrington decerebrate animal in the chronic as well as the acute condition. Brain 45, 185–265.

Beckstead, R.M., Morse, J.R., Norgren, R., 1980. The nucleus of the solitary tract in the monkey: projections to the thalamus and brain stem nuclei. J. Comp. Neurol. 190, 259–382.

Bernard, J.F., Villanueva, L., Carroue, J., Le Bars, D., 1990. Efferent projections from the subnucleus reticularis dorsalis (SRD): a Phaseolus vulgaris leucoagglutinin study in the rat. Neurosci. Lett. 116, 257–262.

Cunningham, E.T., Sawchenko, P.E., 2000. Dorsal medullary pathways subserving oromotor reflexes in the rat: implications for the central neural control of swallowing. J. Comp. Neurol. 417, 448–466.

Davis, P.J., Zhang, S.P., Bandler, R., 1996. Midbrain and medullary regulation of respiration and vocalization. In: Holstege, G., Bandler, R., Saper, C. (Eds.), The Emotional Motor System. Elsevier, Amsterdam, The Netherlands, pp. 315–325.

Dressnandt, J., Jürgens, U., 1992. Brain stimulation-induced changes of phonation in the squirrel monkey. Exp. Brain Res. 89, 549–559.

Düsterhöft, F., Häusler, U., Jürgens, U., 2004. Neuronal activity in the periaqueductal gray and bordering structures during vocal communication in the squirrel monkey. Neuroscience 123, 53–60.

Egnor, S.E.R., Hauser, M.D., 2004. A paradox in the evolution of primate vocal learning. Trends Neurosci. 27, 649–654.

Eibl-Eibesfeld, I., 1973. The expressive behavior of the deaf and blind born. In: von Cranach, M., Vine, I. (Eds.), Social Communication and Movement. Academic Press, London, UK, pp. 163–194.

Esser, K.H., Schubert, J., 1998. Vocal dialects in the lesser spear-nosed bat *Phyllostomus discolor*. Naturwissenschaften 85, 347–349.

Fay, R.A., Norgren, R., 1997. Identification of rat brainstem multisynaptic connections to the oral motor nuclei in the rat using pseudorabies virus. Brain Res. Rev. 25, 255–311.

Fenzl, T., Schuller, G., 2005. Echolocation calls and communication calls are controlled differentially in the brainstem of the bat *Phyllostomus discolour*. BMC Biology 3, 17.

Fischer, J., Hammerschmidt, K., Todt, D., 1998. Local variation in Barbary macaque shrill barks. Anim. Behav. 56, 623–629.

Gerrits, P.O., Holstege, G., 1996. Pontine and medullary projections to the nucleus retroambiguus: a wheat germ agglutinin horseradish peroxidase and autoradiographic tracing study in the cat. J. Comp. Neurol. 373, 173–185.

Hage, S.R., 2008. Vokale Mustergeneration und audiovokale Integration. Telemetrische Einzelzellableitungen im unteren Hirnstamm am sich frei bewegenden Totenkopfaffen. VDM Verlag, Saarbrücken, Germany.

Hage, S.R., Jürgens, U., 2006a. On the role of the pontine brainstem in vocal pattern generation. A telemetric single-unit recording study in the squirrel monkey. J. Neurosci. 26, 7105–7115.

Hage, S.R., Jürgens, U., 2006b. Localization of a vocal pattern generator in the pontine brainstem of the squirrel monkey. Eur. J. Neurosci. 23, 840–844.

Hammerschmidt, K., Freudenstein, T., Jürgens, U., 2001. Vocal development in squirrel monkeys. Behaviour 138, 1179–1204.

Hannig, S., Jürgens, U., 2006. Projections of the ventrolateral pontine vocalization area in the squirrel monkey. Exp. Brain Res. 169, 92–105.

Holstege, G., 1989. Anatomical study of the final common pathway for vocalization in the cat. J. Comp. Neurol. 284, 242–252.

Jürgens, U., 2000. Localization of a pontine vocalization-controlling area. J. Acoust. Soc. Am. 108, 1393–1396.

Jürgens, U., 2002. Neural pathways underlying vocal control. Neurosci. Biobehav. Rev. 26, 235–258.

Jürgens, U., Kirzinger, A., 1985. The laryngeal sensory pathway and its role in phonation. A brain lesioning study in the squirrel monkey. Exp. Brain Res. 59, 118–124.

Jürgens, U., Richter, K., 1986. Glutamate-induced vocalization in the squirrel monkey. Brain Res. 373, 349–358.

Kirzinger, A., Jürgens, U., 1985. The effects of brainstem lesions on vocalization in the squirrel monkey. Brain Res. 358, 150–162.

Larson, C.R., 1991. On the relation of PAG neurons to laryngeal and respiratory muscles during vocalization in the monkey. Brain Res. 552, 77–86.

Li, Y.Q., Takada, M., Mizuno, N., 1993. Identification of premotor interneurons which project bilaterally to the trigeminal motor, facial or hypoglossal nuclei. A fluorescent retrograde double-labeling study in the rat. Brain Res. 611, 160–164.

Li, Y.Q., Takada, M., Mizuno, N., 1993. Premotor neurons projecting simultaneously to 2 orofacial motor nuclei by sending their branched axons. A study with a fluorescent retrograde double-labeling technique in the rat. Neurosci. Lett. 152, 29–32.

Lu, C.L., Jürgens, U., 1993. Effects of chemical stimulation in the periaqueductal gray on vocalization in the squirrel monkey. Brain Res. Bull. 32, 143–151.

Lüthe, L., Häusler, U., Jürgens, U., 2000. Neuronal activity in the medulla oblongata during vocalization. A single-unit recording study in the squirrel monkey. Behav. Brain Res. 116, 197–210.

Manaker, S., Tischler, L.J., Bigler, T.L., Morrison, A.R., 1992. Neurons of the motor trigeminal nucleus project to the hypoglossal nucleus in the rat. Exp. Brain Res. 90, 262–270.

Mantyh, P.W., 1983. Connections of midbrain periaqueductal gray in the monkey. J. Neurophysiol. 49, 567–594.

Monnier, M., Willi, H., 1953. Die integrative Tätigkeit des Nervensystems beim meso-rhombo-spinalen Anencephalus (Mittelhirnwesen). Monatsschr. Psychiat. Neurol. 126, 239–273.

Odeh, F., Antal, M., 2001. The projections of the midbrain periaqueductal grey to the pons and medulla oblongata in rats. Eur. J. Neurosci. 14, 1275–1286.

Petrides, M., Cadoret, G., Mackey, S., 2005. Orofacial somatomotor responses in the macaque monkey homologue of Broca's area. Nature 435, 1235–1238.

Pistorio, A.L., Vintch, B., Wang, X., 2006. Acoustic analysis of vocal development in a New World primate, the common marmoset (*Callithrix jacchus*). J. Acoust. Soc. Am. 120, 1655–1670.

Scheiner, E., Hammerschmidt, K., Jürgens, U., 2004. The influence of hearing impairment on the preverbal emotional vocalization of infants. Folia Phoniat. Logopaed. 56, 27–40.

Shiba, K., Yoshida, K., Miura, T., 1995. Functional roles of the superior laryngeal nerve afferents in electrically induced vocalization in anesthetized cats. Neurosci. Res. 22, 23–30.

Shiba, K., Umezaki, T., Zheng, Y., Miller, A.D., 1997. The nucleus retroambigualis controls laryngeal muscle activity during vocalization in the cat. Exp. Brain Res. 115, 513–519.

Siebert, S., Jürgens, U., 2003. Vocalization after periaqueductal grey inactivation with the GABA agonist muscimol in the squirrel monkey. Neurosci. Lett. 340, 111–114.

Siebert, S., Schuller, G., 2006. Vocalization in bats: interactions between periaqueductal grey and nucleus of the brachium of the inferior colliculus. FENS Abstr. 3, A040.4.

Thoms, G., Jürgens, U., 1981. Role of the internal laryngeal nerve in phonation: an experimental study in the squirrel monkey. Exp. Neurol. 74, 187–203.

Thoms, G., Jürgens, U., 1987. Common input of the cranial motor nuclei involved in phonation in squirrel monkey. Exp. Neurol. 95, 85–99.

Travers, J.B., Yoo, J.E., Chandran, R., Herman, K., Travers, S.P., 2005. Neurotransmitter phenotypes of intermediate zone reticular formation projections to the motor trigeminal and hypoglossal nuclei in the rat. J. Comp. Neurol. 488, 28–47.

VanderHorst, V.G.J.M., Terasawa, E., Ralston, H.J., Holstege, G., 2000. Monosynaptic projections from the nucleus retroambiguus to motoneurons supplying the abdominal wall, axial, hindlimb, and pelvic floor muscles in the female rhesus monkey. J. Comp. Neurol. 424, 233–250.

Winter, P., Handley, P., Ploog, D., Schott, D., 1973. Ontogeny of squirrel monkey calls under normal conditions and under acoustic isolation. Behaviour 47, 230–239.

Yajima, Y., Larson, C.R., 1993. Multifunctional properties of ambiguous neurons identified electrophysiologically during vocalization in the awake monkey. J. Neurophysiol. 70, 529–540.

Zhang, S.P., Bandler, R., Davis, P.J., 1995. Brain stem integration of vocalization: role of the nucleus retroambigualis. J. Neurophysiol. 74, 2500–2512.

CHAPTER 8.3

Neuronal networks involved in the generation of vocalization

Steffen R. Hage*

Department of Physiological Science, University of California, Los Angeles, CA, USA and Department of Animal Physiology, University of Tübingen, Tübingen, Germany

Abstract: Vocalization is a complex behavioral pattern in which several cortical and subcortical structures are involved to produce proper vocal output. While learned vocal patterns, such as speech, are produced by a complex cortical network, genetically preprogrammed vocalizations, to which most mammalian vocalizations belong, are generated by a complex neuronal network in the brainstem. Structures involved in this network are the periaqueductal gray, ventrolateral parabrachial area, nucleus retroambiguus and the lateral pontine, as well as the lateral and caudal medullary reticular formation. This network has direct access to all motoneuron pools controlling laryngeal, respiratory and articulatory components of vocalization. It receives proprioceptive feedback information from laryngeal, pulmonary and oral mechanoreceptors via the solitary tract nucleus and spinal trigeminal nucleus. The network receives facilitating input from emotional, motivational and sensory structures via the periaqueductal gray, which serves as an integration center and triggering structure for vocalization.

Keywords: limbic system; motor coordination; nucleus ambiguus; nucleus retroambiguus; lateral parabrachial nucleus; periaqueductal gray; reticular formation; vocal pattern generation

Abbreviations: ACC: anterior cingulate cortex; PB: lateral parabrachial region; LRF: lateral reticular formation; NA: nucleus ambiguus; NMV: motor trigeminal nucleus; NRA: nucleus retroambiguus; NVII: facial nucleus; NXII: hypoglossal nucleus; PAG: periaqueductal gray; PFC: prefrontal cortex; pLA: paralemniscal area; lat. ret. form.: lateral pontine and medullary reticular formation; SC: superior colliculus; VH: ventral horn of the medulla; VL: ventrolateral; VPG: vocal pattern generator

I. Introduction

Mammalian vocalization is a complex behavioral pattern which takes place at different levels of the brain involving several neuronal networks. This form of communication can be divided into two main groups. The first is determined by genetically-preprogrammed patterns and produced by a vocal pattern generating network in the brainstem (see also Hage, Chapter 8.2 in this volume). These types of innate vocalizations are fixed in their vocal pattern and might only be modulated slightly during ontogenetic development (Pistorio et al., 2006). Almost all non-human mammalian vocalizations belong to this group, as well as non-verbal vocal utterances of humans, such as laughing, crying and moaning (Hammerschmidt et al., 2001; Scheiner et al., 2004). The other group of vocal behavior is characterized by vocal patterns, which are learned by imitation or invention. While this type of vocalization is prominent in oscine birds, it is only rarely found in mammals. Examples are songs of whales and human speech. This chapter focuses on the neural networks underlying the control of innate vocalization. Nevertheless, it will briefly mention which brain structures are indispensable for human speech and, respectively, are not necessary for the production of innate vocalization. Also, the differences and similarities in neuronal control of innate vocalization and human speech will be discussed.

*Corresponding author. E-mail: hage.steffen@gmail.com

ISBN 978-0-12-374593-4
DOI: 10.1016/B978-0-12-374593-4.00032-2

II. Peripheral structures

Essentially, three components are involved in mammalian vocal behavior, irrespective of its complexity: vocal fold movements (laryngeal activity); respiratory activity (usually expiration); and supralaryngeal movements (articulation). Table 1 lists all phonatory muscles with their corresponding innervating nerves and motoneuron pools.

The vocal fold movements are mainly carried out by eight laryngeal muscles (five intrinsic and three extrinsic). The intrinsic muscles interconnect the different laryngeal cartilages and control the position and tension of the vocal folds. The thyroarytenoid muscle forms the body of the vocal folds and is, therefore, responsible for their stiffness. The interarytenoid and cricoarytenoid (lateral and posterior) muscles are responsible for the adduction (glottis closure) and abduction of the vocal folds, by tilting and rotating the arytenoid cartilages toward or away from the midline. Finally, the cricothyroid muscle pulls the cricoid cartilage up toward the thyroid cartilage, which results in stretching the vocal folds by increasing the distance between the anterior (thyroid) and posterior (arytenoid) fixation points of the vocal folds (for further details see Berke and Long, Chapter 10.1 in this volume). The activity of the cricothyroid muscle is directly correlated with vocal pitch, which is also supported by single-unit recording studies (Sutton et al., 1972; Rübsamen et al., 1986; Hage and Jürgens, 2006a). Deactivation of the cricothyroid muscle causes a complete loss of high-pitched sound (Jürgens et al., 1978). The main function of the extrinsic muscles is to control and stabilize the position of the larynx in the neck by direct and indirect connections with surrounding structures. Only three of them play a major role in vocalization: the thyrohyoid muscle, connecting the thyroid cartilage and the hyoid bone; the sternothyroid muscle, connecting the thyroid cartilage with the sternum; and the sternohyoid muscle, connecting the sternum and the hyoid bone. The main function of these muscles is to stabilize the thyroid within the neck, so that the intrinsic muscles can exert their forces.

A mandatory precondition for phonation is an airflow passing the vocal folds. Therefore, inspiratory

Table 1. Innervation of muscles involved in mammalian vocalization (for a detailed list see Jürgens, 2002).

	Muscles	Innervated by	Location of motoneurons	References
Laryngeal	M. cricothyroideus (intrinsic)	N. laryngealis superior externus	Nucl. ambiguus	Davis and Nail, 1984; West and Larson, 1993
	Other intrinsic muscles (Mm. cricoarytaenoideus, interarytaenideus and thyroarytaenoideus)	N. recurrens	Nucl. ambiguus	Schweizer et al., 1981; Yoshida et al., 1992; Yajima and Larson, 1993
	Extrinsic muscles (Mm. sternothyroideus, sternohyoid and thyrohyoid)	Ansa cervicalis	Medullary reticular formation down to VH of cervical spinal cord	Ueyama et al., 1988; Kirzinger and Jürgens, 1994
Respiratory	Mm. abdominalis and intercostalis	Nn. intercostales, subcostalis, iliohypogastricus and ilioinguinalis; Plexus lumbalis	VH of upper lumbar, thoracic and cervical spinal cord	Schriever and Jürgens, 1989; Holstege et al., 1987
	Diaphragm	N. phrenicus	VH of the cervical spinal cord	Duron et al., 1979
Facial	Mm. orbicularis, mentalis, buccinators, zygomaticus major and depressor angulis oris	N. facialis	Facial nucleus	Leanderson et al., 1971; Welt and Abbs, 1990; Horta-Junior et al., 2004
Jaw	Mm. digastricus, masseter, pterygoid., temporalis, geniohyoideus and mylohyoideus	Nn. facialis, mandibularis and hypoglossus	Facial nucleus; motor trigeminal nucleus; VH of the cervical spinal cord	Mizuno et al., 1981; Matsuda et al., 1979
Tongue	Several intrinsic and extrinsic muscles	N. hypoglossus	Hypoglossal nucleus	Uemura-Sumi et al., 1981

muscles have to ensure that sufficient air volume is available in the lungs. On the other hand, expiratory muscles have to produce enough subglottal pressure for proper vocalization. This respiratory component of vocalization is controlled by abdominal and intercostal muscles, as well as the diaphragm. The abdominal muscles, such as the internal and external oblique muscles, as well as the transversus and rectus abdominis muscles and the intercostal muscles, are mainly involved in the expiratory phase of vocalization (Estenne et al., 1990; Jürgens and Schriever, 1991). In addition to building air pressure, the diaphragm is mainly involved in the inspiratory phase of vocalization (West and Larson, 1993).

The sound generated by the vocal folds is modulated in its spectral parameters by the resonance characteristics of the supralaryngeal tract, which is mainly determined by its length and shape. Change in the length of the vocal tract can be obtained in humans and some primates by changing the position of the lips on the one hand, and of the larynx on the other (for some other mammals, see Frey and Gebler, Chapter 10.3 in this volume). Muscles involved in controlling the shape and position of the lips are the orbicularis and mentalis muscles, as well as the buccinators, zygomaticus major and depressor angulis oris muscles. The position of the larynx is controlled by the sternothyroideus and sternohyoideus muscles, as well as the thyrohyoideus and mylohyoideus muscles (Shipp, 1975). The shape of the supralaryngeal tract is mainly modulated by the position of the jaw and the tongue. Muscles involved in jaw movements are the digastricus (opener) and masseter (closer) muscles, as well as the pterygoideus, temporalis, geniohyoideus and mylohyoideus muscles (Matsuda et al., 1979; Mizuno et al., 1981). The significant role of the facial and motor trigeminal nucleus is also supported by single-unit recordings in the squirrel monkey, showing that the majority of recorded neurons fire in correlation with vocal output (Hage and Jürgens, 2006a). Finally, the position of the tongue is controlled by several intrinsic and extrinsic muscles (McClung and Goldberg, 2000).

To the knowledge of the author, only one study has so far reported pre-onset times of vocalization-correlated activity in more than one motoneuron pool (Hage and Jürgens, 2006a). In this study, we were able to record from neurons of the nucleus ambiguus, as well as the motor trigeminal and facial nucleus of spontaneously vocalizing squirrel monkeys. We showed that these cranial motoneuron pools differed significantly in their vocal pre-onset times (see also Hage, Chapter 8.2 in this volume). The longest latencies were found in the facial nucleus, the shortest in the nucleus ambiguus. This result suggests – assuming equal transduction times in motor axons – that, at least in squirrel monkey vocalization, articulation starts before laryngeal activation, and activity of the facial muscles starts before activity of the jaw muscles.

III. Brainstem mechanisms

III.A. Vocal–motor structures

The phonatory motoneuron pools are controlled by a vocal pattern generating network consisting of parts of the lateral reticular formation and the nucleus retroambiguus (NRA) (see Fig. 1). Anatomical, pharmacological and single-unit recording studies indicate that the ventrolateral reticular formation is coordinating the cranial motoneuron pools, thus producing the vocal pattern (Jürgens, 2000; Lüthe et al., 2000; Hannig and Jürgens, 2006, 2006a). The lateral reticular formation has connections to all phonatory motoneuron pools and contains neurons which show correlated activity to specific vocal parameters, such as duration and frequency modulation. Additionally, we showed that the neurons' vocal pre-onset times are in a range that allows them to control the phonatory motoneuron pools during vocalization (Hage and Jürgens, 2006a).

The lateral reticular formation is reciprocally connected with the NRA, which seems to represent a crucial relay station for the laryngeal expiratory component of vocalization (for greater detail on the vocal pattern generator, see Hage, Chapter 8.2 in this volume). In theory, the vocal pattern generator should be reciprocally connected with other pattern generators, as for example those for mastication, respiration or swallowing, since all are using, in part, the same motoneurons (i.e., Rübsamen and Betz, 1986; Yajima and Larson, 1993; Barlow and Estep, 2006; Hage and Jürgens, 2006a; see also Shiba, Chapter 9.1 in this volume). Inhibitory interconnections could prevent two or three patterns happening at the same time. To the knowledge of the author, no studies have yet been performed on this issue with special reference to vocalization. Nevertheless, mastication-correlated neurons found in the region of the masticatory pattern generator were inhibited during vocalization (Hage and Jürgens, 2006b).

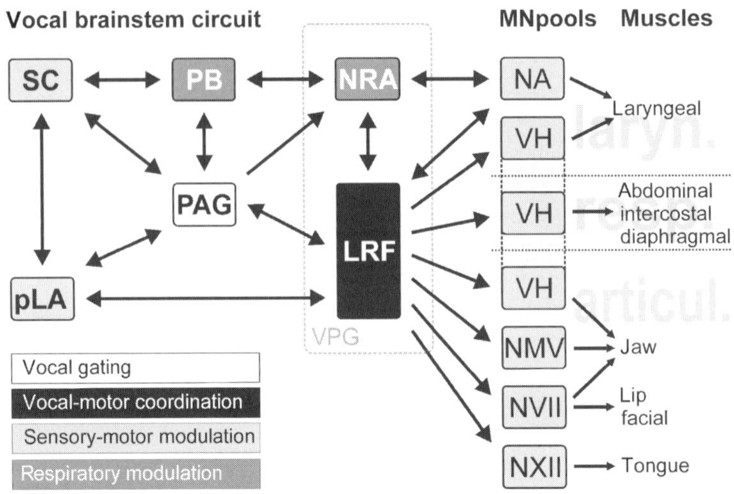

Fig. 1. Brainstem mechanisms involved in the production of vocal motor output. Structures are coded by differential shading according to their suggested role in vocalization (vocal gating, vocal-motor coordination, sensory-motor modulation and respiratory modulation, respectively). Interconnections between structures are based on current anatomical literature (arrows). Not all anatomically verified interconnections of the structures are shown, only the most relevant ones. Sensory feedback reaches almost all structures of the vocal brainstem circuit, but it was omitted for clarity and will be discussed in the text in greater detail. Larynegeal (laryn.), respiratory (resp.), and articulatory (articul.) components of the system are labeled in gray in the background. For abbreviations, see list.

The vocal pattern generating network receives strong input from all regions of the periaqueductal gray (PAG) (Meller and Dennis, 1991; Cameron et al., 1995; Van der Horst et al., 2000; Hannig and Jürgens, 2006). Electrical and chemical stimulation of the PAG has been reported to produce vocalizations in several species, for example rat (Yajima et al., 1980), bat (Suga et al., 1973; Schuller and Radtke-Schuller, 1990), cat (Bandler and Carrive, 1988) and squirrel monkey (Jürgens and Ploog, 1970; Lu and Jürgens, 1993). These vocalizations generally resembled natural, species-specific calls. In squirrel monkeys, the complete vocal repertoire can be elicited from the PAG. Destruction of the PAG resulted in mutism in all animals, including humans (Adametz and O'Leary, 1959; Jürgens and Pratt, 1979; Esposito et al., 1999). Single-unit recordings revealed vocalization-related neuronal activity in the dorsal and lateral part of the caudal third of the PAG (Larson and Kistler, 1984; Suga and Yajima, 1988; Düsterhöft et al., 2004). In the squirrel monkey, about half of the vocalization-related neurons showed an increase in neuronal activity exclusively before vocal onset. The other half of these neurons changed their activity before and during, or only during vocalization. However, none of them exhibited frequency correlated activity (Düsterhöft et al., 2004). In addition, most of the neuronal activity was correlated to several types of vocalizations. These data indicate that the PAG serves as a vocal trigger for the vocal pattern generator, rather than as part of the vocal pattern generator itself (for greater detail see also Gruber-Dujardin, Chapter 8.1 and Hage, Chapter 8.2 in this volume).

The PAG, as well as the vocal pattern generating network, receives modulatory input from the lateral parabrachial region (PB), corresponding to the Kölliker-Fuse nucleus in rats, cats and dogs (Schuller and Radtke-Schuller 1990; Jürgens, 2002). However, the main projections of PB go to the NRA of the vocal patterning network (Smith et al., 1989; Gerrits and Holstege, 1996). The PB itself has a potent influence over respiratory rhythms (Feldman and Gautier, 1976; Feldman et al., 2003; Smotherman et al., Chapter 9.2 in this volume). It also has a significant influence on the mammalian vocal motor network (Jürgens, 2002). Its significant function seems to be the respiratory-vocal coupling that embeds vocal utterances properly in between inspiratory phases. More specifically, the PB seems to be involved in the gating of vocal onset on the basis of the current respiratory status. Its role in the modulation of the respiratory component of vocalization is underlined by the fact that most of the vocalization-correlated neurons in PB were rhythmically active during quiet breathing

(Farley et al., 1992). However, studies in bats have shown that intracerebral injections of glutamatergic and GABA-ergic drugs were changing vocal pitch, duration and interpulse intervals (Smotherman et al., 2003, 2006). Injection of bicuculline, a GABAA antagonist, could even cancel the bat's ability to adjust the pitch of the echolocation pulses so that the returning echo frequencies no longer fall in the optimal hearing range (Doppler shift compensation) (Smotherman et al., 2003). Destruction of PB canceled the bat's ability to change its frequency to compensate for Doppler shifts (Smotherman et al., 2003). Vocal pitch is controlled by the cricothyroid muscle, which alters the tension of the vocal muscles, causing a change in glottal resistance and therefore in glottal airflow. This may explain why, particularly, the control of vocal pitch has to be strongly coordinated with the respiratory system. Recent data indicate that PB is a candidate region for such a control structure.

Another important structure with interconnections to the PAG and the vocal pattern generating network is the paralemniscal area (pLA). Although the whole vocal brainstem circuit receives input from several auditory structures (see below), it seems that the paralemniscal area and the caudally adjacent pontine reticular formation receive the strongest input from the auditory system (Metzner, 1996). Single unit recordings in bats and squirrel monkeys have shown that this area contains neurons that are changing their activity before and during vocalization, as well as during reception of external acoustical stimuli (Metzner, 1993; Hage et al., 2006). In bats, electrical stimulation of this area produced echolocation calls with lower threshold and shorter latency than in the PAG (Schuller and Radtke-Schuller, 1990). Therefore, it has been proposed that in the bat the pLA plays a similar role in vocal motor control as does the PAG in non-echolocating animals.

A further structure involved in sensory–motor modulation of the vocal motor system is the superior colliculus (SC). It is reciprocally interconnected with PAG, PB and pLA (Edwards et al., 1979; Mantyh, 1983; Metzner, 1996; Dujardin and Jürgens, 2005). The SC functions as a sensorimotor interface, linking sensory-based spatial information to orienting responses, such as rapid, directed saccadic movements of the eyes, ears, head and body (Valentine et al., 2002). In bats, electrical microstimulation in SC yielded echolocation calls (Schuller and Radtke-Schuller, 1990) and echolocation calls combined with pinna and head movements (Valentine et al., 2002). In a recent bat study, single units in the SC of freely behaving bats increased their activity prior to vocalizations. This activity showed distinct changes in temporal dynamics related to the duration of the echolocation calls, which foraging bats adjust with target distance (Sinha and Moss, 2007). These findings indicate that, in bat vocalization, the SC plays a specific role in range-dependent adjustments of echolocation call duration. However, its role in the modulation of vocal output of non-echolocating mammals remains unclear.

III.B. Sensory innervation

The vocal motor system gets extensive input from several sensory systems on various hierarchical levels (for review see Smotherman, 2007). Most of these sensory inputs serve feedback purposes, and lack of such feedback leads to distorted vocalization. Another function of the sensory input is that it can serve as a stimulus to trigger/initiate vocal output. In the following paragraph I will focus on the former function, the next paragraph will discuss the vocal initiating structures.

All three subsystems of the phonatory apparatus, that is, the laryngeal, respiratory and articulatory ones, receive somatosensory innervation. This innervation provides central structures with information from corpuscular mechanoreceptors, unmyelinated nerve endings, stretch receptors and muscle spindles in the respective subsystems (Jürgens, 2002). The vocalization-correlated proprioceptive feedback of the larynx is transmitted via the recurrent nerve and the internal branch of the superior laryngeal nerve (Yoshida et al., 1992). Transection of the internal branch of the superior laryngeal nerve, which is exclusively sensory, leads to dysphonia (hoarseness) and a drop in fundamental frequency in cats and squirrel monkeys (Thoms and Jürgens, 1981; Shiba et al., 1995). With respect to the respiratory subsystem, the somatosensory feedback from the lungs, which is mediated by the vagal nerves, is essential for vocal output (Nakazawa et al., 1997). If both vagal nerves are cut, vocalization cannot be elicited any more from the PAG. This reveals that the vocal system relies on information about the current air volume in the lungs to utter vocalizations. Finally, somatosensory feedback of the articulatory subsystem is mediated by the trigeminal, glossopharyngeal and vagal nerves.

While supralaryngeal feedback plays a major role in human speech, its role in innate vocalizations is not well-understood (Smotherman, 2007). All sensory fibers of these subsystems project mainly into the solitary tract nucleus and the spinal trigeminal nucleus (Kalia and Mesulam, 1980; Kalia and Richter, 1985; Altschuler et al., 1989; Travers and Norgren, 1995; Hayakawa et al., 2001). Lesions of one or both of these structures cause severe deterioration of vocalization (Jürgens and Kirzinger, 1985). The vocal motor system is extensively interconnected with at least one of these two sensory nuclei. These structures are the PAG lateral parabrachial region and the vocal pattern generating network, as well as the motoneuron pools such as the nucleus ambiguus, facial nucleus and the motor trigeminal nucleus (Jürgens and Kirzinger, 1985; Shigenaga et al., 1988; Moga and Caper, 1990; Dauvergne et al., 2001; Dujardin and Jürgens, 2005; Yoshida et al., 2005). Besides somatosensory feedback, the vocal motor system receives intense input from several structures of the auditory system, for example from the cochlear nucleus, superior olivary complex, the inferior colliculus, or auditory cortex (Metzner, 1996; Cant and Benson, 2003; Dujardin and Jürgens, 2005). The paralemniscal area receives the strongest auditory input, although the PAG and the lateral reticular formation also receive strong input from auditory structures. While it is well-established that acoustic feedback plays an important role in the echolocation behavior of bats (see also Metzner and Schuller, Chapter 9.4 in this volume) and learned vocal patterns such as human speech (see also Liu et al., Chapter 9.3 in this volume), its role in innate vocalization is not completely understood.

Current data indicate two main functions of auditory feedback on the vocal motor system. This feedback plays a specific role in basic auditory reflexes, for example the Lombard reflex (Lombard, 1911; Brumm et al., 2004), which is the involuntary rising of vocal intensity as a response to an increase in environmental noise. This reflex is still present in decerebrated cats, indicating that the auditory system interacts with the vocal motor system at the brainstem level (Nonaka et al., 1997). Furthermore, differences in the fine vocal structure could be detected in bats and monkeys between different groups of the same species ("dialects"), indicating a direct modulation of innate vocalization by auditory input (Esser and Schubert, 1998; Fischer et al., 1998; Egnor and Hauser, 2004).

IV. Vocal initiating structures

The vocal brainstem circuit, as outlined above, receives direct input from various sensory and motivation-controlling limbic structures mostly via the PAG. Motivation-controlling limbic structures with strong input to the PAG are the anterior cingulate cortex, amygdala, hypothalamus, midline thalamus, nucleus accumbens, nucleus striae terminalis, preoptic area, septum and the subcallosal gyrus (see Fig. 2) (Jürgens and Pratt, 1979; Beitz, 1982; Mantyh, 1982; Marchand and Hagino, 1983; Meller and Dennis, 1986; Dujardin and Jürgens, 2005).

In the squirrel monkey, no qualitative differences were found in the projections of limbic structures to PAG sites that control different call types, (Dujardin and Jürgens, 2005); however, there were quantitative differences (Dujardin and Jürgens, 2006). In most of the structures, projection strength depended on the aversive/hedonic quality of the periaqueductal vocalization-eliciting input site. As an example, the more aversive the emotional state accompanying the vocalization elicited from a specific PAG site, the stronger the input this site received from the posterior hypothalamus. Other studies corroborated this finding by indicating that specific parts of the hypothalamus were involved in the expression of aversive emotional states (Kyuhou and Gemba, 1998). Other limbic structures project more heavily to one or to other periaqueductal site, irrespective of the aversive/hedonic quality of the stimulated PAG vocalization site. One of these limbic structures is the anterior cingulate cortex (Dujardin and Jürgens, 2006), which does not project preferentially to PAG call sites of highly aversive or hedonic quality, but rather to sites producing calls (cackle) associated with higher voluntary control than other vocalizations. This finding is of special interest in light of the results of lesioning studies in macaques. Macaques with lesions in the anterior cingulate cortex still responded to threatening

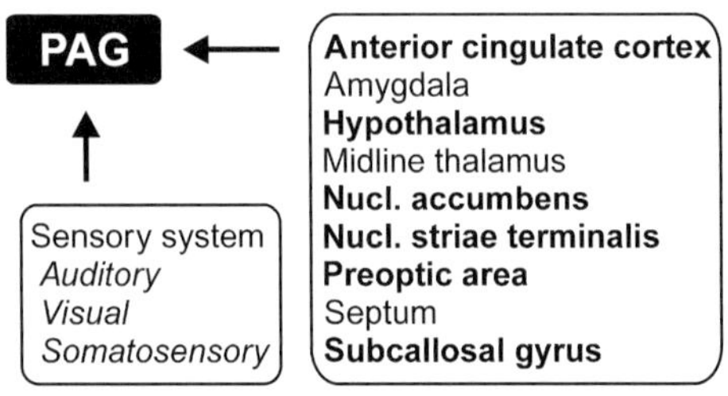

Fig. 2. Limbic and sensory input structures to the periaqueductal gray (PAG), which are known to have a direct influence on vocal initiation, based on lesion, stimulation, single-unit recording and brain imaging studies. Structures highlighted in bold print show quantitative differences in their input to distinct subregions of the PAG, controlling different types of vocalization. Dujardin and Jürgens (2006).

stimuli with warning calls, but were no longer able to produce vocalizations voluntarily as an operant behavioral response when receiving food rewards or avoiding an electroshock (Sutton et al., 1974; Aitken, 1981). These findings underline the important role of the anterior cingulate cortex in the voluntary initiation of monkey vocalization.

After destruction of all above-mentioned limbic structures, vocalizations can still be elicited from specific PAG sites by stimulation, but are no longer uttered spontaneously (Jürgens and Pratt, 1979). This finding indicates that those structures are not involved in the production of the vocal pattern itself, but in the emotional and motivational initiation of the vocal output.

As discussed above, the PAG receives strong input from the auditory, visual and somatosensory systems (Fig. 2). Input from sensory structures to the PAG probably serves as vocal trigger, which initiates or suppresses vocal behavior. An example for vocal initiation via the somatosensory system is the pain shrieking in human infants which is elicited by mild air puffs to the chest. Also, auditory input can induce vocal responses with short latency. Squirrel monkeys, for instance, respond to trill vocalizations of group mates as early as 200 ms after vocal onset (Hage, 2008). This short latency suggests a rather direct vocal initiation pathway. An example for inhibitory vocal control by visual input is the suppression of vocalization during the presence of a predator. These findings of different limbic and sensory pathways to the PAG corroborate the proposed highly integrative mechanism in the PAG and its gating function for different call types.

Apart from their strong projections to the PAG, the vocal initiating structures are interconnected with each other at the cerebral level, forming a complex network. For example, the anterior cingulate cortex is reciprocally connected with the amygdala, hypothalamus, nucleus accumbens, midline thalamus and septum, as well as with the auditory and visual cortices (Müller-Preuss and Jürgens, 1976; Jürgens and Müller-Preuss, 1977; Swanson and Cowan, 1979; Müller-Preuss et al., 1980; Pandya et al., 1981; Vogt and Miller, 1983; Musil and Olson, 1988; Montaron et al., 1996; Hsu and Shyu, 1997). In addition, the nucleus accumbens forms reciprocal connections with the septum and the thalamus (Powell and Leman, 1976; Su and Bentivoglio, 1990). It is not yet clear, however, how this cerebral network affects the vocal initiating process in the PAG.

V. Speech-related structures not necessary for innate vocalization

As reported earlier, most mammalian vocalizations are genetically preprogrammed and determined in their structure. Vocal patterns, which are learned by imitation or invention, are rarely found. The most prominent example of learned vocal utterances is human speech. A complex neuronal network of cortical and subcortical structures is involved in the development and production of human speech (see Fig. 3). Apart from several brainstem structures, which are involved in innate vocalization and in speech, there are other structures that are involved in human speech only, and are not necessary for innate vocalization, such as the facial motor cortex, Broca's area, supplementary motor area, putamen, substantia nigra, pontine gray, cerebellum and the ventrolateral thalamus. Lesions in all these areas had no significant effect on the production of innate vocalizations in monkeys (Jürgens, 2002). These results make clear that innate vocalization and vocal behavior involving sensory-motor learning use, in part, different cortical and subcortical networks.

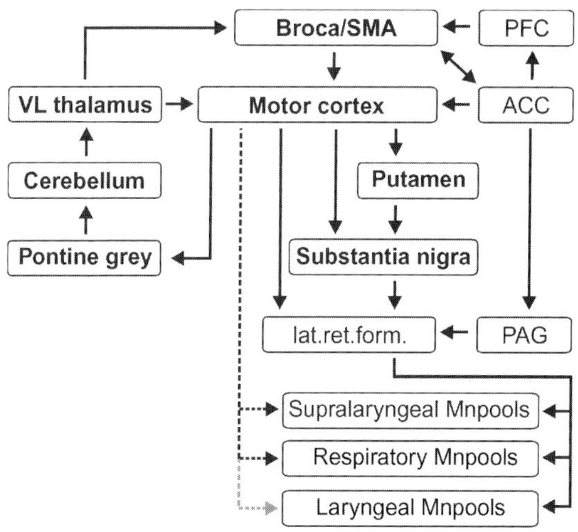

Fig. 3. Circuit diagram summarizing the most important structures for which there is evidence for involvement in speech production from lesion, stimulation, single-unit recording and brain imaging studies. Interconnections between structures are based on current anatomical literature. Dashed interconnections are exclusively found in primates and humans. The interconnection between motor cortex and the laryngeal motoneuron pools (dashed gray arrow) is only present in humans. Lesions in the structures highlighted in bold print showed no significant effect on the production of innate vocalizations. Mnpools: motoneuron pools; SMA: supplementary motor area. For other abbreviations, see list.

VI. Conclusions

The central control of vocalization consists of a complex neuronal network of cortical and subcortical structures. This chapter attempts to explain how innate vocalization is produced by this network in "real time." A complex input from sensory, motivational and emotional structures to the PAG initiates a specific vocalization. The PAG integrates this input and "decides" which vocalization will be uttered and, therefore, which vocal pattern generator will be activated. The system is closely connected with and modulated by PB, pLA and SC. The main function of PB is to coordinate the vocal output with, and embed it properly within, the respiratory cycle. The system receives information about the auditory environment mainly via the pLA. The system might use this input, for instance, to increase vocal amplitude in a noisy environment, if necessary. SC provides information about the position of the head and body of the animal. This information may be important to determine whether the animal is physically able to utter a vocalization without affecting any parameters of normal vocalization. Finally, the vocal pattern generating network, consisting of the lateral reticular formation and the nucleus retroambiguus, innervates the phonatory motoneuron pools and it may inhibit, if necessary, other pattern generators (i.e., for mastication or swallowing) which control, in part, the same muscles. Present data suggest – at least for monkey vocalizations – that the facial muscles are activated prior to the jaw muscles, which themselves are activated prior to the laryngeal muscles. Thus, the articulatory component of vocalization is prepared first, before the sound production is initiated.

Acknowledgments

This work was supported by the Deutsche Forschungsgemeinschaft (Ha 5400/1-1). I thank Walter Metzner and Uwe Jürgens for their support, fruitful discussions and helpful comments on the manuscript.

References

Adametz, J., O'Leary, J.L., 1959. Experimental mutism resulting from periaqueductal lesions in cats. Neurology 9, 636–642.

Aitken, P.G., 1981. Cortical control of conditioned and spontaneous vocal behaviour in rhesus monkeys. Brain Lang. 13, 171–184.

Altschuler, S.M., Bao, X.M., Bieger, D., Hopkins, D.A., Miselis, R.R., 1989. Viscerotopic representation of the upper alimentary tract in the rat: sensory ganglia and nuclei of the solitary and spinal trigeminal tracts. J. Comp. Neurol. 283, 248–268.

Bandler, R., Carrive, P., 1988. Integrated defence reaction elicited by excitatory amino acid microinjection in the midbrain periaqueductal grey region of the unrestrained cat. Brain Res. 439, 95–106.

Barlow, S.M., Estep, M., 2006. Central pattern generation and the motor infrastructure for suck, respiration, and speech. J. Comm. Disord. 39, 366–380.

Beitz, A.J., 1982. The organization of afferent projections to the midbrain periaqueductal gray of the rat. Neuroscience 7, 133–159.

Brumm, H., Voss, K., Köllmer, I., Todt, D., 2004. Acoustic communication in noise: regulation of call characteristics in a New World monkey. J. Exp. Biol. 207, 443–448.

Cameron, A.A., Khan, I.A., Westlund, K.N., Willis, W.D., 1995. The efferent projections of the periaqueductal gray in the rat: a Phaseolus vulgaris-leucoagglutinin study. II. Descending projections. J. Comp. Neurol. 351, 585–601.

Cant, N.B., Benson, C.G., 2003. Parallel auditory pathways: projection patterns of the different neuronal populations in the dorsal and ventral cochlear nuclei. Brain Res. Bull. 60, 457–474.

Dauvergne, C., Pinganaud, G., Buisseret, P., Buisseret-Delmas, C., Zerari-Mailly, F., 2001. Reticular premotor neurons projecting to both facial and hypoglossal nuclei receive trigeminal afferents in rats. Neurosci. Lett. 311, 109–112.

Davis, P.J., Nail, B.S., 1984. On the location and size of laryngeal motoneurons in the cat and rabbit. J. Comp. Neurol. 230, 13–32.

Düsterhöft, F., Häusler, U., Jürgens, U., 2004. Neuronal activity in the periaqueductal gray and bordering structures during vocal communication in the squirrel monkey. Neuroscience 123, 53–60.

Dujardin, E., Jürgens, U., 2005. Afferents of vocalization-controlling periaqueductal regions in the squirrel monkey. Brain Res. 1034, 114–131.

Dujardin, E., Jürgens, U., 2006. Call type-specific differences in vocalization-related afferents to the periaqueductal gray of squirrel monkeys (*Saimiri sciureus*). Behav. Brain Res. 168, 23–36.

Duron, B., Marlot, D., Macron, J.M., 1979. Segmental motor innervation of the cat diaphragm. Neurosci. Lett. 15, 93–96.

Edwards, S.B., Ginsburgh, C.L., Henkel, C.K., Stein, B.E., 1979. Sources of subcortical projections to the superior colliculus in the cat. J. Comp. Neurol. 184, 309–329.

Egnor, S.E.R., Hauser, M.D., 2004. A paradox in the evolution of primate vocal learning. Trends Neurosci. 27, 649–654.

Esposito, A., Demeurisse, G., Alberti, B., Fabbro, F., 1999. Complete mutism after midbrain periaqueductal gray lesion. Neuroreport 10, 681–685.

Esser, K.H., Schubert, J., 1998. Vocal dialects in the lesser spear-nosed bat *Phyllostomus discolor*. Naturwissenschaften 85, 347–349.

Estenne, M., Zocchi, L., Warl, M., Macklem, P.T., 1990. Chest wall motion and expiratory muscle use during phonation in normal humans. J. Appl. Physiol. 68, 2075–2082.

Farley, G.R., Barlow, S.M., Netsell, R., 1992. Factors influencing neural activity in parabrachial regions during cat vocalizations. Exp. Brain Res. 89, 341–351.
Feldman, J.L., Gautier, H., 1976. Interactions of pulmonary afferents and pneumotaxic center in control of respiratory pattern in cats. J. Neurophysiol. 39, 31–44.
Feldman, J.L., Mitchell, G.S., Nattie, E.E., 2003. Breathing: rhythmicity, plasticity, chemosensitivity. Annu. Rev. Neurosci. 26, 239–266.
Fischer, J., Hammerschmidt, K., Todt, D., 1998. Local variation in Barbary macaque shrill barks. Anim. Behav. 56, 623–629.
Gerrits, P.O., Holstege, G., 1996. Pontine and medullary projections to the nucleus retroambiguus: a wheat germ agglutinin horseradish peroxidase and autoradiographic tracing study in the cat. J. Comp. Neurol. 373, 173–185.
Hage, S.R., 2008. Vokale Mustergeneration und audiovokale Integration. Telemetrische Einzelzellableitungen im unteren Hirnstamm am sich frei bewegenden Totenkopfaffen. VDM Verlag, Saarbrücken, Germany.
Hage, S.R., Jürgens, U., 2006a. On the role of the pontine brainstem in vocal pattern generation. A telemetric single-unit recording study in the squirrel monkey. J. Neurosci. 26, 7105–7115.
Hage, S.R., Jürgens, U., 2006b. Localization of a vocal pattern generator in the pontine brainstem of the squirrel monkey. Eur. J. Neurosci. 23, 840–844.
Hage, S.R., Jürgens, U., Ehret, G., 2006. Audio–vocal interaction in the pontine brainstem during self-initiated vocalization in the squirrel monkey. Eur. J. Neurosci. 23, 3297–3307.
Hammerschmidt, K., Freudenstein, T., Jürgens, U., 2001. Vocal development in squirrel monkeys. Behaviour 138, 1179–1204.
Hannig, S., Jürgens, U., 2006. Projections of the ventrolateral pontine vocalization area in the squirrel monkey. Exp. Brain Res. 169, 92–105.
Hayakawa, T., Takanaga, A., Maeda, S., Seki, M., Yajima, Y., 2001. Subnuclear distribution of afferents from the oral, pharyngeal and laryngeal regions in the nucleus tractus solitarii of the rat: a study using transganglionic transport of cholera toxin. Neurosci. Res. 39, 221–232.
Holstege, G., vanNeerven, J., Evertse, F., 1987. Spinal cord location of the motoneurons innervating the abdominal, cuteanous maximus, latissimus dorsi and longissimus dorsi muscles in the cat. Exp. Brain Res. 67, 179–194.
Horta-Junior, J.A., Tamega, O.J., Cruz-Rizzolo, R.J., 2004. Cytoarchitecture and musculotopic organization of the facial motor nucleus in *Cebus apella* monkey. J. Anat. 204, 175–190.
Hsu, M.M., Shyu, B.C., 1997. Electrophysiological study of the connection between medial thalamus and anterior cingulated cortex in the rat. Neuroreport 8, 2701–2707.
Jürgens, U., 2000. Localization of a pontine vocalization-controlling area. J. Acoust. Soc. Am. 108, 1393–1396.
Jürgens, U., 2002. Neural pathways underlying vocal control. Neurosci. Biobehav. Rev. 26, 235–258.
Jürgens, U., Ploog, D., 1970. Cerebral representation of vocalization in the squirrel monkey. Exp. Brain Res. 10, 532–554.
Jürgens, U., Müller-Preuss, P., 1977. Convergent projections of different limbic vocalization areas in the squirrel monkey. Brain Res. 29, 75–83.
Jürgens, U., Pratt, R., 1979. Role of the periaqueductal grey in vocal expression of emotion. Brain Res. 167, 367–378.
Jürgens, U., Kirzinger, A., 1985. The laryngeal sensory pathway and its role in phonation. A brain lesioning study in the squirrel monkey. Exp. Brain Res. 59, 118–124.
Jürgens, U., Schriever, S., 1991. Respiratory muscle activity during vocalization in the squirrel monkey. Folia Primatol. 56, 121–132.
Jürgens, U., Hast, M., Pratt, R., 1978. Effects of laryngeal nerve transection on squirrel monkey calls. J. Comp. Physiol. 123, 23–29.
Kalia, M., Mesulam, M.M., 1980. Brainstem projections of sensory and motor components of the vagus complex in the cat. J. Comp Neurol. 193, 435–508.
Kalia, M., Richter, D., 1985. Morphology of physiologically identified slowly adapting lung stretch receptor afferents stained with intra-axonal horseradish peroxidase in the nucleus of the tractus solitarius of the cat. I. A light microscopic analysis. J. Comp. Neurol. 241, 503–520.
Kirzinger, A., Jürgens, U., 1994. Motoneuronal location of external laryngeal and hyoid muscles involved in primate phonation. J. Hirnforsch. 35, 559–565.
Kyuhou, S., Gemba, H., 1998. Two vocalization-related subregions in the midbrain periaqueductal gray of the guinea pig. Neuroreport 9, 1607–1610.
Larson, C.R., Kistler, M.K., 1984. Periaqueductal gray neuronal activity associated with laryngeal EMG and vocalization in the awake monkey. Neurosci. Lett. 46, 261–266.
Leanderson, R., Persson, A., Ohman, S., 1971. Electromyographic studies of facial muscle activity in speech. Acta Otolaryngol. 72, 361–369.
Lombard, E., 1911. Le signe de l'élévation de la voix. Ann. Mal. Oreil. Larynx 37, 101–119.
Lu, C.L., Jürgens, U., 1993. Effects of chemical stimulation in the periaqueductal gray on vocalization in the squirrel monkey. Brain Res. Bull. 32, 143–151.
Lüthe, L., Häusler, U., Jürgens, U., 2000. Neuronal activity in the medulla oblongata during vocalization. A single-unit recording study in the squirrel monkey. Behav. Brain Res. 116, 197–210.
Mantyh, P.W., 1982. Forebrain projections to the periaqueducal gray in the monkey, with observations in the cat and rat. J. Comp. Neurol. 206, 146–158.
Mantyh, P.W., 1983. Connections of midbrain periaqueductal gray in the monkey. I. Ascending efferent projections. J. Neurophysiol. 49, 567–581.
Marchand, J.E., Hagino, N., 1983. Afferents to the periaqueductal gray in the rat. A horseradish peroxidase study. Neuroscience 9, 95–106.
Matsuda, K., Uemura, M., Takeuchi, Y., Kume, M., Matsushima, R., Mizuno, N., 1979. Localization of motoneurons innervating the posterior belly of the digastric muscle: a comparative anatomical study by the HRP method. Neurosci. Lett. 12, 47–52.
McClung, J.R., Goldberg, S.J., 2000. Functional anatomy of the hypoglossal innervated muscles of the rat tongue: a model for elongation and protrusion of the mammalian tongue. Anat. Rec. 260, 378–386.

Meller, S.T., Dennis, B.J., 1986. Afferent projections to the periaqueductal gray in the rabbit. Neuroscience 19, 927–964.

Meller, S.T., Dennis, B.J., 1991. Efferent projections of the periaqueductal gray in the rabbit. Neuroscience 40, 191–216.

Metzner, W., 1993. An audio–vocal interface in echolocating horseshoe bats. J. Neurosci. 13, 1899–1915.

Metzner, W., 1996. Anatomical basis for audio–vocal integration in echolocating horseshoe bats. J. Comp. Neurol. 368, 252–269.

Mizuno, N., Matsuda, K., Iwahori, N., Uemura-Sumi, M., Kume, M., Matsushima, R., 1981. Representation of the masticatory muscles in the motor trigeminal nucleus of the macaque monkey. Neurosci. Lett. 21, 19–22.

Moga, H.H., Caper, C.B., 1990. Connections of the parabrachial nucleus with the nucleus of the solitary tract and the medullary reticular formation in the rat. J. Comp. Neurol. 293, 540–580.

Montaron, M.F., Deniau, J.M., Menetrey, A., Glowinski, J., Thierry, A.M., 1996. Prefrontal cortex inputs of the nucleus accumbens–nigro–thalamic circuit. Neuroscience 71, 371–382.

Müller-Preuss, P., Jürgens, U., 1976. Projections from the "cingular" vocalization area in the squirrel monkey. Brain Res. 103, 29–43.

Müller-Preuss, P., Newman, J.D., Jürgens, U., 1980. Anatomical and physiological evidence for a relationship between the "cingular" vocalization area and the auditory cortex of the squirrel monkey. Brain Res. 202, 307–315.

Musil, S.Y., Olson, C.R., 1988. Organization of cortical and subcortical projections to anterior cingulated cortex in the cat. J. Comp. Neurol. 272, 203–218.

Nakazawa, K., Shiba, K., Satoh, I., Yoshida, K., Nakajima, Y., Konno, A., 1997. Role of the pulmonary afferent inputs in vocal on-switch in the cat. Neurosci. Res. 29, 49–54.

Nonaka, S., Takahashi, R., Enomoto, K., Katada, A., Unno, T., 1997. Lombard reflex during PAG-induced vocalization in decerebrate cats. Neurosci. Res. 29, 283–289.

Pandya, D.N., Van Hoesen, G.W., Mesulam, M.M., 1981. Efferent connections of the cingulate gyrus in the rhesus monkey. Exp. Brain Res. 42, 319–330.

Pistorio, A.L., Vintch, B., Wang, X., 2006. Acoustic analysis of vocal development in a New World primate, the common marmoset (*Callithrix jacchus*). J. Acoust. Soc. Am. 120, 1655–1670.

Powell, E.W., Leman, R.B., 1976. Connections of the nucleus accumbens. Brain Res. 105, 389–403.

Rübsamen, R., Betz, M., 1986. Control of echolocation pulses by neurons of the nucleus ambiguus in the rufous horseshoe bat, *Rhinolophus rouxi*. I. Single unit recordings in the ventral motor nucleus of the laryngeal nerves in spontaneously vocalizing bats. J. Comp. Physiol. 159, 675–687.

Scheiner, E., Hammerschmidt, K., Jürgens, U., 2004. The influence of hearing impairment on the preverbal emotional vocalization of infants. Folia Phoniat. Logopaed. 56, 27–40.

Schriever, S., Jürgens, U., 1989. Location of respiratory motoneurons involved in phonation. An HRP study in the squirrel monkey. J. Hirnforsch. 30, 521–527.

Schuller, G., Radtke-Schuller, S., 1990. Neural control of vocalization in bats: mapping of brainstem areas with electrical microstimulation eliciting species-specific echolocation calls in the rufous horseshoe bat. Exp. Brain Res. 79, 192–206.

Schweizer, H., Rübsamen, R., Ruhle, C., 1981. Localization of brain stem motoneurones innervating the laryngeal muscles in the rufous horseshoe bat, *Rhinolophus rouxi*. Brain Res. 230, 41–50.

Sinha, S.R., Moss, C.F., 2007. Vocal premotor activity in the superior colliculus. J. Neurosci. 27, 98–110.

Shiba, K., Yoshida, K., Miura, T., 1995. Functional roles of the superior laryngeal nerve afferents in electrically induced vocalization in anesthetized cats. Neurosci. Res. 22, 23–30.

Shigenaga, Y., Yoshida, A., Mitsuhiro, Y., Tsuru, K., Doe, K., 1988. Morphological and functional properties of trigeminal nucleus oralis neurons projecting to the trigeminal motor nucleus of the cat. Brain Res. 461, 143–149.

Shipp, T., 1975. Vertical laryngeal position during continuous and discrete vocal frequency change. J. Speech Hear. Res. 18, 707–718.

Smith, J.C., Morrison, D.E., Ellenberger, H.H., Otto, M.R., Feldman, J.L., 1989. Brainstem projections to the major respiratory neuron populations in the medulla of the cat. J. Comp. Neurol. 281, 69–96.

Smotherman, M.S., 2007. Sensory feedback control of mammalian vocalizations. Behav. Brain Res. 182, 315–326.

Smotherman, M., Zhang, S., Metzner, W., 2003. A neural basis for auditory feedback control of vocal pitch. J. Neurosci. 23, 1464–1477.

Smotherman, M., Kobayasi, K., Ma, J., Zhang, S., Metzner, W., 2006. A mechanism for vocal–respiratory coupling in the mammalian parabrachial nucleus. J. Neurosci. 26, 4860–4869.

Su, H.S., Bentivoglio, M., 1990. Thalamic midline cell populations projecting to the nucleus accumbens, amygdala, and hippocampus in the rat. J. Comp. Neurol. 297, 582–593.

Suga, N., Yajima, Y., 1988. Auditory-vocal integration in the midbrain of the mustached bat: periaqueductal gray and reticular formation. In: Newman, J.D. (Ed.), The Physiological Control of Mammalian Vocalization. Plenum, New York, NY, pp. 87–107.

Suga, N., Schlegel, P., Shimozawa, T., Simmons, J., 1973. Orientation sounds evoked from echolocating bats by electrical stimulation of the brain. J. Acoust. Soc. Am. 54, 793–797.

Sutton, D., Larson, C., Farrell, D.M., 1972. Cricothyroid motor units. Acta Otolaryngol. 74, 145–151.

Sutton, D., Larson, C., Lindeman, R.C., 1974. Neocortical and limbic lesion effects on primate phonation. Brain Res. 71, 61–75.

Swanson, L.W., Cowan, W.M., 1979. The connections of the septal region in the rat. J. Comp. Neurol. 186, 621–656.

Thoms, G., Jürgens, U., 1981. Role of the internal laryngeal nerve in phonation: an experimental study in the squirrel monkey. Exp. Neurol. 74, 187–203.

Travers, S.P., Norgren, R., 1995. Organization of orosensory responses in the nucleus of the solitary tract of the rat. J. Neurophysiol. 73, 2144–2162.

Uemura-Sumi, M., Mizuno, N., Nomura, S., Iwahori, N., Takeuchi, Y., Matsushima, R., 1981. Topographical representation of the hypoglossal nerve branches and tongue muscles in the hypoglossal nucleus of macaque monkeys. Neurosci. Lett. 22, 31–35.

Ueyama, T., Satoda, T., Tashiro, T., Mizuno, N., 1988. The distribution of infrahyoid motoneurons in the cat: a retrograde horseradish peroxidase study. Anat. Embryol. 177, 445–450.

Valentine, D.E., Sinha, S.R., Moss, C.F., 2002. Orienting responses and vocalizations produced by microstimulation in the superior colliculus of the echolocating bat, *Eptesicus fuscus*. J. Comp. Physiol. 188, 89–108.

Van der Horst, V.G.J.M., Terasawa, E., Ralston, H.J., Holstege, G., 2000. Monosynaptic synaptic projections from the lateral periaqueductal gray to the nucleus retroambiguus in the rhesus monkey: implications for vocalization and reproductive behavior. J. Comp. Neurol. 424, 251–268.

Vogt, B.A., Miller, M.W., 1983. Cortical connections between rat cingulate cortex and visual, motor, and postsubicular cortices. J. Comp. Neurol. 216, 192–210.

Welt, C., Abbs, J.H., 1990. Musculotopic organization of the facial motor nucleus in *Macaca fascicularis*: a morphometric and retrograde tracing study with cholera toxin B-HRP. J. Comp. Neurol. 291, 621–636.

West, R., Larson, C.R., 1993. Laryngeal and respiratory activity during vocalization in macaque monkeys. J. Voice 7, 54–68.

Yajima, Y., Larson, C.R., 1993. Multifunctional properties of ambiguous neurons identified electrophysiologically during vocalization in the awake monkey. J. Neurophysiol. 70, 529–540.

Yajima, Y., Hayashi, Y., Yoshii, N., 1980. The midbrain central gray substance as a highly sensitive neural structure for the production of ultrasonic vocalization in the rat. Brain Res. 198, 446–452.

Yoshida, Y., Tanaka, Y., Saito, T., Shimazaki, T., Hirano, M., 1992. Peripheral nervous system in the larynx. An anatomical study on the motor, sensory and autonomic nerve fibers. Folia Phoniatr. 44, 194–219.

Yoshida, A., Yamamoto, M., Moritani, M., Fukami, H., Bae, Y.C., Chang, Z., Sugiyo, S., Takemura, M., Park, K.P., Shigenaga, Y., 2005. Bilateral projection of functionally characterized trigeminal oralis neurons to trigeminal motoneurons in cats. Brain Res. 1036, 208–212.

CHAPTER 8.4

Central pattern generators for orofacial movements and speech

Steven M. Barlow[1–4*], James P. Lund[5], Meredith Estep[2] and Arlette Kolta[6]

[1]*Department of Speech-Language-Hearing: Sciences and Disorders*
[2]*Neuroscience*
[3]*Human Biology*
[4]*Bioengineering Programs, University of Kansas, Lawrence, Kansas, USA*
[5]*Faculty of Dentistry, McGill University, Montreal, Quebec, Canada*
[6]*Faculté de médecine dentaire, Université de Montreal, Montreal, Quebec, Canada*

Abstract: Activity of splanchnocranial and orofacial muscles regulates a number of behaviors and plays an important articulatory role in vocalization and speech. This chapter reviews experimental findings on oromotor central pattern generators (CPGs) and associated distributed neural networks in selected animals and humans. These generators control such primary behaviors as respiration, sucking, licking and mastication. A special emphasis is given to features that are relevant for the development and production of speech.

Keywords: microcircuits; unit burster; sensory experience; neuromodulation; body plan; interneurons; plasticity; brain; suck; lick; mastication; respiration; entrainment; neuroimaging

I. Introduction

The somatomotor system provides us with a repertoire of actions that range from movements essential to survival in all vertebrates (respiration, feeding reflexes, locomotion), to those which evolved only in mammals (suckling, mastication), and finally those unique to humans acquired by learning (speech, piano playing, etc.). Most studies of motor control come from animal models, which mean that we know little about the neuronal machinery of the complex learned motor patterns specific to humans. However, we presume that the systems controlling these complex behaviors are built on the more primitive circuits that control the common body parts. For speech, this means the circuits controlling respiration, suckling, mastication and vocalization. The first three are rhythmic motor patterns that are controlled by primitive neuronal circuits known as central pattern generators (CPGs). The existence of CPGs was first postulated by T. Graham Brown in the early 1900s (see Grillner, 1975), and the first proof for movements other than respiration was provided for orofacial movements (Dellow and Lund, 1971). Outputs of CPGs to motoneurons are generated by repetitive changes that take place at the molecular, cellular and network level (Pearson and Gordon, 2000; Grillner, 2003; Marder and Rehm, 2005).

II. Challenges in the study of central pattern generators

The challenges facing scientists interested in pattern generating networks are substantial (Rossignol et al., 1988; Grillner et al., 2005a,b). The first is the issue of experimental accessibility of a neuronal circuit or set of circuits with known and measurable functions (Marder and Rehm, 2005). Early work on CPGs was limited to invertebrates, due to their relatively simple structure and small number of neural elements. Subsequently, great progress has been made in vertebrates from lampreys to a variety of mammals.

*Corresponding author. E-mail: smbarlow@ku.edu

Stefan M. Brudzynski (Ed.)
Handbook of Mammalian Vocalization
ISBN 978-0-123-74593-4

DOI: 10.1016/B978-0-12-374593-4.00033-4

When studying the cellular mechanisms underlying locomotion, mastication and suckling *in vitro*, there is the challenge of activating the quiescent CPG and of assessing the effects of sensory feedback which is vital to their normal operation.

III. Basic forms of pattern generating circuits

Three basic forms of neural circuits that produce patterned output have been identified (Westberg et al., 1998). *Dedicated circuits* generate only one pattern, and when triggered by sensory input may suppress other ongoing behaviors. *Distributed circuits* consist of a single population of neurons to pattern all variations of movement. For example, neurons in primate motor cortex that control the shoulder muscles seem to perform as a distributed circuit during arm reaching. Each fires maximally during movement in a particular direction. A population response vector reflecting the composite neuronal response determines the trajectory of the arm (Georgopoulos et al., 1988; Kalaska and Crammond, 1992). *Reorganizing circuits* generate different patterns when the effectiveness of synaptic connections between members of the total population of neurons changes. The key feature of this form, which includes the masticatory CPG (see below), is that some neurons do not take part in the full repertoire of behaviors generated by the whole population of neurons.

IV. Basic features of central pattern generators

There are a number of features that characterize primitive vertebrate CPGs. First, the essential circuits are located in the brainstem (mastication, suckling, licking, swallowing, respiration and coughing) or spinal cord (locomotion). They are primarily composed of networks of interneurons that activate groups of motor neurons to generate specific motor patterns. They change cycle duration, and the duration and intensity of motoneuron bursts, in response to central nervous system (CNS) and sensory inputs (Grillner, 1991, 2002; Grillner et al., 2005a,b). Grillner (2003) proposed an elaborate continuum of pattern generating networks, ranging from the most basic CPGs which coordinate protective reflexes (cough), through intermittent (e.g., mastication, swallowing) rhythmic movements to respiration. The respiratory CPG is active throughout life, but its output is continuously modulated by changing metabolic demands, and by the specific requirements of vocalization and speech. Although the neurons, synapses and circuit configurations of different vertebrate microcircuits can differ considerably, Grillner et al. (2005b) noted that it is remarkable that all microcircuits identified thus far in the brain display frequency-specific rhythmic outputs.

V. Neurochemical modulation of central pattern generator activity

Central pattern generators are subject to extensive neuromodulation. This generates flexibility in the rhythmic outputs of these neural networks and ultimately results in behavioral flexibility through an expanded motor repertoire. Neuromodulators exert two basic effects on CPGs, including alteration to intrinsic membrane properties and modulation of the strength of synaptic interactions among members of the pattern generating circuit. Combinations of these effects, with each modulator affecting a select subset of CPG neurons, account for many of the observed effects of neuromodulators. Furthermore, the combination of two or more neuromodulators may interact to cause more complex or different effects than they do when released independently (Dickinson, 2006).

Changes in the pattern of neurotransmitter release and in the relative concentrations of neuromodulators can have profound effects on the operational dynamics of a CPG. A clever set of experiments in the crab have demonstrated systematic fractionation of pyloric CPGs with changes in neuromodulator type and its concentration levels. Neuromodulatory substances can alter the cellular and synaptic properties of neurons in CPG pathways (Swensen and Marder, 2001) to regulate phase switching, reflex reversal and reflex gain.

The effects of neuromodulators may also interact with one another, and even modulatory neurons are often subject to modulation, enabling higher order control and indirect interactions among central pattern generators (Dickinson, 2006). In addition, modulators often directly mediate the interactions between functionally-related CPGs. This enabling feature of CPG control has special relevance for the diverse repertoire of ororhythmic outputs and resulting motor behavior.

The generator in the respiratory system, like other CPGs, is subjected to modulation by several different amines and peptides, including 5HT (Pena and

Ramirez, 2002; see also review by Dickinson, 2006), substance P (Pena and Ramirez, 2004; Del Negro et al., 2005), norepinephrine (Viemari and Ramirez, 2006) and acetylcholine (ACh) functioning via muscarinic acetylcholine receptors (mAChRs) (Shao and Feldman, 2005). Each modulator affects the pacemaker types and other neurons differentially, so that, for example, the relative contributions of cadmium (Cd)-sensitive pacemakers might increase in the presence of substance P, whereas other modulators (e.g., 5HT) enhance the contributions of the non-Cd-sensitive pacemakers (Pena and Ramirez, 2002; Tryba et al., 2006). The important role of neuromodulation in the respiratory pattern was recently shown in a series of experiments examining the role of $5HT_{2A}$ receptors. This is the same receptor type that is considered crucial in the spinal locomotor CPG function in mammals. Blocking $5HT_{2A}$ receptors significantly reduced eupneic activity, but did not eliminate it (Pena and Ramirez, 2002; Tryba et al., 2006). Thus, multiple pacemaker neuron types in the pre-Bötzinger complex (pre-BötC) of the respiratory system interact to produce rhythmic output (Dickinson, 2006). Neuromodulators provide a mechanism to alter the relative contributions of the different pacemakers, leading to substantial changes in motor output and to different forms of respiratory behavior.

VI. Ontogenesis of sucking and chewing

The development of an *in vitro* isolated brainstem preparation (Katakura et al., 1995) was essential for the study of the perinatal rhythmical motor patterns in cranial nerves. In the resting state, rhythmical respiratory activity can be recorded from trigeminal (V), facial (VII) and hypoglossal (XII) motoneurons of neonatal rats (Katakura et al., 1995; Koizumi et al., 2002), but a second much faster rhythm appears if the glutamate agonist N-methyl-D-aspartate (NMDA) is added to the bath, with or without other neurotransmitters or their antagonists (Katakura et al., 1995; Kogo et al., 1996; Kogo et al., 1998; Tanaka et al., 1999; Koizumi et al., 2002; Enomoto et al., 2002). Since all neonatal mammals begin to feed on milk, we presume that the rhythmic non-respiratory oromotor patterns are related to suckling. Indeed, mastication is not seen in young rats before postnatal day (P) 12 (Westneat and Hall, 1992).

The isolated sections of the brainstem of neonates containing the V, VII and XII motor nuclei are each capable of generating the sucking rhythm when isolated from the other motor nuclei (Nakamura et al., 2004), as are the left and right halves of the trigeminal section (Tanaka et al., 1999). These data suggest that there could be at least six separate rhythm generators (two V, two VII and two XII) that are coupled together to coordinate suckling at birth. How these work together, and whether they all persist into adulthood is not known. However, it is possible that the caudal pair is dominant by birth, because the V pair is tonically inhibited by the caudal brainstem (Tanaka et al., 1999). They are also modulated by descending inputs from the "sucking area" of the motor cortex (Nozaki et al., 1986; Iriki et al., 1988).

Cycle duration and other features of the rhythmical trigeminal activity (RTA) recorded at embryonic ages (E) 20–21 in the guinea pig are similar to those of neonatal rats, but RTA is not observed before E18 (Ishihama et al., 2003). This suggests that the basic networks required for RTA generation become functional between E18 and E20. As in older animals, these oromotor rhythmic activities depend on the activation of NMDA receptors.

The transition from suckling to chewing occurs gradually over a period that can vary enormously between species. In rats, the first masticatory movements appear around P12, and the adult pattern is established between P18 and P21 (Westneat and Hall, 1992). Human infants begin to chew after several months, and mastication continues to evolve until the secondary dentition is in place. Mastication and suckling both involve jaw opening and closing, but the power stroke for the suction phase is provided by the jaw depressors during suckling and by the elevators during mastication. Nutritive suckling also includes an expression phase that typically lags the suction phase by 100 ms or more. This involves a stripping "peristaltic" motion of the tongue tip along the length of the nipple or teat. This action requires activation of the intrinsic tongue muscles via the hypoglossal nerve (Lau, 2006). Clearly, many motoneurons are active during all of the oromotor patterns, but we do not yet know if the sucking CPG evolves to also control mastication, or if a mastication CPG emerges during the weaning period. However, there is evidence that some neurons that may form the core of the masticatory CPG in rats (see below) undergo a rapid maturation during weaning. These neurons, which are found in the anterodorsal region of the trigeminal principal sensory nucleus (NVsnpr) of rats, have pacemaker

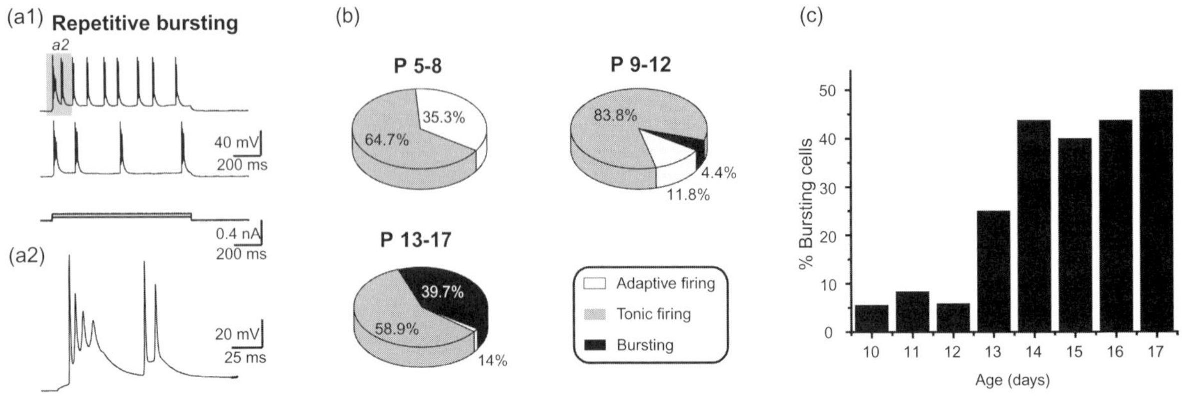

Fig. 1. (a1) Repetitive bursting of a neuron recorded in dorsal NVsnpr was induced by two depolarizing current pulses (superimposed below). The shaded area is enlarged in (a2). (b) Percentages of NVsnpr neurons that showed adaptive, tonic and repetitive bursting during depolarization at different postnatal ages. (c) Percentage of bursting cells at different ages. Modified from Brocard et al. (2006).

properties in young adults (Brocard et al., 2006). Over the period of transition from suckling to adult mastication, the electrical properties of NVsnpr neurons recorded from brainstem slices *in vitro* change dramatically. When current is injected into very young neurons (younger than P8) they can only fire briefly or in short trains. By P12, a few fire in non-repetitive bursts, but the proportion of "bursters" rises quickly to about 45% at P14. Furthermore, many of these neurons burst repetitively (Fig. 1).

The age-related changes in membrane properties that underlie bursting in dorsal NVsnpr neurons (Brocard et al., 2006) are similar to those that occur in the more ventral whisker region of NVsnpr (Lo and Erzurumlu, 2001), the spinal trigeminal nucleus (Guido et al., 1998) and in several other brainstem regions of the rat (Nunez-Abades et al., 1993; Bao et al., 1995; Tsuzuki et al., 1995; Berger et al., 1996; Tanaka et al., 2003; Nguyen et al., 2004). This suggests that the maturation process of neurons in many brainstem nuclei is similar.

VII. Role of spontaneous movements in ontogenesis

Some spontaneous rhythmic movements are present in the embryo before they are needed for behavior. These movements are considered important for circuit formation, early pathfinding decisions and terminal phases of motor axon guidance, pathfinding cues, synaptic tuning and refinement within a nuclear target, and expression of neuronal guidance molecules in locomotor CPGs (Hanson and Landmesser, 2004).

Sucking is a precocious motor skill that begins *in utero* between 15 and 18 weeks gestational age (GA) (Miller et al., 2003). It can be differentiated into non-nutritive and nutritive modes. Infants' readiness to feed is often evaluated by their display of non-nutritive sucking (Lau, 2006). The non-nutritive suck (NNS) is defined as any repetitive mouthing activity on a blind nipple or pacifier that does not deliver a liquid stimulus (Wolff, 1968; Goldson, 1987). Patterned ororhythmic behavior appears between 28 and 33 weeks GA and is remarkably stable by 34 weeks (Hack et al., 1985). As shown in Fig. 2, the NNS pattern observed in the preterm infant is organized into alternating epochs of burst and pause periods. A typical NNS burst consists of 6 to 12 suck cycles that occur at approximately 2 Hz followed by pause "rest" periods to accommodate respiration (Wolf, 1968; Finan and Barlow, 1996; Estep et al., 2008a; Popescu et al., 2008; Stumm et al., 2008). The maturation and coordination of the NNS precedes the suck–swallow–breathe pattern associated with the nutritive suck (Lau and Schanler, 1996; Lau and Hurst, 1999; Gewolb et al., 2001; Medoff-Cooper, 2005). Nutritive sucking (NS) is characterized by a lower cycle rate (1 Hz) and fewer inter-burst pauses (Wolff, 1968). Furthermore, NS differs from NNS because the expression of milk requires coordination of sucking with swallowing and airway protection, and therefore with respiration. The state of maturation of sucking is also of considerable clinical value as a biomarker for the integrity of the developing central nervous system (Mizuno and Ueda, 2005; Barlow and Estep, 2006; Barlow et al., 2008). This has led

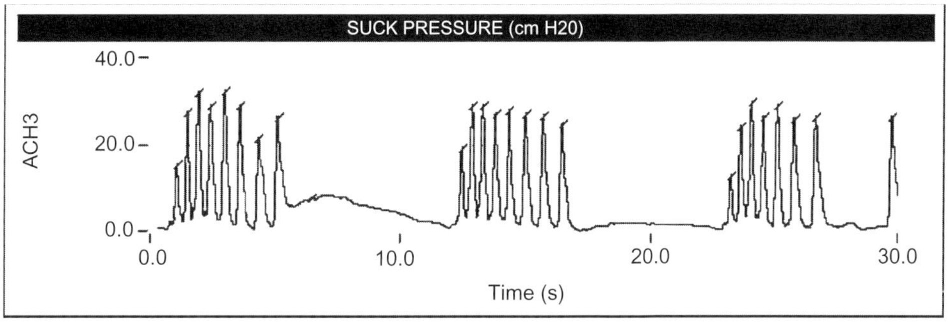

Fig. 2. Waveform display of the classic "burst-pause" non-nutritive suck CPG patterning of a healthy preterm neonate. Within-burst suck rates vary from 3 to 1.5 Hz with inter-burst pause periods of several seconds. Tick marks indicate peak-picking algorithm tags for NNS pressure peaks during burst production. These tick marks are used for automatic pressure amplitude (cm H_2O) and within-burst suck cycle (ms) period measurements. Modified from Stumm et al. (2008).

to the idea of early interventions to restore patterned neural activity if sensory deprivation and reduced motor activity is an issue (i.e., obstructive life saving interventions to support respiration of human preterm infants may actually compromise the development of oromotor control during what is considered to be a critical period of neural development).

VIII. Changes in body plan

Animals which undergo metamorphic changes (e.g., tadpoles) also lead to major reconfiguration of CPGs (Combes et al., 2004). We do not know if variations in body size and training modify CPGs in mammals, but this is a reasonable assumption. Masticatory rate seems to be fast in small species and slow in large ones. When they are eating hard foods, mice chew at about 5.3 Hz (Okayasu et al., 2003), rabbits at about 4.1 Hz (Naganuma et al., 2001) and humans at 1.5 Hz (Peyron et al., 2002). The human facial skeleton and vocal tract change considerably during development (Kent and Vorperian, 1995; Vorperian and Kent, 2007). The facial skeleton expands along inferior and anterior growth trajectories and in the process transforms the human infant from a sucking machine into an organism with an expanded oral cavity suitable for a dimensional vowel space, which makes possible the production of a variety of speech sounds. The acquisition of speech begins with the appearance of vocalization, babbling and early word production. By 12 years of age, the craniofacial skeleton has reached 90% of its adult dimension: however, the peripheral components of the masticatory and vocal systems continue to undergo structural changes until the end of puberty.

IX. Genetic-molecular tools

The use of genetic tools is providing new insights and possibilities for studying the assembly and modification of complex neuronal circuits such as CPGs in mammals. Several laboratories are attempting to develop molecular markers of CPG cell types using genetic transcription factors. For example, in one study a genetic knockout was developed to eliminate a class of ventral interneurons in the mouse locomotor CPG. This manipulation altered reciprocity in limb muscle activation between the left and right sides of the body (Lanuza et al., 2004). Genetic methods have also been used to add neurons to a circuit in order to study the process of CPG evolution in zebrafish (Liu et al., 2003).

The gross licking motor pattern is controlled by a brainstem CPG distributed within several subdivisions of the medullary reticular formation (RF) (Travers et al., 1997). Descending inputs to the lick CPG project to widely-distributed targets of both the medial and lateral reticular formation. Most projections originating from brainstem orosensory nuclei terminate primarily within the lateral reticular formation. Preoromotor neurons appear concentrated in the intermediate zone of the reticular formation, and receive convergent inputs from the lateral and medial RF sites to control this oromotor function (Travers, 2009). A recent study used inbred, F1 and BXD recombinant mice to determine the genetic basis of certain features of the lick CPG (Boughter et al., 2007). Study results suggested a fundamental difference in the rhythmic rate of licking among inbred mice, and the existence of several quantitative trait genes that underlie this phenotypic variation. Analysis of quantitative trait loci (QTL) of licking in the BXD R1 strain indicates significant linkages on chromosomes 1 and 10. Genetic manipulations have also proven useful in examining the role of the cerebellum (Crus I/II and lobus simplex) in regulating the timing or rhythm of the licking CPG in mice (Bryant et al., 2007). Nearly 63% of all Purkinje cells recorded modulated their firing pattern during licking behavior, and 45% of these cells were rhythmically phase-locked to the licking movement. Two strains of ataxic mice, including Lurcher mutants and Cbln1 knockout mice were compared to normal wild-type controls. Somewhat surprisingly, temporal analysis of the licking rhythm showed a significantly increased temporal precision (i.e., invariance) in ataxic mice compared to normal controls that manifested considerable variability between consecutive licks. The authors concluded that the cerebellum coordinates the proper timing of licking with swallowing and respiratory movements.

Continued work on the classic preparations used to study mechanisms of central pattern generation will inform new studies on genetic and molecular development of CPGs, and lead to a better understanding of these circuits in mammals with large numbers of neurons (Marder and Rehm, 2005).

X. The generation of oromotor rhythms

In order to study the fundamental properties of CPGs, it is necessary to work *in vitro* or in paralyzed anesthetized animals in which the motor output is recorded from motoneurons ("fictive movements"). During fictive mastication evoked by cortical stimulation, the CPG evokes long-lasting rhythmical depolarizing potentials in motoneurons of jaw opening muscles (digastrics, mylohyoid) which then fire bursts of spikes at very high frequency ($\leq$250 Hz). Similar plateau potentials depolarize jaw closing motoneurons (masseter, temporalis and medial pterygoid muscles) during the opposite phase, but these usually cause only weak firing. Jaw closing motoneurons are strongly hyperpolarized during the opening phase by glycine (Enomoto et al., 1987), but the membrane potential of opening motoneurons returns to resting levels during the jaw closing phase, because they are not hyperpolarized (Kubo et al., 1981). These important differences between motoneuron pools probably reflect the fact that feedback from their numerous muscle spindles is strong in jaw closing motoneurons, but not in motoneurons supplying jaw opening muscles which contain few, if any, spindles.

The essential parts of the masticatory CPG are found between the rostral poles of motor nuclei V and VII, and each hemisection side can generate a rhythm, although they are normally synchronized by commissural axons (Kolta et al., 2000). Many neurons among several nuclei in this region project not only to V motoneurons, but also to the VII and XII nuclei, and they are strongly and reciprocally interconnected (Bourque and Kolta, 2001; Dal Bo et al., 2005; McDavid et al., 2006). Furthermore, many of them fire bursts during fictive mastication in phase with either the jaw closing or jaw opening motoneurons (Westberg et al., 1998; Tsuboi et al., 2003). Additionally, most of these "masticatory" neurons are activated at very short latency from the masticatory area of sensorimotor cortex and by sensory inputs.

The pattern of human mastication varies greatly between foods and changes systematically during a chewing sequence. Although we presume that much of the variability is caused by changes in sensory feedback (see below), the basic patterns of mastication are also represented centrally. Unlike locomotion, rhythmic orofacial movements are represented in the cerebral cortex. Coordinated movements of the jaw, tongue, lips and cheeks can be evoked in anesthetized animals by repetitive electrical stimulation of a large region of the sensorimotor cortex, and other forebrain and midbrain structures. Some of these movements strongly resemble natural mastication; indeed, the different patterns used by rabbits such as incisal gnawing, left and right molar chewing are represented at specific cortical sites (Lund et al., 1984). Furthermore, the different patterns are represented topographically. In the rabbit (Fig. 3), small rapid vertical movements powered by the digastrics (sucking) are located rostrally, vertical mastication is located behind these, and patterns with strong lateral swings are located in the posterolateral zone of the "masticatory" cortex (Lund et al., 1984). Note that stimulation of the right cortex causes a swing to the left during jaw closure, while a mirror image is generated from the right cortex. During the movement back to the

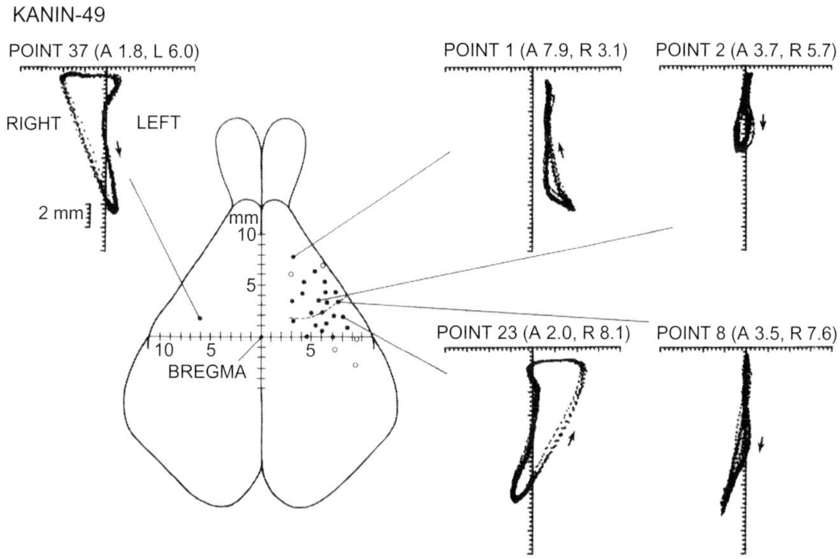

Fig. 3. Drawing of the dorsal surface of the cerebral cortex of the rabbit showing sites from which rhythmical jaw movements could be evoked by repetitive electrical stimulation. The inserts show recordings of the movement of the lower incisor teeth viewed from the front. Several cycles are superimposed. Modified from Lund et al. (1984).

midline the molar teeth are in contact, and the teeth can be heard grinding together.

By comparing activity generated by stimulating several of these sites in paralyzed animals, we now know that the masticatory CPG contains subpopulations of neurons linked to specific motor patterns. Few brainstem masticatory neurons are active during more than two motor patterns. For instance, only half of the neurons in the lateral reticular formation and adjacent V spinal nucleus that fire rhythmically during stimulation of a right posterior cortical site fire during stimulation of a site on the left cortex (Westberg et al., 1998). Those that participate in two or more patterns usually change their burst pattern. This shows that at least two mechanisms contribute to the genesis of different patterns of mastication: addition and subtraction of neurons from the active pool; and changes in firing pattern of common elements

XI. Bursting central pattern generator microcircuits and a termination mechanism

The general organization of brainstem and spinal pattern generating microcircuits has been studied using pharmacological, neurophysiological and computational techniques. Grillner and colleagues have proposed that the core of these networks is formed by pools of excitatory interneurons that generate burst activity and reciprocal activation without the need for inhibitory glycinergic or GABA-ergic neurons (Grillner et al., 2005b). This suggests the presence of a core of multiple unit burst generators described by Grillner and colleagues as unit central pattern generators, or unit CPGs, that can operate without inhibitory mechanisms. Furthermore, individual neurons of the dorsal NVsnpr are unit CPGs *par excellence*, because their pacemaker properties allow them to generate bursts in the absence of synaptic inputs (Brocard et al., 2006). Similar neurons have recently been described in the locomotor CPG (Tazerart et al., 2007, 2008). Inclusion of inhibitory mechanisms are important for coordinating different unit CPGs, such as those involved with coordinating flexor and extensor muscle groups for respiratory control, and for oromotor behaviors such as suckling and mastication (see Section X, The generation of oromotor rhythms).

Spike train modulation during unit burster activity and termination is a key feature of certain locomotor patterns and the human non-nutritive suck CPG. Grillner posited a mechanism to account for locomotor burst generation and termination in the lamprey. In this model, descending inputs drive both excitatory and inhibitory interneurons and output motoneurons. This command drive signal activates the CPG and regulates the burst rate, which is generally in the range of 0.2 to

10.0 Hz. Connectivity patterns among pools of excitatory interneurons result in further excitation within the neuronal pool. Both descending inputs and excitatory interneurons release glutamate to activate NMDA and AMPA receptors, to further enhance excitation. On the question of how does burst activity terminate in a predictable manner, Grillner stated that the principal mechanism is spike-train adaptation (Grillner et al., 2005b). This adaptation results from activation of Ca^{2+}-dependent K^+ channels (K_{Ca}), which generate the post-spike after-hypolarization potential (sAHP), the main determinant of firing frequency. Summation of sAHP events results in spike-train frequency adaptation. In addition, Ca^{2+} entry through NMDA channels leads to a general activation of K_{Ca} channels, leading to burst termination (El Manira et al., 1994). There are good indications that these currents may also play a role in limiting burst duration in NVsnpr neurons (Brocard et al., 2006).

XII. Sensory modulation of central pattern generators

The essential characteristic of CPGs is that they can produce a rhythmic motor pattern in the absence of any sensory feedback. However, the outputs of all known CPGs are modulated by sensory inputs, leading to dynamic reassembly of the neuronal networks to produce adapted forms of activity (Grillner, 1991; Rossignol et al., 1998). Sensory signals are important for error correction and to counteract unexpected perturbations in the environment. As with any control system, loop conduction times determine whether such sensory information can be used to correct ongoing movements, develop predictions for future movement events, or to establish the initial state of the musculoskeletal system (i.e., determining the position of the feet for initiation of locomotion, or determining the initial position of speech articulators in planning a sequence of vocal formants).

Signals from sensory receptors are important for normal motor patterning in intact behaving animals. They supply much of the general excitation that drives CPGs for generation and maintenance of rhythmic activity, but phasic sensory signals also initiate major phase transitions in intact motor systems (Rossignol et al., 1988). For example, elongation of pulmonary stretch receptors and diaphragmatic proprioceptors contribute to the termination of the inspiratory phase of respiration (Speck et al., 1993). Sensory signals serve to regulate the magnitude of ongoing motor activity and dynamically adjust the sensitivity of somatic reflexes, thereby providing an adaptive and flexible neural substrate with changes in task dynamics and environmental conditions.

Most neurons of the masticatory CPG that have been tested receive excitatory inputs from mechanoreceptors in the oral mucosa, periodontal ligament, or jaw muscle spindle afferents (Westberg et al., 1988; Tsuboi et al., 2003), and these have profound effects on the motor program. During experimentally-evoked mastication, the duration and amplitude of the electromyography (EMG) bursts of the jaw closing muscles is strongly dependent on inputs from muscle spindle afferents and from periodontal pressoreceptors (Lavigne et al., 1987; Morimoto et al., 2001). As the teeth engage the food, pressure builds within the periodontal ligament and the velocity of closure drops rapidly, leading to activation of periodontal receptors (Olsson et al., 1988) and muscle spindle afferents (Goodwin and Luschei, 1975). Both provide very short latency feedback excitation of jaw closing motoneurons (about 10 ms in humans). Their output is proportional to the resistance of the food being broken or crushed, as is the size of the EMG bursts. In addition, both inputs cause the EMG activity of the next cycle to increase *before* contact with the food (feedforward activation: Morimoto et al., 2001).

XIII. Entrainment

There are a variety of sensorimotor processes that either drive or modify CPG output. Entrainment is one such phenomenon, defined as the synchronization of an endogenous oscillator to external periodic events (Pavlidis, 1973; Glass and Mackey, 1988; Kriellaars et al., 1994). A stable phase relationship between the stimulus and oscillator must exist to satisfy the conditions for entrainment. The ability of an oscillator to synchronize to an external periodic signal provides adaptive and predictive control that allows fast and reliable responses to external changes (Pavlidis, 1973). Entrainment techniques have been effective in a variety of animal preparations to regulate and modify the rhythm of CPGs that control cyclic motor behaviors such as swimming (Grillner, 2003), stepping and locomotion (Conway et al., 1987; Pearson et al., 1992; Pearson, 2000), and mastication (Rossignol et al., 1988; see also review in Lund, 1991). Stimulation of intraoral tissues is also effective in modulating the masticatory cycle in humans (Hannam and Lund, 1981).

The delivery of synthetically-patterned orocutaneous inputs to healthy term infants up to six months of age can produce harmonic oromotor entrainment (1:1) to drive NNS (Finan and Barlow, 1998). A similar orocutaneous patterned input has been used therapeutically to entrain suckling in premature infants who have endured prolonged periods of orosensory deprivation secondary to respiratory distress syndrome (Barlow et al., 2008; Poore et al., 2008). The synchronous pattern of sensory flow is encoded by trigeminal primary afferents, modulate the firing patterns of trigeminal, facial and hypoglossal lower motor neurons, and relayed centrally along the trigeminal lemniscus to the ventroposteromedial nucleus of the thalamus (VP_m) and the orofacial regions of the sensorimotor cortex (Estep et al., 2007).

XIV. Respiratory central pattern generators and entrainment

In a seminal set of experiments by Jeffrey Smith and colleagues, whole-cell patch-clamp recording techniques were used to study the isolated respiratory pattern generator located in the pre-Bötzinger complex (pre-BötC) of the ventrolateral medulla in rat (Smith et al., 1991, 1992; Butera et al., 1999; Koshiya and Smith, 1999). As shown in Fig. 4, spatially-distributed populations of interneurons, premotoneurons (preMNs), and motoneurons (MNs) in the brainstem and spinal cord serve distinct functional roles in mammalian respiration (Rybak et al., 2007). Interacting populations of interneurons generate temporal features of the motor pattern including network rhythms; preMNs (defined as cells with axonal connections to MNs) function as pattern formation elements and substrates for rhythmic drive transmission, whereas MNs generate motor output. Inspiratory rhythm, generated by interneurons in the pre-BötC of the ventrolateral medulla (Smith et al., 1991; Feldman and Smith, 1995; Rekling and Feldman, 1998), propagates through preMN transmission circuits to spinal and cranial MNs. Topographical features of bulbospinal transmission circuits have been described (Feldman et al., 1988; Smith et al., 1991; Feldman and Smith, 1995; Ellenberger, 1999; Guyenet et al., 2002; Stornetta et al., 2003a,b), but there is less information on cranial motor transmission circuits (Dobbins and Feldman, 1995). Koizumi and colleagues used neural tracing and activity imaging with fluorescent Ca^{2+}-sensitive dyes in conjunction with high-resolution structural imaging (Koshiya and Smith, 1999; Koizumi

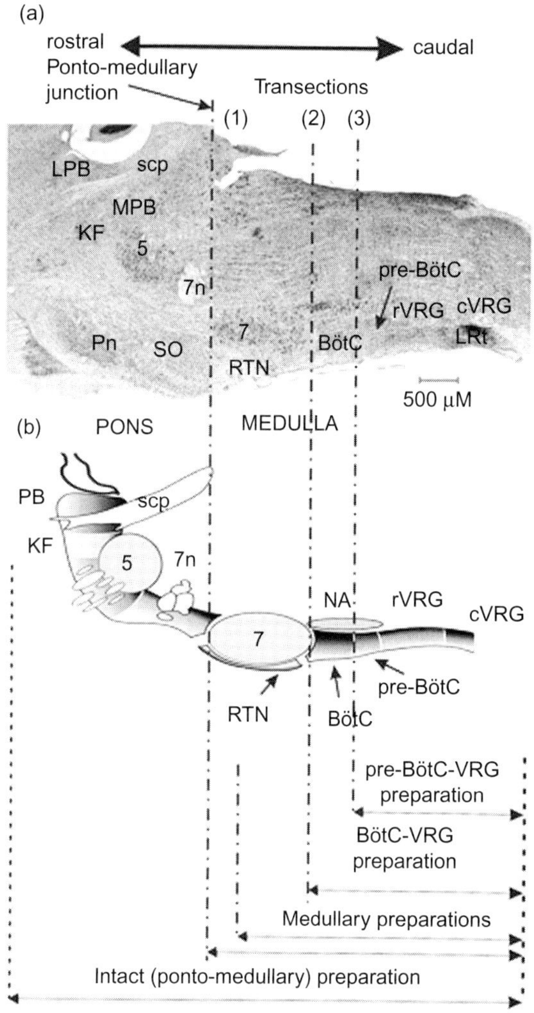

Fig. 4. Parasagittal view of rodent brainstem (section through the level of the compact part of nucleus ambiguus) and spatially arrayed compartments of respiratory CPG network. (a) Respiratory related ponto-medullary regions in the mature rat brainstem with several transection planes (dot-dashed lines) used in experimental studies. (b) Corresponding schematic diagram of respiratory-related brainstem compartments in parasagittal section of rat brain (created by George Alheid and used with permission) with transactions and resultant reduced preparations indicated at the bottom. Abbreviations: 5: trigeminal nucleus; 7: facial nucleus; 7n: facial nerve; BötC: Bötzinger complex; cVRG: caudal ventral respiratory group; KF: Kölliker–Fuse nucleus; LPB: lateral parabrachial nucleus; LRt: lateral reticular nucleus; MPB: medial parabrachial nucleus; NA: nucleus ambiguus; PB: parabrachial nuclei; Pn: pontine nuclei; pre-BötC: pre-Bötzinger complex; RTN: retrotrapezoid nucleus; rVRG: rostral ventral respiratory group; scp: superior cerebellar peduncle; SO: superior olive. Reprinted from Rybak, I.A., Abdala, A.P.L., Markin, S.N., Paton, J.F.R. and Smith, J.C. (2007) Spatial organization and state-dependent mechanisms for respiratory rhythm and pattern generation. Progress in Brain Research, 165: 201–220, with permission from Elsevier.

et al., 2008) to map inspiratory hypoglossal (XII) preMNs and MNs in the circuit transmitting rhythmic inspiratory drive from the pre-BötC to the musculature of the tongue. Rat neonatal medullary slice preparations, containing active rhythm-generating pre-BötC neurons and circuits transmitting rhythmic inspiratory drive to XII MNs, were used for analysis of inspiratory rhythm and pattern generation mechanisms (Smith et al., 1991; Koshiya and Smith, 1999; Del Negro et al., 2001). This combined imaging and electrophysiological approach yielded an architectonic organization whereby many XII preMNs were found to be spatially segregated from pre-BötC rhythm-generating neurons. This pattern of organization has also been found for bulbospinal circuits *in vivo*, and may facilitate downstream integration of convergent inputs for premotor pattern formation. Koizumi et al. (2008) suggested that preMNs and MNs share a similar set of synaptic and intrinsic electrophysiological properties, which they hypothesized to represent a common electrophysiological foundation for neurons functioning as rhythmic drive transmission elements.

Thus, two types of pacemaker neurons that are members of the inspiratory pattern generator have been identified in the pre-BötC, with bursting sustained by different currents. The first type includes a persistent sodium current (cadmium [Cd]-insensitive pacemakers) and the second type is a non-specific Ca^{2+}-dependent cation current (CAN current; Cd-sensitive pacemakers) (Ramirez and Viemari, 2005). Similar to invertebrate models, extensive synaptic interactions contribute to pattern formation. Both the pacemaker properties and the synaptic connections are targets of extensive modulation, with the relative contributions of different pacemaker types and of synaptic interactions being a dynamic function of modulation (Dickinson, 2006). This allows the same neural network to produce several types of breathing patterns: normal breathing (eupnea); sighs; and gasps. The mechanism(s) by which this circuitry is reconfigured for the distinct dynamics of speech breathing is unknown.

Mammalian vocalizations require precise coordination between intrinsic laryngeal muscle adjustments and respiration. Successful call production means integration of central vocal motor commands within the ongoing respiratory CPG, and incorporation of multimodal sensory feedback (stretch receptors, cutaneous, joint, baroreceptors, auditory inputs) onto both respiratory and laryngeal motor patterns. Results in horseshoe bats indicate that GABA-ergic synaptic activity in the ventrolateral parabrachial nucleus is necessary to coordinate vocal and respiratory CPGs during call production (Smotherman et al., 2005; Smotherman et al., Chapter 9.2 in this volume). A similar mechanism is thought to occur in humans during speech production.

Medullary respiratory pattern generators (mRGs) can be driven by periodic electrical stimulation to the ventral part of the spinal gray at the C5 level in newborn rats (Dubayle and Viala, 1996). These authors developed phase-response curves to determine the stimulus parameters that would produce a stable 1:1 harmonic entrainment. The authors concluded that ascending neural connections from the C5 segment of the spinal cord in rat might be involved in locomotor-respiratory coupling. Another line of study that has shown the potency of entraining inputs has been provided by a "breathing" teddy bear (BrBr), which provides optional rhythmic stimulation to irregularly-breathing neonates (Ingersoll and Thoman, 1994). Premature infants of 33 weeks GA were given a BrBr, while a nonbreathing bear served as a control. Respiratory patterning was sampled at 35 and 45 weeks GA in both groups. BrBr babies showed significantly more quiet sleep, less active sleep and increased respiratory regularity. This study demonstrated that the irregular respiratory oscillator of a premature infant will entrain to that of a breathing bear (a regular oscillator, or *zeitgeber*) if the BrBr's rate of breathing reflects that of the individual infant. This suggests that entrainment techniques may be used to facilitate neurobehavioral development of the infant's own biological rhythm.

XV. Relevance to speech production

Distinct cortical areas (e.g., dorsal premotor areas, anterior insula) that project to motor nuclei and peri-aqueductal gray of the brainstem reticular formation may be linked to central pattern generating circuits coordinating mouth and tongue movements associated with speech production (Dronkers, 1996; Wise et al., 1999; Hickok, 2001). Most of the efferent output associated with speech (Bitan et al., 2005; Olthoff et al., 2008) and respiration (McKay et al., 2003, 2008) in humans engages networks of cortical interneurons. In order to control individual speech gestures, cortical neurons may access motoneurons directly via the monosynaptic corticobulbar and corticospinal connections that only exist in humans (Jenny and Saper, 1987; Liscic et al., 1998; see also Eberl, Chapter 11.1 in this volume). However, the great majority of human corticobulbar synapses are made with other neurons and in non-human species activation of orofacial, laryngeal, and respiratory motoneurons occurs only through brainstem and spinal cord interneurons. Furthermore, many neurons that receive cortical inputs form part of orofacial CPGs (see above).

In agreement with what has been demonstrated in studies of modulating centrally-generated behaviors such as locomotion (Bem et al., 2003; Mentel et al., 2006, 2008) and mastication (Westberg et al., 1998), it has been hypothesized that different coordination patterns could be generated via common premotor (i.e., internuncial) networks responsible for speech (Estep and Barlow, 2007) and certain nonspeech behaviors, such as sucking (Barlow and Estep, 2006) and mastication (Lund and Kolta, 2006a,b). The alternating movements of synergistic muscles driving certain speech movements could benefit from dynamically assembled brainstem CPGs, and thereby reduce cortical load (Barlow and Estep, 2006; Lund and Kolta, 2006a).

XVI. Functional neuroimaging of orofacial networks involved in speech and nonspeech

Functional and anatomical imaging of speech and nonspeech neural correlates has recently provided an approach for better understanding of neural networks subserving oromotor behaviors. Distributed networks are composed of functionally connected cortico–cortical (Bressler, 1995), cortico–subcortical (Grillner et al., 2005b), and thalamocortical–brainstem motor areas (Hikosaka et al., 2006). These areas exert influence over one another in a variety of combinations and contexts (Friston et al., 1993a,b; Buchel and Friston, 2001; McIntosh et al., 2003). Variable connectivity patterns are manifest through dynamic links. As a result, distributed networks are robust and provide a large repertoire of adaptive functional interactions among remote brain regions (Tononi et al., 1999). Dysfunction in the links *between* various components of the central motor system, such as the frontoponto–cerebellar pathways (Kent et al., 2001; Ackermann, 2008), corticobulbar tracts (Lo and Ratnagopal, 2007) and basal ganglia interconnections (Ciucci et al., 2008), and loops between basal ganglia and cerebellum (Middleton and Strick, 2000; Garraux et al. 2005; Taniwaki et al., 2006) rather than abnormalities *within* areas may give rise to different articulatory system deficits in different patient populations (Ackermann et al., 1993, 1995, 1997).

Timing and sequencing of orofacial movements are encoded by a bilateral neural network including cerebellum, thalamus, basal ganglia and brainstem (Garraux et al., 2005; Soros et al., 2006). Activity in pontomedullary nuclei has been functionally correlated to the sensorimotor encoding of speech and nonspeech behaviors (Corfield et al., 1999; Komisaruk et al., 2002; Hesselman et al., 2004; Dresel et al., 2005; Ozdemir et al., 2006; Soros et al., 2006; Zhang et al., 2006; Mainero et al., 2007). The feasibility of fMRI to visualize specific lower brainstem sensory and motor nuclei during production of orofacial movements has been demonstrated by Komisaruk and colleagues, who found significant changes in blood oxygenation level-dependent (BOLD) activity within the hypoglossal, facial and trigeminal main sensory nuclei during tongue movement, face/lip movement and stimulation of the malar area, at a field strength of 1.5 T and voxel size of $3.75 \times 3.75 \times 4.00\,mm^3$, respectively (Komisaruk et al., 2002). Other fMRI studies have revealed distributed oromotor networks activated by speech and nonspeech tasks, such as speaking a single syllable (Riecker et al., 2000b), syllable trains produced at different frequencies (Riecker et al., 2005), whistling (Dresel et al., 2005) and opening the mouth while protruding the lips (Soros et al., 2006). Also, increasing the complexity of a stimulus or task enlarges the speech production system by recruiting areas beyond the primary sensorimotor cortices into regions known to be involved in nonspeech motor sequencing. These include the left hemisphere inferior frontal sulcus and posterior parietal cortex, bilateral anterior insula and frontal operculum, the basal ganglia, thalamus and cerebellum (Bohland and Guenther, 2006).

XVII. Do adult speech and nonspeech oromotor behaviors share neural circuitry?

Speech and patterned nonspeech (e.g., suck and mastication) orofacial behaviors are subserved by functionally-specific ensembles of muscles and joints (Kelso and Tuller, 1983; Tuller and Kelso, 1984; Moore et al., 1988). It has been suggested that there are feedback and feedforward mechanisms that are unique for speech or nonspeech motor acts (Ziegler, 2003; Weismer, 2006). These investigators argue that knowledge of nonspeech orofacial kinetics provides little insight into the pathophysiology of neuromotor speech disorders, such as dysarthria, and that training on nonspeech oral behaviors may not be effective in enhancing speech production (Solomon et al., 2000; McAuliffe et al., 2005; Weismer, 2006). Alternatively, clinical applications of nonspeech tasks that closely mimic actual speech movements for assisting in diagnosis of neurologic disease have been supported (Folkins et al., 1995; Ballard et al., 2003) on the basis that speech production relies, in part, on a broad network of brain regions that also subserve nonspeech behaviors (Donnan et al., 1999; Clark, 2003; Ray, 2003). Imaging research of speech and nonspeech central motor control may provide insight as to why non-verbal therapies often do not facilitate speech recovery or inform the bases of motor speech disorders (Ziegler, 2003).

Functional neuroimaging studies demonstrate interdependency of widely separated brain regions correlated to speech and nonspeech behaviors (Kimura and Watson, 1989; Bonilha et al., 2006; Terumitsu et al., 2006) and show that interregional temporal correlations can be modulated with changes in task (Haughton and Biswal, 1998; Lowe et al., 2000; Greicius et al., 2009). However, the functional role of a brain region may depend on the context given by the set of other regions with which it interacts. In this way, a brain region may be involved in different behaviors without significant differential alterations in its gross metabolic activity (Bressler and Kelso, 2001; McIntosh et al., 2001, 2003). How an oromotor network is modulated by task demands and what functions the changing interregional interactions might subserve can only partially be deduced from animal experiments. At the brainstem level, it is known that different patterns of mastication are generated by dynamically-constructed subsets of a larger motor control network (Westberg et al., 1998), although separate, task-specific circuitries are likely to include higher levels of the nervous system.

Although it is unclear to what extent specific oromotor tasks share central neural circuitry, recordings of human orofacial sensory afferents show that many are active during speech and nonspeech tasks (Trulsson and Johansson, 2002). This strongly suggests activation or common circuitry during different tasks. Studies of central activation patterns during the encoding of speech and nonspeech movements may provide insight into why some lesions and diseases differentially affect oromotor performance. Lesions within or near the left face representation of primary motor cortex can result in severe dysarthria (slow, effortful speech, lacking normal prosody), with only mild deficits in tongue movement (Terao et al., 2007). Apraxia can cause a slowing of speech, while nonspeech faculties can remain relatively or completely unimpaired (McNeil and Kent, 1990; Hageman et al., 1994; Ziegler, 2002; Ricci et al., 2008). Such differences may be explained by the fact that, unlike speech, the tongue is bilaterally represented in the cortex (Terumitsu et al., 2006; Terao et al., 2007). From a neuroimaging viewpoint, these findings suggest that functional cortical activation during speech could differ from that during nonspeech orofacial movements.

Neuroimaging studies in healthy adult humans are currently under way to improve our knowledge of the functional interrelationships between cortical and subcortical (including brainstem) regions during changes in speech gestures and repetition rate or task-specific effects (Wildgruber et al., 2001; Riecker et al., 2005, 2006, 2008; Soros et al., 2006; Basho et al., 2007; Christoffels et al., 2007) and in nonspeech behaviors (Corfield et al., 1999; Komisaruk et al., 2002; Onozuka et al., 2002; Dresel et al., 2005; Takahashi et al., 2007). Barlow and colleagues have recently developed an fMRI-compatible silicone pacifier instrumented with a pressure transducer to record participant performance in real-time during functional data collection sessions (Estep et al., 2008b). This has been combined with a protocol for sparse-sampling acquisition of the BOLD response following oromotor task performance, including speech, nonspeech (suck) and rest conditions. This technique minimizes movement artifacts associated with oromotor task performance and susceptibility artifacts associated with changes of the vocal tract configuration during overt oromotor tasks (Munhall, 2001; Bohland and Guenther, 2006; Soltysik and Hyde, 2006). Analyses are currently under way to establish appropriate movement interval duration for use with sparse sampling acquisition of ororhythmic performance correlates optimized for the time course of the hemodynamic response. Preliminary findings suggest nonspeech (i.e., suck) behaviors activate a distributed network of cerebral regions and brainstem loci that also become active during elevated rates (e.g., 3 Hz) of speech production (Estep et al., 2008b).

XVIII. Conclusions

Extensive interactions exist among modulatory neurons, the pattern generators they control and the neuromuscular systems in which they function. Oromotor pattern generators themselves do not exist in isolation from one another, but frequently must interact, either directly or through modulators, to produce coordinated outputs from multiple systems (Dickinson, 2006). This is especially true for oromotor CPGs in the mammalian brain which consist of complex assemblies of interneurons and unit bursters. These may be dynamically reconfigured into subnetworks to generate often simultaneous forms of oromotor activity, including sucking, mastication, licking, swallowing and breathing. In humans, fractionation of these CPG-mediated networks may support the rhythmic features of vocalization and speech production (i.e., canonical babbling, etc.). This adds to the complexity of functions that can be ascribed to modulators. The continued use of reductionist approaches, including genetic isolation techniques, are necessary to understand the fundamental structure of oromotor CPGs, but the modulation of ororhythmic systems must also be examined in a wider context. The exact role of oromotor CPGs and their distributed networks in the production of speech and voice remains unknown and controversial.

The proliferation and continual refinement in the resolution and speed of functional magnetic resonance imaging (fMRI) of the brain with multichannel head coils has greatly increased our ability to study localized cortical, deep brain and now brainstem activity in humans (Logothetis and Wandell, 2004; Logothetis, 2008). The application of fMRI with jittered data acquisition schemes will increase our understanding of functional connectivity among these oromotor networks. The apparent linkages between multiple CPGs which populate the brainstem, via sensory modulation and entrainment, and descending inputs provides scientists and clinicians many new opportunities for exploring cross-modal and cross-CPG facilitation for new therapeutic interventions aimed squarely at the prohabilitation of the late gestation preterm infant, or rehabilitation of sensorimotor networks following a stroke.

Acknowledgments

This review was supported by Grants NIH R01 DC003311 (SM Barlow), NIH P30 HD02528, NIH P30 DC005803, and CIHR 15176.

References

Ackermann, H., 2008. Cerebellar contributions to speech production and speech perception: psycholinguistic and neurobiological perspectives. Trends Neurosci. 31 (6), 265–272.

Ackermann, H., Hertrich, I., Hehr, T., 1995. Oral diadochokinesis in neurological dysarthrias. Folia Phoniatr. Logop. 47, 15–23.

Ackermann, H., Konczak, J., Hertrich, I., 1997. The temporal control of repetitive articulatory movements in Parkinson's disease. Brain Lang. 56, 312–319.

Ackermann, H., Grönem, B.F., Hoch, G., Schönle, P.W., 1993. Speech freezing in Parkinson's disease: a kinematic analysis of orofacial movements by means of electromagnetic articulography. Folia Phoniat. 45, 84–89.

Ballard, K.J., Robin, D.A., Folkins, J.W., 2003. An integrative model of speech motor control: a response to Ziegler. Aphasiology 17, 37–48.

Bao, H., Bradley, R.M., Mistretta, C.M., 1995. Development of intrinsic electrophysiological properties in neurons from the gustatory region of rat nucleus of solitary tract. Brain Res. Dev. Brain Res. 86, 143–154.

Barlow, S.M., Estep, M., 2006. Central pattern generation and the motor infrastructure for suck, respiration, and speech. J. Com. Dis. 39, 366–380.

Barlow, S.M., Finan, D.S., Chu, S., Lee, J., 2008. Patterns for the premature brain: synthetic orocutaneous stimulation entrains preterm infants with feeding difficulties to suck. J. Perinatol. 28, 541–548.

Basho, S., Palmer, E.D., Rubio, M.A., Wulfeck, B., Muller, R-A., 2007. Effects of generation mode in fMRI adaptations of semantic fluency: paced production and overt speech. Neuropsychologia 45, 1697–1706.

Bem, T., Cabelquen, J.M., Ekeberg, O., Grillner, S., 2003. From swimming to walking: a single basic network for two different behaviors. Biol. Cybernetics 88, 79–90.

Berger, A.J., Bayliss, D.A., Viana, F., 1996. Development of hypoglossal motoneurons. J. Appl. Physiol. 81, 1039–1048.

Bitan, T., Booth, J.R., Choy, J., Burman, D.D., Gitelman, D.R., Mesulam, M.M., 2005. Shifts of effective connectivity within a language network during rhyming and spelling. J. Neurosci. 25, 5397–5403.

Bohland, J.W., Guenther, F.H., 2006. An fMRI investigation of syllable sequence production. Neuroimage 32, 821–841.

Bonilha, L., Moser, D., Rorden, C., Baylis, G.C., Fridriksson, J., 2006. Speech apraxia without oral apraxia: can normal brain function explain the physiopathology? Neuroreport 17, 1027–1031.

Boughter, J.D., Bajpai, T., St. John, S.J., Williams, R.W., Lu, L., Heck, D.H., 2007. Genetic analysis of oromotor movements in inbred and BXD recombinant inbred mice. Abstr. Soc. Neurosci 407, 12.

Bourque, M.J., Kolta, A., 2001. Properties and interconnections of trigeminal interneurons of the lateral pontine reticular formation in the rat. J. Neurophysiol. 86 (5), 2583–2596.

Bressler, S.L., 1995. Large-scale cortical networks and cognition. Brain Res. Rev. 20, 288–304.

Bressler, S.L., Kelso, J.A., 2001. Cortical coordination dynamics and cognition. Trends Cogn. Sci. 5, 26–36.

Brocard, F., Verdier, D., Arsenault, I., Lund, J.P., Kolta, A., 2006. Emergence of intrinsic bursting in trigeminal sensory neurons parallels the acquisition of mastication in weanling rats. J. Neurophysiol. 96, 2410–2424.

Bryant, J.L., Roy, S., Boughter, J.D., Goldowitz, D., Swanson, D., Morgan, J.I., Heck, D.H., 2007. A proposed new function of the mouse cerebellum: temporal modulation of brain stem pattern generator activity. Abstr. Soc. Neurosci 78, 17.

Buchel, C., Friston, K., 2001. Interactions among neuronal systems assessed with functional neuroimaging. Revue Neurologique (Paris) 157 (8–9 Pt 1), 807–815.

Butera, R.J., Rinzel, J., Smith, J.C., 1999. Models of respiratory rhythm generation in the pre-Bötzinger complex. I. Bursting pacemaker neurons. J. Neurophysiol. 82, 382–397.

Christoffels, I.K., Formisano, E., Schiller, N.O., 2007. Neural correlates of verbal feedback processing: an fMRI study employing overt speech. Hum. Brain Mapp. 28, 868–879.

Ciucci, M.R., Barkmeier-Kraemer, J.M., Sherman, S.J., 2008. Subthalamic nucleus deep brain stimulation improves deglutition in Parkinson's disease. Mov. Disor. 23, 676–683.

Clark, H.M., 2003. Neuromuscular treatments for speech and swallowing: a tutorial. Am. J. Speech Lang. Path. 12, 400–415.

Combes, D., Merrywest, S.D., Simmers, J., Sillar, K.T., 2004. Developmental segregation of spinal networks driving axial- and hindlimb-based locomotion in metamorphosing *Xenopus laevis*. J. Physiol. 559, 17–24.

Conway, B.A., Hultborn, H., Kiehn, O., 1987. Proprioceptive input resets central locomotor rhythm in the spinal cat. Exp. Brain Res. 68, 643–656.

Corfield, D.R., Murphy, K., Josephs, O., Fink, G.R., Frackowiak, R.S.J., Guz, A., Adams, L., Turner, R., 1999. Cortical and subcortical control of tongue movement in humans: a functional neuroimaging study using fMRI. J. Appl. Physiol. 86, 1468–1477.

Dal Bo, G., Lund, J.P., Verdier, D., Kolta, A., 2005. Inputs to nucleus pontis caudalis from adjacent trigeminal areas. Eur. J. Neurosci. 22 (8), 1987–1996.

Del Negro, C.A., Morgado-Valle, C., Hayes, J.A., Mackay, D.D., Pace, R.W., Crowder, E.A., Feldman, J.L., 2005. Sodium and calcium current mediated pacemaker neurons and respiratory rhythm generation. J. Neurosci. 25, 446–453.

Dellow, P.G., Lund, J.P., 1971. Evidence for central timing of rhythmical mastication. J. Physiol. (London) 215, 1–13.

Dickinson, P.S., 2006. Neuromodulation of central pattern generators in invertebrates and vertebrates. Curr. Opin. Neurobiol. 16, 604–614.

Dobbins, E.G., Feldman, J.L., 1995. Differential innervation of protruder and retractor tongue muscles of rat. J. Comp. Neurol. 357, 376–394.

Donnan, G.A., Carey, L.M., Saling, M.M., 1999. More (or less) on Broca. Lancet 353 (9158), 1031–1032.

Dresel, C., Castrop, F., Haslinger, B., Wohlschlaeger, A.M., Hennenlotter, A., Ceballos-Baumann, A.O., 2005. The functional neuroanatomy of coordinated orofacial movements: sparse sampling fMRI of whistling. Neuroimage 28, 588–597.

Dronkers, N.F., 1996. A new brain region for coordinating speech articulation. Nature 384, 159–161.

Dubayle, D., Viala, D., 1996. Interactions between medullary and spinal respiratory rhythm generators in the *in vitro* brainstem spinal cord preparation from newborn rats. Exp. Brain Res. 109, 1–8.

El Manira, A., Tegnér, J., Grillner, S., 1994. Calcium-dependent potassium channels play a critical role for burst termination in the locomotor network in lamprey. J. Neurophysiol. 72 (4), 1852–1861.

Ellenberger, H.H., 1999. Nucleus ambiguous and bulbospinal ventral respiratory group neurons in the neonatal rat. Brain Res. Bull. 50, 1–13.

Enomoto, E., Kogo, M., Koizumi, K., Ishihama, K., Yamanishi, T., 2002. Localization of premotoneurons for an NMDA-induced repetitive rhythmical activity to TMNs. Neuroreport 13, 2303–2307.

Enomoto, S., Katakura, N., Sunada, T., Katayama, T., Hirose, Y., Ishiwata, Y., Nakamura, Y., 1987. Cortically-induced masticatory rhythm in masseter motoneurons after blocking inhibition by strychnine and tetanus toxin. Neurosci. Res. 4 (5), 396–412.

Estep, M., Barlow, S.M., 2007. Modulation of the trigeminofacial pathway during speech. Brain Res. 1171, 67–74.

Estep, M., Barlow, S.M., Vantipalli, R., Lee, J., Finan, D., 2008a. Non-nutritive suck burst parametrics in preterm infants with RDS and oral feeding complications. J. Neonat. Nursing 14 (1), 28–34.

Estep, M., Barlow, S.M., Auer, E.T., Kieweg, D., Wang, J., Lee, S-P., Savage, C.R., 2008b. Functional neuroimaging correlates of speech and suck ororhythmic activity. Abstr. Soc. Neurosci. 473, 13.

Estep, M., Popescu, M., Auer, E., Popescu, A., Gustafson, K., Barlow, S.M., 2007. MEG registration of hand and face somatosensory inputs in human using patterned pneumatic-cutaneous pulse trains. Abstr. Soc. Neurosci. 74, 12.

Feldman, J.L., Smith, J.C., 1995. Neural control of respiratory pattern in mammals: an overview. In: Dempsey, J.A., Pack, A.I. (Eds.), Regulation of Breathing. Dekker, New York, NY, pp. 39–69.

Feldman, J.L., Smith, J.C., McCrimmon, D.R., Ellenberger, H.H., Speck, D.F., 1988. Generation of respiratory pattern in mammals. In: Cohen, A.M., Grillner, S., Rossignol, S. (Eds.), The Neural Control of Rhythmic Movements in Vertebrates. Wiley, New York, NY, pp. 73–100.

Finan, D.S., Barlow, S.M., 1996. The actifier and neurophysiological studies of orofacial control in neonates. J. Speech Hear. Res. 39, 833–838.

Finan, D.S., Barlow, S.M., 1998. Mechanosensory modulation of non-nutritive sucking in human infants. J. Early Hum. Dev. 52 (2), 181–197.

Folkins, J.W., Moon, J.B., Luschei, E.S., Robin, D.A., Tye-Murray, N., Moll, K.L., 1995. What can nonspeech tasks tell us about speech motor disabilities? J. Phonetics 23, 139–147.

Friston, K.J., Frith, C.D., Frackowiak, R.S.J., 1993b. Time-dependent changes in effective connectivity measured with PET. Hum. Brain Mapp. 1, 69–80.

Friston, K.J., Frith, C.D., Liddle, P.F., Frackowiak, R.S.J., 1993a. Functional connectivity: the principal component analysis of large (PET) data sets. J. Cereb. Blood Flow Metab. 13, 5–14.

Garraux, G., McKinney, C., Wu, T., Kansaku, K., Nolte, G., Hallett, M., 2005. Shared brain areas but not functional connections controlling movement timing and order. J. Neurosci. 25, 5290–5297.

Georgopoulos, A.P., Kettner, R.E., Schwartz, A.B., 1988. Primate motor cortex and free arm movements to visual targets in three-dimensional space. II. Coding of the direction of movement by a neuronal population. J. Neurosci. 8, 2928–2937.

Gewolb, I.H., Vice, F.L., Schweitzer-Kenney, E.L., Taciak, V.L., Bosma, J.F., 2001. Developmental patterns of rhythmic suckle and swallow in preterm infants. Dev. Med. Child Neurol. 43, 22–27.

Glass, L., Mackey, M.C., 1988. From Clocks to Chaos: The Rhythms of Life. Princeton University Press, Princeton, NJ.

Goldson, E., 1987. Nonnutritive sucking in the sick infant. J. Perinatol. 7 (1), 30–34.

Goodwin, G.M., Luschei, E.S., 1975. Discharge of spindle afferents from jaw-closing muscles during chewing in alert monkeys. J. Neurophysiol. 38 (3), 560–571.

Greicius, M.D., Supekar, K., Menon, V., Dougherty, R.F., 2009. Resting-state functional connectivity reflects structural connectivity in the default mode network. Cerebral Cortex 19, 72–78.

Grillner, S., 1975. Locomotion in vertebrates: central mechanisms and reflex interaction. Physiol. Rev. 55 (2), 247–304.

Grillner, S., 1991. Recombination of motor pattern generators. Simple neuronal networks combine to produce complex versatile motor patterns. Curr. Biol. 1, 231–233.

Grillner, S., 2002. Fundamentals of motor systems. In: Zigmond, M.J., Bloom, F.E., Landis, S.C., Roberts, J.L., Squire, L.R. (Eds.), Fundamental Neuroscience. Academic Press, San Diego, CA, pp. 753–766.

Grillner, S., 2003. The motor infrastructure: from ion channels to neuronal networks. Nat. Rev. Neurosci. 4, 573–586.

Grillner, S., Hellgren, J., Menard, A., Saitoh, K., Wikstrom, M.A., 2005a. Mechanisms for selection of basic motor programs – roles for the striatum and pallidum. Trends Neurosci. 28, 364–370.

Grillner, S., Markram, H., De Schutter, E., Silberberg, G., LeBeau, F.E.N., 2005b. Microcircuits in action – from CPGs to neocortex. Trends Neurosci. 28 (10), 525–533.

Guido, W., Gunhan-Agar, E., Erzurumlu, R.S., 1998. Developmental changes in the electrophysiological properties of brain stem trigeminal neurons during pattern (barrelette) formation. J. Neurophysiol. 79, 1295–1306.

Guyenet, P.G., Sevigny, C.P., Weston, M.C., Stornetta, R.L., 2002. Neurokinin-1 receptor-expressing cells of the ventral respiratory group are functionally heterogeneous and predominantly glutamatergic. J. Neurosci. 22, 3806–3816.

Hack, M., Estabrook, M.M., Robertson, S.S., 1985. Development of sucking rhythm in preterm infants. Early Hum. Dev. 11, 133–140.

Hageman, C.F., Robin, D.A., Moon, J.B., Folkins, J.W., 1994. Oral motor tracking in normal and apraxic speakers. Clin. Aphasiol. 22, 219–229.

Hannam, A.G., Lund, J.P., 1981. The effect of intra-oral stimulation on the human masticatory cycle. Arch. Oral Biol. 26, 865–870.

Hanson, M.G., Landmesser, L.T., 2004. Normal patterns of spontaneous activity are required for correct motor axon guidance and the expression of specific guidance molecules. Neuron 43, 687–701.

Haughton, V., Biswal, B., 1998. Clinical application of basal regional cerebral blood flow fluctuation measurements by fMRI. Adv. Exp. Med. Biol. 454, 583–590.

Hesselman, V., Sorger, B., Lasek, K., Guntinas-Lichius, O., Krug, B., Sturm, V., Goebel, R., Lackner, K., 2004. Discriminating the cortical representation sites of tongue and lip movement by functional MRI. Brain Topogr. 16, 159–167.

Hickok, G., 2001. Functional anatomy of speech perception and speech production: psycholinguistic implications. J. Psycholing. Res. 30, 225–235.

Hikosaka, O., Nakamura, K., Nakahara, H., 2006. Basal ganglia orient eyes to reward. J. Neurophysiol. 95, 567–584.

Ingersoll, E.W., Thoman, E.B., 1994. The breathing bear: effects on respiration in premature infants. Physiol. Behav. 56 (5), 855–859.

Iriki, A., Nozaki, S., Nakamura, Y., 1988. Feeding behavior in mammals: corticobulbar projection is reorganized during conversion from sucking to chewing. Dev. Brain Res. 44, 189–196.

Ishihama, K., Kogo, M., Koizumi, H., Nomura, K., Tanaka, S., Yamanishi, T., Enomoto, A., 2003. Oral-motor patterns of rhythmic trigeminal activity generated in fetal rat brainstem *in vitro*. Dev. Brain Res. 145, 163–166.

Jenny, A.B., Saper, C.B., 1987. Organization of the facial nucleus and corticofacial projection in the monkey: a reconsideration of the upper motor neuron facial palsy. Neurology 37, 930–939.

Kalaska, J.F., Crammond, D.J., 1992. Cerebral cortical mechanisms of reaching movements. Science 255, 1517–1523.

Katakura, N., Jia, L., Nakamura, Y., 1995. NMDA-induced rhythmical activity in XII nerve of isolated CNS from newborn rats. Neuroreport 6, 601–604.

Kelso, J.A.S., Tuller, B., 1983. Compensatory articulation: under reduced conditions of afferent information: a dynamic formulation. J. Speech Hear. Res. 26, 217–224.

Kent, R.D., Vorperian, H.K., 1995. Anatomic development of the craniofacial-oral-laryngeal systems: a review. J. Med. Speech Lang. Path. 3, 145–190.

Kent, R.D., Duffy, J.R., Slama, A., Kent, J.F., Clift, A., 2001. Clinicoanatomic studies in dysarthria: review, critique, and directions for research. J. Speech Lang. Hear. Res. 44, 535–551.

Kimura, D., Watson, N., 1989. The relation between oral movement control and speech. Brain Lang. 37, 565–590.

Kogo, M., Funk, G.D., Chandler, S.H., 1996. Rhythmical oral-motor activity recorded in an *in vitro* brainstem preparation. Somatosensory Motor Res. 13, 39–48.

Kogo, M., Tanaka, S., Chandler, S.H., Matsuya, T., 1998. Examination of the relationships between jaw opener and closer rhythmical muscle activity in an *in vitro* brainstem jaw-attached preparation. Somatosensory Motor Res. 15, 200–210.

Koizumi, H., Ishihama, K., Nomura, K., Yamanishi, T., Kogo, M., Matsuya, T., 2002. Differential discharge patterns of rhythmical activity in trigeminal motorneurons during fictive mastication and respiration *in vitro*. Brain Res. Bull. 58, 129–133.

Koizumi, H., Wilson, C.G., Wong, S., Yamanishi, T., Koshiya, N., Smith, J.C., 2008. Functional imaging, spatial reconstruction, and biophysical analysis of a respiratory motor circuit isolated *in vitro*. J. Neurosci. 28 (10), 2353–2365.

Kolta, A., Westberg, K.G., Lund, J.P., 2000. Identification of brainstem interneurons projecting to the trigeminal motor nucleus and adjacent structures in the rabbit. J. Chem. Neuroanat. 19 (3), 175–195.

Komisaruk, B.R., Mosier, K.M., Liu, W-C., Criminale, C., Zaborszky, L., Whipple, B., Kalnin, A., 2002. Functional localization of brainstem and cervical spinal cord nuclei in humans with fMRI. Am. J. Neuroradiol. 23, 609–617.

Koshiya, N., Smith, J.C., 1999. Neuronal pacemaker for breathing visualized *in vitro*. Nature 400, 360–363.

Kriellaars, D.J., Brownstone, R.M., Noga, B.R., Jordan, L.M., 1994. Mechanical entrainment of fictive locomotion in the decerebrate cat. J. Neurophysiol. 71 (6), 2074–2086.

Kubo, Y., Enomoto, S., Nakamura, Y., 1981. Synaptic basis of orbital cortically induced rhythmical masticatory activity of trigeminal motoneurons in immobilized cats. Brain Res. 230 (1–2), 97–110.

Lanuza, G.M., Gosgnach, S., Pierani, A., Jessell, T.M., Goulding, M., 2004. Genetic identification of spinal interneurons that coordinate left–right locomotor activity necessary for walking movements. Neuron 42, 375–386.

Lau, C., 2006. Oral feeding in the preterm infant. Neoreviews 7, 19–27.

Lau, C., Schanler, R.J., 1996. Oral motor function in the neonate. Clinics in Perinatol. 23 (2), 161–178.

Lau, C., Hurst, N., 1999. Oral feeding in infants. Curr. Probl. Pediatr. 29, 105–124.

Lavigne, G., Kim, J.S., Valiquette, C., Lund, J.P., 1987. Evidence that periodontal pressoreceptors provide positive feedback to jaw closing muscles during mastication. J. Neurophysiol. 58, 342–358.

Liscic, R.M., Zidar, J., Mihelin, M., 1998. Evidence of direct connection of corticobulbar fibers to orofacial muscles in man: electromyographic study of individual motor unit responses. Muscle & Nerve 21, 561–566.

Liu, K.S., Gray, M., Otto, S.J., Fetcho, J.R., Beattie, C.E., 2003. Mutations in *deadly seven/notch1a* reveal developmental plasticity in the escape response circuit. J. Neurosci. 23, 8159–8166.

Lo, F.S., Erzurumlu, R.S., 2001. Neonatal deafferentation does not alter membrane properties of trigeminal nucleus principalis neurons. J. Neurophysiol. 85, 1088–1096.

Lo, Y.L., Ratnagopal, P., 2007. Cortico-hypoglossal and corticospinal conduction abnormality in Bickerstaff's brainstem encephalitis. Clin. Neurol. Neurosurg. 109, 523–525.

Logothetis, N.K., 2008. What we can do and what we cannot do with fMRI. Nature 453, 869–878.

Logothetis, N.K., Wandell, B.A., 2004. Interpreting the BOLD signal. Annu. Rev. Physiol. 66, 735–769.

Lowe, M.J., Dzemidzic, M., Lurito, J.T., Mathews, V.P., Phillips, M.D., 2000. Correlations in low-frequency BOLD fluctuations reflect cortico-cortical connections. Neuroimage 12, 582–587.

Lund, J.P., 1991. Mastication and its control by the brain stem. Crit. Rev. Oral Biol. Med. 2, 33–64.

Lund, J.P., Kolta, A., 2006a. Brainstem circuits that control mastication: do they have anything to say during speech? J. Com. Dis. 39 (5), 381–390.

Lund, J.P., Kolta, A., 2006b. Generation of the central masticatory pattern and its modification by sensory feedback. Dysphagia 21 (3), 167–174.

Lund, J.P., Sasamoto, K., Murakami, T., Olsson, K.A., 1984. Analysis of rhythmical jaw movements produced by electrical stimulation of motor-sensory cortex of rabbits. J. Neurophysiol. 52, 1014–1029.

Mainero, C., Zhang, W.T., Kuman, A., Rosen, B.R., Sorensen, A.G., 2007. Mapping the spinal and supraspinal pathways of dynamic mechanical allodynia in the human trigeminal system using cardiac-gated fMRI. Neuroimage 15, 1201–1210.

Marder, E., Rehm, K.J., 2005. Development of central patterning generating circuits. Curr. Opin. Neurobiol. 15, 86–93.

McAuliffe, M.J., Ward, E.C., Murdoch, B.E., Farrell, A.M., 2005. A nonspeech investigation of tongue function in Parkinson's disease. J. Gerontol., Ser. A, Biol. Sci. Med. Sci. 60, 667–674.

McDavid, S., Lund, J.P., Auclair, F., Kolta, A., 2006. Morphological and immunohistochemical characterization of interneurons within the rat trigeminal motor nucleus. Neuroscience 139 (3), 1049–1059.

McIntosh, A.R., Fitzpatrick, S.M., Friston, K.J., 2001. On the marriage of cognition and neuroscience. Neuroimage 14, 1231–1237.

McIntosh, A.R., Rajah, M.N., Lobaugh, N.J., 2003. Functional connectivity of the medial temporal lobe relates to learning and awareness. J. Neurosci. 23, 6520–6528.

McKay, L.C., Evans, K.C., Frackowiak, R.S., Corfield, D.R., 2003. Neural correlates of voluntary breathing in humans. J. Appl. Physiol. 95, 1170–1178.

McKay, L.C., Adams, L., Frackowiak, R.S., Corfield, D.R., 2008. A bilateral cortico-bulbar network associated with breath holding in humans, determined by functional magnetic resonance imaging. Neuroimage 40, 1824–1832.

McNeil, M.R., Kent, R.D., 1990. Motoric characteristics of adult aphasic and apraxic speakers. In: Hammond, G.E. (Ed.), Cerebral Control of Speech and Limb Movements. Elsevier Science Publishers, Amsterdam, The Netherlands, pp. 349–386.

Medoff-Cooper, B., 2005. Nutritive sucking research: from clinical questions to research answers. J. Perinat. Neonat. Nurs. 19, 265–272.

Mentel, T., Cangiano, L., Grillner, S., Buschges, A., 2008. Neuronal substrates for state-dependent changes in coordination between motoneuron pools during fictive locomotion in the lamprey spinal cord. J. Neurosci. 28, 868–879.

Mentel, T., Krause, A., Pabst, M., El Manira, A., Buschges, A., 2006. Activity of fin muscles and fin motoneurons during swimming motor pattern in the lamprey. Eur. J. Neurosci. 23, 2012–2026.

Middleton, F.A., Strick, P.L., 2000. Basal ganglia and cerebellar loops: motor and cognitive circuits. Brain Res. Rev. 31, 236–250.

Miller, J.L., Sonies, B.C., Macedonia, C., 2003. Emergence of oropharyngeal, laryngeal and swallowing activity in the developing fetal upper aerodigestive tract: an ultrasound evaluation. Early Hum. Dev. 71, 61–87.

Mizuno, K., Ueda, A., 2005. Neonatal feeding performance as a predictor of neurodevelopmental outcome at 18 months. Dev. Med. Child Neurol. 47, 299–304.

Moore, C.A., Smith, A., Ringel, R.L., 1988. Task-specific organization of activity in human jaw muscles. J. Speech Hear. Res. 31, 670–680.

Morimoto, T., Yamamoto, Y., Yamatodani, A., 2001. Brain histamine and feeding behavior. Behav. Brain Res. 124, 145–150.

Munhall, K.G., 2001. Functional imaging during speech production. Acta Psychol. (Amst.) 107, 95–117.

Naganuma, K., Inoue, M., Yamamura, K., Hanada, K., Yamada, Y., 2001. Tongue and jaw muscle activities during chewing and swallowing in freely behaving rabbits. Brain Res. 915, 185–194.

Nakamura, Y., Katakura, N., Nakajima, M., Liu, J., 2004. Rhythm generation for food-ingestive movements. Prog. Brain Res. 143, 97–103.

Nguyen, Q.T., Wessel, R., Kleinfeld, D., 2004. Developmental regulation of active and passive membrane properties in rat vibrissa motoneurones. J. Physiol. 556, 203–219.

Nozaki, S., Iriki, A., Nakamura, Y., 1986. Localization of central rhythm generator involved in cortically induced rhythmical masticatory jaw-opening movement in the guinea pig. J. Neurophysiol. 55 (4), 806–825.

Nunez-Abades, P.A., Speielmanm, J.M., Barrionueno, G., Cameron, W.E., 1993. *In vitro* electrophysiology of developing genioglossal motoneurons in the rat. J. Neurophysiol. 70, 1401–1411.

Okayasu, I., Yamada, Y., Kohno, S., Yoshida, N., 2003. New animal model for studying mastication in oral motor disorders. J. Dent. Res. 82, 318–321.

Olsson, K.A., Lund, J.P., Valiquette, C., Veilleux, D., 1988. Activity during mastication of periodontal pressoreceptive neurons of the trigeminal subnucleus oralis of the rabbit. J. Neurophysiol. 59, 341–356.

Olthoff, A., Baudewig, J., Kruse, E., Dechent, P., 2008. Cortical sensorimotor control in vocalization: a functional magnetic resonance imaging study. Laryngoscope 118, 2091–2096.

Onozuka, M., Fujita, M., Watanabe, K., Hirano, Y., Niwa, M., Nishiyama, K., Saito, S., 2002. Mapping brain region activity during chewing: a functional magnetic resonance imaging study. J. Dent. Res. 81, 743–746.

Ozdemir, E., Norton, A., Schlaug, G., 2006. Shared and distinct neural correlates of singing and speaking. Neuroimage 33, 628–635.

Pavlidis, T., 1973. Entrainment of oscillators by external inputs. In: Segel, L.A. (Ed.), Biological Oscillators: Their Mathematical Analysis. Academic Press, London, UK, pp. 71–98.

Pearson, K.G., 2000. Neural adaptation in the generation of rhythmic behavior. Annu. Rev. Physiol. 62, 723–753.

Pearson, K.G., Gordon, J., 2000. Locomotion. In: Kandel, E.R., Schwartz, J.H., Jessell, T.M. (Eds.), Principles of Neural Science. McGraw-Hill, New York, NY, pp. 737–755.

Pearson, K.G., Ramirez, J.M., Jiang, W., 1992. Entrainment of the locomotor rhythm by group Ib afferents from ankle extensor muscles in spinal cats. Exp. Brain Res. 90, 557–566.

Pena, F., Ramirez, J.M., 2002. Endogenous activation of serotonin-2A receptors is required for respiratory rhythm generation *in vitro*. J. Neurosci. 22, 11055–11064.

Pena, F., Ramirez, J.M., 2004. Substance P-mediated modulation of pacemaker properties in the mammalian respiratory network. J. Neurosci. 24, 7549–7556.

Peyron, M.A., Lassauzay, C., Woda, A., 2002. Effects of increased hardness on jaw movement and muscle activity during chewing of visco-elastic model foods. Exp. Brain Res. 142, 41–51.

Poore, M., Zimmerman, E., Barlow, S.M., Wang, J., Gu, F., 2008. NTrainer therapy increases suck spatiotemporal stability in preterm infants. Acta Paediatrica 97, 920–927.

Popescu, E.A., Popescu, M., Wang, J., Barlow, S.M., Gustafson, K.M., 2008. Non-nutritive sucking recorded *in utero* via fetal magnetography. Physiol. Measur. 29, 127–139.

Ramirez, J.M., Viemari, J.C., 2005. Determinants of inspiratory activity. Respir. Physiol. Neurobiol. 147, 145–157.

Ray, J., 2003. Effects of orofacial myofunctional therapy on speech intelligibility in individuals with persistent articulatory impairments. Int. J. Orofacial Myol. 29, 5–14.

Rekling, J.C., Feldman, J.L., 1998. Pre-Bötzinger complex and pacemaker neurons: hypothesized site and kernel for respiratory rhythm generation. Annu. Rev. Physiol. 60, 385–405.

Ricci, M., Magarelli, M., Todino, V., Bianchini, A., Calandriello, E., Tramutoli, R., 2008. Progressive apraxia of speech presenting as isolated disorder of speech articulation and prosody: a case report. Neurocase 14, 162–168.

Riecker, A., Kassubek, J., Groschel, K., Grodd, W., Ackermann, H., 2006. The cerebral control of speech tempo: opposite relationship between speaking rate and BOLD signal changes at striatal and cerebellar structures. Neuroimage 29, 46–53.

Riecker, A., Brendel, B., Ziegler, W., Erb, M., Ackermann, H., 2008. The influence of syllable onset complexity and syllable frequency on speech motor control. Brain Lang. 107, 102–113.

Riecker, A., Ackermann, H., Wildgruber, D., Meyer, J., Dogil, G., Haider, H., Grodd, W., 2000. Articulatory/phonetic sequencing at the level of the anterior perisylvian cortex: a functional magnetic resonance imaging (fMRI) study. Brain Lang. 75, 259–276.

Riecker, A., Mathiak, K., Wildgruber, D., Erb, M., Hertrich, I., Grodd, W., Ackermann, H., 2005. fMRI reveals two distinct cerebral networks subserving speech motor control. Neurology 64, 700–706.

Rossignol, S., Lund, J.P., Drew, T., 1988. The role of sensory inputs in regulating patterns of rhythmical in higher vertebrates. In: Cohen, A., Rossignol, S., Grillner, S. (Eds.), Neural Control of Rhythmic Movements in

Vertebrates. John Wiley and Sons, New York, NY, pp. 201–283.

Rybak, I.A., Abdala, A.P.L., Markin, S.N., Paton, J.F.R., Smith, J.C., 2007. Spatial organization and state-dependent mechanisms for respiratory rhythm and pattern generation. Prog. Brain Res. 165, 201–220.

Shao, X.M., Feldman, J.L., 2005. Cholinergic neurotransmission in the pre-Bötzinger complex modulates excitability of inspiratory neurons and regulates respiratory rhythm. Neurosci. 130, 1069–1081.

Smith, J.C., Ballanyi, K., Richter, D.W., 1992. Whole-cell patch-clamp recordings from respiratory neurons in neonatal rat brainstem *in vitro*. Neurosci. Lett. 134, 153–156.

Smith, J.C., Ellenberger, H.H., Ballanyi, K., Richter, D.W., Feldman, J.L., 1991. Pre-Bötzinger complex: a brainstem region that may generate respiratory rhythm in mammals. Science 254, 726–729.

Smotherman, M., Kobayasi, K., Ma, J., Metzner, W., 2005. Central mechanisms of vocal-respiratory coupling in echolocating bats. Abstr. Soc. Neurosci. 1003, 6.

Solomon, N.P., Robin, D.A., Luschei, E.S., 2000. Strength, endurance, and stability of the tongue and hand in Parkinson disease. J. Speech Lang. Hear. Res. 43, 256–267.

Soltysik, D.A., Hyde, J.S., 2006. Strategies for block-design fMRI experiments during task-related motion of structures of the oral cavity. Neuroimage 29, 1260–1271.

Soros, P., Sokoloff, L.G., Bose, A., McIntosh, A.R., Graham, S.J., Stuss, D.T., 2006. Clustered functional MRI of overt speech production. Neuroimage 32, 376–387.

Speck, D.F., Karius, D.R., Ling, L., 1993. Respiratory afferents and the inhibition of inspiration. In: Speck, D.F., Dekin, M.S., Revelette, W.R., Frazier, D.T. (Eds.), Respiratory Control: Central and Peripheral Mechanisms. University Press of Kentucky, Lexington, KY, pp. 100–103.

Stornetta, R.L., Sevigny, C.P., Guyenet, P.G., 2003a. Inspiratory augmenting bulbospinal neurons express both glutamatergic and enkephalinergic phenotypes. J. Comp. Neurol. 455, 113–124.

Stornetta, R.L., Rosin, D.L., Wang, H., Sevigny, C.P., Weston, M.C., Guyenet, P.G., 2003b. A group of glutamatergic interneurons expressing high levels of both neurokinin-1 receptors and somatostatin identifies the region of the pre-Bötzinger complex. J. Comp. Neurol. 455, 499–512.

Stumm, S., Barlow, S.M., Estep, M., Lee, J., Cannon, S., Carlson, J., Finan, D., 2008. Respiratory distress syndrome degrades the fine structure of the non-nutritive suck in preterm infants. J. Neonatal Nurs. 14 (1), 9–16.

Swensen, A.M., Marder, E., 2001. Modulators with convergent cellular actions elicit distinct circuit outputs. J. Neurosci. 21, 4050–4058.

Takahashi, T., Miyamoto, T., Terao, A., Yokoyama, A., 2007. Cerebral activation related to the control of mastication during changes in food hardness. Neurosci. 145, 791–794.

Tanaka, S., Kogo, M., Chandler, S.H., Matsuya, T., 1999. Localization of oral–motor rhythmogenic circuits in the isolated rat brainstem preparation. Brain Res. 821, 190–199.

Tanaka, S., Wu, N., Hsaio, C.F., Turman, J., Chandler, S.H., 2003. Development of inward rectification and control of membrane excitability in mesencephalic V neurons. J. Neurophysiol. 89, 1288–1298.

Taniwaki, T., Okayama, A., Yoshiura, T., Togao, O., Nakamura, Y., Yamasaki, T., Ogata, K., Shigeto, H., Ohyagi, Y., Kira, J-I., Tobimatsu, S., 2006. Functional network of the basal ganglia and cerebellar motor loops *in vivo*: different activation patterns between self-initiated and externally triggered movements. Neuroimage 31, 745–753.

Tazerart, S., Vinay, L., Brocard, F., 2008. The persistent sodium current generates pacemaker activities in the central pattern generator for locomotion and regulates the locomotor rhythm. J. Neurosci. 28 (34), 8577–8589.

Tazerart, S., Viemari, J.C., Darbon, P., Vinay, L., Brocard, F., 2007. Contribution of persistent sodium current to locomotor pattern generation in neonatal rats. J. Neurophysiol. 98 (2), 613–628.

Terao, Y., Ugawa, Y., Yamamoto, T., Sakurai, Y., Masumoto, T., Abe, O., Masutani, Y., Aoki, S., Tsuji, S., 2007. Primary face motor area as the motor representation of articulation. J. Neurol. 254, 442–447.

Terumitsu, M., Fujii, Y., Suzuki, K., Kwee, I.L., Nakada, T., 2006. Human primary motor cortex shows hemispheric specialization for speech. Neuroreport 17, 1091–1095.

Tononi, G., Sporns, O., Edelman, G.M., 1999. Measures of degeneracy and redundancy in biological networks. Proc. Nat. Acad. Sci. USA 96, 3257–3262.

Travers, J.B., 2009. Motor control of feeding and drinking. In: Squire, L. (Ed.), Encyclopedia of Neuroscience. Elsevier Ltd, San Diego, CA, pp. 1001–1007.

Travers, J.B., DiNardo, L.A., Karimnamazi, H., 1997. Motor and premotor mechanisms of licking. Neurosci. Biobehav. Rev. 21 (5), 631–647.

Trulsson, M., Johansson, R.S., 2002. Orofacial mechanoreceptors in humans: encoding characteristics and responses during natural orofacial behaviors. Behav. Brain Res. 135, 27–33.

Tryba, A.K., Pena, F., Ramirez, J.M., 2006. Gasping activity *in vitro*: a rhythm dependent on 5-HT2A receptors. J. Neurosci. 26, 2623–2634.

Tsuboi, A., Kolta, A., Chen, C.C., Lund, J.P., 2003. Neurons of the trigeminal main sensory nucleus participate in the generation of rhythmic motor patterns. Eur. J. Neurosci, 17, 229–238.

Tsuzuki, S., Yoshida, S., Yamamoto, T., Oka, H., 1995. Developmental changes in the electrophysiological properties of neonatal rat oculomotor neurons studied *in vitro*. Neurosci. Res. 23, 389–397.

Tuller, B., Kelso, J.A., 1984. The timing of articulatory gestures: evidence for relational invariants. J. Acoust. Soc. Am. 76, 1020–1036.

Viemari, J.C., Ramirez, J.M., 2006. Norepinephrine differentially modulates different types of respiratory pacemaker and nonpacemaker neurons. J. Neurophysiol. 95, 2070–2082.

Voperian, H.K., Kent, R.D., 2007. Vowel acoustic space development in children: a synthesis of acoustic and anatomic data. J. Speech Lang. Hear. Res. 50, 1510–1545.

Weismer, G., 2006. Philosophy of research in motor speech disorders. Clin. Linguist. Phonet. 20, 315–349.

Westberg, K.G., Clavelou, P., Sandstrom, G., Lund, J.P., 1998. Evidence that trigeminal brainstem interneurons form subpopulations to produce different forms of mastication in the rabbit. J. Neurosci. 18 (16), 6466–6479.
Westneat, M.W., Hall, W.G., 1992. Ontogeny of feeding motor patterns in infant rats: an electromyographic analysis of suckling and chewing. Behav. Neurosci. 106 (3), 539–554.
Wildgruber, D., Ackermann, H., Grodd, W., 2001. Differential contributions of motor cortex, basal ganglia, and cerebellum to speech motor control: effects of syllable repetition rate evaluated by fMRI. Neuroimage 13, 101–109.
Wise, R.S.J., Green, J., Buchel, C., Scott, S.K., 1999. Brain regions involved in articulation. Lancet 353, 1057–1061.
Wolff, P.H., 1968. The serial organization of sucking in the young infant. Pediatrics 42, 943–956.
Zhang, W.-T., Mainero, C., Kumar, A., Wiggins, C.J., Benner, T., Purdon, P.L., Bolar, D.S., Kwong, K.K., Sorensen, A.G., 2006. Strategies for improving the detection of fMRI activation in trigeminal pathways with cardiac gating. Neuroimage 31, 1506–1512.
Ziegler, W., 2002. Task-related factors in oral motor control: speech and oral diadochokinesis in dysarthria and apraxia of speech. Brain Lang. 80, 556–575.
Ziegler, W., 2003. Speech motor control is task-specific: evidence from dysarthria and apraxia of speech. Aphasiology 17, 3–36.

SECTION 9

Integrative Motor Functions of the Ambiguus, Retroambiguus and Parabrachial Nuclei

The laryngeal muscles are involved in a number of functions in addition to vocalization. They also control breathing, coughing, sneezing and swallowing. The parabrachial nucleus is particularly involved in vocal–respiratory interaction. The vocal output is also influenced by audio–vocal control. Bats have evolved some additional brainstem controlling mechanisms as a special adaptation for echolocation signals.

CHAPTER 9.1

Functions of larynx in breathing, vocalization and airway protective reflexes

Keisuke Shiba[1,2]

[1]*Departments of Otolaryngology, Chiba Medical Center, Chiba, Japan*
[2]*Hikifune Otolaryngology and Day Surgery Clinic, Tokyo, Japan*

Abstract: The larynx serves various functions including respiration, vocalization and airway protection. Central input to laryngeal motoneurons involved in regulation of these behaviors includes a complex combination of excitatory and inhibitory activities characteristic for each behavior. The brain mechanisms that use the same peripheral structure to generate such different behaviors are of central importance for cellular neuroscience. This chapter reviews the laryngeal activity during breathing, vocalization and upper airway reflexes including coughing, sneezing and swallowing at the levels of the vocal folds, laryngeal muscles, and motoneurons. Also discussed is whether laryngeal premotor neurons are shared among the central pattern generators for these behaviors or represent independent networks.

Keywords: larynx; laryngeal motoneuron; premotor neuron; breathing; vocalization; coughing; swallowing; sneezing; central pattern generator; nucleus ambiguus; nucleus retroambiguus

Abbreviations: CPG: central pattern generator; CT: cricothyroid; CTM cricothyroid muscle motoneuron; ELM: expiratory laryngeal motoneuron; ILM: inspiratory laryngeal motoneuron; PAG: periaqueductal gray; PCA: posterior cricoarytenoid; TA: thyroarytenoid

I. Introduction

In the process of evolution the larynx, which had initially contributed merely to maintaining homeostasis by assisting respiration, has eventually become involved in vocal communication. This process caused dramatic functional changes in the airway system, including the larynx itself. As a result, the larynx serves many functions including vocalization, respiration and airway protective reflexes such as coughing, swallowing and sneezing. All of these various glottic behaviors are carried out by activation of intralaryngeal muscles, which are driven by laryngeal motoneurons located in the nucleus ambiguus in the medulla. This chapter describes the multifunctional role of the larynx and different mechanism of its function at the levels of the organ, muscles, motoneurons and premotor neurons.

II. Breathing

II.A. Laryngeal movements and muscle activities during breathing

The respiratory rhythm consists of three respiratory phases: inspiration; stage one of expiration (post-inspiratory phase); and stage two of expiration. During inspiration, in which the diaphragm contracts to pump air into the lung, the glottis is dilated to help the air intake. During the post-inspiratory phase, in which the diaphragm ceases to be active or weakly contract, i.e. post-inspiratory activation, and expiratory muscles such as

*Corresponding author. E-mail: shibak1963@yahoo.co.jp

Stefan M. Brudzynski (Ed.)
Handbook of Mammalian Vocalization
ISBN 978-0-12-374593-4

DOI: 10.1016/B978-0-12-374593-4.00034-6

abdominal muscles do not contract, the air pumped into the lungs is passively exhaled and the glottis is narrowed to brake the passive expiratory airflow, which prevents collapse of the lungs. The post-inspiratory glottic narrowing plays a functional role similar to positive end-expiratory pressure in mechanically ventilated patients for efficient ventilation. It might be a vestige of the air-holding mechanism in amphibians, in which air is forced into the lungs by buccal contraction and held there by glottic closure. During stage two of expiration (active expiration), in which the expiratory muscles contract to pump air out of the lung, the glottis ceases to be narrowed to help exhalation. In addition, the glottic caliber is further changed by hypoxia and hypercapnia for efficient gas exchange (Bartlett, 1986). Therefore, the larynx functions as a gate to the airways and it controls the airflow.

The respiratory-related glottic movements are generated by contractions of intralaryngeal muscles: (1) activation of vocal fold abductor, i.e., posterior cricoarytenoid (PCA) muscle, which dilates the glottis during inspiration by abduction rotation of the arytenoid cartilages; (2) activation of vocal fold adductors, i.e., lateral cricoarytenoid, interarytenoid and thyroarytenoid (TA) muscles, which narrow the glottis during the post-inspiratory phase by adduction rotation of the arytenoid cartilages. The TA muscle also has the function of shortening the vocal folds, and thus its contraction under isometric conditions increases the stiffness of the vocal folds, especially during vocalization. The cricothyroid (CT) muscle, whose constriction tenses the vocal folds, is also one of the intralaryngeal muscles. The CT muscle differs from the other laryngeal muscles, due to its anatomical location and motor nerve innervation. The CT muscle is located ventral to the thyroid cartilage framework and innervated by the superior laryngeal nerve; the other muscles are located dorsal to the thyroid cartilage and are supplied by the recurrent laryngeal nerve. The CT muscle also exhibits respiratory-modulated activity, i.e., inspiratory phasic activation and unstable tonic expiratory activation. However, its physiological role in breathing is minor in comparison to the other intralaryngeal muscles.

II.B. Laryngeal motoneuron activities during breathing

Laryngeal motoneurons that send axons through the recurrent laryngeal nerve, i.e., motoneurons of TA, lateral cricoarytenoid, interarytenoid and PCA muscles, are located in the caudal part of the nucleus ambiguus. On the other hand, CT muscle motoneurons (CTMs) are located in the rostral part of the nucleus ambiguus (Yoshida et al., 1982; Davis and Nail, 1984). Laryngeal motoneurons, whose axons run in the recurrent laryngeal nerve, are classified into expiratory laryngeal motoneurons (ELMs) and inspiratory laryngeal motoneurons (ILMs) according to their expiratory and inspiratory depolarization, respectively (Barillot et al., 1990; Shiba et al., 1999). ELMs and ILMs correspond to adductor and abductor motoneurons, respectively.

ELMs depolarize abruptly at the onset of expiration, and then repolarize slowly throughout expiration, displaying a decreasing firing rate pattern (Fig. 1). They hyperpolarize abruptly at the onset of inspiration and remain hyperpolarized throughout the inspiratory phase. The post-inspiratory phase is defined by ELM activity (Fig. 1). ILMs abruptly depolarize at the onset of inspiration and exhibit a progressive depolarization throughout the inspiratory phase. They usually hyperpolarize sharply at the onset of expiration, and then depolarize gradually throughout expiration. CTMs display depolarization with an augmenting pattern throughout the inspiratory phase, as do the ILMs. They rapidly hyperpolarize at the onset of expiration, and then depolarize again during the post-inspiratory phase. Eupneic activities of laryngeal motoneurons vary according to the respiratory condition. The details and variations of the membrane's potential trajectory in these varying conditions are well-documented by Barillot et al. (1990) and Numasawa et al. (2004).

II.C. Laryngeal premotor neurons for breathing activity

The respiratory-related activity of laryngeal motoneurons results from synaptic inputs from brainstem respiratory neurons and their membrane properties. Fig. 2 summarizes the synaptic connections between medullary respiratory neurons and laryngeal motoneurons in cats. Several studies using the spike-triggered averaging method have unveiled the excitatory input to the ILMs and the inhibitory input to the ELMs and ILMs (Ezure, 1990; Jiang and Lipski, 1990; Ono et al., 2006). Only the excitatory input to the ELMs remains an enigma. The membrane property of ELMs, i.e., post-inhibitory rebound, may contribute to evoking the post-inspiratory depolarization. However, this depolarization cannot be explained solely by post-inhibitory rebound (Barillot et al., 1990; Takeda and Haji, 1993;

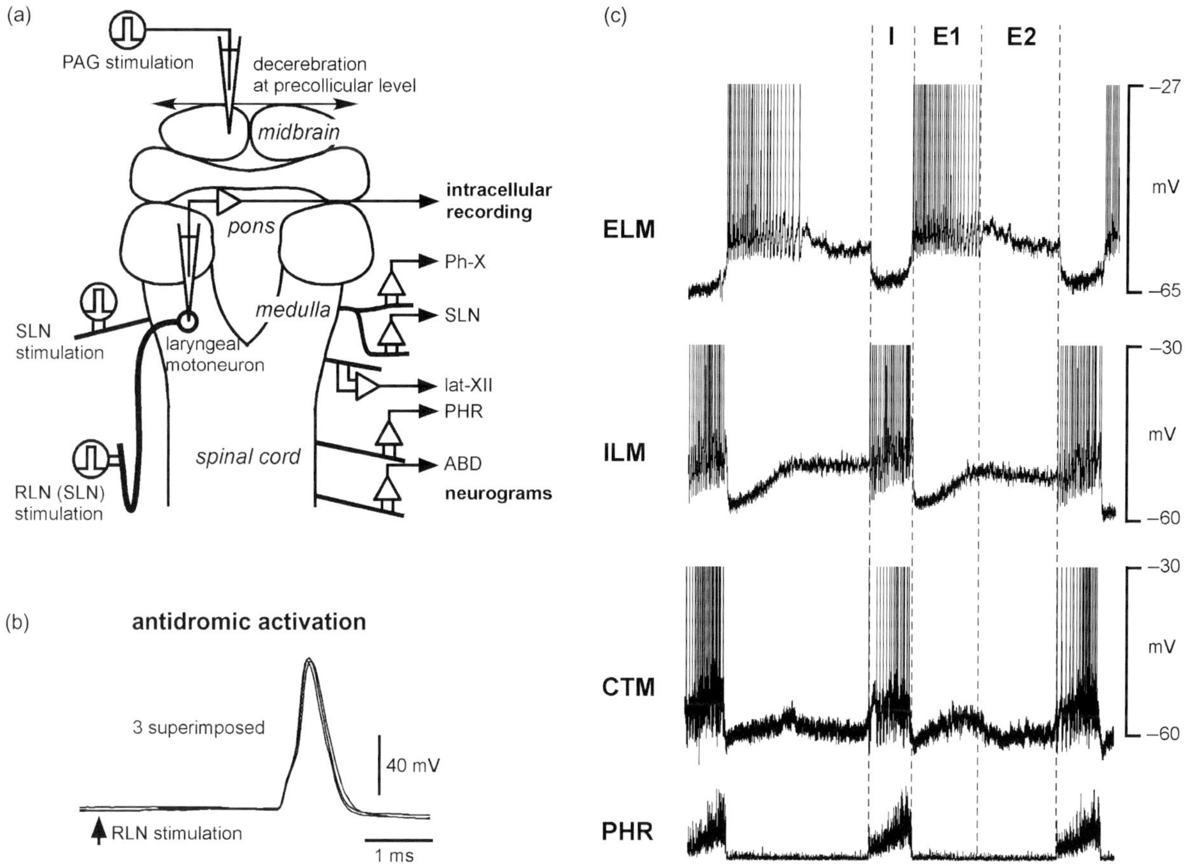

Fig. 1. (a) Scheme of the dorsal view of the cat's brainstem and spinal cord with the arrangement of the stimulating and recording electrodes employed in the experiments providing the data in Figs. 1–5. Membrane potentials of laryngeal motoneurons were recorded during fictive breathing, coughing, sneezing, swallowing and vocalization in decerebrate paralyzed cats. Fictive coughing and sneezing were induced by electrical stimulation of the superior laryngeal nerve (SLN). Fictive sneezing was induced by mechanical stimulation of the nasal mucosa. Fictive vocalization was induced by electrical stimulation of the periaqueductal gray (PAG). These fictive behaviors were identified by the activity patterns of the neurograms recorded from the phrenic (PHR), abdominal (ABD), lateral branch of the hypoglossal (lat-XII), SLN, and pharyngeal branch of the vagus nerves (Ph-X). (b) Antidromic activations from the recurrent laryngeal nerve (RLN) evoked in an expiratory laryngeal motoneuron (ELM), showing constant latency from the stimulus onset. (c) Rhythmic changes in membrane potentials of an ELM, inspiratory laryngeal motoneuron (ILM), and cricothyroid muscle motoneuron (CTM). Inspiration (I), stage one of expiration (E1; post-inspiratory phase), and stage two of expiration (E2; active expiration) were identified by the activities of PHR, ELM, and ABD (not shown in here), respectively. Time duration of respiratory cycles of membrane potential trajectories in Fig. 1–5 were matched using a computer.

Ono et al., 2006). Baekey et al. (2001) have reported that some expiratory neurons in the rostral ventral respiratory group evoked excitation in the recurrent laryngeal nerve. However, we could not identify the expiratory neurons which evoke monosynaptic excitatory postsynaptic potentials (EPSPs) in the ELMs (Ono et al., 2006). It should be noted that, based on the data from chemical stimulation, Dutschmann and Herbert (2006) have argued that the Kölliker-Fuse nucleus contains laryngeal premotor neurons responsible for post-inspiratory activity.

III. Airway protective reflexes

III.A. Coughing and sneezing

III.A.1. Roles of the larynx in coughing and sneezing Coughing and sneezing are the airway's protective reflexes evoked by stimulation of the laryngotracheal and nasal mucosae, respectively, that yield an explosive expiratory airflow to expel foreign bodies and irritants. The glottic movements are the key in the generation of explosive airflow during coughing and sneezing,

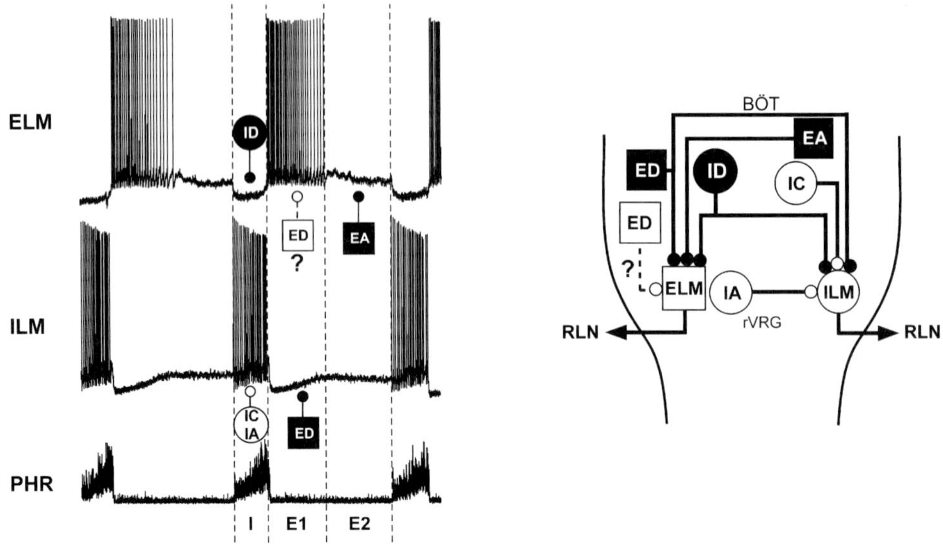

Fig. 2. Activity of ELM, ILM and PHR (left panel), and synaptic connections between medullary respiratory neurons and laryngeal motoneurons (right panel). Inspiratory neurons (circles); expiratory neurons (squares); excitatory neurons (open symbols); inhibitory neurons (filled symbols). BÖT: Bötzinger complex; EA: expiratory neuron with augmenting firing pattern; ED: expiratory neuron with decrementing firing pattern; IA: inspiratory neuron with augmenting firing pattern; IC: inspiratory neuron with constant firing pattern; ID: inspiratory neuron with decrementing firing pattern; rVRG: rostral ventral respiratory group.

and both exhibit the same series of glottic movements. Coughing and sneezing events can be divided into four phases related to changes in glottal caliber: the inspiratory; compressive; expulsive; and narrowing phases (Korpáš and Tomori, 1979). Following the inspiratory phase with dilation of the glottis, it closes in the compressive phase with powerful expiratory muscle activation resulting in an abrupt rise in tracheal pressure. The glottis then dilates transiently at the peak of the tracheal pressure, releasing an explosive expiratory airflow during the expulsive phase. The glottis then constricts again in the narrowing phase.

III.A.2. Differences between coughing and sneezing

Beside the difference in the mucosal stimulation sites, the difference between sneezing and coughing is in the expiratory airflow pathway; in sneezing the forced expiratory airflow through the nose is much higher than that in coughing. This airflow difference is caused by tongue movements. The styloglossus muscle (i.e., the elevator of the back of the tongue) is activated during the expulsion phase of sneezing, but is dormant during the same phase of coughing (Satoh et al., 1998). The elevation of the back of the tongue increases the oral airway resistance, and thus produces a strong airflow through the nose during sneezing. Although it had been thought that the nasopharynx movements are one of the mechanisms responsible for the airflow difference (i.e., opening and closing the nasopharynx might cause the expulsive airflow through the nose in sneezing and through the mouth in coughing, respectively, like a valve between the nose and pharynx), there is not such a big difference in the activities of the pharyngeal constrictors or soft palate elevator between coughing and sneezing (Satoh et al., 1998). Thus, not the nasopharyngeal activity but the tongue movement is the primary factor responsible for the expiratory airflow difference between sneezing and coughing.

III.A.3. Laryngeal motoneuronal activities during coughing and sneezing

Fig. 3 and Fig. 4 show typical activities of laryngeal motoneurons during coughing and sneezing, respectively. The activities of ELMs and ILMs are sharply synchronized with four phases of these reflexes (Shiba et al., 1999). ELMs hyperpolarize during the inspiratory phase (phase I in Fig. 3 and Fig. 4), and then depolarize strongly at the transition from the inspiratory to the expiratory phase (compressive phase; phase II in Fig. 3 and Fig. 4). The membrane potential repolarizes transiently after this depolarization (expulsive phase; phase III in Fig. 3 and Fig. 4), then depolarize again for the remainder of the abdominal burst (narrowing phase; phase IV in Fig. 4). ILMs display the reverse image of

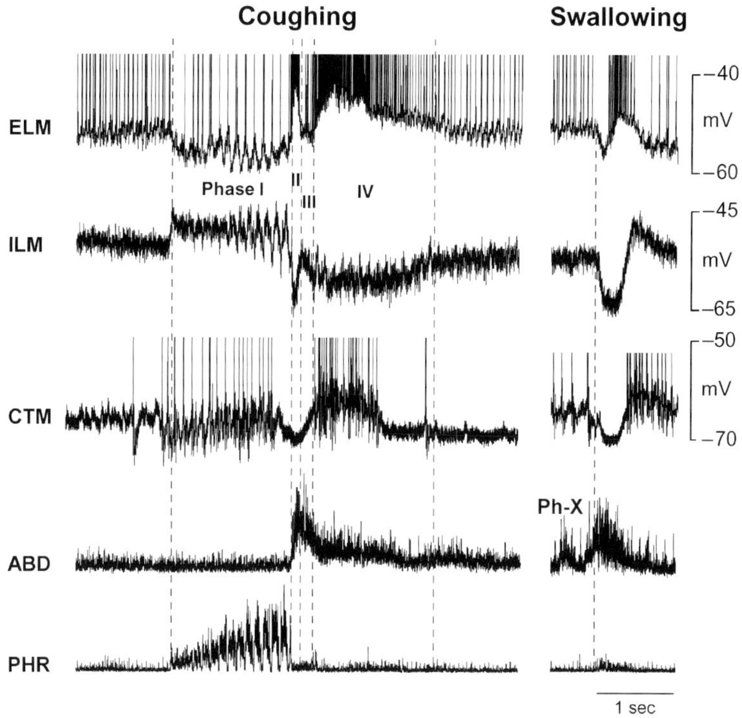

Fig. 3. Membrane potential changes in laryngeal motoneurons during fictive coughing and swallowing induced by SLN stimulation. Fictive coughing under paralyzed conditions is identified by bursting activity in the ABD preceded by increased and prolonged PHR activity. Fictive swallowing is identified by the bursting activity of the Ph-X. Phases I, II, III and IV correspond to the inspiratory, compressive, expulsive and narrowing phases of coughing. A vertical dotted line in the right panel of the figure indicates the onset of ELM hyperpolarization. All pairs of figures in coughing and swallowing in the same rows were recorded from the same motoneurons.

the membrane potential trajectory of ELMs. A coughing event itself or ELMs sometimes lacks the expulsive phase under paralyzed conditions (Shiba et al., 1999; Gestreau et al., 2000). The membrane potential trajectory of CTM during coughing is obviously different from that of ELMs and ILMs. CTMs do not change their membrane potentials in synchrony with the compressive or expulsive phase. CTMs hyperpolarize throughout these phases, and this hyperpolarization starts before the onset of the compressive phase. Because the trajectory is similar to that of the phrenic motoneurons (Grélot et al., 1992), there may be a common synaptic source between the CTMs and phrenic motoneurons during coughing.

III.A.4. Respiratory neuronal network responsible for coughing- and sneezing-related laryngeal movements

The data obtained from the activity of medullary respiratory neurons during coughing and sneezing imply that there is a large overlap between the medullary respiratory circuitries and the central pattern generators (CPGs) of coughing and sneezing (Batsel and Lines, 1978; Jakuš et al., 1985; Orem and Brooks, 1986; Wallois et al., 1992; Oku et al., 1994; Shannon et al., 2000; Pantaleo et al., 2002; Shiba et al., 2007). Baekey et al. (2001) have systematically recorded activities of laryngeal premotor neurons in medullary respiratory neurons during coughing in cats, and argued that these premotor neurons help to shape the motoneuronal activity not only during breathing, but also during coughing. There have been scarcely any studies on the activity of interneurons projecting to the laryngeal premotor neurons, including rhythm generator neurons, although the anatomical locations of the interneurons are suggested by experimental results after the injection of pseudorabies virus, as a trans-synaptic tracer, into intralaryngeal muscles (Fay et al., 1993; Barrett et al., 1994; Waldbaum et al., 2001) and by the expression of the gene c-Fos after coughing and sneezing (Gestreau et al., 1997; Wallois et al., 1995). Further studies of neuronal recordings from these interneurons

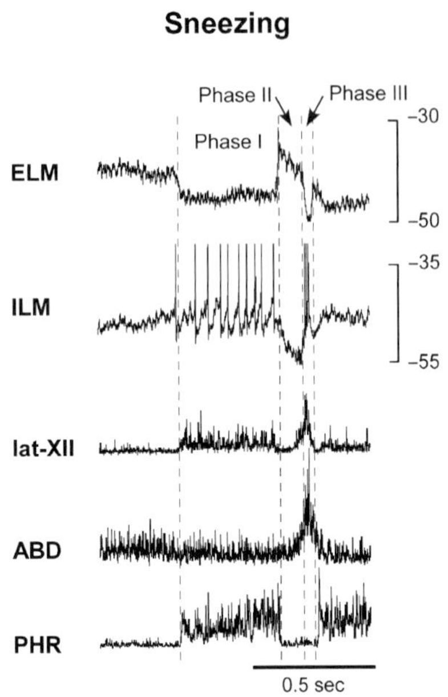

Fig. 4. Membrane potential trajectories of laryngeal motoneurons during fictive sneezing induced by nasal mucosa stimulation. Fictive sneezing is identified by bursting activities in the ABD and lat-XII preceded by increased and prolonged PHR activity. Phases I, II and III correspond to the inspiratory, compressive and expulsive phases of sneezing. Because sneezing was induced repeatedly, the inspiratory phase (I) began just after the expulsive phase (III) without the narrowing phase.

based on these experimental results are necessary for clarifying the function of the CPGs.

III.B. Swallowing

III.B.1. Laryngeal activities during swallowing

The swallowing reflex produces sequential activation of the tongue, pharyngeal and laryngeal muscles to propel the food bolus from the oral cavity to the esophagus without aspiration of food into the airways (Doty and Bosma, 1956; Umezaki et al., 1998). The larynx plays a critical role in swallowing. Glottic closure and laryngeal elevation during the pharyngeal stage of swallowing prevent aspiration of food and facilitate transfer of the bolus from the pharynx to the esophagus. Laryngeal muscle motoneurons receive a complex combination of excitatory and inhibitory synaptic inputs and generate laryngeal muscle activity for swallowing (Zoungrana et al., 1997; Shiba et al., 1999; Gestreau et al., 2000; Saito et al., 2002; Numasawa et al., 2004). As shown in Fig. 3 (on the right), ELMs hyperpolarize briefly at the onset of the pharyngeal stage. Subsequently, they depolarize strongly during the latter part of pharyngeal stage. The preceding inhibition determines the timing of the initiation of ELM excitation (Jean, 2001). ILMs exhibit various responses during swallowing. Some of the ILMs hyperpolarize during the pharyngeal stage, followed by depolarization at the offset of the stage (shown in Fig. 3). The other neurons slightly depolarize or do not change the membrane potential during fictive swallowing. Thus, in contrast to coughing and sneezing, ILMs do not exhibit the reversed image of membrane potential trajectory of ELMs during swallowing. CTMs also show various activities similar to those of the ILMs during swallowing. Depolarization followed by hyperpolarization in the ILMs and CTMs may be partially attributed to the post-inhibitory rebound.

III.B.2. Premotor neurons for swallowing

Because there is a precise coordination of swallowing and respiration-related events, medullary respiratory neurons participate in swallowing. However, considering both their activities during swallowing (Oku et al., 1994; Saito et al., 2003) and their excitatory and inhibitory actions on laryngeal motoneurons (Fig. 2), they are not the main source of swallowing-related laryngeal activity. For example, one class of inhibitory expiratory neurons contributes to the ELM hyperpolarization during breathing and coughing, but does not help in shaping its activity during swallowing (Shiba et al., 2007). Anatomical and neuronal recording studies have suggested that the laryngeal premotor neurons responsible for swallowing are included in the nucleus of the solitary tract and in the reticular formation between it and the nucleus ambiguus (Altschuler, 2001; Jean, 2001). However, recordings of swallowing-related activities of the interneurons identified as laryngeal premotor neurons have rarely been reported.

IV. Vocalization

IV.A. Role of intralaryngeal muscles during vocalization

Although coughing and sneezing, and sometimes breathing, have audible acoustic components, these sounds are neither vocalizations nor intentional communication signals. Generation of vocalization has its own complex mechanisms. Forced expiratory airflow

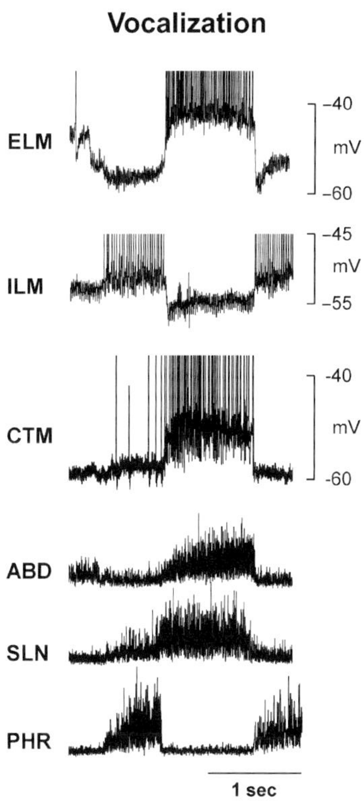

Fig. 5. Intracellular activities of laryngeal motoneurons during fictive vocalization induced by PAG stimulation. Fictive vocalization was identified by the bursting activities of the SLN and ABD.

with glottal narrowing vibrates the vocal fold membrane and evokes vocalization. The vocal folds are adducted by contractions of the interarytenoid, lateral cricoarytenoid and TA muscles during vocalization. These muscles control the degree of glottic narrowing. The TA muscle also increases the stiffness of vocal folds. The CT muscle contraction elongates and tenses the vocal folds, and thus controls vocal pitch. Although the PCA muscle is usually silent during vocalization, it is active during some specific vocalization, such as hissing in cats and during articulation of some consonants in humans (Jürgens, 2002).

IV.B. Laryngeal motoneuron activities during vocalization

Fig. 5 depicts representative membrane potential trajectories of laryngeal motoneurons during vocalization induced by electrical stimulation of the periaqueductal gray (PAG). ELMs rapidly depolarize during vocalization, while ILMs hyperpolarize or slightly depolarize (Shiba et al., 1999). Although CTMs are usually activated during inspiration of normal breathing (Fig. 1), they remarkably depolarize during vocalization (Numasawa et al., 2004). Such drastic depolarization is observed only during vocalization. The activity pattern of the CTM indicates that it plays a main role in vocalization but only a minor role in breathing, coughing and swallowing (Numasawa et al., 2004).

IV.C. Roles of the nucleus retroambiguus and medullary respiratory neurons in vocal-related laryngeal activity

Neurons in the laryngeal motor cortex may directly control laryngeal motoneurons in humans (Kuypers, 1958; Brown et al., 2008). This direct connection is thought to be specific to humans and to be used in voluntary speech. On the other hand, elementary vocalization such as a pain cry is evoked through other pathways, in which the PAG plays an important role (Jürgens, 2002). Although the laryngeal premotor neurons necessary for vocalization have not been well-identified, the nucleus retroambiguus and its surrounding reticular formation are a candidate for anatomical locations of interneurons responsible for PAG-induced vocalization. Based on neuroanatomical tracing studies, Holstege (1989) proposed that PAG-related vocalization is induced by the projections from PAG via the nucleus retroambiguus to motoneurons necessary for vocalization, including laryngeal and abdominal motoneurons. Neurotoxin kainic acid injected into the nucleus retroambiguus and its surrounding reticular formation abolished PAG-induced adductor activation, but preserved the respiratory-related adductor activation and the PAG-induced abductor activation (Shiba et al., 1997). In addition, the lesion also preserved the swallowing-related adductor activity (Umezaki et al., 1997). These findings imply that the nucleus retroambiguus and its surrounding reticular formation do not include the interneurons responsible for the respiratory- or swallowing-related adductor activities, but do include neurons for vocal-related activites. On the other hand, the medullary respiratory neuronal circuitry participates in the vocal-related activity of the respiratory pump muscles (Katada et al., 1996; Sakamoto et al., 1996; Nonaka et al., 1999). Its role in the vocal-related adductor activity is largely unknown.

V. Overlap between central pattern generators for breathing and non-respiratory behaviors

The laryngeal premotor neurons in the medullary respiratory network are a candidate for the neuronal source of laryngeal activities during non-respiratory behaviors, including vocalization and several upper airway reflexes. The question arises whether the same respiratory network can be reconfigured to generate various non-respiratory behaviors, or whether each behavior requires an independent network (Dickinson, 1995; Grélot and Bianchi, 1996). If the former question is answered positively, afferent inputs from the airways may reconfigure the breathing CPG to evoke non-respiratory behaviors, and laryngeal premotor neurons may be shared among the CPGs for breathing and non-respiratory behaviors. Neuronal recordings from medullary respiratory neurons strongly suggest such a mechanism for generation of coughing and sneezing. However, additional premotor neurons, other than those of the respiratory circuitry, are necessary for the excitation of respiratory muscles during coughing (Oku et al., 1994). As mentioned in other sections, there is evidence that some laryngeal premotor neurons are not shared between breathing and vocalization, between breathing and swallowing, or between vocalization and swallowing, although the same medullary respiratory circuitry participates in these behaviors. It thus seems that the CPGs for these behaviors possess features of more than one neuronal circuit, which may function both as a set of hard-wired, independent circuits, and also as a self-reorganizing circuit with a varying degree of overlap (Morton and Chiel, 1994; Dickinson, 1995). Numerous studies using invertebrates have revealed the neuronal circuit reorganization of several types of CPGs result from changes in the intrinsic membrane properties or synaptic connectivity of the circuit, and therefore cause multiple neurons in the circuit to switch from one pattern to another (Nusbaum and Beenhakker, 2002). However, the cellular and circuitry mechanisms of the reorganization among CPGs for respiratory and non-respiratory behaviors are still unknown in mammals. These issues remain to be solved.

VI. Conclusions

The larynx is a key organ not only in vocalization, but also in respiration and airway protective reflexes including coughing, sneezing and swallowing. Laryngeal motoneurons receive excitatory and inhibitory synaptic inputs from the CPGs, integrate neuronal information and then send the motor commands to laryngeal muscles through the recurrent laryngeal nerve. Because premotor neurons are shared among the CPGs of these behaviors, these CPGs overlap to varying degrees.

References

Altschuler, S.M., 2001. Laryngeal and respiratory protective reflexes. Am. J. Med. 111 (Suppl. 8A), 90S–94S.

Baekey, D.M., Morris, K.F., Gestreau, C., Li, Z., Lindsey, B.G., Shannon, R., 2001. Medullary respiratory neurones and control of laryngeal motoneurones during fictive eupnoea and cough in the cat. J. Physiol. 534, 565–581.

Barillot, J.C., Grélot, L., Reddad, S., Bianchi, A.L., 1990. Discharge patterns of laryngeal motoneurones in the cat: an intracellular study. Brain Res. 509, 99–106.

Bartlett, D.J., 1986. Upper airway motor systems. In: Cherniack, N.S., Widdicombe, J.G. (Eds.), Handbook of Physiology: Section 3, The Respiratory System, Vol. II. The American Physiological Society, Bethesda, MD, pp. 223–245.

Barrett, R.T., Bao, X., Miselis, R.R., Altschuler, S.M., 1994. Brain stem localization of rodent esophageal premotor neurons revealed by transneuronal passage of pseudorabies virus. Gastroenterol. 107, 728–737.

Batsel, H.L., Lines, A.J., 1978. Discharge of respiratory neurons in sneezes resulting from ethmoidal nerve stimulation. Exper. Neurol. 58, 410–424.

Brown, S., Ngan, E., Liotti, M., 2008. A larynx area in the human motor cortex. Cerebral Cortex 18, 837–845.

Davis, P.J., Nail, B.S., 1984. On the location and size of laryngeal motoneurons in the cat and rabbit. J. Comp. Neurol. 230, 13–32.

Dickinson, P.S., 1995. Interactions among neural networks for behavior. Curr. Opin. Neurobiol. 5, 792–798.

Doty, B.W., Bosma, J.F., 1956. An electromyographic analysis of reflex deglutition. J. Neurophysiol. 19, 44–60.

Dutschmann, M., Herbert, H., 2006. The Kölliker-Fuse nucleus gates the postinspiratory phase of the respiratory cycle to control inspiratory off-switch and upper airway resistance in rat. Eur. J. Neurosci. 24, 1071–1084.

Ezure, K., 1990. Synaptic connections between medullary respiratory neurons and considerations on the genesis of respiratory rhythm. Prog. Neurobiol. 35, 429–450.

Fay, R., Gilbert, K.A., Lydic, R., 1993. Pontomedullary neurons transsynaptically labeled by laryngeal pseudorabies virus. Neuroreport 5, 141–144.

Gestreau, C., Bianchi, A.L., Grélot, L., 1997. Differential brainstem fos-like immunoreactivity after laryngeal-induced coughing and its reduction by codeine. J. Neurosci. 17, 9340–9352.

Gestreau, C., Grélot, L., Bianchi, A.L., 2000. Activity of respiratory laryngeal motoneurons during fictive coughing and swallowing. Exper. Brain Res. 130, 27–34.

Grélot, L., Bianchi, A.L., 1996. Multifunctional medullary respiratory neurons. In: Miller, A.D., Bianchi, A.L., Bishop, B.P. (Eds.), Neural Control of the Respiratory Muscles. CRC Press, Boca Raton, FL, pp. 297–304.

Grélot, L., Milano, S., Portillo, F., Miller, A.D., Bianchi, A.L., 1992. Membrane potential changes of phrenic motoneurons during fictive vomiting, coughing, and swallowing in the decerebrate cat. J. Neurophysiol. 68, 2110–2119.

Holstege, G., 1989. Anatomical study of the final common pathway for vocalization in the cat. J. Comp. Neurol. 284, 242–252.

Jakuš, J., Tomori, Z., Stransky, A., 1985. Activity of bulbar respiratory neurones during cough and other respiratory tract reflexes in cats. Physiologia Bohemoslov. 34, 127–136.

Jean, A., 2001. Brain stem control of swallowing: neuronal network and cellular mechanisms. Physiol. Rev. 81, 929–969.

Jiang, C., Lipski, J., 1990. Extensive monosynaptic inhibition of ventral respiratory group neurons by augmenting neurons in the Bötzinger complex in the cat. Exper. Brain Res. 81, 639–648.

Jürgens, U., 2002. Neural pathways underlying vocal control. Neurosci. Biobehav. Rev. 26, 235–258.

Katada, A., Sugimoto, T., Utsumi, K., Nonaka, S., Sakamoto, T., 1996. Functional role of ventral respiratory group expiratory neurons during vocalization. Neurosci. Res. 26, 225–233.

Korpáš, J., Tomori, Z., 1979. Cough and Other Respiratory Reflexes. S. Karger, Basel, Switzerland.

Kuypers, H.G., 1958. Corticobular connexions to the pons and lower brain-stem in man: an anatomical study. Brain 81, 364–388.

Morton, D.W., Chiel, H.J., 1994. Neural architectures for adaptive behavior. Trends Neurosci. 17, 413–420.

Nonaka, S., Katada, A., Sakamoto, T., Unno, T., 1999. Brain stem neural mechanisms for vocalization in decerebrate cats. Ann. Otol. Rhinol. Laryngol. Suppl. 108, 12–24.

Numasawa, T., Shiba, K., Nakazawa, K., Umezaki, T., 2004. Membrane potential changes in vocal cord tensor motoneurons during breathing, vocalization, coughing and swallowing in decerebrate cats. Neurosci. Res. 49, 315–324.

Nusbaum, M.P., Beenhakker, M.P., 2002. A small-systems approach to motor pattern generation. Nature 417, 343–350.

Oku, Y., Tanaka, I., Ezure, K., 1994. Activity of bulbar respiratory neurons during fictive coughing and swallowing in the decerebrate cat. J. Physiol. 480, 309–324.

Ono, K., Shiba, K., Nakazawa, K., Shimoyama, I., 2006. Synaptic origin of the respiratory-modulated activity of laryngeal motoneurons. Neuroscience 140, 1079–1088.

Orem, J., Brooks, E.G., 1986. The activity of retrofacial expiratory cells during behavioral respiratory responses and active expiration. Brain Res. 374, 409–412.

Pantaleo, T., Bongianni, F., Mutolo, D., 2002. Central nervous mechanisms of cough. Pulm. Pharmacol. Ther. 15, 227–233.

Saito, Y., Ezure, K., Tanaka, I., 2002. Intracellular activity of superior laryngeal nerve motoneurons during fictive swallowing in decerebrate rats. Brain Res. 956, 262–267.

Saito, Y., Ezure, K., Tanaka, I., Osawa, M., 2003. Activity of neurons in ventrolateral respiratory groups during swallowing in decerebrate rats. Brain Dev. 25, 338–345.

Sakamoto, T., Nonaka, S., Katada, A., 1996. Control of respiratory muscles during speech and vocalization. In: Miller, A.D., Bianchi, A.L., Bishop, B.P. (Eds.), Neural Control of the Respiratory Muscles. CRC Press, Boca Raton, FL, pp. 249–258.

Satoh, I., Shiba, K., Kobayashi, N., Nakajima, Y., Konno, A., 1998. Upper airway motor outputs during sneezing and coughing in decerebrate cats. Neurosci. Res. 32, 131–135.

Shannon, R., Baekey, D.M., Morris, K.F., Li, Z., Lindsey, B.G., 2000. Functional connectivity among ventrolateral medullary respiratory neurones and responses during fictive cough in the cat. J. Physiol. 525, 207–224.

Shiba, K., Umezaki, T., Zheng, Y., Miller, A.D., 1997. The nucleus retroambigualis controls laryngeal muscle activity during vocalization in the cat. Exper. Brain Res. 115, 513–519.

Shiba, K., Satoh, I., Kobayashi, N., Hayashi, F., 1999. Multifunctional laryngeal motoneurons: an intracellular study in the cat. J. Neurosci. 19, 2717–2727.

Shiba, K., Nakazawa, K., Ono, K., Umezaki, T., 2007. Multifunctional laryngeal premotor neurons: their activities during breathing, coughing, sneezing, and swallowing. J. Neurosci. 27, 5156–5162.

Takeda, R., Haji, A., 1993. Mechanisms underlying post-inspiratory depolarization in post-inspiratory neurons of the cat. Neurosci. Lett. 150, 1–4.

Umezaki, T., Zheng, Y., Shiba, K., Miller, A.D., 1997. Role of nucleus retroambigualis in respiratory reflexes evoked by superior laryngeal and vestibular nerve afferents and in emesis. Brain Res. 769, 347–356.

Umezaki, T., Shiba, K., Zheng, Y., Miller, A.D., 1998. Upper airway motor outputs during vomiting versus swallowing in the decerebrate cat. Brain Res. 781, 25–36.

Waldbaum, S., Hadziefendic, S., Erokwu, B., Zaidi, S.I., Haxhiu, M.A., 2001. CNS innervation of posterior cricoarytenoid muscles: a transneuronal labeling study. Resp. Physiol. 126, 113–125.

Wallois, F., Macron, V., Jounieaux, V., Duron, B., 1992. Influence of trigeminal nasal afferents on bulbar respiratory neuronal activity. Brain Res. 599, 105–116.

Wallois, F., Gros, F., Masmoudi, G., Larnicol, N., 1995. C-fos-like immunoreactivity in the cat brainstem evoked by sneeze-inducing air puff stimulation of the nasal mucosa. Brain Res. 687, 143–154.

Yoshida, Y., Miyazaki, T., Hirano, M., Shin, T., Kanaseki, T., 1982. Arrangement of motoneurons innervating the intrinsic laryngeal muscles of cats as demonstrated by horseradish peroxidase. Acta Oto-Laryngol. 94, 329–334.

Zoungrana, O.R., Amri, M., Car, A., Roman, C., 1997. Intracellular activity of motoneurons of the rostral nucleus ambiguus during swallowing in sheep. J. Neurophysiol. 77, 909–922.

CHAPTER 9.2

Vocal–respiratory interactions in the parabrachial nucleus

Michael Smotherman[1*], Christine Schwartz[1] and Walter Metzner[2]

[1]*Biology Department, Texas A&M University, College Station, TX, USA*
[2]*Department of Physiological Sciences, UCLA, Los Angeles, CA, USA*

Abstract: Vocal motor patterning in mammals depends on a network of midbrain and brainstem sensory–motor integration centers that coordinate activity in facial, laryngeal and respiratory muscles. The temporary subjugation of respiratory drive for vocalizing and the entrainment of vocal timing to respiratory rhythms are achieved by a combination of descending motor commands to the respiratory rhythm centers and an overlapping network of somatosensory feedback pathways from the larynx and respiratory system. The parabrachial nucleus is a pontine relay center for the integration of sensory feedback influences on normal breathing, and although its functioning is necessary for normal vocal behaviors the precise nature of its contributions are poorly understood. Here we provide a review of current theories about how the parabrachial nucleus may be facilitating and guiding the expression of vocal motor programs by acting as a state-dependent filter that modulates the normal functioning of respiratory reflexes during vocalization. Understanding how somatosensory feedback influences the dynamic properties of the mammalian vocal motor pathway is an essential step towards grasping the neural representation of speech motor control in the human brain.

Keywords: respiration; vocalizing; sensorimotor; parabrachial nucleus; speech; motor control; larynx; somatosensory feedback; bicuculine

Abbreviations: BMI: bicuculline methiodide; FLI: fos-like immunoreactivity; KF: Kölliker-Fuse nucleus; NST: solitary tract nucleus; NA: nucleus ambiguus; NRA: nucleus retroambiguus; PAG: periaqueductal gray; PB: parabrachial nucleus; PL: paralemniscal area; PRG: pontine respiratory group; VRG: ventral respiratory group

I. Introduction

One of the most challenging goals in neuroscience is to understand how multiple neural networks subserving different motor systems can be coordinated to produce complex behaviors. The execution of a vocal motor command requires the coordinated excitation of many different respiratory and laryngeal muscles activated in precisely defined sequences and with different forces. Coordination of respiratory and laryngeal activity is not unique to vocalizing, but is also essential for a variety of other oral behaviors such as eating, drinking, coughing and vomiting. Each of these behaviors is regulated by a combination of automatic and volitional control mechanisms which must interact with the neural control of breathing. Automatic breathing is controlled by a medullary bulbospinal network that maintains respiratory rhythms (Feldman et al., 2003; Rybak et al., 2004), but this network may be temporarily subjugated by a volitional corticospinal pathway (Butler, 2007). Human speech may differ from other mammalian vocalizations in the extent to which corticospinal pathways circumvent the bulbospinal respiratory control pathways to

*Corresponding author. E-mail: smotherman@tamu.edu

ISBN 978-0-12-374593-4
DOI: 10.1016/B978-0-12-374593-4.00035-8

activate the respiratory musculature during vocalizing (Jarvis, 2004; Jarvis et al., 2005), but ongoing speech is also exquisitely sensitive to somatosensory feedback ascending through the brainstem via a suite of highly-conserved afferent pathways that generally serve all oral and respiratory behaviors. To achieve a complete understanding of how the vocal motor pathways function in humans or mammals generally, it is necessary to understand the dynamic properties of each of the underlying automatic control circuits that may be activated and contribute in real-time to the vocalization process. The parabrachial nucleus, situated at the rostral border of the dorsolateral pons, appears to be a central relay center for many of these circuits, acting as an interface for behavior-specific plasticity in respiratory control (Alheid et al., 2004), and therefore understanding how it functions during vocalizing and how it is interconnected with the rest of the vocal motor pathway is essential.

Respiratory adaptations for vocalizing (or speech sounds) include examples of volitional control of inspiration, such as taking a deeper than normal breath before speaking, or phase shifts in respiratory rhythms such as those that occur naturally during singing or when a speaker postpones inspiration to complete a sentence. Unlike the "simple" vocalizations emitted by most mammals, human speech requires long sequences of sounds, sentences and phrases to be interwoven with periodic rapid inspirations. The timing of inspirations within continuous speech contributes critically to the information content of these complex phonations, and for this reason the ability of the human vocal motor pathway to regulate phase-shifts in respiratory rhythms constitutes a key component of human speech and language. Normal sounding speech also relies on the precise control of subglottic pressure, and the use of subtle fluctuations in loudness to enhance meaning is an important element of speech production. Such examples of vocal–respiratory coordination are difficult to study in animal models, but the neural mechanisms regulating speech breathing are likely to share more than a cursory similarity with other mammals, owing to the conserved nature of the respiratory neurocircuitry in mammals (Farley et al., 1992a; Alheid et al., 2004). For mammals other than humans, vocalizations interact with breathing in ways that are not very different from humans: (1) vocalizing typically requires augmented expiratory force; which (2) may be facilitated by anticipatory augmentation of the preceding inspiration; (3) most mammals appear capable of adjusting the loudness of their call for a variety of reasons, which is achieved by manipulating expiratory force; and (4) extended multisyllabic vocal sequences, although rare among mammals, can be accommodated by phase shifts in respiratory rhythms, especially phase lags created by postponing inspirations. Thus, the respiratory component of human speech may differ from other mammal's vocal behaviors in its precision, but not so much in nature. This is in stark contrast to the laryngeal component of human speech, which apparently reflects a major evolutionary transition from the limbic-based visceromotor vocal pattern generators used by most mammals (Jürgens, 2002) to the more sophisticated corticospinal control of laryngeal motor neurons that underlies human speech (Jarvis, 2004; Jarvis et al., 2005).

II. The functional connectivity of the parabrachial nucleus

The parabrachial complex (PB) is a group of cells surrounding the superior cerebellar peduncle near the junction of the pons and mesencephalon (Fig. 1). It is involved in a variety of sensory integration functions for which it receives and relays ascending afferent inputs, including taste (Cechetto, 1987; DiLorenzo, 1988), nociception (Blomqvist et al., 1989; Ma et al., 1989), cardiopulmonary control (Cechetto, 1987) and most notably regulation of breathing (Cohen, 1971; Alheid et al., 2004). Afferent inputs to the PB have been found to derive from the solitary tract nucleus (NST) (Herbert et al., 1990; Ezure, 2004; Kubin et al., 2006), the nucleus ambiguus (NA) and nucleus retroambiguus (NRA) (Holstege, 1989), and the spinal cord and trigeminal nuclei (Blomqvist et al., 1989; Ma et al., 1989). The PB relays these inputs to several higher brain areas, including the periaqueductal gray (PAG), with which it makes extensive reciprocal connections (Krout et al., 1998), the thalamus (Halsell, 1992; Bester et al., 1999), the dorsal and superior central raphe nuclei (Saper and Loewy, 1980), the hypothalamus (Bester et al., 1997; Saper and Loewy, 1980), the amygdala (Bernard et al., 1993; Saper and Loewy, 1980), the bed nucleus of the stria terminalis (Alden et al., 1994) and several cortical areas, including somatosensory, motor and cingular cortices (Saper and Loewy, 1980). The ascending projections from the PB to the hypothalamic regions and amygdala are believed to mediate the transmission of gustatory and nociceptive stimuli and may be especially important

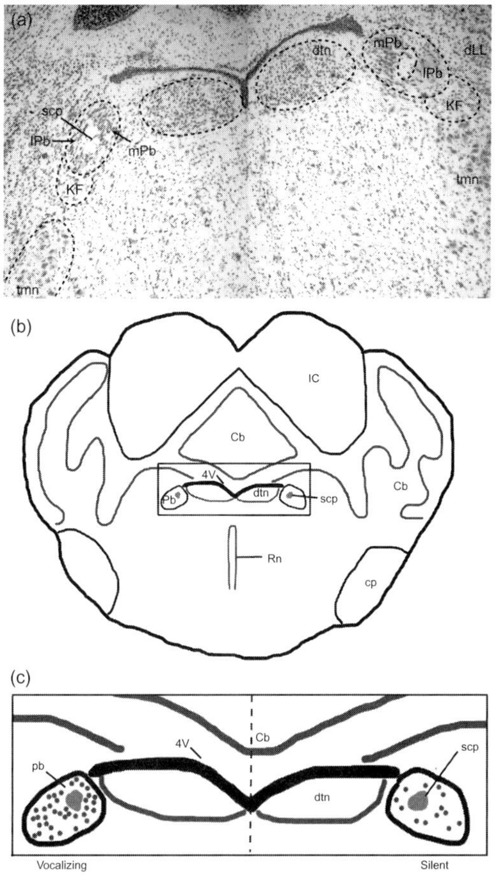

Fig. 1. (a) Cresyl-violet stained coronal section through the midbrain of a free-tailed bat (*Tadarida brasiliensis*) illustrating the position of, and anatomical landmarks associated with, the parabrachial nucleus in mammals. (b) Line drawing presenting the position of the parabrachial nucleus and neighboring brain regions. (c) Enlargement of the boxed region in (b) showing representative c-Fos immunohistochemistry in a vocalizing bat, shown in the left side of the box, versus a silent bat, shown in the right side of the box. Vocalizing bats emitted calls at rates of 5–10 calls per second for thirty minutes and were then isolated in a soundproof room for one hour before sacrifice. Silent bats were inhibited from calling by the presence of the investigator, but otherwise went through the same procedures. Frozen sections were processed with antibodies for c-Fos. Stained nuclei were quantified for each group and a representative animal is shown. Vocalizing bats exhibited significantly more cells within the PB exhibiting Fos-like immunoreactivity (dots) than silent bats (35.4 ± 9.6 cells per PB versus 13.1 ± 5.9, $n = 10$ vocalizing and 10 silent bats, Student t-test, $P < 0.01$), but the increase was not confined to any specific subregion of the PB. Abbreviations: 4V: fourth ventricle; Cb: cerebellum; cp: cerebral peduncle; dLL: dorsal lateral lemniscus; dtn: dorsal tegmental nucleus; IC: inferior colliculus; KF: Kîlliker-Fuse nucleus; lPB: lateral parabrachial nucleus; mPB: medial parabrachial nucleus; PB: parabrachial nucleus; Rn: raphe nucleus; scp: superior cerebellar peduncle; tmn: trigeminal motor nucleus.

for feeding and aversive-defensive behaviors (Bernard et al., 1993; Alden et al., 1994; Bester et al., 1997). Thalamic and cortical projections most likely mediate a diverse array of proprioceptive inputs, including laryngeal and respiratory feedback mechanisms controlling breathing, the larynx and the movements of the mouth associated with vocalizing.

The PB receives somatosensory feedback regarding lung status indirectly from the vagus nerve via the NST, and it also receives somatosensory feedback via the NST regarding the mechanical status of the larynx, including subglottic pressure, tension across the vocal folds and the relative positions of the laryngeal cartilages (Feldman and Gautier, 1976; Feldman, 1986; Ezure et al., 1998). Electrical microstimulation of the internal branch of the superior laryngeal nerve, within which lies the laryngeal sensory afferents, up-regulates expression of the immediate early gene c-Fos in both the PB and the dorsolateral PAG, including vocalization-related areas of the PAG (Ambalavanar et al., 1999). Fos-like immunoreactivity (FLI) was observed in both the medial and lateral portions of the cat PB after experimentally-induced coughing (Gestreau et al., 1997), but in a similar study FLI was only observed in the lateral PB after prolonged bouts of experimentally-induced sneezing (Wallois et al., 1995). Apparently coughing and sneezing were similarly effective in stimulating pulmonary afferents, which were represented by the FLI in the lateral PB, but sneezing did not have a similar impact on the laryngeal afferent activity represented by FLI in the medial PB. Fictive coughing also induced FLI in the NST, the NA, the NRA and several components of the ventral respiratory column which collectively provide the central respiratory drive for the cough reflex. Several studies have provided insight into which regions of the PB are most centrally involved in vocalizing (Jürgens and Ploog, 1970; Jürgens and Pratt, 1979; Schuller and Radtke-Schuller, 1990; Farley et al., 1992a; Krout et al., 1998; Hayward et al., 2004), but it seems likely that several PB subregions participate in different ways. In echolocating bats, FLI is significantly enhanced in both the medial and lateral regions of the PB following long bouts of calling (Fig. 1), presumably because both of these areas are activated by sensory feedback during vocalizing.

The PB receives descending inputs from the forebrain, including the vocalizing regions of the anterior cingulate cortex and the laryngeal region of the motor cortex (Jürgens, 1976; Saper, 1982), the substantia nigra, the hypothalamus and the amygdala (Schneider, 1986; Holstege, 1987). Although it is tempting to predict that direct cortico-parabrachial projections mediate the respiratory components of vocalizing, it is also possible that these projections act indirectly on vocalizing by modulating the efficacy of respiratory reflexes during vocalizing. It has been proposed that the PB acts as a state-dependent filter for ascending sensory pathways, with descending inputs effectively gating sensory feedback in a context-dependent manner (Holstege, 1989; Krout et al., 1998). If so, corticospinal projections may be responsible for generating expiratory force, while cortico-parabrachial projections might mitigate the impact of vagal and laryngeal afferent feedback pathways that might otherwise constrain expression of the vocal motor pattern generator. Normal respiratory reflexes might need to be "short-circuited" during vocalizing, in which case a descending vocal motor command to the PB could be viewed as filtering sensory input to respiratory drive in a behavior-dependent manner. An example of this might be when a speaking human postpones inspiration to complete a sentence, despite passing a threshold of lung deflation that would otherwise trigger an instantaneous inspiration.

The dorsolateral region of the periaqueductal gray (PAG) represents a central convergence of descending forebrain inputs and an output network of pontine and medullary nuclei that comprise the central vocal pattern generating circuitry. Stimulations of the PAG elicit a wide range of natural-sounding, species-specific vocalizations (Schuller and Radtke-Schuller, 1990; Sakamoto et al., 1997; Fenzl and Schuller, 2002; Jürgens, 2002). Vocal motor commands initiated in the PAG generally proceed not directly to phonatory motoneuron pools of the medulla, but rather project to an array of midbrain, pontine and medullary nuclei, which in turn interact directly with the vocal premotor nuclei (Larson, 1991; Sakamoto et al., 1997; Fenzl and Schuller, 2002; Jürgens, 2002). The main PAG projections associated with vocalizing include the midbrain paralemniscal area (PL), the medullary reticular formation (see Hage, Chapter 8.3 in this volume), the PB, the NTS, the NA and the NRA (Sakamoto et al., 1997; Shiba et al., 1997a; Jürgens, 2002). Motoneurons housed within the NA and NRA directly mediate the laryngeal components of vocalization, including closure of the glottis and tension across the vocal membranes (Rübsamen and Schuller, 1981; Schuller and Rübsamen, 1981; Rübsamen and Betz, 1986; Rübsamen and Schweizer, 1986; Neuweiler, 2000; Jürgens, 2002). The PB projects to both laryngeal and respiratory premotor neuronal pools (Moga and Caper, 1990; Chamberlin and Saper, 1994; Fung and St. John, 1994), but the medial PB possesses reciprocal connections with the NA (Saper and Loewy, 1980; Herbert et al., 1990) while the lateral PB is reciprocally connected with the NST and ventral respiratory group (VRG) (Segers et al., 2008). While the lateral PB is most often considered in the context of vocal–respiratory coupling, the medial PB likely contributes to the regulation and maintenance of laryngeal activity during vocalizing. However, it is also likely that intrinsic connections exist between medial and lateral PB neurons (Segers et al., 2008) which would certainly have the potential to mediate the coordination of laryngeal and respiratory activities generally in support of oral behaviors such as vocalizing, eating and coughing, although direct evidence of such a function is lacking (see Shiba, Chapter 9.1 in this volume).

III. The parabrachial complex is endowed with a potent influence on respiration

The parabrachial complex (PB) is composed of several functionally and anatomically distinct subunits, and the various functions of the PB are topographically organized (Chamberlin and Saper, 1994; Krout et al., 1998; Chamberlin, 2004; Ezure, 2004). The lateral and ventrolateral PB subunits appear to be the subregions most particularly involved with the coordination of breathing with vocalizing (Jürgens, 2002; Smotherman et al., 2003), but a large number of cells recorded from within the medial PB were also found to exhibit vocalization-related and respiratory-related firing patterns (Farley et al., 1992a). Electrical stimulations of the PB produced simple vocalizations (Jürgens and Ploog, 1970; Jürgens and Müller-Preuss, 1977; Schuller and Radtke-Schuller, 1990; Farley et al., 1992b), and lesions of the monkey PB can prevent vocalizations normally elicited by forebrain electrical stimulations (Jürgens and Pratt, 1979). In monkeys, lesions of the PB essentially uncoupled respiratory rhythms from laryngeal activity during vocalizations elicited by PAG stimulation, indicating that the PB is essential for vocal–laryngeal coordination (Jürgens, 2002). From both a functional and a neuroanatomical standpoint, the parabrachial complex is well-positioned to coordinate laryngeal and respiratory activity during vocalizations, and has been broadly characterized as a likely candidate for a vocal–respiratory interface (Wild et al., 1990; Farley et al., 1992a; Chamberlin and Saper, 1994; Sakamoto et al., 1997; Jürgens, 2002).

The PB is without doubt an important center for sensory feedback control of respiration control (Bianchi and Pasaro, 1997; Feldman et al., 2003). The ventrolateral PB and neighboring ventral Kölliker-Fuse nucleus (KF) constitute what is currently known as the pontine respiratory group, or PRG (Feldman, 1986). Stimulations of the PRG postpone inspirations and lesions of the PRG caused apneustic breathing characterized by prolonged inspirations (see Alheid et al., 2004 for a review of how the PB and KF regulate breathing). Respiratory-related neuronal activity is most prominent in the Kölliker-Fuse portion of the ventrolateral PB, but it has also been observed in a large percentage of cells extending throughout the lateral edge of the PB (Farley et al., 1992a; Chamberlin and Saper, 1994; Dick et al., 1994). Pharmacological manipulations of the ventrolateral PB caused significant changes in respiratory rhythm (Ling et al., 1993; Chamberlin and Saper, 1994), and electrical stimulations caused immediate respiratory phase switching, either from inspiration to expiration or expiration to inspiration, depending on the precise location of the stimulus (Cohen, 1971; Cohen and Feldman, 1977). Similarly, it was found that respiratory phase shifts triggered by electrical stimulation of afferents within the superior laryngeal nerve were mediated by neurons in the ventrolateral PB (Oku and Dick, 1992). However, the lateral PB is still generally considered non-essential for central respiratory rhythm generation (Feldman, 1986; Oku and Dick, 1992; Dick et al., 1994; Feldman and Del Negro, 2006). The lateral PB plays a key role in the phenomenon known as locomotor–respiratory coupling (Potts et al., 2005), in which somatosensory feedback arising from stretch-activated receptors in skeletal muscle can evoke respiratory phase switching via projections from the lateral PB to the ventral respiratory group. These results suggest that the primary role of the PB in motor control is to modulate respiratory rhythms in support of ongoing motor patterns.

Currently there is little compelling evidence that the PB directly administers the expiratory component of vocalization. Electrical stimulations of the PB elicited vocalizations in monkeys (Jürgens and Ploog, 1970; Jürgens and Muller-Preuss, 1977), and lesions of the PB prevented vocalizing elicited by electrical stimulation of higher vocal control areas (Jürgens and Pratt, 1979), but interpretation of these results is confounded by the general importance of the PB for respiratory control and the overlapping confluence of both respiratory and laryngeal somatosensory afferent pathways in the PB. Jürgens (2002) limited his conclusions to the hypothesis that the lateral PB participated in vocal–respiratory coordination perhaps by gating vocal onset based on current respiratory conditions, which is consistent with the larger theme that the PB generally behaves as a state-dependent filter for somatosensory feedback.

IV. Pharmacological manipulations in the parabrachial complex disrupt vocal–respiratory coupling in spontaneously vocalizing bats

Iontophoretic injections of the $GABA_A$ antagonist bicuculline methiodide (BMI) into the lateral PB caused a reduction in call durations and concurrently changed both the range and distribution of call durations and interpulse intervals used by actively echolocating horseshoe bats (Smotherman et al., 2006). Immediately following BMI injections, we presented the bats with artificial echoes that normally provoked the bats to emit two calls per breath instead of one (a behavioral adaptation to increase call emissions in response to ambiguous acoustic cues), and we discovered that following the BMI injections the bats no longer emitted more than one call per breath (percentage of calls emitted as doublets dropped from 37% to 9%). This effect was quantified as a significant reduction in the relative number of interpulse intervals less than 40 ms following BMI injections (40 ms is the mean duration of a single inspiration during active calling, and interpulse intervals less than this are indicative of multiple calls emitted between succeeding inspirations). Repeating these experiments while monitoring breathing with a custom-built pneumotachograph revealed that the BMI injections caused only a modest increase in quiet breathing rate (from 5.3 to 6.5 Hz, $n = 5$ bats), but had a profound impact on breathing during call emissions; specifically, we found that the shorter calls were associated with shorter than normal expiration durations. We interpreted the shorter calls and absence of doublets as an indication that the bats were no longer able to lengthen expiratory durations in support of normal calling behavior.

Alternatively, we found that application of the $GABA_A$ agonist muscimol increased the mean expiration during vocalizing by almost 10 ms relative to controls (a 20% increase). Mean call duration also increased significantly, but the magnitude of the increase was much less than the increase in expiration duration, which meant that unlike the predrug behavior expirations persisted for several milliseconds after vocalizations terminated. Muscimol injections into the lateral PB caused only minor changes in quiet breathing rates. These results are consistent with the hypothesis that GABA-ergic synapses in the lateral PB are important in the context of behavior-specific respiratory plasticity, and in particular it appears that an inhibitory GABA-ergic mechanism is responsible for regulating the timing of inspirations surrounding vocalizing. Blocking GABA-ergic synapses in the lateral PB may have led to shorter calls because inspirations began sooner than normal, and that the GABA agonist may have extended calls and also increased the likelihood of additional calls being emitted within a single expiration, because it postponed the onset of subsequent inspirations. Based on these observations we can extend Jürgens (2002) hypothesis by stating that GABA-ergic synapses in the lateral PB appear to be gating the onset and duration of vocalizations through their control of inspiratory timing.

V. Vocalization-related neuronal firing patterns in the parabrachial complex

Although the putative function of respiratory neurons in the PB was originally confined to inspiration duration (Cohen and Feldman, 1977), it is now recognized as a topographically heterogeneous population (Dick et al., 1994; Bianchi and Pasaro, 1997). In addition to respiratory-related activity, neurons in the PB may also respond to acoustic stimuli, laryngeal and supralaryngeal somatosensory feedback, and exhibit prevocal and vocal–motor activity (Farley et al., 1992a). Farley et al. (1992a) reported that nearly all the neurons they recorded from in the PB showed some cyclical background activity associated with basic respiratory patterns, a few showed activities related to other oral behaviors such as licking and some showed passive acoustic sensitivities. Vocalization-related neuronal firing patterns fell into three main subtypes, characterized by either: (1) increased neuronal activity during vocalizing (the most common); (2) suppressed activity during vocalizing; or (3) abrupt increases in activity preceding and overlapping with the onset of vocalizing (Farley et al., 1992a). Although very few details are known about the source of the modulatory inputs that created these vocalization-related firing patterns, there is evidence that laryngeal afferent activity can modulate respiratory rhythms and strongly influence vocal motor patterns through connections in the parabrachial complex (Davis et al., 1993; Shiba et al., 1995, 1997b; Clark and Farber, 2001).

In our experiments with horseshoe bats we iontophoresed BMI at multiple injection sites throughout the PB to establish stereotaxic coordinates for localized areas that contributed to vocal–respiratory coordination. We then used tungsten electrodes to monitor and characterize extracellular neuronal activity at these sites while awake, unanesthetized bats spontaneously vocalized. Within an area of approximately 250 μm around our target coordinates in the lateral PB we found a heterogeneous population of neurons that exhibited activity correlated to different phases of respiration, vocalizing and/or auditory stimuli. Similar to the results of Farley et al. (1992a) we found several examples of neurons that fired in phase with respiratory rhythms, and some of these exhibited modulations in firing patterns that followed vocalization-related changes in respiratory drive (Fig. 2). It was impossible to know from these recordings whether changes in firing rates caused either changes in respiration (which would indicate a motor function for the PB), or whether they reflected the effects of somatosensory inputs (which is more consistent with the general functions of the PB). Roughly 40% of 50 neurons recorded in the PB were found to have some vocalization-related activities similar to those found in the cat (Farley et al., 1992a); of these, one subset of neurons ($n = 7$) appeared similar to the vocalization-onset neurons reported in the cat, exhibiting a low background firing rate, generally no respiratory-related activity and spiking just before and through the onset of each vocalization (Fig. 3b). A second and more common neuronal subtype ($n = 24$) resembled the vocalization-inhibition neurons of Farley et al. (1992a), possessing generally high spontaneous activity and both pre- and postvocal periods of excitation separated by vocalization-related inhibition (Fig. 3a). These neurons resemble a class of respiratory phase-switching neurons that have the capacity to trigger or postpone the initiation of inspiration. Based on correlations between the firing patterns of these neurons and changes in inspiratory dynamics surrounding call emission, it is hypothesized that this subset of neurons has the potential to play a significant role in actively regulating the vocalization-related changes in breathing patterns. Activity levels abruptly declined at call onset and remained low throughout the

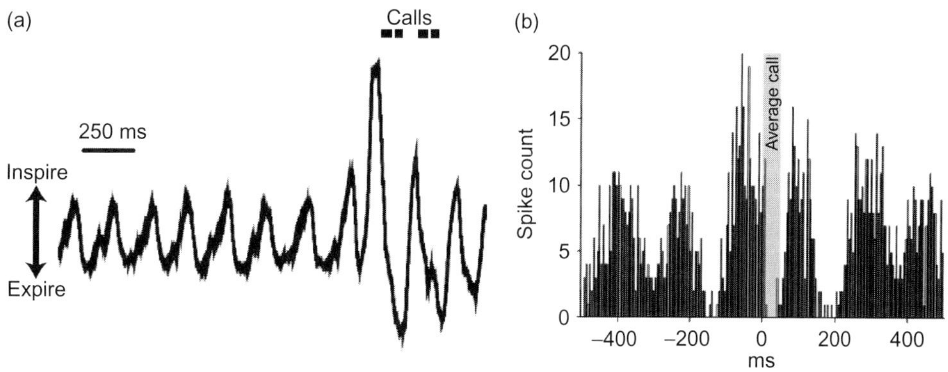

Fig. 2. (a) Pneumotachygraph recording of breathing-related movements of the abdominal cavity before and during spontaneous calling. Upward deflections reflect expansion of the body cavity during inspiration, and downward deflections indicate expirations. The record indicates that the bat took deeper breaths immediately preceding and following call emissions (bars above breathing represent the time-course of calls emitted, in this case two sets of doublets emitted on succeeding breaths). (b) Peri-event histogram of vocalization-related neuronal firing patterns in the lateral PB (n = 20 solitary calls). The neuron exhibited a cyclic respiration-related firing pattern that was exaggerated during calling, with the probability of firing increasing during inspirations that preceded vocalizing. Notice also that firing rates appeared to be suppressed to zero levels during call emissions.

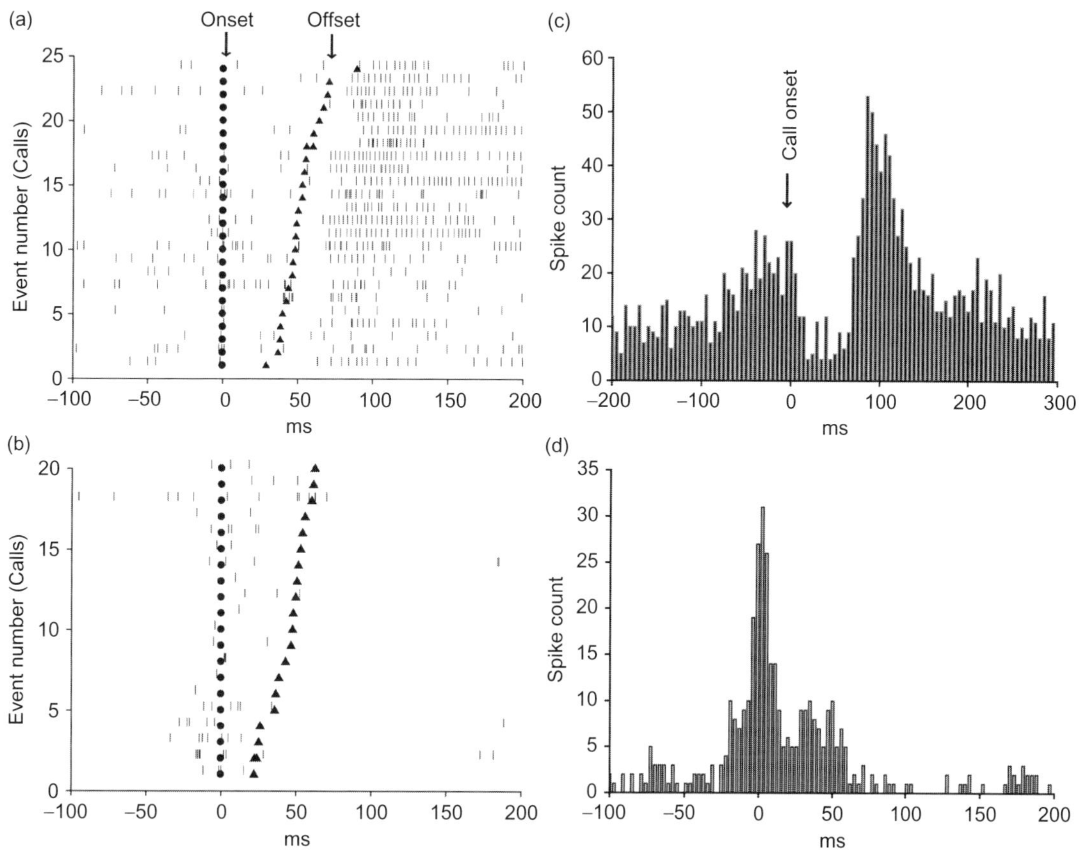

Fig. 3. Examples of two subtypes of vocalization-related activity patterns exhibited by neurons in the lateral PB of the horseshoe bat (*Rhinolophus ferrumequinum*). (a) and (b) are dot-raster representations of neuronal activity surrounding call emissions. Data are ordered by call duration (circles indicate call onset and triangles are offset times). (c) and (d) are peristimulus histograms corresponding to the data in (a) and (b). The neuron shown in (a), (c) was inhibited during vocalizing and exhibited a burst of firing that slightly preceded subsequent inspirations. The neuron in (b), (d) began to fire immediately preceding call onset, and maintained a low level of activity during the call.

vocalization and then rebounded to their highest rates immediately following call offset; this second peak in activity generally preceded the peak of the next inspiratory phase by approximately 15 ms, which indicates that these neurons could have been driving, rather than responding to, changes in inspiration. In three separate trials, BMI applied through a mulit-barreled pipette during extracellular recordings caused the spike rate to rapidly increase and plateau, and no further respiratory or vocal-related modulations in spike rate were observed for the duration of the recordings (approximately two minutes in each case). The BMI applications simultaneously increased breathing rate and altered call temporal patterns, consistent with earlier observations. We speculate that increased activity in these neurons caused by BMI led to premature inspiratory onsets and higher phrenic nerve discharges, which shortened inspirations.

VI. A proposed model of vocal–respiratory rhythm entrainment

It is hypothesized that the putative inspiratory on-switching neurons located in the lateral PB represent part of the previously described population of respiratory phase switching neurons in the lateral PB that govern other forms of locomotor–respiratory coupling (Potts et al., 2005). The region of the PB where we found the vocalization-related activity appears analogous to a site identified as mediating locomotor–respiratory coupling in rodents (Potts et al., 2005) and it appears analogous to a location found to mediate phase shifts in respiratory rhythm caused by direct electrical stimulations of the cat larynx (Oku and Dick, 1992). Based on this background information and the preliminary evidence outlined above, it is possible that the neurons represented in Fig. 3a, and similarly those voc-inhibition neurons identified by Farley et al. (1992a), are inspiratory on-switching neurons, and that the vocalization-related inhibition demonstrated in Fig. 3a,b derives from somatosensory feedback, either from the larynx, pulmonary afferents or diaphragmatic afferents, although a descending inhibitory command cannot be ruled out without further studies. The succeeding spike in activity that follows vocalizing may represent either an inhibitory rebound or a succeeding excitatory input, but in either case serves to link the onset of the next inspiration with the end of a vocalization. Alternatively, extended sequences of vocalizations may postpone inspiratory onset via sustained inhibition of these neurons. Under this model, descending motor commands to the PB are not necessary for phase shifts in respiratory rhythm, but are also not necessarily excluded.

VII. Conclusions

The parabrachial nucleus appears to be a main integration center for a distributed network that coordinates the activities of the laryngeal and respiratory musculature in a behavior-specific manner. There is little evidence that vocal motor patterns are directly expressed through the PB, but considerable evidence that the PB is needed for normal vocalizations, and in particular vocalizations that require substantial adjustments in respiratory rhythms. Descending projections of the vocal motor pathway to the PB likely modulate the dynamic properties of respiratory–laryngeal interactions in ways that facilitate vocalizing. Ascending projections from the PB may contribute to gating of the vocal motor pathway at the level of the PAG or higher and thereby may help regulate of vocal timing. Somatosensory feedback from the larynx and lungs is relayed via the PB to the thalamus and basal ganglia, and may play a critical role in both gating mechanisms and the ongoing maintenance of the acoustic properties of vocalizations. While much is known about the functional connectivity of the PB in the context of respiratory control, surprisingly little is known about the origins, nature and functional significance of descending forebrain projections to the PB, especially those derived from vocal control areas of the cortex and basal ganglia.

References

Alden, M., Besson, J.M., Bernard, J.F., 1994. Organization of the efferent projections from the pontine parabrachial area to the bed nucleus of the stria terminalis and neighboring regions: a PHA-L study in the rat. J. Comp. Neurol. 341, 289–314.

Alheid, G.F., Milsom, W.K., McCrimmon, D.R., 2004. Pontine influences on breathing: an overview. Respir. Physiol. Neurobiol. 143, 105–114.

Ambalavanar, R., Tanaka, Y., Damirjan, M., Ludlow, C.L., 1999. Laryngeal afferent stimulation enhances Fos immunoreactivity in periaqueductal gray in the cat. J. Comp. Neurol. 409, 411–423.

Bernard, J.F., Alden, M., Besson, J.M., 1993. The organization of the efferent projections from the pontine parabrachial complex to the amygdaloid complex: a Phaseolus vulgaris leucoagglutinin study in the rat. J. Comp. Neurol. 329, 201–229.

Bester, H., Besson, J.M., Bernard, J.F., 1997. Organization of efferent projections from the parabrachial area to the hypothalamus: a Phaseolus vulgaris-leucoagglutinin study in the rat. J. Comp. Neurol. 383, 245–281.

Bester, H., Bourgeais, L., Villanueva, L., Besson, J.M., Bernard, J.F., 1999. Differential projections to the intralaminar and gustatory thalamus from the parabrachial area: a PHA-L study in the rat. J. Comp. Neurol. 405, 421–449.

Bianchi, A.L., Pasaro, R., 1997. Organization of central respiratory neurons. In: Miller, A.D., Bianchi, A.L., Bishop, B.P. (Eds.), Neural Control of the Respiratory Muscles. CRC Press, Boca Raton, FL, pp. 77–89.

Blomqvist, A., Ma, W., Berkley, K.J., 1989. Spinal input to the parabrachial nucleus in the cat. Brain Res. 480, 29–36.

Butler, J.E., 2007. Drive to the human respiratory muscles. Respir. Physiol. Neurobiol. 159, 115–126.

Cechetto, D.F., 1987. Central representation of visceral function. Fed. Proc. 46, 17–23.

Chamberlin, N.L., 2004. Functional organization of the parabrachial complex and intertrigeminal region in the control of breathing. Respir. Physiol. Neurobiol. 143, 115–125.

Chamberlin, N.L., Saper, C.B., 1994. Topographic organization of respiratory responses to glutamate microstimulation of the parabrachial nucleus in the rat. J. Neurosci. 14, 6500–6510.

Clark, K.F., Farber, J.P., 2001. Effects of recurrent laryngeal nerve paralysis on superior laryngeal nerve afferents during evoked vocalization. Ann. Otol. Rhinol. Laryngol. Suppl. 187, 18–31.

Cohen, M.I., 1971. Switching the respiratory phases and evoked phrenic responses produced by rostrol pontine electrical stimulation. J. Physiol. 217, 133–158.

Cohen, M.I., Feldman, J.L., 1977. Models of respiratory switching. Fed. Proc. 36, 2367–2374.

Davis, P.J., Zhang, S.P., Bandler, R., 1993. Pulmonary and upper airway afferent influences on the motor pattern of vocalization evoked by excitation of the midbrain periaqueductal gray of the cat. Brain Res. 607, 61–80.

Dick, T.E., Bellingham, M.C., Richter, D.W., 1994. Pontine respiratory neurons in anesthetized cats. Brain Res. 636, 259–269.

DiLorenzo, P., 1988. Taste responses in the parabrachial pons of decerebrate rats. J. Neurophysiol. 59, 1871–1887.

Ezure, K., 2004. Respiration-related afferents to parabrachial pontine regions. Respir. Physiol. Neurobiol. 143, 167–175.

Ezure, K., Tanaka, I., Miyazaki, M., 1998. Pontine projections of pulmonary slowly adapting receptor relay neurons in the cat. Neuroreport 9, 411–414.

Farley, G.R., Barlow, S.M., Netsell, R., 1992a. Factors influencing neural activity in parabrachial regions during cat vocalizations. Exp. Brain Res. 89, 341–351.

Farley, G.R., Barlow, S.M., Netsell, R., Chmelka, J.V., 1992b. Vocalizations in the cat: behavioral methodology and spectrographic analysis. Exp. Brain Res. 89, 333–340.

Feldman, J.L., 1986. Neurophysiology of breathing in mammals. In: Brookhart, J.M., Mountcastle, V.B. (Eds.), Handbook of Physiology; The Nervous System, Intrinsic Regulatory Systems of the Brain. Am. Physiol. Soc. Bethesda, MD.

Feldman, J.L., Del Negro, C.A., 2006. Looking for inspiration: new perspectives on respiratory rhythm. Nat. Rev. Neurosci. 7, 232–242.

Feldman, J.L., Gautier, H., 1976. Interactions of pulmonary afferents and pneumotaxic center in control of respiratory pattern in cats. J. Neurophysiol. 39, 31–44.

Feldman, J.L., Mitchell, G.S., Nattie, E.E., 2003. Breathing: rhythmicity, plasticity, chemosensitivity. Ann. Rev. Neurosci. 26, 239–266.

Fenzl, T., Schuller, G., 2002. Periaquaductal gray and the region of the paralemniscal area have different functions in the control of vocalization in the neotropical bat, *Phyllostomus discolor*. Eur. J. Neurosci. 16, 1974–1986.

Fung, M.L., St. John, W.M., 1994. Separation of multiple functions in the ventilatory control of pneumotaxic mechanisms. Respir. Physiol. 96, 83–98.

Gestreau, C., Bianchi, A.L., Grélot, L., 1997. Differential brainstem Fos-like immunoreactivity after laryngeal-induced coughing and its reduction by codeine. J. Neurosci. 17, 9340–9352.

Halsell, C.B., 1992. Organization of parabrachial nucleus efferents to the thalamus and amygdala in the golden hamster. J. Comp. Neurol. 317, 57–78.

Hayward, L.F., Castellanos, M., Davenport, P.W., 2004. Parabrachial neurons mediate dorsal periaqueductal gray evoked respiratory responses in the rat. J. Appl. Physiol. 96, 1146–1154.

Herbert, H., Moga, M.M., Saper, C.B., 1990. Connections of the parabrachial nucleus with the nucleus of the solitary tract and the medullary reticular formation in the rat. J. Comp. Neurol. 293, 540–580.

Holstege, G., 1987. Some anatomical observations on the projections from the hypothalamus to brainstem and spinal cord: an HRP and autoradiographic tracing study in the cat. J. Comp. Neurol. 260, 98–126.

Holstege, G., 1989. Anatomical study of the final common pathway for vocalization in the cat. J. Comp. Neurol. 284, 242–252.

Jarvis, E.D., 2004. Learned birdsong and the neurobiology of human language. Ann. NY Acad. Sci. 1016, 749–777.

Jarvis, E.D., Gunturkun, O., Bruce, L., Csillag, A., Karten, H., Kuenzel, W., Medina, L., Paxinos, G., Perkel, D.J., Shimizu, T., Striedter, G., Wild, J.M., Ball, G.F., Dugas-Ford, J., Durand, S.E., Hough, G.E., Husband, S., Kubikova, L., Lee, D.W., Mello, C.V., Powers, A., Siang, C., Smulders, T.V., Wada, K., White, S.A., Yamamoto, K., Yu, J., Reiner, A., Butler, A.B., 2005. Avian brains and a new understanding of vertebrate brain evolution. Nat. Rev. Neurosci. 6, 151–159.

Jürgens, U., 1976. Projections from the cortical larynx area in the squirrel monkey. Exp. Brain Res. 25, 401–411.

Jürgens, U., 2002. Neural pathways underlying vocal control. Neurosci. Biobehav. Rev. 26, 235–258.

Jürgens, U., Ploog, D., 1970. Cerebral representation of vocalization in the squirrel monkey. Exp. Brain Res. 10, 532–554.

Jürgens, U., Müller-Preuss, P., 1977. Convergent projections of different limbic vocalization areas in the squirrel monkey. Exp. Brain Res. 29, 75–83.

Jürgens, U., Pratt, R., 1979. The cingular vocalization pathway in the squirrel monkey. Exp. Brain Res. 34, 499–510.

Krout, K.E., Jansen, A.S., Loewy, A.D., 1998. Periaqueductal gray matter projection to the parabrachial nucleus in rat. J. Comp. Neurol. 401, 437–454.

Kubin, L., Alheid, G.F., Zuperku, E.J., McCrimmon, D.R., 2006. Central pathways of pulmonary and lower airway vagal afferents. J. Appl. Physiol. 101, 618–627.

Larson, C.R., 1991. On the relation of PAG neurons to laryngeal and respiratory muscles during vocalization in the monkey. Brain Res. 552, 77–86.

Ling, L., Karius, D.R., Speck, D.F., 1993. Pontine-evoked inspiratory inhibitions after antagonism of NMDA, GABAA, or glycine receptor. J. Appl. Physiol. 74, 1265–1273.

Ma, W., Blomqvist, A., Berkley, K.J., 1989. Spino-diencephalic relays through the parabrachial nucleus in the cat. Brain Res. 480, 37–50.

Moga, H.H., Caper, C.B., 1990. Connections of the parabrachial nucleus with the nucleus of the solitary tract and the medullary reticular formation in the rat. J. Comp. Neurol. 293, 540–580.

Neuweiler, G., 2000. The Biology of Bats. Oxford University Press, New York, NY.

Oku, Y., Dick, T.E., 1992. Phase resetting of the respiratory cycle before and after unilateral pontine lesion in cat. J. Appl. Physiol. 72, 721–730.

Potts, J.T., Rybak, I.A., Paton, J.F.R., 2005. Respiratory rhythm entrainment by somatic afferent stimulation. J. Neurosci. 25, 1965–1978.

Rübsamen, R., Betz, M., 1986. Control of echolocation pulses by neurons of the nucleus ambiguus in the rufous horseshoe bat, *Rhinolophus rouxi*. I. Single unit recordings in the ventral motor nucleus of the laryngeal nerves in spontaneously vocalizing bats. J. Comp. Physiol. A 159, 675–687.

Rübsamen, R., Schuller, G., 1981. Laryngeal nerve activity during pulse emission in the CF-FM bat, *Rhinolophus ferrumequinum*. II. The recurrent laryngeal nerve. J. Comp. Physiol. A 143, 323–327.

Rübsamen, R., Schweizer, H., 1986. Control of echolocation pulses by neurons of the nucleus ambiguus in the rufous horseshoe bat, *Rhinolophus rouxi*. II. Afferent and efferent connections of the motor nucleus of the laryngeal nerves. J. Comp. Physiol. A 159, 689–699.

Rybak, I.A., Shevtsova, N.A., Paton, J.F., Pierrefiche, O., St. John, W.M., Haji, A., 2004. Modeling respiratory rhythmogenesis: focus on phase-switching mechanisms. Adv. Exp. Med. Biol. 551, 189–194.

Sakamoto, T., Nonaka, S., Katada, A., 1997. Control of respiratory muscles during speech and vocalization. In: Miller, A.D., Bianchi, A.L., Bishop, B.P. (Eds.), Neural Control of the Respiratory Muscles. CRC Press, New York, NY, pp. 249–258.

Saper, C., 1982. Reciprocal parabrachial-cortical connections in the rat. Brain Res. 242, 33–40.

Saper, C.B., Loewy, A.D., 1980. Efferent connections of the parabrachial nucleus in the rat. Brain Res. 197, 291–317.

Schneider, J.S., 1986. Interactions between the basal ganglia, the pontine parabrachial region, and the trigeminal system in the cat. Neuroscience 19, 411–425.

Schuller, G., Radtke-Schuller, S., 1990. Neural control of vocalization in bats: mapping of brainstem areas with electrical microstimulation eliciting species-specific echolocation calls in the rufous horseshoe bat. Exp. Brain Res. 79, 192–206.

Schuller, G., Rübsamen, R., 1981. Laryngeal nerve activity during pulse emission in the CF-FM bat, *Rhinolophus ferrumequinum*. I. Superior laryngeal nerve. J. Comp. Physiol. 143, 317–321.

Segers, L.S., Nuding, S.C., Dick, T.E., Shannon, R., Baekey, D.M., Solomon, I.C., Morris, K.F., Lindsey, B.G., 2008. Functional connectivity in the pontomedullary respiratory network. J. Neurophysiol. 100, 1749–1769.

Shiba, K., Yoshida, K., Miura, T., 1995. Functional roles of the superior laryngeal nerve afferents in electrically induced vocalization in anesthetized cats. Neurosci. Res. 20, 23–30.

Shiba, K., Umezaki, T., Zheng, Y., Miller, A.D., 1997a. The nucleus retroambigualis controls laryngeal muscle activity during vocalization in the cat. Exp. Brain Res. 115, 513–519.

Shiba, K., Yoshida, K., Nakazawa, K., Konno, A., 1997b. Influences of laryngeal afferent inputs on intralaryngeal muscle activity during vocalization in the cat. Neurosci. Res. 27, 85–92.

Smotherman, M., Zhang, S., Metzner, W., 2003. A neural basis for auditory feedback control of vocal pitch. J. Neurosci. 23, 1464–1477.

Smotherman, M., Kobayasi, K., Ma, J., Zhang, S., Metzner, W., 2006. A mechanism for vocal–respiratory coupling in the mammalian parabrachial nucleus. J. Neurosci. 26, 4860–4869.

Wallois, F., Gros, F., Masmoudi, K., Larnicol, N., 1995. C-fos-like immunoreactivity in the cat brainstem evoked by sneeze-inducing air puff stimulation of the nasal mucosa. Brain. Res. 687, 143–154.

Wild, J.M., Arends, J.J., Zeigler, H.P., 1990. Projections of the parabrachial nucleus in the pigeon (*Columba livia*). J. Comp. Neurol 293, 499–523.

CHAPTER 9.3

Audio–vocal interactions in the mammalian brain

Hanjun Liu[2], Roozbeh Behroozmand[1] and Charles R. Larson[1*]

[1]*Department of Communication Sciences and Disorders, Northwestern University, Evanston, IL, USA*
[2]*Department of Rehabilitation Medicine, The First Affiliated Hospital, Sun Yat-sen University, Guangzhou, PR China*

Abstract: The ability to hear and recognize vocalizations of others has enabled mammals to develop sophisticated communication systems. The study of neural mechanisms underlying audio–vocal integration has shown that auditory feedback may be used for updating internal representations of mappings between voice feedback and speech motor control. Work on humans and animals have shown that auditory feedback operates reflexively to stabilize voice fundamental frequency (F_0) in vocalization, speech and song. Although precise neural networks of the corrective motor adjustments are still unknown, preliminary processing of auditory feedback takes place in the auditory cortex and posterior parts of the superior temporal gyrus. Comparison of voice feedback and formulation of the corrected vocal response may take place in the frontal cortex, including the anterior cingulate cortex and dorsal premotor cortex. The inferior frontal gyrus may be involved in the generation of vocalizations and efference copy back to the auditory cortex.

Keywords: audio–vocal; auditory cortex; auditory feedback; vocalization; efferent copy; internal model; motor induced suppression; feedforward control

Abbreviations: ACC: anterior cingulate cortex; DAF: delayed auditory feedback; DIVA: directions into velocities of articulators; Fgain: frequency gain; FTc: frequency time constant; F_0: fundamental frequency; Lgain: loudness gain; LTc: loudness time constant; MEG: magnetoencephalography; MIS: motor induced suppression; PAG: periaqueductal gray; pSTG: posterior superior temporal gyrus; SSM: speech sound map; STG: superior temporal gyrus

I. Introduction

The integration of audition and vocalization has been appreciated for centuries, although little was known about the mechanisms and functions of this process until recently. It is now known that these mechanisms are closely related functionally, behaviorally and neurally. We will refer to these mechanisms as the audio–vocal system. There are several functions of the audio–vocal system, including communication in the broad sense, and learning and development of species-specific vocalizations and speech, recognition of conspecifics, and regulation of vocalization and speech. In the quest to understand the audio–vocal mechanisms, it has also been learned that neural interactions between the auditory system and the vocal motor system exist both at the brainstem and cortical levels. Research on these interactions has been intensive for the past 15 years and, although we are learning more about how auditory feedback is used for the control of vocalization and speech, a complete understanding of the underlying neural mechanisms remains speculative.

Auditory-based communication between members of the same species or across species entails sound production by one individual and sound reception, perception and recognition by another member. The goal of communication is varied and includes cohesion within a group (troop, family, clan, etc.),

*Corresponding author.
Charles R. Larson, Dept. of Communication Sciences and Disorders, Northwestern University, 2240 Campus Dr., Evanston, IL 60208

Stefan M. Brudzynski (Ed.)
Handbook of Mammalian Vocalization
ISBN 978-0-12-374593-4

DOI: 10.1016/B978-0-12-374593-4.00036-X

isolation (maintaining boundaries) between unrelated groups, mating behavior, food acquisition and general survival. Within each of these categories there are nuances in the vocal signal that may convey social hierarchy, sexual dominance, threat, aggression, fear and other affect. The vocal signals produced for each situation must be uniquely identifiable by conspecifics. Vocal signals used across species boundaries, such as in aggression, may be readily identifiable, solely because of the common acoustical properties shared by many animal species. It is therefore necessary that all members of a group share "knowledge" and neural mechanisms responsible for sound recognition and production. Research has taught us that many of these neural mechanisms are inherited, part of the basic neural machinery we are born with, and some are learned. The basic mechanisms of the auditory and vocal motor systems are described in other portions of this book (see Section 8 in this volume).

II. Auditory mechanisms of vocal perception and recognition

One of the most widely studied aspects of audio–vocal interaction is that of auditory mechanisms involved in recognition of species-specific vocalizations, including human speech. Beginning in the 1970s there were several studies that investigated the role of the auditory cortex in recognition of vocalization among conspecifics. In both cats and monkeys, it was found that lesions of the auditory cortex caused animals to produce vocalizations that were louder than normal (Talmage-Riggs et al., 1972; Buchwald and Shipley, 1985). This observation provided physiological support for a much earlier observation that the presence of noise causes people to vocalize louder than in a quiet situation (Lombard, 1911; Lane and Tranel, 1971). Others recorded from single cells in auditory cortex and found that neurons were sensitive to changing temporal properties of sounds in general, as well as species-specific vocalizations (Wollberg and Newman, 1972; Winter and Funkenstein, 1973; Steinschneider et al., 1995; Bieser and Müller-Preuss, 1996). These studies were supported by the observation that lesions of auditory cortex led to deterioration in discrimination of species-specific vocalizations (Heffner and Heffner, 1984). A more specific organizational picture of auditory cortex for detection of conspecific vocalization was shown by Rauschecker et al. (1995) in the rhesus monkey and Bieser et al. (1996) in the squirrel monkey. The identification of cells in multiple areas of auditory cortex that are responsive to relatively specific vocalizations raises the issue that these areas may reflect an organizational pattern that has been elaborated in humans for speech perception. Recently, Eliades and Wang (2008) have shown that neurons in the auditory cortex of the marmoset change their activity with pitch-shifted voice feedback. This important finding suggests that the auditory cortex of monkeys may be sensitive to conspecific vocalizations not only for purposes of group cohesion, but that they may also be involved in monitoring one's own vocalization for regulation, a property that has recently been investigated in humans.

The human auditory cortex is endowed with mechanisms that are specifically sensitive to the human voice (Fecteau et al., 2004). Identification of vocalizations takes place in different stages and locations. The first stage of voice detection appears to occur in Heschl's gyrus (Gunji et al., 2003; Schonwiesner et al., 2007), where potentials have been recorded as early as 11 ms after a sound onset (Steinschneider et al., 1999). These early neural potentials are responsive to the onset of vocalization, and thus may be actively involved in detection of voice onset time in speech (Steinschneider et al., 1999). More laterally, cells are more responsive to the sound frequency of vocalization and less so to clicks (Schönweisner and Zatorre, 2008). Along the length of the superior temporal gyrus (STG), the right posterior part is responsive to temporal variations of voices and sounds, and the anterior part is involved in discriminating speaker identity (Belin and Zatorre, 2003; Kriegstein and Giraud, 2004). The left STG is more sensitive to frequency transitions than the right, a property that may also make the left hemisphere more sensitive to sounds of speech than the right hemisphere (Belin et al., 1998; Binder et al., 2000).

III. Regulation of vocalization and speech by auditory feedback

Auditory feedback provides information necessary to implement speech motor goals to correct for errors to guide learning during speech development. One of the earliest demonstrations of the role of auditory feedback in voice control is the Lombard effect, where people raise their voice amplitude to overcome environmental noise (Lombard, 1911; Lane and Tranel, 1971). A related phenomenon is side-tone amplification, in which people increase their voice loudness when their self-perceived loudness feedback is too quiet to accomplish the vocal task, and *vice versa* (Lane and Tranel, 1971). Thus, people rely on auditory feedback to adjust their voice amplitude to achieve a communication goal.

The importance of auditory feedback for voice control is also shown by the delayed auditory feedback (DAF) technique, in which the timing of voice feedback is delayed when individuals hear themselves speaking. It has been demonstrated that DAF of 100 to 200 ms can have severe disruptive effects on speech articulation, resulting in disfluencies in the speech of normal speakers (Fairbanks and Guttman, 1958). On the other hand, severe stutters become more fluent with DAF of 50 to 200 ms (Nessel, 1958; Lotzmann, 1961; Burke, 1975), showing different impacts of DAF on speech production for different populations.

Generally, the importance of auditory feedback changes during the course of speech development (Lane and Tranel, 1971; Borden, 1979). Namely, auditory feedback is crucial to speech–language acquisition for children learning to speak (Siegel et al., 1984), but its importance is significantly reduced after language acquisition (Van Riper, 1971). For example, deafness that occurs prelingually can severely affect a child's articulation or prevent the normal development of speech production (Oller and Eilers, 1988). In contrast, the deleterious effects on speech are greatly reduced if hearing loss occurs after language acquisition (Waldstein, 1990; Cowie and Douglas-Cowie, 1992). However, the reduction in importance of auditory feedback in speech production after language acquisition has been challenged by studies on postlingually deafened adults showing degradation in phonatory, articulatory and suprasegmental control during speech production (Binnie et al., 1982; Lane and Webster, 1991). Furthermore, experimental studies have demonstrated that if voice auditory feedback is missing, masked or altered, voice F_0 control is diminished (Elliott and Niemoeller, 1970; Svirsky et al., 1992; Mürbe et al., 2002). It has also been shown that changes in kinesthetic feedback through anesthetization procedures reduce fine control of voice F_0 in pitch-masking tasks (Mallard et al., 1978; Sundberg et al., 1995). Therefore, these studies suggest that auditory feedback is involved in monitoring speech production and maintaining normal speech motor control over the long term (Singh, 1998).

A promising technique for quantifying the role of auditory feedback in vocal control is the perturbation technique (Löfqvist and Lindblom, 1994). Various investigators have presented perturbations in voice pitch or loudness over headphones to subjects speaking or sustaining vowel phonations (Fig. 1) (Kawahara, 1994; Burnett et al., 1997, 1998; Hain et al., 2000; Natke and Kalveram, 2001; Donath et al., 2002; Natke et al., 2003; Heinks-Maldonado and Houde, 2005; Bauer et al., 2006; Liu and Larson, 2007; Chen et al., 2007; Larson et al., 2007, 2008; Russo et al., 2008). The response magnitudes vary as a function of stimulus magnitude (Liu and Larson, 2007), stimulus direction (Chen et al., 2007), stimulus duration (Burnett et al.,

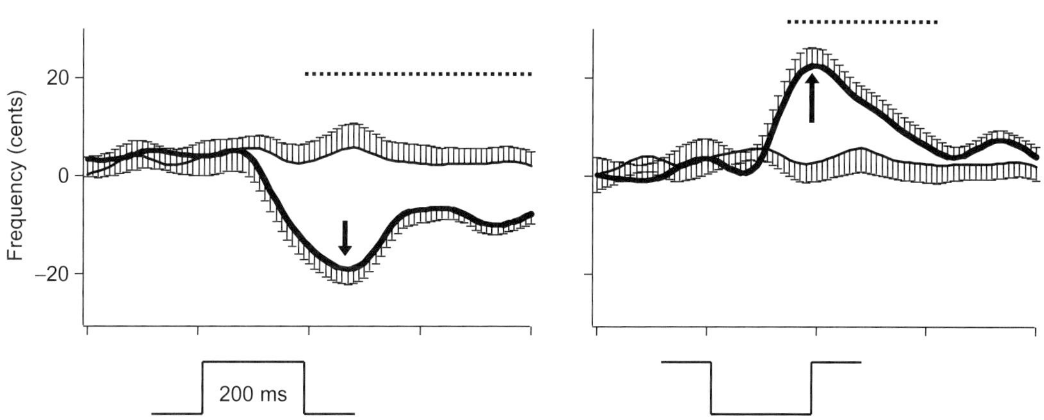

Fig. 1. Illustration of voice fundamental frequency (F_0) responses to pitch-shifted voice feedback. Left side shows downward F_0 response (thick line, with standard error bars in one direction) and control average (thin line with SE bars) to upward pitch-shift stimulus (square bracket at bottom). Right side shows voice F_0 response to downward pitch shift stimulus. Y-axis, frequency in cents (100 cents = 1 semitone); X-axis is time in milliseconds.

1998), stimulus onset velocity (Larson et al., 2000) and vocal pitch level (Liu and Larson, 2007). During speech production, the responses are modulated in a task-dependent manner to facilitate accurate and timely adjustments in voice control, and thus maintain accuracy of speech articulation (Xu et al., 2004; Liu et al., 2007). Larger vocal responses were also observed in singing compared to speaking (Natke et al., 2003), in speech compared to vowels (Chen et al., 2007) and in stressed conditions compared to unstressed conditions (Natke and Kalveram, 2001). The fact that the response latencies range from about 80 to 200ms and can be reduced in some speaking tasks (Xu et al., 2004) is a further indication of the task-dependent nature of the system. Similar effects were reported in a series of sensorimotor adaptation experiments in which subjects modified their F_0 or formant frequencies opposite in direction to the shifted feedback during vowels and phrases (Houde and Jordan, 1998; Jones and Munhall, 2000; Houde and Jordan, 2002; Jones and Munhall, 2002, 2005; Purcell and Munhall, 2006; Villacorta et al., 2007; Jones and Keough, 2008; Tourville et al., 2008). The compensatory nature of these responses suggests that the system can recognize the direction of pitch, loudness or formant perturbations and can help to maintain the voice F_0, intensity or formants at a desired level during speech and nonspeech vocal tasks. It is also suggested that vocal–auditory feedback may be used to update an internal representation of the mapping between auditory feedback and the motor control systems.

IV. Neural mechanisms of auditory feedback

The neural mechanisms underlying many of the above observations have not been worked out in detail; however, recent observations have provided important information on some of the neural pathways likely to be involved in these processes. Generally, the STG or the intraparietal sulcus have been shown to be activated in response to changes in voice auditory feedback (Hirano et al., 1997; Curio et al., 2000; Gunji et al., 2000, 2001; Toyomura et al., 2007; Aliu et al., 2008). Moreover, Müller-Preuss demonstrated that activity levels of single neurons in the squirrel monkey STG were reduced during vocalization compared to listening to previously recorded vocalizations (Muller-Preuss et al., 1980; Muller-Preuss and Ploog, 1981), a phenomenon often referred to as motor-induced suppression (MIS). A similar effect has also been investigated using magnetoencephalographic (MEG) (Numminen and Curio, 1999; Numminen et al., 1999; Houde et al., 2002; Heinks-Maldonado et al., 2006) and electroencephalographic (EEG) recordings (Heinks-Maldonado et al., 2005, 2006) in human auditory cortex, showing that the neural responses to voice feedback were attenuated during speaking compared to listening. It was proposed that this suppression reflected a feedforward model (Wolpert, 1997; Blakemore et al., 1998) in which vocal motor control reafference, presumably in auditory cortex, is compared with voice auditory feedback. When the feedback closely matches the vocal efferent copy, the feedback signal is recognized as self-vocalization, and is suppressed relative to listening to the same signal. However, if the feedback signal is altered, as in a change in frequency or delay in time, there is not a close match with efference copy, and the feedback signal is not suppressed. It was suggested that this process might be important for identifying vocalizations as belonging either to self or to someone else. Through detailed studies of MIS in marmosets, Eliades and Wang (2008) reported that while many neurons reduce their activity levels during vocalization, others increase their level of activity. Furthermore, those neurons showing a reduction in firing rate also showed increased activity when voice auditory feedback was unexpectedly shifted in pitch. This observation led to the suggestion that the MIS, as reported by others, may actually function to enhance neural sensitivity by increasing the dynamic range of the auditory cortex neurons during feedback alteration. Thus, vocalization primes auditory cortex to be more sensitive to self-produced vocalization, which could be related to a regulatory mechanism of vocal control through auditory feedback.

Although most of these studies were focused on the auditory cortex, it has recently been found that considerable processing occurs at the level of the midbrain. The frequency following response is thought to originate in the inferior colliculus and it registers the fundamental frequency of vocalization, as well as formant frequencies of vowels in human speech (Wible et al., 2004; Johnson et al., 2005). These observations raise the point that some of the processing that was thought to occur at the level of the cerebral cortex may in fact take place in the brainstem (Papanicolaou et al., 1986; Galbraith et al., 1995; Pieper and Jürgens, 2003). Another possibility is that cortical mechanisms of pitch decoding have descending projections to the brainstem that modulate activity at that level (Musacchia et al., 2006). Brainstem mechanisms may also be responsible in part for the Lombard effect, as Nonaka et al. (1997), using decerebrate cats, reported that an increase in vocal loudness elicited by electrical stimulation of the midbrain periaquedutal gray (PAG) could be induced by added environmental noise. It has been shown that monkeys also produce louder vocalizations in the presence of noise (Sinnott et al., 1975). Neural connections between the inferior colliculus and PAG are known to be capable of modulating vocalization, as has been shown in studies on bats (see Metzner and Schuller, Chapter 9.4 in this volume). Similarly, single neurons in the PAG, inferior colliculus and adjacent reticular formation of vocalizing monkeys were shown to be modulated by auditory stimulation (Larson and Kistler, 1984; Pieper and Jürgens, 2003).

V. Models of auditory feedback control of the voice

In order to improve our understanding of the audio–vocal system, several models of the system have been proposed. Such models generally incorporate both feedback and feedforward mechanisms of control. Internal models have been proposed to model the control of speech and based on feedforward control (Wolpert et al., 1995; Perkell et al., 1997; Kawato, 1999), representing the relationship between motor commands and their acoustic output. It has been suggested that the neural system learns how to generate motor commands that can produce the desired acoustic output, and to continuously fine-tune the internal representation of such motor–acoustic relationships (Wolpert et al., 1998; Jones and Munhall, 2000). On the other hand, sensory feedback plays a role in updating internal models and correcting for errors in execution of a task. Speech production is thought to involve both feedback and feedforward control. Several recent studies have shown that auditory feedback impacts both control processes (Houde and Jordan, 1998; Jones and Munhall, 2005; Bauer et al., 2006; Purcell and Munhall, 2006).

Directions into velocities of articulators (DIVA) is a neural network-based model for movement control of the speech articulators that provides a computational framework of neural processing involved in the production of phonemes, syllables and words (Guenther, 2006; Guenther et al., 2006) (see Fig. 2). This model combines feedforward- and feedback-based controlling strategies to provide a realistic way of investigating the role of sensory (somatosensory and auditory) feedback in speech production and has been widely used to account for issues in child development (Guenther, 1995), hearing impairment and stuttering (Guenther et al., 2006).

The feedforward subsystem of DIVA hypothesizes the initiation of speech by activation of speech sound map (SSM) cells that are located in the left ventral premotor areas, including Broca's area in the opercular portion of the inferior frontal gyrus (Guenther et al., 2006). The projection from cells in the premotor area is associated with activation of the stored motor cortex program of the corresponding syllabus that is learned during network training (learning phase). Three distinct maps of motor cortical cells representing positional commands to the speech articulators, velocity commands originating from the feedback control subsystem and from the feedforward control subsystem, respectively, are thought to be involved in the feedback and feedforward process. A simple function transforms the motor commands to a speech synthesizer parameter to produce the intended speech. The auditory representation of the produced speech sounds includes a vector that contains the information of the first three formants. This property of DIVA has made it a suitable model for investigating the effect of formant shifts in auditory perturbation studies (Villacorta et al., 2007).

Projections from the SSM cells to primary motor and auditory cortical areas have been proposed to be a part of the neural circuitry involved in feedforward initiation and auditory feedback-based control of voice (Guenther, 2006; Guenther et al., 2006). Based on the proposed "reafference hypothesis" (feedforward model), DIVA posits that the "efference copy" of the motor commands transmits the sensory neural representation of an intended vocal output (corollary discharges) to the auditory cortical areas where the error neurons encode disparity between the intended and actual output (Houde et al., 2002; Heinks-Maldonado et al., 2005; Guenther et al., 2006). The neural pathways between different components also include projections from auditory error cells in posterior superior temporal gyrus (pSTG) to motor correction cells in anterior cingulate cortex (ACC) of the right prefrontal area (Tourville et al., 2008). This pathway is hypothesized to be involved in fine tuning of motor commands during the learning period and feedback monitoring of voice. Zarate and Zatorre (2008) and Warren et al. (2005) have proposed that

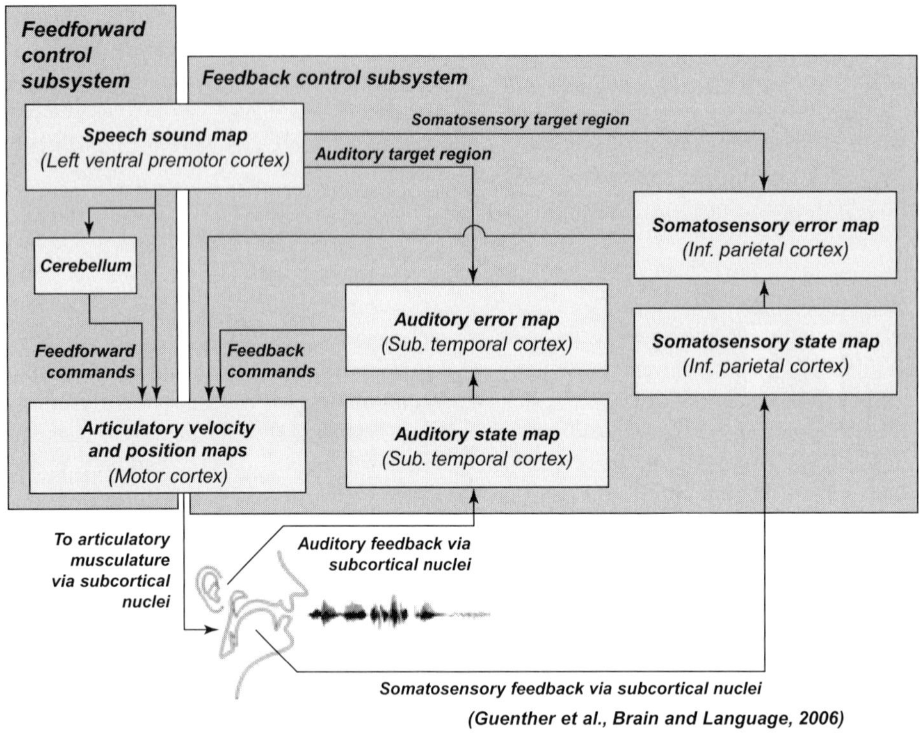

Fig. 2. Model of speech feedback control system. From Guenther et al. (2006) published with permission of the author and Elsevier Publishers.

for untrained singers, the posterior temporal area and intraparietal sulcus are involved in auditory–motor transformations with projections to the dorsal premotor cortex (Zarate and Zatorre, 2008) where adjustment of vocal output occurs. For trained singers, they argue that the ACC and putamen become more involved in making vocal corrections to errors in auditory feedback. It was also suggested that with trained singers there is greater involvement in the left hemisphere, while in untrained singers the right hemisphere is more important. In support of the suggested role of the ACC in this process, Muller-Preuss et al. (Muller-Preuss et al., 1980; Muller-Preuss and Ploog, 1981) reported that stimulation of the vocalization-related neurons in cingulate cortex resulted in attenuation of the auditory cortex neural activity in squirrel monkeys.

In a series of studies to investigate different aspects of sensory perturbation during speech, Hain et al. (2000) proposed and implemented a model to illustrate a possible arrangement of neural circuitry that explains a large body of experimental observations during voice F_0 compensation and other speech-related tasks (Hain et al., 2000; Larson et al., 2000; Hain et al., 2001; Bauer et al., 2006; Larson et al., 2007) (see Fig. 3). The execution of this model is based on the comparison of intended F_0 with the perceived F_0, or a connection between auditory input and motor output. Namely, the model uses voice pitch and loudness feedback that are lowpass filtered and delayed, to account for neural processing and transmission delays, and then multiplexed. The demultiplexed signals are then compared with intended voice F_0 and amplitude to generate corrected output. An important feature of this model is that voice pitch and loudness are processed independently, as this allows for independent adjustments of voice F_0 and amplitude. This model does not say whether voice pitch and/or loudness are processed in the brainstem or at the cortical level, but given the literature discussed above, it is more likely that some features of this model represent cortical events. This model is designed in a way that accounts for the intrinsic delays involved in signal transmission and neuromuscular reproduction that occurs during speech initiation, presenting a feasible manner in which neural systems can correct for errors in both voice F_0 and amplitude by using negative closed-loop feedback

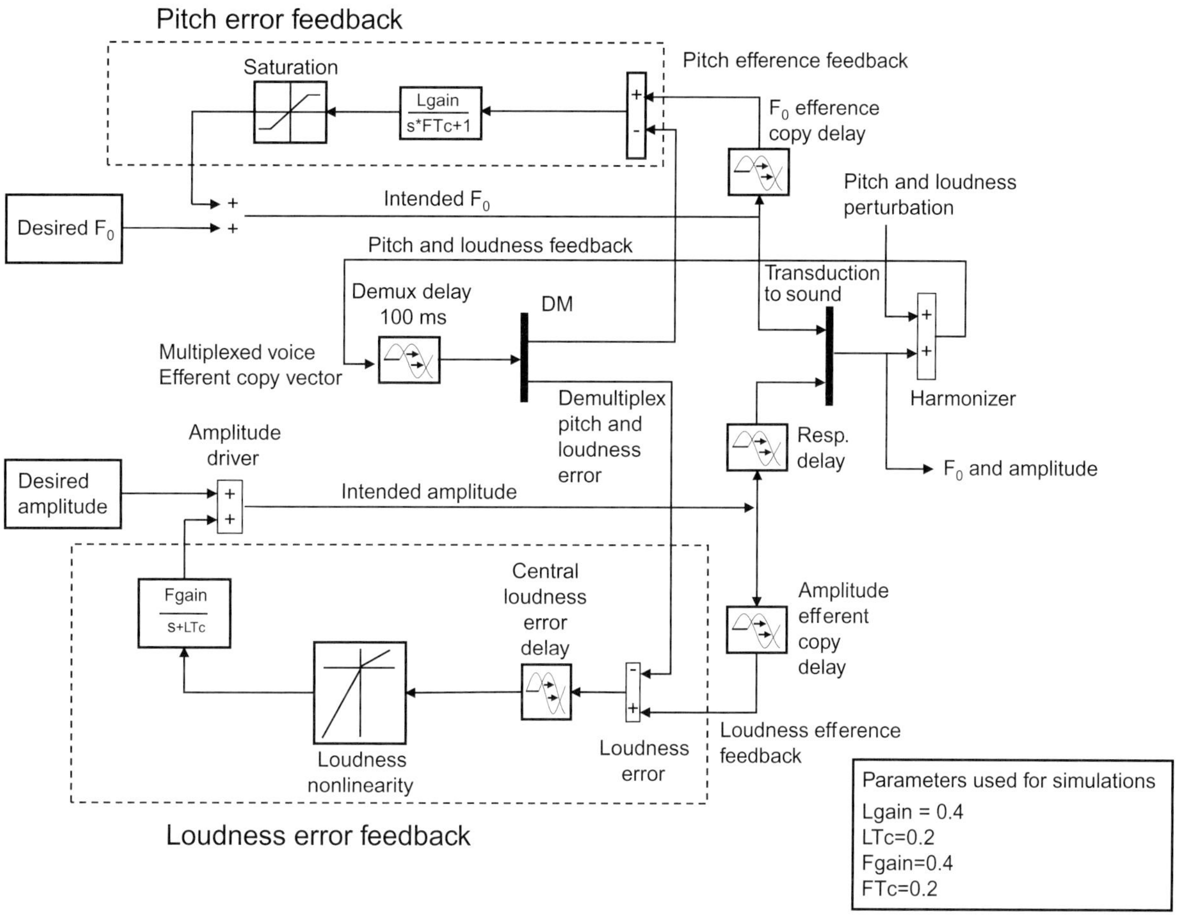

Fig. 3. Model of circuitry that reproduces voice F_0 and amplitude responses to perturbations in pitch and loudness feedback. Reprinted with permission of the Acoustical Society of America.

control. It should be noted that Hain's model contains the same topology as Guenther's model (2006). The primary difference is that Guenther's model is a neural network model containing many nonlinearities and free parameters corresponding to individual simulated neurons, while Hain's model is a mathematical model containing a number of simple parameters that are sufficient to simulate the processing of voice F_0 and amplitude, including the perturbed auditory feedback (Larson et al., 2007).

VI. Conclusion

Significant progress has been made in the last 30 years on the understanding of neural mechanisms underlying audition and vocalization. We now know that audio–vocal interactions occur at both the brainstem and cortical levels. The precise details of how brainstem processing affects vocalization in mammals are poorly understood. Moreover, it also appears that cortical circuits can modulate the brainstem. The auditory cortex is organized into several different components which are responsible for decoding different acoustical properties of the voice and speech. This information is used for recognition of conspecifics, communication and voice regulation. From auditory cortex, the decoding process proceeds in a posterior direction where error between planned and actual vocalization takes place for regulatory purposes. From these temporal/parietal areas, information proceeds to the frontal lobe where more direct planning of vocalizations, based on auditory feedback, takes place. Resulting vocal output takes place in motor cortex. At some undefined location within the temporal, parietal and frontal circuit, efference copy of vocal output is routed to the inferior frontal region and then back to auditory cortex. Many details of this proposed network are unknown, but future studies using neural imaging and modeling combined with perturbed auditory feedback holds promise for defining this network in greater detail.

References

Aliu, S.O., Houde, J.F., Nagarajan, S.S., 2008. Motor-induced suppression of the auditory cortex. J. Cogn. Neurosci. 21 (4), 791–802.

Bauer, J.J., Mittal, J., Larson, C.R., Hain, T.C., 2006. Vocal responses to unanticipated perturbations in voice loudness feedback: an automatic mechanism for stabilizing voice amplitude. J. Acoust. Soc. Am. 119, 2363–2371.

Belin, P., Zatorre, R.J., 2003. Adaptation to speaker's voice in right anterior temporal lobe. Neuroreport 14, 2105–2109.

Belin, P., Zilbovicius, M., Crozier, S., Thivard, L., Fontaine, A., Masure, M.C., Samson, Y., 1998. Lateralization of speech and auditory temporal processing. J. Cog. Neurosci. 10, 536–540.

Bieser, A., Müller-Preuss, P., 1996. Auditory responsive cortex in the squirrel monkey: neural responses to amplitude-modulated sounds. Exp. Brain Res. 108, 273–284.

Binder, J.R., Frost, J.A., Hammeke, P.S.F., Bellgowan, J.A., Springer, J.N., Possing, E.T., 2000. Human temporal lobe activation by speech and nonspeech sounds. Cereb. Cortex 10, 512–528.

Binnie, C.A., Daniloff, R.G., Buckingham, H.W., 1982. Phonetic disintegration in a five-year-old following sudden hearing loss. J. Speech Hear. Dis. 47, 181–189.

Blakemore, S.J., Rees, G., Frith, C.D., 1998. How do we predict the consequences of our actions? A functional imaging study. Neuropsychologia 36, 521–529.

Borden, G.J., 1979. An interpretation of research of feedback interruption in speech. Brain Lang. 7, 307–319.

Buchwald, J.S., Shipley, C., 1985. A comparative model of infant cry. In: Lester, B.M., Zachariah Boukydis, C.F. (Eds.), Infant Crying. Plenum Press, New York, NY.

Burke, B.D., 1975. Variables affecting stutterer's initial reactions to delayed auditory feedback. J. Comm. Dis. 8, 141–155.

Burnett, T.A., Senner, J.E., Larson, C.R., 1997. Voice F_0 responses to pitch-shifted auditory feedback: a preliminary study. J. Voice 11, 202–211.

Burnett, T.A., Freedland, M.B., Larson, C.R., Hain, T.C., 1998. Voice F_0 responses to manipulations in pitch feedback. J. Acoust. Soc. Am. 103, 3153–3161.

Chen, S.H., Liu, H., Xu, Y., Larson, C.R., 2007. Voice F_0 responses to pitch-shifted voice feedback during English speech. J. Acoust. Soc. Am. 121, 1157–1163.

Cowie, R. and Douglas-Cowie, E. (1992) Postlingually acquired deafness: speech deterioration and the wider consequences. Trends in Linguistics, Studies and Monographs, 62.

Curio, G., Neuloh, G., Numminen, J., Jousmaki, V., Hari, R., 2000. Speaking modifies voice-evoked activity in the human auditory cortex. Hum. Brain Mapp. 9, 183–191.

Donath, T.M., Natke, U., Kalveram, K.T., 2002. Effects of frequency-shifted auditory feedback on voice F_0 contours in syllables. J. Acoust. Soc. Am. 111, 357–366.

Eliades, S.J., Wang, X., 2008. Neural substrates of vocalization feedback monitoring in primate auditory cortex. Nature 453, 1102–1106.

Elliott, L., Niemoeller, A., 1970. The role of hearing in controlling voice fundamental frequency. Int. Aud. IX, 47–52.

Fairbanks, G., Guttman, N., 1958. Effects of delayed auditory feedback upon articulation. J. Speech Hear. Res. 1, 12–22.

Fecteau, S., Armony, J.L., Joanette, Y., Belin, P., 2004. Is voice processing species-specific in human auditory cortex? An fMRI study. Neuroimage 23, 840–848.

Galbraith, G.C., Arbagey, P.W., Branski, R., Comerci, N., Rector, P.M., 1995. Intelligible speech encoded in the human brain stem frequency-following response. Neuroreport 6, 2363–2367.

Guenther, F.H., 1995. Speech sound acquisition, coarticulation, and rate effects in a neural network model of speech production. Psych. Rev. 102, 594–621.

Guenther, F.H., 2006. Cortical interactions underlying the production of speech sounds. J. Comm. Dis. 39, 350–365.

Guenther, F.H., Ghosh, S.S., Tourville, J.A., 2006. Neural modeling and imaging of the cortical interactions underlying syllable production. Brain Lang. 96, 280–301.

Gunji, A., Hoshiyama, M., Kakigi, R., 2000. Identification of auditory evoked potentials of one's own voice. Clin. Neurophysiol. 111, 214–219.

Gunji, A., Hoshiyama, M., Kakigi, R., 2001. Auditory response following vocalization: a magnetoencephalographic study. Clin. Neurophysiol. 112, 514–520.

Gunji, A., Koyama, S., Ishii, R., Levy, D., Okamoto, H., Kakigi, R., Pantev, C., 2003. Magnetoencephalographic study of the cortical activity elicited by human voice. Neurosci. Lett. 348, 13–16.

Hain, T.C., Burnett, T.A., Kiran, S., Larson, C.R., Singh, S., Kenney, M.K., 2000. Instructing subjects to make a voluntary response reveals the presence of two components to the audio-vocal reflex. Exp. Brain Res. 130, 133–141.

Hain, T.C., Burnett, T.A., Larson, C.R., Kiran, S., 2001. Effects of delayed auditory feedback (DAF) on the pitch-shift reflex. J. Acoust. Soc. Am. 109, 2146–2152.

Heffner, H.E., Heffner, R.S., 1984. Temporal lobe lesions and perception of species-specific vocalizations by Macaques. Science 226, 75–76.

Heinks-Maldonado, T.H., Houde, J.F., 2005. Compensatory responses to brief perturbations of speech amplitude. Acoust. Res. Lett. Online 6, 131–137.

Heinks-Maldonado, T.H., Nagarajan, S.S., Houde, J.F., 2006. Magnetoencephalographic evidence for a precise forward model in speech production. Neuroreport 17, 1375–1379.

Heinks-Maldonado, T.H., Mathalon, D.H., Gray, M., Ford, J.M., 2005. Fine-tuning of auditory cortex during speech production. Psychophysiology 42, 180–190.

Hirano, S., Kojima, H., Naito, Y., Honjo, I., Kamoto, Y., Okazawa, H., Ishizu, K., Yonekura, Y., Nagahama, Y., Fukuyama, H., Konishi, J., 1997. Cortical processing mechanism for vocalization with auditory verbal feedback. Neuroreport 8, 2379–2382.

Houde, J.F., Jordan, M.I., 1998. Sensorimotor adaptation in speech production. Science 279, 1213–1216.

Houde, J.F., Jordan, M.I., 2002. Sensorimotor adaptation of speech I: Compensation and adaptation. J. Speech Hear. Res. 45, 295–310.

Houde, J.F., Nagarajan, S.S., Sekihara, K., Merzenich, M.M., 2002. Modulation of the auditory cortex during speech: an MEG study. J. Cog. Neurosci. 14, 1125–1138.

Johnson, K., Nicol, T., Kraus, N., 2005. The brainstem response to speech: a biological marker. Ear Hearing 25, 424–433.

Jones, J.A., Munhall, K.G., 2000. Perceptual calibration of F_0 production: evidence from feedback perturbation. J. Acoust. Soc. Am. 108, 1246–1251.

Jones, J.A., Munhall, K.G., 2002. The role of auditory feedback during phonation: studies of Mandarin tone production. J. Phon. 30, 303–320.

Jones, J.A., Munhall, K.G., 2005. Remapping auditory-motor representations in voice production. Curr. Biol. 15, 1768–1772.

Jones, J.A., Keough, D., 2008. Auditory-motor mapping for pitch control in singers and nonsingers. Exp. Brain Res. 190, 279–287.

Kawahara, H., 1994. Interactions between speech production and perception under auditory feedback perturbations on fundamental frequencies. J. Acoust. Soc. Jpn. 15, 201–202.

Kawato, M., 1999. Internal models for motor control and trajectory planning. Curr. Opin. Neurobiol. 9, 718–727.

Kriegstein, K.V., Giraud, A.L., 2004. Distinct functional substrates along the right superior temporal sulcus for the processing of voices. Neuroimage 22, 948–955.

Lane, H., Tranel, B., 1971. The Lombard sign and the role of hearing in speech. J. Speech Hear. Res. 14, 677–709.

Lane, H., Webster, J.W., 1991. Speech deterioration in postlingually deafened adults. J. Acoust. Soc. Am. 89, 859–866.

Larson, C.R., Kistler, M.K., 1984. Periaqueductal gray neuronal activity associated with laryngeal EMG and vocalization in the awake monkey. Neurosci. Lett. 46, 261–266.

Larson, C.R., Sun, J., Hain, T.C., 2007. Effects of simultaneous perturbations of voice pitch and loudness feedback on voice F_0 and amplitude control. J. Acoust. Soc. Am. 121, 2862–2872.

Larson, C.R., Burnett, T.A., Kiran, S., Hain, T.C., 2000. Effects of pitch-shift onset velocity on voice F_0 responses. J. Acoust. Soc. Am. 107, 559–564.

Larson, C.R., Altman, K.W., Liu, H., Hain, T.C., 2008. Interactions between auditory and somatosensory feedback for voice F(0) control. Exp. Brain Res. 187, 613–621.

Liu, H., Larson, C.R., 2007. Effects of perturbation magnitude and voice F_0 level on the pitch-shift reflex. J. Acoust. Soc. Am. 122, 3671–3677.

Liu, H., Zhang, Q., Xu, Y., Larson, C.R., 2007. Compensatory responses to loudness-shifted voice feedback during production of Mandarin speech. J. Acoust. Soc. Am. 122, 2405–2412.

Löfqvist, A., Lindblom, B., 1994. Speech motor control. Curr. Opin. Neurobiol. 4, 823–826.

Lombard, E., 1911. Le signe de l'évélation de la voix. Ann. Mal. Oreil. Larynx 37, 101–119.

Lotzmann, G., 1961. On the use of varied delay times in stammerers. Folia Phoniatr. Logo. 13, 276–310.

Mallard, A.R., Ringel, R.L., Horii, Y., 1978. Sensory contributions to control of fundamental frequency of phonation. Folia Phoniat. 30, 199–213.

Muller-Preuss, P., Ploog, D., 1981. Inhibition of auditory cortical neurons during phonation. Brain Res. 215, 61–76.

Muller-Preuss, P., Newman, J.D., Jürgens, U., 1980. Anatomical and physiological evidence for a relationship between the "cingular" vocalization area and the auditory cortex in the squirrel monkey. Brain Res. 202, 307–315.

Mürbe, D., Pabst, F., Hofmann, G., Sundberg, J., 2002. Significance of auditory and kinesthetic feedback to singers' pitch control. J. Voice 16, 44–51.

Musacchia, G., Sams, M., Nicol, T., Kraus, N., 2006. Seeing speech affects acoustic information processing in the human brainstem. Exp. Brain Res. 168, 1–10.

Natke, U., Kalveram, K.T., 2001. Effects of frequency-shifted auditory feedback on fundamental frequency of long stressed and unstressed syllables. J. Speech Lang. Hear Res. 44, 577–584.

Natke, U., Donath, T.M., Kalveram, K.T., 2003. Control of voice fundamental frequency in speaking versus singing. J. Acoust. Soc. Am. 113, 1587–1593.

Nessel, E., 1958. Delayed auditory feed-back in stutterers (Lee effect). Folia Phoniatr. Logo. 10, 199–204.

Nonaka, S., Takahashi, R., Enomoto, K., Katada, A., Unno, T., 1997. Lombard reflex during PAG-induced vocalization in decerebrate cats. Neurosci. Res. 29, 283–289.

Numminen, J., Curio, G., 1999. Differential effects of overt, covert and replayed speech on vowel-evoked responses of the human auditory cortex. Neurosci. Lett. 272, 29–32.

Numminen, J., Salmelin, R., Hari, R., 1999. Subject's own speech reduces reactivity of the human auditory cortex. Neurosci. Lett. 265, 119–122.

Oller, D., Eilers, R., 1988. The role of audition in infant babbling. Child Develop. 51, 441–449.

Papanicolaou, A.C., Raz, N., Loring, D.W., Eisenberg, H.M., 1986. Brain stem evoked response suppression during speech production. Brain Lang. 27, 50–55.

Perkell, J.M.M., Lane, H., Guenther, F., Wilhelms-Tricarico, R., Wozniak, J., Guiod, P., 1997. Speech motor control: acoustic goals, saturation effects, auditory feedback and internal models. Speech Comm. 22, 227–249.

Pieper, F., Jürgens, U., 2003. Neuronal activity in the inferior colliculus and bordering structures during vocalization in the squirrel monkey. Brain Res. 979, 153–164.

Purcell, D.W., Munhall, K.G., 2006. Compensation following real-time manipulation of formants in isolated vowels. J. Acoust. Soc. Am. 119, 2288–2297.

Rauschecker, J.P., Tian, B., Hauser, M., 1995. Processing of complex sounds in the macaque nonprimary auditory cortex. Science 268, 111–114.

Russo, N., Larson, C., Kraus, N., 2008. Audio-vocal system regulation in children with autism spectrum disorders. Exp. Brain Res. 188, 111–124.

Schönweisner, M., Zatorre, R., 2008. Depth electrode recordings show double dissociation between pitch processing in lateral Heschl's gyrus and sound onset processing in medial Heschl's gyrus. Exp. Brain Res. 187, 97–105.

Schonwiesner, M., Novitski, N., Pakarinen, S., Carlson, S., Tervaniemi, M., Naatanen, R., 2007. Heschl's gyrus, posterior superior temporal gyrus, and mid-ventrolateral prefrontal cortex have different roles in the detection of acoustic changes. J. Neurophysiol. 97, 2075–2082.

Siegel, G., Pick, H., Garber, S., 1984. Auditory feedback and speech development. In: Rees, H. (Ed.), Advances in Child Development and Behavior. Academic Press, Orlando, FL.

Singh, S.A. (1998) The Influence of Delayed Auditory Feedback on the Speech Movement Coordination of the Lips, Jaw and Larynx. Communication Sciences and Disorders. Northwestern University, Evanston, IL.

Sinnott, J.M., Stebbins, W.C., Moody, D.B., 1975. Regulation of voice amplitude by the monkey. J. Acoust. Soc. Am. 58, 412–414.

Steinschneider, M., Schroeder, C.E., Arezzo, J.C., Vaughan, H.G., 1995. Physiologic correlates of the voice onset time boundary in primary auditory cortex (A1) of the awake monkey: temporal response patterns. Brain Lang. 48, 326–340.

Steinschneider, M., Volkov, I.O., Noh, M.D., Garell, P.C., Howard, 3rd, M.A., 1999. Temporal encoding of the voice onset time phonetic parameter by field potentials recorded directly from human auditory cortex. J. Neurophysiol. 82, 2346–2357.

Sundberg, J., Iwarsson, J., Billstrom, A.H., 1995. Significance of mechanoreceptors in the subglottal mucosa for subglottal pressure control in singers. J. Voice. 9, 20–26.

Svirsky, M.A., Lane, H., Perkell, J.S., Wozniak, J., 1992. Effects of short-term auditory deprivation on speech production in adult cochlear implant users. J. Acoust. Soc. Am. 92, 1284–1300.

Talmage-Riggs, G., Winter, P., Ploog, D., Mayer, W., 1972. Effect of deafening on the vocal behavior of the squirrel monkey (*Saimiri sciureus*). Folia Primatol. 17, 404–420.

Tourville, J.A., Reilly, K.J., Guenther, F.H., 2008. Neural mechanisms underlying auditory feedback control of speech. Neuroimage 39, 1429–1443.

Toyomura, A., Koyama, S., Miyamaoto, T., Terao, A., Omori, T., Murohashi, H., Kuriki, S., 2007. Neural correlates of auditory feedback control in human. Neurosci. 146, 499–503.

Villacorta, V.M., Perkell, J.S., Guenther, F.H., 2007. Sensorimotor adaptation to feedback perturbations of vowel acoustics and its relation to perception. J. Acoust. Soc. Am. 122, 2306–2319.

Waldstein, R., 1990. Effects of postlingual deafness on speech production: implications for the role of auditory feedback. J. Acoust. Soc. Am. 88, 2099–2114.

Warren, J.E., Wise, R.J.S., Warren, J.D., 2005. Sounds do-able: auditory–motor transformations and the posterior temporal plane. Trends Neurosci. 28, 636–643.

Wible, B., Nicol, T., Kraus, N., 2004. Atypical brainstem representation of onset and formant structure of speech sounds in children with language-based learning problems. Biol. Psychol. 67, 299–317.

Winter, P., Funkenstein, H.H., 1973. The effect of species-specific vocalization on the discharge of auditory cortical cells in the awake Squirrel monkey. Exp. Brain Res. 18, 489–504.

Wollberg, Z., Newman, J.D., 1972. Auditory cortex of Squirrel monkey: response patterns of single cells to species-specific vocalizations. Science 175, 212–214.

Wolpert, D.M., 1997. Computational approaches to motor control. Trends Cogn. Sci. 1, 209–216.

Wolpert, D.M., Ghahramani, Z., Jordan, M.I., 1995. An internal model for sensorimotor integration. Science 269, 1880–1882.

Wolpert, D.M., Miall, R.C., Kawato, M., 1998. Internal models in the cerebellum. Trends Cogn. Sci. 2, 338–347.

Xu, Y., Larson, C., Bauer, J., Hain, T., 2004. Compensation for pitch-shifted auditory feedback during the production of Mandarin tone sequences. J. Acoust. Soc. Am. 116, 1168–1178.

Zarate, J.M., Zatorre, R.J., 2008. Experience-dependent neural substrates involved in vocal pitch regulation during singing. Neuroimage 40, 1871–1887.

CHAPTER 9.4

Vocal control in echolocating bats

Walter Metzner[1*] and Gerd Schuller[2*]

[1]*Department of Physiological Science, UCLA, Los Angeles, CA, USA*
[2]*Department Biology II, Ludwig-Maximilians-University Munich, Planegg-Martinsried, Germany*

Abstract: Call production in the bat larynx follows the general mammalian pattern. However, several adaptations in the design of the larynx and vocal tract enable bats to produce precisely timed, high-intensity ultrasonic echolocation calls. Whereas the laryngeal innervation by the vocal motor nucleus, the nucleus ambiguus, also follows the common mammalian scheme, certain brainstem areas feeding into the nucleus ambiguus play a peculiar role in bats and operate in parallel to the descending connections from the periaqueductal gray to the final common vocal motor pathway commonly described in other mammals. Some brainstem areas exclusively control echolocation pulses, but not social calls. A similar separated involvement in either echolocation or communication may also occur at higher levels of vocal control, such as in the anterior cingulate cortex, and involve differential gene expression. Most brainstem areas involved in vocal control also receive auditory inputs, providing the audio–vocal feedback quintessential for bat echolocation.

Keywords: vocal motor pathway; larynx biomechanics; brainstem; anterior cingulate cortex; FoxP2; auditory feedback; bat

Abbreviations: ACC: anterior cingulate cortex; AMPA: (±)-α-amino-3-hydroxy-5-methylisoxazole-4-propionic acid (excitatory transmitter agonist); AP: pretectal area; CUN: nucleus cuneiformis; DMN: deep mesencephalic nucleus; DSC: Doppler-shift compensation; GABA: γ-amino-n-butyric acid (inhibitory transmitter agonist); Icrp: inferior colliculus, rostral pole; Icx: inferior colliculus, external portion; NBIC: nucleus of the brachium of the inferior colliculus; NCAT: nucleus of the central acoustic tract; PAG: periaqueductal gray; PB: parabrachial nucleus; PLA: paralemniscal area; RET: reticular formation; RF: resting frequency; SCdp: superior colliculus, deep layers

I. Introduction

Bats constitute more than 20% of all mammalian species and populate almost every type of terrestrial habitat. Common to all bats is their ability to fly and their mostly nocturnal lifestyle. Although all mammals, including bats, use acoustic signals for passive listening to ambient sounds and for intra- and interspecies communication, the acoustic sense has gained considerable importance in bats compared to most other more visually or olfactory oriented mammals. In addition, most bats have evolved an active sonar system which uses ultrasonic calls for acoustically probing the environment. Such sophisticated laryngeal echolocation behavior, however, is only present in the suborder microchiroptera; it is lacking in megachiropteran bats (flying foxes), which instead rely on visual and olfactory cues for orientation and foraging (Schuller and Moss, 2004; for general overview see Thomas et al., 2004).

Communication on one side and acoustical imaging of the environment on the other necessitate different temporal and spectral structures of calls used for these distinct tasks. Information transmitted in

*Corresponding author. E-mail: metzner@ucla.edu; gschuller@lmu.de

ISBN 978-0-12-374593-4
DOI: 10.1016/B978-0-12-374593-4.00037-1

communication calls is generally encoded with complex time-varying modulations of amplitude and frequency during ongoing calls that are long enough to transmit sufficient spectrotemporal information about the intended "meaning" of these calls (see Fig. 1) (for overview see Kanwal et al., 2006). Acoustical signals used for imaging obstacles that a bat may encounter in its flight path or imaging its prey, however, demand a different spectrotemporal design. Echolocation pulses (see Fig. 1) are therefore generally shorter to optimize (distance resolution) target range accuracy, and contain higher energies to yield sufficiently strong echoes even from small objects. Hence, their spectrotemporal characteristics are rather simple and stereotypical, with a well-defined time course. This facilitates the detection and discrimination of target-induced modulations in the echo that are imprinted on the carrier signal, i.e., the echolocation pulse, by being easily recognizable as deviations from the call template. At the extremes of this range of echolocation call types (Fig. 1) are, on one end, very short, broadband signals composed of several,

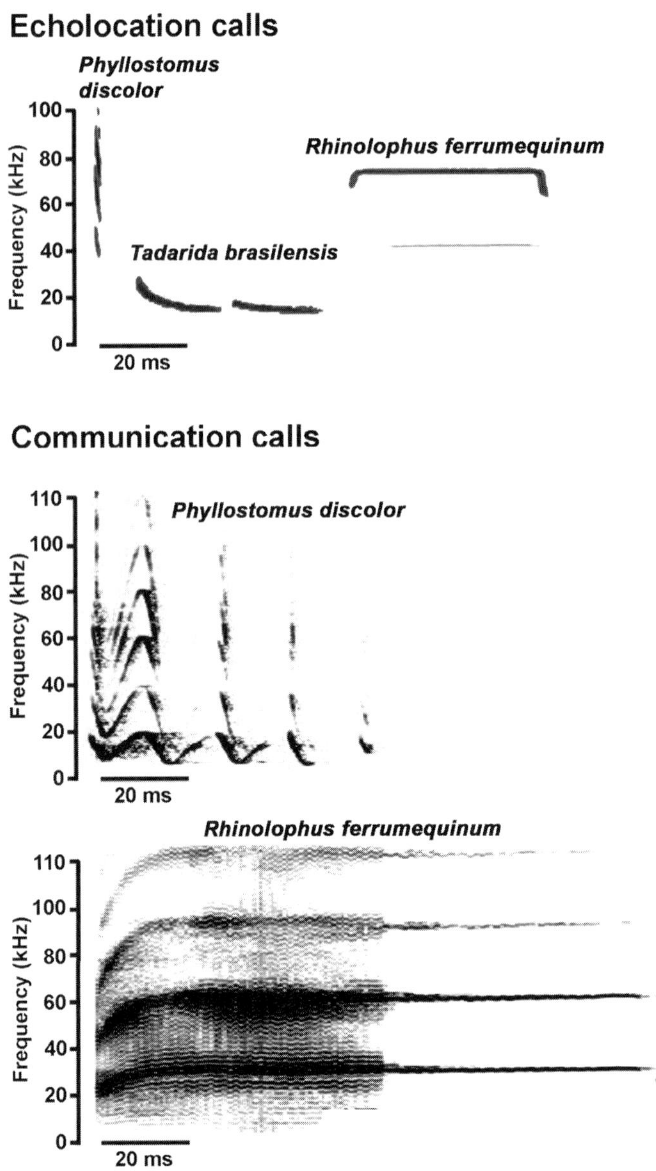

Fig. 1. Sonographic examples for different types of echolocation pulses (top) and communication calls (center and bottom).

often overlapping harmonic components (Fig. 1, *Phyllostomus discolor*) and, on the other end, long-lasting, constant frequency calls with an extraordinarily small bandwidth. Whereas the former yield high temporal resolution and range accuracy, the latter produce long ranging echolocation calls, especially for low-frequency signals (Fig. 1, *Tadarida brasilensis*), or in high-frequency, narrowband pulses, which encode small frequency deviations induced by moving targets ("flutter detection;" Fig. 1, *Rhinolophus ferrumequinum*).

Because such extremely different demands on vocal production and control are not common in other terrestrial mammals, one may wonder whether echolocating bats show peculiar adaptations in the biomechanical properties of the larynx, its motor innervation, as well as the premotor and sensory feedback control through the central nervous system. During echolocation, vocal design obviously constrains the auditory feedback provided by the returning echo signals and, conversely, in order to optimize signal detection, vocal production changes depending on the target-induced modulations in the auditory feedback. This suggests that sensory feedback mechanisms at a premotor level are present or more pronounced in the bat brain than in the brain of other mammals. It also raises the question of how specialized the vocal periphery, as well as neuronal circuits and mechanisms involved in call production, actually are in echolocating bats.

II. Motor periphery: the bat larynx and its neural innervation

II.A. Laryngeal morphology and functions

Echolocating bats, like all other mammals, produce all their vocalizations in the larynx during expiration. The supralaryngeal space (throat, mouth, nostrils and noseleaf) acts as an acoustic filter, shaping the emitted sound spectrum as well as its directional characteristics, without, however, shaping call design in an articulation-like manner. Because echolocation calls in bats show several distinct features when compared with vocalizations in other mammals, such as very high call frequencies, short durations and high repetition rates, as well as high intensities, one wonders if the bat larynx exhibits any morphological adaptation or specialization in the muscular physiology that relate to these unusual call designs. Although morphological data are relatively sparse, they do not indicate any particular design principle that differs from that of the general mammalian structural plan of the larynx (Robin, 1881; Elias, 1907; Sprague, 1943; Fischer, 1961; Fischer and Gerken, 1961; Denny, 1976; Griffith, 1983). Important features of the bat larynx that are likely to affect its function include its large size (see Frey and Gebler, Chapter 10.3 in this volume), the level of ossification of the laryngeal cartilages, rather rigid arytenoid cartilages, the relative strength and ultrastructural design of the internal laryngeal muscles, the configuration of the laryngeal membranes and the design of tracheal cavities and their filter characteristics.

First, the relatively high degree of ossification that occurs early during the development of many bats reinforces the laryngeal scaffolding, a prerequisite for tolerating the high tension of the vocal folds needed to produce high-pitched glottal pulses. It is unclear whether laryngeal ossifications observed in mammals other than bats are merely age-dependent alterations or morphologically constitutive traits.

Another morphological adaptation for generating high-frequency calls is the fact that the cricothyroid muscle of most echolocating bats is rather hypertrophied, which during contraction allows it to exert strong tension on the ventral insertion point of the vocal fold. At its opposite end, the vocal fold is attached to the arytenoid cartilages, which are relatively rigid in bats. As a result, the cricothyroid muscle, a very fast twitching muscle, is virtually the only muscle affecting the tension of the vocal fold.

Bats emit calls with extraordinarily large variations in duration and emission rates, even within one individual; they can range from 0.5 ms for the shortest echolocation pulses up to several hundred milliseconds in many communication calls. To cope with these extreme physiological demands, laryngeal muscles in bats, most notably the cricothyroid muscle, seem to have developed ultrastructural features that are characteristic for high-speed excitation–contraction coupling seen in other fast twitching muscles. Although studies are sparse, they indicate, among other characteristics, an extremely well-developed sarcoplasmatic reticulum and an extremely large number of mitochondria (Revel, 1962; Reger, 1978).

Another morphological specialization found in many bat larynges are vocal membranes or "vocal lips," which are membranous rostral extensions of the vocal fold (Griffin, 1958). The vocal membranes are delicate and lightweight structures, and their low mass allows them to be excited (i.e., vibrate) at the high frequencies characteristic for bat vocalizations. Laryngeal airflow might directly drive vibrations of the vocal membranes, or they may be biomechanically coupled to the oscillations of the underlying vocal folds. This mechanism enables bats to emit high-frequency calls at elevated sound pressure levels of more than 100 dB SPL (when measured 15 cm in front of the bat), with only moderately high subglottic pressures (Fattu and Suthers, 1981). Apart from boosting call emission, the coupling of the activity of the vocal membranes to that of the vocal fold can also produce nonlinear by-products in the sound production, such as non-harmonically related call components or chaotic features in calls (see more details below).

Nonlinear phenomena in vocal production are also known in other mammals, for example in dogs, monkeys and humans (Fitch et al., 2002), and yield such call features as subharmonics and biphonation (i.e., two independent pitches). In cases where such nonlinear vocal features have been investigated they were most often related to coupled oscillations of vocal membranes adjacent to the vocal fold. Modeling approaches (Mergell et al., 1999) corroborated that vocal membranes can indeed account for nonlinear laryngeal properties and cause many of the complex spectrotemporal features seen in a wide range of mammalian vocalizations. The presence of pronounced "vocal lips" in bats, however, represents a prime example of how purely biomechanical adaptations in the chiropteran larynx can generate the particularly rich variation of call features observed in bat echolocation and communication signals (e.g., Kanwal et al., 1994; Ma et al., 2006). Some aspects are outlined below.

Laryngeal vocalizations are transmitted through the supralaryngeal tract and emitted either through the mouth, as in other mammals, or through the nostrils. Approximately one-third of all microchiropteran bats emit their calls through the nose, such as Old World Rhinolophids, Hipposiderids, Megadermatids and various New World Phyllostomids (*Carollia* spp., *Phyllostomus* spp.). Transmission through the mouth allows for spectral as well as directional shaping, which is quite variable and depends on the opening constellation of the mouth (Strother and Mogus, 1970; Mogensen and Møhl, 1987; Ghose et al., 2007). Call emission through the nostrils and its resulting filtering properties are spectrally and directionally less variable, due to the rather fixed structure of the transmitting ducts. However, they dramatically alter the frequency composition of the emitted calls and protrusions ("noseleaves") surrounding the nostrils in these bats, acting as acoustic beamforming baffles that influence the shape of the emitted biosonar beam (Hartley and Suthers, 1987; Henze and O'Neill, 1991; Zhuang and Mueller, 2006, 2007).

The trachea of horseshoe bats exhibits a peculiar morphological specialization not found in other bats (Elias, 1907; Möhres, 1953; Suthers et al., 1988). It is enlarged below the larynx by three tracheal pouches (two lateral pouches and one medial pouch) that are connected by small openings with the trachea. These tracheal enlargements were initially believed to possess resonant properties that aid in the creation of the unusually long-lasting, constant frequency portion of horseshoe bat echolocation calls (see Fig. 1). Total or partial filling of the hollows that connect the pouches with the trachea, however, failed to affect the emitted call frequency, but instead affected the intensity of the calls; the tracheal chambers appear to suppress the fundamental frequency component in the trachea and prevent it from being reflected from the lungs back towards the cochlea. This could prevent cochlear stimulation along multiple indirect pathways (via tissue conduction), which would severely smear clear reception of sound (Suthers et al., 1988).

II.B. Biomechanical properties of the isolated bat larynx

In an isolated larynx preparation obtained from freshly-prepared greater horseshoe bats, *Rhinolophus ferrumequinum*, we investigated how much biomechanical properties of the larynx contribute to call frequency control (Kobayasi et al., 2008). Altering tracheal airflow within the physiological range resulted in a combination of linear and nonlinear changes of the spectral composition of sounds (Fig. 2). Low tracheal air pressure above excitation threshold initially produced sounds at high frequencies (and lower amplitudes), reflecting the main frequency component of echolocation pulses. Sound frequencies linearly increased with air pressure until further increases in air pressure suddenly caused the sound frequency to jump to that of the lower main frequency component of most communication signals (Ma et al., 2006). An even greater rise in air pressure yielded, in addition to the partially prevailing main frequency components, transitions from periodic to chaotic excitation that occurred within one

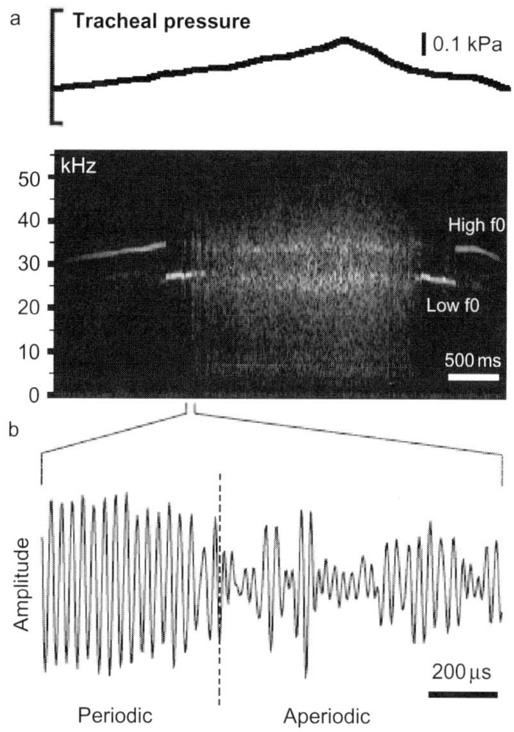

Fig. 2. Sound emitted by an isolated horseshoe bat larynx for varying tracheal air pressures. (a) Tracheal pressure and corresponding sonogram of emitted sounds; (b) enlarged portion (see brackets) of the waveform of the sound signal illustrating the sudden change from periodic (tonal) to aperiodic (chaotic) emissions (vertical hatched line) within only a few sound cycles. For additional details, see text.

or two signal cycles (<20 μs). These spectral features ("noise components") are observed in many horseshoe bat communication signals.

When simulating contractions of the cricothyroid muscle (the sole laryngeal muscle involved in call frequency control in bats) by varying the antero-posterior tension exerted on the thyroid, call frequencies changed in a highly linear fashion over a range corresponding to those observed naturally in both communication and echolocation signals. Sinusoidal frequency modulations, which represent another prominent feature of many communication calls, could also sometimes be elicited by varying the tracheal airflow (exact conditions unknown), without simulated periodic contractions of the cricothyroid muscle. Nonlinear interactions of intrinsic biomechanical properties of the larynx most likely account for such modulations developing independently from externally applied manipulations. These results can help one to better understand which features of bat vocalizations (pitch, amplitude and/or frequency modulations) are based exclusively on intrinsic biomechanical properties of the larynx, and which are controlled or at least modulated by neuronal input from the brain.

II.C. Neuronal control of the bat larynx

The spectrotemporal features of the briefest echolocation calls, the characteristics of "noisy" broadband echolocation calls, or the frequency–time course of upward or downward frequency sweeps are most likely determined by the biomechanical properties of the larynx, together with trigger-like muscular activation and relaxation, respectively. In contrast, most of the spectrotemporal fine control in longer calls (tens of milliseconds), as well as call-by-call changes in the frequency–time course, duration and rate of emission, are governed by neural control from the brain. Such changes in call design, which are usually caused by changing echolocation demands, are most likely achieved by the temporal delay through neuronal feedback loops, rather than by the laryngeal reaction time (Schuller, 1977).

Similar to other mammals (Holstege 1989; Gerrits, 1996; Vanderhorst and Holstege, 1996), the brainstem exerts this motor control through the external branch of the superior laryngeal nerve and via the recurrent or inferior laryngeal nerve. The former exclusively innervates the cricothyroid muscle, which controls the tension of the vocal fold and thus the fundamental frequency of the calls (Schuller and Suga, 1976). The inferior laryngeal nerve is the motor input to all other laryngeal muscles and is primarily responsible for the temporal fine structuring of the calls (Rübsamen and Schuller, 1981).

Finally, unlike in most other mammals, the vocalization behavior of echolocating bats is embedded into a complex mesh with other behavioral patterns that typically involve other motor acts, such as active flight and highly-directional movements of head and pinnae (Ghose and Moss, 2003). The emission of echolocation calls is locked to distinct phases of respiration, which itself is temporally tightly linked to wing stroke cycles during active flight (Lancaster and Speakman, 2001). Although the functional organization and anatomical interconnection of motor subsystems, such as those for locomotion or vocalization, with respiratory control are not uniquely present in bats, they are nevertheless more apparent and pronounced, and therefore more accessible to investigation than in other mammals (Smotherman et al., 2006). The neuronal connections of these functional subsystems are discussed separately (see Smotherman et al., Chapter 9.2 in this volume).

III. Premotor control and feedback from auditory structures

Following the general mammalian organization (Holstege 1989; Gerrits, 1996; Vanderhorst and Holstege, 1996), the motor somata of the superior and the recurrent laryngeal nerves are located within the nucleus ambiguus (NA), a rostrocaudally elongated structure in the medulla oblongata (e.g., *Rhinolophus*: Rübsamen and Schweizer, 1986). Similar to other mammals, such as the rabbit (Kitamura, 1993), motor neurons also appear to be distributed in a myotopical manner in bats (*Pteronotus*: Kobler, 1983; *Rhinolophus*: Schweizer et al., 1986). Somata giving rise to the superior laryngeal nerve are located more rostrally, and those of the recurrent laryngeal nerve occupy a slightly larger area and are distributed over the central and more caudal portions of the nucleus ambiguus. This myotopic organization of the bat nucleus ambiguus represents a functional topography of motor neurons in a rostral-to-caudal direction, controlling spectral and temporal parameters of echolocation calls, respectively (Schuller and Rübsamen, 1981; Rübsamen and Schuller, 1981).

The nucleus ambiguus is part of a final common network of motor pathways in which coordination of all subsystems for vocalization is taken over by an area in the reticular formation which receives major inputs from the periaqueductal gray. This area extends from just above the superior olivary complex at its anterior end through the lateral reticular formation at the level of the facial nucleus and nucleus ambiguus posteriorly to the caudal-most medulla, including the nucleus retroambiguus (Vanderhorst et al., 2000; see also Hage, Chapter 8.3 in this volume). So far, however, no study has addressed the cytoarchitecture or anatomical connectivity of a nucleus retroambiguus in bats.

III.A. Control of call frequency and call temporal parameters by nucleus ambiguus: pharmacology and single cell recordings

Most animals constantly adjust the spectrotemporal composition of their vocalizations depending on the information content intended to be conveyed. Little is known, however, about the neuronal networks acting as pattern generators in the vertebrate brain to adjust such spectral or temporal call parameters (see Hage, Chapter 8.2 in this volume). In order to investigate how activity of the nucleus ambiguus affected frequency, duration and interpulse interval (IPI) of horseshoe bat vocalizations, the synaptic activity of the nucleus ambiguus was manipulated (Kobayasi et al., 2009, submitted). For this purpose, various transmitter agonists and antagonists were injected into the nucleus ambiguus of spontaneously vocalizing bats while they adjusted their calls in response to electronically altered playbacks of their own calls (Doppler-shift compensation behavior, DSC). The frequency of echolocation calls could only be affected in anterior portions of the nucleus ambiguus, where only $GABA_A$-receptor specific, but not glutamate-receptor specific, agents were effective. Increasing GABA-ergic inhibition by application of muscimol lowered the emitted call frequency; conversely, the GABA-inhibitor bicuculline methiodide led to an increase in call frequency. Drug injections into the posterior nucleus ambiguus affected only the timing parameters of the calls (call duration and call rate), and in a manner characteristic for DSC (see, e.g., Smotherman and Metzner, 2005). Excitation induced by bicuculline and a glutamate agonist (AMPA) increased the proportion of short duration calls and of short IPIs, whereas inhibition with muscimol and kynurenic acid, a glutamate receptor-antagonist, decreased it. GABA-ergic drugs affected call durations within the intermediate and ventral portions of the nucleus ambiguus and IPIs in the more caudodorsal nucleus ambiguus. Glutamatergic drug effects on both duration and IPI were found throughout the posterior nucleus ambiguus. The main topography of functional involvement for frequency and temporal control in the nucleus ambiguus is in line with neurophysiological findings of Rübsamen and Betz (1986) and anatomical projection studies from Schweizer et al. (1981) in the Sri Lankan horseshoe bat (*R. rouxi*), and results by Kobler (1983) in *Pteronotus*.

The pharmacological investigations represent an important step towards identifying the details of the neural basis for the production of complex vocal signals in a mammal. The data support a novel hypothesis for the control of call frequency by the laryngeal motoneurons of the anterior nucleus ambiguus. The lack of glutamatergic influence in this nucleus ambiguus area suggests an intrinsic spontaneous activity in these neurons that is only modulated by $GABA_A$, giving rise to the range of call frequency responses encountered in these bats.

Fig. 3 illustrates this idea. When the bat is silent, the spontaneous activity is constantly suppressed by GABA-ergic input to cricothyroid motoneurons from premotor structures (Fig. 3, left column). During emission of a call, premotor input releases this inhibition, and the amount of release from inhibition is directly proportional to call frequency. When bats are calling at rest, and therefore not compensating for flight-induced shifts in the echo frequency (resting frequency, RF; Fig. 3, left column), the premotor input to the nucleus ambiguus yields more release from inhibition during call emission than when calls are emitted at lower frequencies (Fig. 3, center column), such as during DSC when compensating for echo frequencies that are shifted above RF. Muscimol injections during call emission mimicked this effect by increasing inhibition, equivalent to a reduction of release from inhibition. The activity of neurons predicted by this hypothesis is consistent with neurophysiological responses found by Rübsamen in the rostral nucleus ambiguus (Rübsamen and Betz, 1986). Conversely, when shifts in echo frequency below RF cause pulse frequencies to rise above RF (Metzner et al., 2002), premotor input releases cricothyroid motoneurons from inhibition even more (Fig. 3, right column), much as do bicuculline injections. The fact that only a single transmitter, GABA, is sufficient to control call frequencies in horseshoe bats is consistent with the rather straightforward control of this call parameter by a single phonatory pool of motoneurons innervating only a single laryngeal muscle, the cricothyroid muscle.

A combination of single unit recordings and pharmacological manipulations of nucleus ambiguus motoneurons revealed evidence for two cell types that may be elements of such a frequency control circuit. The most common type of nucleus ambiguus neurons exhibited prevocal excitatory activity which was directly proportional to call frequency. The prevocal activity was suppressed by muscimol and slightly elevated by bicuculline, whereas glutamatergic drugs had no effect. Another relatively common type of neuron, which exhibited relatively high spontaneous firing rates, was inhibited during call emission and showed an inverse correlation between spike count and call frequency. Their activity was increased by application of glutamate (or AMPA) and reduced by GABA (or muscimol). This type of neuron, presumably an interneuron, may represent the source of GABA-ergic inhibition onto nucleus ambiguus motoneurons controlling call frequency (Hage et al., 2008).

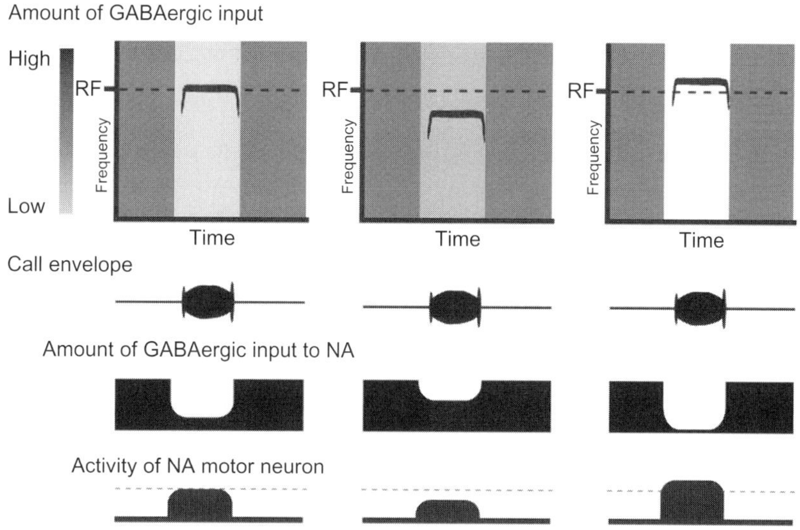

Fig. 3. Hypothesis on how changes in call frequency (top row) could result from GABA-ergic modulation of activity in nucleus ambiguus motoneurons (bottom row; neuronal activity expressed by the height of black bars). The amount of GABA-ergic inhibition of nucleus ambiguus motoneurons is given in the second row from the bottom (expressed by the depth of the indentations in the black bar and as gray shading in the top row. Calls emitted at RF (left column), call frequency below RF (center column) and above RF (right column). Hatched lines in top row indicate RF, and in bottom row the resting activity.

III.B. Premotor vocal areas in the brainstem of bats

The concept of a strictly hierarchical, almost exclusive, dominance of the periaqueductal gray (PAG) as the controlling input to the final common motor pathway for vocalization did not prove advantageous for the investigation of the descending vocal control system in bats. Electrical or pharmacological stimulation in the PAG elicits vocalizations in several bat species (Suga et al., 1973; Fenzl and Schuller, 2007) as it does in other mammals (e.g., Jürgens, 1994). In bats, PAG stimulation yields echolocation calls as well as communication calls, although at slightly different locations (Suga et al., 1973; Fenzl and Schuller, 2002). Echolocation calls, but not communication calls, could be elicited at low threshold stimulating currents or pharmacologically in several well-defined areas in the brainstem below the level of the PAG (Schuller and Radtke-Schuller, 1990). Stimulations yielded natural echolocation calls in a one-to-one manner in the paralemniscal area, the deep layers of the superior colliculus, the deep mesencephalic nucleus, the nucleus of the brachium of the inferior colliculus, the pretectal area and in marginal zones of the inferior colliculus (rostral pole and external part) (Schuller, 1998; Moss and Sinha, 2003; Schuller and Moss, 2004). Communication calls could not be elicited in these brain regions. These vocally active regions are not considered to directly control the distinct spectrotemporal structure of echolocation pulses similar to a pattern generator, as calls were elicited without distortions (see Hage, Chapter 8.2 in this volume).

Stimulation of some areas within the reticular formation and caudal brainstem of bats, however, yielded clearly distorted calls (Schuller and Radtke-Schuller, 1990; Behrend and Schuller, 2000). In these brain areas the electrical microstimulation not only triggers the vocalization, but also interferes with functional processes organizing the spectrotemporal pattern of the calls. Similar brainstem sites have been reported by Hage and Jürgens (2006) in the squirrel monkey. They suggest that these reticular formation networks represent the pattern generator for distinct squirrel monkey calls that consist of pronounced frequency modulations (see Hage, Chapter 8.2 in this volume).

In bats, vocalizations caused by stimulation in the paralemniscal area (PLA) and in the nucleus of the brachium of the inferior colliculus (NBIC) (other regions not tested so far) do not depend on a functioning PAG. This is evident from reversible inactivations of the PAG with kynurenic acid (Siebert, unpublished results). Kynurenic acid-based inactivation of PAG eliminated the ability to elicit any type of calling with microelectrical PAG stimulation, even at elevated current levels. In contrast, microelectrical stimulations of the two brainstem regions, the paralemniscal area and nucleus of the brachium of the inferior colliculus, after PAG inactivation yielded vocalizations at the same low current settings as the control without any distortions or alterations in the elicited calls. On the other hand, the ability to evoke vocalization in the PAG greatly depended on an intact paralemniscal area and nucleus of the brachium of the inferior colliculus (Fenzl and Schuller, 2002, 2005). When the paralemniscal area was lesioned bilaterally, echolocation calls were no longer elicited in the PAG (Fenzl and Schuller, 2005), whereas communication calls could still be generated.

The premotor vocal control of echolocation pulses and communication calls must therefore be differently organized. The descending vocalization pathway, as described for mammals other than bats (Holstege, 1989; see also Hage, Chapter 8.2 in this volume), represents the condition for the control of bat echolocation calls only insufficiently. Instead, it appears that a network of various brainstem regions interacting in parallel with the proposed pathway from PAG to the ventral reticular formation (including nucleus ambiguus and nucleus retroambiguus) portrays the condition in bats more appropriately. Further neurophysiological and pharmacological investigations will aid in elucidating the functional significance of the complicated interconnections of brain areas involved in triggering or controlling bat echolocation calls. This will, first, clarify their integration within the descending vocal control pathway, and second, determine their role as recipients of auditory information from nuclei of the ascending auditory system.

III.C. Audio–vocal interactions and vocal areas in the brainstem of bats

Echolocation in bats relies heavily on adaptive feedback between hearing and vocal control, because the characteristics of the vocal signal determine the efficiency of the spectral and temporal resolution achieved by the echo analysis. Changes in the acoustic environment therefore require constant adjustments in call design, in order to optimize echo signal detection. Echolocation is a fast-reacting system necessitating adjustments within less than a few tens of milliseconds and is not tolerant to long neural delays. To avoid lengthy feedback loops, audio–vocal feedback may take place at rather low levels in the brain. Neurons that are situated in vocal areas not belonging to the ascending auditory pathway, but responding to acoustic stimuli, were found in the PLA (*Rhinolophus rouxi*: Metzner, 1993, 1996), the NCAT and the immediately adjacent reticular formation (*Pteronotus parnellii*: Casseday et al., 1989; *Rhinolophus rouxi*: Schuller et al., 2001), the nucleus cuneiformis (*Rhinolophus rouxi*: Schuller and Wenstrup, unpublished results), the deep layers of the superior colliculus (*Rhinolophus rouxi*: Reimer, 1991; Sinha and Moss, 2007), the rostral pole of the inferior colliculus (Metzner, unpublished results; Prechtl, unpublished results; Pieper and Jürgens, 2003), the pretectal area (Nixdorf, 2003) and the NBIC (ferret: Schnupp and King, 1997).

In bats, only neurons in the paralemniscal area and the pretectal area were tested simultaneously during active vocalizations and acoustical stimulation, whereas acoustically-activated neurons were not tested during active vocalization in other vocally active brain regions. Metzner (1993) has demonstrated in the paralemniscal area that neuronal activity before and during vocalization, as well as acoustically-induced activity can occur in the same neuron. This indicates that audio–vocal interfacing can occur directly at the level of single neurons. Neurons active before and during self-produced vocalization and responding also to acoustic stimulation are not restricted to bats, but have also been recorded in the brainstem of squirrel monkeys (Hage et al., 2006). It appears, however, that this area cannot be homologized with the more dorsally located paralemniscal area in the bat's brain, which differs both in its vocal motor and auditory connectivity from the connections found in the monkey (Hannig and Jürgens, 2005).

In several mammals, neuroanatomical connection studies of brainstem nuclei, in which stimulation or inactivation experiments suggested that they are functionally involved in vocalization, focus mostly on the final common vocal motor pathway. Tracer studies combined with neurophysiological characterization of the injection sites as being involved in vocal control, and with the objective of determining the connectivity that mediates audio–vocal feedback in mammals are sparse and limited mostly to bats and monkeys (e.g., Metzner, 1996; Hannig and Jürgens, 2005). The audio–vocal interconnections are complex and the main results from work in bats are compiled in Table 1 (without claiming to be complete).

IV. FoxP2 expression in the brain of echolocating and non-echolocating bats

FoxP2 is a transcription factor implicated in the development and neural control of vocalization. Recently it has been shown that it underwent accelerated evolution in bats employing different sonar systems (high-duty cycle, constant frequency calls with at least partial Doppler shift compensation, such as *Rhinolophus* and *Hipposideros* versus low-duty cycle bats, such as *Myotis*), thus possibly playing a role in the evolution and development of echolocation (Li et al., 2007). We investigated the expression of FoxP2 in different brain areas of bats using different echolocation systems (Yin et al., 2008). We compared three microchiropteran species with well-developed echolocation with two species of megachiropteran bats lacking laryngeal echolocation. The first group comprised *Rhinolophus ferrumequinum* and *Hipposideros armiger*, two species emitting constant frequency pulses and exhibiting Doppler shift compensation (DSC) behavior, and *Myotis ricketti*, a species producing short frequency modulated sonar pulses. Megachiropteran bats studied were *Rousettus leschenaulti*, which uses a very rudimentary sonar system based on tongue-clicks, and *Cynopterus sphinx*, which lacks any echolocation. The most striking differences in FoxP2 expression were seen in the suprageniculate nucleus and the anterior cingulate cortex (ACC), which both exhibited significant expression in bats with laryngeal echolocation (*Rhinolophus*, *Hipposideros* and *Myotis*), but not in the group of megachiropteran bats (*Rousettus* and *Cynopterus*). Similarly dramatic expression disparities were found in the olfactory tubercle. Here, however, the labeling was more than twice as dense in both fruitbats, *Rousettus* and *Cynopterus*, than in any of the echolocating microchiropteran bats. No significant differences in FoxP2 expression were observed between the species within each of the micro- and megachiropteran bat groups.

The significantly stronger expression of FoxP2 in ACC in echolocating bats prompted us to investigate its behavioral significance for echolocation in a bat with a highly-sophisticated echolocation behavior, such as *Hipposideros armiger*. Bats of this species very precisely adjust their call parameters, such as the

Table 1. Synopsis of functional involvement of brainstem nuclei in vocal control in bats

Brainstem area involved in vocal control	Vocal response (a)	Recordings (b)	Connectivity (c)	Literature
NCAT	*ElStim*: E-calls, no C-calls	Rec_audit	WGA: extralemniscal pathway	(a) Behrend and Schuller, 2000; (b, c) Schuller et al., 1991; (c) Casseday et al., 1989
	PhStim: E-calls, no C-calls			
	Lesion: DSC disturbed			
RET	*ElStim*: E-calls (distorted)			(a) Behrend and Schuller, 2000; (a) Schuller and Radtke-Schuller, 1990
PLA	*ElStim*: E-calls, no C-calls	Rec_audit, Rec_vocal	WGA: auditory afferences, motor efferences	(a) Schuller and Radtke-Schuller, 1990; (a) Pillat and Schuller, 1998; (a) Fenzl and Schuller, 2002, 2005; (b) Metzner, 1993; (c) Metzner, 1996
	PhStim: E-calls, no C-calls			
	Lesion: DSC unimpaired, E-calls impaired, not C-calls			
CUN	*ElStim*: E-calls, arousal		WGA: to Ncl. Ambiguus, afferences from "vocal" areas	(a) Schuller and Radtke-Schuller, 1990; (c) Schuller, 1998
DMN	*ElStim*: E-calls			(a) Schuller and Radtke-Schuller, 1990
PB	*PhApplic*: vocal-respiratory coupling; DSC affected			(a) Smotherman et al., 2003, 2006
ICrp, ICx	*ElStim*: E-calls	Rec_audit, Rec_vocal		(a, b) Prechtl (pers. communication)
PAG	*ElStim*: E-calls, C-calls (different locations); *lesion*: E-calls unimpaired			(a) Fenzl and Schuller, 2002, 2005
SCdp	*ElStim*: E-calls, no C-calls	Rec_audit, Rec_vocal		(a) Schuller and Radtke-Schuller, 1990; Sinha and Moss, 2007; (b) Reimer, 1991
	PhStim: E-calls, no C-calls			
NBIC	*ElStim*: E-calls, no C-calls			(a) Schuller, 1998; Schuller and Sripathi, 1999
	PhApplic: DSC suppressed			
AP	*Estim*: E-calls, no C-calls	Rec_audit, Rec_vocal	WGA: auditory afferences, motor efferences	(a) Nixdorf, 2003

Effectiveness of microstimulation for vocal activation, consequences of lesions or pharmacological modulation for vocal behavior and selective involvement in the control of echolocation versus communication calls are indicated in column (a). Availability of recordings to acoustical stimuli or to ongoing vocalization is given in (b). Tracer experiments are mentioned in column (c) with general indications of main connectivity. The last column gives relevant publications for the different items.

Abbreviations: AP: pretectal area; CUN: nucleus cuneiformis; DMN: deep mesencephalic nucleus; Icrp: ICx inferior colliculus rostral pole/external; NBIC: nucleus of the brachium of the inferior colliculus; NCAT: nucleus of the central acoustic tract; PAG: periaqueductal gray; PB: parabrachial nucleus; PLA: paralemniscal area; RET: reticular formation; SCdp: superior colliculus deep layers; C-call: communication call; DSC: Doppler shift compensation in CF-FM bats; E-call: echolocation call; ElStim: electrical stimulation; PhApplic: application of transmitter agonists or antagonists; PhStim: pharmacological stimulation; Rec_audit: recording of auditory responses; Rec_vocal: recording during active vocalization; WGA: wheat germ agglutinin (HRP associated).

frequency and emission rate, depending on the auditory feedback provided by the returning echo signal (Doppler shift compensation, DSC).

Knocking-down FoxP2 expression in the ACC of *Hipposideros* had marked consequences for echolocation call control (Yin et al., 2008). Whereas it had no effect on the bat's resting frequency (i.e., the call frequency emitted when not performing DSC), it did significantly alter the DSC behavior by reducing the amount of frequency compensation and the call activity.

This suggests that FoxP2 expression in the ACC of a bat with a highly-sophisticated echolocation behavior, such as *Hipposideros*, is involved in controlling call parameters during more complex echolocation tasks (DSC), whereas it has no significant effect on the control of more basic vocalization parameters, such as the call frequency emitted when not performing DSC (i.e., the resting frequency). This supports the notion that ACC is likely to be involved in high-level, probably voluntary initiation of calling behavior, and it underlines the importance of FoxP2 in the control of echolocation calls. The difference in expression patterns in ACC between echolocating and non-echolocating bats may therefore indicate that the same higher-order structure (ACC), although present in both groups of bats, can play a significantly different role in call production. Based on knowledge from other mammals (see Hage, Chapter 8.3 in this volume), the ACC may ultimately affect different vocal patterns through its direct connection with the PAG, but in addition also via several cortical and cortico–thalamic routes, such as through the amygdala or the nucleus accumbens.

V. Summary and conclusions

Call production in the larynx of bats follows the general mammalian pattern. However, several adaptations in the design of the larynx and vocal tract allow bats to generate and emit precisely timed high-frequency, ultrasonic echolocation calls at high intensities. Such adaptations are, for example, a hypertrophied cricothyroid muscle that allows for high-speed excitation–contraction coupling, the presence of vocal membranes ("vocal lips"), and in bats with nasal call emission, tracheal and nasal cavities as well as "noseleaves" that alter the filter characteristics of the vocal tract and increase the directionality of call emission. The innervation of the different laryngeal muscles by the nucleus ambiguus also follows the common mammalian scheme. At the level of premotor control by structures feeding into the final common vocal motor pathway, however, some areas and connections appear to play a peculiar role in echolocating bats. Certain brainstem areas, such as different portions of the periaqueductal gray, the paralemniscal area and the superior colliculus operate in parallel to the descending connections from the PAG to the ventral reticular formation (encompassing the nucleus ambiguus and nucleus retroambiguus) commonly described in mammals other than bats (see Hage, Chapter 8.3 in this volume). Some of these brainstem areas are devoted exclusively to the control of echolocation pulses, but not communication calls. Similar differences in the involvement of brain structures in either echolocation or the production of other vocal signals may also occur at higher levels of vocal motor control, such as in the anterior cingulate cortex, and involve differential expression of certain genes, such as FoxP2, involved in the control of vocal motor functions. Finally, most of the brainstem areas involved in vocal control are also targeted by inputs from different levels of the ascending auditory system, and therefore represent potential links within the audio–vocal feedback loops that are quintessential for echolocation in bats.

Acknowledgments

We wish to thank M. Smotherman and S. Hage for comments on this manuscript. Supported by NIH DC 5400 (WM) and Deutsche Forschungsgemeinschaft (DFG) Schu390/7-1 (GS).

References

Behrend, O., Schuller, G., 2000. The central acoustic tract and audio–vocal coupling in the horseshoe bat, *Rhinolophus rouxi*. Eur. J. Neurosci. 12, 4268–4280.

Casseday, J.H., Kobler, J.B., Isbey, S.F., Covey, E., 1989. Central acoustic tract in an echolocating bat: an extralemniscal auditory pathway to the thalamus. J. Comp. Neurol. 267, 247–259.

Denny, S.P., 1976. The bat larynx. In: Hinchcliffe, R., Harrison, D.F. (Eds.), Scientific Foundations of Otolaryngology. Heinemann Medical Books Ltd, London, UK, pp. 346–370.

Elias, H., 1907. Zur Anatomie des Kehlkopfes der Microchiropteren. Morphol. Jahrbuch, 37, 70–118.

Fattu, J.M., Suthers, R.A., 1981. Subglottic pressure and the control of phonation by the echolocating bat, *Eptesicus*. J. Comp. Physiol. A, 143, 465–475.

Fenzl, T., Schuller, G., 2002. Periaqueductal gray and the region of the paralemniscal area have different functions in the control of vocalization in the neotropical bat, *Phyllostomus discolor*. Eur. J. Neurosci. 16, 1974–1986.

Fenzl, T., Schuller, G., 2005. Echolocation calls and communication calls are controlled differentially in the brainstem of the bat *Phyllostomus discolor*. BMC Biology 3, 17. doi: 10.1186/1741-7007-3-17.

Fenzl, T., Schuller, G., 2007. Dissimilarities in the vocal control over communication and echolocation calls in bats. Behav. Brain Res. 182, 173–179.

Fischer, H., Gerken, H., 1961. [The larynx of the bat (*Myotis myotis*) and the human larynx]. Ann. Otolaryngol. Chir. Cervicofac. 78, 577–585 in French.

Fitch, W.T., Neubauer, J., Herzel, H., 2002. Calls out of chaos: the adaptive significance of nonlinear phenomena in mammalian vocal production. Anim. Behav. 63, 407–418.

Ghose, K., Moss, C.F., 2003. The sonar beam pattern of a flying bat as it tracks tethered insects. J. Acoust. Soc. Am. 114, 1120–1131. doi:10.1121/1.1589754.

Ghose, K., Moss, C.F., Horiuchi, T.K., 2007. Flying big brown bats emit a beam with two lobes in the vertical plane. J. Acoust. Soc. Am. 122, 3717–3724. doi: 10.1121/1.2799491.

Griffin, D.R., 1958. Listening in the Dark. Yale University Press, New Haven, CT.

Hage, S., Jürgens, U., 2006. Localization of a vocal pattern generator in the pontine brainstem of the squirrel monkey. Eur. J. Neurosci. 23, 840–844.

Hage, S.R., Jürgens, U., Ehret, G., 2006. Audio–vocal interaction in the pontine brainstem during self-initiated vocalization in the squirrel monkey. Eur. J. Neurosci. 23, 3297–3307.

Hage, S., Kobayasi, K., Metzner, W., 2008. Call frequency control by neurons in the vocal motor nucleus in echolocating horseshoe bats. 38th Annual Meeting of the Society for Neuroscience, Washington, DC.

Hannig, S., Jürgens, U., 2005. Projections of the ventrolateral pontine vocalization area in the squirrel monkey. Exp. Brain Res. 169, 92–105. doi: 10.1007/s00221-005-0128-5.

Hartley, D.J., Suthers, R.A., 1987. The sound emission pattern and the acoustical role of the noseleaf in the echolocating bat, *Carollia perspicillata*. J. Acoust. Soc. Am. 82, 1892–1900.

Henze, D., O'Neill, W.E., 1991. The emission pattern of vocalizations and directionality of the sonar system in the echolocating bat, *Pteronotus parnelli*. J. Acoust. Soc. Am. 89, 2430–2434.

Holstege, G., 1989. Anatomical study of the final common pathway for vocalization in the cat. J. Comp. Neurol. 284, 242–252.

Jürgens, U., 1994. The role of the periaqueductal grey in vocal behaviour. Behav. Brain Res. 62, 107–117.

Kobayasi, K., Hage, S., Berquist, S., Metzner, W., 2008. How laryngeal dynamics affect call frequency in an echolocating bat. Progr. No. 796.14 38th Annual Meeting of the Society for Neuroscience, Washington, DC.

Kobayasi, K., Ma, J., Smotherman, M., Zhang, S., Feng, J., Metzner, W., 2009. Segregated control of call features in the mammalian vocal motor nucleus, the nucleus ambiguus (submitted).

Kobler, J.B., 1983. The nucleus ambiguus of the bat, *Pteronotus parnellii*. Peripheral targets and central inputs. Ph.D. Thesis, University of North Carolina at Chapel Hill, NC.

Lancaster, W.C., Speakman, J.R., 2001. Variations in respiratory muscle activity during echolocation when stationary in three species of bat (Microchiroptera: *Vespertilionidae*). J. Exp. Biol. 204, 4185–4197.

Li, G., Wang, J., Rossiter, S.J., Jones, G., Zhang, S., 2007. Accelerated FoxP2 evolution in echolocating bats. PLoS ONE, 2: e900.

Ma, J., Kobayasi, K., Zhang, S., Metzner, W., 2006. Vocal communication in adult greater horseshoe bats, *Rhinolophus ferrumequinum*. J. Comp. Physiol. A, 192, 535–550.

Mergell, P., Fitch, W.T., Herzel, H., 1999. Modeling the role of nonhuman vocal membranes in phonation. J. Acoust. Soc. Am. 105, 2020–2028.

Metzner, W., 1993. An audio–vocal interface in echolocating horseshoe bats. J. Neurosci. 13, 1899–1915.

Metzner, W., 1996. Anatomical basis for audio–vocal integration in echolocating horseshoe bats. J. Comp. Neurol. 368, 252–269.

Metzner, W., Zhang, S., Smotherman, M., 2002. Doppler-shift compensation behavior in horseshoe bats revisited: auditory feedback controls both a decrease and an increase in call frequency. J. Exp. Biol. 205, 1607–1616.

Möhres, F.P., 1953. Über die Ultraschallorientierung der Hufeisennasen (Chiroptera: *Rhinolophinae*). Z. Vergl. Physiol. 34, 547–588.

Moss, C.F., Sinha, S.R., 2003. Neurobiology of echolocation in bats. Curr. Opin. Neurobiol. 13, 751–758.

Nixdorf, A., 2003. Die prätektale Region bei der Fledermaus Phyllostomus discolor: Eine audio-motorische Schnittstelle? Ph.D. Dissertation, LMU München, Faculty of Biology. <http://edoc.ub.uni-muenchen.de/1114/>

Pieper, F., Jürgens, U., 2003. Neuronal activity in the inferior colliculus and bordering structures during vocalization in the squirrel monkey. Brain Res. 979, 153–164.

Pillat, J., Schuller, G., 1998. Audiovocal behavior of Doppler-shift compensation in the horseshoe bat survives bilateral lesion of the paralemniscal tegmental area. J. Exp. Brain Res. 119, 17–26.

Reger, J.F., 1978. A comparative study on the fine structure of tongue and cricothyroid muscle of the bat, *Myotis grisescens*, as revealed by thin section and freeze-fracture techniques. J. Ultrastruct. Res. 63, 275–286.

Reimer, K., 1991. Auditory properties of the superior colliculus in the horseshoe bat, *Rhinolophus rouxi*. J. Comp. Physiol. A, 169, 1351–1432. doi:10.1007/BF00194900.

Revel, J.P., 1962. The sarcoplasmic reticulum of the bat cricothyroid muscle. J. Cell Biol. 12, 571–688.

Robin, H.A., 1881. Recherches anatomiques sur les mammiferes de l'ordre des chiroptères. Ann. Sci. Nat., 6th serie, Zool. 12, 1–180.

Rübsamen, R., Schuller, G., 1981. Laryngeal nerve activity during pulse emission in the CF-FM bat, *Rhinolophus ferrumequinum*: J. Comp. Physiol. A, 143, 323–327.

Rübsamen, R., Betz, M., 1986. Control of echolocation pulses by neurons of the nucleus ambiguus in the rufous horseshoe bat, *Rhinolophus rouxi*. J. Comp. Physiol. A, 159, 675–687.

Rübsamen, R., Schweizer, H., 1986. Control of echolocation pulses by neurons of the nucleus ambiguus in the rufous horseshoe bat, *Rhinolophus rouxi*. II. Afferent and efferent connections of the motor nucleus of the laryngeal nerves. J. Comp. Physiol. A, 159, 689–699.

Schnupp, J.W.H., King, A.J., 1997. Coding for auditory space in the nucleus of the brachium of the inferior colliculus in the ferret. J. Neurophysiol. 78, 2717–2731.

Schuller, G., 1977. Echo delay and overlap with emitted orientation sounds and Doppler-shift compensation in the bat, *Rhinolophus ferrumequinum*. J. Comp. Physiol. A, 4, 103–114.

Schuller, G., 1998. Neural control of vocalization in bats. On-line presentation, 5th Internet World Congress of Biomedical Sciences, INABIS 1998. <http://www.mcmaster.ca/inabis98/brudzynski/schuller0511/four.html.>

Schuller, G., Suga, N., 1976. Laryngeal mechanisms for the emission of CF-FM sounds in the Doppler-shift compensating bat, *Rhinolophus ferrumequinum*. J. Comp. Physiol. A, 107, 253–262.

Schuller, G., Rübsamen, R., 1981. Laryngeal nerve activity during pulse emission in the CF-FM bat, *Rhinolophus ferrumequinum*. J. Comp. Physiol. 143, 317–321.

Schuller, G., Radtke-Schuller, S., 1990. Neural control of vocalization in bats: mapping of brainstem areas with electrical microstimulation. Exp. Brain Res. 79, 192–206. doi: 10.1007/BF00228889.

Schuller, G., Sripathi, K., 1999. Audio-motor control in horseshoe bats. In: Shakunthala, S. (Ed.), Contributions to the XXVI Intern. Ethol. Conference. Advances in Ethology 34, Suppl. Blackwell, London, UK.

Schuller, G., Moss, C.F., 2004. Vocal control and acoustically guided behavior in bats. In: Thomas, J.A., Moss, C.F., Vater, M. (Eds.), Echolocation in Bats and Dolphins. The University of Chicago Press, Chicago, IL and London, UK, pp. 3–17.

Schuller, G., Covey, E., Casseday, J.H., 1991. Auditory pontine grey: connections and response properties in the horseshoe bat. Eur. J. Neurosci. 3, 648–662.

Schweizer, H., Rübsamen, R., Rühle, C., 1981. Localization of brain stem motoneurons innervating the laryngeal muscles in the rufous horseshoe bat, *Rhinolophus rouxi*. Brain Res. 230, 41–50.

Sinha, S.R., Moss, C.F., 2007. Vocal premotor activity in the superior colliculus. J. Neurosci. 27, 98–110. doi: 10.1523/JNEUROSCI.2683-06.2007.

Smotherman, M.S., Metzner, W., 2005. Auditory-feedback control of temporal call patterns in echolocating horseshoe bats. J. Neurophysiol. 93, 1295–1303.

Smotherman, M., Zhang, S., Metzner, W., 2003. A neural basis for auditory feedback control of vocal pitch. J. Neurosci. 23, 1464–1477.

Smotherman, M., Kobayasi, K., Ma, J., Zhang, S., Metzner, W., 2006. A mechanism for vocal–respiratory coupling in the mammalian parabrachial nucleus. J. Neurosci. 26, 4860–4869.

Sprague, J.M., 1943. The hyoid region of placental mammals with special reference to the bats. Am. J. Anat. 72, 385–472.

Strother, G.K., Mogus, M., 1970. Acoustical beam patterns for bats: some theoretical considerations. J. Acoust. Soc. Am. 48, 1430–1432.

Suga, N., Schlegel, P., Shimozawa, T., Simmons, J., 1973. Orientation sounds evoked from echolocating bats by electrical stimulation of the brain. J. Acoust. Soc. Am. 54, 793–797.

Suthers, R.A., Hartley, D.J., Wenstrup, J.J., 1988. The acoustic role of tracheal chambers and nasal cavities in the production of sonar pulses by the horseshoe bat, *Rhinolophus hildebrandti*. J. Comp. Physiol. A, 162, 799–813.

Thomas, J.A., Moss, C.F., Vater, M. (Eds.), 2004. Echolocation in Bats and Dolphins. University Press, Chicago, IL.

Vanderhorst, V.G.A., Holstege, G., 1996. A concept for the final common pathway of vocalization and lordosis behavior in the cat. Prog. Brain Res. 107, 327–342.

Vanderhorst, V.G.A., Terasawa, E., Ralston III, H.J., Holstege, G., 2000. Monosynaptic projections from the lateral periaqueductal gray to the nucleus retroambiguus in the rhesus monkey: implications for vocalization and reproductive behavior. J. Comp. Neurol. 424, 251–268.

Yin, J., Ma, J., Zhang, S., Metzner, W., 2008. FoxP2 expression in the brain of echolocating and non-echolocating bats and its possible role in vocalization. Progr. No. 796.13 38th Annual Meeting of the Society for Neuroscience, Washington, DC.

Zhuang, Q., Mueller, R., 2006. Noseleaf furrows in a horseshoe bat act as resonance cavities shaping the biosonar beam. Phys. Rev. Lett. 97 (21), 218701.

Zhuang, Q., Mueller, R., 2007. Numerical study of the effect of the noseleaf on biosonar beamforming in a horseshoe bat. Phys. Rev. E, 76, 051902.

SECTION 10

Sound Production by Larynx

Larynx represents a complex organ with internal cartilages and muscular elements and it functions as a principal sound generator in mammals. Vocalization is produced by vibration of vocal folds that form a highly efficient oscillator. Macromolecular structure of vocal folds supports their oscillatory function. Acoustic parameters of vocalizations may also be regulated by extralaryngeal mechanisms, such as size and anatomical position of larynx, and length and volume of the articulatory apparatus. The vocalization-producing system has developed additional modification as adaptation to underwater life in marine mammals, enabling vocalization underwater.

CHAPTER 10.1

Functions of the larynx and production of sounds

Gerald S. Berke* and Jennifer L. Long*

Laboratory of Laryngeal Physiology and Dynamics,
University of California-Los Angeles, David Geffen School of Medicine,
Division of Head and Neck Surgery, CHS, Los Angeles, CA, USA

Abstract: Vocalization in most mammalian species occurs in the larynx. This organ has developed a degree of complexity commensurate with its role in communication and as the source of self-expression through language in humans. However, the larynx also performs more primitive actions relating to respiration and swallowing. Laryngeal functions and functional anatomy are considered in this section. The myoelastic aerodynamic theory, describing how sound is produced by the vocal folds, is outlined. Illustrative examples of laryngeal pathology and laryngeal control mechanisms are discussed.

Keywords: larynx; evolution of larynx; vocal folds; airway; anatomy; phonation; phonatory control; voice; laryngospasm; spasmodic dysphonia; laryngeal nerves

I. Introduction

Vocalization requires a vibratory source to transform airflow into a series of acoustic pressure waves. This physiological sound production is, at its core, the same in all mammals in utilizing an energy source and an oscillator. Air pressure from the lungs supplies the energy source to an oscillating glottal opening to produce sound. A modifiable vocal tract shape, consisting of concatenary resonators such as the oral cavity, pharynx and nasal cavity, then alters the sound to produce characteristic harmonic amplitudes or spectra. The ear of mammals can deduce minute rapid changes in harmonic spectra produced by vocal tract movement, the culmination of which is the recognition of sound differences and spoken language in humans. We focus here on the central sound generator, the larynx.

II. Functions of the larynx through evolution

Laryngeal physiology encompasses other functions more fundamental to survival than phonation. The larynx is often called the gatekeeper of the upper respiratory tract, in order to emphasize its importance in the regulation of respiration and deglutition in humans and other mammals. For the original treatise on comparative physiology of the larynx, based on anatomical study of hundreds of species, the reader is referred to Negus, 1949 (see also Shiba, Chapter 9.1 in this volume). Harrison (1995) updated the field with high-quality photomicrographs and current concepts. Key points with regard to conserved laryngeal functions are reviewed here.

The first evolutionary appearance of a primitive larynx was as a protective organ in the pulmonary system to prevent water entry into air-breathing fish lungs. This original sphincteric function is evident in some species of lung fish as a single constrictor muscle sealing the lung inlet. When breathing air, relaxation of the sphincter allows air entry into the lungs. Later, dilator muscles and cartilaginous attachments evolved to facilitate inspiration. In marine mammals, the larynx no longer serves to exclude water as seen in the air-breathing fish, pointing to a possible de-evolutionary path as land-dwelling mammals returned to the sea (Gordon and Tyack, 2002). Instead, water exclusion in those orders is accomplished by strong closure of the nares or blowhole. The larynx can remain open underwater, which allows rebreathing of inspired air within the large nasal chambers (see Reidenberg and Laitman, Chapter 10.4 in this volume).

*Corresponding authors. E-mails: GBerke@mednet.ucla.edu; JLong@mednet.ucla.edu

Stefan M. Brudzynski (Ed.)
Handbook of Mammalian Vocalization
ISBN 978-0-12-374593-4

DOI: 10.1016/B978-0-12-374593-4.00038-3

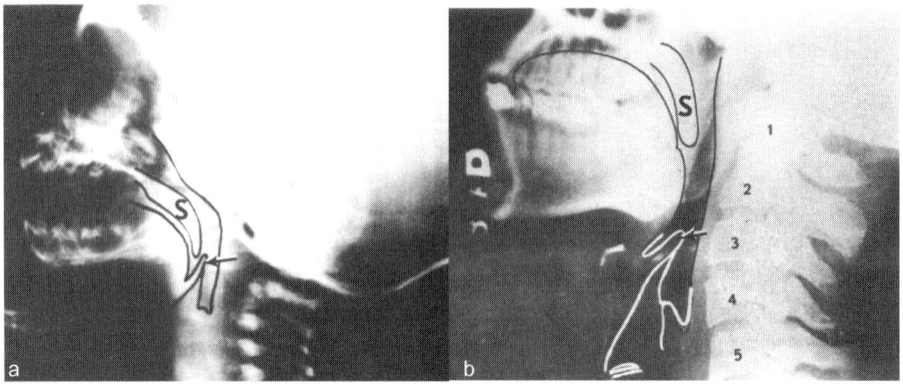

Fig. 1. Lateral radiographs of the upper aerodigestive tract in: (a) a newborn human; and (b) an adult. Note the close apposition of the soft palate (S) and epiglottis (arrow) in the newborn, effectively separating the larynx from the oral cavity. Reprinted from Crelin (1987).

In adult humans and some carnivorous mammals, sphincteric closure of the larynx shields the lungs from food aspirated during swallowing. This action is part of the complex coordinated sequence of voluntary and involuntary movements during swallowing, all of which are under central control (Logemann, 2006). Other mammals including newborn humans and ruminants keep the glottis open during swallowing. They can therefore simultaneously respire and swallow, which grants an evolutionary advantage considering the large amount of time herbivores and human infants spend eating. In those animals, airway protection is accomplished by more complete separation of the alimentary and respiratory tracts by the epiglottis, a flap-like structure above the glottis (Fig. 1). The epiglottis and soft palate touch, forcing liquids to pass laterally around the larynx into the esophagus. This anatomy also improves olfaction, by routing the inspired airstream over the nasal sensory epithelium even when the mouth is open for feeding, another important evolutionary advantage particularly in grazing species to warn of approaching predators. The epiglottis and lateral epiglottic folds provide variable degrees of aspiration protection in other mammals, with adult humans relying relatively little on such anatomic features to direct food into the esophageal inlet (Fig. 2).

In all mammals, an additional laryngeal closure reflex protects against aspiration when food or a foreign body actually enters the immediate supraglottic area. This action differs from the coordinated glottic closure during swallowing in that it functions as an emergency response. The reflex is triggered by superior laryngeal nerve stimulation, and completed by recurrent laryngeal nerve action, but still maintains some degree of central control (Sasaki et al., 2001, 2005). Oversensitivity in this system is disadvantageous since it interrupts respiration; the dangerous condition of laryngospasm does seem to be an uncontrolled glottic closure reflex (Ikari and Sasaki, 1980). Laryngospasm producing apnea and bradycardia has occurred in humans and many domestic and experimental animals in response to a variety of chemical and mechanical stimuli throughout the respiratory tract (Rex, 1970).

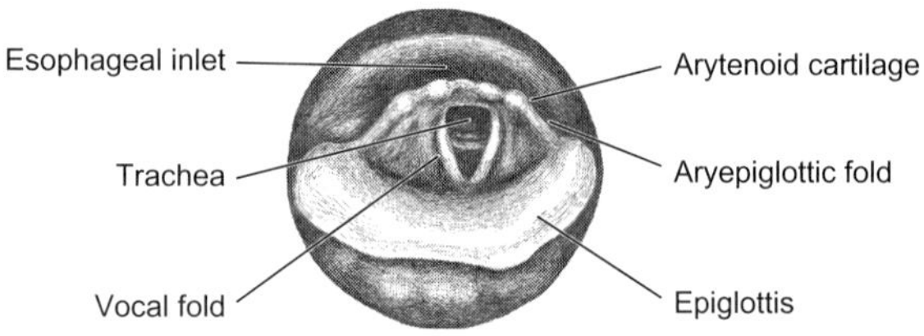

Fig. 2. Endoscopic view of adult human larynx from above. The epiglottis (anteriorly) and bilateral aryepiglottic folds provide modest protection to the trachea by directing food into the esophagus posteriorly. The upper part of the trachea, leading to the lungs, can be seen between the two open vocal folds. To view this figure in color please visit. http://www.elsevierdirect.com/companions/9780123745934.

As a bidirectional valve with fine control of constrictor and dilator functions, the human larynx has acquired the secondary function of controlling lung volume during expiration. The partial expiratory obstruction posed by the glottis controls pulmonary expiratory resistance, allowing adequate time for pulmonary gas exchange. It also prevents wasting of air during phonation by minimizing air escape. By closing tightly during breath-holding, the larynx permits efficient use of intercostal musculature during exercise without loss of intrathoracic volume. Similarly, the closed glottis stabilizes the thorax and upper body during lifting in bipeds.

Most recently in evolution, the larynx has gained its characteristic function in sound generation in some phyla, including mammals. This function is utilized by many species for communication. It is interesting to note that not only does the larynx maintain almost all of the primitive regulatory functions discussed above in humans, but it also contributes the central role in sound generation. It is this most advanced function that arguably separates hominoids most distinctly from other species, namely, the ability to communicate ideas, thoughts and even emotions through spoken language. The larynx's dense neural connections to the brain's emotive and language areas indicate the important role this organ plays in primates and particularly in humans. Furthermore, when one considers the degrees of freedom that mammals have over the laryngeal musculature in altering the organ's biomechanics, one can only marvel at the millions of years of evolution that were required to achieve its current operational state.

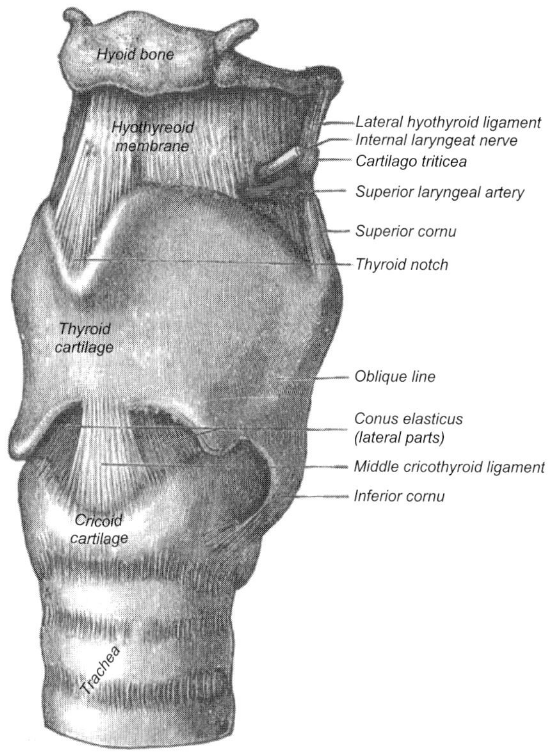

Fig. 3. Laryngeal anatomy, anterolateral view. From Gray (1918).

III. Laryngeal anatomy

While it is beyond the scope of this chapter to discuss all of the physiological processes known about the larynx, some anatomic features are essential to understand how the human larynx performs its many roles. The essential structure is similar in most mammals (Fig. 3). The larynx consists of a cartilaginous framework (the *cricoid* and *thyroid cartilages*), muscles connecting the framework to its surroundings in the body (*extrinsic laryngeal muscles*) and smaller muscles found entirely within the larynx (*intrinsic laryngeal muscles*). The cricoid cartilage, named for its ring shape, sits at the top of the trachea. The shield-shaped thyroid cartilage, known colloquially as the Adam's apple, attaches to the cricoid ring posteriorly and with a few degrees of articulation. Sitting on top of the ring are two smaller pyramidal or funnel-shaped cartilages (*arytenoid cartilages*), which each articulate in a complex rotational movement. Two sets of paired intrinsic muscles attach the arytenoid cartilages to the cricoid and can move the tips of the arytenoids apart or together, along with the attached ligaments. Opening (abduction) occurs during inspiration when the posterior cricoarytenoid muscles on the back of the cricoid contract and rotate the arytenoid cartilages laterally and outward. Closure (adduction) occurs during swallowing or phonation, when the lateral cricoarytenoid muscles attached to the front of the cricoid contract, rotating the cartilages inward. A single muscle passing between the arytenoid cartilages, the interarytenoid muscle, provides additional assistance with closure (Fig. 4).

Another important muscle pair runs from the arytenoid cartilages anteriorly to insert into the thyroid cartilage. These so-named thyroarytenoid muscles are covered with several layers of loose connective tissue, unlike the other intrinsic muscles. The paired structures of muscle and loose connective tissue are known as the vocal folds, and ultimately produce voice by their vibration against each other and modulation of the airstream. (The common term "vocal cords" has been abandoned, because the tissues neither have the shape of a cord nor produce sound by the vibratory mechanism of a cord or string.) Abduction and adduction of the arytenoid cartilages open and close the attached vocal folds, as seen in Fig. 4. Contraction of the thyroarytenoid muscles themselves can further adduct the vocal folds, as well as shortening them by rotating the arytenoid cartilages inward. The rocking articulation

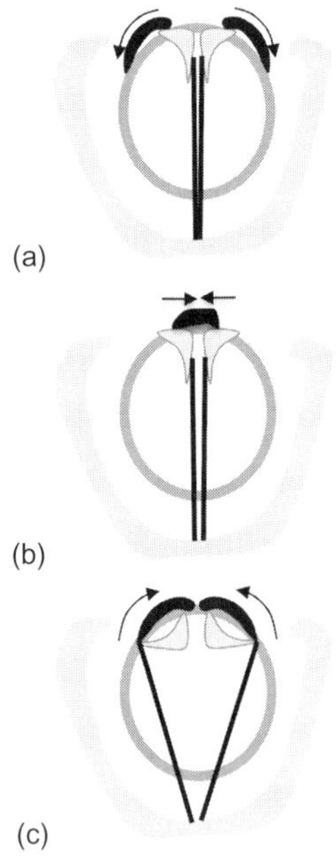

Fig. 4. Functions of intrinsic laryngeal muscles in arytenoid movement, schematic endoscopic view from above. Arytenoid cartilages are seen resting on top of the posterior aspect of the round cricoid ring. Vocal folds seen as two strings connect the arytenoid cartilages with the thyroid cartilage anteriorly. The muscles of interest are drawn in thick lines, with adjacent arrows indicating their direction of contractile force. Resulting arytenoid tip movement causes vocal fold abduction or adduction. (a) Lateral cricoarytenoid muscle contraction produces adduction or closure of the arytenoid cartilage tips and attached vocal folds. (b) Interarytenoid muscle contraction augments adduction. (c) Posterior cricoarytenoid muscle contraction opens or abducts the arytenoids and vocal folds.

of the thyroid and cricoid cartilages, controlled by the curtain-like cricothyroid muscle, lengthens the vocal folds (Fig. 5). Thus, a total of five muscles contribute to the movement or lengthening of each vocal fold, providing a remarkable degree of control over a structure less than an inch long in an adult human.

IV. Laryngeal innervation

The extensive muscular control of the vocal folds is completely supplied by the vagus nerve. Efferent lower motor neural fibers arise from the nucleus ambiguus; afferent sensory fibers also traverse there. Each vagus sends two laryngeal branches, the superior laryngeal nerve (SLN) with both sensory and motor components, and the purely motor recurrent laryngeal nerve (RLN) (Fig. 6).

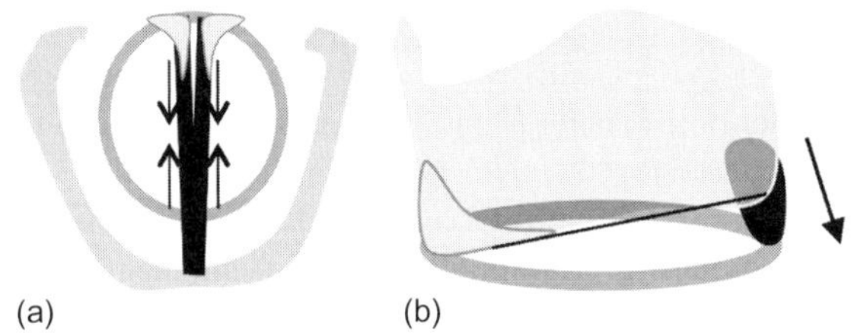

Fig. 5. Thyroarytenoid and cricothyroid muscle actions. (a) Thyroarytenoid muscle contracts and shortens the overlying vocal folds. (b) Lateral cutaway view of laryngeal cartilages, showing the thyroid cartilage resting above the cricoid ring. The anterior cricothyroid muscle pivots the two cartilages, elongating the attached vocal folds.

The SLN takes a direct route from the vagus across the neck to the larynx, and divides into two branches near the larynx. The "external" branch supplies motor signals to the cricothyroid muscle on the external laryngeal framework surface. The "internal" branch enters the larynx via the thyrohyoid membrane to carry afferent sensory fibers from the mucosa lining the larynx, epiglottis and part of the tongue.

Both RLNs branch from the vagus low in the neck and reverse course to supply their target tissue. This awkward "recurrent" nerve arrangement is a result of the embryological development of the branchial arches. The larynx arises as condensation of mesenchymal tissue from the fourth and sixth branchial arches, associated with the vagus nerve and the fourth aortic arch artery. The artery associated with the fourth branchial arch ultimately forms part of the left-sided aortic arch and the right subclavian artery. As the embryo grows, the motor branches of the bilateral vagus nerves supplying the developing larynx are trapped beneath these blood vessels in the thorax. Both nerves assume long and recurrent courses back up into the neck, with the left RLN dipping further into the chest than the right. The tortuous path occurs in most mammals, and may contribute to the idiopathic dysfunction that has been recognized in several species. Large racehorses are especially susceptible to a degenerative, demyelinating neuropathy affecting the left RLN almost exclusively, with the length of that nerve postulated to be a predisposing factor (Cook, 1965; Delahunta and Glass, 2008). Human idiopathic vocal fold paralysis also favors the left side, but is less severe than in the horse. Unlike the equine disease, airway obstruction is not the dominant feature in humans, and about 40% of humans recover some function (Sulica, 2008). The etiology in humans is suspected to include viral

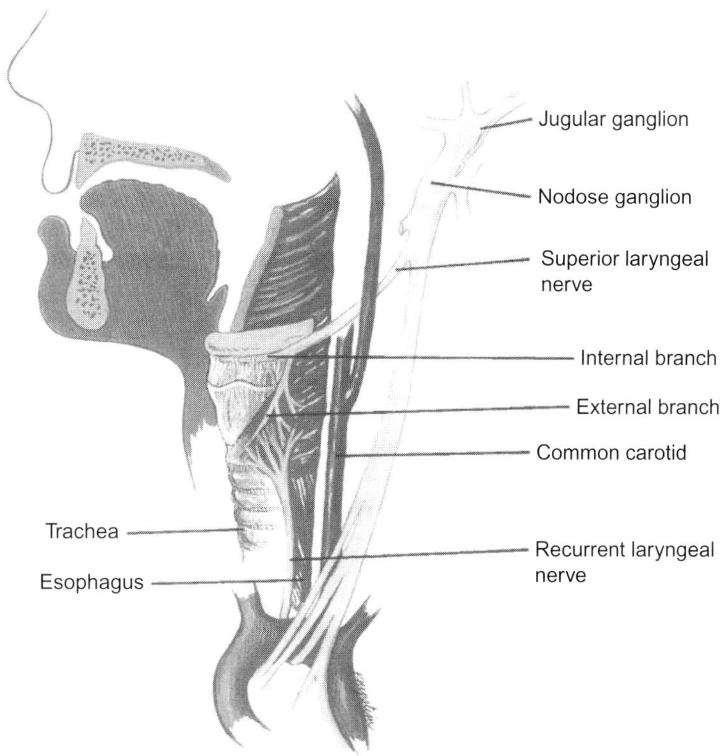

Fig. 6. Pathways of vagal nerve branches to the larynx. The superior laryngeal nerve travels directly from the vagus, while the recurrent laryngeal nerve loops under the subclavian artery. From Benninger et al. (2003).

neuritis, but postmortem pathological studies of the nerve involved are lacking.

After reaching the larynx, the RLN branches and sends terminal branches to the posterior cricoarytenoid muscle, the interarytenoid muscle, the lateral cricoarytenoid muscle and the thyroarytenoid muscle (Damrose et al., 2003). The precise branching pattern varies between subjects, but the order is conserved. Recognizing this order of innervation can localize injury, and is useful in surgical reinnervation treatments for laryngeal paralysis. It is interesting to note that each muscle can be individually controlled by the human motor cortex. Certainly, these neural connections and the resulting degrees of neuromuscular control must have conferred a significant evolutionary advantage on any species manifesting them.

It is quite conceivable that, just as in a morass of tangled telephone lines acquired through years of additions to a service area, the human larynx today must have some higher neural connections which are no longer primarily active but, nevertheless, still exist. Residual primitive pathways could emerge during limbic system activities producing, for example, fear or anger, or could be abnormally active in pathologic conditions. For example, a disorder known as spasmodic dysphonia presents as frequent voice breaks which impact speech fluency. To the listener, it may sound similar to a person being so surprised that they become speechless several times in a phrase. Laryngeal spasms, usually in the adductor muscles, abruptly halt phonation midphoneme. The voice breaks usually improve temporarily with botulinum toxin injection into the spastic intrinsic laryngeal muscles. However, the disorder is not purely a muscular dystonia, because phonation during laughter, anger or singing is at times quite normal, indicating that alternate limbic pathways exist in tandem with the speech neural projections that can also modulate laryngeal motor function (Berke and Blumin, 2000).

V. Mechanism of phonation

When the larynx closes under appropriate control of the recurrent laryngeal nerve and the vocal cords approximate, a very special function can be observed. Air from the lungs forced through the vocal folds can set the loose connective tissue layer into self-sustained oscillations. Sound is then produced due to the compression and rarefaction of the air molecules above and below the vocal cords. The vocal fold oscillation is explained by the myoelastic aerodynamic theory of phonation, which has been accepted as the mechanism of laryngeal sound production since van den Berg's defense of it in 1958 (see also Finck and Lejeune, Chapter 10.2 in this volume).

According to the myoelastic aerodynamic theory, voice is achieved by a complex repeating cycle in which glottal opening and closing modulates the transglottic airstream at anywhere from 50 to 300 cycles per second. Each cycle begins with increasing subglottic pressure pushing against the under-surface of the closed vocal folds. The aerodynamic pressure from the lungs eventually overcomes the muscular and elastic recoil forces holding the folds closed. The inferior-most part of the folds open first, and the tissue is progressively compressed as an "air bubble" rises to the superior surface of the folds. This compression from inferior to superior is called a traveling wave or mucosal wave. When the "bubble" or wave reaches the superior portion of the folds, they start to unzip and pressurized air begins to escape from the folds as a jet. The ensuing airflow leads to loss of pressure holding the folds open, and they start to close. The vocal folds contact each other again at the most inferior points, and then progress to complete closure and the end of a single cycle. As subglottic pressure builds against the closed glottis, the cycle repeats (Fig. 7).

Sound is produced during vocal fold oscillation by the inertial interaction of the pulsating jet against the air columns above and below the folds. The resulting air pressure wave interacts with the resonators of the vocal tract to shape the harmonic spectrum before exiting the lips. Of course, many partially-collapsed elastic tubes in nature vibrate when air or fluid passes through, causing sounds such as snoring, nose blowing, flatulence and borborygmi. One must recall that, unlike in those simple vibrating tubes, mammals have individual control over the intrinsic laryngeal muscles. As we will continue to discuss in the next section, alteration in laryngeal muscle activity alters the phonatory sound quality. Thus, through central neuromuscular control, voice quality is altered.

VI. Control of sound qualities

Detailed laryngeal control mechanisms have been outlined by selectively stimulating the terminal nerve branches in animal models. The cricothyroid muscle,

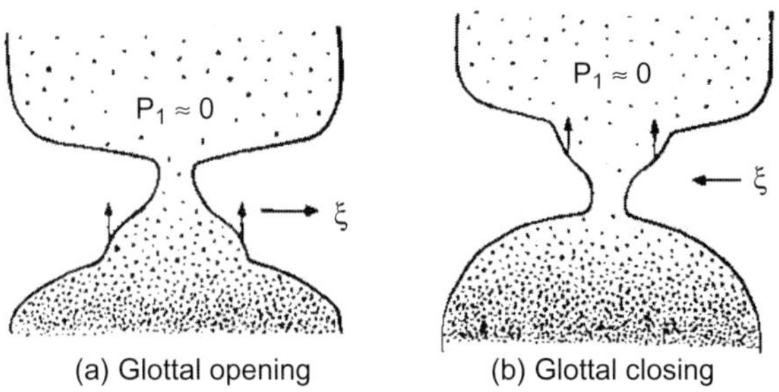

Fig. 7. Schematic drawing of the mucosal wave of the medial vocal fold surfaces, in a coronal plane. The hourglass shape represents the airway, with the bilateral vocal folds seen as the constriction. (a) During the glottal opening phase, subglottic pressures are relatively larger (indicated by dot density) and elastic recoil forces ξ are directed laterally. Paired arrows indicate the direction of vocal fold opening, from inferior to superior. (b) During the glottal closure phase, intraglottal pressure is lower and elastic recoil forces are directed medially. Vocal fold closing also occurs from inferior to superior, indicated by paired arrows. Reprinted from Titze (1988a).

which rotates the major laryngeal cartilages, in turn passively stretches and tightens the vocal folds. As they lengthen and become stiffer, the fundamental frequency of vocal fold vibration increases and a higher-pitched sound is produced. The thyroarytenoid muscle, conversely, shortens the vocal fold by contracting. While this could be expected to have the opposite effect and lower the pitch, canine studies have instead found that isolated thyroarytenoid activation actually raises pitch, although less steeply than cricothyroid activation does (Choi et al., 1993). How similar voice effects can be produced from seemingly opposite actions on the vocal folds is a matter of ongoing investigation, and likely relates to complex control of the stiffness and bulk of the vibrating structures, as suggested by the cover-body model of phonation. In that mathematical model, intrinsic stiffness and effective vibratory mass influence the input energy required and the output pitch and amplitude (Titze et al., 1988b).

In actual phonation, the interplay of cricothyroid and thyroarytenoid muscle activation controls not only pitch, but also register. Humans (and canines) phonate in at least two different voice regimes, the "chest" or modal voice and the "head" or falsetto voice. While head voice generally occurs at a higher pitch than chest voice, some overlap exists so that a given pitch can be produced in either register by one speaker. Registers are recognizable by their different timbres, with the head voice sounding breathier due to more airflow exiting the nasopharynx. Although the vocal tract certainly plays a role in actual phonation, register changes can be produced solely by laryngeal stimulation in canine models. If the cricothyroid muscle is maximally contracted, engaging the thyroarytenoid initially has no effect on pitch. Once a threshold level of thyroarytenoid stimulation is reached, a rapid pitch decrease occurs, analogous to the register change from head to chest voice in humans (Choi, 1993). Conflicting data on human register change mechanisms, as determined by electromyography, have been published (Faaborg-Anderson, 1957; Hirano et al., 1970; Shipp and McGlone, 1971; Roubeau et al., 1997). Regardless of the mechanism, smoothing the transition between registers is a primary goal of voice training for singers, indicating that voluntary control can modify laryngeal function.

Sound amplitude is the other key phonatory feature controlled at the level of the larynx. To produce a louder voice, speakers may either increase the peak airflow rate or raise subglottic pressure via vocal fold adduction force. One may postulate a sensory mechanism and feedback loop for either of these variables. However, classic human studies demonstrated that phonation was essentially unaffected by blocking afferent laryngeal sensory responses (Gould and Okamura, 1974; Gould and Tanabe, 1975). Thus, under normal operating circumstances most laryngeal servo control is a learned response, independent of sensory input. Homeostatic sensory mechanisms may come into play only during perturbation of normal conditions, or while learning how to manipulate the voice. That learning is largely achieved during infancy and childhood, but may undergo continued refinements with training. For example, Sundberg (1987) demonstrated that to increase amplitude, experienced singers increased airflow more than normal subjects, who relied more on increased subglottic pressure.

Consideration of phonatory control illustrates the complexity of the larynx, and reinforces the imperfection of current experimental techniques. Animal preparations with selected nerve sectioning or stimulation allow specific study of nerve branches, but are only beginning to capture the graded nerve stimulation which is typical of actual phonation. Human EMG studies detect neurologic function during actual phonatory tasks, but can only minimally manipulate the system to study individual influences. Human subjects also display remarkable variation in their approach to phonatory tasks, demonstrating the redundancy in the larynx producing similar output from multiple muscular mechanisms (Hillel, 2001).

VII. Vocal tract modification of source sound

This section has explored the function of the larynx in generating sound. Downstream structures, including the pharynx, oral cavity, lips and even nasal cavities, modify the sound waves to produce recognizable speech waves, as was described in detail by Fant (1960). Much as the shape of a trumpet produces a trumpet sound rather than a trombone sound, even when the two instruments play the same note, the vocal tract is as important as the larynx in producing the characteristic timbres of different speakers or even different species. Laryngeal vibrations produce a complex sound spectrum with multiple harmonics; the vocal tract amplifies some of those harmonics and dampens others during reflections of the sound waves off the walls of the vocal tract. This process is controlled by the muscular actions of the entire vocal tract, and involves learned behaviors and feedback with the larynx, auditory system and language centers. It is this finely coordinated control of the vocal tract and larynx that has produced complex spoken language in humans.

VIII. Conclusions

The larynx is essential for airway protection in most vertebrate species. It has secondarily developed into the principal sound generator for mammalian vocal communication by vocal fold oscillation, as predicted by the myoelastic aerodynamic theory. Control of all the laryngeal functions is mediated by the vagus nerve with complex cortical inputs. Pathologic states arise when either the distal nerves or the governing cortical pathways are deranged, and the precise mechanisms of these pathologies remain a focus of basic and clinical research.

References

Benninger, M.S., Gardner, G.M., Schwimmer, C., 2003. Laryngeal neurophysiology. In: Rubin, J.S., Sataloff, R.T., Korovin, G.S. (Eds.), Diagnosis and Treatment of Voice Disorders, 2nd edn. Thomson Delmar Learning, Clifton Park, NY, pp. 107–112.

Berke, G., Blumin, J., 2000. Spasmodic dysphonia: therapeutic options. Curr. Opin. Otol. Head Neck Surg. 8, 509–513.

Choi, H., Ye, M., Berke, G., Kreiman, J., 1993. Function of the thyroarytenoid muscle in a canine laryngeal model. Ann. Otol. Rhinol. Laryngol. 102, 769–776.

Cook, W.R., 1965. A comparison of idiopathic laryngeal paralysis in man and horse. J. Laryngol. Otol. 84, 819–835.
Crelin, E.S., 1987. The Human Vocal Tract: Anatomy, Function, Development, and Evolution. Vantage Press, New York, NY.
Damrose, E.J., Huang, R.Y., Ye, M., Berke, G.S., Sercarz, J.A., 2003. Surgical anatomy of the recurrent laryngeal nerve: implications for laryngeal reinnervation. Ann. Otol. Rhinol. Laryngol. 112, 434–438.
Delahunta, A., Glass, E., 2008. Veterinary Neuroanatomy and Clinical Neurology, 3rd edn. Elsevier Health Sciences, Philadelphia, PA, pp. 134–167.
Faaborg-Anderson, K., 1957. Electromyographic investigation of intrinsic laryngeal muscles in humans. Acta Physiol. 41 (Suppl. 140), 1–149.
Fant, G., 1960. Acoustic Theory of Speech Production. Mouton & Co., The Hague, The Netherlands.
Gordon, J., Tyack, P., 2002. Sound and cetaceans. In: Raga, J.A. (Ed.), Marine Mammals: Biology and Conservation. Plenum Publishers, New York, NY, pp. 139–196.
Gould, W.J., Okamura, H., 1974. Interrelationships between voice and laryngeal mucosal reflexes. In: Wyke, B. (Ed.), Ventilatory and Phonatory Control Systems. Oxford University Press, London, UK, pp. 347–360.
Gould, W.J., Tanabe, M., 1975. The effect of anesthesia of the internal branch of the superior laryngeal nerve upon phonation: an aerodynamic study. Folia Phoniat p. 27, 337–349.
Gray, H., 1918. Anatomy of the Human Body. Lea and Febiger, Philadelphia, PA.
Harrison, D.F., 1995. The Anatomy and Physiology of the Mammalian Larynx. Cambridge University Press, Cambridge, UK.
Hillel, A.D., 2001. The study of laryngeal muscle activity in normal human subjects and in patients with laryngeal dystonia using multiple fine-wire electromyography. Laryngoscope 111, 1–47.
Hirano, M., Vennard, W., Ohala, J., 1970. Regulation of register, pitch, and intensity of voice. An electromyographic investigation of intrinsic laryngeal muscles. Folia Phoniatr. (Basel), 22, 1–20.
Ikari, T., Sasaki, C.T., 1980. Glottic closure reflex: control mechanisms. Ann. Otol. Rhinol. Laryngol. 89, 220–224.
Logemann, J.A., 2006. Upper digestive tract anatomy and physiology. In: Bailey, B.J., Johnson, J.T. (Eds.), Head and Neck Surgery – Otolaryngology, 4th edn. Lippincott Williams and Wilkins, Philadelphia, PA, pp. 685–692.
Negus, V.E., 1949. The Comparative Anatomy and Physiology of the Larynx. Grune and Stratton, Inc., New York, NY.
Rex, M.A.E., 1970. A review of the structural and functional basis of laryngospasm and a discussion of the nerve pathways involved in the reflex and its clinical significance in man and animals. Brit. J. Anaesth. 42, 891–899.
Roubeau, B., Chevrie-Muller, C., Lacau St. Guilly, J., 1997. Electromyographic activity of strap and cricothyroid muscles in pitch change. Acta Otolaryngol. 117, 459–464.
Sasaki, C.T., Ho, S., Kim, Y.H., 2001. Critical role of central facilitation in the glottic closure reflex. Ann. Otol. Rhinol. Laryngol. 110, 401–405.
Sasaki, C.T., Hundal, J.S., Kim, Y.H., 2005. Protective glottic closure: biomechanical effects of selective laryngeal denervation. Ann. Otol. Rhinol. Laryngol. 114, 271–275.
Shipp, T., McGlone, R.E., 1971. Laryngeal dynamics associated with voice frequency change. J. Speech Hear. Res. 14, 761–768.
Sulica, L., 2008. Natural history of idiopathic unilateral vocal fold paralysis. Laryngoscope 118, 1303–1307.
Sundberg, J., 1987. The Science of the Singing Voice. Northern Illinois University Press, DeKalb, IL, pp. 49–91.
Titze, I., 1988a. The physics of small-amplitude oscillation of the vocal folds. J. Acoust. Soc. Am. 83, 1536–1553.
Titze, I., Jiang, J., Drucker, D.G., 1988b. Preliminaries to the body-cover theory of pitch control. J. Voice 1, 314–319.
Van den Berg, J., 1958. Myoelastic-aerodynamic theory of voice production. J. Speech Hear. Res. 1, 227–244.

CHAPTER 10.2

Structure and oscillatory function of the vocal folds

Camille Finck[1*] and Lionel Lejeune

Service d'Oto-Rhino-Laryngologie, Centre Hospitalier Universitaire, Sart Tilman, Liège, Belgium

Abstract: Voice can be produced by the vibration of the vocal folds that together form a highly efficient oscillator. Sustained phonation and self-protection against vibratory stresses depend on the histological organization and molecular composition of the folds. Their connective tissue, the lamina propria, has a multilayered structure in humans and other mammalian species. The layered organization determines its biomechanical features. Laryngeal muscles can modify the biomechanical characteristics of the vocal folds, thus permitting the emission of a wide variety of sounds. Laminar organization of human and animal vocal folds is described and their biomechanical properties and relationship with the oscillatory function is explained.

Keywords: structure of vocal folds; oscillation of vocal folds; lamina propria; Reinke's space; vibratory stress; extracellular matrix; collagen; elastin; hyaluronic acid; vocal fold damage

Abbreviations: VFLP: vocal folds' lamina propria; SLLP: superficial layer of the lamina propria; ILLP: intermediate layer of the lamina propria; DLLP: deep layer of the lamina propria; ECM: extracellular matrix; GAGs: glycosaminoglycans; HA: hyaluronic acid

I. Introduction

Vocal folds are intrinsic parts of the composite laryngeal structure. They are two tissue folds stretched between the internal anterior angle of the thyroid cartilage and the vocal process of the arytenoid cartilages. They appear as two white and shiny ribbons which, in phonation, are adducted, i.e., the free edge of both folds come into contact (see Fig. 1). During respiration, they abduct in order to permit the free flow of inspiratory and expiratory air volumes. The varying space between the folds is the glottis. Structures located above the folds are supraglottal, while structures located under the folds are infraglottal (for anatomical details, see Berke and Long, Chapter 10.1 in this volume).

The length of the vocal folds in the human species is between 10 mm for females and 16 mm for males (Hirano et al., 1981). Both the folds consist of a muscle (the vocal muscle or internal thyroarytenoid muscle), a ligament (the vocal ligament) and a covering mucosa (see Fig. 2). The vocal ligament is the medial edge of the laryngeal triangular membrane (also called conus elasticus). The vocal muscle is the most internal portion of the large thyroarytenoid muscle lying externally to the laryngeal quadrangular

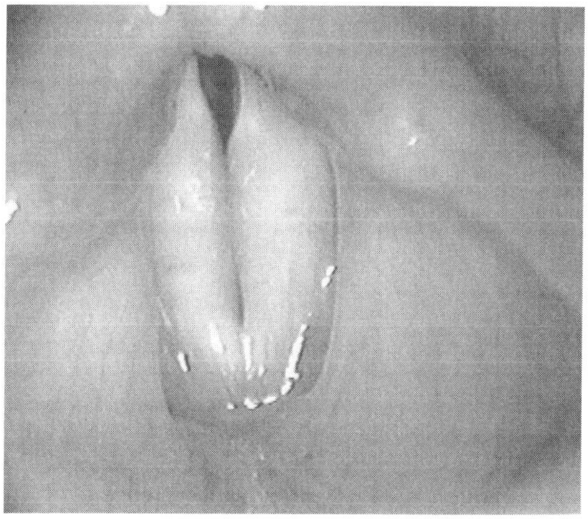

Fig. 1. Normal vocal folds in their adducted phonatory position in the human larynx. Glottis is closed, except in the posterior, interarytenoid, region.

*Corresponding author. E-mail: Camille.finck@chu.ulg.ac.be or Camillefinck@scarlet.be

Stefan M. Brudzynski (Ed.)
Handbook of Mammalian Vocalization
ISBN 978-0-12-374593-4

DOI: 10.1016/B978-0-12-374593-4.00039-5

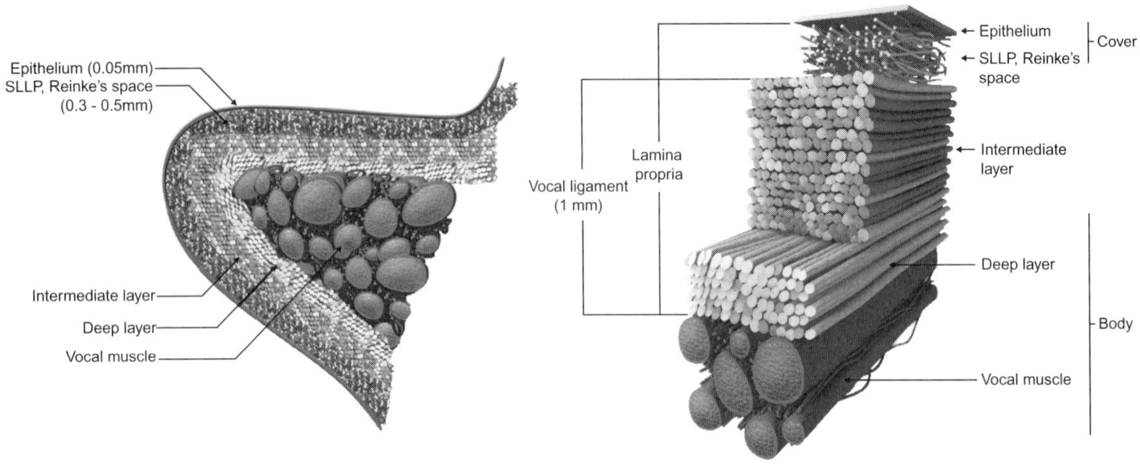

Fig. 2. Laminar structure of the vocal folds. Coronal cross-section of a vocal fold, on the left, and three-dimensional model of the layers, on the right. Blue: epithelium; yellow: collagen fibers; green: elastic fibers; red: vocal muscle fibers. Adapted from Hirano (1975) Phonosurgery – basic and clinical investigation. Otologia, 21: 239 (To see the full color version of this figure please refer to the color plate in the back of the book. Copies produced via our print on demand service do not contain color plates. If your copy does not have the color plate, please go to this website to view the figure in color http://www.elsevierdirect.com/companions/9780123745934).

membrane. The muscle fibers are parallel and external to the vocal ligament. The mucosa covering the underlying vocal ligament and the muscle is made of a squamous stratified epithelium and of an underlying layer of loose connective tissue called Reinke's space. Reinke's space (Reinke, 1895, 1897; see Fig. 2) takes its name from the anatomist who first described it, has well-defined vertical and anteroposterior limits, and plays a major role in the natural oscillatory function of the vocal folds (Hirano et al., 1980, 1981). Reinke's space and the underlying vocal ligament together constitute the vocal folds' lamina propria (VFLP).

Despite this quite simple macroscopic anatomy, vocal folds form a highly efficient vibrator. In their adducted and phonatory position (Fig. 1) they exhibit rapid and complex oscillatory movements that depend on the molecular composition and unique tissue organization of the lamina propria (Finck, 2005).

When observed at a microscopic level, the connective tissue of the vocal folds appears to be organized in successive layers differing in the composition of the extracellular matrix (ECM) filled with a variety of macromolecules (Alberts et al., 2008). The composition of the ECM in the lamina propria is particularly important because it determines the tissue's physical properties (such as shape, thickness and mass) and also the tissue's biomechanical characteristics, such as its rigidity, viscosity and pliability. The vibratory capability of the vocal folds depends on these biomechanical features, which appear to be substantially different in the three layers of the lamina propria, depending on the molecular composition of their ECM (Hirano, 1981; Gray et al., 1995, 1999; Chan et al., 2001; Ward et al., 2002; Thibeault, 2005; Jiang, 2006; Maragos, 2006).

The ECM is made up of two main classes of molecules: the fibrous proteins, which include collagen and elastin; and the interstitial molecules including the glycosaminoglycans (GAGs), the proteoglycans and the glycoproteins. The first description of the multilayered structure of the VFLP was based on three distinct distributions of the fibrous proteins, collagen and elastin (Hirano, 1974, 1981) (Fig. 2). The most superficial layer of the lamina propria (SLLP) possesses loose fibrous scaffolding with a few collagen and elastic fibers. SLLP corresponds to the formerly described Reinke's space and its thickness is between 0.3 mm and 0.5 mm.

The intermediate layer of the lamina propria (ILLP) is rich in elastic fibers, while the deep layer of the lamina propria (DLLP) is rich in collagen fibers. The intermediate and the deep layer of the VFLP together constitute the vocal ligament, and its mean thickness is between 1 mm and 2 mm. In his description of the VFLP, Hirano did not take into account macromolecules of the ECM other than fibers. More recent research refined the description and demonstrated that the interstitial molecules also show a layered organization inside the VFLP.

In this chapter, we will focus on the molecular composition and organization of the lamina propria of the human and other studied mammalian species, and on the oscillatory process of the vocal folds.

II. Laminar organization of the extracellular matrix of human lamina propria

II.A. *The fibrous proteins: collagens and elastin*

II.A.1. Collagens

Collagens are the dominant fibrous proteins of all connective tissues. Their structural role is mainly to maintain the shape of the tissue and to resist elongation stresses. Thus, collagenous fibers are flexible but offer resistance to any pulling force (Sato, 1998; Gray et al., 2000). The primary structure of collagen is a triple-stranded helical structure in which three polypeptide chains, called alpha (α) chains, are wound around each other in a rope-like helix. Twenty-five different types of α chains have been identified, all coded by a separate gene. Collagens are present in the three layers of the VFLP, in the epithelial basal membrane and in the connective tissue surrounding the muscle fibers of the vocal muscle. Collagen represents 43% of the total protein content of the lamina propria, which is approximately 60% to 70% of that in the human dermis (Hahn et al., 2006b). The most common collagen types of the human body (types I, II and III) have been identified in the VFLP (Gray, 1993, 2000; Sato, 1998; Lodish et al., 2008; Madruga de Melo et al., 2003; Tateya et al., 2006, 2007).

Fibrils made up of collagen type I possess an unusual tensile strength, and this collagen is as strong as steel (Lodish et al., 2000; Tateya et al., 2006). Collagen type III is usually present in tissues containing collagen type I, but in smaller amounts. Type III is less rigid (Sato, 1998) and tends to be abundant in organs requiring flexibility and elasticity, such as the walls of blood vessels (Tateya et al., 2006).

In the vocal folds, type III collagen is found throughout the lamina propria, whereas type I is observed in the SLLP, close to the epithelial basal membrane and in the DLLP (Tateya et al., 2006). The size of the collagen fibers in respect to the collagen type is still a subject of discussion, but the three-dimensional organization of these fibers appears to be an intertwined network arrangement in the form of a "wicker basket" (Madruga de Melo et al., 2003; Tateya et al., 2006).

II.A.2. Elastin

Elastin represents 8.5% of the total protein of the VFLP. In comparison, the human dermis contains only 3% of elastin (Hahn et al., 2006a). Elastin is secreted in the extracellular space as a soluble precursor, the tropoelastin. After secretion, tropoelastin becomes highly cross-linked and generates a network capable of stretch and resilience. Elastic fibers are not solely made up of elastin; the elastin core is covered by a sheet of microfibrils composed of glycoproteins (Alberts et al., 2008). Three types of elastin structures can be distinguished by their elastin-to-microfibril ratio: oxytalan; elaunin; and mature elastic fibers. All of them exist in the ECM of the VFLP (Hammond et al., 1997; Gray et al., 2000; Hahn et al., 2006a). Oxytalan and elaunin are immature forms of elastin. Oxytalan is purely microfibrillar in structure and, in fact, does not contain any elastin. Elaunin contains only a small amount of elastin, while mature elastic fibers have a rich core of elastin molecules (Hammond et al., 1997). Because of their poor content of elastin, the immature forms exhibit low elasticity and oxytalan does not elongate under mechanical stress (Hammond et al., 1997). On the contrary, mature elastic fibers can be stretched up to two times their resting length (Hahn et al., 2006a). In the SLLP, one can only observe the immature forms, oxytalan and elaunin (Hammond et al., 1997; Hahn et al., 2006a). Mature elastic fibers are present in both ILLP and DLLP, but their higher concentrations were only found in the ILLP (Hammond et al., 1997).

II.B. *Interstitial molecules*

Between the cells and the fibrous proteins, the ECM is composed of an amorphic substance that forms a highly hydrated gel composed of polysaccharide and protein molecules: the glycosaminoglycans (GAGs); the proteoglycans; and the glycoproteins. These interstitial molecules control hydration, volume and viscosity of the tissue, and they influence the size and density of the collagen fibers (Gray et al., 1999, 2000; Alberts et al., 2008). They also resist compressive stress.

GAGs are unbranched polysaccharide chains composed of repeating disaccharide units. Four main groups of these molecules are identified, depending on the sugar type and number and location of sulfate groups: keratan sulfate; dermatan sulfate; heparan sulfate; and hyaluronic acid (HA). Unlike HA, keratan, dermatan and heparan sulfate do not exist in isolation in the ECM and are covalently attached to a core protein in the form of proteoglycans (Pawlak et al., 1996; Gray et al., 1999; Alberts et al., 2008).

The ECM also contains a number of glycoproteins, which are noncollagen proteins. As described in the

VFLP, they include fibronectin and laminin glycoproteins and they play an important role in organizing the ECM and helping cells to attach to it (Nicolai et al., 1990; Pawlak et al., 1996; Gray et al., 1999; Alberts et al., 2008). Glycoproteins have a much lower carbohydrate content than the proteoglycan family in the form of relatively short and branched oligosaccharide chains (Alberts et al., 2008). Some of these interstitial molecules in the VFLP appear to be layer-specific, and all of them, especially HA, play an important biomechanical role.

II.B.1. Hyaluronic acid

Hyaluronic acid (HA) is a large glycosaminoglycan which is part of the ECM in all vertebrates. It represents 0.82% of the human VFLP relative to tissue total protein (Hahn et al., 2006a), which is equivalent to that of the human dermis. HA is made up of repeating disaccharide structures of glucuronic acid and N-acetylglucosamin (up to 50,000 repeats) (Meyer and Palmer, 1934). Depending on the number of repeats, the molecular weight of the HA molecule varies from 10,000 to 4,000,000 Da (Lodish et al., 2000).

The macromolecule of HA is frequently referred to as hyaluronan, a polyanion, reflecting the fact that HA is not present in the "acid" form *in vivo*. HA is both hydrophilic and hydrophobic (axial hydrogens create a hydrophobic patch). Negatively charged, HA attracts cations, mostly Na^+, that are osmotically active, attracting large amounts of water to the ECM.

Because of its remarkable molecular and physical properties, HA occupies a large space and creates a turgor inside the ECM and the ability of the connective tissue to resist compressive stress (Gray et al., 1999; Lodish et al., 2000; Alberts et al., 2008), these "swelling" forces are necessary to maintain the volume of three-dimensional structures created by other proteoglycans and proteins in the ECM.

Although HA is present in all the three layers of VFLP, it is not distributed uniformly. HA is more concentrated in the ILLP and SLLP than in the DLLP, where its concentration is the lowest (Hammond et al., 1997; Gray et al., 1999; Chan et al., 2001; Ward et al., 2002; Hahn et al., 2006a) (see Fig. 3). HA is produced by fibroblasts and macrophages, and its turnover is very rapid, with a 3–5 day half-life inside the VFLP (Ward et al., 2002; Hallen et al., 1998; Alberts et al., 2008).

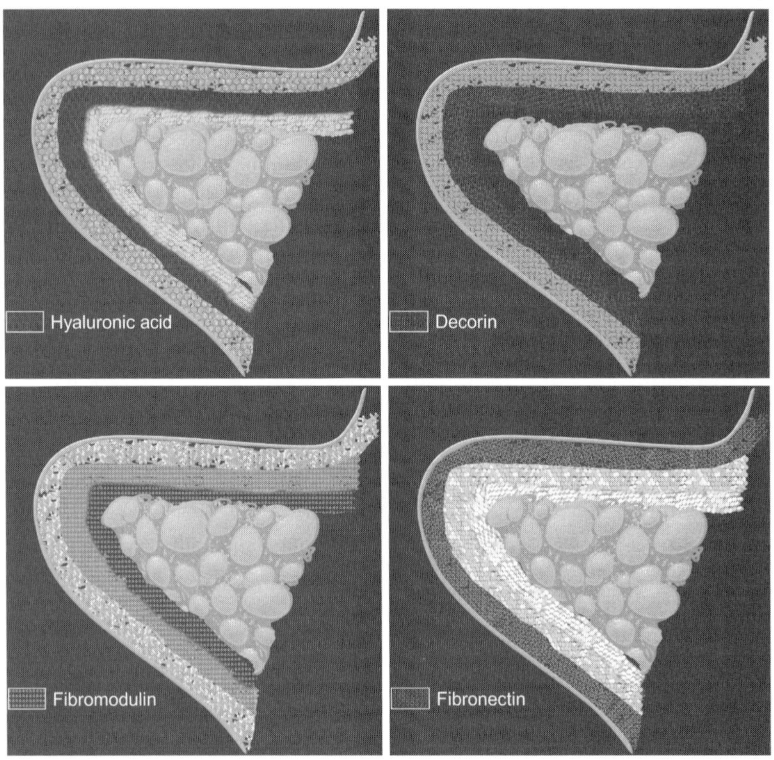

Fig. 3. Distribution of hyaluronic acid (purple), decorin (green), fibromodulin (blue) and fibronectin (red) in the three layers of the VFLP (To see the full color version of this figure please refer to the color plate in the back of the book. Copies produced via our print on demand service do not contain color plates. If your copy does not have the color plate, please go to this website to view the figure in color http://www.elsevierdirect.com/companions/9780123745934).

HA degradation is mostly initiated by binding to CD44 receptor on the plasma membrane, cellular internalization and degradation by lysosomal hyaluronidase in its monosaccharidic components (Ward et al., 2002).

HA plays an important biological role associated with potential microdamage caused by the vocal folds vibration, and it is a regulator of morphogenic processes and tissue repair. It influences cell attachment, cell proliferation, cell migration and cell differentiation through binding to HA receptors (hyaladherins) located intracellularly and on cell membranes (Lodish et al., 2000; Petreaca, 2008). HA reduces the amount of collagen deposit during the remodeling phase of the healing process, although the precise mechanisms of this process are not fully understood.

II.B.2. Lamina propria proteoglycans

Two proteoglycans appear to be layer-specific: decorin and fibromodulin. Decorin (50–200 kDa) is a small proteoglycan only present in the SLLP that is able to attach to collagen type I and II, modifying the assembly of these collagen fibers and reducing their size (Pawlak et al., 1996; Gray et al., 1999) (see Fig. 3).

Fibromodulin (59 kDa) is structurally very similar to decorin, but is present only in the intermediate and deep layer of the lamina proria (Fig. 3). Fibromodulin also binds to collagen type I and II, and it can inhibit fibril formation *in vitro* (Thibeault et al., 2003). The association of fibromodulin with the ligament zone may be related to a lubricating role of this proteoglycan (Gray et al., 1999; Gray, 2000).

Very little is known about other proteoglycans in the lamina propria. Versican, a high molecular weight molecule (>1,000 kDa) has been found in the macrophages and fibroblasts of the VFLP. Versican is, like HA, a space-filling molecule, and is also able to bind to HA molecules (Pawlak et al., 1996).

II.B.3. Glycoproteins

The ECM contains non-fibrous proteins that play multiple roles. They constitute binding sites for other matrix macromolecules and for cell-surface receptors, thus helping in organizing the matrix and permitting cell-matrix attachment (Alberts et al., 2008).

Fibronectin is a glycoprotein found in all vertebrates and is composed of two large polypeptide chains joined by disulfide bonds. In the matrix, 20 isoforms of fibronectin can be found, all of them being encoded by a single large gene (Alberts et al., 2008). In the VFLP, fibronectin is present in all three layers (Gray et al., 1999), but immunohistochemical studies have also shown intense staining for fibronectin in the SLLP and the ILLP (Fig. 3) (Gray et al., 1995). In the VFLP, the exact role of the fibronectin isoforms is still unknown.

Fibronectin possesses strong adhesion properties and it binds to collagen type I, II, III, V, to HA, heparin and the cell surface. It was postulated that fibronectin mediates the adhesion of fibroblasts to matrix collagen and influences the collagen fibril formation (Obara and Yoshizato, 1997; Lodish et al., 2000; Hirschi et al., 2002; Alberts et al., 2008).

II.C. The epithelial basal membrane as a specialized extracellular matrix

The epithelial basal lamina is a specialized ECM securing the vocal fold epithelium to the underlying lamina propria. Its molecular components are secreted by the epithelial cells that rest on it. Basal lamina is composed of collagen type I and III–VI (Tateya et al., 2006), two glycoproteins (laminin and entactin) and the proteoglycan perlecan (Nicolai et al., 1990; Benninger et al., 1996; Alberts et al., 2008). The basal lamina is a flexible and thin mat (40–120 nm thick) tethered to the SLLP by specialized anchoring fibrils made of type VII collagen molecules. Together, those anchoring fibrils and the basal lamina form the basal membrane zone. The basal lamina separates epithelial cells from fibroblasts of the connective tissue, and influences cell survival, proliferation and differentiation (Alberts et al., 2008; Gray, 1991; Gray et al., 1994).

II.D. The cellular population of the vocal folds

II.D.1. Epithelium of the vocal folds

The epithelium of the vocal folds is a squamous, stratified epithelium of approximately eight layers of cells, and is different from the ciliated and pseudostratified epithelia lining most of the laryngeal cavities (Graney and Flint, 1998).

II.D.2. Cell population of the vocal folds' lamina propria

Three types of cells are present in the connective tissue of the folds: fibroblasts; myofibroblasts; and macrophages. Fibroblasts, observed in all three layers of the vocal folds lamina propria (VFLP), are the most numerous cells, while macrophages are the least numerous. SLLP contains very few cells, but it is in this layer that macrophages and myofibroblasts are more frequently seen. The presence of macrophages in the most superficial layer of the folds is probably due to their protective role against germs and mucosal irritants. Myofibroblasts are differentiated fibroblasts and are important in tissue repair. They contain actin and are connected to the extracellular matrix by a transmembrane complex between intra- and extracellular fibronectin called a fibronexus.

II.D.3. The maculae flavae

Located at the anterior and posterior ends of each vocal fold in the human species, maculae flavae are small yellow elliptical formations composed of dense masses of extracellular matrix and cells approximately 1.5 × 1.5 × 1 mm in size (Hirano et al., 1981; Sato et al., 2003). The dense and fibrous ECM is mainly composed of collagen, elastin and HA. Collagen type I and type III, and elastin are denser than in the adjacent VFLP, do not form distinct fibrous structures, and are not organized in a specific direction (Sato et al., 2001, 2003; Tateya et al., 2007). Both maculae are connected to the adjacent structures of the larynx. Many authors suggest that maculae flavae play a role in the growth and development of the vocal ligament (Campos Banales et al., 1995; Sato et al., 2003; Fayoux et al., 2004). At birth vocal ligament is nonexistent, but immature maculae flavae are present. They grow and produce ECM during intrautero life. After birth, the ECM molecules migrate from the maculae toward the middle portion of the membranous vocal fold to initiate the formation of the vocal ligament (Fayoux et al., 2004). At adult age, the maculae flavae continue to produce ECM and are thought to take part in the maintenance of the vocal ligament (Sato et al., 2003).

Maculae flavae are also present at both ends of the membranous portion of the vocal folds of different animal species. Studies in the dog and rat have revealed, however, that they are less developed than in the human species, and their cell populations and composition of the surrounding ECM are also different (Sato et al., 2000; Tateya and Bless, 2006). These differences could explain the existence of a well-developped vocal ligament and a unique multilayered structure of the VFLP in humans (Sato et al., 2003).

III. Laminar organization of the extracellular matrix of the lamina propria in different mammalian species

The VFLP of other mammalian species differs from the VFLP of the human species. Some species do not exhibit the same three-layer structure as that in humans. Pigs, rabbits and sheep have only a two-layer VFLP, while cats possess a single layer VFLP (Nagata, 1982). However, a three-layer VFLP, like in humans, has been observed in dogs, horses, monkeys, guinea pigs, oxen and rats, but the layers differ greatly from those in the human species (Nagata, 1982; Hahn and Langer, 2005).

Quantitative and comparative studies of the ECM of the VFLP of human, pig, ferret and dog show very different distributions of collagen, elastin, hyaluronic acid and proteoglycans among the studied species. Porcine collagen and elastin distributions appear to be more similar to that of humans. Both species possess the highest concentration of collagen in the deep layer of the VFLP, but in the pig, elastin is evenly distributed throughout the VFLP. By contrast, collagen and elastin distributions in canine and ferret VFLP are mirror images of the distribution in humans (Hahn et al., 2006a,b). HA is more abundant in the VFLP of studied species and the mean concentration of HA is 3–4 times higher than that in humans. The highest concentration is located in the ILLP of the human, the canine and ferret VFLP, while porcine HA concentration is high in the most superficial and deepest layer of the VFLP (Hahn et al., 2006a). The vocal folds of the Japanese monkey have also been studied. Their lamina propria is thinner and less elastin is present throughout the VFLP (Garrett et al., 2000). The density of elastin and collagen also decrease in this species with increasing depth of the VFLP (Ishii et al., 1999).

IV. Biomechanical properties and oscillatory function of the vocal folds

Phonation refers to the physical and physiological processes of vocal fold vibration that produces the glottic sound. A common misconception is to believe that vocal folds vibrate like guitar strings and that the sound produced at their level is a direct result of their vibration. In fact, sound is produced by the periodic escape of air through the glottis while the vocal folds vibrate. Voice production requires interaction between the expiratory airflow and the adducted vocal folds that transform aerodynamic power into acoustic power.

Structural and biomechanical properties of the vocal folds allow them not only to sustain oscillation, but also to ensure self-protection against mechanical vibratory stresses. Sustained oscillation depends on favorable biomechanical properties, in terms of viscosity, rigidity and elasticity. Elasticity refers to the ability of the tissue to undergo an elongation stress, to store energy and to revert to its initial length. Viscosity represents the speed of deformation of the tissue submitted to a mechanical stress. Rigidity is representative of the tissue's resistance to deformation. Stiffness is different from rigidity; it is a mechanical characteristic of composite structures which increases with the rigidity and viscosity, and which is directly proportional to the tension applied to the tissue.

By varying the geometry of the glottis and the tension in the layers of the lamina propria, laryngeal muscles modify the mechanical properties of the laryngeal vibrator and allow the production of a wide range of sound frequencies and intensities.

IV.A. Sustaining oscillation

Many theories of sound production by the larynx have been described. In 1958, Van den Berg suggested a myoelastic–aerodynamic theory of phonation (Van den Berg, 1958) which takes into consideration tissue elasticity, tissue collision, subglottic pressure under perfectly closed vocal folds and the Bernoulli effect. According to this theory, vocal folds need to be closed by the laryngeal adductor muscles to allow the building of the subglottal air pressure which opens the glottis. Cyclic closure of the folds relies on tissue elasticity (or resilience) and on a negative pressure inside the glottis because of the Bernoulli effect. Bernoulli's energy conservation law states that the total energy of a gas streaming down a pipe is a constant. In the glottis in phonation, air velocity increases because of the constricted region, while pressure decreases. The decreased pressure draws the folds together. The subglottal pressure builds up again under the collapsed folds, allowing the cycle to repeat. But Van den Berg's theory is inadequate to explain the self-sustained oscillation of the folds; in this purely aerodynamic theory, Bernoulli's effect is insensitive to the direction of the tissue displacement (inward versus outward), leading to a rapid extinguishing of the vibration. To maintain oscillation, the inertia of the supraglottal air column plays a necessary additional role. When the vocal folds open, the supraglottal air condenses and this creates an area of high pressure above the folds, which helps to drive them apart (Fig. 4). When the glottis then closes, the supraglottal air continues to move upward, creating an area of low pressure which helps to draw the folds further together. Thus, cyclic supraglottal pressure variations help to maintain the cyclic opening and closing of the vocal folds (see Fig. 4) (Titze, 1994c).

Because the vocal folds are vibrating too rapidly for the naked eye to see, special techniques, such as stroboscopy or high speed cinematography, are necessary to observe the vibratory displacements of the vocal folds. The observer can then see back-and-forth movements of the free edge of the vocal fold, moving in the medial and lateral direction. The slow motion observation of the vocal folds also shows that the tissue is never displaced in a uniform manner, i.e., the vocal fold does not move as a homogeneous solid bar. A wavelike motion of the pliable mucosal cover can be seen traveling from the inferior lip of the vocal fold to the superior aspect of the fold (Fig. 5). Hirano, in his body-cover theory of vibration, explains this vibratory behavior by different viscoelastic properties of the layers of the VFLP. The SLLP has a low viscosity (Hirano, 1974, 1981; Titze, 1994c). This biomechanical characteristic is linked to its ECM composition;

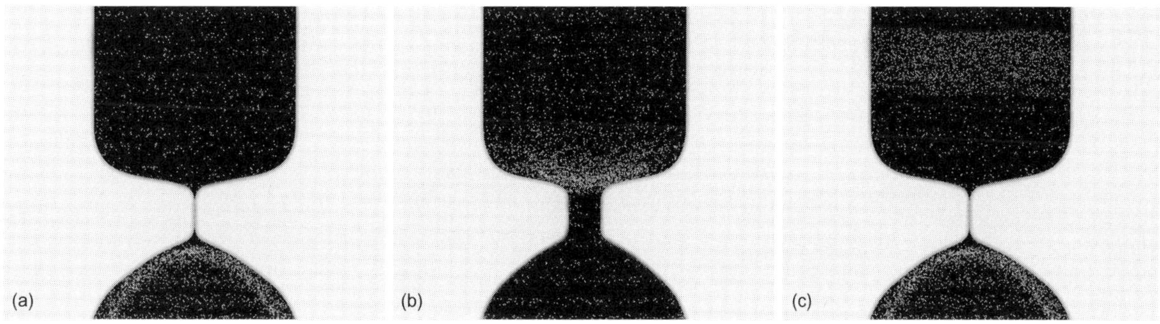

Fig. 4. Cyclic infra- and supraglottal air pressure variations during vocal folds' vibration. (a) Closed vocal folds: subglottal pressure is building up; (b) vocal folds open, condensation of supraglottal air. Subglottal pressure diminishes; (c) vocal folds close: rarefaction of supraglottal air. Subglottal pressure builds up again.

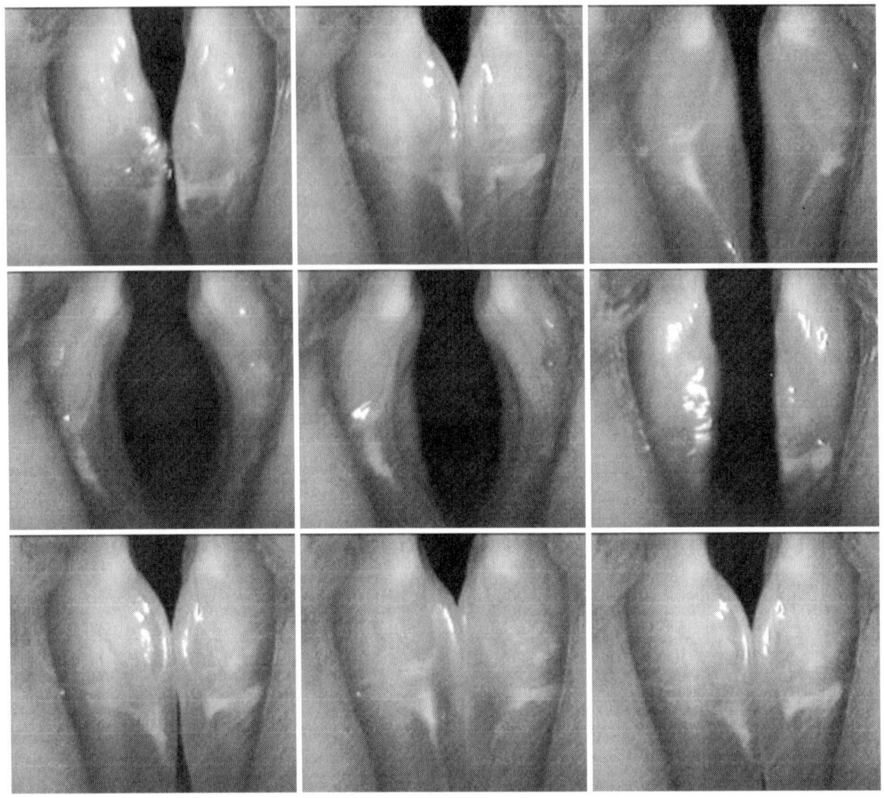

Fig. 5. Still images from high-speed cinematography of vocal folds in vibration (2,000 frames per second). Images from KayPentax Corp., Lincoln Park, NJ, USA (reproduced with permission).

a low content in fibrous proteins and a high concentration of HA. Rheologic studies on vocal folds in human cadavers showed that enzymatic removal of HA from the SLLP led to a two- to four-fold increase in the tissue viscosity (Gray et al., 1999; Chan et al., 2001).

Fibronectin and HA contained in the SLLP give to this layer the optimal stiffness necessary to stabilize the frequency of vibration. Removal of the HA from the vocal fold's cover leads to a 35% decrease in the tissue's stiffness (Chan et al., 2001).

Together with the epithelium, SLLP forms the "mucosal cover" that is responsible for the mucosal wave propagating on a more rigid "body." The "body" of the fold is made up of the vocal ligament (ILLP and DLLP) and vocal muscle. ILLP and DLLP are more rigid, because of their high content of fibrous proteins. Thus, the cover is inherently more mobile than the deeper portions of the VFLP, and therefore constitutes vocal fold's main vibrator.

Mucosal wave is necessary to sustain oscillation; the upward propagating mucosal wave creates a difference of phase between the inferior and the superior lip of the folds. During oscillation, the bottom of the folds is always ahead of the top. Different glottal shapes are then obtained during the different parts of the vibratory cycle. During the opening phase (outward movement), the glottis is convergent. During closing phase, the glottis is divergent. Pressure inside the glottis is higher during outward movement than during the inward movement, modulating Bernoulli's effect and enabling a sustained oscillation (Fig. 6) (Ishizaka and Matsudaira, 1972; Scherer and Titze, 1983, 1994c).

Although possessing a different layered organization of the VFLP, a mucosal wave has been observed in the dog. Limited data showed that no mucosal wave or vertical phase difference were observed during videostroboscopy of pigs' vocal folds (Garrett et al., 2000).

In the human species, the existence of a pliable cover is essential to voice production. Lesions localized to that vocal cord's cover (e.g., intracordal cysts), destroying it (e.g., scars and sulci), or deforming it (e.g., Reinke's edema), have a deleterious effect on the voice, because of the alterations of the mechanical and viscoelastic properties of the vocal folds (Finck, 2005; Finck and Lefebvre, 2005).

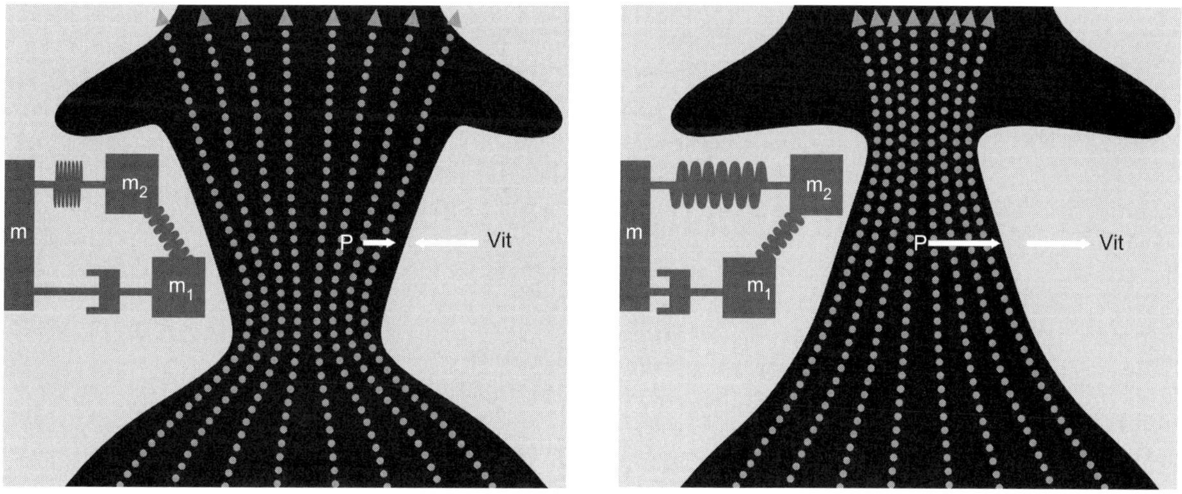

Fig. 6. Frontal section of the laryngeal vibrator showing the phase opposition between inferior and superior lips of the vibrating folds, creating intraglottal pressure differences. Divergent glottis (closing phase of the vibratory cycle, on the left) and convergent glottis (opening phase of the vibratory cycle, on the right). Symbols: m: mass of the body; m1: mass of the inferior lip of the cover; m2: mass of the superior lip of the cover; P: intraglottal pressure; vit: tissue displacement. Adapted from Titze, I.R. (1994b) Vocal fold oscillation. In: Titze, I.R (Ed.), Principles of Voice Production. Prentice-Hall Inc, Englewood Cliffs, NJ, p. 95 (reproduced with permission).

IV.B. Modification of biomechanical properties of the laryngeal vibrator

The mechanical properties of the laryngeal vibrator (length, mass, position and tension of the tissue in vibration) can be modified by the contraction of the intrinsic and extrinsic laryngeal muscles. In the body-cover model of vibration, a fine tuning of the frequency of vibration can be obtained, knowing that frequency is directly proportional to tension of the vibrating structure and inversely proportional to the mass in vibration. Two muscles play important roles in controlling the frequency: the vocal muscle (the body of the vocal fold); and the cricothyroid muscle. The isolated contraction of the cricothyroid muscle leads to an increased tension both in the cover and the body of the laryngeal vibrator; viscosity and stiffness are high, amplitude of vibration is small and frequency increases with increased tension. The isolated contraction of the vocal muscle leads to an increased "active" stiffness in the body of the laryngeal vibrator, but to a decreased tension in the cover of the vibrator. In this situation, when the vocal fold shortens, viscosity decreases in the cover, but increases in the body of the vibrator. Depending on the depth of the layers in vibration, this could lead either to a lowering of the frequency of the vibration (if only the cover vibrates), or to an increase of the frequency of vibration (if some part of the body vibrates).

When both muscles (vocal muscle and cricothyroid muscle) contract together, which is the usual muscular setting for most of the speech phonation situations, the effect is difficult to predict, because it depends on the balance between the two muscles and again on the depth of vibration of the folds (Hirano, 1974; Titze, 1994b).

IV.C. Self-protection against vibratory stresses

To be a performing oscillator, vocal folds also need to develop a self-protection against vibratory stresses. There are many mechanical stresses associated with vibration: tensile stress applied to the vocal folds by contraction of the cricothyroid muscle; contractile stress in muscles; collision stress between the vibrating folds; inertial stress due to tissue acceleration and deceleration; shear stresses; aerodynamic stress by intraglottal pressure; and arytenoids contact stress. Many benign vocal fold lesions observed in the human species are the consequence of excessive voice load. The SLLP is the more fragile layer, and vocal fold lesions usually occur in this layer. They are the consequence of both tissue destruction and tissue response to the mechanical trauma (Titze, 1994a).

The VFLP possesses molecular and tissue characteristics to be able to withstand vibratory stresses. In a vibrating tissue, such as the vocal folds, collagen type III provides flexibility to the tissue, but also resistance to excessive deformation, while type I collagen provides the tensile strength particularly needed close to the surface and the muscle fibers. The "wicker basket" organization creates the necessary mechanical features for forces acting in all directions, which is required for a structure vibrating in nonuniform modes.

Vocal ligament is probably as solid as the knee ligament and can absorb the very high tensile stress developed during emission of high-frequency sounds, therefore protecting the SLLP from excessive stress (Titze, 1994a). During vibration of the vocal fold, mature elastic fibers permit elongation of the structures and store the energy necessary for the system to return to the resting state (Madruga de Melo et al., 2003). In the SLLP, however, no mature elastic fibers are observed, but there is a high level of oxytalan. This immature elastic molecule is known to exist in tissues where compressive and shear stresses are common (as in the superficial layer of a tendon and in the cartilage). This might be one of the reasons for the presence of these molecules in Reinke's space (Hammond et al., 1997). Thus, oxytalan could represent an element of the tissue's self-protection mechanism against compressive and shear stresses (Gunter, 2004).

In the VFLP, HA is essential for the protection of the vibrating tissue against compressive and collisional stresses (Hammond et al., 1997; Gray et al., 1999). HA content influences the thickness of the lamina propria, as suggested by the fact that HA is two (Butler et al., 2001) to three times (Hammond et al., 1997) more abundant in the lamina propria of men (thicker lamina) than women (Hammond et al., 1997). The lower amount of HA in women's SLLP, as compared to men's SLLP, provides less tissue protection against compressive stress and this could be a possible cause for the more frequent development of vocal fold lesions in women (Butler et al., 2001). In addition to HA, fibronectin brings stiffness and resistance to deformation to the tissue, because of its adhesion properties (Gray et al., 1995, 1999; Thibeault et al., 2002a,b, 2003).

The basal membrane zone is essential to the vibrating role of the vocal folds and it secures the epithelium to the SLLP. Without a solid basal membrane zone, the inertial, compressive and collisional stresses existing during normal vibration of the tissue would tear the epithelium apart (Titze, 1994a). Lesions of the basal membrane zone have been demonstrated in the vocal fold nodules, observed in the human species, in subjects who overwork their vocal structures (Gray et al., 1995; Courey et al., 1996).

Myofibroblasts, differentiated fibroblasts important in tissue repair, are present in the SLLP. This suggests that they could act as repairing cells in vocal fold areas enduring normal and daily vibrating stress (Catten et al., 1998; Hinz et al., 2001; Alberts et al., 2008; Petreaca, 2008).

It is important to notice that many of these protective molecular and tissue mechanisms are present in the main vibrator, the SLLP. The SLLP therefore possesses not only the necessary pliability for a sustained phonation, but also the biological and structural features for self-protection.

V. Conclusions

The vocal folds' microanatomy has raised considerable interest, because its molecular composition and tissue organization determine the oscillatory function.

Production of the human voice, with its large variety of sounds, relies on the unique laminar organization of its connective tissue. No other mammalian species possesses such a complex vocal fold structure. The rapid vibration of the human vocal folds can be observed thanks to special techniques, such as stroboscopy or high speed cinematography. Very little is known about the oscillatory behavior of the vocal folds of other mammalians, because of the lack of direct observation of their folds when they phonate.

References

Alberts, B., Johnson, A., Walter, P., Lewis, J., Raff, M., Roberts, K., 2008. Cell junction, cell adhesion and the extracellular matrix. In: McDonald Institute for Archaeological Research (Ed.), Molecular Biology of the Cell. Garland Science, New York, NY and London, UK, pp. 1131–1204.

Benninger, M.S., Alessi, D., Archer, S., Bastian, R., Ford, C., Koufman, J., Sataloff, R.T., Spiegel, J.R., Woo, P., 1996. Vocal fold scarring: current concepts and management. Otolaryngol. Head Neck Surg. 115, 474–482.

Butler, J.E., Hammond, T.H., Gray, S.D., 2001. Gender-related differences of hyaluronic acid distribution in the human vocal fold. Laryngoscope 111, 907–911.

Campos Banales, M.E., Perez Pinero, B., Rivero, J., Ruiz Casal, E., Lopez Aguado, D., 1995. Histological structure of the vocal fold in the human larynx. Acta Otolaryngol. (Stockholm) 115, 701–704.

Catten, M., Gray, S.D., Hammond, T.H., Zhou, R., Hammond, E., 1998. Analysis of cellular location and concentration in vocal fold lamina propria. Otolaryngol. Head Neck Surg. 118, 663–667.

Chan, R.W., Gray, S.D., Titze, I.R., 2001. The importance of hyaluronic acid in vocal fold biomechanics. Otolaryngol. Head Neck Surg. 124, 607–614.

Courey, M.S., Shohet, J.A., Scott, M.A., Ossoff, R.H., 1996. Immunohistochemical characterization of benign laryngeal lesions. Ann. Otol. Rhinol. Laryngol. 105, 525–531.
Fayoux, P., Devisme, L., Merrot, O., Chevalier, D., Gosselin, B., 2004. Histologic structure and development of the laryngeal macula flava. Ann. Otol. Rhinol. Laryngol. 113, 498–504.
Finck, C., 2005. [Vocal fold structure and speech pathologies]. Rev. Laryngol. Otol. Rhinol. (Bord.) 126, 295–300 (in French).
Finck, C., Lefebvre, P., 2005. Implantation of esterified hyaluronic acid in microdissected Reinke's space after vocal fold microsurgery: first clinical experiences. Laryngoscope 115, 1841–1847.
Garrett, C.G., Coleman, J.R., Reinish, L., 2000. Comparative histology and vibration of the vocal folds: implications for experimental studies in microlaryngeal surgery. Laryngoscope 110, 814–824.
Graney, D.O., Flint, P.W., 1998. Anatomy of the larynx. In: Cummings, C.W., Fredrickson, J.M., Harker, L.A., Krause, C.J., Richardson, M.A., Schuller, D.E. (Eds.), Otolaryngology and Head Neck Surgery. Mosby, St. Louis, MO, pp. 1824–1833.
Gray, S.D., 1991. Basement membrane zone injury in vocal nodules. In: Gauffin, J., Hammarberg, B. (Eds.), Vocal Fold Physiology: Acoustic, Perceptual, and Physiological Aspects of Voice Mechanisms. Whurr Publishers, Singular Publishing Group, San Diego, CA, pp. 21–27.
Gray, S.D., 2000. Cellular physiology of the vocal folds. Otolaryng. Clin. of North Am. 33, 679–697.
Gray, S.D., Hirano, M., Sato, K., 1993. Molecular and cellular structure of vocal fold tissue. In: Titze, I.R. (Ed.), Vocal Fold Physiology: Frontiers in Basic Science. Singular Publishing Group, San Diego, CA, pp. 1–34.
Gray, S.D., Pignatari, S.S., Harding, P., 1994. Morphologic ultrastructure of anchoring fibers in normal vocal fold basement membrane zone. J. Voice 8, 48–52.
Gray, S.D., Hammond, E., Hanson, D.F., 1995. Benign pathologic responses of the larynx. Ann. Otol. Rhinol. Laryngol. 104, 13–18.
Gray, S.D., Titze, I.R., Chan, R., Hammond, T.H., 1999. Vocal fold proteoglycans and their influence on biomechanics. Laryngoscope 109, 845–854.
Gray, S.D., Titze, I.R., Alipour, F., Hammond, T.H., 2000. Biomechanical and histologic observations of vocal fold fibrous proteins. Ann. Otol. Rhinol. Laryngol. 109, 77–85.
Gunter, H.E., 2004. Modeling mechanical stresses as a factor in the etiology of benign vocal fold lesions. J. Biomech. 37, 1119–1124.
Hahn, M.S., Langer, R., 2005. Midmembranous vocal fold lamina propria proteoglycans across selected species. Ann. Otol. Rhinol. Laryngol. 114, 451–462.
Hahn, M.S., Kobler, J.B., Zeitels, S.M., Langer, R., 2006. Quantitative and comparative studies of the vocal fold extracellular matrix II: collagen. Ann. Otol. Rhinol. Laryngol. 115, 225–232.
Hahn, M.S., Kobler, J.B., Starcher, B.C., Zeitels, S.M., Langer, R., 2006a. Quantitative and comparative studies of the vocal fold extracellular matrix. I: Elastic fibers and hyaluronic acid. Ann. Otol. Rhinol. Laryngol. 115, 156–164.
Hallen, L., Dahlqvist, A., Laurent, C., 1998. Dextranomeres in hyaluronan (DiHA): a promising substance in treating vocal cord insufficiency. Laryngoscope 108, 393–397.
Hammond, T.H., Zhou, R., Hammond, E.H., Pawlak, A., Gray, S.D., 1997. The intermediate layer: a morphologic study of the elastin and hyaluronic acid constituents of normal human vocal folds. J. Voice 11, 59–66.
Hinz, B., Mastrangelo, D., Iselin, C.E., Chaponnier, C., Gabbiani, G., 2001. Mechanical tension controls granulation tissue contractile activity and myofibroblast differentiation. Am. J. Pathol. 159, 1009–1020.
Hirano, M., 1974. Morphological structure of the vocal cord as a vibrator and its variations. Folia Phoniatr. (Basel) 26, 89–94.
Hirano, M., 1975. Phonosurgery – basic and clinical investigation. Otologia 21, 239.
Hirano, M., 1981. The structure and control of the vocal fold as a vibrator. In: Arnold, G.E., Winckel, A.F., Wycke, B.D. (Eds.), Clinical Examination of Voice. Springer-Verlag, Wien, Germany, New York, NY, pp. 4–7.
Hirano, M., Kurita, S., Nakashima, T., 1981. The structure of the vocal folds. In: Stevens, K.N., Hirano, M. (Eds.), Vocal Fold Physiology. University of Tokyo Press, Tokyo, Japan, pp. 33–41.
Hirano, M., Gould, W.J., Lambiase, A., Kakita, Y., 1980. Movements of selected points on a vocal fold during vibration. Folia Phoniatr. (Basel) 32, 39–50.
Hirschi, S.D., Gray, S.D., Thibeault, S.L., 2002. Fibronectin: an interesting vocal fold protein. J. Voice 16, 310–316.
Ishii, K., Yamashita, K., Akita, M., 1999. Fibrous structure of connective tissue in the vocal fold of the Japanese monkey. Okajimas Folia Anat. Jpn. 76, 107–115.
Ishizaka, K., Matsudaira, M., 1972. Fluid mechanical considerations of vocal cord vibration. SCRL Monograph no. 8. Speech Communication Research Laboratory, Santa Barbara, CA.
Jiang, J., 2006. Physiology of voice production: how does voice work? In: Benninger, M.S., Murry, T. (Eds.), The Performer's Voice. Plural Publishing, San Diego, CA, pp. 23–32.
Lodish, H., Berk, A., Kaiser, C.A., Krieger, M., Scott, M.P., Bretscher, A., Ploegh, H. (Eds.), 2008. Integrating cells into tissues. In: Molecular Cell Biology, 6th edn, Chapter 19. WH Freeman & Co., New York, NY, pp. 801–846.
Madruga de Melo, E.C., Lemos, M., Ragao Ximenes, F.J., Sennes, L.U., Nascimento Saldiva, P.H., Tsuji, D.H., 2003. Distribution of collagen in the lamina propria of the human vocal fold. Laryngoscope 113, 2187–2191.
Maragos, N.E., 2006. Anatomy of the vocal mechanism: structure of the voice. In: Benninger, M., Murry, T. (Eds.), The Performer's Voice. Plural Publishing, San Diego, CA, pp. 15–32.
Meyer, K., Palmer, J.W., 1934. The polysaccharide of the vitreous humor. J. Biol. Chem. 107, 629–634.
Nagata, K., 1982. A comparative study of the layer structure of the vocal fold: a morphological investigation of 11 mammalian species. Otologia (Fukuoka) 28, 699–738.
Nicolai, P., Cappiello, J., Peretti, G., Antonelli, A.R., Parolini, S., Rosa, D., Favret, M., Maroccolo, D., Molinari Tosatti, M.P., 1990. [Distribution of laminin, type IV collagen and fibronectin in normal, dysplastic and neoplastic laryngeal tissue]. Acta Otorhinolaryngol. Ital. 10, 139–149 (In Italian).

Obara, M., Yoshizato, K., 1997. A novel domain of fibronectin revealed by epitope mapping of a monoclonal antibody which inhibits fibroblasts-mediated collagen gel contraction. FEBS Lett. 412, 48–52.

Pawlak, A.S., Hammond, T., Hammond, E., Gray, S.D., 1996. Immunocytochemical study of proteoglycans in vocal folds. Ann. Otol. Rhinol. Laryngol. 105, 6–11.

Petreaca, M., Martins-Green, M., 2008. Cell-ECM interactions in repair and regeneration. In: Atala, A., Lanza, R., Thomson, J.A., Nerem, R.M. (Eds.), Principles of Regenerative Medicine. Academic Press, Elsevier, Amsterdam, The Netherlands, pp. 66–99.

Reinke, F., 1895. Untersuchungen ueber das menschliche Stimmband. Fortschr. Med. 13, 469–478.

Reinke, F., 1897. Über die funktionelle Struktur der menschlichen Stimmlippe mit besonderer Berücksichtigung des elastischen Gewebes. Ant. Hefte 9, 103–117.

Sato, K., 1998. Reticular fibers in the vocal fold mucosa. Ann. Otol. Rhinol. Laryngol. 107, 1023–1028.

Sato, K., Hirano, M., Nakashima, T., 2000. Comparative histology of the maculae flavae of the vocal folds. Ann. Otol. Rhinol. Laryngol. 109, 136–140.

Sato, K., Hirano, M., Nakashima, T., 2001. Stellate cells in the human vocal fold. Ann. Otol. Rhinol. Laryngol. 110, 319–325.

Sato, K., Hirano, M., Nakashima, T., 2003. 3D structure of the macula flava in the human vocal fold. Acta Otolaryngol. 123, 269–273.

Scherer, R., Titze, I., 1983. Pressure-flow relationships in a model of laryngeal airway with diverging glottis. In: Bless, D.M., Abbs, J.H. (Eds.), Vocal Fold Physiology: Contemporary Research and Clinical Issues. College-Hill Press, San Diego, CA, pp. 179–193.

Tateya, T., Bless, D.M., 2006. Postnatal development of rat vocal folds. Ann. Otol. Rhinol. Laryngol. 115, 215–224.

Tateya, T., Tateya, I., Bless, D.M., 2006. Collagen subtypes in human vocal folds. Ann. Otol. Rhinol. Laryngol. 115, 469–476.

Tateya, T., Tateya, I., Bless, D.M., 2007. Immuno-scanning electron microscopy of collagen types I and III in human vocal fold lamina propria. Ann. Otol. Rhinol. Laryngol. 116, 156–159.

Thibeault, S.L., 2005. Advances in our understanding of the Reinke space. Curr. Opin. Otolaryngol. Head Neck Surg. 13, 148–151.

Thibeault, S.L., Bless, D.M., Gray, S.D., 2003. Interstitial protein alterations in rabbit vocal fold with scar. J. Voice 17, 377–383.

Thibeault, S.L., Gray, S.D., Bless, D.M., Chan, R.W., Ford, C.N., 2002a. Histologic and rheologic characterization of vocal fold scarring. J. Voice 16, 96–104.

Thibeault, S.L., Gray, S.D., Li, W., Ford, C.N., Smith, M.E., Davis, R.K., 2002b. Genotypic and phenotypic expression of vocal fold polyps and Reinke's edema: a preliminary study. Ann. Otol. Rhinol. Laryngol. 111, 302–309.

Titze, I.R., 1994a. Mechanical stress in phonation. J. Voice 8, 99–105.

Titze, I.R., 1994b. Control of fundamental frequency. In: Titze, I.R. (Ed.), Principles of Voice Production. Prentice-Hall, Englewood Cliffs, NJ, pp. 191–217.

Titze, I.R., 1994c. Vocal fold oscillation. In: Titze, I.R. (Ed.), Principles of Voice Production. Prentice-Hall, Englewood Cliffs, NJ, pp. 80–111.

Van den Berg, J.W., 1958. Myoelastic-aerodynamic theory of voice production. J. Speech Hear. Res. 1, 227–244.

Ward, P.D., Thibeault, S.L., Gray, S.D., 2002. Hyaluronic acid: its role in voice. J. Voice 16, 303–309.

CHAPTER 10.3

Mechanisms and evolution of roaring-like vocalization in mammals

Roland Frey[1*] and Alban Gebler[2]

[1]*Leibniz Institute for Zoo and Wildlife Research (IZW), Berlin, Germany*
[2]*Institute for Theoretical Biology, Humboldt University of Berlin, Germany*

Abstract: Typical examples for roaring are provided by rutting red deer stags and male elephant seals, but also by both sexes of lions. Roaring is characterized by a low fundamental frequency and lowered vocal tract resonances (formants). The fundamental frequency depends on the mass and length of the vocal folds and, thereby, correlates with larynx size. A lowered dispersion of formants is decisively determined by vocal tract length that correlates with body size. Several species evolved pronounced larynx retraction causing vocal tract elongation and, therefore, acoustically simulating a larger body size. Roaring, however, remains an honest signal and it may prevent conflict escalation by acoustic assessment of a rival's quality. If roaring is restricted to the males of a species, it is directly linked to their reproductive success. If roaring occurrs in both sexes, it may improve intergroup spacing, territory defense, or communication in a densely forested habitat.

Keywords: vocal system anatomy; roaring; roaring-like vocalization; formants; larynx retraction; nasal vestibulum; vocal tract; resonance volumes; sexual selection

Abbreviations: Ad ventric lar: entrance to laryngeal ventricle; App hyo: hyoid apparatus; Art tempmand: temporomandibular joint; Basih: basihyoid; Bul cric: cricoid bulla; Bul thyr: thyroid bulla; Bul tymp: tympanic bulla; Burs pal (dex): (right) palatine pouch; Burs vest oris: cheek pouch, extending dorsocaudally towards the ear; Cart aryt: arytenoid cartilage; Cart cric: cricoid cartilage; Cart intaryt: interarytenoid cartilage; Cart thyr: thyroid cartilage; Cartt nasi: nasal cartilages; Cav infrglott: infraglottic cavity; Costa 1: first rib; Diaphr: diaphragm; Epigl: epiglottis; FEP: fibroelastic pad; Lam cric: lamina of cricoid cartilage; Lar: larynx; Lig thyroh: thyrohyoid ligament; Ling: tongue; nasvt: nasal portion of vocal tract; nostr: nostril; M hyoepigl: hyoepiglottic muscle; M inc sup: superior incisive muscle, main protractor of nasal vestibulum; Oesoph: esophagus; oralvt: oral portion of vocal tract; Ost intrphar: intrapharyngeal opening, between nasal and laryngeal portion of pharynx; Palat mol: soft palate (palatine velum); Phar: pharynx; Plic voc: vocal fold; Proc corn: corniculate process of arytenoid cartilage; Proc med: medial process of arytenoid cartilage; Proc musc: muscular process of arytenoid cartilage; Proc voc: vocal process of arytenoid cartilage; Prom hyo: hyoid prominence; Prom lar: laryngeal prominence; Rec lat: lateral recess of laryngeal vestibulum; Spin proc corn: spine of corniculate process; stern: sternum; Trach: trachea; Vestib lar: laryngeal vestibulum; Vestib nasi: nasal vestibulum

I. Introduction

The term "roaring" is commonly used for the vocalizations of several species, e.g., lion, leopard, jaguar, elephant seals, red deer and musk ox (Fig. 1). In contrast to some species-specific definitions, Weissengruber et al. (2002) suggested a more general definition of roaring. According to their suggestion, a roar is specified as a call type with a low fundamental frequency and lowered formant frequencies. In this chapter not only roaring vocalizations which conform strictly to this definition, but also other similar roaring-like vocalizations are considered. Roaring-like vocalizations are

*Corresponding author. E-mail: frey@izw-berlin.de

ISBN 978-0-12-374593-4
DOI: 10.1016/B978-0-12-374593-4.00040-1

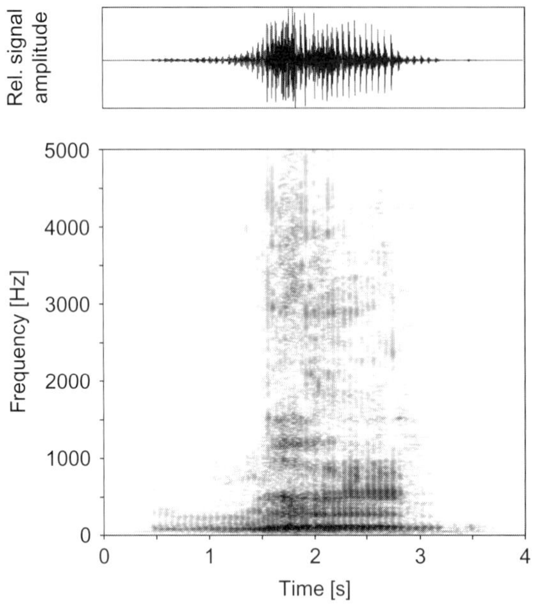

Fig. 1. Spectrogram of a male muskox roar (*Ovibos moschatus*). The acoustic energy is concentrated to fundamental frequency and formants in the lower part of the spectrum. The pulsed structure of the roar presumably relates to the bulky vocal fold of the musk ox, but is not an inherent feature of roaring in general.

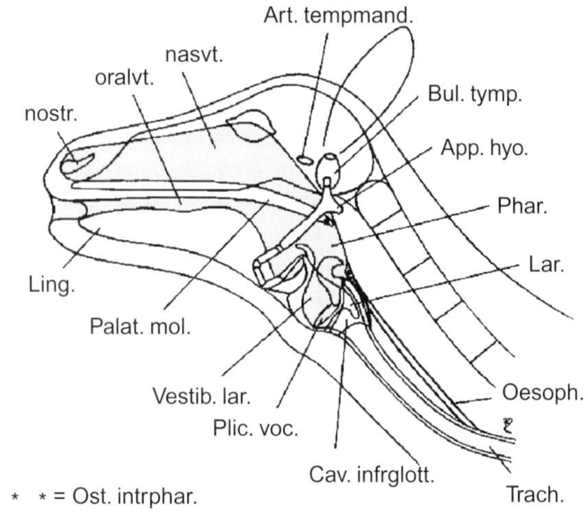

Fig. 2. General anatomical overview of the mammalian vocal tract (shaded). Teeth and lower jaw omitted for the sake of clarity, left lateral view. Mouth opened and larynx slightly retracted as in typical oral exhalatory vocalization; soft palate touches dorsal pharyngeal wall to guide the entire airflow through the mouth opening. Intrapharyngeal portion of esophagus closed by contraction of caudal pharynx sphincters, remaining esophagus collapsed. The vocal folds (source) are forced to oscillate by the exhalatory air stream. On its passage towards the mouth opening, the source signal is transformed by the vocal tract (filter). Abbreviations: App hyo: hyoid apparatus; Art tempmand: temporomandibular joint; Bul tymp: tympanic bulla; Cav infrglott: infraglottic cavity; Lar: larynx; Ling: tongue; nasvt: nasal portion of vocal tract; nostr: nostril; Oesoph: esophagus; oralvt: oral portion of vocal tract; Ost intrphar: intrapharyngeal opening, between nasal and laryngeal portion of pharynx; Palat mol: soft palate (palatine velum); Phar: pharynx; Plic voc: vocal fold; Trach: trachea; Vestib lar: laryngeal vestibulum.

those in which one of the two components (fundamental frequency or formant frequencies) is not as low as in typical roaring.

In a simplified model of sound production, the vocal folds in the larynx oscillate during vocalizing and produce a primary signal composed of the fundamental frequency and its integer multiples, the harmonics or overtones (see Berke and Long, Chapter 10.1 in this volume). This primary signal does not necessarily have to sound such as a listener will hear. In the source-filter model of vocalization (Fant, 1960), the audible acoustic output is restricted to those frequency ranges where the formants, i.e., the resonances of the vocal tract, and the laryngeal source signal overlap.

The vocal tract is formed by all anatomical structures of the upper respiratory pathway, e.g., laryngeal cavity, throat, oral and nasal cavity, lips and nostrils, and possible additional air-filled volumes such as laryngeal ventricles or air sacs (Fig. 2). The formants depend not only on the stiffness of vocal tract boundary tissues. Formant dispersion, i.e., the spectral position of formants, is mainly determined by the geometrical shape and, particularly, the length of the vocal tract. The vocal tract length correlates with skull size, which is in turn closely correlated with body size in mammals (Fitch, 1997). Body size in general is linked to fighting ability and, ultimately, to dominance rank. As a consequence, formant dispersion has been assumed to transmit reliable cues to body size, fighting ability and dominance rank in mammals. Social dominance rank is an important aspect of male quality and is particularly important for reproductive success. This has been demonstrated in red deer and fallow deer (Reby et al., 2005; Charlton et al., 2007; Vannoni and McElligott, 2008).

Since the fundamental frequency is a function of vocal fold length and mass, it is also correlated with the size of the larynx that determines vocal fold size. In contrast to the vocal tract length, the size of the larynx is potentially more independent of body size. Fitch (1997) states that the fundamental frequency on its own is an unreliable indicator of body size.

However, there are several species with sexually dimorphic and conspicuously enlarged larynges in the males, e.g., the hammer-headed fruit bat (Schneider et al., 1967; Zeller, 1984) and the Mongolian gazelle (Kleinschmidt, 1961; Frey and Gebler, 2003; Frey and Riede, 2003). Considering the polygynous mating systems involved, the enlarged larynx is probably the result of sexual selection towards longer vocal folds and lower fundamental frequency. In these cases an increase in larynx size (fundamental frequency decrease) appears to contribute to the honest signaling of male quality.

In the case of evolutionary vocal tract elongation, a low fundamental frequency and, therefore, dense harmonics are necessary to transmit formant-related information. In general, the accuracy of formant dispersion as an acoustic cue to body size is best with noisy, low-pitched vocalizations such as roaring.

It should be noted that it may not only be sexual selection which leads to the evolution of a roaring vocalization; in some polygynous species vocal tract specializations suitable for roaring may also occur in females, indicative of other reasons for roaring, e.g., habitat requirements (see IV). Different evolutionary pathways may lead to lower frequency and dark-timbred phonations which are perceived by humans as roaring-like vocalizations.

II. Evolution of roaring along different pathways

II.A. Larynx size increase

The fundamental frequency is dependent on the length and the mass of the vocal folds. Thus, one evolutionary option to lower the fundamental frequency is to increase the size of the entire larynx, because this will entail an anatomical elongation of the vocal folds. A larger larynx may have evolved by ontogenetically decoupling advanced larynx growth from body growth. As a consequence, the final larynx size would exceed the dimensions expected from comparisons with conspecifics or other species of comparable body size.

An extreme larynx size is observed in the African adult male hammer-headed fruit bat (*Hypsignathus monstrosus*, Epomophorinae, Megachiroptera), which has a larynx almost half the length of its vertebral column and occupying most of the thoracic cavity (Fig. 3). There is a pronounced sexual dimorphism in this species. The larynx is almost three times larger in males than in females (Lang and Chapin, 1917; Schneider et al., 1967; Zeller, 1984). The extreme enlargement in the male could only occur at the expense of the displacement and changes in shape of other organs. Linked with the increase in size of the larynx, very large vocal folds evolved (Schneider et al., 1967; Kingdon, 1984). In all probability, the disproportionate larynx size in the male hammer-headed fruit bat has evolved by sexual selection in the context of a polygynous lek mating system (Bradbury, 1977).

In a comparison with two other species of nearly-related epauletted fruit bats, *Epomophorus wahlbergi* and *Epomops franqueti*, the evolution of an increased larynx size, most pronounced in the male hammer-headed fruit bat, clearly correlated with a decrease in fundamental frequencies (Wickler and Seibt, 1976), thus confirming the interdependency between larynx size and vocal fold length.

A disproportionate larynx size increase has also occurred in the male Mongolian gazelle (*Procapra gutturosa*, Bovidae). The larynx, including the large epiglottis, is approximately double the size relative to female conspecifics and to other bovid species of comparable body size (Kleinschmidt, 1961; Frey and Riede, 2003).

As in the males of the hammer-headed fruit bat, the remarkable larynx size in male Mongolian gazelles is assumed to have arisen by sexual selection and the evolution of a polygynous mating system (Frey and Gebler, 2003; Frey and Riede, 2003; Frey et al., 2008).

II.B. Mass increase of the vocal folds

A second evolutionary option to lower the fundamental frequency is an increase in the mass of the vocal folds which, typically, consist of thin mucous membrane duplications enclosing specific vocal musculature and a vocal ligament. Such a mass increase may be achieved by incorporating fibroelastic pads, consisting of fat and fibrous connective tissue, into the vocal folds. As a consequence, the vocal folds increase in bulk and their oscillation rates slow down. In addition, high vocal fold masses may lead to a pulsed source signal (Fig. 1).

A mass increase in the vocal folds is found not only in the pantherine felids, the so-called "roaring cats:" lion, tiger, leopard, jaguar (Hast, 1986, 1989), but also in the takin (*Budorcas taxicolor*), a stoutly-built south-Asian bovid (Frey and Hofmann, 2000), in the male Mongolian gazelle (Fig. 4) (Frey and Gebler,

2003; Frey and Riede, 2003), in the male Arctic musk ox (*Ovibos moschatus* – Frey et al., 2006), in the central-Asian saiga antelope (*Saiga tatarica* – Frey et al., 2007) and in the male hammer-headed fruit bat (Schneider et al., 1967; Kingdon, 1984).

II.C. Larynx retraction

A crucial factor determining vocal tract resonance frequencies (formants) is vocal tract length, i.e., the distance between the lips and the vocal folds. As vocal tract length correlates with neck length, and neck length with body size, vocal tract resonances are used for signaling body size, e.g., by male red deer (*Cervus elaphus*) during their loud roaring displays in male–male competition (Reby et al., 2005) and in female choice contexts (Charlton et al., 2007).

Vocal tract elongation in mammals can be achieved in three ways: (1) a permanent way; (2) a seasonal way; or (3) a call-synchronous way. The first way is represented by the pantherine felids in which the larynx, assisted by an elongated and ligamentous epihyoid and an enlarged pharynx, is subjected to a permanent caudal shift. The second way is represented by males of red and fallow deer (*Dama dama*) and males of the Mongolian gazelle in which the larynx, assisted by an extensible ligament-like thyrohyoid connection and an extended pharynx, descends to a lower resting position during the rut (Fig. 5). The third way is represented by all taxa mentioned in (1) and (2). It consists of a pronounced retraction of the larynx for the duration of the roar, and a subsequent return of the larynx to its resting position. Retraction is effected by so-called strap muscles connecting the larynx to the sternum. Return of the larynx is caused by the elastic recoil of tissues involved, and by the action of certain muscles connecting the larynx to the hyoid apparatus that is itself connected by muscles to the tongue, the lower jaw and neck vertebrae, and by cartilage and muscles to the skull base (Frey et al., 2008).

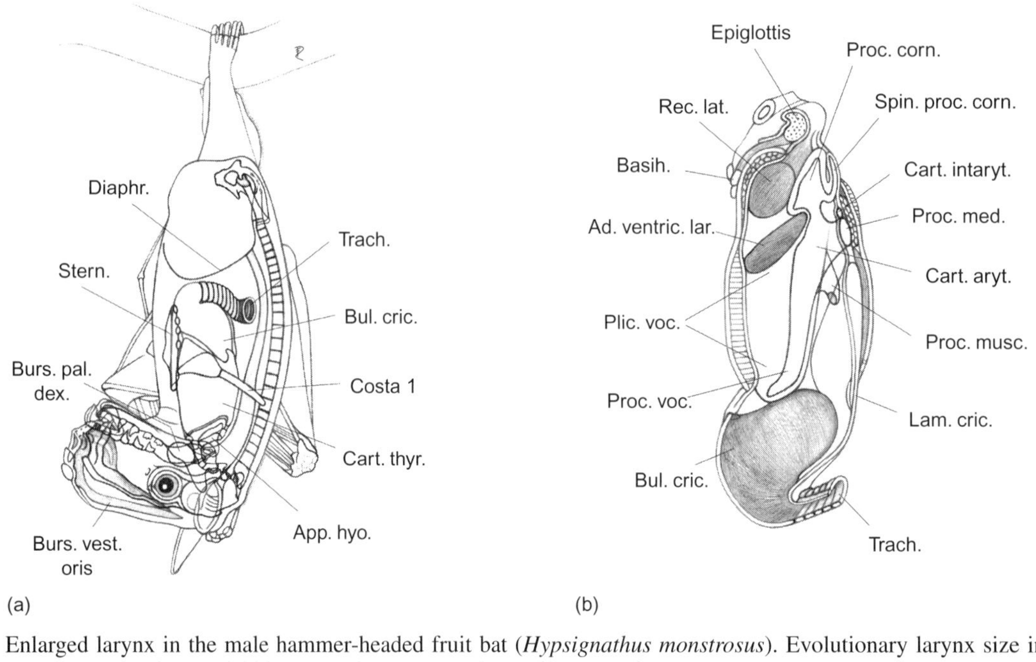

Fig. 3. Enlarged larynx in the male hammer-headed fruit bat (*Hypsignathus monstrosus*). Evolutionary larynx size increase implicates an increase of vocal fold length and, thus, an option to lower the fundamental frequency. Male in its natural upside-down hanging position (a), right half of isolated larynx, upright, with its very large vocal fold, medial view (b). (a) After Schneider et al., 1967; Zeller, 1984; photos of Pavel Zuber and one anonymus photographer in www.animalpicturesarchive.com, Minnie Visions SWD. (b) After Schneider et al., 1967 (redrawn). Abbreviations: Ad ventric lar: entrance to laryngeal ventricle; App hyo: hyoid apparatus; Basih: basihyoid; Bul cric: cricoid bulla; Burs pal dex: right palatine pouch; Burs vest oris: cheek pouch, extending dorsocaudally towards the ear; Cart aryt: arytenoid cartilage; Cart intaryt: interarytenoid cartilage; Cart thyr: thyroid cartilage; Costa 1: first rib; Diaphr: diaphragm; Epigl: epiglottis; Lam cric: lamina of cricoid cartilage; Plic voc: vocal fold; Proc corn: corniculate process of arytenoid cartilage; Proc med: medial process of arytenoid cartilage; Proc musc: muscular process of arytenoid cartilage; Proc voc: vocal process of arytenoid cartilage; Rec lat: lateral recess of laryngeal vestibulum; Spin proc corn: spine of corniculate process; Stern: sternum; Trach: trachea.

All three ways serve to decrease formant frequencies by a considerable increase of vocal tract length (Luckhaus, 1969; Peters and Hast, 1994; Fitch and Reby, 2001; Weissengruber et al., 2002). As there are anatomical constraints to further larynx retraction, the acoustic exaggeration of body size by elongation of the vocal tract remains an honest signal (Reby and McComb, 2003).

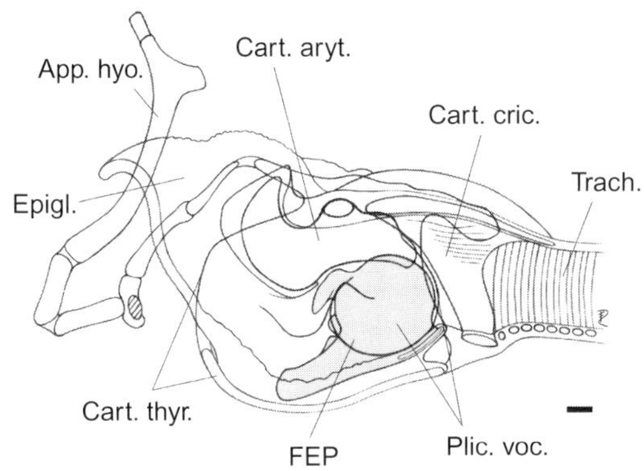

Fig. 4. Right half of the hyoid apparatus and larynx of an adult male Mongolian gazelle (*Procapra gutturosa*), disclosing the large and bulky right vocal fold, medial view. The vocal fold is supported by a cymbal-like fibroelastic pad (shaded) that is dorsally connected to the vocal process of the arytenoid cartilage and ventrally to the thyroid cartilage. A flexible mucous membrane fold and a string-like vocal ligament, as in a typical mammalian larynx, are lacking. Scale bar = 10 mm. Abbreviations: App hyo: hyoid apparatus; Cart aryt: arytenoid cartilage; Cart cric: cricoid cartilage; Cart thyr: thyroid cartilage; Epigl: epiglottis; FEP: fibroelastic pad; Plic voc: right vocal fold; Trach: trachea.

In red and fallow deer and in the Mongolian gazelle larynx, the vocal tract and associated structures are all sexually dimorphic. The much more pronounced features in males are assumed to have evolved by sexual selection in the context of a polygynous mating system.

II.D. Nasal vestibulum extension

In almost the same manner as the retraction of the larynx, a rostrally-directed extension of an inflatable nasal vestibulum causes an elongation of the vocal tract. Accordingly, this offers an additional evolutionary option for decreasing formant frequencies and signaling large body size.

In contrast to red and fallow deer stags and male Mongolian gazelle, which produce loud oral roars or barks, rutting saiga males produce loud nasal roars, i.e., they keep the mouth closed when displaying acoustically in male–male competition and female choice contexts. In rutting saiga males, vocal tract elongation is achieved by forward nose extension, in particular by protraction of the vestibular floor and pronounced S-shaping of the nasal vestibulum, both controlled by evolutionarily transformed musculature (Fig. 6). A visual comparison between vocal tract lengths achieved with extended or relaxed noses among saiga males suggests a mean vocal tract elongation of about 20% during the production of rutting calls. An acoustic comparison between rutting and non-rutting calls of saiga males confirmed formant depression during vocal tract extension (Frey et al., 2007b).

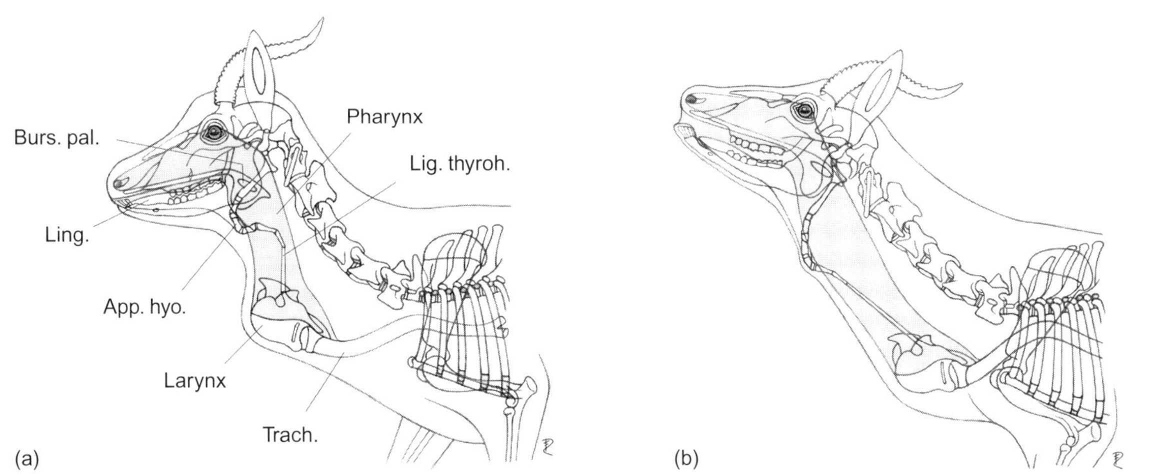

Fig. 5. Adult male Mongolian gazelle, larynx in resting position during the rut (a) and maximally retracted during a head-up roar (b). Vocal tract shaded. Abbreviations: App hyo: hyoid apparatus; Burs pal: palatinal pouch; Lig thyroh: thyrohyoid ligament; Ling: tongue; Trach: trachea.

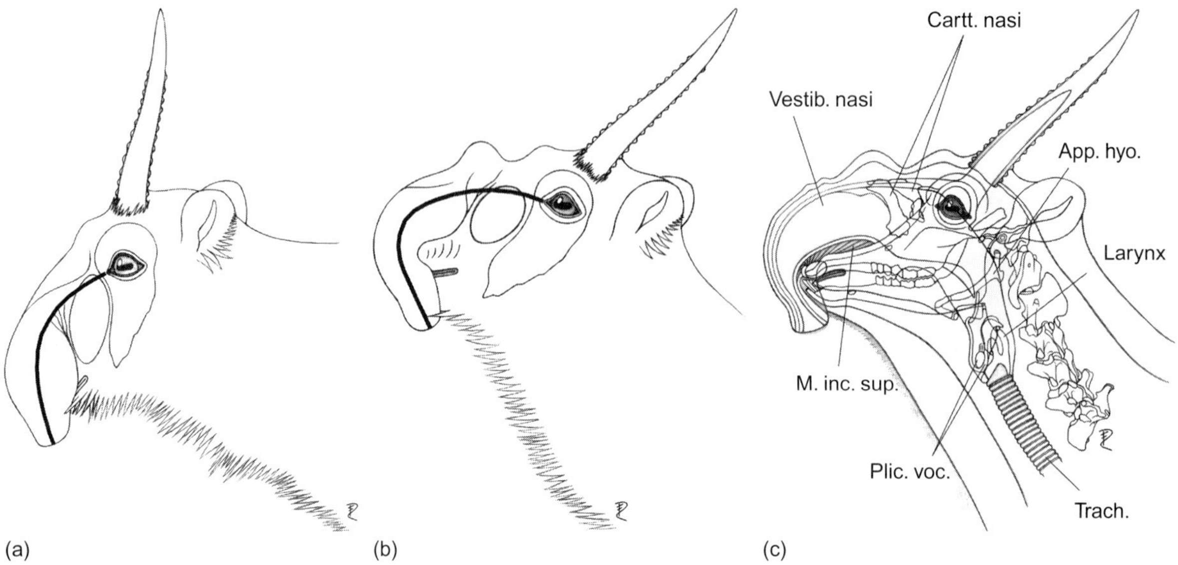

Fig. 6. Forward extension of the nasal vestibulum in rutting saiga (*Saiga tatarica*) male, synchronously to a loud nasal roar. Vocal tract elongation by nose extension is about 20% in saiga. Nose in resting position (a), nose extension during a head-up nasal roar (b) and overlay of structures involved in nasal roaring (c). Black line indicates forward extension of nasal vestibulum. (a), (b) and contour of (c) after video single frames recorded in 2007 by Elena and Ilya Volodin, Moscow. Abbreviations: App hyo: hyoid apparatus; Cartt nasi: nasal cartilages; Lar: larynx; M inc sup: superior incisive muscle, main protractor of nasal vestibulum; Plic voc: vocal fold with its egg-shaped fat pad; Trach: trachea; Vestib nasi: nasal vestibulum.

As in the males, female saigas also possess an evolutionarily enlarged nasal vestibulum. However, the size of the male nasal vestibulum, particularly during the rut, is sexually dimorphic and considerably larger than in females, even after correction for body size. Probably, this dimorphism was induced by sexual selection in the context of a polygynous mating system. As with the vocal tract elongation of pantherine felids, certain cervids and Mongolian gazelle, anatomical constraints can be expected to set a limit to the rostral pulling of the saiga nasal vestibulum, so that acoustic display of body size remains an honest signal (Frey et al., 2007b).

As in the retraction of the larynx in red and fallow deer and in Mongolian gazelle, the protraction of the nasal vestibulum in saiga can be assumed to have evolved by sexual selection in the context of a polygynous mating system.

II.E. Vocal postures and vocal gestures

In addition to anatomical adaptations of the larynx itself or of adjacent structures, vocalizing with an extended throat region and elevated head tends to decrease formant frequencies by a slight elongation of vocal tract length and is, thus, well-suited for making vocalizations sounding deeper (Hauser and Schön-Ybarra, 1994). Therefore, there is a typical posture for roaring-like vocalizations demonstrated by red deer stags and male Mongolian gazelles (Fig. 7). Saiga males during their nasal roaring adopt a similar vocal posture in which the head is held up in an almost horizontal position and the ventral neck region extended. Male elephant seals (*Mirounga* spp.) likewise erect the forepart of their body, extend the ventral neck region and raise their head during a roar (Fig. 8).

In addition to these vocal postures there are vocal gestures which modify the vocal tract shape, thereby influencing formants. Vocal gestures comprise, e.g., oral roaring with an oval contour of the mouth opening as in red deer stags; or stereotypic nose configuration while producing nasal roars as in saiga males (Frey et al., 2007b). Further vocal gestures are lip and tongue configurations and movements of the lower jaw as, e.g., in the roaring lion (*Panthera leo*). Comparable active controlling of vocal tract shape by articulatory gestures has been reported in the South-American howler monkeys (*Alouatta* spp. – Schön-Ybarra, 1988; Whitehead, 1995).

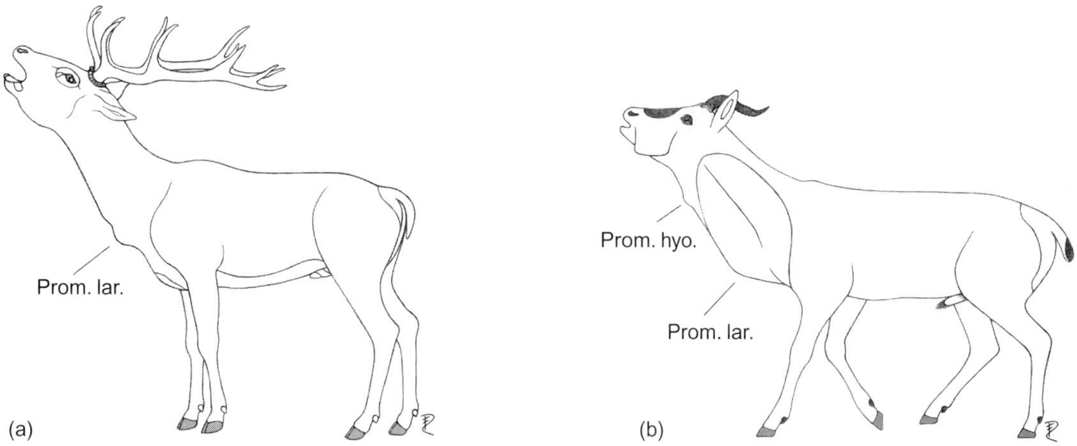

Fig. 7. Typical roaring postures of red deer (*Cervus elaphus*) stag (a) and male Mongolian gazelle (b) with its whiskers-like cheek hair coat. Abbreviations: Prom hyo: hyoid prominence; Prom lar: laryngeal prominence (a) and (b) not to scale.

Fig. 8. Additional resonance volume in male northern elephant seal (*Mirounga angustirostris*). Roaring is produced with inflated proboscis and widely opened mouth. Lateral view of the head. After a photo by Phillip Colla, 2008, Oceanlight.com, CA.

II.F. Additional vocal tract resonance volumes

Inflatable or rigid resonance volumes which are connected to the vocal tract represent another evolutionary option to influence the acoustic characteristics of vocalizations. Such volumes are, e.g., laryngeal ventricles, air sacs and thyroid, cricoid and hyoid bullae (see Frey et al., 2007a). Additional resonance volumes, among other effects, can increase the dynamic range of sound emission and shift the formants of the main vocal tract, particularly affecting the first formant (Gautier, 1971; Riede et al., 2008).

Inflatable resonance volumes involved in roaring-like vocalizations are, e.g., a greatly enlarged nasopharynx that connects to paired palatinal pouches, plus paired cheek pouches which cover the frontoparietal region in a hood-like shape up to the rostrum in the male hammer-headed fruit bat (Kingdon, 1984), paired two-chambered laryngeal ventricles in the male Mongolian gazelle (Frey and Gebler, 2003; Frey and Riede, 2003), an unpaired rostroventral laryngeal air sac in the musk ox (Frey et al., 2006), an unpaired dorsal pharyngeal air sac in bears (Ursidae) (Ganzberger et al., 1995; Weissengruber et al., 2001), and an unpaired rostroventral laryngeal air sac covered by a rigid hyoid bulla and additional paired lateral laryngeal ventricles in howler monkeys (Kelemen and Sade, 1960; Starck and Schneider, 1960). For a more comprehensive list of laryngeal saccules in mammals see Frey et al. (2007a).

Male elephant seals (*Mirounga angustirostris* and *M. leonina*) have a greatly enlarged, proboscis-like inflatable nasal vestibulum that increases in size during the rut, dangling in front of the mouth as in male saiga. By contrast to saiga, however, male elephant seals roar with their mouth widely opened (Fig. 8). Accordingly, despite the elongated nasal portion of the vocal tract, the main phonative exhalation occurs through the mouth (Sanvito et al., 2007a,b). Therefore, it is inferred that the proboscis may act more as an additional inflatable resonance volume rather than as an elongated vocal tract (see Riede et al., 2008). Nose configuration during a roar, i.e., during inflation, is different from saiga. Apart from a conspicuous dorsal mid-nose depression, the distal nostril portion of the male elephant seal proboscis is strongly bent caudally towards the widely-opened mouth.

Dominant males produce extremely loud roars when displaying in male–male competition or in a female choice context. The proboscis is inflated synchronously with respective roars and then appears to be used as a resonator that increases the amplitude of the roars, which can be heard over a 500–1,000 m range (Bartholomew and Collias, 1962; Sanvito and Galimberti, 2000; Southall et al., 2003). The increase of amplitude in male roars may result from a temporarily-inflated nasal portion of the main vocal tract functioning as a resonator, similar to the inflatable air sacs connected as side branches to the main vocal tract in other species, e.g., in reindeer (*Rangifer tarandus*), in the Siamang gibbon (*Symphalangus syndactylus*) and in the great apes (Hill and Booth, 1957; Starck and Schneider, 1960; Schneider, 1964; Frey et al., 2007a; Riede et al., 2008).

In addition, the proboscis of male elephant seals might serve as a visual signal (Sanvito et al., 2007b) as has been discussed for the hood-and-nasal-septum display of rutting male hooded seals (*Cystophora cristata*) (Terhune and Ronald, 1973; Ballard and Kovacs, 1995).

The nasal region is highly sexually dimorphic in elephant seals. Formation of the excessive inflatable proboscis in male elephant seals can be assumed to have evolved by sexual selection in the context of a polygynous mating system. It develops gradually in the course of male ontogeny (Sanvito et al., 2007b).

Interestingly, the male hammer-headed fruit bat also inflates its large dorsally extended hood-like cheek pouches while flapping its wings and emitting its roars. The roars comprise two distinct components: at close quarters they sound guttural, explosive and blaring; but with increasing distance they are perceived as a ringing chink (Kingdon, 1984).

III. Roaring and reproductive success

III.A. Mating systems associated with roaring

It appears that roaring and roaring-like vocalizations are particularly associated with polygynous female defense or lek mating systems. Dominant males will monopolize females (collect a harem) and exclude other males from their respective harems, or dominant males will gather at often traditional court grounds (leks) where they defend small non-resource territories which females enter for mating.

An advantage of lek mating is that males pool their advertising and courtship signals. In the hammer-headed fruit bat, males gather in large groups at traditional lek sites and make use of their giant larynx to produce a chorus of high-amplitude and, compared to related species of fruit bats, low-frequency advertisement calls which are performed uninterruptedly for several hours, thereby attracting female mating partners. Similarly, fallow deer stags, assembled at a lek, produce remarkably high groaning rates to induce females to approach the court ground and their individual lek territory for mating. In polygynous mating systems, the successful acoustic advertisement, mostly by roaring-like vocalizations, is closely linked to reproductive success.

Acoustic displays are not restricted to terrestrial mating systems. They may also occur underwater, e.g., in polygynous aquatically-mating harbor seals (*Phoca vitulina*) with a lek-type mating system (Hayes et al., 2004; Boness et al., 2006). Dominant males defend underwater non-resource territories and produce underwater low-frequency broadband roars (Hanggi and Schusterman, 1994). Large older males produce roars of lower frequency and longer duration than younger males. Underwater roars serve an intra-sexual function in male–male competition to assess each other and to contingently ward off rival male, as well as attract female mating partners (Hayes et al., 2004; Boness et al., 2006). As in terrestrial mammals, underwater roars of harbor seals are energetically expensive to produce and can be assumed to be honest signals of male quality (Zahavi and Zahavi, 1997; Hayes et al., 2004).

III.B. Intra- and inter-sexual selection

A decisive feature of polygynous mating systems is that male mating success is highly skewed, i.e., few males of high social rank get most of the matings and most males get few or none. This provides a basis for sexual selection on both an intra- and inter-sexual level. Intra-sexually, males will compete for access to estrous females and inter-sexually, females have the opportunity to make a mate choice (Andersson, 1994).

In the well-studied red deer, there is a clear evidence that formants are used as acoustic cues to body size, both in male–male competition (Reby et al., 2005) and in a female choice context (Charlton et al., 2007b). This appears to have induced a selection pressure towards the evolution of acoustic exaggeration of body size by vocal tract elongation, as observed in red deer (Fitch and Reby, 2001), in saiga (Frey et al., 2007) and in the Mongolian gazelle (Frey and Gebler, 2003; Frey et al., 2008).

Apparently, the evolved mechanisms of larynx retraction or nasal extension have some anatomical constraint. This constraint, applying to all conspecific males, ensures maintaining the honesty of the signal (Reby and McComb, 2003).

Apart from formants, an acoustic estimate of fighting strength or male quality might be signaled by the repetition rates of roars. Roaring repetition has been shown to be highest in the highest-ranking red and fallow deer stags, which have the highest mating success (McComb, 1991; McElligot and Hayden, 1999). Similarly, the roaring of dominant male musk oxen and the barking of dominant male Mongolian gazelles during the rut are described as frequently repeated over longer periods of time (Smith, 1976; Gray, 1984; Frey et al., 2008).

III.C. Roaring-like vocalizations as a handicap

The accuracy of formant dispersion as an acoustic cue to body size is best with noisy, low-pitched vocalizations (Fitch, 1997). Simultaneously, high roaring rates correlate positively with male mating success (McComb, 1991; McElligot and Hayden, 1999). Thus, the darkest timbred roars emitted at the highest possible rate indicate the largest male, the best fighter and, altogether, the highest-quality male.

These general conditions have important implications for male–male competition. The emission of noisy, low-pitched roaring-like vocalizations is intrinsically tied to a release of acoustic energy over a broad spectrum and, simultaneously, to the exhalation of large air volumes. This will entail a substantial energy loss to the environment. Muscular action required for retraction and protraction of the larynx, protraction and retraction of the nasal vestibulum, or the inflation of a laryngeal air sac will further increase the energetic costs of roaring-like vocalizations. And finally, high roaring rates, several hundred times per day, will further increase the energetic costs of the acoustic display. As a consequence, we may conclude that those males which have the highest reproductive success must be capable of tolerating the high-energy demands imposed on them by roaring-like vocalizations, so fitter males can better stand the handicap of acoustic display than less fit ones (see Zahavi, 1975, 1977, 1981, 1997).

IV. Additional reasons to evolve roaring

In the above examples, sexual selection was the driving force for the evolution of roaring-like phonation in males. However, other reasons appear to exist which promote the evolution of roaring. Howler monkeys (*Alouatta* spp.) invest much time and energy in roaring (da Cunha and Byrne, 2006). Roaring bouts are typically initiated by the alpha male (Fig. 9) and females regularly join in (da Cunha and Byrne, 2006). The highly-specialized vocal apparatus of both sexes involves a greatly enlarged, thin-walled basihyoid bulla, presumably functioning as the main resonator organ (Schön, 1970). In contrast to other primates, the fragile basihyoid is shifted rostrally and becomes located between the mandibular bodies which provide lateral protection by gaining dorsoventral height (Kelemen and Sade, 1960; Starck and Schneider, 1960; Watanabe, 1982). In addition, the accommodation of the enlarged vocal apparatus requires remodeling of the skull shape to allow for a permanent extension of the ventral neck region (Starck and Schneider, 1960; Starck, 1995; Koppe et al., 2005).

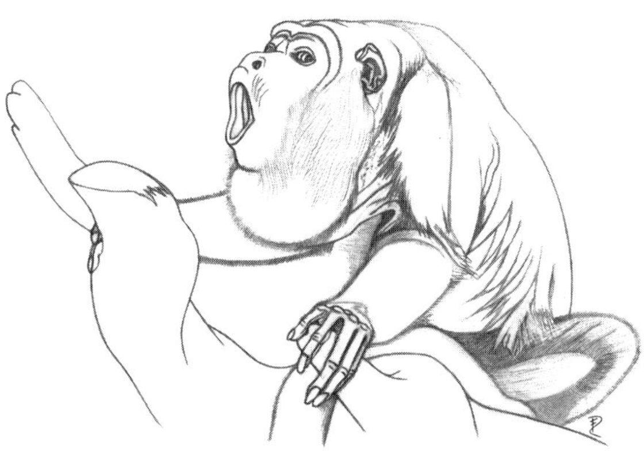

Fig. 9. Roaring posture of a male red howler monkey (*Alouatta seniculus*). Accommodation of the enlarged vocal apparatus in the throat region entails a permanent extension of the ventral neck region and facilitates transition to the roaring posture. After a photo of Wilfried Berns (2006), www.tiermotive.de.

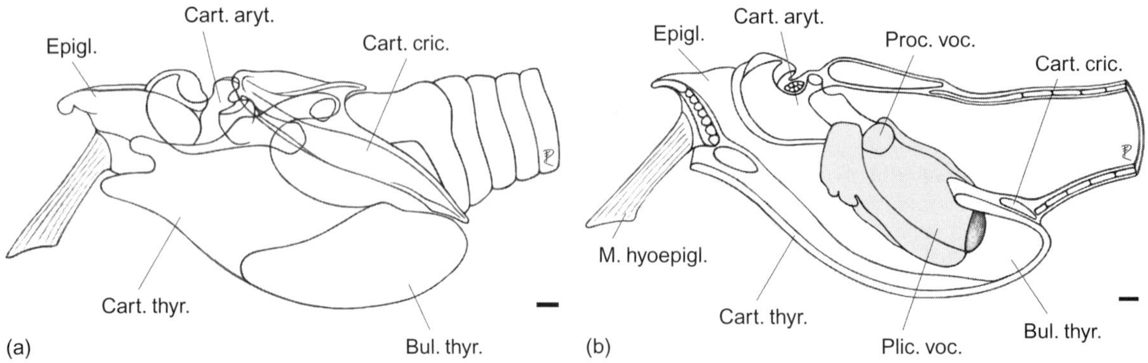

Fig. 10. Enlarged larynx and vocal folds in the takin (*Budorcas taxicolor*) occur in both sexes and probably evolved by natural selection. Outward appearance of an adult male's larynx with its large thyroid bulla and overlay of laryngeal cartilages (a) and right half of larynx for demonstration of the semi-cylindrical, long and bulky right vocal fold (shaded), medial view (b). Scale bars = 10 mm, respectively. Abbreviations: Bul thyr: thyroid bulla; Cart aryt: arytenoid cartilage; Cart cric: cricoid cartilage; Cart thyr: thyroid cartilage; Epigl: epiglottis; M hyoepigl: hyoepiglottic muscle; Plic voc: vocal fold; Proc voc: vocal process of arytenoid cartilage.

The loud and prolonged roaring in different species of howler monkeys has been suggested to function mainly in intergroup spacing (da Cunha and Byrne, 2006; da Cunha and Jalles-Filho, 2007). Intergroup spacing and territory defense by males and females suggest that natural selection is responsible for vocal tract specializations in both sexes. Apart from the above, the roaring of howler monkeys has been attributed a wide range of additional functions, e.g., assessment of relative fighting ability of intruders (Chiarello, 1995; Kitchen, 2004) or repelling potentially infanticidal males (Holzmann, 2006).

In the lion (*Panthera leo*) both males and females are social and territorial and use roars to advertise and defend their territories (Packer et al., 1988; McComb et al., 1994). The species exhibits obvious sexual dimorphism. However, sexual selection has not affected the acoustic parameters of the lion's roar, as it is not directly related to reproductive success. Differences between the sexes can be explained by differences in body size (Pfefferle et al., 2007). As adult males are larger than adult females, conspecific listeners can be assumed to recognize acoustically not only the number of individuals roaring, but also the sex of a roaring individual (McComb et al., 1993, 1994). Other evidence additionally suggests that the primary function of roars in lions is the advertisement and defense of the territory or coordination of hunting (Schaller 1972; McComb et al., 1994; Grinnel and McComb, 2001).

Under specific habitat conditions, roars might be more advantageous than vocalization with higher frequencies or more tonal characteristics. Concerning sound propagation, absorption and scattering will attenuate higher frequencies more than lower ones. Especially in habitats with dense vegetation, and therefore with a distinctive influence of scattering, the use of low frequencies might increase the range of the acoustic communication (Martens and Marler, 1977; Wiley and Richards, 1978; Crocker, 1998).

An example of a habitat-dependent evolution of roaring-like vocalization are the calls of the takin (*Budorcas taxicolor*) which lives most of the year in dense scrub. Evolutionary transformations of the larynx, such as an impressive thyroid bulla and long, massive vocal folds occur in both sexes (Fig. 10) (Frey and Hofmann, 2000). It can be assumed that in the takin an additional intralaryngeal resonance space (thyroid bulla) and a down-shifting of the fundamental frequency (large vocal folds) have evolved by natural selection. Probably, the semicylindrical, bulky vocal folds are also responsible for the takin's pulsed source signal.

V. Conclusions

Contrary to common belief, the morphological database on vocal tract anatomy of wild living species is sparse. Most species of mammals, not to mention other land vertebrates, have not yet been subjected to detailed investigation of their respective vocal apparatus. With further studies, we can expect to find more evolutionary specializations.

In order to achieve an improved understanding of animal vocalizations, an integrative approach is needed. Therefore, attempts should be made to cross-link morphophysiological, anatomical, ethological, bioacoustic and neurobiological results as closely as possible.

Using dissection data and measured anatomical features as a starting point, new insights can be expected to arise from advanced computational modeling of "prototypical" vocal tracts and, ultimately, vocal tracts of individual species. In an optimistic perspective, this kind of research might pave the way for predicting acoustic output from knowledge of anatomical structure. Finally, the application of a comparative approach will increase our understanding of the causes and courses of the evolutionary transformation of vocal anatomies.

References

Andersson, M., 1994. Sexual Selection. Monographs in Behavior and Ecology. Princeton University Press, Princeton, NJ.

Ballard, K.A., Kovacs, K.M., 1995. The acoustic repertoire of hooded seals (*Cystophora cristata*). Can. J. Zool. 73, 1362–1374.

Bartholomew, G.A., Collias, N.E., 1962. The role of vocalization in the social behaviour of the northern elephant seal. Anim. Behav. 10, 7–14.

Boness, D.J., Bowen, W.D., Buhleier, B.M., Marshall, G.J., 2006. Mating tactics and mating system of an aquatic-mating pinniped: the harbour seal, *Phoca vitulina*. Behav. Ecol. Sociobiol. 61, 119–130.

Bradbury, J.W., 1977. Lek mating behavior in the hammer-headed bat. Z. Tierpsychol. 45, 225–255.

Charlton, B.D., Reby, D., McComb, K., 2007. Female red deer prefer the roars of larger males. Biol. Lett. 3, 382–385.

Chiarello, A.G., 1995. Role of loud calls in brown howlers, *Alouatta fusca*. Am. J. Primatol. 36, 213–222.

Crocker, M.J., 1998. Handbook of Acoustics. J. Wiley and Sons, Inc., New York, NY.

Da Cunha, R.G.T., Byrne, R.W., 2006. Roars of black howler monkeys (*Alouatta caraya*): evidence for a function in inter-group spacing. Behaviour 143, 1169–1199.

Da Cunha, R.G.T., Jalles-Filho, E., 2007. The roaring of southern brown howler monkeys (*Alouatta guariba clamitans*) as a mechanism of active defence of borders. Folia Primatol. 78, 259–271.

Fitch, W.T., 1997. Vocal tract length and formant frequency dispersion correlate with body size in rhesus macaques. J. Acoust. Soc. Am. 102, 1213–1222.

Fitch, W.T., Reby, D., 2001. The descended larynx is not uniquely human. Proc. R. Soc. Lond. B. 268, 1669–1675.

Frey, R., Hofmann, R.R., 2000. Larynx and vocalization of the takin (*Budorcas taxicolor* Hodgson, 1850 – Mammalia, Bovidae). Zool. Anz. 239, 197–214.

Frey, R., Gebler, A., 2003. The highly specialized vocal tract of the male Mongolian gazelle (*Procapra gutturosa* Pallas, 1777 – Mammalia, Bovidae). J. Anat. 203, 451–471.

Frey, R., Gebler, A., Fritsch, G., 2006. Arctic roars – laryngeal anatomy and vocalization of the muskox (*Ovibos moschatus* Zimmermann, 1780, Bovidae). J. Zool. 268, 433–448.

Frey, R., Gebler, A., Fritsch, G., Nygrén, K., Weissengruber, G.E., 2007a. Nordic rattle: the hoarse vocalization and the inflatable laryngeal air sac of reindeer (*Rangifer tarandus*). J. Anat. 210, 131–159.

Frey, R., Gebler, A., Olson, K.A., Odonkhuu, D., Fritsch, G., Batsaikhan, N., Stuermer, I.W., 2008. Mobile larynx in Mongolian gazelle – retraction of the larynx during rutting barks in male Mongolian gazelle (*Procapra gutturosa* PALLAS, 1777). J. Morphol. 269, 1223–1237.

Frey, R., Riede, T., 2003. Sexual dimorphism of the larynx of the Mongolian gazelle (*Procapra gutturosa* Pallas, 1777) (Mammalia, Artiodactyla, Bovidae). Zool. Anz. 242, 33–62.

Frey, R., Volodin, I., Volodina, E., 2007b. A nose that roars: anatomical specializations and behavioural features of rutting male saiga. J. Anat. 211, 717–736.

Ganzberger, K., Henninger, W., Kübber-Heiss, A., Forstenpointner, G., 1995. Das Diverticulum pharyngeum beim Eisbären (*Thalarctos maritimus*). Zool. Garten. N.F. 65, 237–244.

Gautier, J.-P., 1971. Étude morphologique et fonctionelle des annexes extra-laryngées des Cercopithecinae; liaisons avec les cris d'espacement. Biol. Gabon. 7, 230–267.

Gray, D.R., 1984. Dominance fighting in muskoxen of different social status. Biol. Pap. Univ. Alaska Spec. Rep. 4, 118–122.

Hauser, M.D., Schön-Ybarra, M., 1994. The role of lip configuration in monkey vocalizations: experiments using xylocain as a nerve block. Brain Lang. 46, 232–244.

Hayes, S.A., Kumar, A., Costa, D.P., Mellinger, D.K., Harvey, J.T., Southall, B.L., Le Boeuf, B.J., 2004. Evaluating the function of the male harbour seal, *Phoca vitulina*, roar through playback experiments. Anim. Behav. 67, 1133–1139.

Hill, W.C.O., Booth, A.H., 1957. Voice and larynx in African and Asiatic Colobidae. J. Bombay. Nat. Hist. Soc. 54, 309–321.

Kelemen, G., Sade, J., 1960. The vocal organ of the howling monkey (*Alouatta palliata*). J. Morphol. 10, 123–140.

Kingdon, J., 1984. East African mammals, an atlas of evolution in Africa. Insectivores and Bats, Vol. II, Part A. University of Chicago Press, Chicago, IL.

Kitchen, D.M., 2004. Alpha male black howler monkey responses to loud calls: effect of numeric odds, male companion behaviour and reproductive investment. Anim. Behav. 67, 125–139.

Kleinschmidt, A., 1961. Über die große mongolische Kropfgazelle (*Procapra gutturosa* Pallas, 1777). Stuttgarter Beiträge zur Naturkunde 79, 1–24.

Koppe, T., Moormann, T., Wallner, C.-P., Röhrer-Ertl, O., 2005. Extensive enlargement of the maxillary sinus in *Alouatta caraya* (Mammalia, Primates, Cebidae): an allometric approach to skull pneumatization in Atelinae. J. Morphol. 263, 238–246.

Lang, H., Chapin, J.P., 1917. Notes on the distribution and ecology of Central African Chiroptera. Bull. Am. Mus. Nat. Hist. 37, 479–563.

Luckhaus, G., 1969. Vergleichend-anatomische Betrachtungen über den Pharynx des Löwen. Zbl. Vet. Med., A. 16, 240–256.

Martens, K., Marler, P., 1977. Sound transmission and its significance for animal vocalization. I. Temperate habitats. II. Tropical forest habitats. Behav. Ecol. Sociobiol. 2, 271–302.

McComb, K.E., 1991. Female choice for high roaring rates in red deer, *Cervus elaphus*. Anim. Behav. 41, 79–88.

McComb, K.E., Packer, C., Pusey, A., 1994. Roaring and numerical assessment in contests between groups of female lions, *Panthera leo*. Anim. Behav. 47, 379–387.

McComb, K.E., Pusey, A., Packer, C., Grinnell, J., 1993. Female lions can identify potentially infanticidal males from their roars. Proc. R. Soc. London B, 252, 59–64.

McElligot, A.G., Hayden, T.J., 1999. Context-related vocalization rates of fallow bucks, *Dama dama*. Anim. Behav. 58, 1095–1104.

Packer, C., Herbst, L., Pusey, A.E., Bygott, J.D., Hanby, J.P., Cairns, S.J., Borgerhoff-Mulder, M., 1988. Reproductive success of lions. In: Clutton-Brock, T.H. (Ed.), Reproductive Success. University of Chicago Press, Chicago, IL, pp. 363–383.

Peters, G., Hast, M.H., 1994. Hyoid structure, laryngeal anatomy, and vocalization in felids (Mammalia: Carnivora: Felidae). Z. Säugetierk. 59, 87–104.

Pfefferle, D., West, P.M., Grinnell, J., Packer, C., Fischer, J., 2007. Do acoustic features of lion, *Panthera leo*, roars reflect sex and male condition? J. Acoust. Soc. Am. 121, 3947–3953.

Reby, D., McComb, K., 2003. Anatomical constraints generate honesty: acoustic cues to age and weight in the roars of red deer stags. Anim. Behav. 65, 519–530.

Reby, D., McComb, K., Cargnelutti, B., Darwin, C., Fitch, W.T., Clutton-Brock, T., 2005. Red deer stags use formants as assessment cues during intrasexual agonistic interactions. Proc. R. Soc. London 272, 941–947.

Riede, T., Tokuda, I.T., Munger, J.B., Thomson, S.L., 2008. Mammalian laryngeal air sacs add variability to the vocal tract impedance: physical and computational modeling. J. Acoust. Soc. Am. 124, 634–647.

Sanvito, S., Galimberti, F., 2000. Bioacoustics of southern elephant seals. I. Acoustic structure of male aggressive vocalisations. Bioacoustics 10, 259–285.

Sanvito, S., Galimberti, F., Miller, E.H., 2007a. Vocal signalling of male southern elephant seals is honest but imprecise. Anim. Behav. 73, 287–299.

Sanvito, S., Galimberti, F., Miller, E.H., 2007b. Having a big nose: structure, ontogeny, and function of the elephant seal proboscis. Can. J. Zool. 85, 207–220.

Schaller, G.B., 1972. The Serengeti Lion: A Study of Predator–Prey Relations. Chicago University Press, Chicago, IL.

Schneider, R., 1964. Der Larynx der Säugetiere. In: Helmcke, J.-G., Lengerken, H.V., Starck, D., Wermuth, H. (Eds.), Handbuch der Zoologie, 8. Band, 35. Lieferung, 5(7). de Gruyter, Berlin, Germany, pp. 1–128.

Schneider, R., Kuhn, H.-J., Kelemen, G., 1967. Der Larynx des männlichen *Hypsignathus monstrosus* Allen, 1861 (Pteropodidae, Megachiroptera, Mammalia). Z. Wiss. Zool. 175, 1–53.

Schön, M.A., 1970. On the mechanism of modulating the volume of the voice in howling monkeys. Acta Otolaryng. 70, 443–447.

Schön-Ybarra, M.A., 1988. Morphological adaptations for loud phonations in the vocal organ of howling monkeys. Primate Rep. 22, 19–24.

Smith, T.E., 1976. Reproductive behavior and related social organization of the muskox on Nunivak Island. Master's thesis, University of Alaska, Fairbanks, AK.

Southall, B.L., Schusterman, R.J., Kastak, D., 2003. Acoustic communication ranges for northern elephant seals (*Mirounga angustirostris*). Aquat. Mamm. 29, 202–213.

Starck, D., 1995. Lehrbuch der Speziellen Zoologie. G. Fischer, Jena, Stuttgart, New York.

Starck, D., Schneider, R., 1960. Respirationsorgane, A. Larynx. Primatologia 3, 423–587.

Terhune, J.M., Ronald, K., 1973. Some hooded seal (*Cystophora cristata*) sounds in March. Can. J. Zool. 51, 319–321.

Vannoni, E., McElligott, A.G., 2008. Low frequency groans indicate larger and more dominant fallow deer (*Dama dama*) males. PLoS One 3, e3113.

Watanabe, T., 1982. Mandible/basihyal relationships in red howler monkeys (*Alouatta seniculus*): a craniometrical approach. Primates 23, 105–129.

Weissengruber, G.E., Forstenpointner, G., Peters, G., Kübber-Heiss, A., Fitch, W.T., 2002. Hyoid apparatus and pharynx in the lion (*Panthera leo*), jaguar (*Panthera onca*), tiger (*Panthera tigris*), cheetah (*Acinonyx jubatus*) and domestic cat (*Felis silvestris* f.catus). J. Anat. 201, 195–209.

Weissengruber, G.E., Forstenpointner, G., Peters, G., Kübber-Heiss, A., Riedelberger, H., Ganzberger, K., 2001. Occurrence and structure of epipharyngeal pouches in bears (Ursidae). J. Anat. 198, 309–314.

Wickler, W., Seibt, U., 1976. Field studies on the African fruit bat *Epomophorus wahlbergi* (Sundevall), with special reference to male calling. Z. Tierpsychol. 40, 345–376.

Wiley, R.H., Richards, D.G., 1978. Physical constraints on acoustic communication in the atmosphere: implications for the evolution of animal vocalizations. Behav. Ecol. Sociobiol. 3, 69–94.

Zahavi, A., 1975. Mate selection: a selection for a handicap. J. Theoret. Biol. 53, 205–214.

Zahavi, A., 1977. The cost of honesty (further remarks on the handicap principle). J. Theoret. Biol. 67, 603–605.

Zahavi, A., 1981. Natural selection, sexual selection and the selection of signals. In: Scudder, G.G.E., Reveals, J.H. (Eds), Evolution Today. Carnegie-Mellon University Press, Pittsburgh, PA, pp. 133–138.

Zahavi, A., Zahavi, A., 1997. The Handicap Principle, a Missing Piece of Darwin's Puzzle. Oxford University Press, New York, NY.

Zeller, U., 1984. Zur Kenntnis des Stimmapparates der Epauletten-Flughunde Epomophorini, Pteropodidae, Megachiroptera). Z. Säugetierk. 49, 207–220.

CHAPTER 10.4

Generation of sound in marine mammals

Joy S. Reidenberg* and Jeffrey T. Laitman

Center for Anatomy and Functional Morphology, Mount Sinai School of Medicine, New York, NY, USA

Abstract: The mechanisms for sound production in marine mammals utilize pneumatically-driven vibrations of anatomical structures derived from those of terrestrial mammals. As marine mammals inhabit a liquid environment, these structures have evolved to accommodate new protective and sound generating/transmitting functions in this medium. For example, while two marine mammal species (polar bears and sea otters) produce sounds in air only, members of all other marine mammal orders (Pinnipedia, Sirenia, and Cetacea) can produce sounds underwater. Marine mammals use the vocal folds of the larynx for generating most or all of the initial sound vibrations. One cetacean group (toothed whales) generates echolocation sounds by vibrating structures in the nasal region. Many marine mammals vocalize underwater without releasing air by capturing airflow from the lungs into collapsible/expandable internal reservoirs. These reservoirs recycle air back to the lungs, allowing multiple or long vocalizations without needing to resurface to breathe. The various tracheal, pharyngeal, laryngeal and nasal air sacs probably evolved as diving adaptations, enabling availability of extra air reserves at depth.

Keywords: sound production; sound generation; vocalization; marine mammal; pinniped; sirenian; cetacean; odontocete; mysticete; larynx; nasal; air sacs; phonic lip

I. Introduction

Marine mammals exist at an evolutionary crossroads: they bear traits inherited from their terrestrial ancestors, while expressing new traits that adapt them to living in water. Communication is one of many behaviors that were affected by the evolutionary transition from a terrestrial to an aquatic lifestyle. Communication may occur through various modes (e.g., visual, tactile, olfactory, acoustic). Some modes work well underwater, while others work better in air.

Visual communication only works well under conditions of good visibility (i.e., transparent media, daylight). On land, visual communication is generally limited to daylight hours and between individuals within a relatively close range. In water, vision is even more restricted, because it cannot easily occur under conditions in which light penetration is reduced due to depth or suspended particulate matter. Tactile communication is only possible if the subjects stay in close proximity to each other (e.g., between slowly swimming individuals or when clustered together on a beach). Unfortunately, touch communication is not effective in conditions where marine mammals are moving at great speeds and are spaced apart. Olfactory communication can be a primary means of conveying a host of information, ranging from territorial boundaries to mating status. However, as most mammals are adapted for smelling in air, olfaction has little or no use in water and is thought to be greatly reduced or even absent in many marine mammals (Lowell and Flanigan, 2008). Terrestrial mammals can use the tongue for liquid chemoreception (taste), but it is also unclear how widespread this sense is in marine mammals (Lowell and Flanigan, 2008). Perhaps taste sensation is reduced in marine mammals that generally swallow their food whole. That may leave only one sense that is actually enhanced in water: acoustic communication (vocalizing and hearing). This is because sound travels faster and farther in water than in air, due to the denser medium of liquid compared with gas.

*Corresponding author. E-mail: joy.reidenberg@mssm.edu

ISBN 978-0-12-374593-4

DOI: 10.1016/B978-0-12-374593-4.00041-3

The evolutionary transition from land to sea involved modifications of both the vocalizing and hearing apparati of marine mammals. Structures originally fit for acoustic communication in air during a terrestrial existence were subjected to evolutionary selection in a semiaquatic or fully-aquatic environment. While some marine mammals have retained the ability to produce and receive sound in air, others evolved changes that allow sound production and reception underwater. This chapter will focus on the various mechanisms of sound generation in marine mammals.

II. Evolution from the terrestrial to the aquatic habitat

Aquatic sound production in marine mammals is the result of evolutionary modifications to the respiratory tract, including the larynx. It is imperative to realize that the primary function of the larynx is not, as is commonly held, as a "voice box," but rather as the protective guardian of the lungs (Reidenberg and Laitman, 2005). Indeed, the vocal folds (Fig. 1), a key element as the initial generator of sounds (fundamental frequencies) used in vocal communication, evolved as the primary protector against the incursion of foreign material into the trachea and lungs (Laitman and Reidenberg, 1997; Lipan et al., 2006). Additional roles evolved for these folds, including changing and maintaining intrathoracic and intra-abdominal pressures, activities essential for physiologic functions ranging from defecation, to lifting an object, to childbirth (Laitman and Reidenberg, 2009). Indeed, the complex central and peripheral neurosensory and neuromotor feedback loops to/from the larynx have evolved for precise monitoring of this organ. Second – or perhaps in tandem – with these basic biological functions, emerged the "role" of the vocal folds, its laryngeal housing and extra-laryngeal surroundings, in sound generation (Laitman and Reidenberg, 1993). While sound production is often cited as the major function of the vocal folds, it is prudent to also remember their evolutionarily older functions in protection and intrathoracic/abdominal pressure regulation.

Increased protection from aquatic incursions was likely a major factor in the evolution of the marine mammal respiratory tract. As air breathing was preserved in the transition from a fully-terrestrial to an amphibious or fully-aquatic habitat, respiratory modifications that conferred additional protection were likely retained through natural selection in the evolution of marine mammals. For example, the nostrils changed in shape and/or position to increase protection from drowning and facilitate breathing while swimming (e.g., closable slit-shaped nasal openings or nasal plugs, nasal opening positioned at the top of the head in whales).

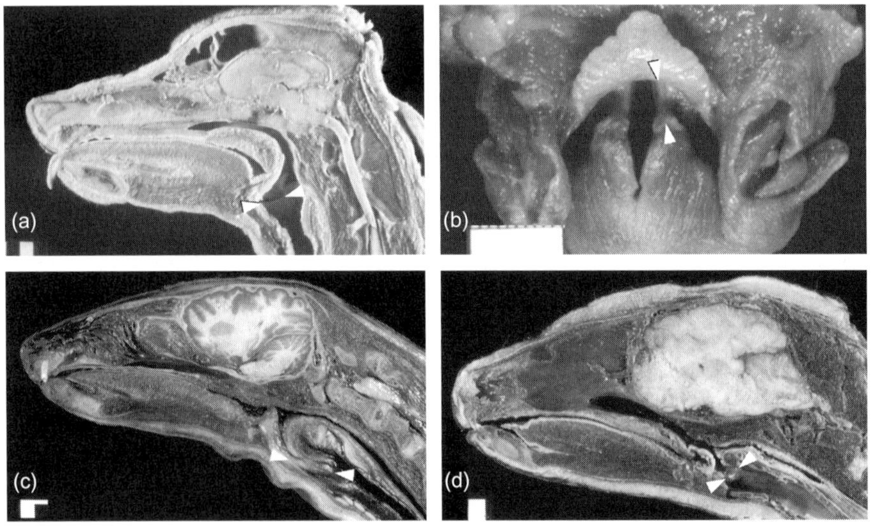

Fig. 1. (a) Midsagittal section through the head of an adult dog, *Canis familiaris*, showing the right side of the head. (b) Dorsal view of the larynx of an adult sea otter, *Enhydra lutris*. The rostral aspect is at the top of the image. (c) Midsagittal section through the head of a juvenile California sea lion, *Zalophus californianus*, showing the right side of the head. (d) Midsagittal section through the head of a sub-adult harbor seal, *Phoca vitulina*, showing the right side of the head. The vocal fold of the right side is indicated between the two white arrowheads for each specimen. White square or rectangle indicates 1 cm in each figure. (To view this figure in color please visit http://www.elsevierdirect.com/companions/9780123745934)

The pneumatic sound generating mechanism present in terrestrial mammals is also used in marine mammals, indicating this is a shared (and therefore ancestral) characteristic. Essential laryngeal functions (e.g., protection, pressure maintenance) were also preserved, and in some species (e.g., toothed whales) the anatomy is modified to provide increased laryngeal protection during swallowing (Reidenberg and Laitman, 1987). Vocal fold function in sound generation is retained in some marine mammals, particularly those that vocalize in air. Underwater vocalizations, however, are accomplished by a variety of unique respiratory tract modifications found in marine mammals that also appear to relate to diving adaptations (Reidenberg and Laitman, 2008).

Understanding the variety of sound production mechanisms employed begins with appreciating that not all marine mammals have the same behavioral lifestyle. Some marine mammals, such as polar bears and pinnipeds (seals, fur seals, sea lions, walruses) have an amphibious lifestyle (i.e., they regularly spend considerable amounts of time out of the water on a dry surface). Other marine mammals, such as sea otters, sirenians (manatees and dugongs) and cetaceans (porpoises, dolphins and whales) have a fully-aquatic lifestyle. These differences affect whether they produce sounds in air, in water, or both.

III. Non-vocal sound production

Marine mammals generate sound in several ways, many of which are unrelated to true "vocalizations," i.e., sounds generated via a pneumatic mechanism involving vibration of the vocal folds of the larynx. Non-respiratory tract sound generation is accomplished often in conjunction with visual displays, such as locomotor movements on land or in water, striking a body part (flippers, flukes) against the water, or breaching clear of the water and crashing the body against the surface during re-entry. Some displays do make use of the respiratory tract, but the signals are largely visual in the form of air releases. For example, displays of bubbles are released underwater from the nose or mouth by pinnipeds (Ray et al., 1969; Hewer, 1974; Møhl et al., 1975; Hanggi and Schusterman, 1992, 1994), humpback whales (*Megaptera novaeangliae*) (Tyack and Whitehead, 1983; Reidenberg and Laitman, 2007a) and dolphins (Caldwell and Caldwell, 1972; McCowan and Reiss, 1995; Herzing, 1996). It is not clear whether sounds (e.g., whistles from dolphins) associated with these air releases are coincidental or an essential acoustic component of the display.

IV. Vocalizations in air by "amphibious" marine mammals

All of the "amphibious" marine mammals (i.e., those that regularly spend time on land) and some of the exclusively aquatic marine mammals are capable of vocalizing in air. Most of these vocalizations are calls between pups and mothers, or the aggressive sounds of competing males. For example, California sea lions (*Zalophus californianus*) make a variety of laryngeal sounds including a loud bark (Poulter, 1965). Harbor seals (*Phoca vitulina*) make roars and bubbly growls (Hanggi and Schusterman, 1992). Monk seals (*Monachus schauinslandi*) can produce threat roars, vocalizations called "bubble sounds" and a guttural sound on expiration (Miller and Job, 1992; Tyack and Miller, 2002). Harp seals (*Pagophilus groenlandicus*), gray seals (*Halichoerus grypus*) and hooded seals (*Cystophora cristata*) can make roars, growls and moans (Ballard, 1993 cited in Thomas and Golladay, 1995; Miller and Murray, 1995). Walrus (*Odobenus rosmarus*) pups produce in-air vocalizations (Kastelein et al., 1995) and the adult walrus can whistle (Verboom and Kastelein, 1995). Polar bears (*Ursus maritimus*) generate sounds that, as in many terrestrial mammals, are emitted orally (Wemmer et al., 1976). Sea otters (*Enhydra lutris*) have been described as capable of emitting a range of in-air vocalizations, including cry, scream, whistle/whine, coo, snarl/growl, hiss, grunt and bark (Kenyon, 1969).

Vocalizations in air by pinnipeds, polar bears and sea otters are generated by tissue vibrations caused by air movements within the respiratory tract. The source of most of the orally-emitted and some of the nasally-emitted vocalizations appears to be vocal fold vibrations in the larynx (see Miller and Murray, 1995 for review of glottal pulse rates recorded for various seal species). These fundamental frequencies are then filtered through the supralaryngeal vocal tract (pharyngeal, oral and nasal cavities). Movements of the pharyngeal walls, soft palate, tongue, lips and in some species the flexible nasal membranes, all affect the qualities of the sound produced in air. Both pinnipeds and polar bears can also generate chuffs or snorts. These sounds may be the result of forceful exhalations against narrow nasal passageways.

Pinniped vocalizations in air are enhanced by supralaryngeal modifications. An expanded nasal chamber, such as is found in the male elephant seal (*Mirounga angustirostris*), may add additional resonance while providing a prominent visual display.

In the male hooded seal, the membranous nasal septum is expanded beyond the nasal chamber and extruded out through the nostril (Berland, 1966; Ballard and Kovacs, 1995). This not only dramatically increases the resonant space of the vocal tract, but it also provides a dramatic visual breeding display. The sac may also function as a drum, transmitting vibrations in air. Larger larynges with longer vocal folds produce lower sounds and larger resonant spaces may amplify sounds (see Frey and Gebler, Chapter 10.3 in this volume). Loud low-frequency sounds travel farther, and thus these traits likely advertise larger overall body size in males. Females, however, may produce higher frequency and softer vocalizations to be used for close-range communication with pups.

Polar bears, sea otters, seals, fur seals and sea lions all have vocal folds in their larynx (Fig. 1). The vocal folds vary in their length, thickness and orientation, and these differences may correlate with different voice qualities (e.g., variation in frequency ranges). It is interesting to note that while the vocal folds of a sea otter or a harbor seal are very similar to those of a typical land mammal such as a dog, those of a California sea lion are placed ventrally within the laryngeal lumen. The sea lion, however, has large arytenoid cartilages. These paired cartilages may oppose each other and function essentially as the vocal folds in forming a thick and strong valve, thereby regulating airflow. It is unclear whether the thickness of these cartilages contributes to voice quality (thicker folds usually result in lower frequencies), as it is not known if the cartilages are capable of vibrating separately from the vocal folds (or at all) to generate a fundamental frequency.

V. Pinniped vocalizations underwater

California sea lions are able to make multiple simultaneous sounds, which indicate that they may have more than one sound generator (Brauer et al., 1966). Interestingly, the laryngeal lumen is expanded in the sea lion, compared with seals, and may function as a resonant space for vibrating laryngeal cartilages or vocal folds. Sea lions make a variety of vocalizations that include pulsed sounds underwater (Richardson et al., 1995). It was thought that these pulses might represent evidence of echolocation in sea lions (Shaver and Poulter, 1967), but behavioral evidence and controlled experimental studies showed that they did not use sound to navigate (Schevill, 1968; Schusterman et al., 2000).

Otariid pinnipeds (sea lions, fur seals), polar bears and sea otters use sounds mainly in air, while phocid pinnipeds (seals) primarily vocalize underwater (Schusterman et al., 2000; Tyack and Miler, 2002). However, all pinnipeds are capable of vocalizations both in air and underwater. While the mechanism of sound generation in air is probably similar to terrestrial mammals (i.e., pneumatic processes involving laryngeal vibration and air release), the mechanism and mode of transmission of underwater sounds is still elusive, as the anatomy of the sound producing/transducing organs is not well-understood (Miller and Murray, 1995).

Many marine mammals can vocalize underwater without releasing air, and therefore must have collapsible/expandable internal air reservoirs. Sea lions, for example, can bark underwater without releasing air (Brauer et al., 1966). This indicates that there may be supralaryngeal modifications of the upper respiratory tract that allow air recapturing and recycling, such as an expandable supralaryngeal reservoir. Without a flexible reservoir to receive the air used in pneumatic sound generation, a "closed" vocal tract (i.e., no air release) with rigid walls would quickly pressurize. Supralaryngeal pressurization would prevent airflow through the larynx, and therefore cause sound production to stop. Capturing this air, and subsequently contracting the reservoir to recycle it back to the lungs, would allow multiple vocalizations and the ability to vocalize for a long period without needing to resurface to breathe. The pathway for underwater sound transmission is unknown. It is possible that laryngeal vibrations are transmitted through the surrounding neck tissues. This may be a sexually dimorphic ability, as the girth of the neck is much thicker in males than in females. If the cervical fat layer has a density similar to seawater, then it may function as a transducer, allowing propagation of laryngeal vibrations to water. The ability to clearly transmit a loud bark underwater through the neck fat may allow males to advertise their size vocally.

The walrus also vocalizes underwater (Schevill et al., 1966; Stirling et al., 1987). Walruses have a specialized pharyngeal pouch located lateral and superior to the larynx (Fay, 1960). While it is unclear how this sac functions, it is possible that it is used for air recapture and recycling during laryngeal vocalizations to prevent air release underwater. The pharyngeal pouch may also provide resonance, perhaps facilitating an unusual gong or bell sound that is unique to walruses (Fay, 1960). The thick fatty neck may function similarly as proposed for sea lions, perhaps enabling an underwater sound transmission pathway for laryngeal vibrations to reach seawater.

Seals are perhaps the most vocal of the pinnipeds in producing underwater sounds. The Weddell seal (*Leptonychotes weddelli*), for example, has been studied for its elaborate underwater vocal repertoire, which are often referred to as "songs" (Green and Burton, 1988; Morrice et al., 1994). The leopard seal (*Hydrurga leptonyx*) has a large repertoire of underwater calls (Rogers et al., 1995). Harp seals are also particularly vocal underwater (Møhl et al., 1975). It is unclear whether seal vocal folds function for vocalizations only above the water's surface, or if they also provide the fundamental frequencies for underwater vocalizations. It is interesting to note that some seal species have expanded infralaryngeal air spaces. The trachea is often widened, with expanded membranes and flexible (compressible) tracheal cartilages that may facilitate rapid inspiration/expiration at the surface (Kooyman and Andersen, 1969; Bryden and Felts, 1974; Ray, 1981; Kooyman, 1981). The bearded seal (*Erignathus barbatus*) has an enlarged dorsal membrane in the lower tracheal region (Solokov et al., 1970). In the ribbon seal (*Histriophoca fasciata*), the trachea expands into one very large, extracostal (outside the ribcage) sac on the right side that is regulated by a valve near the carina (Abe et al., 1977; Burns, 1981).

The expanded seal trachea, enlarged dorsal membrane and sac expansions are probably diving adaptations, providing additional air reservoirs. Diving to extreme depths causes compression of the volume of the air in the respiratory tract. The ability to carry additional air reserves would allow the seal to dive longer before depleting oxygen stores, and would also prevent complete lung collapse as this additional air is added to the total lung volume at depth.

These tracheal specializations may subserve additional functions besides diving. Vibration of the dorsal tracheal membrane or tracheal sac may either generate sound or facilitate transmission of sounds underwater (Pierard, 1969; Ray et al., 1969; Bryden and Felts, 1974; Ray, 1981). It is unclear, however, how seals generate and regulate vibration of these structures. It is possible that vibrations of the vocal folds are passed from the larynx to the dorsal tracheal membrane through a soft tissue connection. This mechanism might explain the unusual "comb calls" of leopard seals (Thomas and Golladay, 1995). Alternatively, pulsatile movements of air from rapid diaphragm contractions might cause the dorsal membrane or tracheal sac to vibrate like a drum. The valve that regulates the opening into the tracheal sac may also generate vibrations. The enlarged tracheal spaces may facilitate the production of louder sounds. Tracheal sac size is sexually dimorphic in ribbon seals, with the sac being larger in males. It is therefore likely that the tracheal sac is used for mating calls or advertising size to other competing males. The tracheal sac probably remains expanded near the surface, where ambient pressure is low, thus providing a large resonating chamber.

The various parapharyngeal and paratracheal air reservoirs of pinnipeds are likely the key to underwater sound production without air release. The ability to keep air flowing during sound production is problematic while submerged, because the nose and mouth are generally closed to prevent drowning. The length of a vocalization is dependent on the movement of air causing structures to vibrate, but would be quickly arrested in a closed system of fixed volume with rigid walls as this arrangement provides little or no air movement. The flexible walls of the air sacs can allow air to flow from a region of high pressure (the contracting lungs) to a region of low pressure (the expanding sacs). However, as the pressure gradient between the lungs and sacs equalizes, air ceases to flow and sound production stops. A quick reversal of airflow may enable sound production to continue nearly uninterrupted, as long as vibrations could be generated by both ingressive (towards the lungs) and egressive (towards the sacs) air movement (Reidenberg and Laitman, 2008).

VI. Sirenian vocalizations

Sirenians are fully-aquatic marine mammals. As such, they do not appear to make sounds in air, but they are capable of underwater vocalizations (Schevill and Watkins, 1965). Sirenian vocalizations are likely made with the larynx (Reynolds and Odell, 1991). The West Indian manatee (*Trichechus manatus*) has a relatively limited vocal repertoire (Nowacek and Casper, 2003), which may be related to the poorly-developed vocal folds (Fig. 2). No air is emitted during underwater vocalizations (Hartman, 1971). This indicates that the sound transmission pathway is likely through the surrounding neck tissues. Manatees have an area of increased cervical blubber fat along the ventral neck (Fig. 2). It is likely that this fat serves as an acoustic window through which laryngeal sounds may be transferred to the surrounding water (Reidenberg and Laitman, 2002).

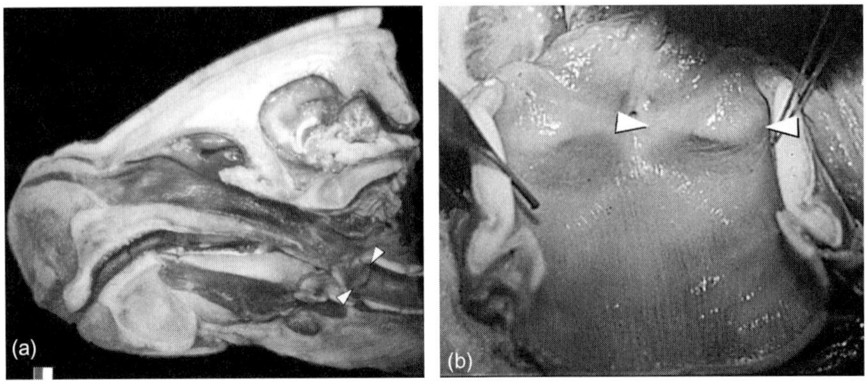

Fig. 2 (a) Midsagittal section through the head of an adult West Indian manatee *Trichechus manatus*, showing the right side of the head. The right vocal fold of the larynx is indicated between the two white arrowheads. The fat pad located ventral to the larynx (lower right corner of image) is thicker than the dorsal blubber fat. White square indicates 1 cm. (b) Dorso-caudal view of the larynx of an adult West Indian manatee *Trichechus manatus*. The rostral aspect is at the top of the image. The larynx is cut dorsally through the cricoid cartilage, and the cut edges are spread laterally by forceps to reveal lumen. Note the two blunt-edged vocal folds. The vocal fold of the right side is indicated between the two white arrowheads for each specimen. (To view this figure in color please visit http://www.elsevierdirect.com/companions/9780123745934)

VII. Cetacean sound generation

Cetaceans, like sirenians, are also fully-aquatic marine mammals. Unlike sirenians, however, cetaceans make very complex underwater sounds. As in other marine mammals, cetaceans generate sound through pneumatically driven vibrations of upper respiratory tract structures (Cranford et al., 1996; Cranford, 1999; Aroyan et al., 2000; Reidenberg and Laitman, 2007). An air space adjacent to phonic tissues, such as an air sac, allows vibration of these structures (Reidenberg and Laitman, 2008). In terrestrial mammals, the tissue-conducted vibrations attenuate, while the air-borne pressure waves are what translate into the audible sounds in air. Although airflow also drives the system in cetaceans, it is the energy of the vibrating tissues that must be transferred to water as sound rather than the pressure waves in air. Cetaceans have multiple potential sites for sound generation, which differ between mysticetes (baleen whales) and odontocetes (toothed whales, including dolphins and porpoises).

The two cetacean suborders, Mysticeti and Odontoceti, exhibit divergent anatomy of their upper respiratory tract structures. Although both groups possess elements of the basic terrestrial mammalian configuration, similarities and differences in their upper respiratory tract anatomy support the hypothesis that cetaceans are monophyletic, but diverged along different evolutionary paths (Uhen, 2007). Cetacean larynges share some unique characteristics, particularly a ventrally-incomplete cricoid cartilage, rostrally-elongated corniculate cartilages and ventrocaudally elongated vocal processes of the arytenoid cartilages (Reidenberg and Laitman, 1987, 1988, 1994, 2007a,b). Mysticetes continue to use the larynx for sound generation, but have specializations that enable long, loud, low-frequency sound generation and transmission. Odontocetes, however, exhibit unique modifications of both the laryngeal and nasal regions, and are known for emitting a variety of sounds including whistles and high-frequency pulsed sounds used for echolocation.

VII.A. Mysticete sound generation

Mysticetes are well-known for their loud low-frequency calls. The humpback whale, in particular, is famous for the complexity of its songs (Payne and McVay, 1971; Payne et al., 1983; Payne and Payne, 1985). Unfortunately, the anatomy and physiology of mysticete sound production is very difficult to study, due to the impracticality of keeping a live specimen in captivity. Most information is derived from postmortem dissections and hydrophone recordings that capture the range and complexities of mysticete vocalizations. A recent study by Potter et al. (2007) showed that the source of emitted vocalizations in live singing male humpback whales originates in the ventral and lateral throat region. This is consistent with the location of the mysticete larynx and supports it as the source of underwater vocalizations (Reidenberg and Laitman, 2007b).

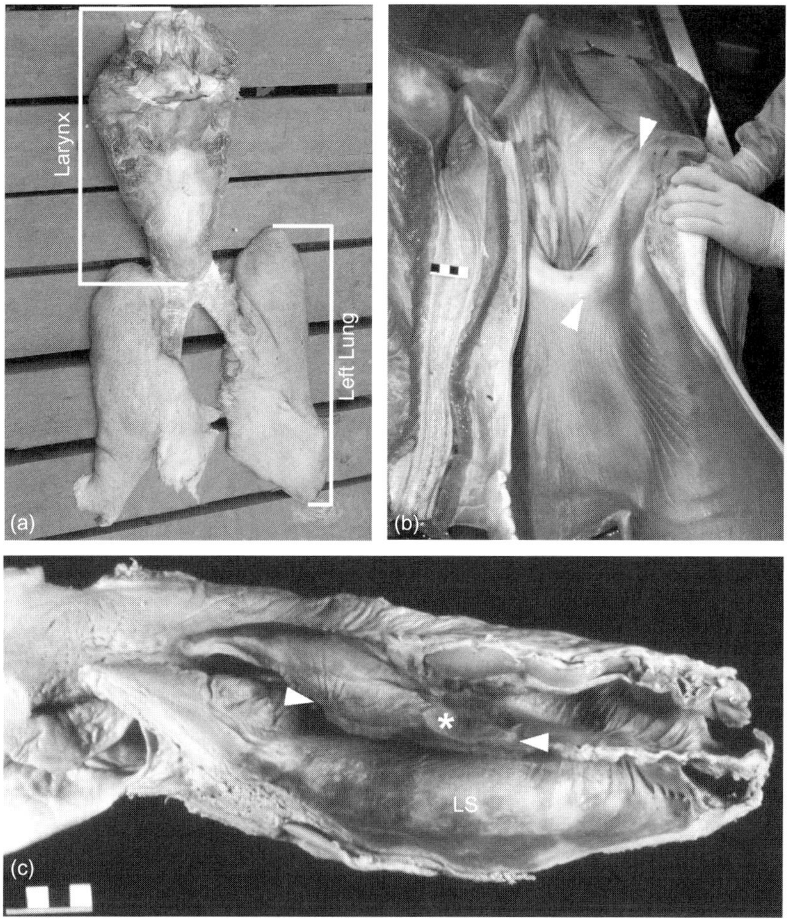

Fig. 3 (a) Ventral view of the larynx, trachea and lungs of a juvenile minke whale, *Balaenoptera acutorostrata*. Note that the larynx (top structure) is larger than either lung (bottom two structures). Most of the mass of the larynx, in this view, is comprised of the laryngeal sac. For size reference, each beam of wood in the background is approximately 30 cm wide. (b) Dorsocaudal view of the larynx of a juvenile humpback whale, *Megaptera novaeangliae*. The rostral aspect is at the top of the image. The larynx is cut dorsally through the cricoid cartilage, and the right side is retracted by the gloved hands to reveal the lumen. Note the right portion of the U-shaped fold, indicated between the white arrowheads. Each white or black square on the scale bar = 1 cm. (c) Midsagittal section through the larynx of a juvenile minke whale, *Balaenoptera acutorostrata*. The cut edge of the U-shaped fold (cut at the bottom of the "U") is indicated by the asterisk. The right portion of the U-shaped fold is indicated between the white arrowheads. LS: laryngeal air sac. Each white or black square on the scale bar = 1 cm. (To view this figure in color please visit http://www.elsevierdirect.com/companions/9780123745934)

The size of the mysticete larynx is absolutely and relatively large, compared with that of other mammals. Even in a juvenile specimen, it is noted to be larger than either one of the whale's lungs (Fig. 3). The overall size is an indication of the function of this organ as a producer of loud and low-frequency sounds, as increased size of vocal folds and resonant spaces correspond to longer wavelengths and increased amplification (Aroyan et al., 2000). As a larger whale would likely have a larger larynx, the quality of the sound emitted may be an advertisement of body size and therefore of fitness for breeding (Tyack and Clark, 2000). Vocal advertising may be an important component of baleen whale behavior, particularly for solitary males that may be separated from other conspecifics by long distances, and therefore out of range for visual assessment.

The mysticete larynx contains a U-shaped fold that is homologous to the vocal folds of terrestrial mammals (Reidenberg and Laitman, 2007b). The U-shaped fold is a very large structure comprised of two parallel bar-like portions that meet caudally at the bottom of the "U." These long and thick bars are each supported by an arytenoid cartilage. The U-shaped fold is controlled by muscular contractions that move the arytenoids at their synovial joint with the cricoid cartilage. The bars of the U-shaped fold may be opposed towards each other to regulate airflow through the gap between them. These bars may vibrate in a manner similar to terrestrial vocal folds, causing generation of a fundamental frequency. Their length and thickness likely corresponds to the low-frequency range in which most mysticete sounds are generated.

Adduction/abduction and elevation/depression of the U-shaped fold may control the rate and direction of airflow, as well as tension of the vibrating tissue. Interestingly, the U-shaped fold is oriented parallel to the long axis of the larynx and trachea, rather than perpendicular to it as in other mammals (Reidenberg and Laitman, 2007b). This means that air may be diverted from the inspiratory/expiratory path to flow through the gap in the U-shaped fold. Air flowing through the gap in the U-shaped fold enters a large diverticulum, called the laryngeal air sac.

The mysticete's laryngeal air sac appears to be derived from midline fusion of the paired, lateral laryngeal ventricles, which are present above the vocal folds in most terrestrial mammals (Reidenberg and Laitman, 2007b). The laryngeal sac has a flexible wall and is surrounded by a large layer of muscle. The mysticete laryngeal sac may be used for diving and sound production needs in a manner similar to that of the pharyngeal and tracheal air sacs of pinnipeds.

Laryngeal air sacs may provide an important air reservoir for increasing total air volume available during diving (Reidenberg and Laitman, 2008). Air in the lungs is compressed at depth – gas volume is halved for each doubling of ambient pressure. Filling this reservoir at the surface enables baleen whales to add this "extra" air to the total volume in the compressed respiratory spaces at depth, thereby perhaps preventing total lung collapse or ear damage. The extra volume also provides additional oxygen, and therefore may extend breath-hold time (i.e., the time available for diving between breaths).

As many mysticete vocalizations occur at depth, the ability to replenish air used would require a trip back to the surface. Mysticetes can conserve the limited quantity of air in the respiratory tract by vocalizing without air release at depth. The laryngeal sac provides an air recapture and recycling reservoir to facilitate airflow between it and the lungs during vocalizations (Reidenberg and Laitman, 2008). The flexibility of the sac allows a pressure gradient to exist and be reversed, thereby allowing air to flow back and forth from the lungs. The flow (and perhaps reversal of flow) across the U-shaped fold likely causes vibrations, and therefore is a critical component of sound generation. Without this gradient (i.e., if the sac were a rigid space similar to a cranial sinus), pressure in the two spaces would equalize and airflow (and therefore sound production) would stop.

When inflated, the mysticete laryngeal sac and contiguous nasal passageways may also serve as a resonant space, with larger volumes resulting in longer, louder and lower sounds. The ability to expand and maintain a large space, even under high pressure while diving, may be an advertisement of absolute size and strength. Some mysicetes (e.g., blue whales, *Balenoptera musculus*) may use the laryngeal sac and nasal passageways as a Helmholtz resonator (monopole source oscillations of an air-filled cavity). This model proposes that the change in pressure during diving is the force that drives compression of the pliant membranes of the lungs, and forces air to flow through an oscillating laryngeal valve (Aroyan et al., 2000). This implies that a constant frequency is maintained over changing depths, allowing monotonic sound production. This matches the blue whale type "B" calls: 17 Hz, 10 seconds, 185 dB re 1 μPa, assuming a volume displacement of 1,000 liters and a time-averaged depth of 30 meters (Aroyan et al., 2000). Other vocalizations (e.g., pulsive type "A" calls, shorter duration calls and higher frequency calls produced by other mysticete species) could be produced by laryngeal sac contractions alone, with air recycling between calls or call segments.

Laryngeal sac walls may propagate vibrations generated by the oscillating valve (movements of the U-shaped fold). Contraction of the laryngeal sac at depth may change its resonant properties or sound transference abilities. As the sac is located immediately under the flexible throat skin (which has expandable pleats in the rorqual whales), sac vibrations are likely transferred to the skin and then to water with relatively little energy loss (Reidenberg and Laitman, 2007b). A sound transmission pathway through these overlying tissues is supported by triangulated hydrophone and video recordings of singing whales that localized the points of sound emission to the ventral and lateral cervical region (Potter et al., 2007).

VII.B. Odontocete sound generation

The topic of odontocete sound generation has gained much interest since hydrophone recordings and controlled experiments with captive dolphins revealed that they are capable of echolocation (Kellogg, 1958;

Norris et al., 1961). Some researchers championed the larynx as the sound source (Purves, 1966; Purves and Pilleri, 1978, 1983; Reidenberg and Laitman, 1987, 1988), while others claimed the nasal region was involved (see reviews in Mead, 1975; Heyning 1989; Aroyan et al., 2000; Cranford, 2000; Cranford and Amundin, 2003). Controlled experimental work on live captive specimens (Dormer, 1979; Mackay, 1980; Ridgway et al., 1980; Cranford et al., 1997) and exquisite anatomical studies (Cranford et al., 1996; Cranford, 1999, 2000; Cranford and Amundin, 2003; Cranford et al., 2008a) have shown that the nasal region is primarily responsible for odontocete sound generation, and that the path of transmission is through the specialized structures of the forehead region. The "unified hypothesis" of sonar-signal production (Cranford et al., 1996) states that all odontocetes use a similar mechanism of homologous nasal structures to produce echolocation. This mechanism, described in detail by Cranford et al. (1996), is summarized below.

During exhalation, air passes superiorly through the bony nares, is then channeled through the soft tissue component of the external nasal passageways of the odontocete's forehead region and merges into one chamber that exits through a single nostril (blowhole) at the top of the head. The width of the passageway is regulated by contraction of a pair of thick, muscular tissues called nasal plugs. At rest, the nasal plugs cover the external openings of the paired bony nares. Between the nasal plugs and the blowhole are a number of complex structures (Fig. 4), including usually three pairs of air sacs and two pairs of fat-filled structures associated with cartilage, known as "phonic lips" (Cranford, 1999), also known as "museau de singe" (Clarke, 1978) or "MLDB/monkey-lip dorsal bursae complex" (Cranford et al., 1996). Each phonic lip consists of a shelf of tissue that tapers to a free border.

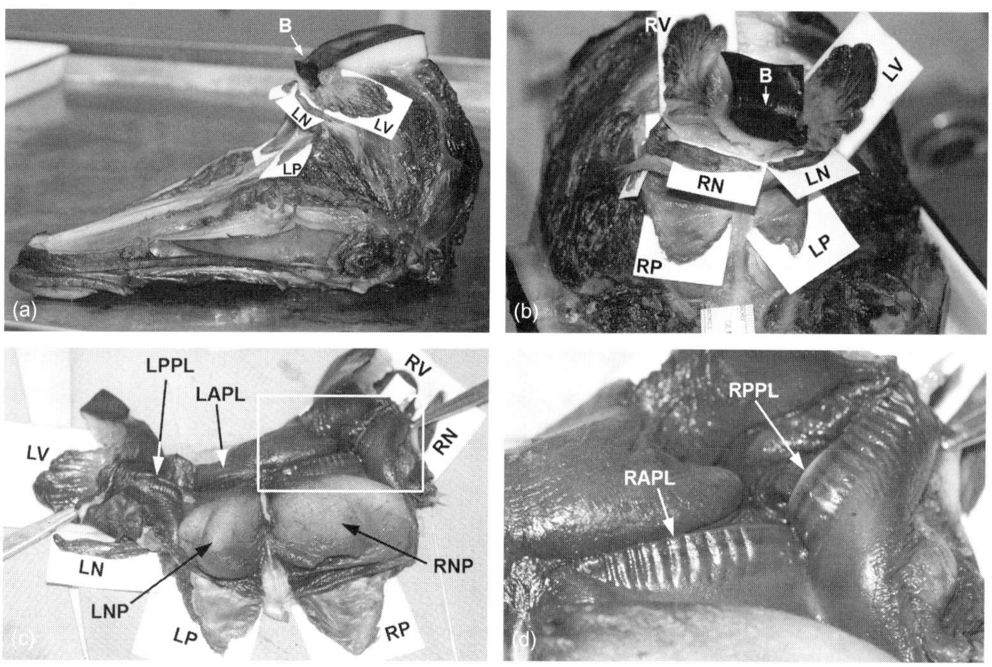

Fig. 4. (a) Lateral-oblique view of left side of the head of an adult Atlantic white-sided dolphin, *Lagenorhynchus acutus*, that has been dissected to reveal the three pairs of nasal air sacs. Each air sac has been highlighted by the white 3×5 inch index card placed under it. (b) Frontal view of the rostral aspect of the same specimen shown in (a). (c) Ventral view of the nasal sac complex, dissected off the skull of the same specimen shown in (a) and (b). Note the two pairs of phonic lips, which have a crinkled appearance. The posterior phonic lip has been separated from the anterior lip at the midline on both sides and each is reflected laterally by a hemostat. (d) Enlargement of region indicated by the white box in image (c). Note the crinkled appearance of the surfaces (likely due to relaxation after severing medial connection of posterior phonic lip). Each phonic lip has a shelf of tissue that tapers to a thin edge at its free border. B: blowhole opening; LAPL: left anterior phonic lip; LN: left nasofrontal air sac; LNP: left nasal plug; LP: left premaxillary air sac; LPPL: left posterior phonic lip; LV: left vestibular air sac; RAPL: right anterior phonic lip; RN: right nasofrontal air sac; RNP: right nasal plug; RP: right premaxillary air sac; RPPL: right posterior phonic lip; RV = right vestibular air sac (To see the full color version of this figure please refer to the color plate in the back of the book. Copies produced via our print on demand service do not contain color plates. If your copy does not have the color plate, please go to this website to view the figure in color http://www.elsevierdirect.com/companions/9780123745934).

It is thought that during underwater sonar signal production, the blowhole is closed while air is directed between the two overlying pairs of phonic lips (one for each nasal passage), causing them to part rapidly and reoppose. The air is captured in the various nasal sacs surrounding the nasal passageway under the blowhole (Fig. 4). Expansion of these sacs allows air to continue to flow past the phonic lips until the volume of air to be used is exhausted, or the nasal region becomes pressurized. The nasal sacs capture the air so that it is not released through the blowhole and lost. Contraction of the muscles covering these sacs may force the air to flow in reverse, causing more vibration of the phonic lips and/or recycling of the air to be used again in another sound production event. During phonic lip oscillations, the posterior phonic lip strikes against the anterior phonic lip. As the anterior phonic lip is attached to the melon (a large fatty structure in the odontocete forehead), these pulsed vibrations are transmitted directly to and through this tissue (Fig. 5). The melon has a density approaching that of seawater, and thus provides a good medium for transferring the pulsed sound energy to water without much reflection or attenuation. The melon may act as a filter or

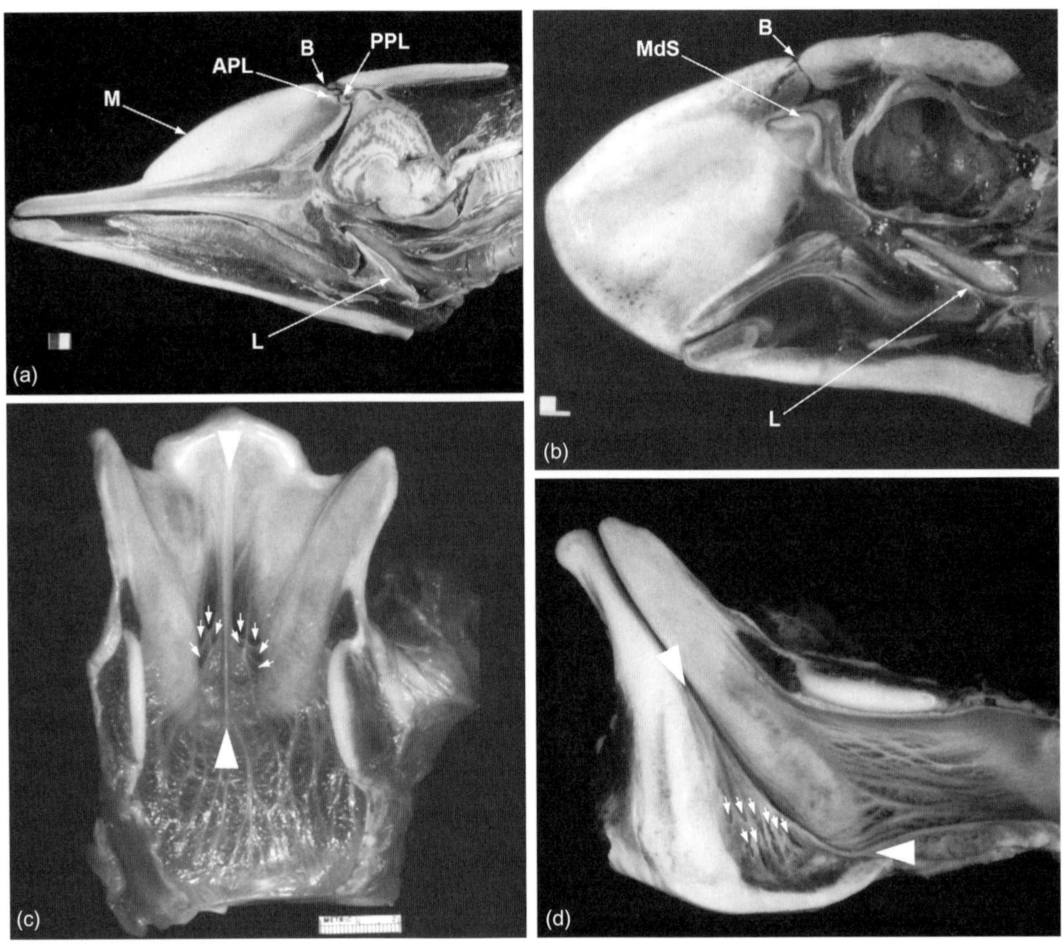

Fig. 5 (a) Midsagittal section through the head of an adult common dolphin, *Delphinus delphis*, showing the right side of the head. (b) Midsagittal section through the head of a juvenile pygmy sperm whale, *Kogia breviceps*, showing the right side of the head. (c) Dorsal view of the larynx of an adult pilot whale, *Globicephala melas*. The larynx is cut dorsally through the cricoid cartilage, and the cut edges are retracted laterally to reveal the lumen. The rostral aspect is at the top of the image. (d) Midsagittal section through the larynx of an adult pilot whale, *Globicephala melas*, showing the right side of the larynx. In images (c) and (d), note the trabeculated ventral surface (at the caudal half of the larynx), the midline fold (indicated between the two larger white arrowheads), and the laryngeal ventricles which are divided into numerous laryngeal air sacs (small white arrows). APL: anterior phonic lip; B: blowhole opening; L: larynx; M: melon (fatty forehead region); MdS: museau de singe; PPL: posterior phonic lip. (To view this figure in color please visit http://www.elsevierdirect.com/companions/9780123745934)

lens, causing changes in the pattern or directionality of sound release (beam formation).

The mechanism described above appears to occur in all odontocetes. Although all odontocetes appear to have the same structures, there is great diversity in their size, shape and position within the head. The greatest variation is seen in three families in particular: great sperm whale (Physeteridae), pygmy/dwarf sperm whales (Kogiidae) and beaked whales (Ziphiidae). In the sperm whales, the phonic lips are greatly enlarged into a structure called the museau de singe and are only found on the right side (Clarke, 1978) (Fig. 5), the spermaceti organ appears to be an enlargement of the right bursae and the "junk" (a fatty tissue in the forehead) is a modification of the melon (Cranford et al., 1996; Cranford, 1999). In the beaked whales, the air sacs are reduced or absent and are replaced by a dense connective tissue theca (Cranford et al., 2008a).

Specific shapes and positions of the air sacs may also serve to direct sound transmission via reflection at the density interface between air and tissue. It is interesting to note that the nasal air sacs are generally larger on the right side and smaller on the left side of the head (Mead, 1975), and appear to be controlled separately. This difference may indicate different rates or volumes of airflow, perhaps enabling simultaneous production of pulses with different frequencies. As with stereovision, in which input from two eyes allows depth perception, the production of two different sonar signals may similarly enable comparisons that help odontocetes determine depth, texture, or density of the object scanned by echolocation. Odontocetes with more asymmetry likely have an increased ability to vary the characteristics of the emitted sound (e.g., generate a standing wave or emphasize certain frequencies within a click), while those with relatively symmetrical configurations probably produce less variable sounds (Cranford et al., 1996).

While production of odontocete high-frequency pulses or clicks (which are used as sonar signals) is explained by the complex nasal anatomy, the origin of whistles and other sounds used in communication is still unknown. While they may be produced in the nasal region, it is also possible that some of these sounds are generated by the larynx.

The larynx of odontocetes is modified rostrally into an elongated, tubular structure that can be circumferentially "gripped" by the palatopharyngeal sphincter (comprised of muscles of the soft palate) (Fig. 5). The anterior edge of the epiglottis is enlarged and shaped like an arrowhead, thus facilitating interlocking with the encircling sphincter. This configuration locks the laryngeal opening into a permanently intranarial position (Reidenberg and Laitman, 1987). The palatopharyngeal sphincter seals the permanently intranarial larynx from the digestive tract, thereby protecting odontocetes from drowning while breathing, swallowing, or vocalizing. In addition, the larynx is often seen positioned asymmetrically towards the left side of the head and neck region, thus displacing the larynx to the left. This results in enlargement of the right piriform sinus (lateral food channel) while maintaining respiratory tract integrity, thereby allowing odontocetes to swallow relatively large prey whole without dislodging the larynx (Reidenberg and Laitman, 1994; see also Mignucci-Giannoni et al., 2009 for the consequences of swallowing prey that is too large).

Laryngeal protection in odontocetes is an exaggeration of a basic mammalian pattern of a larynx placed high in the neck to allow simultaneous breathing (and smelling) while swallowing (Laitman and Reidenberg, 1993, 1998, 2009). This level of protection indicates that the laryngeal portion of the respiratory tract must remain patent during non-digestive activities, such as sound production. Otherwise, a pattern of simple laryngeal closure at the level of the vocal folds would be sufficient to obstruct the larynx during feeding, as occurs in many reptiles.

The internal anatomy of the larynx is also modified (Fig. 5). The laryngeal lumen has a complex appearance, being comprised of ventral trabeculations and numerous small diverticulae called laryngeal air sacs (the latter are not present in sperm whales) (Reidenberg and Laitman, 1988). The presence of these structures indicates that the laryngeal walls are flexible and must undergo considerable changes in volume. A ventrally-incomplete cricoid cartilage facilitates even more flexibility in the tissues along the ventral aspect of the larynx. These features are likely diving adaptations that allow for air volume changes at depth without risking cracking the more rigid cartilages of the laryngeal skeleton. The remainder of the larynx is relatively rigid and is unlikely to collapse at depth. This laryngeal structure appears capable of maintaining its shape near the surface, but contracting ventrally to accommodate decreasing air volume at increasing pressures during diving. It is possible that this anatomy only facilitates a patent larynx at the surface for breathing, but allows laryngeal lumen obstruction without damage at depth. Alternatively, perhaps a patent laryngeal lumen at the surface may be used to produce communication sounds between socializing individuals.

As odontocetes exchange air very rapidly during breathing, it is expected that tissue folds (such as vocal folds) would be reduced to vestigial structures if unnecessary. Thin and delicate membranes projecting into the laryngeal lumen would be shredded with the explosive force of a "blow" (exhalation followed by inhalation at the surface) that, due to its rapidity, resembles a sneeze. It is expected that evolutionary selection would favor elimination of vocal folds to prevent damage and bleeding. However, contrary to expectations, a vocal fold homolog exists and is present in every odontocete species examined to date (Reidenberg and Laitman, 1988, 2007b; Laitman and Reidenberg, 1999).

A midline fold is present along the ventral aspect of the odontocete larynx (Fig. 5). The attachments of this fold are the same as the vocal folds of terrestrial mammals, thus indicating homology (Reidenberg and Laitman, 1988). Interestingly, the fold is reoriented parallel to airflow, but the attachments are reversed from those seen in mysticetes (Reidenberg and Laitman, 2007b). The fold is elongated in the rostrocaudal direction, and extends from the vocal processes of the arytenoid cartilages to the midline of the thyroid cartilage. Movements of the arytenoid can adjust tension of the fold and raise it higher into the laryngeal lumen. The midline fold is a thin tissue oriented in a vertical plane, parallel to airflow direction. Midline fold morphology is variable among different species, appearing as a single fold in the dolphins, a bifurcated fold in sperm whales and a trifurcated fold in the harbor porpoise. Laryngeal ventricles (which are subdivided into laryngeal air sacs) and vestibular folds are also found consistently lateral to the midline fold, further supporting homology with vocal folds. The folds may divide the airstream within the larynx into separate air currents. Air channeled on either side of the fold may cause passive vibrations, much like the oscillating reed in a woodwind musical instrument. The laryngeal air sacs extend the volume of the airspace around the midline fold, possibly facilitating lateral excursions during vibration.

The sound transmission pathway for laryngeal vibrations is not understood. It is possible that fold vibrations are transferred ventrally through the throat tissues, similar to the pathway proposed for mysticetes. It is interesting to note that a gular sound channel has recently been discovered in odontocetes (Cranford et al., 2008b). While this sound channel is thought to function in receiving sounds, it is possible that its purpose is also to transmit sounds. Alternatively, fold vibrations may be transferred to the epiglottic cartilage. Oscillations of this cartilage could, in turn, be transferred to the rostrum and transmitted along its length to be emitted at the front of the head.

VIII. Conclusions

Marine mammals are a very diverse group and encompass a large range of sound-generating morphologies. The least derived conditions exist in the marine mammals that make most of their vocalizations in air (polar bears, sea otters, pinnipeds). This group uses essentially the same apparatus as a terrestrial mammal: vocal fold vibrations in air emitted through the mouth or nose. Pinnipeds, particularly the deep diving seals, are more derived than polar bears and sea otters in their sound producing abilities. They are still capable of emitting sounds in air, but have acquired the ability to vocalize underwater. While some may continue to use the larynx, others have developed specialized structures (dorsal tracheal membrane and air sacs) that facilitate sound generation and/or transmission. Sirenians only generate sounds underwater, but their vocal capabilities are limited. Mysticete cetaceans, however, are vocal athletes, performing some of the loudest, longest, lowest and often very complex songs. Although they use the larynx, their anatomy is quite different from pinnipeds and terrestrial mammals. Their vocal folds are reoriented parallel to airflow and they have developed a large laryngeal sac. Odontocete cetaceans are perhaps the most derived in their sound producing anatomy. Unlike all the other marine mammals, only odontocetes have developed echolocation. Along with this new sense, they have developed complex nasal structures that are capable of generating, modifying and transmitting the pulsed sounds used in echolocation. They may even have dual nasal sound sources, with the larger one located on the right side. While the use of their larynx in sound production remains controversial, the complexity of that organ warrants further investigation as to its function, vocal or not.

References

Abe, H., Hasegawa, Y., Wada, K., 1977. A note on the airsac of ribbon seal. Scientific Reports of the Whales Research Institute 29, 129–135.

Aroyan, J.L., McDonald, M.A., Webb, S.C., Hildebrand, J.A., Clark, D., Laitman, J.T., Reidenberg, J.S., 2000.

Acoustic models of sound production and propagation. In: Au, W., Fay, R., Popper, A. (Eds.), Hearing by Whales and Dolphins. Springer-Verlag, New York, NY, pp. 409–469.

Ballard, K.A., 1993. Comparison of the acoustic repertoires of the grey (*Halichoerus grypus*), harp (*Phoca groenlandica*), hooded (*Cystophora cristata*), and harbor (*Phoca vitulina*) seals. Ph.D. Thesis. University of Waterloo, Ontario, Canada.

Ballard, K.A., Kovacs, K.M., 1995. The acoustic repertoire of hooded seals (*Cystophora cristata*). Canadian J. Zool. 73, 1362–1374.

Berland, B., 1966. The hood and its extrusible balloon in the hooded seal – *Cystophora cristata* Erxl. Norsk Polarinstitutt, Arbok, Oslo 1965, 95–102.

Burns, J.J., 1981. Ribbon seal *Phoca fasciata* Zimmermann, 1783. In: Ridgway, S.H., Harrison, R.J. (Eds.), Handbook of Marine Mammals, Vol. 2. Seals. Academic Press, London, UK, pp. 89–109.

Brauer, R.W., Jennings, R.A., Poulter, T.C., 1966. The effect of substituting helium and oxygen for air on the vocalization of the California sea lion, *Zalophus californianus*. In: Poulter, T.C. (Ed.), Proceedings of the Third Annual Conference on Biological Sonar and Diving Mammals, Abstracts. Stanford Research Institute, Fremont, CA, pp. 68–72.

Bryden, M.M., Felts, W.J.L., 1974. Quantitative anatomical observations on the skeletal and muscular systems of four species of Antarctic seals. J. Anat. 118, 589–600.

Caldwell, D.K., Caldwell, M.C., 1972. The World of the Bottlenose Dolphin. Lippincott, Philadelphia, PA.

Clarke, M.R., 1978. Structure and proportions of the spermaceti organ in the sperm whale. J. Mar. Biol. Assn. UK 58, 1–17.

Cranford, T.W., 1999. The sperm whale's nose: sexual selection on a grand scale? Mar. Mamm. Sci. 15, 1134–1158.

Cranford, T.W., 2000. In search of impulse sound sources in odontocetes. In: Au, W.W.L., Popper, A.N., Fay, R.R. (Eds.), Hearing by Whales and Dolphins. Springer-Verlag, New York, NY, pp. 109–156.

Cranford, T.W., Amundin, M.E., 2003. Biosonar pulse production in odontocetes: the state of our knowledge. In: Thomas, J.A., Moss, C.F., Vater, M. (Eds.), Echolocation in Bats and Dolphins. The University of Chicago Press, Chicago, IL, pp. 27–35.

Cranford, T.W., Amundin, M., Norris, K.S., 1996. Functional morphology and homology in the odontocete nasal complex: implications for sound generation. J. Morphol. 228, 223–285.

Cranford, T.W., Van Bonn, W.G., Chaplin, M.S., Carr, J.A., Kamolnick, T.A., Carder, D.A., Ridgway, S.H., 1997. Visualizing dolphin sonar signal generation using high-speed video endoscopy. J. Acoust. Soc. Am. 102, 3123.

Cranford, T., McKenna, M., Soldeville, M., Wiggins, S., Goldbogen, J., Shadwick, R., Krysl, P., St. Leger, J., Hildebrand, J., 2008a. Anatomic geometry of sound transmission and reception in Cuvier's beaked whale (*Ziphius cavirostris*). Anat. Rec. 291, 353–378.

Cranford, T.W., Krysl, P., Hildebrand, J., 2008b. Sound pathways revealed: simulated sound transmission and reception in Cuvier's beaked whale (*Ziphius cavirostris*). Bioinspir. Biomim. 3, 1–10.

Dormer, K.J., 1979. Mechanism of sound production and air recycling in delphinids: cineradiographic evidence. J. Acoust. Soc. Am. 65, 229–239.

Fay, F.H., 1960. Structure and function of the pharyngeal pouches of the walrus (*Odobenus rosmarus L.*). Mammalia 24, 361–371.

Green, K., Burton, H.R., 1988. Do Weddell seals sing? Polar Biol. 8, 165–166.

Hanggi, E.B., Schusterman, R.J., 1992. Underwater acoustical displays by male harbour seals (*Phoca vitulina*): initial results. In: Thomas, J.A., Kastelein, R.A., Supin, A.Y. (Eds.), Marine Mammal Sensory Systems. Plenum Press, New York, NY, pp. 449–458.

Hanggi, E.B., Schusterman, R.J., 1994. Underwater acoustic displays and individual variation in male harbour seals, *Phoca vitulina*. Animal Behav. 48, 1275–1283.

Hartman, D.S., 1971. Behavior and ecology of the Florida manatee, *Trichechus manatus latirostris* (Harlan), at Crystal River, Citrus County. Ph.D. Thesis, Cornell University, Ithaca, NY.

Herzing, D.L., 1996. Vocalizations and associated underwater behavior of free-ranging Atlantic spotted dolphins, *Stenella frontalis* and bottlenose dolphins, *Tursiops truncatus*. Aq. Mammals 22, 61–79.

Hewer, H.R., 1974. British Seals. William Collins & Sons, Glasgow, UK.

Heyning, J.E., 1989. Comparative facial anatomy of beaked whales (Ziphiidae) and a systematic revision among the families of extant Odontoceti. Contrib. Sci. Los Angeles County Mus. 405, 1–64.

Kastelein, R.A., Postma, J., Verboom, W.C., 1995. Airborne vocalizations of Pacific walrus pups (*Odobenus rosmarus divergens*). In: Kastelein, R.A., Thomas, J.A., Nachtigall, P.E. (Eds.), Sensory Systems of Aquatic Mammals. De Spil Publishers, Woerden, The Netherlands, pp. 265–285.

Kellogg, W.N., 1958. Echo ranging in the porpoise. Science 128, 982–988.

Kooyman, G.L., 1981. Leopard seal *Hydrurga leptonyx* Blainville, 1820. In: Ridgway, S.H., Harrison, R.J. (Eds.), Handbook of Marine Mammals, Vol. 2. Seals. Academic Press, London, UK, pp. 261–274.

Kooyman, G.L., Andersen, H.T., 1969. Deep diving. In: Andersen, H.T. (Ed.), The Biology of Marine Mammals. Academic Press, New York, NY, pp. 65–94.

Laitman, J.T., Reidenberg, J.S., 1993. Specializations of the human upper respiratory and upper digestive tract as seen through comparative and developmental anatomy. Dysphagia 8, 318–325.

Laitman, J.T., Reidenberg, J.S., 1997. The human aerodigestive tract and gastroesophageal reflux: an evolutionary perspective. Am. J. Medicine 103, 2S–8S.

Laitman, J.T., Reidenberg, J.S., 1998. Comparative and developmental anatomy of laryngeal position. In: Bailey, B.J. (Ed.), Head and Neck Surgery – Otolaryngology, 2nd edn., Vol. 1. J. B. Lippincott Co., Philadelphia, PA, pp. 45–52.

Laitman, J.T., Reidenberg, J.S., 1999. Evolution of the vocal folds in cetaceans. Eur. Res. Cetaceans 12, 286–289.

Laitman, J.T., Reidenberg, J.S., 2009. The evolution of the human larynx: nature's great experiment. In: Fried, M.P.A.

Ferlito, A. (Eds.), The Larynx, 3rd edn, vol. 1. Plural Publishing, San Diego, CA, pp. 19–38.

Lipan, M.J., Reidenberg, J.S., Laitman, J.T., 2006. Anatomy of reflux: a growing health problem affecting structures of the head and neck. Anat. Rec. (part B: New Anat.), 289B, 261–270.

Lowell, W.R., Flanigan, W.F., 2008. Marine mammal chemoreception. Mamm. Rev. 10 (1), 53–59.

Mackay, R.S., 1980. Dolphin air sac motion measurements during vocalization by two noninvasive ultrasonic methods. In: Busnel, R., Fish, J. (Eds.), Animal Sonar Systems. Plenum Press, New York, NY, pp. 933–935.

Mead, J.G., 1975. Anatomy of the external nasal passages and facial complex in the Delphinidae (Mammalia: Cetacea). Smithson. Contrib. Zool. 207, 1–72.

McCowan, B., Reiss, D., 1995. Quantitative comparison of whistle repertoires from captive adult bottlenose dolphins (Delphinidae, *Tursiops truncatus*): a re-evaluation of the signature whistle hypothesis. Ethology 100, 193–209.

Mignucci-Giannoni, A.A., Rosario-Delestre, R.J., Alsina-Guerrero, M.M., Falcón-Matos, L., Guzmán-Ramírez, L., Williams, Jr., E.H., Bossart, G.D., Reidenberg, J.S., 2009. Asphyxiation in a Bottlenose dolphin (*Tursiops truncatus*) from Puerto Rico due to choking on a Black Margate (*Anisotremus surinamensis*). Aquatic Mammals 35 (1), 48–54.

Miller, E.H., Job, D.A., 1992. Airborne acoustic communication in the Hawaiian monk seal, *Monachus schauinslandi*. In: Thomas, J.A., Kastelein, R.A., Supin, A.Y. (Eds.), Marine Mammal Sensory Systems. Plenum Press, New York, NY, pp. 485–531.

Miller, E.H., Murray, A.V., 1995. Structure, complexity, and organization of vocalizations in harp seal (*Phoca groenlandica*) pups. In: Kastelein, R.A., Thomas, J.A., Nachtigall, P.E. (Eds.), Sensory Systems of Aquatic Mammals. De Spil Publishers, Woerden, The Netherlands, pp. 237–264.

Møhl, B., Terhune, J.M., Ronald, K., 1975. Underwater calls of the harp seal, *Pagophilus groenlandicus*. Rapports et Proces-Verbaux des Reunions, Conseil International pour l'Exploration de la Mer 169, 533–543.

Morrice, M.G., Burton, H.R., Green, K., 1994. Microgeographic variation and songs in the underwater vocalization repertoire of the Weddell seal (*Leptonychotes weddellii*) from the Vestfold Hills. Antartica Polar Biol. 14, 441–446.

Norris, K.S., Prescott, J.H., Asa-Dorian, P., Perkins, P., 1961. An experimental demonstration of echolocation behavior in the porpoise *Tursiops truncatus* (Montagu). Biol. Bull. 120, 163–176.

Nowacek, D.P., Casper, B.M., 2003. Intraspecific and geographic variation of West Indian manatee (*Trichechus manatus spp.*) vocalizations (L). J. Acoust. Soc. Am. 114 (1), 66–69.

Payne, K., Payne, R., 1985. Large scale changes over 19 years in the songs of humpback whales in Bermuda. Z. Tierpsychol. 68, 89–114.

Payne, K.B., Tyack, P., Payne, R.S., 1983. Progressive changes in the songs of humpback whales. In: Payne, R. (Ed.), Communication and Behavior of Whales. Westview Press, Boulder, CO, pp. 9–59.

Payne, R.S., McVay, S., 1971. Songs of humpback whales. Science 173, 585–597.

Pierard, J., 1969. Le Larynx du phoque de Weddell (*Leptonychotes weddelli*, Lesson, 1926). Canadian J. Zool. 47, 77–87.

Purves, P.E., 1966. Anatomical and experimental observations on the cetacean sonar system. In: Busnel, R. (Ed.), Les Systems Sonars Animaux: Biologie et Bionique/Animal Sonar Systems: Biology and Bionics. Laboratoire de Physiologie Acoustique, Jouy-en-Josas, France, pp. 197–270.

Purves, P.E., Pilleri, G., 1978. The functional anatomy and general biology of *Pseudorca crassidens* (Owen) with a review of the hydrodynamics and acoustics in Cetacea. In: Pilleri, G. (Ed.), Investigations on Cetacea, Vol. 9. Brain Anatomy Institute, Berne, Switzerland, pp. 67–227.

Purves, P.E., Pilleri, G., 1983. Echolocation in Whales and Dolphins. Academic Press, New York, NY.

Potter, J.R., Pack, A.A., Reidenberg, J., Hoffmann-Kuhnt, M., Seekings, P., Chitre, M.A., Koay, T.B., 2007. Humpback song source location in the head, source levels and directionality from *in situ* rebreather diver recordings. 17th Biennial Conference on the Biology of Marine Mammals. Cape Town, South Africa.

Poulter, T.C., 1965. Location of the point of origin of the vocalization of the California sea lion, *Zalophus californianus*. In: Poulter, T.C. (Ed.), Proceedings of the Second Annual Conference on Biological Sonar and Diving Mammals. Stanford Research Institute, Fremont, CA, pp. 41–48.

Ray, G.C., 1981. Ross seal *Ommatophoca rossi* Gray, 1844. In: Ridgway, S.H., Harrison, R.J. (Eds.), Handbook of Marine Mammals, Vol. 2. Seals. Academic Press, London, UK, pp. 237–260.

Ray, G.C., Watkins, W.A., Burns, J.J., 1969. The underwater song of Erignathus (bearded seal). Zoologica (New York) 54, 79–83.

Reidenberg, J.S., Laitman, J.T., 1987. Position of the larynx in Odontoceti (toothed whales). Anat. Rec. 218, 98–106.

Reidenberg, J.S., Laitman, J.T., 1988. Existence of vocal folds in the larynx of Odontoceti (toothed whales). Anat. Rec. 221, 884–891.

Reidenberg, J.S., Laitman, J.T., 1994. Anatomy of the hyoid apparatus in Odontoceti (toothed whales). Specializations of their skeleton and musculature compared with those of terrestrial mammals. Anat. Rec. 240, 598–624.

Reidenberg, J.S., Laitman, J.T., 2002. Proposed mechanisms of pneumatic sound production and transference to water in aquatic mammals: a comparative anatomical study. American Association of Anatomists: Experimental Biology Meeting, FASEB Journal (abstracts published on CD-ROM).

Reidenberg, J.S., Laitman, J.T., 2005. Morphophysiology of the Larynx. In: Van De Water, T., Staecker, H. (Eds.), Basic Science Review for Otolaryngology. Thieme, New York, NY, pp. 505–515.

Reidenberg, J.S., Laitman, J.T., 2007a. Blowing bubbles: an aquatic adaptation that risks protection of the respiratory tract in humpback whales (*Megaptera novaeangliae*). Anat. Rec. 290, 569–580.

Reidenberg, J.S., Laitman, J.T., 2007b. Discovery of a low frequency sound source in mysticeti (baleen whales): anatomical establishment of a vocal fold homolog. Anat. Rec. 290, 745–760.

Reidenberg, J.S., Laitman, J.T., 2008. Sisters of the sinuses: cetacean air sacs. Anat. Rec. 291, 1389–1396.

Reynolds, J.E., Odell, D.K., 1991. Manatees and Dugongs. Facts on File, Inc., New York, NY.

Richardson W.J., Greene, Jr., C.R., Malme, C.I., Thomson, D.H., 1995. Marine Mammals and Noise. Academic Press, New York, NY.

Ridgway, S.H., Carder, D.A., Green, R.F., Gaunt, S.L., Gaunt, A.S., Evans, W.E., 1980. Electromyographic and pressure events in the nasolaryngeal system of dolphins during sound production. In: Busnel, R., Fish, J. (Eds.), Animal Sonar Systems. Plenum Press, New York, NY, pp. 239–249.

Rogers, T., Cato, D.H., Bryden, M.M., 1995. Underwater vocal repertoire of the leopard seal (*Hydrurga leptonyx*) in Prydz Bay, Antartica. In: Kastelein, R.A., Thomas, J.A., Nachtigall, P.E. (Eds.), Sensory Systems of Aquatic Mammals. De Spil Publishers, Woerden, The Netherlands, pp. 223–236.

Schevill, W.E., 1968. Sea lion echo ranging? J. Acoust. Soc. Am. 43, 1458–1459.

Schevill, W.E., Watkins, W.A., 1965. Underwater calls of *Trichechus* (manatee). Nature 205, 373–374.

Schevill, W.E., Watkins, W.A., Ray, G.C., 1966. Analysis of underwater Odobenus calls with remarks on the development of pharyngeal pouches. Zoologica (New York) 51 (3), 103–106.

Schusterman, R.J., Kastak, D., Levenson, D.H., Reichmuth, C.J., Southall, B.L., 2000. Why pinnipeds don't echolocate. J. Acoust. Soc. Am. 107 (4), 2256–2264.

Shaver, H.N., Poulter, T.C., 1967. Sea lion echo ranging. J. Acoust. Soc. Am. 42 (2), 428–437.

Sokolov, A.S., Kosygin, G.M., Shustov, A.P., 1970. Structure of lungs and trachea of Bering Sea pinnipeds. In: Arsen'ev V.A., Panin, K.I. (Eds.), Pinnipeds of the North Pacific [English translation of Vsesoyuznyi Nauchno-Issledovatel'skii Institut Morskogo Rybnogo Khozyaistva I Okeanografii (VINRO), Vol. 68 (1968)]. Israel Program for Scientific translations, Jerusalem, Israel, pp. 250–262.

Stirling, I., Calvert, W., Spencer, C., 1987. Evidence of stereotyped underwater vocalizations of male Atlantic walruses (*Odobenus rosmarus rosmarus*). Can. J. Zool. 65, 2311–2321.

Thomas, J.A., Golladay, C.L., 1995. Geographic variation in leopard seal (*Hydrunga leptonyx*) underwater vocalizations. In: Kastelein, R.A., Thomas, J.A., Nachtigall, P.E. (Eds.), Sensory Systems of Aquatic Mammals. De Spil Publishers, Woerden, The Netherlands, pp. 201–221.

Tyack, P.L., Whitehead, H., 1983. Male competition in large groups of wintering humpback whales. Behaviour 83, 132–154.

Tyack, P.L., Clark, C.W., 2000. Communication and acoustic behavior of dolphins and whales. In: Au, W., Fay, R., Popper, A. (Eds.), Hearing by Whales and Dolphins. Springer-Verlag, New York, NY, pp. 156–224.

Tyack, P., Miller, E.H., 2002. Vocal anatomy, acoustic communication and echolocation. In: Hoelzel, A.R. (Ed.), Marine Mammal Biology: An Evolutionary Approach. Blackwell Science Ltd, Malden, MA, pp. 142–184.

Uhen, M.D., 2007. Evolution of marine mammals: back to the sea after 300 million years. Anat. Rec. 290, 514–522.

Verboom, W.C., Kastelein, R.A., 1995. Rutting whistles of a male Pacific walrus (*Odobenus rosmerus divergens*). In: Kastelein, R.A., Thomas, J.A., Nachtigall, P.E. (Eds.), Sensory Systems of Aquatic Mammals. De Spil Publishers, Woerden, The Netherlands, pp. 287–298.

Wemmer, C., Von Elbers, M., Scow, K., 1976. An analysis of the chuffing vocalization in the polar bear (*Ursus maritimus*). J. Zool. (Lond.) 180, 425–439.

SECTION 11

Vocal Communication Systems in Mammals: Semiotic Codes in Vocalization

Mammalian vocalization signals may be structured to code multidimensional information, and may include motivational and referential signals, signature vocalizations and reflect categorization of auditory perception. Vocal communication may also be reinforced by other media, such as gestures which, particularly in apes, are increasingly interdependent.

CHAPTER 11.1

Control of gestures and vocalizations in primates

Eva Maria Eberl*

Language Development and Cognitive Sciences, Philosophical Faculty, University of Graz, Graz, Austria

Abstract: Primate communication systems are complex and utilize manual gestures and vocalization for the transfer of information. The relationship between these two communicatory media has become increasingly interdependent in apes as specific cerebral regions have gradually assumed a governing role in cognitive communication processes. Over the course of primate evolution, subcortically processed monkey vocalizations have become largely corticalized and lateralized to the left hemisphere in humans. The emergence of handedness has shifted cortical control over manual gestures to the left hemisphere and established a cerebral specialization subserving gesture and vocalization in higher primates.

Keywords: primate gestural communication; corticalization of vocalization; synchrony of gesture and vocalization; evolution of Broca's area; primate motor theory of sound perception

I. Introduction

Gesture and vocalization constitute a unified system for communication in higher primates. Humans and great apes use gestural and vocal signals for denoting intentions and events in a referential way, and thereby use both media of expression for the transferral of information between individuals. Sounds (and combinations of sounds) and manual gestures are intrinsically linked in human speech with the precedents of that behavior already evident in great apes. Chimpanzees often combine hand gestures with a call during their communication (Hopkins and Cantero, 2003). This complex communication system, on which many aspects of social life are based, is deeply-rooted in primate neurobiological mechanisms where gestures of the hands and gestures of the mouth have increasingly come under shared control of cerebral regions that have specifically been adapted for communication.

II. Neurobiology of gesture in primates

II.A. Gesture and the macaque motor cortex

The macaque motor cortex encodes single manual movements and integrates them into unified manual gestures according to ethological action categories (Graziano, 2006). Single movements are simple changes in position of an upper limb without constituting semiotic value; a gesture is a combination of single movements to form a larger, meaningful, unit of action.

Cortical neuronal populations around the precentral gyrus are partly organized according to behaviorally-relevant movements (Graziano, 2006). A specified "map of hand location" in the precentral gyrus of the macaque brain represents hand locations in the upper space (near face), in the midlevel space (in front of the body) and in the lower space, with activation of the relevant regions within the area of the arcuate and central sulcus evoking certain movements (see Fig. 1).

The cortical area responsible for hand-to-mouth movements is located on the border of the F4 and

*Corresponding author. E-mail: eva.eberl@uni-graz.at

Stefan M. Brudzynski (Ed.)
Handbook of Mammalian Vocalization
ISBN 978-0-12-374593-4

DOI: 10.1016/B978-0-12-374593-4.00042-5

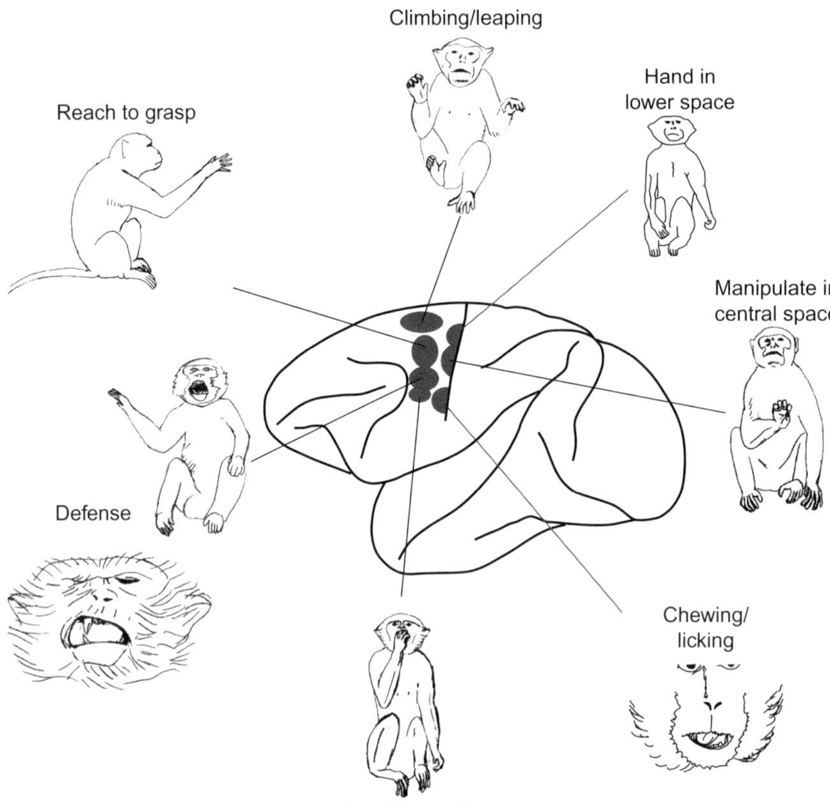

Fig. 1. Ethological organization of movement on the macaque motor cortex. From Graziano and Aflalo (2007) p. 243. Reprinted with permission from Elsevier.

F5 regions (terminology according to Matelli et al., 1985). In the F5 region, Rizzolatti and colleagues (1988) recorded single cells which fired selectively during a combination of grip posture of the hand and movements of the mouth. This result indicates a close-knit network subserving hand and mouth movement coordination and has implications for evolutionary perspectives on the origin of speech. In particular, defensive actions and hand-to-mouth (feeding) actions that facilitate the simultaneous use of hand and mouth are located close to each other in the premotor cortex.

It is believed that the motor cortex computes parameters relevant for a specific behavioral output, rather than specifying a single movement type. This goal-directed encoding specifies actions from a functional point of view (e.g., feeding, defensive actions). The global–functional consequence of an action trajectory, i.e., a sequence of single movements which equals the ethological relevance, can be broken down into its constituent elements (the single movements) which in themselves bear no functional content, but only gain meaning in the context of the global gesture.

Aflalo and Graziano (2006) found that neurons on the macaque precentral gyrus are partially tuned to preferred end postures of a series of single movements. The manual gestures therefore are encoded in a goal-directed way that focuses on the final output action. This *end posture coding* emphasizes the final function of an action and melds chronology of movements with semiotic value (*semiosis*). Since ethological organization of action in the macaque motor cortex reflects the species-typical motor output in the monkey, the semiotic connotation results from, and is at the same time constrained by, the common behavioral repertoire of the animal. The *pars-pro-toto* principle ("one part for the whole") that is inherent in end posture coding establishes a conceptual connection between parts (*partes*), single movements and *toto*, the semiosis where components of a movement trajectory anticipate the final product.

The rule of end-posture coding also becomes evident in speech, where the vocal tract posture is ascribed to approximate a certain "final position," but does not necessarily have to reach it to produce intelligible speech (Fowler et al., 1980). Graziano et al. (2004) hypothesized speech mechanisms: "were built on a pre-existing mechanism for motor control" (p. 225).

While the gestural repertoire is rudimentary in monkeys, with movement chronology and semiotic content strongly restricted and numerically limited, ape gestures show some flexibility, as evidenced by idiosyncrasies and neologisms, as well as the ability to use gestures symbolically (Tomasello and Zuberbühler, 2002; Pika et al., 2005).

II.B. Lateral asymmetry in manual gestures of apes

Cortical control of gestures in great apes shows human-like asymmetric hemispheric specialization. Several studies have reported population-level right-handedness for gorillas and chimpanzees (e.g., Byrne and Byrne, 1991; Hopkins et al., 2002), although there is no universal agreement on whether non-human primates show population-level handedness (see McGrew and Marchant, 1997). Chimpanzees are more proficient at certain hand-grasp types, such as the thumb index-finger "precision grip" when it is performed with the right hand, and, in turn, use of the right precision-grip facilitates the preferred use of the right hand (Hopkins et al., 2002). Hemispheric specialization may be the result of a left-hemisphere asymmetry in the region of the precentral gyrus that governs digit control, in particular the left dorsal area of the precentral gyrus (Hopkins and Pilcher, 2001; Hopkins and Cantalupo, 2004). Hopkins and Cantalupo (2004) demonstrated neuroanatomical differences in the precentral gyrus in chimpanzees, where left- and right-handed subjects showed different degrees of left-hemisphere asymmetry. Neuroanatomical research suggests that handedness may be associated with the left-hemispheric asymmetry in the gyrification processes in chimpanzees (Hopkins et al., 2006). Humans show a similar trend, with one study suggesting greater gyrification for female brains (Luders et al., 2004).

There have been attempts to link handedness to body posture in apes. Hopkins (1993) observed chimpanzees (*Pan troglodytes*) and orang-utans (*Pongo pygmaeus*) displaying right-hand preference only when in the bipedal position, but not in the quadrupedal position. A similar study by Olson et al. (1990) showed a clear left-hand preference of gibbons (*Hylobates*) and right-hand preference for gorillas on bipedal tasks, whereas chimpanzees and orang-utans appeared more ambidextrous. In the wild, gorillas and gibbons display upright walking more often than the other ape species and might therefore have developed handedness. These findings imply a potential connection between handedness and bipedalism.

Regarding communication, gorillas and chimpanzees preferentially use the right hand for intentional and referential gestures (Shafer, 1993, 1997). In gorillas, a left-hemispheric bias for motor control during communicative actions has been suggested. In a pilot study, gorillas showed a bias toward using the right hand synchronous with mouth movements (Forrester, 2008). Hopkins and Cantero (2003) observed an increase in right-hand use in chimpanzees when gestures were accompanied by vocalization. This finding suggests that communicative gestures and vocalizations are both lateralized to the left hemisphere, and linked at some level. While in the macaque brain, the link between mouth and hand gestures is rooted in feeding and defense behavior, chimpanzees (and possibly other apes) connect manual gestures and vocalizations on a communicative level.

III. Vocal communication and the underlying brain organization

Production of vocalizations in non-human primates is generally attributed to the subcortical cerebral regions and therefore regarded as fundamentally different from human language (Steklis and Harnad, 1976). Hauser (1996) suggested that cortical regions might play a role in the perception of primate vocalizations, since monkeys showed ear preferences when listening to conspecific calls, while production of calls remained only weakly linked to the cortex. However, the emotional factor of alarm calls may contribute to right-hemispheric specialization, as is the case with prosody in human speech (Vauclair, 2004). Regarding production of vocalization, Hauser and Akre (2001) reported that rhesus monkeys initiate mouth movements starting with the left side of the face, which may point to a right-hemispheric cerebral specialization for production of vocalization.

III.A. Subcortical organization of vocalization in the non-human primate brain

In monkeys, apes and humans, voluntary learned phonation is associated with activity in specified regions around the Sylvian fissure, including the neocortical larynx area (Jürgens and Zwirner, 2000). Involuntary and emotional vocalizations, as well as timing of phonation, are processed primarily by a second phonatory pathway, the limbic pathway that passes through the periaqueductal gray (PAG) in the midbrain. Damage to this area causes mutism in humans and animals, without damage to the articulatory abilities of the vocal apparatus (Esposito et al., 1999). The subcortical pathway is phylogenetically older and projects more heavily to the laryngeal motor neurons than the cortical pathway, as shown for squirrel monkeys (*Saimiri sciureus*). Vocalizations elicited from the PAG region were shown to be dominant over cortically-elicited phonation as they exerted more influence on the laryngeal muscles than the cortical control (Jürgens and Zwirner, 2000).

Although initially Romanski et al. (2005) found that neuronal activity in the ventrolateral prefrontal cortex was related only to acoustic properties of the perceived vocalization, subsequent results indicate otherwise. Gifford et al. (2005) and Cohen et al. (2006) described neuronal populations in the ventrolateral prefrontal cortex of rhesus monkeys (*Macaca mulatta*) as essential in the analysis of species-specific calls. The authors discovered differences in neuron activity as a function of the semiotic content of the vocalization (e.g., food-related calls caused similar neural responses).

III.A.1. Hemispheric dominance for call analysis

Playback experiments in the rhesus monkeys (*Macaca mulatta*) demonstrated a clear left-hemispheric specialization for the analysis of conspecific calls (Hauser and Andersson, 1994). For perception of heterospecific vocalizations, however, the animals showed a right-hemispheric preference. Japanese macaques (*Macaca fuscata*) have been shown to possess left-hemispheric bias when processing species-specific calls (Petersen et al., 1984).

In a single-case ERP (event-related potential) study on a chimpanzee, Bernston et al. (1993) found right-lateralized processing of vocal stimuli from conspecific vocalizations and human speech, but not for simple nonsignal stimuli. The length of the Sylvian fissure of chimpanzees and rhesus monkeys is asymmetric, as it is longer in the left cerebral hemisphere than in the right hemisphere (Heilbronner and Holloway, 1988). This bears implications for the lateralization of auditory processing during the evolution of human cognition.

III.A.2. Hemispheric asymmetry in vocal fold control

In primates, the larynx motor neurons receive their input from neurons of the reticular formation of the pons and medulla oblongata. There are several classes of cells in the reticular formation that become differentially active during phonation (Lüthe et al., 2000). Some of these cells fire during certain patterns of phonation, while others fire during all types of phonation; a further class of cell fires during the entire length of the phonation, while other cells fire only shortly before the onset of phonation. Some neurons have been found to change their firing rate depending on the fundamental frequency (Jürgens, 2004). These findings suggest that the reticular formation plays an important role in structuring/patterning phonation (see Hage, Chapter 8.2 in this volume).

In addition to the corticoreticular tract present in primates, humans also have a corticoambigual tract which establishes a direct connection between the primary motor cortex and the laryngeal motor neurons in the nucleus ambiguus (Jürgens, 2004). This direct link between the motor neurons and the motor cortex is suggested to have evolved within the last two million years, and it enables complete control over articulation (Jürgens, 2004). This development may have been a crucial step in the evolution of language.

Squirrel monkeys demonstrate hemispheric dominance in the cortical control over their vocal folds (Jürgens and Zwirner, 2000), with equal distribution of left- and right-lateralization in the population. Ipsi- and contralateral stimulation of the phonation areas revealed that only the cortical areas govern hemispheric dominance. Although there is no population-level bias for lateralization, individual animals show preferential hemispheric dominance.

III.B. Corticalization of vocalization in humans

During the course of evolution there was an increase in corticalization where more cognitive processes came under control of the cortex. In primates, there is a direct pathway between the motor cortex and the motor neurons in the brainstem, which enables fine-skilled motor movements of individual muscles and conscious planning of movement actions.

Vocalizations, which are under subcortical control in non-human primates, have been corticalized in humans, rendering them voluntarily controllable (Deacon, 1997). Ploog (1992) found that when the cortical face and larynx areas of non-human primates were electrically stimulated, movements of the vocal folds without phonation occur, whereas in humans actual vocalizations resulted. Ploog (1992) suggested that chimpanzees show more advanced corticalization than the other great apes and therefore they may use cortical areas for a direct control of phonation.

IV. Evolutionary relationship between gesture and vocalization

IV.A. Area BA44: where gesture and vocalization meet

The regions BA44 and BA45, also known as Broca's area in the human brain, have long been associated with speech and language processes. Ever since Paul Broca published the *Tan*-case (Broca, 1861), describing how a lesion on the left inferior frontal gyrus caused inability to speak (the patient's language production was restricted to the syllable "tan"), research over the last century has accumulated further evidence linking Broca's area to language processes, in particular grammatical structuring (Sahin et al., 2006). This cortical area has even been suggested as the seat of Noam Chomsky's "universal grammar," the human inborn set of rules for organizing language (see Musso et al., 2003). In addition to language, Broca's area in humans has also been implicated in the processing of complex hand and finger movements (Bonda et al., 1994; Binkofski and Buccino, 2006).

Finally, Cantalupo and Hopkins (2001) were able to show that a similar structure to Broca's area exists in chimpanzees and gorillas, where the left inferior frontal gyrus exceeds that of the right hemisphere in size. Cantalupo and Hopkins (2001) argue that the neuroanatomical asymmetry in the ape BA44 area is related to right-handed gesture production accompanied by vocalizations. The predominance of speech may have caused the expansion of the adjacent BA45 area in the human brain.

IV.B. Object–mouth synchrony during grasping in chimpanzees and humans

According to Armstrong et al. (1995), shared neural regions for manual and vocal tract gestures (area BA44) generate multiple motor commands that are being sent to the hands and the vocal organs at the same time. Thus, interfering effects can occur between gesture and vocalization which become evident at various levels of language processing, e.g., lexis and syntax (see Willems and Hagoort, 2007, for a review).

In studies on humans, Gentilucci and colleagues (Gentilucci et al., 2001) found mutual influences between grasp gestures performed with the mouth and those performed with the hands which were governed by similar neuronal populations in Broca's area. Grasping movements of the hands can affect syllable pronunciation and lead to spectral changes in the acoustic structure of speech sounds (Gentilucci et al., 2001). The close connection between hand and mouth movements also becomes evident in non-human primates. Chimpanzees have been observed to perform apparently aimless movements of articulatory organs during fine-motor tasks with the use of a finger (Waters and Fouts, 2002). The shared control over fingers and mouth can produce simultaneous activation at times. This close neurophysiological interplay of hand gestures and speech has spurred so-called "gesture-first theory" that see the roots of modern human language in an earlier manual–gestural form of communication (Corballis, 2002).

In macaque monkeys, the homologous area to BA44 houses "mirror neurons" which are associated with gestural behavior and imitation (Rizzolatti et al., 1988). Mirror neurons relate perception of a gesture to actual performance, and therefore define the two processes as interdependent. It has been demonstrated for speech processes that perception and speech production are "two sides of the same coin" that are connected via their motor features (Liberman and Mattingly, 1985, p. 30). Speech sounds are defined by their underlying vocal tract gestures (e.g., "lip rounding"). Perceiving a sound means identifying gestures of the articulatory apparatus where sets of gestural features determine the quality of the different speech sounds.

IV.C. The "primate motor hypothesis of sound perception"

Liberman's sound perception theory, which links production and perception mechanisms at the neurophysiological level, has already been documented for at least one species of non-human primates (Eberl, 2008). Analyses of gibbon (*Nomascus* spp.) song excerpts show that the same sound units, such as

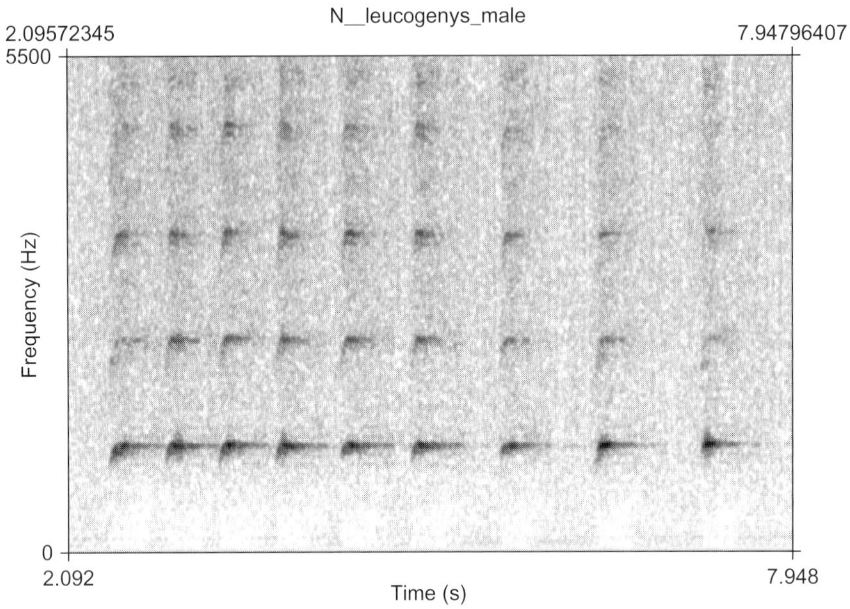

Fig. 2. Sonagram showing sound frequency versus time (y-axis: total scale 5 kHz; x-axis: total scale 15 seconds) of an excerpt from a male gibbon (*Nomascus leucogenys*) song phrase. Each of the nine consecutive staccato notes was spectrally analyzed (see two examples in Fig. 3).

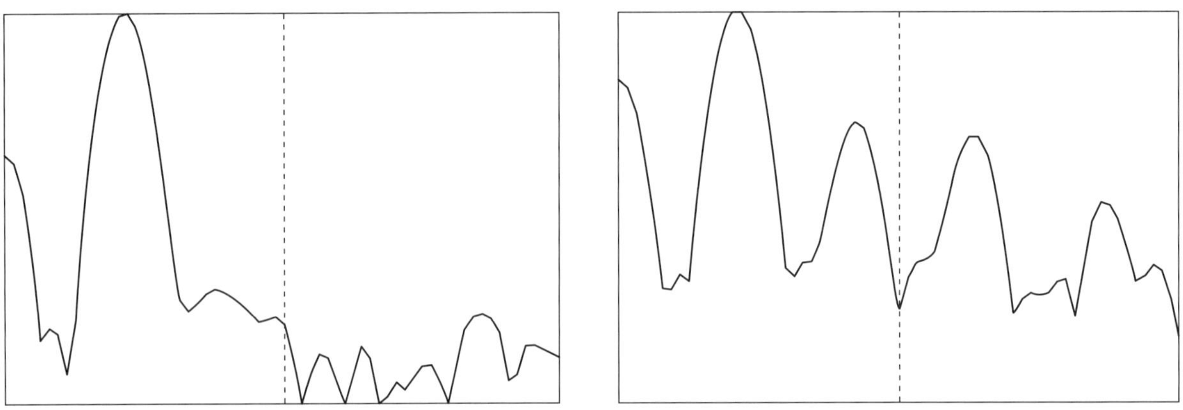

Fig. 3. Notes 3 (on the left hand side) and 8 (on the right hand side) from the sonagram in Fig. 2. Spectral analysis after Fourier transformation – sound intensity versus sound frequency (5 kHz on x-axis; 60 dB on y-axis). The overtone patterns are markedly different in the two notes, even though both belong to the same phonetic class of staccato notes.

notes of a male staccato phrase (see Fig. 2), can have different acoustic realizations. In order to perceive a sound correctly, the receiving animal therefore has to identify articulation action, with the acoustic patterns of the sound indicating the underlying vocal tract gesture. Acoustic features may provide cues to the articulatory gesture that underlies every sound. This hypothesis defines the motor movement of an articulation action (vocal tract gesture) as the actual percept.

The differing acoustic patterns for the same phonetic units (staccato notes), as shown in Fig. 3, identify the acoustic signal as unreliable as there is not a one-to-one relationship between the note (*phoneme*) and acoustic output. Fig. 3 compares the different spectral patterns for the same note. The relationship between vocal tract gesture and sound, however, is of a direct nature, with each sound characterized by only one articulation movement.

The motor perception mechanisms of speech may be evolutionarily older than Liberman and colleagues previously assumed (Liberman and Mattingly, 1985) and may go back to the emergence of the apes.

The indication that sounds may be perceived by apes and humans according to their underlying production features, which are mediated by the acoustic channel, gains meaning in the context of "mirror neuron" functionality, which suggests a constant mapping between production and perception of a gesture (including articulation) as they activate similar brain regions (see Scott et al., Chapter 5.5 in this volume).

As mentioned above, sound perception and production mechanisms are neurophysiologically connected. Groups of neurons in the area F5 of monkeys and BA44 of apes and humans (see Rizzolatti and Craighero, 2004, for a review) possess properties that link production and perception of both manual and articulatory gestures through mirroring processes that generate similar neural activity on the motor cortex of the performer, as well as the recipient. Communication takes place as shared neural activation between sender and receiver of a message, regardless of the communicatory medium, whether it is a manual or articulatory gesture.

V. Conclusions

During the course of primate evolution, manual gesture and articulation have gradually come under shared cortical control (areas BA44 and BA45), resulting in an intricate network subserving gesture and speech in apes and humans. Beginning lateralization of communication, as evident in apes, may constitute a precursor to the evolution of left-hemispheric language areas in the human brain.

Acknowledgment

I thank David Loyall for his valuable help with the figures in this work.

References

Aflalo, T., Graziano, M.S.A., 2006. Partial tuning of motor cortex neurons to final posture in a free-moving paradigm. Proc. Natl. Acad. Sci. USA 103 (8), 2909–2914.

Armstrong, D.F., Stokoe, W.C., Wilcox, S.E., 1995. Gesture and the Nature of Language. Cambridge University Press, Cambridge, UK.

Bernston, G.G., Boysen, T.S., Torello, M.W., 1993. Vocal perception, brain event related potentials in a chimpanzee. Dev. Psychobiol. 26, 305–319.

Binkofski, F., Buccino, G., 2006. The role of ventral premotor cortex in action execution and action understanding. J. Physiol. (Paris) 99, 396–405.

Bonda, E., Petrides, M., Frey, S., Evans, A.C., 1994. Frontal cortex involvement in organized sequence of hand movements: evidence from positron emission tomography study. Soc. Neurosci. Abstr. 20.353.

Broca, P., 1861. Sur le principe des localisations cérébrales. Bulletin de la Société d'Anthropologie 2, 190–204.

Byrne, R.W., Byrne, J.M., 1991. Hand preferences in the skilled gathering tasks of mountain gorillas (*Gorilla gorilla berengei*). Cortex 27, 521–536.

Cantalupo, C., Hopkins, W.D., 2001. Asymmetric Broca's area in great apes: a region of the ape brain is uncannily similar to one linked with speech in humans. Nature 29, 505.

Cohen, Y.E., Hauser, M.D., Russ, B.E., 2006. Spontaneous processing of abstract categorical information in the ventrolateral prefrontal cortex. Biol. Lett. 2, 261–265.

Corballis, M.C., 2002. From Hand to Mouth: The Origins of Language. Princeton University Press, Princeton, NJ.

Deacon, T., 1997. The Symbolic Species: The Co-Evolution of Language and the Human Brain. W. W. Norton & Co., New York, NY.

Eberl, E.M., 2008. A primate "motor theory of sound perception:" three case studies with gibbons. Gibbon J. 4, 64–72.

Esposito, A., Demeurisse, G., Alberti, B., Fabbro, F., 1999. Complete mutism after midbrain periaqueductal gray lesion. Neuroreport 10, 681–685.

Forrester, G.S., 2008. A multidimensional approach to investigations of behaviour: revealing structure in animal communication signals. Anim. Behav. 76, 1749–1760.

Fowler, C.A., Rubin, P., Remez, R.E., Turvey, M.T., 1980. Implications for speech production of a general theory of action. In: Buttersworth, B. (Ed.), Language Production, Vol. 1. Speech and Talk. Academic Press, London, UK, pp. 373–420.

Gentilucci, M., Benuzzi, F., Gangitano, M., Grimaldi, S., 2001. Grasp with hand and mouth: a kinematic study on healthy subjects. J. Neurophysiol. 86, 1685–1699.

Gifford III, G.W., MacLean, K.A., Hauser, M.D., Cohen, Y.E., 2005. The neurophysiology of functionally meaningful categories: macaque ventrolateral prefrontal cortex plays a critical role in spontaneous categorization of species-specific vocalizations. J. Cogn. Neurosci. 17, 1471–1482.

Graziano, M.S.A., 2006. The organization of behavioral repertoire in motor cortex. Annu. Rev. Neurosci. 29, 105–134.

Graziano, M.S.A., Aflalo, T., 2007. Mapping behavioral repertoire onto the cortex. Neuron 56, 239–251.

Graziano, M.S.A., Taylor, C.S.R., Cooke, D.F., Moore, T., 2004. A map of complex movements in motor cortex of primates. In: Humphreys, G.W., Riddoch, M.J. (Eds.), Attention in Action – Advances from Cognitive Neuroscience. Psychology Press, Hove, UK, pp. 211–232.

Hauser, M.D., 1996. The Evolution of Communication. Bradford MIT, Cambridge, MA.

Hauser, M.D., Andersson, K., 1994. Left hemisphere dominance for processing vocalizations in adult, but not infant, rhesus monkeys: field experiments. Proc. Natl. Acad. Sci. USA 91, 3946–3948.

Hauser, M.D., Akre, K., 2001. Asymmetries in the timing of facial and vocal expression in rhesus monkeys: implications for hemispheric specialization. Anim. Behav. 61, 391–408.

Heilbronner, P.L., Holloway, R.L., 1988. Anatomical brain asymmetries in New World and Old World monkeys. Stages of temporal lobe development in primate evolution. Am. J. Phys. Anthropol. 76, 39–48.

Hopkins, W.D., 1993. Posture and reaching in chimpanzees (*Pan troglodytes*) and orang-utans (*Pongo pygmaeus*). J. Comp. Psychol. 17, 162–168.

Hopkins, W.D., Cantalupo, C., 2004. Handedness in chimpanzees (*Pan troglodytes*) is associated with asymmetries of the primary motor cortex but not with homologous language areas. Behav. Neurosci. 118, 1176–1183.

Hopkins, W.D., Cantero, M., 2003. From hand to mouth in the evolution of language: the influence of vocal behavior on lateralized hand use in manual gestures by chimpanzees (*Pan troglodytes*). Dev. Sci. 6, 55–61.

Hopkins, W.D., Pilcher, D.L., 2001. Neuroanatomical localization of the motor hand area using magnetic resonance imaging: the left hemisphere is larger in great apes. Behav. Neurosci. 115, 1159–1164.

Hopkins, W.D., Cantalupo, C., Wesley, M.J., Hostetter, A. B., Pilcher, D.L., 2002. Grip morphology and hand use in chimpanzees (*Pan troglodytes*): evidence of a left hemisphere specialization in motor skill. J. Exp. Psychol. Gen. 131, 412–423.

Hopkins, W.D., Cantalupo, C., Taglialatela, J., 2006. Handedness is associated with asymmetries in gyrification of the cerebral cortex of chimpanzees. Cerebral Cortex 17, 1750–1756.

Jürgens, U., 2004. Vergleichende zentral-motorische Innervation des Larynx. 21. Wissenschaftliche Jahrestagung der DGPP Deutsche Gesellschaft für Phoniatrie und Pädaudiologie. Freiburg: September 10–12, 2004. URL: <http://www.egms.de/en/meetings/dgpp2004/04dgpp08.shtml/>

Jürgens, U., Zwirner, P., 2000. Individual hemispheric asymmetry in vocal fold control of the squirrel monkey. Behav. Brain Res. 109, 213–217.

Liberman, A.M., Mattingly, I.G., 1985. The motor theory of speech perception revised. Cognition 21, 1–36.

Luders, E., Narr, K.L., Thompson, P.M., Rex, D.E., Jancke, L., Steinmetz, H., Toga, A.W., 2004. Gender differences in cortical complexity. Nature Neurosci. 7, 799–800.

Lüthe, L., Häusler, U., Jürgens, U., 2000. Neuronal activity in the medulla oblongata during vocalization. A single-unit recording study in the squirrel monkey. Behav. Brain Res. 116, 197–210.

Matelli, M., Luppino, G., Rizzolatti, G., 1985. Patterns of cytochrome oxidase activity in the frontal agranular cortex of the macaque monkey. Behav. Brain Res. 18, 125–136.

McGrew, W.C., Marchant, L.F., 1997. On the other hand: current issues in and meta-analysis of the behavioral laterality of hand function in nonhuman primates. Yearbook of Physical Anthropology 40, 201–232.

Musso, M., Moro, A., Glauche, V., Rijntjes, M., Reichenbach, J., Büchel, C., Weiller, C., 2003. Broca's area and the language instinct. Nature Neurosci. 6, 774–781.

Olson, D.A., Ellis, J.E., Nadler, R.D., 1990. Hand preferences in captive gorillas, orang-utans, and gibbons. Am. J. Primatol. 20, 83–94.

Petersen, M.R., Beecher, M.D., Zoloth, S.R., Green, S., Marler, P.R., Moody, D.B., Stebbins, W.C., 1984. Neural lateralization of vocalizations by Japanese macaques: communicative significance is more important than acoustic structure. Behav. Neurosci. 98, 779–790.

Pika, S., Liebal, K., Call, J., Tomasello, M., 2005. The gestural communication of apes. Gesture 5, 41–56.

Ploog, D.W., 1992. The evolution of vocal communication. In: Papoušek, H., Jürgens, U., Papoušek, M. (Eds.), Nonverbal Vocal Communication. Cambridge University Press, Cambridge, UK, pp. 6–30.

Rizzolatti, G., Craighero, L., 2004. The mirror-neuron system. Annu. Rev. Neurosci. 27, 169–192.

Rizzolatti, G., Camarda, R., Fogassi, L., Gentilucci, M., Luppino, G., Matelli, M., 1988. Functional organization of inferior area 6 in the macaque monkey. II. Area F5 and the control of distal movements. Exp. Brain Res. 71, 491–507.

Romanski, L.M., Averbeck, B.B., Diltz, M., 2005. Neural representation of vocalization in the primate ventrolateral prefrontal cortex. J. Neurophysiol. 93, 734–747.

Sahin, N., Pinker, S., Halgren, E., 2006. Abstract grammatical processing of nouns and verbs in Broca's area: evidence from fMRI. Cortex 42, 540–562.

Shafer, D.D., 1993. Patterns of hand preference in gorillas and children. In: Ward, J.P., Hopkins, W.D. (Eds.), Primate Laterality: Current Behavioral Evidence of Primate Asymmetries. Springer, New York, NY, pp. 67–283.

Shafer, D.D., 1997. Hand preference behaviors shared by two groups of captive bonobos. Primates 38, 303–313.

Steklis, H.D., Harnad, S., 1976. From hand to mouth: some critical stages in the evolution of language. Ann. N.Y. Acad. Sci. 280, 445–455.

Tomasello, M., Zuberbühler, K., 2002. Primate vocal and gestural communication. In: Bekoff, M., Allen, C., Burghardt, G.M. (Eds.), The Cognitive Animal: Empirical and Theoretical Perspectives on Animal Cognition. MIT Press, Cambridge, MA, pp. 293–299.

Vauclair, J., 2004. Lateralization of communicative signals in nonhuman primates and the hypothesis of the gestural origin of language. Interaction Studies 5, 365–386.

Waters, G.S., Fouts, R.S., 2002. Sympathetic mouth movements accompanying fine motor movements in chimpanzees (*Pan troglodytes*) with implications toward the evolution of language. Neurol. Res. 24, 174–180.

Willems, R.M., Hagoort, P., 2007. Neural evidence for the interplay between language, gesture, and action: a review. Brain Lang. 103, 278–289.

CHAPTER 11.2

The generation of functionally referential and motivational vocal signals in mammals

Marta B. Manser[*]

Zoological Institute, University of Zurich, Zurich, Switzerland

Abstract: Animal vocalizations may be divided into functionally referential calls and motivational calls, depending on whether the calls refer to objects and events in the external environment or whether they are the expression of the internal state of the caller although most call types include probably both kinds of information to a receiver. In this chapter I compare several aspects of these two call categories, including the context in which they are produced and what they refer to, how referential and motivational information is expressed in vocal behavior, and what is their ontogeny and evolution. Functionally referential calls have only been described in the context of encounters with a predator, food resource and in agonistic social interactions. All other animal vocalizations are considered motivational calls, although they may contain some referential information to the receiver regarding the context the caller experiences. Functionally referential signals refer to specific attributes of the eliciting external stimuli and cause an appropriate response in the receivers in the absence of any additional information. Motivational calls alternatively refer not to a specific external event, but to the emotional state of the caller. Both referential and motivational information are expressed in frequency-related acoustic parameters. In addition, referential information, which appears at a later age than motivational information, also seems to be expressed in the combination of different call types, whereas motivational information relates more to temporal aspects of calls. The evolution of functionally referential calls is limited to a small number of contexts where the receiver's responses to calls elicited by different attributes of the external stimuli are of high sociobiological benefit. So far, ecological as well as social constraints have been reported as limiting factors.

Key words: functionally referential calls; motivational calls; alarm calls; food/resource calls; vocalization function; ontogeny; evolution; acoustic structure; call combination

I. Introduction

Depending on the context and the calling individual, animal vocalizations vary in structure and the rate at which they are emitted. Over the last 30 years, evidence has accumulated that, besides the expression of the emotional state of an individual and information on individual physical traits, some calls can refer to external events or objects (for a review see Bradbury and Vehrencamp, 1998). It has been argued for a long time that the variation in animal calls expressed only the emotion of an individual (Darwin, 1872). With the documentation of alarm calls in vervet monkeys (*Cercopithecus aethiops*) referring to the predator type (Struhsaker, 1967; Seyfarth et al., 1980), it became clear that some calls denote external events and receivers responded to the calls as if the eliciting stimuli (predator) were present. These calls have been defined as functionally referential vocalizations (Marler et al., 1992). In contrast, alarm calls that are not predator-specific, and appear to relate to the level of urgency and to express the emotional state of the caller, are referred to as motivational calls (Robinson 1980; Blumstein and Armitage, 1997).

Functionally referential calls are distinguished from motivational calls due to their high production and perception specificities. This subsequently allows receivers to gain accurate information and respond independently without any additional information. Motivational calls, on the other hand, are the expression of the caller's internal state in a defined context (e.g., foraging, leading

[*]Corresponding author. E-mail: marta.manser@zool.uzh.ch

Stefan M. Brudzynski (Ed.)
Handbook of Mammalian Vocalization
ISBN 978-0-12-374593-4

DOI: 10.1016/B978-0-12-374593-4.00043-7

the group, alarm, etc.) and they do not refer to accurate information about an external event or object. As a consequence, the receivers do not gain precise information regarding the external environment, but may still gain some referential information from almost any call type as long as it is closely associated with a given context (Seyfarth and Cheney, 2003a,b). These differences between functionally referential calls and motivational calls in the eliciting stimuli and in the information conveyed by them to the receivers are reflected in several aspects of the generation of these two different call categories.

In this chapter, I review vocal signals according to their referential or motivational information in mammals. In particular, I address the following questions: (1) in what context do we find functionally referential calls and motivational calls; (2) what do functionally referential calls refer to, and what do motivational calls express; (3) are functionally referential calls denotative, imperative, or do they refer to the subsequent behavior of the caller; (4) how are functionally referential and motivational aspects expressed in the vocal behavior; (5) how does referential and motivational information develop; (6) when do functionally referential calls and motivational calls evolve?

II. In what context do we find functionally referential and when are motivational calls used?

Functionally referential calls have only been described in a few situations, where it seems beneficial to emit different call types to specific stimuli in the external environment and specific contexts that allow receivers to show different adaptive responses. Such vocalizations have been described for a few species in the context of predators and also food encounters (Table 1). Functional referents have also been suggested for agonistic screams in primates (Gouzoules and Gouzoules, 1984; Slocombe and Zuberbühler, 2005a,b; Slocombe et al., 2009). These calls can typically be distinguished as discrete acoustic signals, although animals also appear to perceive graded calls as discrete units (Fischer and Hammerschmidt, 2001; Fischer et al., 2001) (Table 3).

Most animal vocalizations, however, are considered to be motivational, as they express the emotional state of the caller and they are not always elicited by specific external objects or events (Table 1). In particular, in social groups that also forage together, almost any behavior is accompanied by vocalizations allowing the coordination of group movement and activity,

Table 1. Studies of contexts where motivational and functionally referential calls have been documented

Context	Call type	Call category	Species	Reference
Antipredator strategies	Alarm calls	Functionally referential	Vervet monkeys	Seyfarth et al., 1980
			Diana monkeys	Zuberbühler et al., 1997
			Baboons	Fischer et al., 2001
			Meerkats	Manser, 2001
			Prairie dogs	Kiriazis and Slobodchikoff, 2006
		Motivational	Belding's ground squirrels	Robinson, 1980
			Cape ground squirrels	Furrer and Manser, 2009
			Marmots	Blumstein, 1995
Foraging	Food calls	Functionally referential	Toque macaque	Dittus, 1984
		Motivational	Capuchin monkeys	Gros-Louis, 2004
Social interactions	Screams	Functionally referential	Rhesus macaques	Gouzoules et al., 1984
			Chimpanzees	Slocombe and Zuberbühler, 2005a,b; Slocombe et al., 2009
	Reconciliation calls	Motivational	Baboons	Cheney et al., 1995
Spatial orientation	Contact calls Lead calls Separation calls	Motivational	Golden tamarins	Boinski et al., 1994
Offspring–parent communication	Begging calls	Motivational	Meerkats	Manser et al., 2008

particularly when visual contact is difficult or impossible due to vegetation. Such calls express underlying motivation of the current behavior of the caller, e.g., contact calls while foraging (Robinson, 1982), lead calls when an individual initiates group movement (Boinski et al., 1994), sentinel calls while on guard (Rasa, 1986), grooming calls while grooming each other (Cheney et al., 1995), etc. In the context of predator encounters, where most of the functionally referential calls have been described, non-specific alarm calls that are given to a variety of different predator types have been considered as being the expression of emotional state with little referential specificity (Macedonia and Evans, 1993). The explanation behind this assumption is that receivers do not gain much information from these non-specific alarm calls and they do not respond in obviously different ways to different sources of danger. Motivational alarm calls have been documented for many sciurid and other rodent species (Robinson, 1980; Blumstein and Armitage, 1997; Litvin et al., Chapter 5.1 in this volume) and also birds (Leavesley and Magrath, 2005). These species use different graded call types and express the corresponding level of urgency in responding to the approaching predator, independent of whether it is an aerial or terrestrial animal.

Recent studies have shown that alarm calls can contain both functionally referential and motivational information, and that the production specificity can vary between call types in the same context (e.g., predator encounter) within the species. In meerkat (*Suricata suricatta*) alarm calls, the acoustic structure differs depending on the predator type and also varies in a consistent way in relation to the distance of the approaching predator, which is likely to reflect response urgency (Manser, 2001; Manser et al., 2001, 2002). In red-fronted lemurs (*Eulemur fulvus rufus*) and white sifakas (*Propitheus verreauxi verreauxi*), a specific alarm call type is given to aerial predators, but the alarm call given to terrestrial predators is also given in other contexts, suggesting that production specificity for the different call types can vary (Fichtel and Kappeler, 2002).

III. What do functionally referential calls refer to and what do motivational calls express?

III.A. Functionally referential calls

To qualify as functionally referential, the vocalization must correlate in some respect with specific attributes of an object or an event in the external environment of the caller. In several monkey species (Seyfarth and Cheney, 1980; Zuberbühler, 2000), other mammals (Manser, 2001) and bird species (Gyger et al., 1987), specific alarm call types are elicited by defined predator types, such as aerial or terrestrial predators. However, in most of these studies it is not clear whether these calls denote the spatial area of the predator approaching, physical properties of the predator, or its behavior (Evans et al., 1993; Manser and Fletcher, 2004; Templeton et al., 2005). Since aerial and terrestrial predators hunt in different ways, and also from different spatial areas – raptors (birds of prey) approach with high speed out of the air, while terrestrial animals approach with moderate speed on the ground – the variation of these call types may refer to the way of approach, or the location of the predator (Manser and Fletcher, 2004; Templeton et al., 2005).

For bird and mammalian species, it has been shown that they emit different call types to the same species of predators depending on their spatial location. Chickadees (*Poecile atricapilla*) produce seet calls to raptors in flights and mobbing calls to perched raptors (Templeton et al., 2005). Siberian jays also emit different calls to the same species of raptor depending on whether it is flying or perched in a tree (Griesser, 2008). In the same way, meerkats bark at perched raptors (Manser, 2001) and do not emit the predator-specific aerial alarm calls given to raptors in the air. Also, a terrestrial predator encountered resting in a bolthole elicits mobbing behavior with recruitment calls in meerkats, while the same predator species walking or standing in the surrounding elicits terrestrial alarm calls (Manser, 2001). Therefore, predator-specific calls in the different species may likely refer to the spatial location of the approaching animal, and only be secondary to a specific predator species or category itself. However, the receivers may still recognize the difference as predator species-specificity. For prairie dogs (*Cynomys gunnisoni*), it has been shown that calls appear to relate to specific predator species within terrestrial predators, and also to different characteristics of humans (Frederiksen and Slobodchikoff, 2007). Yet, in most other studies, physical attributes such as size, color or shape, which correlate to the risk of the encountered stimuli, appear to elicit the object-specific calls in a lesser degree, but more so to affect the arousal of the caller expressed in a graded call system rather than discrete call types.

Call specificity has also been suggested to refer to the behavior of the encountered predator (e.g., hunting behavior) (Templeton et al., 2005; Griesser, 2008), rather than the spatial location of the predator causing the use of different call types. Studies investigating the influence of the spatial position and the speed of predators conducted under controlled conditions on chickens in captivity (Evans et al., 1993) revealed that predators approaching from the air or on the ground elicited acoustically-distinct call types. Speed differences were shown to change the call rate, but not the structure. This result is supported by the pattern observed in meerkats that circling raptors at some distance to the groups elicit aerial alarm calls in the same way as raptors directly approaching them with high speed (Manser, personal observation). The only clear evidence for a call to refer to the behavior of the predators has recently been demonstrated in Siberian jays (Griesser, 2008). The alarm call types in this species were specific to the behavior (perched, prey search and attack) of different species of hawks, and elicited distinct escape responses. In a different way, a call referring to the behavior of another animal than the caller has been shown for a non-specific alarm call type, the "animal moving" call in meerkats (Manser 2009). Animal moving calls are emitted to almost any category of animals in their environment: to low flying non-dangerous birds close to the group, such as vultures or Southern yellow-billed hornbills (*Tockus leucomelas*); to raptors, such as eagles flying close on the ground far away; to perched raptors moving their wings to potentially fly off; to non-dangerous terrestrial animals, such as herbivores; to terrestrial predators, in particular when moving after having interrupted their approach; and also to foreign meerkats close by. As it has been experimentally confirmed, the consistent aspect in all these contexts initiating the call was that the animals were moving.

The functionally referential aspect for food calls and agonistic calls is less clear than for alarm calls. For most studies, the calls given in the context of encountering food appear to be distinct from call types in other contexts (Evans and Evans, 1999; Di Bitetti, 2003). It has also been shown for a few species that the acoustic structure varies when the animal is encountering different food sources (Slocombe and Zuberbühler, 2006). In particular, calls can vary systematically with food preference (Elowson et al., 1991; Hauser and Marler, 1993a,b), food quality (Hauser, 1998) and food divisibility (Hauser et al., 1993a,b). However, within the context of encountering food, for most species, the physical properties of the signal appear to change in a continuously graded manner with food amount or quality related to arousal, or the presence of other conspecifics. Similar agonistic calls described as functionally referential vary depending on the status of the individual involved in the social interaction or the severity of the interaction. In addition to this, receivers show clear differences in their responses by looking for a prolonged period of time towards the caller, more likely to approach the location of the food source. However, it is difficult to distinguish whether these are discrete different responses, or the differences reflect the emotional state of the caller and therefore should be classified as motivational calls, as has been described for many sciurid alarm calls that express the level of response urgency (Blumstein, 2007).

III.B. Motivational calls

Motivational calls have been referred to as call types that are elicited by a wide range of environmental circumstances and vary in their acoustic structure between and within a context, depending on the emotional state of the caller. The acoustic variation between motivational call types likely reflects different valences (e.g., fear, anxiety, comfort, etc.) (Paul et al., 2005; Bugrdorf and Moskal, Chapter 6.2 in this volume), whereas the variation within motivational call types may be due to differences in arousal states (Brecovitch et al., 1995; Fichtel et al., 2001; Blumstein et al., 2006). Valences are predicted to be similar if it is a positive context, such as friendly social interactions (e.g., grooming), or negative, hostile (agonistic) interactions (e.g., aggression: food competition, dominance assertions; fear: appeasement, predator responses; mix of aggression and fear: mobbing of predators) (Morton, 1977). The arousal of a caller seems to relate to the urgency to respond, e.g., to an approaching predator, depending on the distance (Manser, 2001), its speed (Evans et al., 1993) and size (Templeton et al., 2005; Palleroni et al., 2005).

IV. Are functionally referential calls denotative, imperative, or do they have "behavioral referents"?

Whether animal vocalizations are denotative, imperative, or have behavioral referents is of interest because these options suggest different underlying cognitive mechanisms in the production of the calls (Evans, 1997; Hauser, 1998). This question has been discussed, in particular, in the context of functionally referential alarm calls (Cheney and Seyfarth, 1990; Marler et al., 1992; Evans, 1997; Palleroni et al., 2005). If calls have a denotative function, it means they represent labels for stimulus properties (Table 2). If they have an imperative function, it means the signaler instructs other group members on the appropriate response at the given moment. If the calls had behavioral referents by referring to the subsequent behavior of the caller, the acoustic structure should relate to the escape mode of that individual (Smith, 1981; Hauser, 1998). In the case of the denotative function, the call may express a simple association to the assessed threat. If calls had behavioral referents and are related to the escape mode of the signaler, the caller would have to assess the nature of the threat and also a suitable response depending on the context it experiences in that instance. Imperative calls may be cognitively more challenging, because they require the signaler to assess both the threat and the appropriate response to be emitted to the receivers, thereby taking the location and the behavior of the audience into account. Considering it from the receiver side, with the information on stimuli properties, receivers are able to assess the danger of the situation and adjust their response according to their own behavior at the moment of the alarm. In contrast, if the calls were imperative, receivers would have to rely on the assessment of the caller as to the correct escape strategy to be used.

To distinguish whether calls are denotative, imperative, or have behavioral referents may be difficult in most systems, as functionally referential alarm calls appear to have evolved in species where predator-specific calls are linked with the escape mode of the animals (Macedonia and Evans, 1993). Yet, if the animals gave the same alarm call to the same stimulus independent of their momentary behavior/situation (for example, foraging and away from shelter versus being close to a shelter location without need to escape), this would clearly suggest that these calls were denotative. If the calls varied depending on the location of the receiver in relation to shelter, this may suggest they have an imperative function.

The fact that meerkat alarm calls vary with the properties of the predator approach, independent of the spatial position in relation to shelter or the behavior of the signaler or receiver (Manser, 2009), supports the hypothesis that functionally referential alarm calls in this system are denotative. Denotative calls may be more informative than imperative calls to the receivers if individuals within groups are typically dispersed while foraging, in such a way that the trade-off between predation risk and energy gain from foraging may continuously differ among individuals. Each of the receivers is likely to be in a context that may vary in specific attributes, such as distance to the shelter location, distance to the approaching predator, location within the group, or costs of escape due to foraging investment. In particular in meerkats, which depend on mobile prey, the trade-off between risk of predation and food intake may play an important role in their response to alarm calls. Therefore, receivers may benefit more by information from the signaler referring to the stimuli, than information on

Table 2. Vocal parameters expressing referential and motivational information

Vocal parameter	Motivational		Referential
	Valences	Arousal	
Energetic distribution	yes	yes	yes
Call sequence	no	no	yes
Repetition of call	no	yes	no
Call rate	no	yes	no

Table 3. Production and perception specificity in studies reporting functionally referential calls. Examples of species that use distinct or graded call types and show the according distinct or graded response to them

		Perception/response type	
		Distinct	Graded
Production/call type	distinct	e.g., Vervet monkey alarm calls (Seyfarth et al., 1980)	e.g., Gunnison prairie dogs alarm calls (Kiriazis and Slobodchikoff, 2006)
	graded	e.g., Baboons (Fischer et al., 2001)	e.g., Chimpanzee screams (Slocombe et al., in press)

how to escape. This allows receivers to adjust their own escape strategy at a given moment according to the information on the approaching stimuli, rather than on the assessment of the signaler in regard to which escape behavior to employ. However, we cannot generalize that functionally referential alarm call are always denotative until more species have been tested. In species where for a caller a specific receiver or group of receivers may be more important than others, it may preferably to produce the call that takes into account the best response for those individuals. The only evidence that callers take into account the degree of danger experienced by other group members comes from blue monkeys (*Cercopithecus mitis stuhlmanni*) (Papworth et al., 2008). Here, the males produced more alarm calls when their group members were close to the playback stimulus rather than far away, independent of their own spatial position.

V. How are functionally referential and motivational aspects expressed in vocal behavior?

Functionally referential and motivational information of the calls are expressed through different aspects in the vocal production. Both referential and motivational calls within their categories differ in the acoustic structure of the calls. In particular, energy distributions at different frequency and temporal locations cause differences in frequency parameters, such as height of fundamental frequency, modulation of fundamental frequency, bandwidth and pulsed structure (Evans, 1997; Manser, 2001). Recent empirical evidence from acoustic analyses of meerkat alarm calls shows that the referential information is conveyed by different acoustic parameters than the motivational information within the same call type (Manser, 2001). The change in the acoustic structure related to the level of urgency to respond, at least partly, supports the rules put forward by Morton's motivational structure rules (Morton, 1977). These rules attempt to describe specific aspects of the acoustic structure of calls in relation to the motivation of the caller.

In addition to differences in the acoustic structure, functionally referential information also appears to be encoded by the specific combination of call types. In several primate species, consistent differences in how calls are assembled into sequences appear to convey functionally referential information to the receiver in a predator-specific way (Stephan and Zuberbühler, 2008). For example, putty-nosed monkeys (*Cercopithecus nictitans*) produce two alarm call types with predator-specific information and also with information as to the initiation of group travel depending on the combination of the within-the-call sequence (Arnold and Zuberbühler, 2006). Diana monkeys, at two different study sites with different predator pressure, emitted call sequences that differed in composition and number of calls when they were exposed to additional predator species (Stephan and Zuberbühler, 2008). Similarly, King colobus (*Colobus polykomos*) and guereza monkeys (*Colobus guereza*) emitted two alarm call types, which they assemble differently to meaningful predator-specific sequences (Schel et al., 2009, in press).

Motivational information relates more to temporal aspects, such as call rate, call duration, or lengths of a call bout. One parameter, namely the repetition of the call, correlates linearly across all different call types and contexts with the arousal of the animal. In several studies, calls were more frequent and fast-repeated as the urgency of the situation increased (Blumstein and Armitage, 1997; Manser, 2001; Schehka et al., 2007). This is particularly well-documented in chickens, with videos simulating the approach of predators at different speeds (Evans et al., 1993). Furthermore, although the call's acoustic structure varies with the predator-specific call type and varies in a linear fashion with the arousal of the caller, the animal moving call in meerkats does not change in its acoustic structure. This being said, the repetition of the call does increase with the tension of the situation for this call type (Manser, 2009). Only for the call types emitted under immediate threat and extreme time constraints (high urgency aerial call and panic call), meerkats (Manser, 2001) and marmots (Blumstein and Armitage, 1997) emit a single or a few calls and do not repeat them.

VI. The ontogeny of functionally referential and motivational calls

Little is known about how the characteristics encoding referential and motivational information in animal vocalizations develop. Although several studies investigated the development of response to functionally referential signals with age (Seyfarth and Cheney, 1986; Fichtel, 2008), not much information on the ontogeny of the production of these call types exists

(Hollén and Manser, 2007). In human speech, children seem to enter the language system of producing words through the use of vocal forms which are more adult-like in sound structure than in their semantic function, and may stimulate participation in social interactions rather than transmitting information (Locke and Snow, 1997). For functionally referential alarm calls it had been suggested that they might evolve from motivational alarm calls (Macedonia, 1993). There is some support for this from the alarm calls of meerkats that seem to convey the information about the level of urgency along a general common rule, whereas the referential information does not follow a linear change (Manser, 2001; Hollén and Manser, 2007). Furthermore, referential aspects of alarm calls change much more than motivational aspects over time, and predator-specific alarm calls appear much later in the vocal repertoire than non-specific alarm calls (Hollén et al., 2008). Taken together, these findings suggest that producing and correctly articulating calls encoding referential information requires practice and shows similarities to that observed in human speech development, where children learn to attach specific meanings to different words (Locke and Snow, 1997; Hollén and Manser, 2007).

VII. When do functionally referential calls and motivational calls evolve?

The evolution of functionally referential calls has been explained by the need for more than one adaptive way of responding, in the context of alarm calls (Macedonia and Evans, 1993) and food calls (Evans and Evans, 1999; Slocombe and Zuberbühler, 2006). For functionally referential alarm calls, it is beneficial to escape differently from predators with different hunting strategies (Marler, 1967; Cheney and Seyfarth, 1990; Macedonia and Evans, 1993). For example, ring-tailed lemurs (*Lemur catta*) are semi-arboreal and live in open woodlands. In this complex habitat, they are hunted in different ways by different predators, which they escape from using specific strategies, profiting from the structured environment (Macedonia and Evans, 1993). The evolution of predator-specific calls has been explained in the same way for other primate species and also for red squirrels (*Tamiasciurus hudsonicus*) (Greene and Meagher, 1998). In contrast species that live in relatively open habitats, for example some ground-dwelling rodents, run to their burrows in response to any predator type (Blumstein and Armitage, 1997a). For these species, information about the urgency to respond may be more important than about predator type.

However, the habitat structure does not explain why meerkats have evolved functionally referential alarm calls, since they live like many ground-dwelling rodents on relatively open, unstructured plains and use several burrow systems to retreat (Doolan and Macdonald, 1996; Manser and Bell, 2004). Unlike other species that emit referential alarm calls, their only escape from predators is to retreat into a bolt-hole (Manser et al., 2001). The comparison of the anti-predator behavior with the sympatric living Cape ground squirrel (*Xerus inauris*) suggests instead that group coordination may explain why meerkats evolved functionally referential alarm calls, allowing them to increase their foraging efficiency substantially in an open habitat (Furrer and Manser, 2009). This may also explain the evolution of functionally referential alarm calls in the group-foraging dwarf mongoose (Beynon and Rasa, 1989), whereas the solitary-foraging yellow mongoose uses urgency-dependent alarm calls, unrelated to the predator type (Le Roux et al., 2009). Whether group coordination may also explain the evolution of functionally referential alarm calls in Gunnison's prairie dogs (*Cynomys gunnisoni*), another ground dwelling species, is not clear. They emit different alarm calls in response to aerial and terrestrial predators (Placer and Slobodchikoff, 2000, 2001), with receivers showing adaptive subtle differences to calls elicited by different predators (Kiriazis and Slobodchikoff, 2006). In marmot species (*Marmota* spp.), the range of the alarm call repertoire varies with the degree of social complexity (Blumstein and Armitage, 1997b), with more socially complex species producing more alarm call types (Blumstein, 2003). However, the evidence that social complexity triggers the evolution of functional reference is lacking, as none of these marmot species emit functionally referential alarm calls (Blumstein, 2007). Although in some species different adaptive escape strategies due to habitat structure may trigger the evolution of predator-specific calls, social constraints, such as the coordination of group movements for increased foraging efficiency, should also be considered when investigating the evolution of functionally referential alarm calls (Furrer and Manser, 2009).

The evolution of functionally referential food calls (Evans, 1997; Hauser, 1998; Slocombe and Zuberbühler, 2006) is less clear than that of alarm calls. Food calls typically cause the receivers to initiate feeding rather than any other behavior (Evans and Evans, 1999; Di Bitteti, 2003) or approach to the caller (Hauser, 1998), and it is difficult to distinguish any further between different response categories. Only the recent studies on captive chimpanzees (Slocombe and Zuberbühler, 2005b, 2006) have suggested that receivers use the referential information in the call to know where to look for the specific food type. The authors suggested that referential information on food type in the wild may not be as important as the "meaning" of calls. Instead, the information may be taken from a combination of call type and context.

VIII. Conclusions

The generation of functionally referential calls and motivational calls has been of interest since the first alarm calls described to refer to objects and events in the external environment, rather than to express the emotional state of the caller. This is partly due to the fact that functionally referential signals have been suggested as precursors of human speech, and also because they are a useful tool to investigate how animals categorize their external environment and information processes their behavior underlies.

However, despite the rather clear definition of calls to be considered as functional referential calls, several recent studies have shown that the distinction between the two call categories may be more complex. Firstly, animals may categorize the external environment in a different way than predicted by observers from their studies. This may suggest a nonspecific call production to us, although it may be highly specific for the animal. For example, the production of moving animal calls in meerkats elicited by different predator types, but always when the animal moves, is regarded as nonspecific within the context of predator type-specificity. However, within the specificity regarding the behavior of the external object it is defined within a narrow context. This may, in a similar way, explain why aerial alarm calls, but not terrestrial alarm calls in sifakas are highly production-specific. The information of terrestrial calls may have a different function than just alerting to predators, but to this species may still be highly specific. Second, animal species differ in the variation of their vocal repertoire, where some appear to have discrete clear-cut call types, other species use graded calls. Few studies have shown that animals are able to extract discrete information from such graded call systems and behave accordingly. Third, the receivers may not have to show distinct response categories, but behaviors that gradually change from one category to another. For example, whether the response in prarie dogs to run to the shelter and observe the predator is a discrete different response from an escape into a bolthole, or whether this response reflects an increase in the emotional state of the caller, is difficult to say.

Considering the variation in the interpretation of functionally referential information versus motivational information the question arises whether the dichotomy of these two call categories is helpful. I would argue that it still helps to elucidate the biosociological meaning of different call types based on the referential information to events or objects in the external environment used by different species. It also helps to understand the mechanisms underlying the production and perception of a given call category. From this point of view, it seems beneficial to keep this distinction as long as it is clearly documented what aspects of the production and perception side fulfill the criteria of referential information. Studies of referential and motivational aspects of calls will also help to identify how animals categorize their external environment, and may allow identification of the underlying information processes the different behaviors are based on.

Acknowledgments

I would like to thank Stefan M. Brudzynski as editor of this book for his support in writing this chapter. Many thanks to Simon Townsend, Alex Thornton and Tim Clutton-Brock for their comments on previous versions and corrections to language.

References

Arnold, K., Zuberbühler, K., 2006. Language evolution: semantic combinations in primate calls. Nature 441, 303.

Beynon, P., Rasa, A.E., 1989. Do dwarf mongooses have a language – warning vocalizations transmit complex information. S. Afr. J. Sci. 85, 447–450.

Blumstein, D.T., 1995. Golden-marmot alarm calls: the meaning of situationally specific calls. Ethol. 100, 113–125.

Blumstein, D.T., 2003. Social complexity but not the acoustic environment is responsible for the evolution of complex alarm communication. In: Ramousse, R., Allainé, D., Le Berre, M. (Eds.), Adaptive Strategies and Diversity in Marmots. International Marmot Network, Lyon, France, pp. 41–49.

Blumstein, D.T., 2007. The evolution, function, and meaning of marmot alarm communication. Adv. Stud. Behav. 37, 371–401.

Blumstein, D.T., Armitage, K.B., 1997a. Alarm calling in yellow-bellied marmots: I. The meaning of situationally variable alarm calls. Anim. Behav. 53, 143–171.

Blumstein, D.T., Armitage, K.B., 1997b. Does sociality drive the evolution of communicative complexity? A comparative test with ground-dwelling sciurid alarm calls. Am. Nat. 150, 179–200.

Blumstein, D.T., Patton, M.L., Saltzman, W., 2006. Faecal glucocorticoid metabolites and alarm calling in free-living yellow-bellied marmots. Biol. Lett. 2, 29–32.

Boinski, S., Moraes, E., Kleiman, D.G., Dietz, J.M., Bakker, A.J., 1994. Intra-group vocal behaviour in wild golden lion tamarins, *Leontopithecus rosalia*: honest communication of individual activity. Behaviour 130, 53–75.

Bradbury, J.W., Vehrencamp, S.L., 1998. Principles of Animal Communication. Sinauer Associates, Sunderland, MA.

Brecovitch, F.B., Hauser, M.D., Jones, J.H., 1995. The endocrine stress response and alarm vocalizations in rhesus macaques. Anim. Behav. 49, 1703–1706.

Cheney, D.L., Seyfarth, R.M., 1990. How Monkeys See the World. University of Chicago Press, Chicago, IL.

Cheney, D.L., Seyfarth, R.M., Silk, J., 1995. The role of grunts in reconciling opponents and facilitating interactions among adult female baboons. Anim. Behav. 50, 249–257.

Darwin, C., 1872. The Expression of the Emotions in Animals and Man. Appleton & Co., New York, NY.

Di Bitetti, M.S., 2003. Food-associated calls of tufted Capuchin monkeys (*Cebus paella nigritus*) are functionally referential signals. Behaviour 140, 565–592.

Doolan, S.P., Macdonald, D.W., 1996. Diet and foraging behaviour of group-living meerkats, *Suricata suricatta*, in the south-western Kalahari. J. Zool. 239, 697–716.

Elowson, A.M., Tannenbaum, P.L., Snowdon, C.T., 1991. Food associated calls correlate with food preferences in cotton-top tamarins. Anim. Behav. 42, 931–937.

Evans, C.S., 1997. Referential signals. In: Owings, D.H., Beacher, M.D., Thompson, N.S. (Eds.), Perspectives of Ethology. Plenum Press, New York, NY, pp. 99–143.

Evans, C.S., Evans, L., 1999. Chicken food calls are functionally referential. Anim. Behav. 58, 307–319.

Evans, C.S., Macedonia, J.M., Marler, P., 1993. Effects of apparent size and speed on the response of chickens, *Gallus gallus*, to computer-generated simulations of aerial predators. Anim. Behav. 46, 1–11.

Fichtel, C., 2008. Ontogeny of conspecific and heterospecific alarm call recognition in wild verreaux's sifakas (*Propithecus verreauxi verreauxi*). Am. J. Primatol. 70, 127–135.

Fichtel, C., Kappeler, P.M., 2002. Anti-predator behavior of group-living Malagasy primates: mixed evidence for a referential alarm call system. Behav. Ecol. Sociobiol. 51, 262–275.

Fichtel, C., Hammerschmidt, K., Jürgens, U., 2001. On the vocal expression of emotion. A multi-parametric analysis of different states of aversion in the squirrel monkey. Behaviour 138, 97–116.

Fischer, J., Hammerschmidt, K., 2001. Functional referents and acoustic similarity revisited: the case of Barbary macaque alarm calls. Anim. Cogn. 4, 29–35.

Fischer, J., Metz, M., Cheney, D.L., Seyfarth, R.M., 2001. Baboon responses to graded bark variants. Anim. Behav. 61, 925–931.

Frederiksen, J.K., Slobodchikoff, C.N., 2007. Referential specificity in the alarm calls of the black-tailed prairie dog. Ethol. Ecol. Evol. 19, 87–99.

Furrer, R.D., Manser, M.B., 2009. The evolution of urgency-based and functionally referential alarm calls in ground-dwelling species. Am. Nat. 173, 400–410.

Greene, E., Meagher, T., 1998. Red squirrels, *Tamiasciurus hudsonicus*, produce predator-class specific alarm calls. Anim. Behav. 55, 511–518.

Griesser, M., 2008. Referential calls signal predator behavior in a group-living bird species. Curr. Biol. 18, 69–73.

Gouzoules, S., Gouzoules, H., Marler, P., 1984. Rhesus monkey (*Macaca mulatta*) screams: representational signaling in the recruitment of agonistic aid. Anim. Behav. 32, 182–193.

Gyger, M., Marler, P., Pickert, R., 1987. Semantics of an avian alarm call system: the male domestic fowl (*Gallus gallus domesticus*). Anim. Behav. 36, 358–365.

Hauser, M.D., 1998. Functional referents and acoustic similarity: field playback experiments with rhesus monkeys. Anim. Behav. 55, 1647–1658.

Hauser, M.D., Marler, P., 1993a. Food-associated calls in rhesus macaques (*Macaca mulatta*). I. Socio-ecological factors influencing call production. Behav. Ecol. 4, 194–205.

Hauser, M.D., Marler, P., 1993b. Food-associated calls in rhesus macaques (*Macaca mulatta*). II. Costs and benefits of call production and suppression. Behav. Ecol. 4, 206–212.

Hollén, L.I., Manser, M.B., 2007. Motivation before meaning: motivational information encoded in meerkat alarm calls develops earlier than referential information. Am. Nat. 169, 758–767.

Hollén, L.I., Clutton-Brock, T.H., Manser, M.B., 2008. Factors affecting the development of alarm-call production, usage and responses in meerkats (*Suricata suricatta*). Behav. Ecol. Sociobiol. 62, 821–829.

Kiriazis, J., Slobodchikoff, C.N., 2006. Perceptual specificity in the alarm calls of Gunnison's prairie dogs. Behav. Proc. 73, 29–35.

Leavesley, A.J., Magrath, R.D., 2005. Communicating about danger: urgency alarm calling in bird. Anim. Behav. 70, 365–373.

Le Roux, A., Cherry, M.I., Manser, M.B., 2009. The vocal repertoire in a solitary foraging carnivore, *Cynctis penicillata*, may reflect facultative sociality. Naturwissenschaften 96 (5), 575–584.

Locke, J.L., Snow, C., 1997. Social influences on vocal learning in human and nonhuman primates. In: Snow, C.T., Hausberger, M. (Eds.), Social Influences on Vocal Development. Cambridge University Press, Cambridge, UK, pp. 249–273.

Macedonia, J.M., 1993. The vocal repertoire of the ring-tailed lemur (*Lemur catta*). Folia Primatol. 61, 186–217.

Macedonia, J.M., Evans, C.S., 1993. Variation among mammalian alarm call systems and the problem of meaning in animal signals. Ethol. 93, 177–197.

Manser, M.B., 2001. The acoustic structure of suricate alarm calls varies depending on predator type and level of urgency. Proc. R. Soc. Lond. B 268, 2315–2324.
Manser, M.B., 2009. What do functionally referential alarm calls refer to? In: Dukas, R., Ratcliffe, J. (Eds.), Cognitive Ecology. Chicago University Press, Chicago, IL.
Manser, M.B., Bell, M.B., 2004. The spatial representation of shelter locations in meerkats. Anim. Behav. 68, 151–157.
Manser, M.B., Fletcher, L.B., 2004. Vocalise to localise – a test on the functionally referential alarm calls in meerkats. Interact. Stud. 5, 325–342.
Manser, M.B., Fletcher, L.B., Bell, M.B., 2001. The information that receivers extract from alarm calls in suricates. Proc. R. Soc. Lond. B 268, 2485–2491.
Manser, M.B., Cheney, D.L., Seyfarth, R.M., 2002. Suricate alarm calls signal predator class and urgency. Trends Cogn. Sci. 6, 55–57.
Manser, M.B., Madden, J.R., Kunc, H.J., English, S., Clutton-Brock, T., 2008. Signals of need in a cooperatively breeding mammal with mobile offspring. Anim. Behav. 76, 1805–1813.
Marler, P., Evans, C.S., Hauser, M.D., 1992. Animals signals: motivational, referential, or both? In: Papousek, H., Jürgens, U., Papousek, M. (Eds.), Nonverbal Communication: Comparative and Developmental Approaches. Cambridge University Press, Cambridge, UK, pp. 66–86.
Morton, E.S., 1977. On the occurrence and significance of motivation-structural rules in some bird and mammal sounds. Am. Nat. 111, 855–869.
Palleroni, A., Hauser, M., Marler, P., 2005. Do responses of galliform birds vary adaptively with predator size? Anim. Cogn. 8, 200–210.
Papworth, S., Böse, A.S., Barker, J., Schel, A.M., Zuberbühler, K., 2008. Male blue monkeys alarm call in response to danger experienced by others. Biol. Lett. 4, 472–475.
Placer, J., Slobodchikoff, C.N., 2000. A fuzzy-neural system for identification of species-specific alarm calls of Gunnison's prairie dogs. Behav. Proc. 52, 1–9.
Placer, J., Slobodchikoff, C.N., 2001. Developing new metrics for the investigation of animal vocalizations. Intel. Automat. Soft Comput. 7, 249–258.
Paul, E.S., Harding, E.J., Mendl, M., 2005. Measuring emotional processes in animals: the utility of a cognitive approach. Neurosci. Biobehav. Rev. 29, 469–491.
Rasa, O.A.E., 1986. Coordinated vigilance in dwarf mongoose family groups: the "Watchman's Song" hypothesis and the costs of guarding. Ethol. 71, 340–344.
Robinson, J.G., 1982. Vocal systems regulating within-group spacing. In: Snowdon, S.T., Brown, C.H., Petersen, M.R. (Eds.), Primate Communication. Cambridge University Press, New York, NY, pp. 94–116.
Robinson, S.R., 1980. Antipredator behaviour and predator recognition in Belding's ground squirrel. Anim. Behav. 28, 840–852.
Schehka, S., Esser, K.H., Zimmermann, E., 2007. Acoustical expression in conflict situations in tree shrews (*Tupaia belangeri*). J. Comp. Physiol. A 193, 845–852.
Seyfarth, R.M., Cheney, D.L., 1986. Vocal development in vervet monkeys. Anim. Behav. 34, 1640–1658.
Seyfarth, R.M., Cheney, D.L., 2003a. Signalers and receivers in animal communication. Annu. Rev. Psychol. 54, 145–173.
Seyfarth, R.M., Cheney, D.L., 2003b. Meaning and emotion in animal vocalizations. Ann. N.Y. Acad. Sci. 1000, 32–55.
Seyfarth, R.M., Cheney, D.L., Marler, P., 1980. Monkey responses to three different alarm calls: evidence for predator classification and semantic communication. Science 210, 801–803.
Slocombe, K.E., Zuberbühler, K., 2005a. Agonistic screams in wild chimpanzees vary as a function of social role. J. Comp. Psychol. 119 (1), 67–77.
Slocombe, K.E., Zuberbühler, K., 2005b. Functionally referential communication in a chimpanzee. Curr. Biol. 15, 1779–1784.
Slocombe, K.E., Zuberbühler, K., 2006. Food-associated calls in chimpanzees: responses to food types or food preferences? Anim. Behav. 72, 989–999.
Slocombe, K.E., Townsend, S.W., Zuberbühler, K., 2009. Wild chimpanzees (pan troglodytes schweinfurthii) distinguish between different scream types: evidence from a playback study. Anim. Cogn. 12, 441–449.
Smith, W.J., 1981. Referents of animal communication. Anim. Behav. 29, 1273–1274.
Stephan, C., Zuberbühler, K., 2008. Predation increases acoustic complexity in primate alarm calls. Biol. Lett. 4, 641–644.
Struhsaker, T.T., 1967. Auditory communication among vervet monkeys (*Cercopithecus aethiops*). In: Altmann, S.A. (Ed.), Social Communication Among Primates. Chicago University Press, Chicago, IL, pp. 281–324.
Templeton, C.N., Greene, E., Davies, K., 2005. Allometry of alarm calls: black-capped chickadees encode information about predator size. Science 308, 1934–1937.
Zuberbühler, K., 2000. Referential labeling in Diana monkeys. Anim. Behav. 59, 917–927.
Zuberbühler, K., Noe, R., Seyfarth, R.M., 1997. Diana monkey long distance calls: messages for conspecifics and predators. Anim. Behav. 53, 589–604.

CHAPTER 11.3

Auditory categories in the non-human primate

Yale E. Cohen,* Jung Hoon Lee, Joji Tsunada and Brian E. Russ

Department of Otorhinolaryngology-Head and Neck Surgery, University of Pennsylvania, 3400 Spruce, 5 Raydin Philadelphia, PA10104, USA

Abstract: Communication is one of the fundamental components of both human and non-human animal behavior. Whereas the benefits of language in human evolution are obvious, other communication systems have also evolved to convey information that is critical for survival. This chapter focuses on auditory communication signals, specifically species-specific vocalizations and the underlying neural processes that may support their use in guiding goal-directed behavior. We first highlight the fundamental role that species-specific vocalizations play in the socioecology of several species of non-human primates, with a focus on rhesus monkeys (*Macaca mulatta*). Finally, we discuss the role that the ventrolateral prefrontal cortex may play in the categorization of species-specific vocalizations.

Keywords: auditory system; categorization; non-human primate; prefrontal cortex; vocalization

I. Introduction

Categories provide a mechanism to mentally reason, manipulate and respond to objects in our environment. If goal-directed behavior can be thought of as the processes that transform sensory signals into actions, the computations that form the intermediate steps of this transformation involve categorical representations (Ashby and Berretty, 1997; Grinband et al., 2006). These computations are further modulated by context, motivation and other factors that permit behavior to be both flexible and adaptive. In the wild, consider what happens when a lion sees a gazelle. If the lion is hungry, it may categorize the gazelle as prey and form a motor plan that allows it to capture the gazelle. But if the lion is not hungry, the lion may not categorize the gazelle as prey and may disregard the gazelle entirely. In this chapter, we highlight categorization and the factors contributing to categorization with an emphasis on auditory categorization in human and non-human primates. In particular, we highlight the categorization of higher-order features of vocalizations.

*Corresponding author. E-mail: ycohen@mail.med.upenn.edu

II. Types of categories

Any given stimulus (object) can be a member of several different categories, depending on what rules process the stimulus. These different categorical levels are often thought of as hierarchical; categories at the top of the hierarchy are the most general (superordinate categories) whereas those at the bottom (subordinate categories) are the most specific. Basic-level subordinate categories are the terms that people commonly use when encountering an object. For example, Lassie can be categorized as an animal, a dog, or a specific collie. The superordinate category would be "animal," the basic-level category would be "dog," and the subordinate category would be "collie." However, these different categories are not equipotent, in the sense that they require different amounts of neural processing; we categorize objects into basic level categories faster and more accurately than we categorize objects into superordinate or subordinate categories (Rosch et al., 1976). Also, the level at which an object is categorized depends on previous experience and knowledge; a dog expert might classify Lassie, at the basic level, as a "collie" or more specifically a "rough collie" (Gauthier and Logothetis, 2000; Marschark et al., 2004).

Stefan M. Brudzynski (Ed.)
Handbook of Mammalian Vocalization
ISBN 978-0-12-374593-4

DOI: 10.1016/B978-0-12-374593-4.00044-9

Different hypothetical frameworks can be used to describe the relationship between an object's membership in basic and more superordinate categories. One framework posits that superordinate categories contain a set of features that belong to all of the members of the more basic category (Smith et al., 1974; Rosch et al., 1976; Damasio, 1989; Devlin et al., 1998; Martin et al., 2002). For instance, the basic-level of category of "dog" might contain descriptors like "has fur," "has wet nose," "has four legs," "breathes," "is mobile," "can reproduce on its own," etc. On the other hand, the superordinate category of "animal" contains descriptors "breathes," "is mobile," "can reproduce on its own," etc. An alternative view is that the properties of a basic-level category are not omitted from the superordinate category, but are represented as more abstract variable values in this higher-order category (MacNamara, 1982; Macnamara, 1999; Prasada, 2000).

II.A. Perceptual categories

Perceptual similarity is one of the key elements that determine a stimulus' categorical membership (Liberman et al., 1967; Eimas et al., 1971; Kuhl and Miller, 1975; Lasky et al., 1975; Miyawaki et al., 1975; Streeter, 1976; Sandell et al., 1979; Kuhl and Padden, 1982, 1983; Boyton and Olson, 1987, 1990; Wyttenbach et al., 1996; Doupe and Kuhl, 1999). Perceptual categories are based on physical similarities or dissimilarities between auditory objects. For example, we can categorize male and female voices by listening to the pitch of their voices, with female voices characteristically having a higher pitch than males. In another example, listeners can perceive different speech signals as belonging to the same phonemic category, independent of pitch or timbre differences, or viewers can perceive different visual signals as being members of the same color category (Bornstein et al., 1976; Sandell et al., 1979; Boyton and Olson, 1987, 1990).

One prominent feature of perceptual categories is that they are often accompanied by categorical perception. In categorical perception, a subject's perception of an object does not vary smoothly with changes in the physical properties of the object (Liberman et al., 1967; Ashby and Berretty, 1997; Miller et al., 2003). That is, objects on one side of the categorical boundary are treated similarly, despite potentially large differences between their physical properties. At locations near the category boundary, small changes in an object's properties can lead to large changes in perception.

The classic example of categorical perception is the categorization of speech units into phonetic categories (Liberman et al., 1967; Eimas et al., 1971; Mann, 1980; Kuhl and Padden, 1982, 1983; Lotto et al., 1998; Holt, 2006). In a seminal study, Liberman et al. (1967) created morphed versions of two different phonemes and asked subjects to report the phoneme that they heard. Liberman and colleagues found that when subjects were presented with a morphed stimulus that contained more than 50% of a phoneme prototype, the subjects reliably perceived that stimulus as the prototype. That is, even though the presented stimuli varied smoothly in their physical features, subjects perceived the presented stimuli as being either one of the two phoneme prototypes.

Interestingly, the perceptual categorization of phonemes is not strictly a human behavior (Liberman et al., 1967; Kuhl and Padden, 1982, 1983; Kluender et al., 1987; Lotto et al., 1997; Russ et al., 2008). Rhesus macaques, chinchillas and Japanese quail perceive human phonemes in a manner comparable to that of humans. Since the manner in which perceptual categories, at least phonemes, are coded appears to be similar across a wide variety of animal species, it is hypothesized that the mechanisms underlying this perceptual categorization may be a fundamental component of vertebrate auditory processing from which human speech was bootstrapped.

II.B. Abstract categories

Categories are not only formed based on the perceptual (physical) features of stimuli. Categories can also be based on more abstract information. An abstract category is one in which a group of arbitrary stimuli are linked together as a category based on some shared feature, a functional characteristic, or acquired knowledge. For instance, despite vast physical differences, "hammer," "band saw," and "pliers" are all members of the "tool" category. Similarly, a combination of physical characteristics and knowledge about their reproductive processes allow us to categorize "dogs," "cats," and "killer whales" in the category of "mammals." However, if we use different criteria to form a category of "pets," "dogs" and "cats" would be members of the "pet" category but "killer whales" would not.

Non-human primates can also categorize stimuli into abstract categories. Monkeys can be trained to categorize objects as being animals or non-animals (Fabre-Thorpe et al., 1998) or as trees or non-trees (Vogels, 1999). The capacity to represent even more abstract categories, such as ordinal numbers (Orlov et al., 2000; Nieder et al., 2002), is also present. Although these studies provide important insight into how abstract categories are built, their generalization to more ethological, natural conditions is limited by their use of non-ethological stimuli.

Behavioral studies that have used ethological stimuli have shown that non-human primates may form categories "spontaneously." That is, they form categories in the absence of laboratory-based operant training. A classic example is the categorization of food-related species-specific vocalizations by rhesus monkeys (Hauser and Marler, 1993a,b; Hauser, 1998; Gifford III et al., 2003). In rhesus monkeys, information about the discovery of rare, high-quality food is transmitted by two different vocalizations: by a "harmonic arch;" and by a different vocalization called a "warble." Importantly, whereas both harmonic arches and warbles transmit the same type of information, they have distinct spectrotemporal properties (i.e., they sound different). In contrast, "grunts" transmit a different type of information (the discovery of common, low-quality food) and are acoustically distinct from harmonic arches and warbles.

Despite these acoustic differences, rhesus monkeys categorize these food-related calls based on the transmitted information and not their acoustic features. Monkeys do not discriminate between vocalizations that transmit the same referential information (i.e., harmonic arches and warbles) even though these vocalizations have different acoustic features. In contrast, they do discriminate between vocalizations that transmit different types of information (i.e., grunts versus warbles/harmonic arches). That is, rhesus monkeys perceive harmonic arches and warbles as if they belong to a single functional category (based on referential information and not based on acoustics) and treat grunts as a second, distinct category.

The formation of abstract acoustic categories is seen in other monkey species. Female Diana monkeys categorize and respond similarly to a male Diana monkey who is eliciting a leopard-alarm call or to a crested guinea fowl that is eliciting its unique species-specific leopard-alarm call (Züberbuhler and Seyfarth, 1997; Züberbuhler, 2000a,b,c). Diana monkeys also form cross-species categories with putty-nose monkeys, based on the ability of putty-nose monkeys to provide vocal warnings of eagles (Eckardt and Züberbuhler, 2004). These observations suggest that Diana monkeys form abstract categorical representations of vocalizations independent of acoustics and the species generating the signal. Finally, an example of a non-communicative multimodal category is "looming" stimuli; rhesus monkeys treat approaching auditory or visual stimuli in a comparable manner (Schiff et al., 1962; Ghazanfar et al., 2002; Maier et al., 2004).

III. Neurophysiological representations of categories

Traditionally, the auditory cortex has been thought to be involved in feature extraction. However, more recent work has shown that the auditory cortex, particularly the primary auditory cortex, plays a substantive role in more advanced stages of auditory processing, such as auditory-object analysis (Sutter et al., 2000; Fishman et al., 2001, 2004; Miller et al., 2001; Nelken et al., 2003; Petkov et al., 2003; Poremba et al., 2003; Micheyl et al., 2005; Griffiths et al., 2007; Petkov et al., 2008). Consistent with these studies, other work suggests that the auditory cortex may also be involved in the computations underlying category processing in both human and non-human primates (Steinschneider et al., 1995; Guenther et al., 2004; Poeppel et al., 2004; Selezneva et al., 2006). For example, Brosch and colleagues (Selezneva et al., 2006) have shown that auditory-cortex neurons respond categorically to sequences of tone pips that are either increasing or decreasing in frequency. In a related study, the primary auditory cortex has been shown to contain a distributed representation of the voice-onset time of human phonemes (Steinschneider et al., 1995). It is thought that this representation and related representations form the neurophysiological bases for the perceptual categorization of phonemes. However, unlike the Brosch study, this work by Steinschneider emphasizes that categories are represented in population activity and not at the level of the single cell. Finally, visual and multimodal stimuli, such as faces and bimodal looming representations, also appear to have categorical-like representations in different regions of the auditory cortex (Hoffman et al., 2008; Maier et al., 2008).

What types of categorical processing then occur in subsequent areas of the cortical hierarchy? We suggest that there are two major classes of computational processing. First, neurons become increasingly sensitive to more abstract categories. For example, recent work from our group has suggested a role for the ventrolateral prefrontal cortex (vPFC) in categorizing the referential information that a vocalization transmits, as opposed to the vocalization's acoustic properties (Gifford III et al., 2005; Cohen et al., 2006). Using an oddball paradigm (Näätänen and Tiitinen, 1996), we found that the activity of vPFC neurons was not modulated by transitions between presentations of food vocalizations that transmitted the same information (high-quality food), even though these vocalizations had significantly different acoustic structures. The vPFC activity, however, was modulated by transitions between presentations of food vocalizations that transmitted different types of information (low-quality versus high-quality food). These data suggested that, on average, vPFC neurons are modulated preferentially by transitions between presentations of food vocalizations that belong to functionally meaningful and different categories.

Second, the categorical representations in more central areas such as the prefrontal cortex are used to flexibly guide an animal's behavior (Miller, 2000; Miller et al., 2002). That is, categorical information in the prefrontal cortex is critical for both the selection and retrieval of task-relevant information as it relates to the rules of an ongoing task (Asaad et al., 2000; Ashby and Spiering, 2004; Bunge, 2004; Badre et al., 2005; Bunge et al., 2005). Indeed, recent work from our laboratory (Russ et al., 2008) has demonstrated that when monkeys are asked to categorize pairs of the spoken words, vPFC neurons do not reflect the perceptual differences between the spoken words. Instead, vPFC activity reflects the monkeys' behavioral reports (decisions) as to whether they perceive that the pair of spoken words is the same or whether they are different. That is, vPFC activity reflects how the monkeys *actually* respond and does not reflect how they *should* respond (which is based on the actual sensory/perceptual differences between the spoken words). Moreover, this activity seems to play a causal role in the decision-making process; disruption of the vPFC through transcranial magnetic stimulation significantly alters the time it takes the monkeys to report their decision (Russ and Cohen, unpublished observations).

IV. Conclusions

This chapter has emphasized the relationship between categorical perception and the neural correlates underlying these representations. The focus has been on how vocalizations and other communication signals are represented in the auditory pathway. This chapter raises a number of questions. How are perceptual categories transformed into more abstract representations and what are the neural computations underlying this transformation? On a related note, what specific cortical regions are involved in categorization? Are categories processed in distinct cortical hierarchies or do cortical regions code all levels of categories simultaneously? Finally, this chapter emphasized unimodal categorization. However, communication signals are inherently multimodal (Sumby and Pollack, 1954; McGurk and MacDonald, 1976; Stein and Meredith, 1993; Calvert et al., 1997; Partan and Marler, 1999). Thus, a fundamental function of neural processing may be to integrate auditory and visual stimuli that provide complementary information (Hinde and Rowell, 1962; van Hooff, 1962; Maestripieri, 1997; Partan and Marler, 1999; Hauser and Akre, 2001; Partan, 2002; Ghazanfar and Logothetis, 2003; Ghazanfar et al., 2005). Are similar processes and areas involved in the categorization of visual and auditory stimuli or are new processes and cortical areas engaged to process these multimodal stimuli (Ghazanfar and Schroeder, 2006)? Since our world is full of multimodal sensory information, it will be important to investigate how we combine information from multiple domains into a single coherent signal.

Acknowledgments

This work was supported by grants from the NIMH-NIH and the NIDCD-NIH.

References

Asaad, W.F., Rainer, G., Miller, E.K., 2000. Task-specific neural activity in the primate prefrontal cortex. J. Neurophysiol. 84, 451–459.

Ashby, F.G., Berretty, P.M., 1997. Categorization as a special case of decision-making or choice. In: Marley, A.A.J. (Ed.), Choice, Decision, and Measurement: Essays in Honor of R. Duncan Luce. Lawrence Erlbaum Associates, Inc., Mahwah, NJ, pp. 367–388.

Ashby, F.G., Spiering, B.J., 2004. The neurobiology of category learning. Behav. Cogn. Neurosci. Rev. 3, 101–113.

Badre, D., Poldrack, R.A., Pare-Blagoev, E.J., Insler, R.Z., Wagner, A.D., 2005. Dissociable controlled retrieval and generalized selection mechanisms in ventrolateral prefrontal cortex. Neuron 47, 907–918.

Bornstein, M.H., Kessen, W., Weiskopf, S., 1976. The categories of hue in infancy. Science 191, 201–202.

Boyton, R.M., Olson, C.X., 1987. Locating basic color terms in the OSA space. Color Res. Appl. 12, 94–105.

Boyton, R.M., Olson, C.X., 1990. Salience of basic color terms confirmed by three measures. Vis. Res. 30, 1311–1317.

Bunge, S.A., 2004. How we use rules to select actions: a review of evidence from cognitive neuroscience. Cogn. Affect. Behav. Neurosci. 4, 564–579.

Bunge, S.A., Wallis, J.D., Parker, A., Brass, M., Crone, E.A., Hoshi, E., Sakai, K., 2005. Neural circuitry underlying rule use in humans and nonhuman primates. J. Neurosci. 25, 10347–10350.

Calvert, G.A., Bullmore, E.T., Brammer, M.J., Campbell, R., Williams, S.C., McGuire, P.K., Woodruff, P.W., Iversen, S.D., David, A.S., 1997. Activation of auditory cortex during silent lipreading. Science 276, 593–596.

Cohen, Y.E., Hauser, M.D., Russ, B.E., 2006. Spontaneous processing of abstract categorical information in the ventrolateral prefrontal cortex. Biol. Lett. 2, 261–265.

Damasio, A.R., 1989. Time-locked multiregional retroactivation: a systems-level proposal for the neural substrates of recall and recognition. Cogn. 33, 25–62.

Devlin, J.T., Gonnerman, L.M., Andersen, E.S., Seidenberg, M.S., 1998. Category-specific semantic deficits in focal and widespread brain damage: a computational account. J. Cog. Neurosci. 10, 77–94.

Doupe, A.J., Kuhl, P.K., 1999. Birdsong and human speech: common themes and mechanisms. Ann. Rev. Neurosci. 22, 567–631.

Eckardt, W., Züberbuhler, K., 2004. Cooperation and competition in two forest monkeys. Behav. Ecol. 15, 400–411.

Eimas, P.D., Siqueland, E.R., Jusczyk, P., Vigorito, J., 1971. Speech perception in infants. Science 171, 303–306.

Fabre-Thorpe, M., Richard, G., Thorpe, S.J., 1998. Rapid categorization of natural images by rhesus monkeys. Neuroreport 9, 303–308.

Fishman, Y.I., Arezzo, J.C., Steinschneider, M., 2004. Auditory stream segregation in monkey auditory cortex: effects of frequency separation, presentation rate, and tone duration. J. Acoust. Soc. Am. 116, 1656–1670.

Fishman, Y.I., Reser, D.H., Arezzo, J.C., Steinschneider, M., 2001. Neural correlates of auditory stream segregation in primary auditory cortex of the awake monkey. Hear Res. 151, 167–187.

Gauthier, I., Logothetis, N.K., 2000. Is face recognition not so unique after all? Cog. Neuropsych. 17, 125–142.

Ghazanfar, A.A., Logothetis, N.K., 2003. Neuroperception: facial expressions linked to monkey calls. Nature 424, 937–938.

Ghazanfar, A.A., Schroeder, C.E., 2006. Is neocortex essentially multisensory? Trends Cog. Sci. 10, 278–285.

Ghazanfar, A.A., Neuhoff, J.G., Logothetis, N.K., 2002. Auditory looming perception in rhesus monkeys. Proc. Natl. Acad. Sci. USA 99, 15755–15757.

Ghazanfar, A.A., Maier, J.X., Hoffman, K.L., Logothetis, N.K., 2005. Multisensory integration of dynamic faces and voices in rhesus monkey auditory cortex. J. Neurosci. 25, 5004–5012.

Gifford III, G.W., Hauser, M.D., Cohen, Y.E., 2003. Discrimination of functionally referential calls by laboratory-housed rhesus macaques: implications for neuroethological studies. Brain Behav. Evol. 61, 213–224.

Gifford III, G.W., MacLean, K.A., Hauser, M.D., Cohen, Y.E., 2005. The neurophysiology of functionally meaningful categories: macaque ventrolateral prefrontal cortex plays a critical role in spontaneous categorization of species-specific vocalizations. J. Cog. Neurosci. 17, 1471–1482.

Griffiths, T.D., Kumar, S., Warren, J.D., Stewart, L., Stephan, K.E., Friston, K.J., 2007. Approaches to the cortical analysis of auditory objects. Hear. Res. 229, 46–53.

Grinband, J., Hirsch, J., Ferrera, V.P., 2006. A neural representation of categorization uncertainty in the human brain. Neuron 49, 757–763.

Guenther, F.H., Nieto-Castanon, A., Ghosh, S.S., Tourville, J.A., 2004. Representation of sound categories in auditory cortical maps. J. Speech. Lang. Hear. Res. 47, 46–57.

Hauser, M.D., 1998. Functional referents and acoustic similarity: field playback experiments with rhesus monkeys. Anim. Behav. 55, 1647–1658.

Hauser, M.D., Akre, K., 2001. Asymmetries in the timing of facial and vocal expressions by rhesus monkeys: implications for hemispheric specialization. Anim. Behav. 61, 391–400.

Hauser, M.D., Marler, P., 1993a. Food-associated calls in rhesus macaques (*Macaca mulatta*) 1. Socioecological factors influencing call production. Behav. Ecol. 4, 194–205.

Hauser, M.D., Marler, P., 1993b. Food-associated calls in rhesus macaques (*Macaca mulatta*) II. Costs and benefits of call production and suppression. Behav. Ecol. 4, 206–212.

Hinde, R.A., Rowell, T.E., 1962. Communication by postures and facial expressions in the rhesus monkey (*Macaca mulatta*). Proc. Zool. Soc. Lond. 138, 1–21.

Hoffman, K.L., Ghazanfar, A.A., Gauthier, I., Logothetis, N.K., 2008. Category-specific responses to faces and objects in primate auditory cortex. Fron. Sys. Neurosci. 1, 2. doi:10.3389/neuro.06.002.2007.

Holt, L.L., 2006. Speech categorization in context: joint effects of nonspeech and speech precursors. J. Acoust. Soc. Am. 119, 4016–4026.

Kluender, K.R., Diehl, R.L., Killeen, P., 1987. Japanese quail can learn phonetic categories. Science 237, 1195–1197.

Kuhl, P.K., Miller, J.D., 1975. Speech perception by the chinchilla: voiced-voiceless distinction in alveolar plosive consonants. Science 190, 69–72.

Kuhl, P.K., Padden, D.M., 1982. Enhanced discriminability at the phonetic boundaries for the voicing feature in macaques. Percept. Psychophys. 32, 542–550.

Kuhl, P.K., Padden, D.M., 1983. Enhanced discriminability at the phonetic boundaries for the place feature in macaques. J. Acoust. Soc. Am. 73, 1003–1010.

Lasky, R.E., Syrdal-Lasky, A., Klein, R.E., 1975. VOT discrimination by four to six and a half month old infants from Spanish environments. J. Exp. Child. Psychol. 20, 215–225.

Liberman, A.M., Cooper, F.S., Shankweiler, D.P., Studdert-Kennedy, M., 1967. Perception of the speech code. Psychol. Rev. 5, 552–563.
Lotto, A.J., Kluender, K.R., Holt, L.L., 1997. Perceptual compensation for coarticulation by Japanese quail (*Coturnix coturnix japonica*). J. Acoust. Soc. Am. 102, 1134–1140.
Lotto, A.J., Kluender, K.R., Holt, L.L., 1998. Depolarizing the perceptual magnet effect. J. Acoust. Soc. Am. 103, 3648–3655.
MacNamara, J., 1982. Names for Things: A Study of Human Learning. MIT Press, Cambridge, MA.
MacNamara, J., 1999. Through the Rearview Mirror: Historical Reflections on Psychology. MIT Press, Cambridge, MA.
Maestripieri, D., 1997. Gestural communication in macaques: usage and meaning of nonvocal signals. Evol. Comm. 1, 193–222.
Maier, J.X., Chandrasekaran, C., Ghazanfar, A.A., 2008. Integration of bimodal looming signals through neuronal coherence in the temporal lobe. Curr. Biol. 18, 963–968.
Maier, J.X., Neuhoff, J.G., Logothetis, N.K., Ghazanfar, A.A., 2004. Multisensory integration of looming signals by rhesus monkeys. Neuron 43, 177–181.
Mann, V.A., 1980. Influence of preceding liquid on stop-consonant perception. Percept. Psychophys. 28, 407–412.
Marschark, M., Convertino, C., McEvoy, C., Masteller, A., 2004. Organization and use of the mental lexicon by deaf and hearing individuals. Am. Ann. Deaf. 149, 51–61.
Martin, A., Ungerleider, L.G., Haxby, J.V., 2002. Category specificity and the brain: the sensory/motor model of semantic representations of objects. In: Gazzaniga, M.S. (Ed.), The New Cognitive Neurosciences. MIT Press, Cambridge, MA.
McGurk, H., MacDonald, J., 1976. Hearing lips and seeing voices. Nature 264, 746–748.
Micheyl, C., Tian, B., Carlyon, R.P., Rauschecker, J.P., 2005. Perceptual organization of tone sequences in the auditory cortex of awake macaques. Neuron 48, 139–148.
Miller, E.K., 2000. The prefrontal cortex and cognitive control. Nat. Rev. Neurosci. 1, 59–65.
Miller, C.T., Dibble, E., Hauser, M.D., 2001. Amodal completion of acoustic signals by a nonhuman primate. Nat. Neurosci. 4, 783–784.
Miller, E.K., Freedman, D.J., Wallis, J.D., 2002. The prefrontal cortex: categories, concepts, and cognition. Phil. Trans. R. Soc. Lond. B Biol. Sci. 29, 1123–1136.
Miller, E.K., Nieder, A., Freedman, D.J., Wallis, J.D., 2003. Neural correlates of categories and concepts. Curr. Opin. Neurobiol. 13, 198–203.
Miyawaki, K., Strange, W., Verbrugge, R., Liberman, A.M., Jenkins, J.J., Fujimura, O., 1975. An effect of linguistic experience: the discrimination of [r] and [l] by native speakers of Japanese and English. Percept. Psychophys. 18, 331–340.
Näätänen, R., Tiitinen, H., 1996. Auditory information processing as indexed by the mismatch negativity. In: Sabourin, M., Craik, F., Robert, M. (Eds.), Advances in Psychological Science: Biological and Cognitive Aspects, Vol. 2. Taylor and Francis Group, Montreal, Canada, pp. 145–170.
Nelken, I., Fishbach, A., Las, L., Ulanovsky, N., Farkas, D., 2003. Primary auditory cortex of cats: feature detection or something else? Biol. Cyber. 89, 397–406.
Nieder, A., Freedman, D.J., Miller, E.K., 2002. Representation of the quantity of visual items in the primate prefrontal cortex. Science 297, 1708–1711.
Orlov, T., Yakovlev, V., Hochstein, S., Zohary, E., 2000. Macaque monkeys categorize images by their ordinal number. Nature 404, 77–80.
Partan, S., 2002. Single and multichannel signal composition: facial expressions and vocalizations of rhesus macaques (*Macaca mulatta*). Behaviour 139, 993–1027.
Partan, S., Marler, P., 1999. Communication goes multimodal. Science 283, 1272–1273.
Petkov, C.I., O'Connor, K.N., Sutter, M.L., 2003. Illusory sound perception in macaque monkeys. J. Neurosci. 23, 9155–9161.
Petkov, C.I., Kayser, C., Steudel, T., Whittingstall, K., Augath, M., Logothetis, N.K., 2008. A voice region in the monkey brain. Nat. Neurosci. 11, 367–374.
Poeppel, D., Guillemin, A., Thompson, J., Fritz, J., Bavelier, D., Braun, A.R., 2004. Auditory lexical decision, categorical perception, and FM direction discrimination differentially engage left and right auditory cortex. Neuropsychologia 42, 183–200.
Poremba, A., Saunders, R.C., Crane, A.M., Cook, M., Sokoloff, L., Mishkin, M., 2003. Functional mapping of the primate auditory system. Science 299, 568–572.
Prasada, S., 2000. Acquiring generic knowledge. Trends Cog. Sci. 4, 66–72.
Rosch, E., Mervis, C.B., Gray, W.D., Johnson, D.M., Boyes-Braem, P., 1976. Basic objects in natural categories. Cog. Psych. 8, 382–439.
Russ, B.E., Orr, L.E., Cohen, Y.E., 2008. Prefrontal neurons predict choices during an auditory same-different task. Curr. Biol. 18, 1483–1488.
Sandell, J.H., Gross, C.G., Bornstein, M.H., 1979. Color categories in macaques. J. Comp. Phsyiol. Psychol. 93, 626–635.
Schiff, W., Caviness, J.A., Gibson, J.J., 1962. Persistent fear responses in rhesus monkeys to the optical stimulus of "looming. " Science 136, 982–983.
Selezneva, E., Scheich, H., Brosch, M., 2006. Dual time scales for categorical decision making in auditory cortex. Curr. Biol. 16, 2428–2433.
Smith, E.E., Shoben, E.J., Rips, L.J., 1974. Structure and process in semantic memory: a feature model for semantic decisions. Psychol. Rev. 81, 214–221.
Stein, B.E., Meredith, M.A., 1993. The Merging of the Senses. MIT Press, Cambridge, MA.
Steinschneider, M., Schroeder, C.E., Arezzo, J.C., Vaughan, Jr., H.G., 1995. Physiologic correlates of the voice onset time boundary in primary auditory cortex (A1) of the awake monkey: temporal response patterns. Brain Lang. 48, 326–340.
Streeter, L.A., 1976. Language perception of 2-month-old infants shows effects of both innate mechanisms and experience. Nature 259, 39–41.
Sumby, W.H., Pollack, I., 1954. Visual contribution to speech intelligibility in noise. J. Acoust. Soc. Am. 26, 212–215.
Sutter, M.L., Petkov, C., Baynes, K., O'Connor, K.N., 2000. Auditory scene analysis in dyslexics. Neuroreport 11, 1967–1971.

van Hooff, J.A.R.A.M, 1962. Facial expressions in higher primates. Symp. Zool. Soc. London, 8, 97–125.
Vogels, R., 1999. Categorization of complex visual images by rhesus monkeys. Part 1: behavioural study. Eur. J. Neurosci. 11, 1223–1238.
Wyttenbach, R.A., May, M.L., Hoy, R.R., 1996. Categorical perception of sound frequency by crickets. Science 273, 1542–1544.
Züberbuhler, K., 2000a. Causal cognition in a non-human primate: field playback experiments with Diana monkeys. Cogn. 76, 195–207.
Züberbuhler, K., 2000b. Interspecies semantic communication in two forest primates. Proc. R. Soc. Lond. B Biol. Sci. 267, 713–718.
Züberbuhler, K., 2000c. Referential labelling in Diana monkeys. Anim. Behav. 59, 917–927.
Züberbuhler, K., Seyfarth, R.M., 1997. Diana monkey long-distance calls: messages for conspecifics and predators. Anim. Behav. 53, 589–604.

CHAPTER 11.4

Recognition of individuals within the social group: signature vocalizations

Ari Daniel Shapiro*

Biology Department, Woods Hole Oceanographic Institution, Woods Hole, MA, USA

Abstract: Signature vocalizations contain sufficiently unique spectral and/or temporal identity to allow conspecifics to differentiate between individuals. Experimental assessment of signature content relies on: (1) accurate discrimination of vocalizations according to individual callers using human observers and statistical methods; (2) playback experiments with conspecifics in captivity or in the wild. Signature vocalizations are often acquired through social and vocal learning. Functions for such signals include individual recognition (e.g., between parents and offspring, mating pairs, territorial rivals) and the mediation of social interactions. Signature cues are likely to be subtle, and their analysis is likely to benefit from the use of new data collection technologies and human speech recognition techniques.

Keywords: signature vocalizations; vocal learning; signature whistles; playback experiments; individual recognition; playback

I. Introduction

Animals can use a variety of sensory modalities, including visual, olfactory and auditory cues, to communicate information about individual identity. These signals contain signature content when their inter-individual variability exceeds their intra-individual variability. Olfactory signals often require conspecifics to approach the site where a scent is left and they tend to be carried by wind or air currents. Visual signals can be obscured by physical obstacles or darkness and can be attenuated in marine habitats. Acoustic methods of encoding individual identity, however, do not have these disadvantages and offer animals the opportunity to communicate rapidly and coordinate behavior and interaction over a range of distances. Such individually-distinctive signals are called signature vocalizations. Certain animals incorporate additional spectral or temporal cues into their calls to supply sufficient acoustic variation to allow differentiation between individuals.

This chapter is divided into three main sections. The first section will focus on what constitutes a signature vocalization; namely, the different acoustic features that might endow a signal with signature information, how these signals are defined experimentally and how they are acquired by an animal and subsequently developed. The second part is concerned with the functional aspects of signature signals and how the consequence of individual recognition may impact both affiliative and aversive social interactions. Finally, recommendations for future research are offered.

II. What is a signature vocalization?

II.A. Features that embed signature content

Two general sets of acoustic properties might be used by an animal to encode signature information. Firstly, spectral features can differ between individuals, including such basic features as a call's fundamental frequency and harmonics. A variety of species of birds and mammals produce contact calls that can be differentiated spectrally, including pairs of orange-fronted parakeets (*Aratinga canicularis*) from the Guanacaste province of Costa Rica (Fig. 1a) (Cortopassi and Bradbury, 2000) and mother–pup pairs of South American fur seals

*Corresponding author. E-mail: ashapiro@whoi.edu
www.aridanielshapiro.com

ISBN 978-0-12-374593-4
DOI: 10.1016/B978-0-12-374593-4.00045-0

(*Arctocephalus australis*, Phillips and Stirling, 2000), Galapagos fur seals (*Arctocephalus galapagoensis*) and Galapagos sea lions (*Zalophus californianus wollebaeki*, Trillmich, 1981). Features of spectral variability, such as relative harmonic energies and the mean, modulation and sonographic shape of the fundamental frequency, are used in different combinations by individuals of numerous species to encode

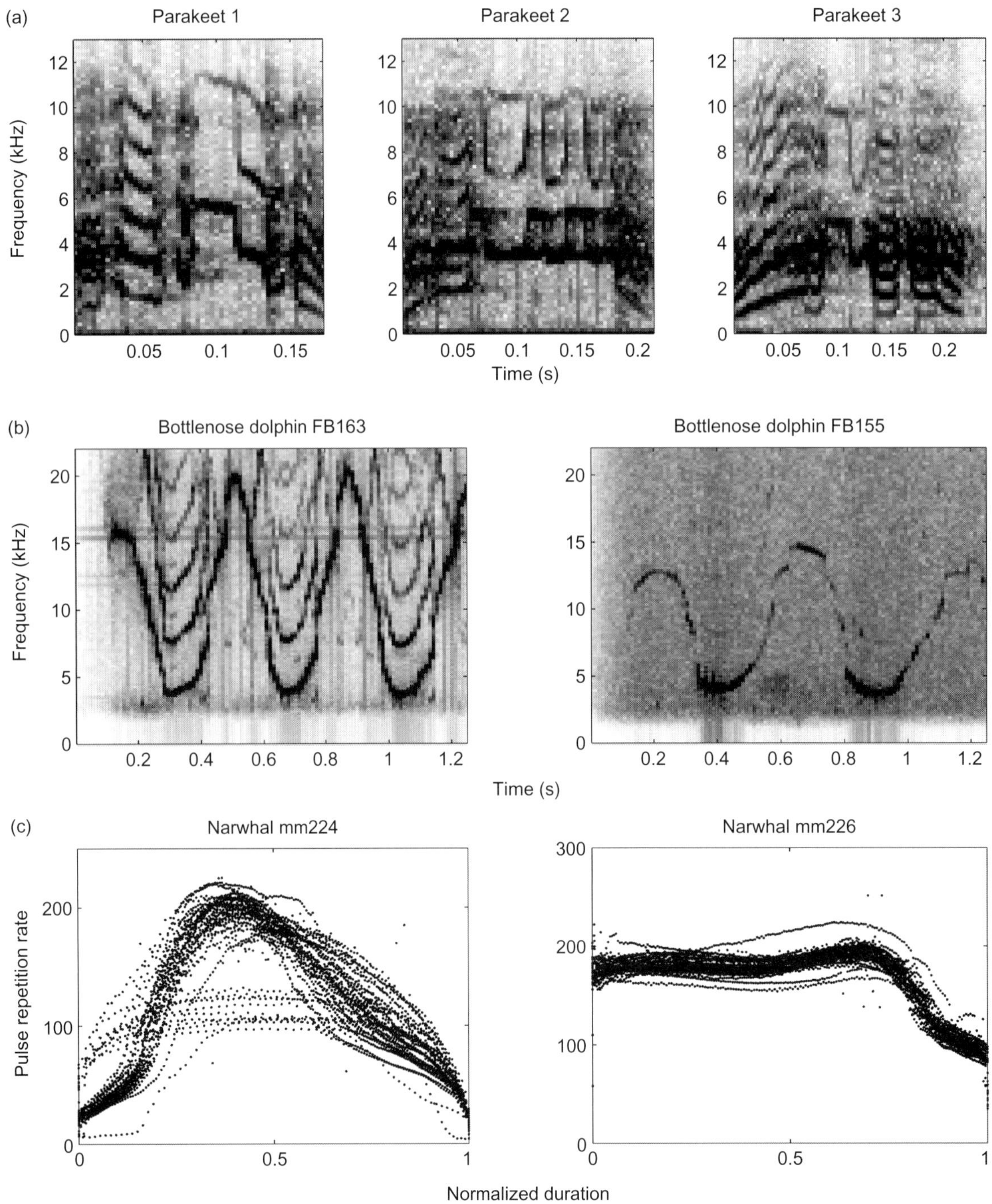

Fig. 1. Examples of signature vocalizations. (a) Orange-fronted parakeet contact calls, courtesy of J.W. Bradbury. Each subplot corresponds to a different individual (labeled 1–3 here). (b) Signature whistles of female bottlenose dolphin individuals FB163 and FB5, courtesy of L.S. Sayigh. (c) Pulse repetition rates of combined tonal/pulsed signals for individual narwhals mm224 and mm226, respectively, reprinted courtesy of the Journal of the Acoustical Society of America.

signature information (e.g., squeaks of adult timber wolves (*Canis lupus*, Goldman et al., 1995); calls of subantarctic fur seal pups (*Arctocephalus tropicalis*, Charrier et al., 2002); signature whistles of bottlenose dolphins (Fig. 1b) (*Tursiops truncatus*, Caldwell and Caldwell, 1965; Janik et al., 2006)).

Secondly, temporal elements can contain signature information. For example, the sonar signals of big brown bats (*Eptesicus fuscus*) and the isolation calls of pallid bats (*Antrozous pallidus*) showed individual differences in the duration and temporal arrangement of certain frequency elements, respectively (Masters et al., 1995; Brown, 1976). In another example, the relative timing and repetition rate of the pulses of combined tonal/pulsed signals of two Arctic narwhals (*Monodon monoceros*) off Baffin Island, Canada, showed individual variation suitable for individual signatures (Fig. 1c) (Shapiro, 2006).

II.B. Experimental determination of signature vocalizations

Two general stages are commonly followed to demonstrate that a particular signal constitutes a signature vocalization. Because it is often straightforward to identify vocalizations of individual animals in captivity, the first step of recordings often occurs in the laboratory. Once the individual recordings are acquired, it is necessary to determine whether there is a sufficient combination of inter-individual variability and intra-individual consistency to sort the signals according to the individual animals that produced them. One common approach is to ask a group of naïve human observers to categorize a set of signals, usually by sorting spectrograms visually into different classes (e.g., Janik, 1999). This technique is straightforward and useful as a first approach, since human classifiers have good skills in matching patterns and evaluating signals in a general way. These results can indicate whether more advanced analyses are likely to be fruitful. However, the use of human subjects can be time-consuming and may not reveal the precise acoustic features used to discriminate individuals.

Several analytical approaches have been developed, most of which require extracting a set of temporal and spectral features from all or part of the signal to classify the calls. Such approaches tend to require an *a priori* assessment of which characteristics to compare across a set of signals. Tonal signals are often represented as contour traces to facilitate comparison. It is straightforward to measure such parameters as a contour's initial, final, mean, minimum and maximum frequencies, temporal duration and the degree of its frequency modulation. In addition, the relative energies between different harmonics can be determined from the original recording. A signal composed of discrete pulses might be characterized by its duration, number of pulses and average inter-pulse interval (or, inversely, the pulse repetition rate). Once these features have been extracted from vocalizations produced by known individuals, statistical methods can be used to assess the significance of inter-individual differences (Falls, 1982; Sokal and Rohlf, 1995).

The coefficient of variation (i.e., the ratio between the standard deviation and the mean) of each feature is often also computed. Its value should be greater between than within individuals if the selected vocalizations are to qualify as potential signature signals (Lengagne, 2001; Charrier et al., 2002). Discriminant function analysis (Jones et al., 1993) and principal coordinates analysis (Cortopassi and Bradbury, 2000) have also been successfully applied to differentiate among the signals produced by multiple individuals. Other, nonparametric approaches can also be useful. For example, Shapiro (2006) applied a modified cross-correlation metric to compare the whistles produced by two narwhals.

The second essential step is a playback experiment to test whether conspecifics can discern between vocalizations produced by different individuals. Once demonstrated, the manipulation or removal of certain acoustic features in a playback can then indicate which parameters in the signal contain the relevant signature information for other conspecifics. For example, Charrier et al. (2001) used playback experiments to demonstrate that subantarctic fur seal pups on Amsterdam Island in the Indian Ocean responded to the calls of any female a few hours after birth. Within 2–5 days, however, this response became specific only to the vocalizations of their own mother. The pups stopped responding to other similarly-aged, possibly familiar reproductive females from the same study colony (Charrier, personal communication). Subsequent experimental manipulations of the maternal vocal signals showed that the pups relied on the harmonic structure and the initial ascending frequency modulation to recognize their mothers, but did not use amplitude modulation (Charrier et al., 2003). Charrier et al. (2002) also studied the question from the mothers' perspective and found that mothers used the fundamental frequency and the next two harmonics of their pup calls to recognize the pup successfully.

Similar results confirming individual discrimination based on signature vocalizations were obtained from bottlenose dolphins (Sayigh et al., 1998, discussed in greater detail below). Results of these studies go beyond the recognition of familiar or unfamiliar signals (e.g., Bee and Gerhardt, 2002) which can arise from simple habituation. Instead, they provide evidence for individual recognition capabilities among conspecifics, a demanding discrimination task.

II.C. Ontogeny of signature vocalizations

The development and acquisition of a signature vocalization can reveal how animals learn and at least some of the salient features of an animal's acoustic and/or social environment. Numerous terrestrial mammals can distinguish among vocalizations of individuals because of the unique physical differences characterizing their vocal tracts (see Tyack, 2000; and also Frey and Gebler, Chapter 10.3 in this volume). Such morphological differences represent inherited physical features, which cannot be learned. They likely account for certain aspects of the individually characteristic signals in some species. Often, however, animals can control the articulation and modulation of these signals to further modify certain acoustic features.

In a variety of species, animals acquire signature vocalizations through learning. Although physical constraints limit the temporal and spectral space in which the signature vocalization can be generated, learning allows an animal to differentiate and mimic species-specific acoustic patterns that it perceives in its environment. Vocal learning, a form of production learning in which signals are rendered more or less similar to those produced by other conspecific, heterospecific or synthetic sound models (Janik and Slater, 1997, 2000), can explain the ontogeny of a signature vocalization. Bottlenose dolphins provide an example. Early observations suggested that bottlenose dolphins produce signature whistles (Caldwell and Caldwell, 1965) and that the structure of these whistles may be influenced by their acoustic environment. Tyack (1986) reported that two captive animals produced their own signature whistles most of the time and sometimes produced the whistle of the other companion dolphin, suggesting that they could mimic one another. Demonstrations of vocal imitation in bottlenose dolphins (Richards et al., 1984) and observations of matching artificial flat tone whistles (Miksis et al., 2002) support the conclusion that these animals are capable of vocal learning. The studies on signature whistle ontogeny collectively reinforce the notion that bottlenose dolphins use vocal learning to imitate or sometimes eschew an acoustic model.

In some situations, an individual animal learns from itself. Human beings (*Homo sapiens*) are the most striking mammalian example of this phenomenon and show numerous parallels with vocal development among songbirds (Doupe and Kuhl, 1999). These similarities include learning difficult vocalizations early in life, using specialized brain structures to help integrate motor and auditory processes, possessing intrinsic perception and learning mechanisms, relying on acoustic models, and practicing through listening to themselves (i.e., auditory feedback).

III. Functional aspects of signature vocalizations

It is important to address the various functions of signature vocalizations to understand how these signals operate in both captive and natural settings.

III.A. Parent–offspring recognition

Recognition between parents and offspring is one of the primary functions of some signature vocalizations (Falls, 1982). It appears to drive signature vocalizations in numerous gregarious species of birds and mammals. Most studies on this topic have been conducted on bird species. As the signature vocalizations in birds are likely similar to those in mammals, a broad comparative perspective is useful. Colonial species generally encounter selective pressure for parents and offspring to recognize one another in a flurry of other individuals, a situation that is important but less confusing among non-colonial species. In birds, swallows demonstrate parent–offspring recognition convincingly. Parents and young of colonial species such as bank (*Riparia riparia*) and cliff swallows (*Hirundo pyrrhonota*) both produce and recognize signature vocalizations (Beecher et al., 1981, 1985; Sieber, 1985; Stoddard and Beecher, 1983), which ensures parental care is directed toward the correct offspring. In addition, some bat species live in very large colonies where recognition of individuals within the social group, particularly infants, is vital for their survival (Boughman, 1997; Bohn et al., 2007; Kazial et al., 2008). On the other hand, among non-colonial barn (*Hirundo rustica*) and tree swallows (*Tachycineta bicolor*) and solitary rough-winged swallows (*Stelgidopteryx ruficollis*), recognition between parents and offspring is weaker or absent altogether (Beecher, 1981; Medvin and Beecher, 1986; Leonard et al., 1997). In these non-colonial species, parents and offspring can locate one another more easily using location cues, since fewer conspecifics are nearby.

As expected, parent–offspring vocal recognition is common among colonial penguin species. The complexity of their signature vocalizations varies, however, according to whether the penguins do or do not build nests. Since nests provide a helpful visual landmark for individuals attempting to locate one another, the signature content of parental calls of certain nest-building species tends to be simpler. For example, Adélie (*Pygoscelis adeliae*) and gentoo (*P. papua*) penguins rely primarily on the basic feature of pitch of the vocalization to distinguish individuals (Jouventin and Aubin, 2002). In contrast, non-nest-building species such as king and emperor penguins must rely more heavily on acoustic discrimination and their resulting two-voice signature systems are more complex (Robisson, 1990; Jouventin et al., 1999; Aubin et al., 2000; Lengagne et al., 2000, 2001).

Although learning can require a long period of time, sometimes the survival needs can facilitate a quick and early learning. For example, as described above, fur seal pups learn to recognize their mothers' voices within 2–5 days after birth (Charrier et al., 2001). The reason for this early and fast learning is related to the fact that fur seal mothers must go to sea to feed and to build sufficient nutrient reserves for lactation while their newborn pups remain ashore. Because the mother must reunite with her pup once she returns to land, the two animals need to recognize each other in a crowd of conspecifics. The situation leads to both the pup and the mother acquiring vocal signatures (Charrier et al., 2002, 2003). The timing of this recognition always occurs before the mother's departure (2–10 days following birth). The importance of such recognition is evidenced by the fairly widespread synchrony between when parents and offspring first separate and the development of the capacity to recognize one another (Falls, 1982).

III.B. Individual recognition among adults

Within the context of acoustic communication, Falls (1982) pointed out that individual variation, or the differences in the vocal signals between individuals, form a precondition for individual recognition by using these acoustic cues to discriminate between conspecifics. Such recognition can arise in both affiliative and aversive (e.g., aggressive) situations.

Mate recognition is one of the contexts in which animals need to identify each other reliably, a scenario that can benefit from employing signature vocalizations. Mated pairs of colonial seabird species, for example, can encounter the same difficulties of locating one another in large, noisy colonies as the parents and offspring of the species described earlier. For instance, Manx shearwaters (*Puffinus puffinus*) entered their burrows at night when visual landmarks were less helpful. Males arrived at a burrow first and responded generically to the sounds of any vocalizing females that arrived later in the season. Playback experiments demonstrated that the female was the member of the pair bond that discriminated among the male calls to locate her mate and the appropriate burrow (Brooke, 1978).

Vocalizations are used by some mammalian species for advertising and maintaining their territories (e.g., in elephants, see Garstang, Chapter 3.2 in this volume). Among certain species of songbirds, males also sing to each other to maintain and defend their territories. These contexts serve an aggressive function. The responses to an intruding male with unfamiliar song may result from a simple discrimination between known neighbors and unknown strangers (e.g., indigo buntings (*Passerina cyanea*), Emlen, 1971; little owls (*Athene noctua*), Hardouin et al., 2006), a dichotomous classification system that does not demand individual recognition. Sometimes, however, individuals respond vocally (i.e., aggressively) to the song of a familiar neighbor if that song originates from a different and unexpected location (e.g., white-throated sparrows (*Zonotrichia albicollis*) (Falls and Brooks, 1975); hooded (*Wilsonia citrina*) and Kentucky warblers (*Oporornis formosus*) (Godard and Wiley, 1995). This suggests that for some species, birds are not simply habituating to the songs produced repeatedly by their neighbors. Rather, they might be recognizing conspecifics insofar as they associate particular songs with specific individuals whose geographical distribution generally remains fixed. More work is required to rigorously demonstrate individual recognition, but such a scenario would help these animals conserve their energy resources by targeting their vocal behavior at, and establishing their territories in response to, intruders or new arrivals in specific defensive contexts.

III.C. Mediating social interactions

Signature vocalizations can greatly facilitate social interactions between animals where individual-specific relationships (evidenced by distinctive behavioral responses to different conspecifics) are important. Bottlenose dolphins provide a useful case study in this regard, since their social and vocal behavior has been studied extensively in both the laboratory and the wild. Their social structure is characterized by a fission–fusion dynamic, a situation in which animals fluidly and repeatedly associate with, separate from, and rejoin one another (Wells et al., 1987). This fluidity of social structure, coupled with individual-specific relationships in which dolphins treat different animals as distinct entities, provides a scenario for which their individually characteristic signature whistles are well-suited (Tyack and Sayigh, 1997).

In Sarasota, Florida, temporarily restrained mother bottlenose dolphins and their independent offspring responded more strongly to one another's signature whistles than to those of other individuals of the same age (Sayigh et al., 1998). The results of this work demonstrate that signature whistles are likely used for individual recognition and they might play a role in reuniting mothers with their calves when they separate from one another. Similarly, free-ranging adult, allied male dolphins produced their signature whistles most frequently when they separated voluntarily from their partners and least frequently when they were swimming together (Watwood et al., 2005). In addition, temporarily restrained allied males that were separated involuntarily from one another produced more signature whistles than during periods when they were swimming freely (Watwood et al., 2005). These findings suggest that allied male dolphins, while separated, might also use signature whistles as contact calls to help them find one another and stay in contact.

In addition to the vocal learning described above, bottlenose dolphins are capable of vocal labeling (Richards et al., 1984) in which a vocalization is used to identify an object or individual acoustically. This has particularly interesting implications when animals become separated. Janik and Slater (1998) studied the social context of whistle usage of a group of four bottlenose dolphins (including one juvenile, one sub-adult and two adult animals) at a captive facility consisting of two connected pools of unequal size. Whenever an individual voluntarily left the group by swimming to the smaller pool apart from the remaining animals, it produced its signature whistle at a much higher rate than when all four dolphins swam together in the main pool. These results suggested that the animals were using their signature whistles to facilitate group cohesion (Janik and Slater, 1998). Because group composition is likely to change in the wild, a common group call would be insufficient to maintain this cohesion. Rather, given the vocal mimicry abilities of these animals (see Richards et al., 1984; Tyack, 1986; Miksis et al., 2002), they could accommodate an individual arriving at, or departing from, the group by simply learning its signature whistle. Taken together, these studies provide evidence that bottlenose dolphins (at least in captivity) produce their own signature whistles in certain contexts but are also capable of copying the signature whistles of their companions in other situations.

An example of such whistle copying may have been observed in the Moray Firth, Scotland, where whistles were localized to the free-ranging bottlenose dolphins (Janik, 2000). For each pair of consecutive whistles, both the distance between their localized origins and the inter-whistle time interval were computed. If a dolphin could not have traveled this distance within the calculated time interval when swimming at its maximum speed, the two whistles were considered to have been produced by different individuals. Among 176 such whistle interactions between dolphins, 39 (22%) involved matched exchanges of the same whistle between different individuals separated by up to 579 m ($\bar{x}$ = 179 m). These matched interactions may have functioned as a mechanism for animals to address one another, indicate alliance membership, or coordinate interactions over large distances (Janik, 2000). This may have been a demonstration of signature whistles being produced and then copied, although additional work on identifying the whistler and the social and behavioral context is required.

In summary, bottlenose dolphin fission–fusion social structure is characterized by individual specific social relationships that benefit from an ability to detect and recognize individuals by their signature whistles. The production of these individually specific vocal signals is not uncommon among other species that share a similarly fluid social dynamic in environments that are at least occasionally visually occluded. Furthermore, signature vocalizations often accompany situations where: (1) individual contact between animals is biologically important (e.g., parents and offspring, mated pairs, social allies, or aggressors); and (2) separation due to visual barriers or confusion among numerous conspecifics is present.

IV. Future directions

A couple of areas of research related to signature vocalizations stand out as especially promising for future work.

The signature vocalizations described in this chapter often differ from one another in a somewhat obvious manner. However, it is likely that animals are capable of distinguishing each other using far more subtle acoustic cues than those we have described. Only recently, for example, the stereotyped calls of resident, fish-eating killer whales (*Orcinus orca*) off the coast of Vancouver Island, Canada, were shown to contain sex-specific spectral features relating to the relative energy levels of different harmonics (Miller et al., 2007). A towed hydrophone array (Miller and Tyack, 1998) allowed the determination of which killer whale was vocalizing, a technical advancement that greatly facilitated the collection of the data. Data collection technologies, which also include digital archival tags (e.g., Johnson and Tyack, 2003) and both terrestrial and marine recorder arrays (e.g., Miller et al., 2004; Mennill and Vehrencamp, 2008), will improve the reliability of assigning vocalization sequences to the individuals producing them. The application of these technologies towards deepening our understanding of how and why individual animals produce and use signature vocalizations in the wild should be encouraged.

A second area for future work should involve investigation of the detailed acoustic features that animals are using to identify one another. Animals use a rich array of feature detectors to detect signals buried in background noise. This problem is especially important in colonial situations, in which, with the increasing number of individuals, the noise generally occupies a nearly identical temporal and spectral bandwidth as the signal being sought. However, even young animals overcome the masking easily and probably rely on subtle features to make the discrimination possible. In addition, many terrestrial species rely on the morphological differences of the vocal tract and/or vocal production organ to differentiate between speakers (see Tyack, 2000). This has been harder to assess for marine mammals. All of the members of a killer whale acoustic clan, for example, produce a common set of stereotyped vocalizations that are difficult for humans to discriminate.

It is possible, however, that these animals are embedding barely noticeable (to us) signature information into their vocalizations that a more detailed investigation might detect. For human speech recognition, extensive research has been conducted by analyzing voices of human speakers and comparing subtle features of their verbal utterances (e.g., Reynolds, 1995). One approach worth exploring involves taking the techniques that have been developed to differentiate between human speakers and using them to classify similar-sounding animal vocalizations according to the individuals that produced them. This may lead to the discovery of additional features that animals use to encode their individual identity acoustically.

V. Conclusions

Signature vocalizations are attributes of individual animals containing spectral and temporal acoustic features that allow conspecifics to perform individual recognition. Signature information is best demonstrated by coupling computer- or human-based classification schemes in the laboratory with playback experiments in the wild. The ability to recognize other individuals offers an adaptive advantage in such situations as parents searching for their young, social companions attempting to reestablish contact and male aggressors asserting their territorial claims. Signature vocalizations also offer a way for researchers to single out both relevant social partnerships in the midst of a gregarious colony and meaningful encounters in the mist of a communicative flurry.

Acknowledgments

I am grateful to P.L. Tyack for his review of this chapter and for financial support. Special thanks to J.W. Bradbury and L. Sayigh for providing me with their audio recordings.

References

Aubin, T., Jouventin, P., Hildrebrand, C., 2000. Penguins use the two-voice system to recognize each other. Proc. Royal Soc. Lond. Ser. B-Biol. Sci. 267, 1081–1087.

Bee, M.A., Gerhardt, H.C., 2002. Individual voice recognition in a territorial frog (*Rana catesbeiana*). Proc. Royal Soc. Lond. Ser. B-Biol. Sci. 269, 1443–1448.

Beecher, M.D., 1981. Development of parent–offspring recognition in birds. In: Aslin, R.N., Alberts, J.R., Petersen, M.R. (Eds.), Development of Perception: Psychobiological Perspectives. Academic Press, New York, NY, pp. 45–66.

Beecher, M.D., Beecher, I.M., Hahn, S., 1981. Parent–offspring recognition in bank swallows (*Riparia riparia*): II. Development and acoustic basis. Anim. Behav. 29, 95–101.

Beecher, M.D., Stoddard, P.K., Loesche, P., 1985. Recognition of parents' voices by young cliff swallows. The Auk 102, 600–605.

Bohn, K.M., Wilkinson, G.S., Moss, C.F., 2007. Discrimination of infant isolation calls by female greater spear-nosed bats, *Phyllostomus hastatus*. Anim. Behav. 73, 423–432.

Boughman, J.W., 1997. Greater spear-nosed bats give group-distinctive calls. Behav. Ecol. Sociobiol. 40, 61–70.

Brooke, M.D.L., 1978. Sexual differences in the voice and individual vocal recognition in the Manx Shearwater (*Puffinus puffinus*). Anim. Behav. 26, 622–629.

Brown, P., 1976. Vocal communication in the pallid bat, *Antrozous pallidus*. Z. Tierpsychol. 41, 34–54.

Caldwell, M.C., Caldwell, D.K., 1965. Individualized whistle contours in bottlenose dolphins (*Tursiops truncatus*). Nature 207, 434–435.

Charrier, I., Mathevon, N., Jouventin, P., 2001. Mother's voice recognition by seal pups. Nature, 412–873.

Charrier, I., Mathevon, N., Jouventin, P., 2002. How does a fur seal mother recognize the voice of her pup? An experimental study of *Arctocephalus tropicalis*. J. Exp. Biol. 205, 603–612.

Charrier, I., Mathevon, N., Jouventin, P., 2003. Vocal signature recognition of mothers by fur seal pups. Anim. Behav. 65, 543–550.

Cortopassi, K.A., Bradbury, J.W., 2000. The comparison of harmonically rich sounds using spectrographic cross-correlation and principal coordinates analysis. Bioacoustics 11, 89–127.

Doupe, A.J., Kuhl, P.K., 1999. Birdsong and human speech: common themes and mechanisms. Annu. Rev. Neurosci. 22, 567–631.

Emlen, S.T., 1971. The role of song in individual recognition in the Indigo Bunting. Z. Tierpsychol. 28, 241–246.

Falls, J.B., 1982. Individual recognition by sounds in birds. In: Kroodsma, D.E., Miller, E.H. (Eds.), Acoustic Communication in Birds. Academic Press, New York, NY.

Falls, J.B., Brooks, R.J., 1975. Individual recognition by song in White-throated Sparrows. II. Effects of location. Can. J. Zool. 53, 1412–1420.

Godard, R., Wiley, R.H., 1995. Individual recognition of song repertoires in two wood warblers. Behav. Ecol. Sociobiol. 37, 119–123.

Goldman, J.A., Phillips, D.P., Fentress, J.C., 1995. An acoustic basis for maternal recognition in timber wolves (*Canis lupus*). J. Acoust. Soc. Am. 97, 1970–1973.

Hardouin, L.A., Tabel, P., Bretagnolle, V., 2006. Neighbour-stranger discrimination in the little owl, *Athene noctua*. Anim. Behav. 72, 105–112.

Janik, V.M., 1999. Pitfalls in the categorization of behaviour: a comparison of dolphin whistle classification methods. Anim. Behav. 57, 133–143.

Janik, V.M., 2000. Whistle matching in wild bottlenose dolphins (*Tursiops truncatus*). Science 289, 1355–1357.

Janik, V.M., Slater, P.J.B., 1997. Vocal learning in mammals. Adv. Study Behav. 26, 59–99.

Janik, V.M., Slater, P.J.B., 1998. Context-specific use suggests that bottlenose dolphin signature whistles are cohesion calls. Anim. Behav. 56, 829–838.

Janik, V.M., Slater, P.J.B., 2000. The different roles of social learning in vocal communication. Anim. Behav. 60, 1–11.

Janik, V.M., Sayigh, L.S., Wells, R.S., 2006. Signature whistle shape conveys identity information to bottlenose dolphins. Proc. Nat. Acad. Sci. USA, 103, 8293–8297.

Johnson, M.P., Tyack, P.L., 2003. A digital acoustic recording tag for measuring the response of wild marine mammals to sound. IEEE J. Oceanic Eng. 28, 3–12.

Jones, B.S., Harris, D.H.R., Catchpole, C.K., 1993. The stability of the vocal signature in phee calls of the common marmoset, *Callithrix jacchus*. Am. J. Primatol. 31, 67–75.

Jouventin, P., Aubin, T., 2002. Acoustic systems are adapted to breeding ecologies: individual recognition in nesting penguins. Anim. Behav. 64, 747–757.

Jouventin, P., Aubin, T., Lengagne, T., 1999. Finding a parent in a king penguin colony: the acoustic system of individual recognition. Anim. Behav. 57, 1175–1183.

Kazial, K.A., Kenny, T.L., Burnett, S.C., 2008. Little brown bats (*Myotis lucifugus*) recognize individual identity of conspecifics using sonar calls. Ethology 114, 469–478.

Lengagne, T., 2001. Temporal stability in the individual features in the calls of eagle owls (*Bubo bubo*). Behaviour 138, 1407–1419.

Lengagne, T., Aubin, T., Jouventin, P., Lauga, J., 2000. Perceptual salience of individually distinctive features in the calls of adult king penguins. J. Acoust. Soc. Am. 107, 508–516.

Lengagne, T., Lauga, J., Aubin, T., 2001. Intra-syllabic acoustic signatures used by the king penguin in parent–chick recognition: an experimental approach. J. Exp. Biol. 204, 663–672.

Leonard, M.L., Horn, A.G., Brown, C.R., Fernandez, N.J., 1997. Parent–offspring recognition in tree swallows, *Tachycineta bicolor*. Anim. Behav. 54, 1107–1116.

Masters, W.M., Raver, K.A.S., Kazial, K.A., 1995. Sonar signals of big brown bats, *Eptesicus fuscus*, contain information about individual identity, age and family affiliation. Anim. Behav. 50, 1243–1260.

Medvin, M.B., Beecher, M.D., 1986. Parent–offspring recognition in the barn swallow (*Hirundo rustica*). Anim. Behav. 34, 1627–1639.

Mennill, D.J., Vehrencamp, S.L., 2008. Context-dependent functions of avian duets revealed by microphone-array recordings and multispeaker playback. Curr. Biol. 18, 1314–1319.

Miksis, J.L., Tyack, P.L., Buck, J.R., 2002. Captive dolphins, *Tursiops truncatus*, develop signature whistles that match acoustic features of human-made model sounds. J. Acoust. Soc. Am. 112, 728–739.

Miller, P.J., Tyack, P.L., 1998. A small towed beamforming array to identify vocalizing resident killer whales (*Orcinus orca*) concurrent with focal behavioral observations. Deep Sea Res. II, 45, 1389–1405.

Miller, P.J.O., Samarra, F.I.P., Perthuison, A.D., 2007. Caller sex and orientation influence spectral characteristics of "two-voice" stereotyped calls produced by free-ranging killer whales. J. Acoust. Soc. Am. 121, 3932–3937.

Miller, P.J.O., Shapiro, A.D., Tyack, P.L., Solow, A.R., 2004. Call-type matching in vocal exchanges of free-ranging resident killer whales, *Orcinus orca*. Anim. Behav. 67, 1099–1107.

Phillips, A.V., Stirling, I., 2000. Vocal individuality in mother and pup South American fur seals, *Arctocephalus australis*. Mar. Mammal Sci. 16, 592–616.

Reynolds, D.A., 1995. Speaker identification and verification using Gaussian mixture speaker models. Speech Communic. 17, 91–108.

Richards, D.G., Woltz, J.P., Herman, LM., 1984. Vocal mimicry of computer-generated sounds and vocal labeling of objects by a bottlenosed dolphin (*Tursiops truncatus*). J. Comp. Psychol. 98, 10–28.

Robisson, P., 1990. The importance of the temporal pattern of syllables and the syllable structure of display calls for individual recognition in the genus Aptenodytes. Behav. Proc. 22, 157–163.

Sayigh, L.S., Tyack, P.L., Wells, R., Solow, A.R., Scott, M. D., Irvine, A.B., 1998. Individual recognition in wild bottlenose dolphins: a field test using playback experiments. Anim. Behav. 57, 41–50.

Shapiro, A.D., 2006. Preliminary evidence for signature vocalizations among free-ranging narwhals (*Monodon monoceros*). J. Acoust. Soc. Am. 120, 1695–1705.

Sieber, O.J., 1985. Individual recognition of parental calls by bank swallow chicks (*Riparia riparia*). Anim. Behav. 33, 107–116.

Sokal, R.R., Rohlf, F.J., 1995. Biometry: The Principles and Practices of Statistics in Biological Research. W. H. Freeman and Co., New York, NY.

Stoddard, P.K., Beecher, M.D., 1983. Parental recognition of offspring in the cliff swallow. The Auk 100, 795–799.

Trillmich, F., 1981. Mutual mother–pup recognition in Galapagos fur seals and sea lions: cues used and functional significance. Behaviour 78, 21–42.

Tyack, P., 1986. Whistle repertoires of two bottlenosed dolphins, *Tursiops truncatus*: mimicry of signature whistles? Behav. Ecol. Sociobiol. 18, 251–257.

Tyack, P.L., 2000. Dolphins whistle a signature tune. Science 289, 1310–1311.

Tyack, P.L., Sayigh, L.S., 1997. Vocal learning in cetaceans. In: Snowdon, C., Hausberger, M. (Eds.), Social Influences on Vocal Development. Cambridge University Press, Cambridge, UK, pp. 208–233.

Watwood, S.L., Owen, E.C.G., Tyack, P.L., Wells, R.S., 2005. Signature whistle use by temporarily restrained and free-swimming bottlenose dolphins, *Tursiops truncatus*. Anim. Behav. 69, 1373–1386.

Wells, R.S., Scott, M.D., Irvine, A.B., 1987. The social structure of free-ranging bottlenose dolphins. In: Genoways, H.H. (Ed.), Current Mammalogy. Plenum, New York, NY, pp. 247–305.

CHAPTER 11.5

Evolution of mammalian vocal signals: development of semiotic content and semantics of human language

Klaus Zuberbühler*

School of Psychology, University of St Andrews, St Andrews, Scotland, UK

Abstract: Human language is the product of a specialized cognitive architecture that interacts with a uniquely adapted vocal apparatus. Current evidence suggests that many of the necessary core components, mental concepts, social awareness, cooperative motivation and vocal control, are also present in non-human animals, in various rudimentary forms. First, the ability to represent the world in terms of mental concepts appears to have evolved early, although it is difficult to determine how animal concepts compare with those of humans. Second, apes have demonstrated a high degree of social awareness and cooperative motivation, which are other key components of the language faculty, although these features are also found in other social mammals. Vocal control, finally, is well-developed in humans, but relevant precursors are also seen in non-human animals. The emerging picture is one of a patchwork of adaptations, uniquely combined and elaborated in humans, suggesting that many of the critical language components are latently present in the mammalian class. Cooperation and acoustic control have evolved through convergent evolution, suggesting that these capacities are latently present in a wide range of species. Mental concepts and social awareness have a more clearly visible phylogenetic history, with social awareness exhibiting the most limited distribution.

Keywords: semantic; semiotic; alarm calls; social intelligence; referential; mental concept; intention; triadic awareness; cooperation; gestures

I. Introduction

From a biological perspective, human language is just another example of the diverse ways by which mammals communicate. Although it is a complex and flexible behavior, it also has shared most of its evolutionary past with other mammals. Only very recently, some hundred thousand years according to one theory (Enard et al., 2002), did humans begin to diverge significantly from the rest of the animal world in terms of their communication skills and how they managed to construct and understand meaning. This process coincided with other important changes, including a rapid increase in brain size, which makes it difficult to decide whether language is a specially adapted behavior, or the by-product of other transitions. Language is the product of a range of adaptations and first needs to be "carved by its joints," before any comparative investigations can be instigated (Hurford, 2007). One difficult problem is to decide what the relevant components are exactly, and to distinguish the core ones from their derivatives.

Here, I will focus on some components that are widely considered as basal: mental concepts; social awareness; motivation to cooperate; and ability to control the sound production apparatus. All are crucial for the semiotic features and semantic force that characterize human communication. The main points, in brief, are the following. First, language relies on the cognitive ability to organize the world in discrete and mentally

*Corresponding author. E-mail: kz3@st-and.ac.uk

Stefan M. Brudzynski (Ed.)
Handbook of Mammalian Vocalization
ISBN 978-0-12-374593-4

DOI: 10.1016/B978-0-12-374593-4.00046-2

represented actions and categories. Many of these natural mental concepts appear to emerge more or less automatically in human beings, without any specific training, and they do not require language. Without such a biological predisposition, it is difficult to imagine how language, a largely arbitrary system, could ever become grounded in human cognition. Another important component of language concerns the ability of speakers to be socially aware of the current situation, the composition of the audience, their behaviors and relationships, motivations and knowledge. This is relevant because a primary way by which humans generate meaning is by taking into account what others know about the world or how they assess the situation. A third component has to do with a general ability and willingness to liaise with others and to cooperate with them in everyday activities. Cooperatively-motivated individuals are more likely to share attention and information with each other, a fundamental psychological skill, especially during language acquisition (Tomasello, 2008). Finally, human communication is primarily an acoustic behavior, which relies on strong and precise motor control of a number of articulators in the supralaryngeal vocal tract, particularly the tongue. Vocal learning and imitation is also important, but they may be a mere by-product of advanced oro-facial motor skills.

II. Mental concepts

II.A. Mental concepts in animals

In primates, considerable research has been devoted to the question of what types of mental concepts individuals are able to entertain. In one classic study, Dasser (1988) trained long-tailed macaques to select images of group members with a special social relationship, the mother–offspring dyad. In the test conditions, monkeys were able to transfer to untrained pictures of group members, demonstrating that they were able to utilize this social concept to organize their world. In other captive work, baboons were tested on the ability to make same/different judgements and to apply this ability to compare categories of objects (Bovet and Vauclair, 2001). In apes, it was found that individuals could discriminate natural concepts at various levels of abstractness, such as discriminating between pictures of animals and non-animals (Vonk and MacDonald, 2002, 2004).

In non-primates, bottle-nosed dolphins could be trained to classify pairs of objects as same or different, which they could transfer to two-dimensional representations of novel items and use to discriminate between categories of items (Mercado et al., 2000). Domestic dogs could be trained to discriminate between 200 different items following a verbal command ("name"). If forced to choose from an array containing an unfamiliar item, the subject spontaneously associated this item with a new name and remembered this link, even after a considerable delay (Kaminski et al., 2004). The dog's performance was based on something akin to reasoning by exclusion, further demonstrated by other experiments with dogs (Erdohegyi et al., 2007). Some of these concepts are modality independent. In one case, dogs were able to generate a visual representation of the owner's face after hearing their voice (Adachi et al., 2007).

II.B. Mental concepts in the wild

One critique of many of these studies on concept formation in animals is that they typically rely on a considerable training effort and it is often unclear whether this is handled by general learning mechanisms. Studies with free-ranging animals, particularly their communicative behavior in response to external events, can thus be more relevant to address the question of mental concepts in animals. Such research in the wild has relied considerably on alarm calls, mainly because there is often a strong and direct link between a call type and a specific biologically relevant event in the world (Zuberbühler, 2006).

If a monkey consistently produces one type of alarm call in response to raptors, then this contingency typically becomes meaningful to others. Many animals possess sizeable repertoires of alarm calls, produced to signal different types of dangers in relatively specific ways (e.g., ground squirrels: Leger et al., 1979; marmots: Blumstein, 1999; prairie dogs: Kiriazis and Slobodchikoff, 2006) (suricates: see Manser, Chapter 11.2 in this volume).

In primates, there are considerable differences in how species convey information about external events. A common finding is that one acoustically-distinct alarm call type is given to a narrow range of events, such as dangerous raptors, while a second type is given to other types of danger, usually terrestrial predators or other disturbances. In baboons, however, alarm barks contain considerable acoustic variation, with some variants produced in predation and others in non-predation contexts (Fischer et al., 2001). Some other primates convey information by assembling a small repertoire of calls into more complex sequences (Schel et al., 2009). White-handed gibbons use their calls not only in pair bonding, but also to indicate the presence of predators, such as clouded leopards or tigers. These predator-induced calls consist of the same types of notes as normal calls, but they are assembled in subtly different ways (Clarke et al., 2006). In putty-nosed monkeys, adult males produce two basic types of alarm calls to predators, but the calls are then assembled into sequences specific to the presence of eagles or leopards and sequences that indicate group travel (Arnold and Zuberbühler, 2008) (Fig. 1).

One interpretation of these results is that, when producing and responding to each other's alarm calls, monkeys generate semiotically diverse routes that are part of their phylogenetic history and which tap into the same basic mental representations of the different predator classes. However, one relevant difference to human language is that, by responding to specific external events with species-specific biologically predetermined calls, individuals have little control over the relationship between the acoustic features of their signals and the triggering of external events. In other words, the mental concepts generated alongside, are in some sense a coincidental side-product, rather than a necessary prerequisite for successful communication.

III. Social awareness

Humans are generally aware of what their receivers understand and know about a situation, and their communications are heavily influenced by these factors. Meaning is shared between individuals, rather than merely being a product of a stimulus-response contingency. Even though social awareness that includes the receiver is pivotal for language, Darwinian common sense predicts that it is unlikely for this cognitive skill to have emerged *de novo*, and socially living mammals are the primary place to start a comparative investigation.

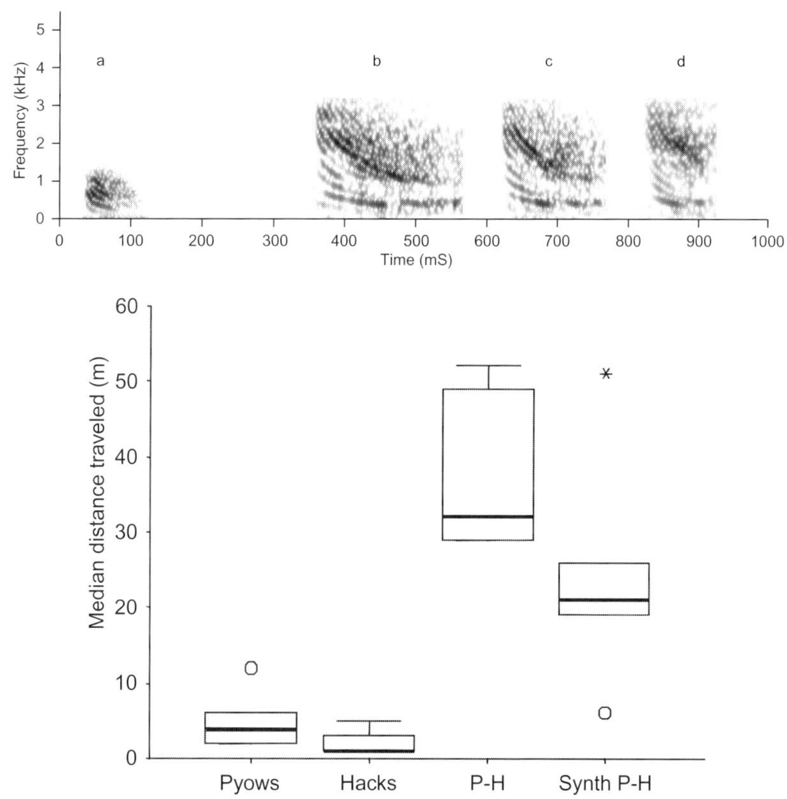

Fig. 1. Top: Spectrographic illustration of representative exemplars of the two types of male putty-nosed load alarm calls (a) hack, (b–d) pyows. Bottom: Responses of female putty-nosed monkeys to playbacks of their own male's pyows, hacks, natural pyow-hack sequence, or artificial pyow-hack sequence composed of natural pyows and hacks. Arnold and Zuberbühler (2008).

Somewhat surprisingly, sophisticated social skills appear to evolve relatively easily in animals. Research using species as diverse as cleaner fish, rooks, or chimpanzees have revealed that complex social behaviors can evolve readily, provided individuals interact with each other on a regular basis and in intricate ways (Zuberbühler and Byrne, 2006). Advanced social skills are often also revealed during agonistic interactions.

In primates, various peace-making behaviors have been observed following aggression, such as "reconciliation" between former opponents, or kin reconciling on behalf of victims or aggressors (Wittig et al., 2007). Other third-party behavior can involve "consolation" by which a bystander intervenes to support the victim after a fight (de Waal, 1982). In some species, high-ranking individuals sometimes intervene to terminate aggression between two group members, a behavior sometimes termed "policing." A recent example comes from free-ranging chimpanzees in the Budongo Forest, Uganda, where high-ranking males have attempted to stop a group of resident females from committing coalitionary lethal attacks on the infants of recent immigrant mothers (Townsend et al., 2007).

There is controversy about the cognitive sophistication required to process third-party interactions and other social interactions that require a certain degree of awareness. In baboons, individuals appear to understand the hierarchical organization of their group at both the individual and matrilineal levels (Bergman et al., 2003), while in sooty mangabeys egocentric heuristics may be at work when individuals solicit third-party aid (Range and Noë, 2005). Like most social mammals, chimpanzees produce acoustic signals during conflicts, most likely to address their opponent, to alert nearby group members and to recruit aid (Gouzoules et al., 1984). In the Budongo Forest, Uganda, chimpanzees produce variable screams that convey information about the role of the caller during the conflict (victim or aggressor) and the nature of the conflict, e.g., whether it involved direct physical contact or mere threats (Slocombe and Zuberbühler, 2007). Victims of aggression modified the acoustic structure to increase the chances of eliciting help from high-ranking individuals. If high-ranking individuals were present, victims of moderately severe aggression produced screams that were acoustically identical to screams given in cases of very severe aggression, suggesting that they understand who is able to help them effectively and how to obtain this support (Fig. 2).

Some relevant evidence is also available for monkeys. In Thomas langurs, males only emitted alarm calls if there was an audience (Wich and de Vries, 2006), a common principle in many species. Primates and other animals are sensitive to the composition of their audience, particularly the presence of kin or mates, but an unresolved issue is whether callers can alter their calling behavior in systematic ways to affect the behavior of nearby individuals (Zuberbühler, 2008). In the Thomas langur study, surprisingly, males continued to give alarm calls to a model predator until all other group members had given at least one alarm call themselves, as if the males were trying to keep track of which group members had and had not called (Wich and de Vries, 2006). In another field study, male blue monkeys were exposed to playbacks of a neighboring male's eagle alarm calls. It was found that subjects produced significantly more of their own eagle alarm calls if their own group members were close to the presumed eagle, compared to when they were further away, regardless of the calling male's own position relative to the eagle (Papworth et al., 2008).

IV. Cooperative motivation: cooperation in animals

Cooperation is relatively common in the animal world, and reports of cooperative behaviors are widely available, including species such as cleaner fish, unlikely carriers of high intelligence (Bshary et al., 2002). Cooperative behavior as such appears to be an unreliable indicator of advanced cognition.

One way to study the cognitive sophistication underlying a cooperative propensity is during social games with objects, as this requires coordinating attention to task, the object and to one another. Children commonly engage in such triadic games, but similar observations have also been reported from dogs and dolphins interacting with humans (Mitchell and Thompson, 1993; Kuczaj and Highfill, 2005) and, more interestingly, with each other (McBride and Hebb, 1948). In non-human primates, social games are rarely triadic or cooperative. In one systematic study, Warneken and Tomasello (2006) compared the cooperative skills of children and chimpanzees during triadic social games (Warneken et al., 2006). The key experimental manipulation was to engage the subject in a social game, but then abruptly stop the interaction. The subject's response

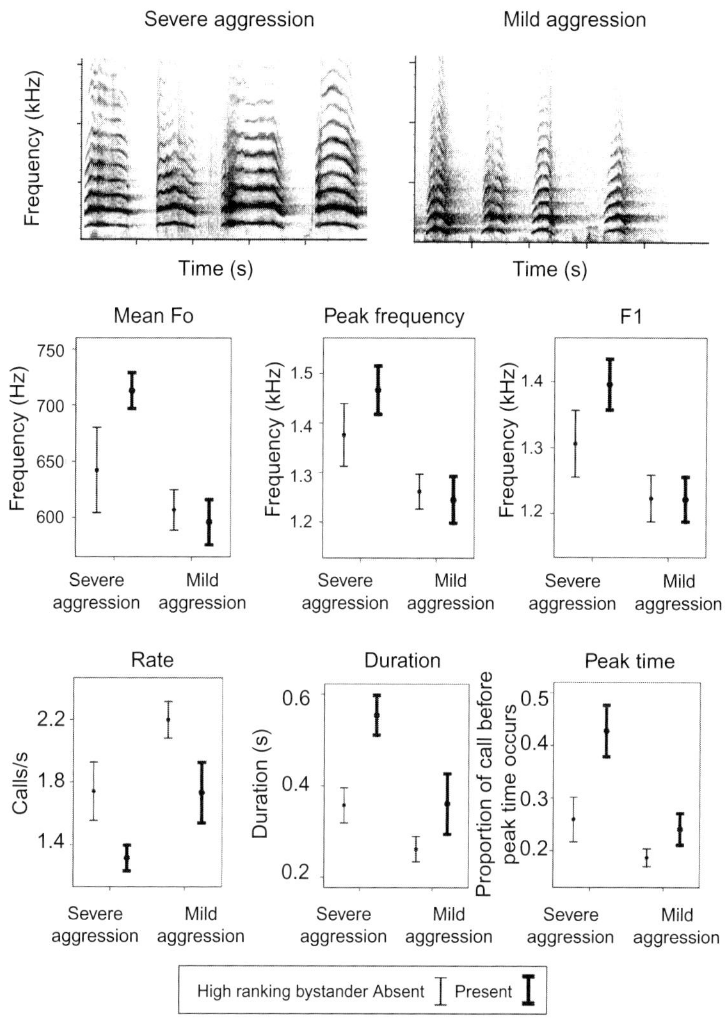

Fig. 2. Top: Chimpanzee victim screams given to severe and mild aggression. Bottom: Victims of severe aggression tend to shift the acoustic features of their screams if high ranking bystanders are nearby, who are potentially able to interfere. Slocombe and Zuberbühler (2007).

was used to assess his understanding of the game's cooperative nature and the different roles it required. Results were clear-cut. While children tried to encourage the experimenter to reuptake his role, by using communicative signals, this was not observed in chimpanzees.

In a recent study with bonobos, an experimenter encouraged subjects to play one of several games with an object. As in the study by Warneken et al. (2006), the experimenter then suddenly stopped the interaction, and the subject's response was monitored. Unlike chimpanzees, these apes behaved more like children by gesturing to the experimenter, an apparent attempt to re-engage the suddenly passive partner (Pika and Zuberbühler, 2007). Crucially, during the game, subjects were not interested in gaining possession of the object *per se*, but used it to maintain the joint activity.

Results suggest that apes are also able to form joint intentions with a partner towards a common goal. Why chimpanzees behaved differently is difficult to assess, but the current evidence does not support a fundamental cognitive difference between humans and apes in terms of the ability to share intention (Tomasello, 2008; Gomez and Martin-Andrade, 2002). Clearly, the "re-engaging a reluctant partner" paradigm is only one observational criterion for shared intentionality. In particular, children appear to go a step further by coordinating their plans of action and joint intentions to allow role reversal. In the bonobos study, individuals appeared to understand and attend to both roles of the game, although they never attempted to take on the role of the human playmate.

V. Motor control of sound production

V.A. *Sound production and articulation*

Human speech is largely the product of an extremely agile tongue that can cause constrictions in the vocal tract to generate the various speech sounds. Humans can alter the constrictions during sound production, which generates formant transitions. In contrast, the default assumption for mammalian vocal tracts has long been that they resemble a uniform tube with no relevant constrictions, and no ability to generate frequency modulations in this way (Lieberman, 1984). Humans were also thought to differ from animals in terms of the permanently low resting position of the larynx.

More recent research has shown that these views are not correct. Frequency modulations caused by articulation can be observed in animal calls and many species are capable of lowering of the larynx, including chimpanzees (Nishimura et al., 2003). Using lateral radiographs and postmortem dissection, Riede and colleagues determined the geometry of the Diana monkey's vocal tract and identified a number of constrictions between vocal folds and lips (Riede et al., 2005). From these data, it was possible to generate a computer model which was capable of replicating the sound properties of natural Diana monkey eagle and leopard alarm calls to an astonishingly accurate degree. The study demonstrated that some nonhuman animals can use flexible constrictions in their vocal tract to convey semantic information (Fig. 3).

V.B. *Vocal imitation*

In a recent review, Tyack (2008) argued that a case for vocal learning is made if an individual imitates a sound that is atypical for the species' normal repertoire. Examples are elephants (Poole et al., 2005), harbor seals (Ralls et al., 1985), bottle-nosed dolphins (Richards et al., 1984) and other species. In the elephant study, the calls of one African elephant were unlike any of the normal calls made by other African elephants, because they sounded like moving trucks. Another African elephant, which had been raised with two Asian elephants, produced calls that were more similar to Asian than African elephants (Poole et al., 2005).

In primates, the evidence for vocal flexibility and vocal control is weak. The default position is illustrated by a study conducted by Hammerschmidt et al. (2001) who reared squirrel monkeys in normal social conditions with a mute mother, or congenitally deaf. Age-related acoustic changes occurred throughout the vocal repertoire, but the calls of both individuals deprived of hearing adult calls fell within the range of calls produced by normally raised monkeys. In this primate, auditory input had little effect on call production, while age-related variations in calls were primarily caused by physical growth. In another more controlled experiment, rhesus macaques were cross-fostered with Japanese macaques, but despite this radical social intervention, subjects showed only minor developmental effects in their call structure (Owren et al., 1993). Cross-fostered Japanese macaques adhered to their species-typical behavior, while cross-fostered rhesus subjects also exhibited species-typical behavior in many contexts, although some call rates were affected. A number of exceptions exist, however. Recently, Wich and colleagues reported that a captive orang-utan could be brought to imitate whistling sounds made by humans, but little was known about the subject's past history and training (Wich et al., 2009).

Another major road of vocal learning is by means of vocal convergence, which is much more widespread (Tyack, 2008). In a typical case, the acoustic properties of calls from different individuals converge in a number of acoustic parameters, especially when they form social bonds. Vocal convergence has been demonstrated for animals as diverse as songbirds, bats, elephants, cetaceans and primates. Call convergence may function as a group identifier, or an indication of strong social bonds.

Clearly, humans differ substantially from the other primates in terms of their vocal repertoire. Although vocal development in human infants also starts with the emergence of calls, such as crying or laughter, this is soon followed by babbling sounds, a behavior not normally seen in primates, which paves the road for phonetic development. However, even in humans the phonetic repertoire is limited and, in a sense, part of the species-specific vocal repertoire. What is unusual in humans is not the ability to incorporate new sounds, as for example seen in dolphins, but the capacity to

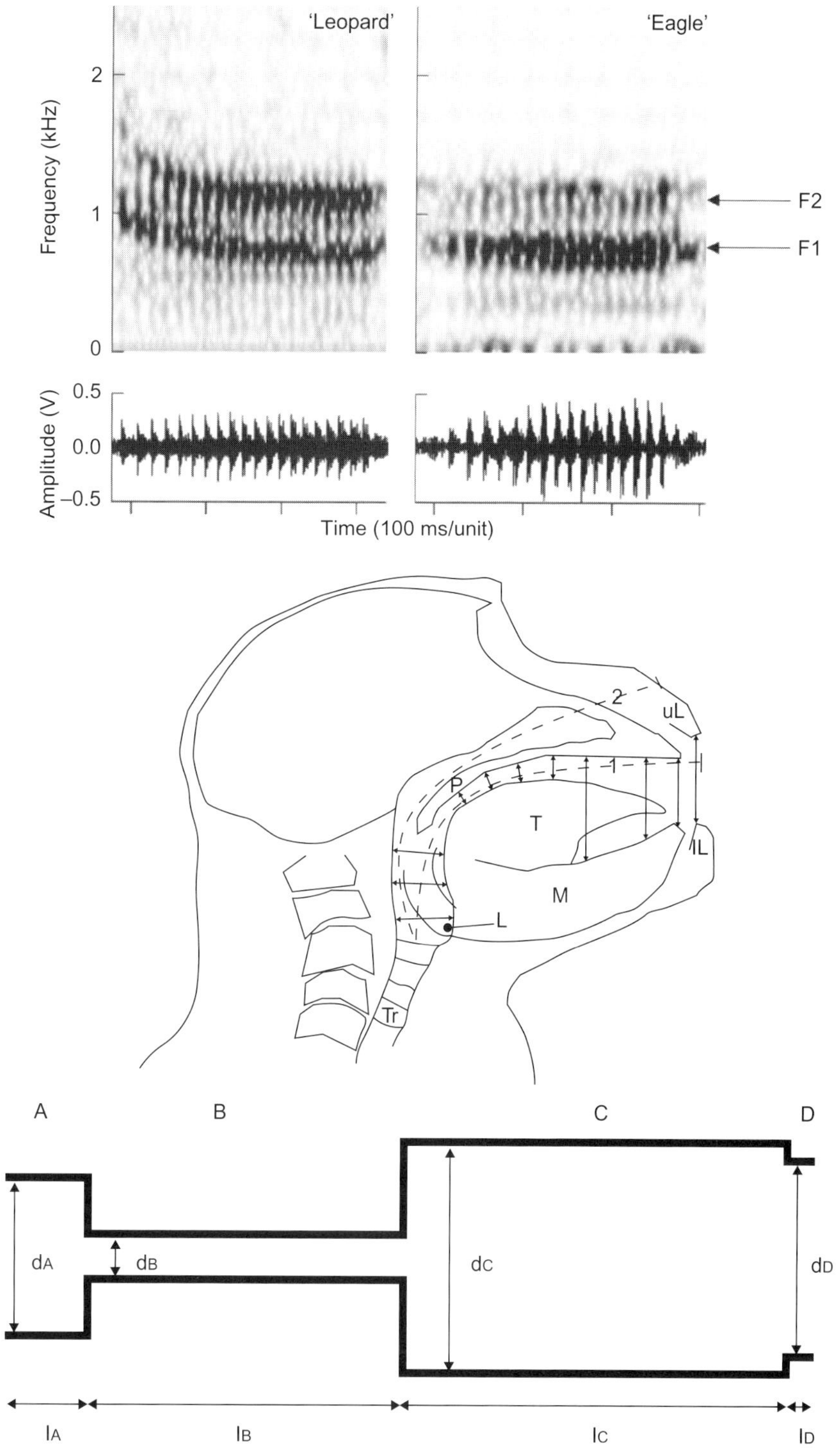

Fig. 3. Top: Diana monkey alarm calls to leopards and crowned eagles contain two formant frequencies with differing transitions. Middle: Schematic drawing of the head-neck region of a Diana monkey, as seen in dissection and lateral x-rays (T – tongue, Tr – trachea, uL – upper lip, lL – lower lip, L – larynx, P – palate, dashed line 1 – oral vocal tract length, dashed line 2 – nasal vocal tract length, arrows indicate the dorso-ventral distances of the oral vocal tract). Bottom: A multi-tube model of a Diana monkey's vocal tract can generate the full range of formant patterns given by live animals. Tube C–D simulates the mouth opening with the lip aperture (D). Riede et al. (2005).

combine the species-specific phonetic units rapidly and precisely to generate more complex utterances, usually to initiate some type of social interaction, a behavior not observed in other mammals.

V.C. Rigid callers: flexible gesturers?

The previous sections have suggested that the core components necessary for human language can be found, at least in their roots, throughout the mammalian order. Whether a communicative ability is expressed in the vocal or, as some propose, in the gestural domain is a somewhat irrelevant point. Once the necessary cognitive powers are established, individuals will seek ways to express them, and if one production channel is blocked, the next best one will come into effect. The fact that all human languages are auditory-based is indicative of an evolutionary adaptation for acoustic communication, and this appears to have a long evolutionary history. Although research with gestural communication has shown that apes are able of producing imperative-like statements (Tomasello, 2008), similar effects can also be found in the vocal domain, although the necessary controlled empirical studies have yet to be done (e.g., Slocombe and Zuberbühler, 2007). Non-human primates may be constrained by species-specific vocal repertoires, but it is also the case that there is often significant acoustic flexibility within some call types which is largely determined by the callers' ontogenetic experiences (e.g., Stephan and Zuberbühler, 2008). The occasionally proposed dichotomy "rigid caller–flexible gesturer" is not really based on much empirical support (see Eberl, Chapter 11.1 in this volume).

VI. Conclusions

The main purpose of this chapter was to review the mammalian evidence for language precursors, and to understand their phylogenetic distribution and origins. Human language is a patchwork of abilities, uniquely combined and elaborated, with deep roots in mammalian evolutionary history. From the available evidence, it is difficult to see clear evolutionary patterns. The ability to represent the world in mental concepts appears to have evolved early, and it may even be a cognitive feature of most higher animals. However, it is often not clear whether the natural mental concepts experienced by humans are similar to or differ from the ones available to other animals. Cooperation has a widespread presence in the animal world, and seems to evolve independently, although in many species cooperative acts are restricted to very specific behavioral contexts, such as hunting or breeding. Evidence of social awareness is more restricted, and mainly available for apes, suggesting that it may be a hallmark of the more encephalized primates. Vocal control, finally, is well-developed in humans, but precursor abilities are found in a number of non-human primates, with no clear phylogenetic patterns. Vocal control appears to emerge relatively easily, although the underlying production mechanisms vary widely. In humans control of the tongue is crucial, and is perhaps the result of recent genetic changes (Enard et al., 2002).

Acknowledgments

I am grateful to the Leverhulme Trust, the BBSRC, the EU FP6 Pathfinder Initiative, and the Royal Zoological Society of Scotland for supporting much of the research reviewed in this chapter.

References

Adachi, I., Kuwahata, H., Fujita, K., 2007. Dogs recall their owner's face upon hearing the owner's voice. Anim. Cogn. 10, 17–21.

Arnold, K., Zuberbühler, K., 2008. Meaningful call combinations in a non-human primate. Curr. Biol. 18, R202–R203.

Bergman, T.J., Beehner, J.C., Cheney, D.L., Seyfarth, R.M., 2003. Hierarchical classification by rank and kinship in baboons. Science 30, 1234–1236.

Blumstein, D.T., 1999. Alarm calling in three species of marmots. Behaviour 136, 731–757.

Bovet, D., Vauclair, J., 2001. Judgment of conceptual identity in monkeys. Psychon. Bull. Rev. 8, 470–475.

Bshary, R., Wickler, W., Fricke, H., 2002. Fish cognition: a primate's eye view. Anim. Cogn. 5, 1–13.

Clarke, E., Reichard, U., Zuberbühler, K., 2006. The syntax and meaning of wild gibbon songs. PLoS One 1, e73.

Dasser, V., 1988. A social concept in Java monkeys. Anim. Behav. 36, 225–230.

de Waal, F.B.M., 1982. Chimpanzee Politics. Harper, New York, NY.

Enard, W., Przeworski, M., Fisher, S.E., Lai, C.SL., Wiebe, V., Kitano, T., Monaco, A.P., Paabo, S., 2002. Molecular evolution of FOXP2, a gene involved in speech and language. Nature 418, 869–872.

Erdohegyi, A., Topal, J., Viranyi, Z., Miklosi, A., 2007. Dog-logic: inferential reasoning in a two-way choice task and its restricted use. Anim. Behav. 74, 725–737.

Fischer, J., Hammerschmidt, K., Cheney, D.L., Seyfarth, R.M., 2001. Acoustic features of female chacma baboon barks. Ethology 107, 33–54.

Gomez, J.C., Martin-Andrade, B., 2002. Possible precursors of pretend play in nonpretend actions of captive gorillas (*Gorilla gorilla*). In: Mitchell, R.W. (Ed.), Pretending and imagination in animals and children. Cambridge University Press, Cambridge, UK, pp. 255–268.

Gouzoules, S., Gouzoules, H., Marler, P., 1984. Rhesus monkey (*Macaca mulatta*) screams: representational signalling in the recruitment of agonistic aid. Anim. Behav. 32, 182–193.

Hammerschmidt, K., Freudenstein, T., Jürgens, U., 2001. Vocal development in squirrel monkeys. Behaviour 138, 1179–1204.

Hurford, J.R., 2007. The Origins of Meaning. Oxford University Press, Oxford, UK.

Kaminski, J., Call, J., Fischer, J., 2004. Word learning in a domestic dog: evidence for "fast mapping". Science 304, 1682–1683.

Kiriazis, J., Slobodchikoff, C.N., 2006. Perceptual specificity in the alarm calls of Gunnison's prairie dogs. Behav. Proc. 73, 29–35.

Leger, D.W., Owings, D.H., Boal, L.M., 1979. Contextual information and differential responses to alarm whistles in California ground squirrels. Zeitschr. Tierpsychol. 49, 142–155.

Lieberman, P., 1984. The Biology and Evolution of Language. Harvard University Press, Cambridge, MA.

McBride, A.F., Hebb, D.O., 1948. Behavior of the captive bottle-nose dolphin, *Tursiops truncatus*. J. Comp. Physiol. Psychol. 41, 111–123.

Mercado, E., Killebrew, D.A., Pack, A.A., Macha, I.V.B., Herman, L.M., 2000. Generalization of "same–different" classification abilities in bottlenosed dolphins. Behav. Proc. 50, 79–94.

Mitchell, R.W., Thompson, N.S., 1993. Familiarity and the rarity of deception: two theories and their relevance to play between dogs (*Canis familiaris*) and humans (*Homo sapiens*). J. Comp. Psychol. 107, 291–300.

Nishimura, T., Mikami, N., Suzuki, J., Matsuzawa, T., 2003. Descent of the larynx in chimpanzee infants. Proc. Natl. Acad. Sci. 100, 6930–6933.

Owren, M.J., Dieter, J.A., Seyfarth, R.M., Cheney, D.L., 1993. Vocalizations of rhesus (*Macaca mulatta*) and Japanese (*M. fuscata*) macaques cross-fostered between species show evidence of only limited modification. Dev. Psychobiol. 26, 389–406.

Papworth, S., Böse, A.-S., Barker, J., Zuberbühler, K., 2008. Male blue monkeys alarm call in response to danger experienced by others. Biol. Lett. 4, 472–475.

Pika, S., Zuberbühler, K., 2007. Social games between bonobos and humans: evidence for shared intentionality? Am. J. Primatol. 69, 1–6.

Poole, J.H., Tyack, P.L., Stoeger-Horwath, A.S., Watwood, S., 2005. Elephants are capable of vocal learning. Nature 434, 455–456.

Ralls, K., Fiorelli, P., Gish, S., 1985. Vocalizations and vocal mimicry in captive harbor seals, *Phoca vitulina*. Can. J. Zool. (Rev. Can. Zool.) 63, 1050–1056.

Range, F., Noë, R., 2005. Can simple rules account for the pattern of triadic interactions in juvenile and adult female sooty mangabeys? Anim. Behav. 69, 445–452.

Richards, D.G., Wolz, J.P., Herman, L.M., 1984. Vocal mimicry of computer-generated sounds and vocal labeling of objects by a bottlenosed dolphin, *Tursiops truncatus*. J. Comp. Psych. 98, 10–28.

Riede, T., Bronson, E., Hatzikirou, H., Zuberbühler, K., 2005. Vocal production mechanisms in a non-human primate: morphological data and a model. J. Human Evol. 48, 85–96.

Schel, A.M., Tranquilli, S., Zuberbühler, K., 2009. The alarm call system of two species of black-and-white colobus monkeys (*Colobus polykomos* and *Colobus Quereza*). J. Comp. Psychol. 123 (2), 136–150.

Slocombe, K.E., Zuberbühler, K., 2007. Chimpanzees modify recruitment screams as a function of audience composition. Proc. Nat. Acad. Sci. USA 104, 17228–17233.

Stephan, C., Zuberbühler, K., 2008. Predation increases acoustic complexity in primate alarm calls. Biol. Lett. 4, 641–644.

Tomasello, M., 2008. Origins of Human Communication. MIT Press, Cambridge, MA.

Townsend, S.W., Slocombe, K.E., Thompson, M.E., Zuberbühler, K., 2007. Female-led infanticide in wild chimpanzees. Curr. Biol. 17, R355–R356.

Tyack, P.L., 2008. Convergence of calls as animals form social bonds, active compensation for noisy communication channels, and the evolution of vocal learning in mammals. J. Comp. Psychol. 122, 319–331.

Vonk, J., MacDonald, S.E., 2002. Natural concepts in a juvenile gorilla (*Gorilla gorilla gorilla*) at three levels of abstraction. J. Exp. Anal. Behav. 78, 315–332.

Vonk, J., MacDonald, S.E., 2004. Levels of abstraction in orangutan (*Pongoabelii*) categorization. J. Comp. Psychol. 118, 3–13.

Warneken, F., Tomasello, M., 2006. Altruistic helping in human infants and young chimpanzees. Science 311, 1301–1303.

Wich, S.A., de Vries, H., 2006. Male monkeys remember which group members have given alarm calls. Proc. Roy. Soc. B-Biol. Sci. 273, 735–740.

Wich, S.A., Swartz, K.B., Hardus, M.E., Lameira, A.R., Stromberg, E., Shumaker, R.W., 2009. A case of spontaneous acquisition of a human sound by an orang-utan. Primates 50, 56–64.

Wittig, R.M., Crockford, C., Wikberg, E., Seyfarth, R.M., Cheney, D.L., 2007. Kin-mediated reconciliation substitutes for direct reconciliation in female baboons. Proc. Roy. Soc. B-Biol. Sci. 274, 1109–1115.

Zuberbühler, K., 2006. Alarm calls – evolutionary and cognitive mechanisms. In: Brown, K. (Ed.), Encyclopedia of Language and Linguistics. Elsevier, Oxford, UK, pp. 143–155.

Zuberbühler, K., 2008. Audience effects. Curr. Biol. 18, R189–R190.

Zuberbühler, K., Byrne, R.W., 2006. Social cognition. Curr. Biol. 16, R786–R790.

Index

An asterisk before a page span indicates mentions on each page covered rather than unbroken discussion.

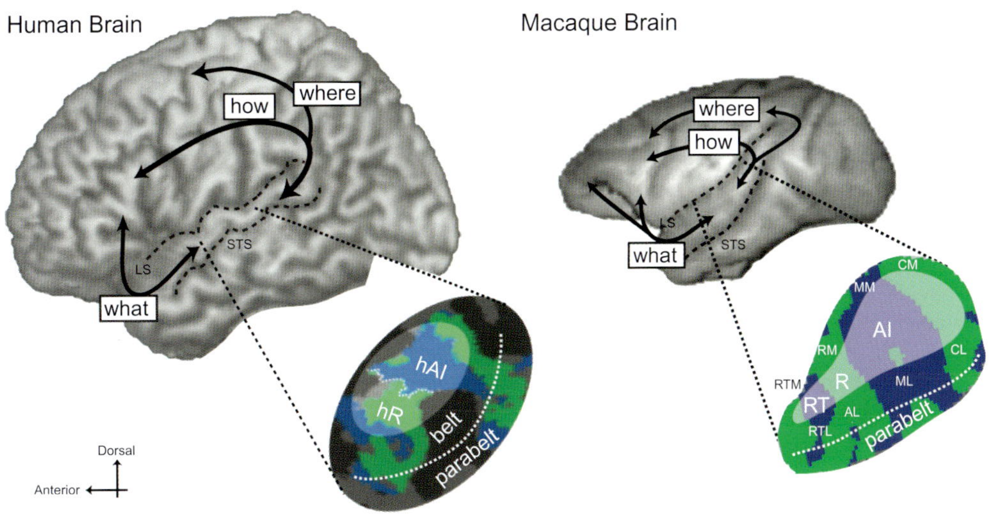

Plate 4.4.1 Pathways supporting vocal communication in humans and monkeys. The pathways for parallel brain processing are labeled with black arrows. These are termed "what," "where" and "how," reflecting their respective contribution in evaluating what a sensory object is, where it is located in sensory space and how it was generated. Although the figure suggests a close correspondence between the human and monkey brain, it remains unclear how many pathways there are, their precise roles and the extent of the homologies between the species. The inserts illustrate the functional organization of the auditory cortex. Cortical processing begins with the primary auditory cortex (see the lighter region in the enlarged inserts, which includes field AI in monkeys and its presumed homolog hAI in humans). The colors of the insert reflect the direction of the tonotopic gradients of the auditory fields (dark blue: fields with high to low frequency preference toward the anterior direction; light green: fields with low to high frequency preference in the anterior direction). Data such as these can be used to functionally parcellate the auditory cortex and delineate borders between fields with mirror reversed tonotopic gradients. More detailed processing of sound follows in the hierarchically higher auditory "belt" and "parabelt" regions of auditory cortex. See the text for further details and references. This figure contains a rendered human brain image kindly contributed by J. Obleser and an example of the mapping of human auditory cortical fields contributed by E. Formisano. Abbreviations: AL: antero-lateral; CL: caudo-lateral; CM: caudo-medial; CPB: caudal parabelt; LS: lateral sulcus; ML: medio-lateral MM: medio-medial; STS: superior temporal sulcus; RM: rostro-medial; RPB: rostral parabelt; RL: rostro-lateral; RT: rostro-temporal; RTL: rostro-temporal lateral; RTM: rostro-temporal medial.

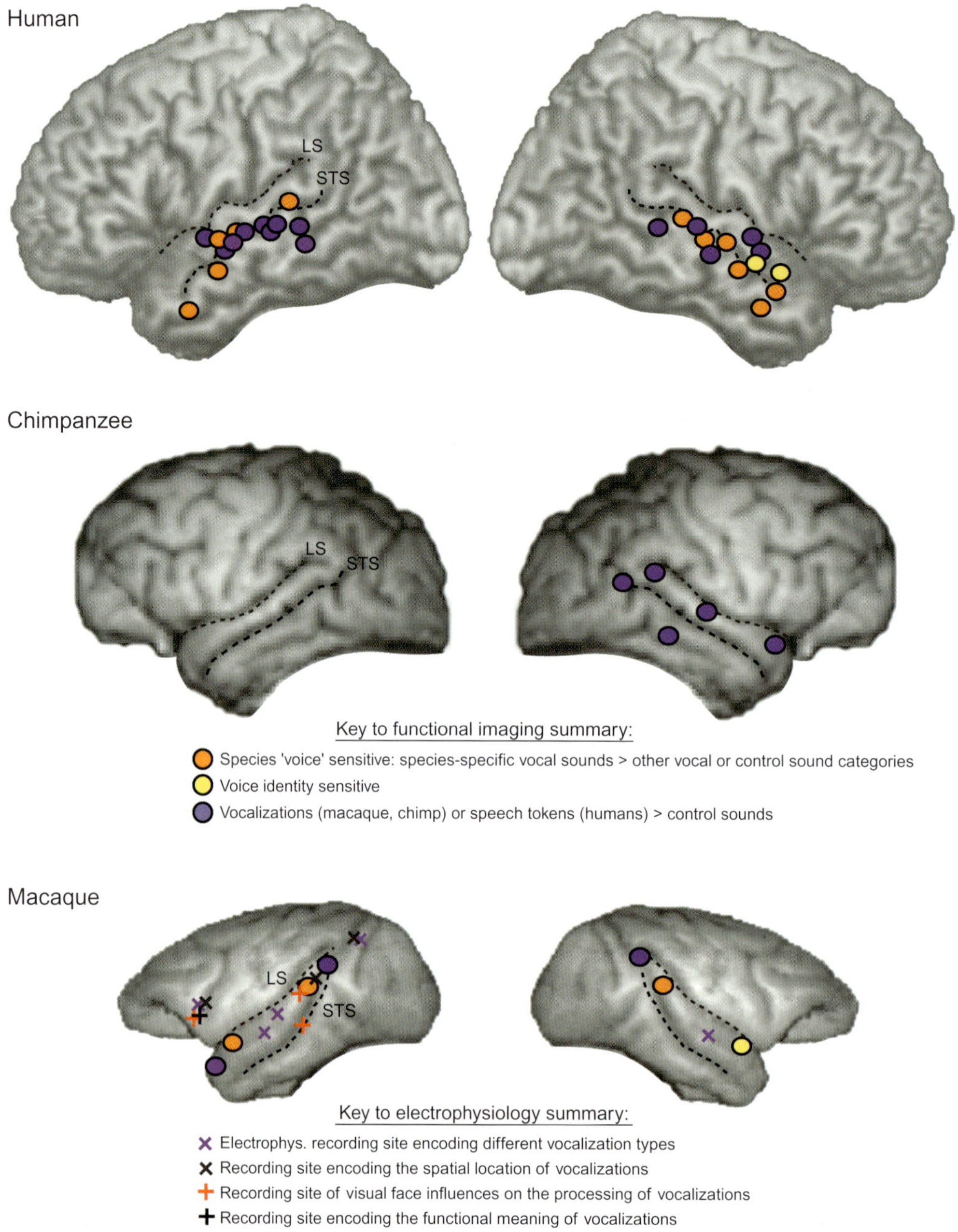

Plate 4.4.2 Comparative summary of human, chimpanzee and macaque processing of species-specific communication sounds. Colored circles summarize several functional imaging results (see key) focusing on the stimulus-bound processing of vocal signals in the temporal lobe. For humans, we summarize the peaks of activity reported in studies of voice sensitivity (Belin et al., 2000; Belin and Zatorre, 2003; von Kriegstein et al., 2003), voice identity (Belin and Zatorre, 2003; von Kriegstein et al., 2003) and the sublexical or stimulus-bound aspects of speech (Dehaene-Lambertz et al., 2005; Liebenthal et al., 2005; Rimol et al., 2005; Obleser et al., 2006, 2007); for the exact coordinates of the summaries in humans and monkeys see Petkov et al., 2009 (in press). For chimpanzees we summarize a recent study evaluating chimp vocal sound processing in these great apes (Taglialatela et al., 2009). For the macaque brain we show the sensitivity to macaque vocalizations from both PET (Poremba et al., 2004; Gil-da-Costa et al., 2006) and fMRI (Petkov et al., 2008) studies. The monkey voice sensitive regions (orange circles) and voice-identity selective regions (yellow circle) identified in the monkey fMRI study (Petkov et al., 2008) can be directly compared with the human studies on human voice sensitivity and selectivity (compare the orange and yellow circles in the human and macaque). For macaques, we also identify electrophysiological recording sites (see the lower key), from the temporal (Rauschecker et al., 1995; Tian et al., 2001; Ghazanfar et al., 2005, 2008; Russ et al., 2008), parietal (Gifford and Cohen, 2005) and prefrontal cortices (Romanski and Goldman-Rakic, 2002; Cohen et al., 2004; Gifford et al., 2005; Sugihara et al., 2006). This figure contains a rendered chimpanzee brain image kindly contributed by J. Taglialatela. For abbreviations, see Plate 4.4.1.

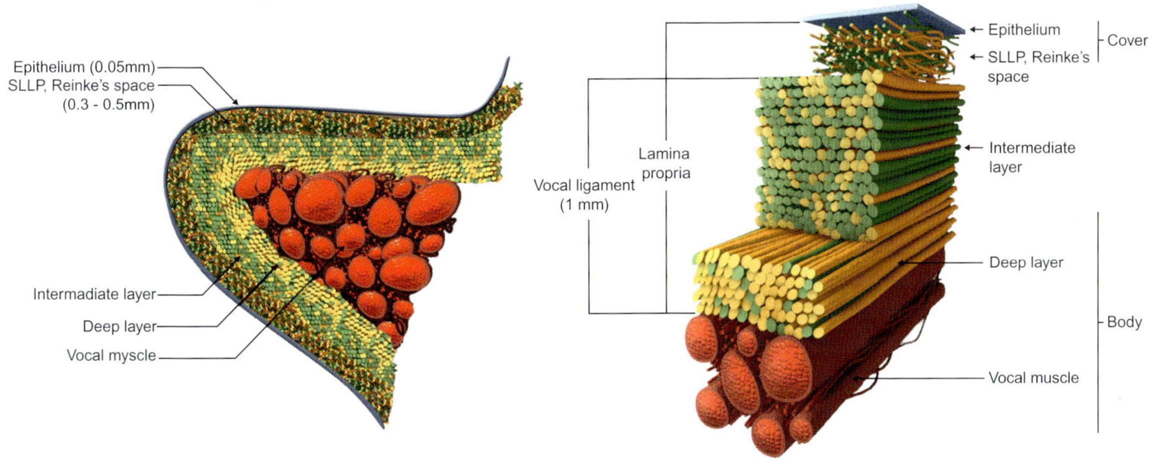

Plate 10.2.2 Laminar structure of the vocal folds. Coronal cross-section of a vocal fold, on the left, and three-dimensional model of the layers, on the right. Blue: epithelium; yellow: collagen fibers; green: elastic fibers; red: vocal muscle fibers. Adapted from Hirano (1975) Phonosurgery – basic and clinical investigation. Otologia, 21: 239.

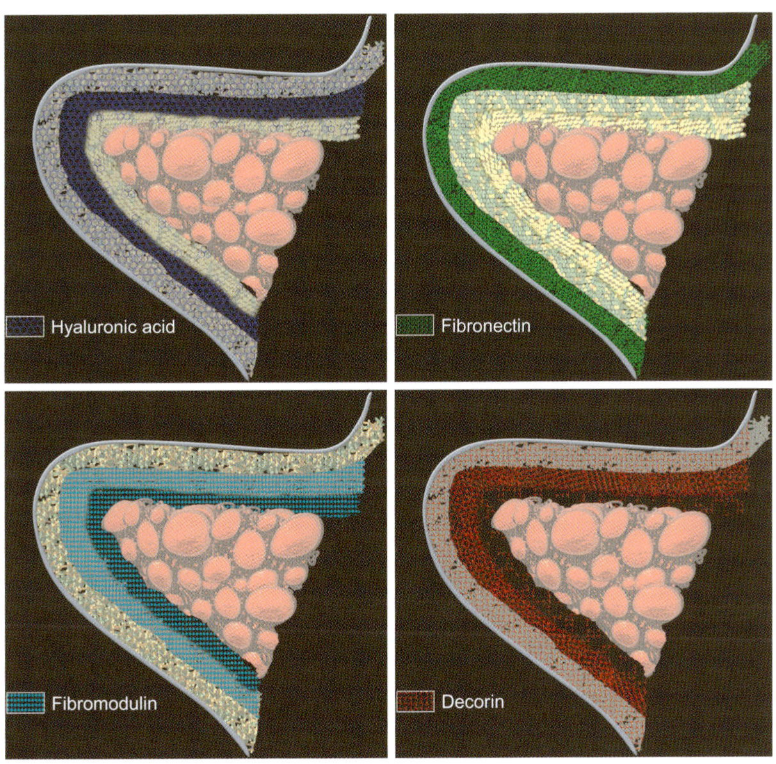

Plate 10.2.3 Distribution of hyaluronic acid (purple), decorin (green), fibromodulin (blue) and fibronectin (red) in the three layers of the VFLP.

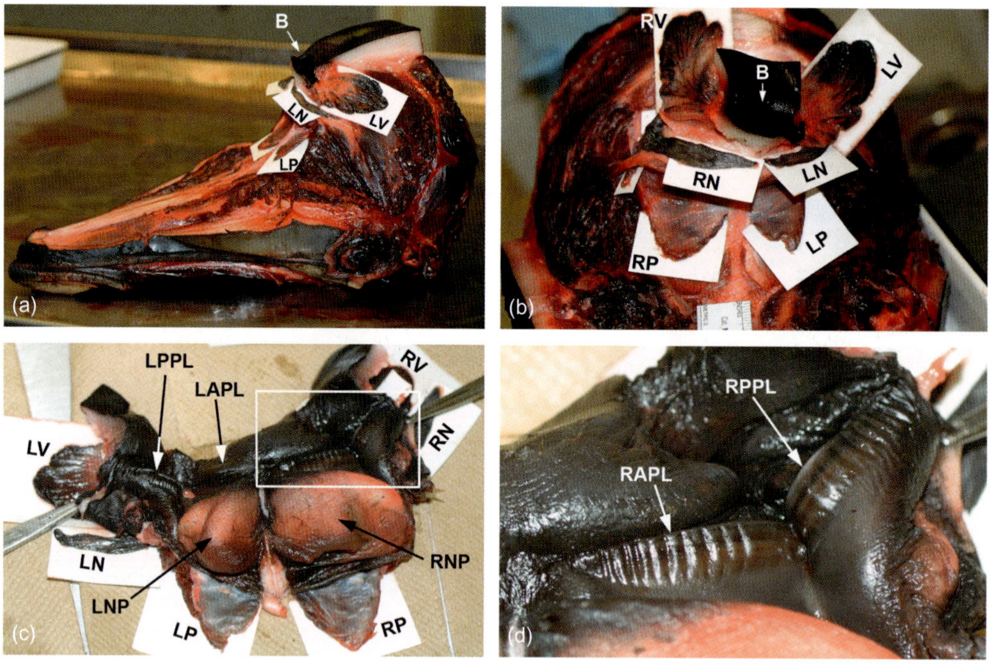

Plate 10.4.4 (a) Lateral-oblique view of left side of the head of an adult Atlantic white-sided dolphin, *Lagenorhynchus acutus*, that has been dissected to reveal the three pairs of nasal air sacs. Each air sac has been highlighted by the white 3 × 5 inch index card placed under it. (b) Frontal view of the rostral aspect of the same specimen shown in (a). (c) Ventral view of the nasal sac complex, dissected off the skull of the same specimen shown in (a) and (b). Note the two pairs of phonic lips, which have a crinkled appearance. The posterior phonic lip has been separated from the anterior lip at the midline on both sides and each is reflected laterally by a hemostat. (d) Enlargement of region indicated by the white box in image (c). Note the crinkled appearance of the surfaces (likely due to relaxation after severing medial connection of posterior phonic lip). Each phonic lip has a shelf of tissue that tapers to a thin edge at its free border. B: blowhole opening; LAPL: left anterior phonic lip; LN: left nasofrontal air sac; LNP: left nasal plug; LP: left premaxillary air sac; LPPL: left posterior phonic lip; LV: left vestibular air sac; RAPL: right anterior phonic lip; RN: right nasofrontal air sac; RNP: right nasal plug; RP: right premaxillary air sac; RPPL: right posterior phonic lip; RV = right vestibular air sac.